Dictionary of Environmental Science and Engineering

ENGLISH–SPANISH | SPANISH–ENGLISH
INGLÉS–ESPAÑOL | ESPAÑOL–INGLÉS

Diccionario de Ciencia e Ingeniería Ambiental

About the Authors

Until 1994 **Howard Headworth** was Managing Director of a British environmental company which provided consultancy and laboratory services. He is a geologist by training and hydrogeologist by profession. During his career he has been involved in all aspects of groundwater resources management and protection and latterly led multidisciplinary scientific and environmental teams. He is a European Geologist and Fellow of the Institution of Water and Environmental Management. He now lives in southern Spain with his co-author wife, Sarah.

After obtaining a bachelors degree and masters degree in Applied Biology and Biological Computation respectively, **Sarah Steines** worked for fifteen years as an environmental scientist. She has expertise over a range of disciplines including ecology, chemistry, waste treatment and disposal, aquatic and terrestrial pollution and mathematical modelling. She has spent several years working with engineers, scientists and town planners on environmental impact assessments for a range of construction projects.

Sobre los Autores

Hasta 1994 **Howard Headworth** fue Director Gerente de una compañía británica que provee servicios de consultoría y laboratorio. Es geólogo por formación e hidrogeólogo por profesión. Durante su carrera ha trabajado en todos los aspectos de la dirección y protección de recursos acuíferos y posteriormente ha dirigido equipos multidisciplinarios científicos y ambientales. Es Geólogo Europeo y Socio de la Chartered Institution of Water and Environmental Management. Ahora vive en el sur de España con su esposa y co-autora del libro, Sarah.

Después de obtener su licenciatura y master en Biología Aplicada y Computación Biología respectivamente, **Sarah Steines** trabajó durante quince años como científica medioambiental. Tiene experiencia en muchos temas como ecología, química, tratamiento y vertido de desechos, contaminación acuática y terrestre y modelización matemática. Ha pasado varios años trabajando con ingenieros, científicos y urbanistas en evaluaciones de impactos ambientales para numerosos proyectos de construcción.

Dictionary of Environmental Science and Engineering

ENGLISH–SPANISH | SPANISH–ENGLISH
INGLÉS–ESPAÑOL | ESPAÑOL–INGLÉS

Diccionario de Ciencia e Ingeniería Ambiental

HOWARD HEADWORTH AND SARAH STEINES
Almería, Spain

with the assistance of | con la asistencia de

Manuel Regueiro y Gonzáles-Barros
Instituto Geológico y Minero de España, Madrid, Spain

and | y

Alberto Bustani Adem
Instituto Tecnológico y de Estudios Superiores de Monterrey, Mexico

JOHN WILEY & SONS
Chichester • New York • Weinheim • Brisbane • Singapore • Toronto

Other Wiley Editorial Offices

John Wiley & Sons, Inc., 605 Third Avenue,
New York, NY 10158-0012, USA

WILEY-VCH Verlag GmbH, Pappelallee 3,
D-69469 Weinheim, Germany

Jacaranda Wiley Ltd, 33 Park Road, Milton,
Queensland 4064, Australia

John Wiley & Sons (Asia) Pte Ltd, 2 Clementi Loop #02-01,
Jin Xing Distripark, Singapore 129809

John Wiley & Sons (Canada) Ltd, 22 Worcester Road,
Rexdale, Ontario M9W 1L1, Canada

Library of Congress Cataloging-in-Publication Data

Headworth, Howard.
    Spanish/English & English/Spanish dictionary of environmental science and
engineering / Howard Headworth and Sarah Steines ; with the assistance of
Manuel Regueiro y González-Barros and Alberto Bustani Adem.
        p.   cm.
    ISBN 0-471-96273-2
    1. Environmental sciences—Dictionaries—Spanish. 2, Spanish language—
Dictionaries—English. 3. Environmental sciences—Dictionaries. 4. English
language—Dictionaries—Spanish. 5. Environmental engineering—Dictionaries—
Spanish. 6. Environmental engineering—Dictionaries. I. Steines, Sarah.
II. Title. III Title: Diccionario español/inglés e inglés/español de ciencia e
ingeniería ambiental
GE10.H43 1997
363.7'003—dc21   97-28550
                                                                        CIP

British Library Cataloguing in Publication Data

A catalogue record for this book is available from the British Library

ISBN 0-471-96273-2

Typeset in 8/9pt Times by Mayhew Typesetting, Rhayader, Powys
Printed and bound in Great Britain by Biddles Ltd, Guildford and King's Lynn

This book is printed on acid-free paper responsibly manufactured from sustainable forestation,
for which at least two trees are planted for each one used for paper production.

Contents

The dictionary includes words and expressions from the following component subjects of environmental science and environmental engineering which the practitioner is likely to need in the course of his or her work:

archaeology
biology
chemistry
civil engineering and construction
computing
contaminated land
ecology
environmental engineering
environmental science
finance and mangement
geography

geology
geomorphology
hydrogeology
hydrology
landfill
law and regulation
mathematics and modelling
meteorology
sanitary engineering
water and wastewater engineering

Presentation

The dictionary conforms with traditional format. The first half comprises English/Spanish and the second half comprises Spanish/English. Each side of the dictionary is made up of some 10000 principal entries which between them contain 14000 associated expressions. The principal entries are given alphabetically and are given in bold italic type. Of the principal entries, 6500 contain at least one associated expression and these are presented alphabetically within each principal entry. In these cases the leading word is abbreviated.

For example: *park,* n. parque (m); **national p.,** parque nacional.

Grammatical function and gender are given throughout.

English verbs, nouns or adjectives which share the same spelling are distinguished within a single entry.

Alternative translations of words are given in order of most common use.

(US) after an English word denotes the spelling in the USA. It refers to the single word which it follows.

For example: *litro,* m. litre, liter (US).

In cases where words can be translated in several ways, the meaning of the translated words are shown in brackets. These bracketed meanings apply to all translated words contained within semi-colon separators.

For example: *pasaje,* m. passage, passing (acción); voyage, crossing (Mar); fare (tarifa).

The dictionary follows normal convention with Spanish principal entries starting with 'ch', 'll' and 'ñ'. (That is, words starting with 'ch' follow those starting with 'c', words starting with 'll' follow those starting with 'l' and words starting with 'ñ' follow those starting with '*n*'.) Likewise, where the letters '*ch*', '*ll*' and '*ñ*' are contained with Spanish entries than these words follow those containing '*c*', '*l*' and '*n*'.

For example: *nicho* follows *nicotina*: *callampa* follows *calzada*: *estaño* follows *estanqueidad*.

Registered products are followed by (R).

Spanish principal entries having a Latin American origin have their country of origin show in brackets before the English translation.

For example: **desmenuzador,** m. (LAm) shredder (máquina).
        huayco, f. (Perú, Ch) landslide.

In all other cases, the country of origin appears after the English translation.

For example: **barata,** f. cockroach (Ch); bargain (Méx).

Similarly, subject abbreviations follow the translated term.

For example: **pod,** n. vaina (f) (Bot); manada (f) (Zoo); ranura (f) (Eng, groove).

Abbreviations

Grammatical Abbreviations

a.	adjective	pf.	prefix
adv.	adverb	pl.	plural
e.g.	for example	sf.	suffix
n.	noun	v.	verb
p.	preposition		

Subject Abbreviations

Anat	anatomy	Fig	figurative use
Agr	agriculture	Geog	geography, geomorphology and landuse
Arch	archaeology		
Biol	biology	Geol	geology
Bot	botany	Ind	industry
Chem	chemistry	Jur	law
Com	commerce, finance, economics	Mat	mathematics
Comm	communications, radio and television, printing	Med	medicine and pharmacy
		Met	meteorology
Comp	computer science, information technology	Mil	military
		Min	mining
Con	construction, building and architecture	Mar	marine, maritime, nautical
		Phys	physics
Ecol	ecology	R	registered product
Elec	electricity	Reg	regulation and government, public administration
Eng	engineering, mechanics, technical		
		Treat	waste treatment
Fam	familiar, colloquial use	Zoo	zoology

National Abbreviations

Afr	Africa	ElS	El Salvador
Ar	Arabia	EurU	European Union
Austr	Australia	Fr	France
Arg	Argentina	GB	Great Britain
Bol	Bolivia	Ger	Germany
Br	Brazil	Guat	Guatemala
Can	Canada	Hond	Honduras
Carib	Caribbean (Cuba, Dominican Republic, Puerto Rico)	Ir	Ireland
		LAm	Latin America
CAm	Central America (Guatemala, El Salvador, Honduras, Costa Rica, Nicaragua, Panama)	Mex	Mexico
		Nic	Nicaragua
		Pan	Panama
Ch	Chile	Par	Paraguay
Col	Colombia	Peru	Peru
CRica	Costa Rica	PRico	Puerto Rico
CSur	Southern Cone (Argentina, Uruguay, Paraguay, Chile)	Sp	Spain
		Ur	Uruguay
Cuba	Cuba	US	United States
DomR	Dominican Republic	Ven	Venezuela
Ec	Ecuador		

Contenido

Este diccionario incluye palabras y expresiones de los siguientes temas que componen la ciencia ambiental y la ingeniería que el usario encontrará o necesitará usar en el curso de su empleo:

arqueología
biología
ciencia ambiental
ecología
finanzas y administración
geografía
hidrogeología
hidrología
informática
ingeniería ambiental

ingeniería civil y contrucción
ingeniería sanitaria
ley y regulación
matemática y modelización
meteorología
química
suelos contaminados
tratamiento de aguas potables y
   residuales
vertederos

Presentación

Este diccionario sigue un formato tradicional. La primera mitad es inglés/español y la segunda mitad es español/inglés. Cada mitad abarca 10.000 registros principales que contienen 14.000 expresiones asociadas. Aparecen en itálica las palabras principales y en orden alfabético. 6.500 de los registros principales contienen al menos una expresión asociada y éstas se presentan alfabéticamente dentro cada entrada principal. En estos casos la palabra principal está abreviada.

Ejemplo: *park*, n. parque (m); **national p.**, parque nacional.

Se da la función gramatical y el género en todas las partes.

Verbos, sustantivos o adjectivos ingleses que comparten la misma ortografía se distingues dendro una solar entradas.

Se dan traducciones alternativas de palabras por orden de uso más común.

(US) despues de una palabra ingelsa denota la ortografía en los Estados Unidos. Se refiere a la única palabra a la que sigue.

Ejemplo: *litro*, m. litre, liter (US).

En los casos en que las palabras pueden ser traducidas de varias maneras, se muestra entre paréntesis el sentido de las palabras traducidas. Estos sentidos entre paréntesis se aplican a todas las palabras traducidas que hay entre dos puntos y comas.

Ejemplo: *pasaje*, m. passage, passing (acción); voyage, crossing (Mar); fare (tarifa).

El diccionario sigue la convención de las entradas principales españolas que empiezan con '*ch*', '*ll*' y '*ñ*' apareciendo por separado (es decir, las palabras que comienzan por '*ch*' siguen a las que comienzan con '*c*', las palabras que comienzan por '*ll*' sigue a las que comienzan con '*l*' y las palabras que comienzan con '*ñ*' siguen a las comienzan con '*n*').

De la misma forma, si las letras '*ch*', '*ll*' y '*ñ*' van contenidas en las palabras españolas, seguirán a aquellas palabras que contengan '*c*', '*l*' y '*n*'.

Ejemplo: *nicho* sigue a *nicotina*: *callampa* sigue a *calzada*: *estaño* sigue a *estanqueidad*.

Los productos registrados van seguido por (R).

Se muestra entre paréntesis el país de origen de las principales entradas españolas que tienen un origen Latinoamericano antes de la traducción inglesa.

Ejemplo: **desmenuzador**, m. (LAm) shredder (máquina).
      huayco, m. (Perú, Ch) landslide.

Aparece en todos los demás casos el país de origen después de la traducción inglesa.

Ejemplo: **barata**, f. cockroach (Ch), bargain (Méx).

De forma semejante a la entrada traducida siguen abreviaturas temáticas.

Ejemplo: **pod**, n. viana (f) (Bot); manada (f) (Zoo); ranura (f) (Eng, groove).

Abreviaturas

Abreviaturas Gramaticales

a.	adjetivo	p.	preposición
adv.	adverbio	p.e.	por ejemplo
f.	femenino	pf.	prefijo
f.pl.	femenino plural	pl.	plural
m.	masculino	sf.	sufijo
m.pl.	masculino plural	véase	véase
m.y f.	masculino y femenino		

Abreviaturas de Temas

Anat	anatomía	Geog	geografía, geomorfología, ocupación de terreno
Agr	agricultura		
Arq	arqueología	Geol	geología
Biol	biología	Ing	ingeniería, mecánica, técnica
Bot	botánica	Jur	ley
Com	comercio, finanzas, economía	Mar	marítima, náutica
		Med	medicina y farmacia
Comp	informática	Met	meteorología
Con	construcción, arquitectura	Mil	militar
Ecol	ecología	Min	minería
Elec	electricidad	R	producto registrado
Fam	uso familiar	Quím	química
Fig	uso figurativo	Reg	regulación, gobierno, administración pública
Fís	física		
Gen	uso general	Zoo	zoología

Abreviaturas Nacionales

Ale	Alemania	DomR	República Dominicana
Afr	Africa	Ec	Ecuador
Ar	Arabia	ElS	El Salvador
Arg	Argentina	Esp	España
Austr	Australia	EU	Estados Unidos
Bol	Bolivia	EurU	Unión Europea
Br	Brasil	Fr	Francia
CAm	Centroamérica (Guatemala, El Salvador, Honduras, Costa Rica, Nicaragua, Panama)	GB	Gran Bretaña
		Guat	Guatemala
		Hond	Honduras
Can	Canadá	Ir	Irlanda
Carib	Caribe (Cuba, República Dominicana, Puerto Rico)	LAm	Latinoamérica
		Méx	México, Méjico
Ch	Chile	Nic	Nicaragua
Col	Colombia	Pan	Panamá
CRica	Costa Rica	Par	Paraguay
CSur	Cono Sur (Paraguay, Argentina, Uruguay, Chile)	Péru	Péru
		PRico	Puerto Rico
Cuba	Cuba	Ur	Uruguay
		Ven	Venezuela

Foreword

While Spanish to English and multi-language dictionaries and glossaries for environmental science exist, most concentrate on the biological sciences and avoid the wider spectrum of the subject covering the developed and contaminated environment in which we are obliged to live. This dictionary seeks to remedy this by encompassing all the science subjects pertaining to the environment and by including, in addition, the wide field of environmental engineering. This book is primarily aimed at the practising environmental scientist and engineer, whether in government, regulating agency, industry or consultancy who is working in the English/Spanish domain. Others with an involvement or interest in these subjects will also find the dictionary of value.

Acknowledgements

The authors have drawn on a large number of books, manuals, technical journals, magazines and newspaper articles in preparing this dictionary and they would like to express their particular thanks to the Argentina Geological Association for permission to draw on their Diccionario Geológico Español–Inglés y Inglés–Español, to Professor Emilio Custodio at the Polytechnic University of Catalonia in Barcelona for his approval to make use of the index in his classic textbook on Hydrogeology, Hidrológia Subterránea, to José Manuel Reyero for permission to draw on the book Terminología Popular de los Humedales by Fernando Gonzáles Bernáldez, to the Panamerica Centre of Sanitary Engineering and Environmental Sciences (CEPIS) in Lima for making available their Vocabulary in Spanish, English, Portuguese, German and French of Sanitary and Environmental Engineering, to the International Association of Hydrogeologists for permission to use its six-language List of Terms in Hydrogeology, Geochemistry and Geothermals of Mineral and Thermal Waters (copyright Verlag Heinz Heise) and to the Institution of Water and Environment Management in Britain for permission to draw on their second edition glossary of Handbooks of UK Wastewater Practice. They express their appreciation to Professor Stephen Foster and Rob Evans of the British Geological Survey and Rick Brassington, consultant hydrogeologists, for their help and encouragement during the preparation of this dictionary. Lastly and principally, the authors thank their two advisors, EuroGeol Professor Manual Regueiro y Gonzáles-Barros at the Instituto Geológico y Minero de España in Madrid and Dr Alberto Bustani Adem, Director of the Centro de Calidad Ambiental, Instituto Tecnológico y de Estudios Superiores de Monterrey, Monterrey, Mexico, for correcting and commenting on the final draft manuscript. Their thanks also go to Natalia García in Madrid and Ana Capel in Almería for their valuable assistance.

Prologo

Aunque ya existen diccionarios y glosarios Inglés/Español y diccionarios y glosarios políglotos para las ciencias ambientales muchos de ellos se centran en las ciencias biológicas y evitan el espectro más amplio del tema que abarca el medioambiente desarrollado y contaminado en cual tenemos que vivir. Este diccionario intenta remediar esto, abarcando todos los temas científicos relacionados con el medio ambiente y además, el amplio campo de la ingeniería ambiental. El diccionario está destinado principalmente al científico e ingeniero ambiental que trabaja para el gobierno, agencias de regulación industria o consultarías en los idiomas Inglés/Español. Tambien encontrarán útil este diccionario otras personas interesadas o relacionadas con estos temas.

Reconocimiento

Al mismo tiempo que los autores han utilizado numeroso libros, manuales, periódicos, técnicos, revistas y artículos populares en la preparación del diccionario, les gustaría dar las gracias en particular a la Asociación Geológica Argentina por su permiso para usar su Diccionario Geológico Español-Inglés y Inglés-Español, al profesor Emilio Custodio de la Universitat Politècnica de Catalunya en Barcelona por su aprobación de hacer uso de la tabla de índices de temas de su clásico libro de hidrogeología, Hidrología Subterránea, a José Manuel Reyero por su permiso para usar el libro Terminología Popular de los Humedales de Fernando Gonzalés Bernaldéz, al Centro Panamericano de Ingenería Sanitaria y Ciencias del Ambiente en Lima por poner a disposición su Vocabulario en Español, Inglés, Portugués, Alemán y Francés de Ingeniería Sanitaria y Ambiental, a la Asociación Internacional de Hidrogeólogos por su Lista de Terminos de Hidrogeología, Hidroquímica y Geotermia de Aguas Minerales y Termales (derechos de autor Verlag Heinz Heise), y al Institution of Water and Environmental Management en Gran Bretaña por su permiso para usar la segunda edición de su glosario de Handbook of UK Wastewater Practice. Expresan sus agradecimentos al profesor Stephen Foster, y Rob Evans del British Geological Survey y a Rick Brassington, consultor hidrogeológico, por la ayuda y estímulo que han dado durante la preparación de este trabajo. Finalmente y principalmente, los autores agradecen a sus dos consejeros, el geólogo europeo profesor Manuel Regueiro y Gonzalés-Barros del Instituto Geológico y Minero de España en Madrid y al Dr Alberto Bustani Adem del Centro de Calidad Ambiental, Instituto Tecnológico y de Estudios Superiores de Monterrey, Monterrey, México, por corregir y comentar el borrador. Tambien quieren expresar sus agradacimientos a Natalía Garcia en Madrid y Ana Capel en Almería.

Part 1

ENGLISH–SPANISH
INGLÉS–ESPAÑOL

A

Aalenian, n. Aaleniano (m).
abandon, v. abandonar; **to a. a claim**, abandonar una pertenencia.
abandonment, n. abandono (m).
abatement, n. disminución (f), moderación (f).
abattoir, n. matadero (m).
abdomen, n. abdomen (m), vientre (m).
abdominal, a. abdominal.
abiotic, a. abiótico/a.
ablation, n. ablación (f).
abnormality, n. anormalidad (f).
aboral, a. aboral (furthest from).
abrasion, n. abrasión (f), rozadura (f); **eolian (wind) a.**, abrasión eólica; **fluvial a.**, abrasión fluvial; **glacial a.**, abrasión glaciar, abrasión glacial; **marine a.**, abrasión marina.
abrasive, a. abrasivo/a.
abrupt, a. brusco/a, repentino/a (sudden); cortado/a (style).
abscisin, n. abscisina (f).
abscissa, n. abscisa (f).
abscissae, n.pl. abscisas (f.pl).
abscission, n. abscisión (f).
absence, n. ausencia (f).
absenteeism, n. ausentismo (m), absentismo (m).
absolute, a. absoluto/a; **a. zero**, cero (m) absoluto (0 deg. K); **degrees a.**, grados (m.pl) absolutos.
absorb, v. absorber.
absorbency, n. absorbencia (f).
absorbent, n. absorbente (m).
absorption, n. absorción (f), empapamiento (m).
absorptive, a. absortivo/a; **a. capacity**, capacidad (f) de absorción.
absorptivity, n. absorbencia (f).
abstract, 1. v. abstraer, extraer (to extract), resumir (to summarize); 2. n. resumen (m), sumario (m) (e.g. of a report).
abstraction, n. abstracción (f), extracción (f).
abundance, n. abundancia (f), abundamiento (m).
abundant, a. abundante.
abuse, n. abuso (m) (misuse); **environmental a.**, abuso ambiental.
abut, v. lindar, confinar (to border), apoyarse (to lean against).
abutment, n. contrafuerte (m) (buttress); estribo (m) de puente (of bridge).
abyss, n. abismo (m).
abyssal, a. abisal, abismal.
AC, n. CA; **alternating current**, corriente (f) alterna.
acacia, n. acacia (f), guaje (m) (LAm), huizache (LAm).

academic, a. académico/a.
Acadian, n. Acadiense (m).
acari, n.pl. ácaros (m.pl).
Acarina, n.pl. ácaros (m.pl), el orden Acarina (Zoo).
acarus, n. ácaro (m).
accelerate, v. acelerar.
acceleration, n. aceleración (f); **a. of gravity**, aceleración de la gravedad.
accelerometer, n. acelerómetro (m).
acceptor, n. aceptante (m) (Chem).
accident, n. accidente (m), casualidad (m); **road a.**, siniestro (m), accidente.
acclimatization, n. aclimatación (f).
acclimatize, v. aclimatizar.
accomplish, v. realizar.
account, n. cuenta (f) (bill), informe (f) (report), estado (m) (statement); **bank a.**, cuenta bancaria; **current a.**, cuenta corriente; **management accounts**, cuentas de gestión.
accountant, n. contable (m, f), contador (m) (LAm).
accounting, n. contabilidad (f).
accreditation, n. acreditación (f).
accretion, n. acreción (f), acrecentamiento (m).
accumulate, v. acumular.
accumulation, n. acumulación (f); **alluvial a.**, acumulación aluvial; **glacial a.**, acumulación glaciar, acumulación glacial.
accuracy, n. precisión (f), exactitud (f).
accused, a. acusado/a.
acetaldehyde, n. acetaldehído (m).
acetamide, n. acetamida (f).
acetate, n. acetato (m).
acetic, a. acético/a.
acetone, n. acetona (f).
acetyl, n. acetilo.
acetylene, n. acetileno (m).
ache, n. dolor (m).
achievement, n. realización (f), ejecución (f) (action).
acid, n. ácido (m); **abscisic a.**, ácido abscísico; **acetic a.**, ácido acético; **ascorbic a.**, ácido ascórbico; **boric a.**, ácido bórico; **carbonic a.**, ácido carbónico; **carboxylic a.**, ácido carboxílico; **chlorosulphonic a.**, ácido clorsulfónico; **citric a.**, ácido cítrico; **deoxyribonucleic a. (abbr. DNA)**, ácido desoxirribonucleico (abbr. ADN); **fatty a.**, ácido graso; **formic a.**, ácido fórmico; **humic a.**, ácido húmico; **hydrobromic a.**, ácido bromhídrico; **hydrochloric a.**, ácido clorhídrico; **hydrofluoric a.**, ácido fluorhídrico; **maleic a.**, ácido maléico; **nitric a.**, ácido nítrico; **oxalic a.**, ácido oxálico; **phosphoric a.**, ácido fosfórico; **ribonucleic a. (abbr. RNA)**, ácido

ribonucleico (abbr. ARN); **salicylic a.**, ácido salicílico; **sulphuric a.**, ácido sulfúrico; **uric a.**, ácido úrico.

acidification, n. acidificación (f).

acidimetry, n. acidimetría (f).

acidity, n. acidez (f).

acidize, v. acidificar.

acidogeneisis, n. acidogénesis (m).

acidophile, a. acidófilo/a.

acmite, n. acmita (f).

acoustic, a. acústico/a; **a. materials**, materiales (m.pl) acústicos.

acoustics, n. acústica (f).

acquisition, n. adquisición (f).

acre, n. acre (m) (equiv.to 0.4047 ha).

acreage, n. número (m) de acres, superficie (f) en acres.

acre-foot, n. acre-pie (m) (unit of water volume, 271,200 imperial gallons, 226,100 US gallons or 1233.5 cubic metres).

acre-inch, n. acre-pulgada (m) (unit of water volume, 22,600 imperial gallons, 18,850 US gallons or 102.8 cubic metres).

acropolis, n. acrópolis (m).

acrylic, a. acrílico/a; **a. resin**, resina (f) acrílica.

act, n. acto (m), acción (f); decreto (m), ley (f) (Jur); **a. of Congress (US)**, ley del Congreso; **a. of Parliment (GB)**, ley del parlamento, ley del parlamento británico; **Water A. (GB)**, Ley de aguas.

actinide, n. actínido (m).

actinium, n. actinio (m) (Ac).

actinolite, n. actinolita (f).

action, n. acción (f), actuación (f); **hydraulic a.**, acción hidráulica; **wind a.**, acción eólica.

activated, a. activado/a; **a. sludge**, lodos (m.pl) activados.

activation, n. activación (f).

active, a. activo/a; vigente (law); vivo/a (of interest); **a. volcano**, volcán (m) en actividad.

activist, n. activista (f).

activity, n. actividad (f); **commercial a.**, actividad comercial; **human a.**, actividad humana; **ionic a.**, actividad iónica; **manufacturing a.**, actividad fabril; **seismic a.**, actividad sísmica.

actuarial, a. actuarial.

actuary, n. actuario/a (m, f) de seguros.

actuator, n. actuador (m); **pneumatic a.**, actuador neumático.

acyclic, a. acíclico/a.

ad, n. anuncio (Fam) (abbr. of advertisement).

adaptability, n. adaptabilidad (f), flexibilidad (f).

adaptation, n. adaptación (f).

add, v. añadir; suma (Mat).

addition, n. adición (f).

additive, n. aditivo (m); **anti-incrustation a.**, aditivo antiincrustante.

address, n. dirección (f), señas (f.pl).

adductor, n. aductor (m).

adenosine, n. adenosina (f); **a. triphosphate** (abbr. ATP), trifosfato (m) de adenosina (abbr. ATP).

adherence, n. adherencia (f), adhesión (f).

adhesion, n. adhesión (f), adherencia (f).

adiabatic, a. adiabático/a; **a. lapse rate (Met)**, velocidad (f) de caída adiabática.

adit, n. galería (f), socavón (m); qanat (m) (Arab).

adjacent, a. adyacente.

adjustment, n. adaptación (f); ajuste (m) (action); **graphical a.**, ajuste gráfico; **least squares a.**, ajuste por mínimos cuadrados.

administration, n. administración (f); **a. of water resources**, administración de recursos hidráulicos.

admiralty, n. almirantazgo (m); **A. charts (GB)**, cartas (f.pl) náuticas de Ministerio de Marina.

admissible, a. admisible.

admixture, n. admixtura (f), comezcla (f) (Agr), ingrediente (m).

adobe, n. adobe (m).

adrenalin, n. adrenalina (f).

adsorb, v. adsorber.

adsorption, n. adsorción (f).

advance, n. avance (m), progreso (m).

advection, n. advección (f).

adverse, a. adverso/a; **a. effect**, efecto (m) adverso.

advertise, v. anunciar.

advertisement, n. anuncio (m).

advice, n. consejo (m); asesoramiento (m) (action); **a piece of a.**, un consejo; **technical a.**, asesoramiento técnico.

advisor, n. asesor (m, f); **biological a.**, biólogo asesor; **legal a.**, asesor jurídico.

adze, n. azuela (f) (axe).

AEA, n. JEN (f); **Atomic Energy Authority (GB)**, Junta (f) de Energía Nuclear (Sp).

AEC, n. JEN (f); **Atomic Energy Commission (US)**, Junta (f) de Energía Nuclear (Sp).

aeolian, a. eólico/a.

aeon, n. eón (m).

aerate, v. airear, gasificar.

aerated, a. aireado/a, aerado/a, gaseoso/a; **a. water**, agua (f) gaseosa.

aeration, n. aireación (f) (ventilation); **bubble a.**, aireación con burbujas.

aerator, n. ventilador (m); **cone a.**, ventilador cónico.

aerial, 1. n. antena (f) (Comm); **receiving a.**, antena receptora; **transmitting a.**, antena emisora; 2. a. aéreo/a; **a. photography**, fotografía (f) aérea.

aerobic, a. aerobio/a, aeróbico/a; **a. respiration**, respiración (f) aerobia.

aerodrome, n. aeródromo (m).

aerodynamic, a. aerodinámico/a.

aerodynamics, n. aerodinámica (f).

aeronautical, a. aeronáutico/a.

aerosol, n. aerosol (m).

aesthetic, a. estético/a.

aestivation, n. estivación (f).
affinity, n. afinidad (f).
affluent, a. afluente, copioso/a; **a. tributary**, tributario (m) afluente.
afforestation, n. repoblación (f) forestal, plantación (f) de bosques; forestación (f), aforestación (esp. LAm).
aflux, n. afluencia (f).
Africa, n. África (f).
African, a. africano/a.
after-effect, n. consecuencias (f.pl).
afteruse, n. uso (m) posterior.
agar, n. agar (m).
agar-agar, n. agar-agar (m).
agaric, n. agárico (m).
agate, n. ágata (f).
agave, n. agave (m), henequén (m), pita (f).
age, n. edad (f); **actual a.**, edad real; **Bronze A.**, Edad de Bronce; **Copper A.**, Edad de Cobre; **geological a.**, edad geólogica; **Ice A.**, Edad del Hielo; **Iron A.**, Edad de Hierro; **Iron-A. settlement**, castro (m), asentamiento (m) neolítico; **Stone A.**, Edad de Piedra, Paleolítica.
ageing, n. envejecimiento (m).
agency, n. agencia (f); **development a.**, agencia de desarrollo; **Environmental Protection A. (abbr. EPA) (US)**, agencia de protección del medio ambiente.
agenda, n. agenda (f), orden (m) del día.
agent, n. agente (m); **a. of contamination**, agente contaminante; **a. of erosion**, agente erosivo; **a. of mineralization**, agente mineralizante; **antiknocking a.**, inhibidor (m) antidetonante; **brightening a.**, abrillantador (m) (for laundry); **disinfecting a.**, desinfectante (m), agente de desinfección; **dispersive a.**, dispersante (m,f); **land a.**, corredor (m) de fincas rurales, administrador (m) territorial, valorador (m); **whitening a.**, blanqueador (m).
agglomerate, n. aglomerado (m); **volcanic a.**, aglomerado volcánico.
agglomeration, n. aglomeración (f); **urban a.**, aglomeración urbana.
aggrade, v. elevarse (Geol).
aggregate, n. agregado (m).
aggregation, n. agregación (f), agregado (m).
aggression, n. agresión (f).
aggressive, a. agresivo/a; **a. waters**, aguas (f.pl) agresivas.
aggressiveness, n. agresividad (f); **a. of water**, agresividad del agua.
agitate, v. agitar.
agitation, n. agitación (f).
agitator, n. agitador (m); vibrador (m) (vibrator).
Agnatha, n.pl. agnatos (m.pl), la superclase Agnatha (Zoo).
agoraphobia, n. agorafobia (f).
agrarian, a. agrario/a; **a. reform**, reforma (f) agraria.

agreement, n. acuerdo (m); contrato (m) (contract), convenio (m).
agribusiness, n. industria (f) agropecuaria, agronegocios (m.pl).
agricultural, a. agrícola; **a. college**, escuela (f) de agricultura; **a. engineer**, ingeniero (m) agrónomo; **a. foodstuffs**, agroalimentaría (f).
agriculture, n. agricultura (f); **a. under plastic**, plasticultura (f); **sustainable a.**, agricultura sostenible.
agrochemical, n. producto (m) agroquímico.
agroforestry, n. agrosilvicultura (f).
agronomist, n. agrónomo/a (m. y f.).
agronomy, n. agronomía (f).
aid, n. ayuda (f) (Gen).
AIDS, n. SIDA (m); **acquired immune deficiency syndrome**, síndrome (m) de inmunodeficiencia adquirida.
air, n. aire (m); **a. cleaner**, purificador (m) de aire; **a. conditioned**, aire climatizado, aire acondicionado; **a. conditioning**, aire acondicionado, climatización (f); **a. contamination**, contaminación (f) atmosférica; **a. current**, corriente (f) de aire; **a. duct**, conducto (m) de aire; **a. entrainment**, inclusión (f) de aire; **a. filtration**, filtración (f) de aire; **a. lock**, bolsa (f) de aire; **a. monitoring**, vigilancia (f) de la contaminación atmosférica; supervisión (m) del aire; **a. pollution**, contaminación (f) del aire, contaminación atmosférica; **a. pressure**, presión (f) atmosférica, presión del aire; **a. purification**, purificación (f) del aire; **a. quality control**, control (m) de la calidad del aire; **a. quality standards**, normas (f.pl) de la calidad del aire; **a. space**, cámara (f) de aire; **a. terminal**, terminal (f) aérea; **a. vessel**, recipiente (m) de aire; **compressed a.**, aire comprimido.
airbus, n. aerobús (m).
airfield, n. campo (m) de aviación.
airline, n. línea (f) aérea.
airlock, n. esclusa (f) de aire, bolsa (f) de aire.
airmail, n. aeropostal (m), correo (m) aéreo.
airplane, n. avión (f).
airport, n. aeropuerto (m); **a. terminal (air terminal)**, terminal (f) aérea.
airship, n. aeronave (m).
airstrip, n. pista (f) de aterrizaje.
airtight, a. hermético/a.
airtightness, n. hermeticidad (f).
alabaster, n. alabastro (m).
albedo, n. albedo (m).
Albian, n. Albiense (m).
albinism, n. albinismo (f).
albino, n. albino (m).
albite, n. albita (f).
albumen, n. albumen (m).
albumin, n. albúmina (f).
albuminoid, a. albuminoideo/a; **a. nitrogen**, nitrógeno (m) albuminoideo.
alburnum, n. alburno (m).
alchemy, n. alquimia (f).

alcohol, n. alcohol (m).
alcoholic, a. alcohólico/a.
alcoholism, n. alcoholismo (m).
alcoholometer, n. alcoholímetro (m).
alcove, n. alcoba (f).
aldehyde, n. aldehído (m).
alder, n. aliso (m).
aldrin, n. aldrín (m).
ale, n. cerveza (f) inglesa.
alert, n. alerta (f); **a. system**, sistema (m) de alerta.
alfalfa, n. alfalfa (f), mielga (f).
alga, n. alga (f).
algae, n.pl. algas (f.pl); **blue-green a.**, algas verde-azuladas; **brown a.**, algas pardas; **green a.**, algas verdes; **red a.**, algas rojas.
algal, a. algal; **a. bloom**, florecimiento (m) de algas, floración (f) de algas.
algebra, n. álgebra (f) (Mat); **Boolean a.**, álgebra booleana.
algebraic, a. algebraico/a, algébrico/a.
algicide, n. alguicida (m).
algology, n. algología (f), ficología (f).
algorithm, n. algoritmo (m).
alicyclic, a. alicíclico/a.
align, v. alinear.
alignment, n. alineamiento (m), alineación (f); **borehole a.**, alineación de la perforación; **to be out of a. (v)**, estar fuera de alineación.
alimentary, a. alimentario/a, alimenticio/a; **a. canal**, tubo (m) digestivo.
aliphatic, a. alifático/a.
aliquot, n. alícuota (f).
alive, a. vivo/a.
alkali, n. álcali (m), álcalis (m).
alkaline, a. alcalino/a; **a. earths**, alcalinotérreos (m.pl); **calc-a. series**, serie (f) calcoalcalina, serie alcalino cálcica.
alkaline-earth, n. alcalinotérreo (m).
alkalinity, n. alcalinidad (f), basicidad (f).
alkalinization, n. alcalinización (f).
alkaloid, n. alcaloide (m).
alkane, n. alcano (m).
alkene, n. alqueno (m).
alkibenzene, n. alquibenzol (m).
alkyl, n. alquíl (m).
alkyne, n. alquino (m).
allegation, n. alegación (f).
allele, n. alelo (m).
allelomorph, n. alelo (m).
allergic, a. alérgico/a; **a. reaction**, manifestación (f) alérgica.
allergy, n. alergia (f).
alleviate, v. aliviar, paliar; mitigar (to mitigate).
alleviation, n. alivio (m); mitigación (f) (mitigation).
alley, n. callejón (m).
alliance, n. alianza (f).
allied, a. conexo/a (connected with), aliado/a (joined in alliance).
allo-, pf. alo-.

allocation, n. asignación (f), reparto (m) (distribution); **a. of funds**, asignación de fondos.
allochthonous, a. alóctono/a.
allophane, n. halofana (f); alofana (f) (LAm).
allotrophy, n. alotropía (f).
alloy, n. aleación (f).
alluvial, a. aluvial; **a. fan**, abanico (m) aluvial.
alluviation, n. aluvionamiento (m).
alluvium, n. aluvión (m).
almandite, n. almandino (m).
almond, n. almendra (f) (nut); almendro (m) (tree).
alomorphism, n. alomorfismo (m).
alpha, n. alfa (f).
alter, v. alterar.
alteration, n. alteración (f).
altered, a. alterado/a.
alternate, 1. v. alternar; 2. a. alterno/a.
alternating, a. alterno/a.
alternative, n. alternativa (f); **viable a.**, alternativa viable.
alternator, n. alternador (m).
altimeter, n. altímetro (m).
altitude, n. altitud (f).
altocumulus, n. altocúmulo (m).
altostratus, n. altoestrato (m).
alum, n. alumbre (m).
alumina, n. alúmina (f), óxido (m) de aluminio.
aluminate, n. aluminato (m).
aluminium, n. aluminio (m) (Al).
aluminosis, n. aluminosis (f).
alveoli, n.pl. alveolos (m.pl).
alveolus, n. alveolo (m), alvéolo (m), alveolo pulmonar.
amalgam, n. amalgama (f).
amalgamation, n. amalgamación (f).
amber, n. ámbar (m).
ambient, a. ambiente; **a. temperature and pressure**, temperatura (f) y presión (f) ambiente.
ambiguity, n. ambigüedad (f).
ameliorate, v. mejorar.
amelioration, n. mejora (f), mejoría (f), mejoramiento (m).
amend, v. enmendar (e.g. law), corregir (to correct).
amendment, n. enmienda (f); corrección (f) (correction).
amenity, n. comodidad (f), esparcimiento (m) (for recreation); amabilidad (f) (of a person), amenidad (f) (pleasantness).
America, n. América (f) (Note: The Spanish word América on its own often refers solely to Latin America though it can also mean the United States); **South A.**, América del Sur, Suramérica.
American, a. americano/a, estadounidense, norteamericano/a (of the United States); **South A.**, suramericano/a.
Amerindian, n. amerindio/a (m, f).
Amerindian, a. amerindio/a.

amethyst, n. amatista (f).
amide, n. amida (f).
amine, n. amina (f).
amino acid, n. aminoácido (m).
aminophenol, n. aminofenol (m).
ammonia, n. amoníaco (m).
ammoniacal, a. amoniacal; **a. nitrogen,** nitrógeno (m) amoniacal.
ammonification, n. amonificación (f).
ammonite, n. amonita (f).
ammonium, 1. n. amonio (m); 2. a. amónico/a; **a. carbonate,** carbonato (m) amónico.
amnion, n. amnios (m).
amoeba, n. ameba (f), amiba (f).
amoebae, n.pl. amibas (f.pl) (see: amoeba).
amoeboid, a. ameboide, amiboide; **a. movement,** movimiento (m) ameboide.
amorphous, a. amorfo/a.
amortization, n. amortización (f).
amp, n. amperio (m).
ampere, n. amperio (m), amp (m) (abbr. amp).
Amphibia, n.pl. anfibios (m.pl), la clase Amphibia (Zoo).
amphibian, 1. n. anfibio (m); 2. a. anfibio/a.
amphibole, n. anfíbol (m).
amphibolite, n. anfibolita (f).
amphiphilic, a. amfifílico/a.
amphitheater, n. anfiteatro (m).
amphitheatre, n. anfiteatro (m).
ampholyte, n. anfólito (m).
amphora, n. ánfora (f).
amplification, n. amplificación (f), ampliación (f) (enlargement).
amplitude, n. amplitud (f).
AMU, n. UMA (f); **atomic mass unit,** unidad (f) de masa atómica.
amyl, n. amilo (m).
amylene, n. amileno (m).
anabolic, a. anabólico/a.
anabolism, n. anabolismo (m).
anaemia, n. anemia (f).
anaerobe, n. anaerobio (m).
anaerobic, a. anaerobio/a, anaeróbico/a; **a. zone,** zona (f) anaerobia.
anal, a. anal.
analog, n. análoga (f).
analogue, 1. n. análogo/a (m, f); **electrical a.,** análogo eléctrico; 2. a. analógico/a; **a. computer,** ordenador (m) analógico.
analogy, n. analogía (f).
analyse, v. analizar.
analysis, n. análisis (m); **a. of variance (abbr. ANOVA),** análisis de varianza; **bacterial a., bacteriological a.,** análisis bacteriológico; **canonical a.,** análisis canónico; **chemical a.,** análisis químico; **cluster a.,** análisis de conglomerados/grupos/agrupamientos; **cost-benefit a.,** análisis de costos-beneficios; **critical path a.,** análisis de paso crítico; **data a.,** análisis de datos; **discriminant a.,** análisis discriminante; **factorial a.,** análisis factorial; **grain-size a.,** granulometría (f),

análisis granulométrico; **harmonic a.,** análisis armónico; **hydrological a.,** análisis hidrológico; **laboratory a.,** análisis de laboratorio; **market a.,** análisis de mercados; **mechanical a.,** análisis mecánico; **microscopic a.,** análisis microscópico; **parasitological a.,** análisis parasitológico; **physicochemical a.,** análisis fisicoquímico; **principal components a. (abbr. PCA),** análisis de componentes principales; **risk a.,** análisis de riesgos; **sample a.,** análisis de muestras; **spectral a.,** análisis espectral; **statistical a.,** análisis estadístico; **volumetric a.,** análisis volumétrico.
analyst, n. analista (m, f).
analytical, a. analítico/a; **a. technique,** método (m) analítico.
analyzer, n. analizador (machine).
anatase, n. anatasa (f), octaedrita (f).
anatomic, a. anatómico/a.
anatomy, n. anatomía (f).
anchor, n. ancla (f) (Mar); ancora (f) (Con), anclaje (m) (anchorage); **a. block,** bloque (m) de anclaje; **a. bolt,** perno (m) de anclaje; **ground a. (Eng),** anclaje del terreno; **weighing a. (acción),** zarpa (f).
anchorage, n. fondeo (m).
anchovy, n. boquerón (m) (fresh fish), anchoa (f).
ancillary, a. auxiliar, subordinado/a, anexo/a, secundario/a.
andalucite, n. andalucita (f).
Andean, a. andino/a; **A. high plateau,** altiplano (m) (LAm), altiplanicie (f).
andesite, n. andesita (f).
andro-, pf. andro-.
androecium, n. androceo (m).
androgen, n. andrógeno (m).
anemia, n. anemia (f).
anemometer, n. anemómetro (m).
aneroid, a. aneroide; **a. barometer,** barómetro (m) aneroide.
angiosperm, n. angiosperma (f).
Angiospermae, n.pl. angiospermas (f.pl), la división Angiospermae (Bot).
angle, n. ángulo (m); **a. bracket,** ménsula (f) en escuadra; **a. iron,** angular (m); **a. of dip,** ángulo de inclinación; **a. of repose,** ángulo de reposo; **a. of slip,** ángulo de desplazamiento.
angler, n. pescador (m) de caña; **a.-fish,** rape (m), pejesapo (m).
angling, n. pesca (f) con caña.
anglosaxon, n. anglosajón (m).
anglosaxon, a. anglosajón, anglosajona.
angstrom, n. angstrom (m) (abbr. Å, unit of measurement of atomic size); **a. unit (Phys),** angstromio (m).
angular, a. angular.
angularity, n. angulosidad (f).
anhedral, a. anhedral.
anhydride, n. anhídrido (m).
anhydrite, n. anhidrita (f).

anhydrous, a. anhidro/a.

anilin, n. anilina (f); **a. blue**, azul (m) de anilina.

aniline, n. anilina (f).

animal, n. animal (m); **domestic a.**, animal doméstico; **lower animals**, animales inferiores; **pack a.**, animal de carga; **wild a.**, animal salvaje; **a. kingdom**, reino (m) animal.

anion, n. anión (m).

Anisian, n. Anisiense (m).

anisotropic, a. anisotrópico/a, anisotrópo/a.

anisotropy, n. anisotropía (f).

ankle, n. tobillo (m).

annealing, n. recocido (m).

annelid, n. anélido (m).

Annelida, n.pl. anélidos (m.pl), el filum Annelida (Zoo).

announcment, n. anuncio (m) (Gen), declaración (f) (declaration), anuncios (m.pl) (e.g. in newspapers).

annual, n. planta (f) anual, especie (f) anual (Bot); anuario (m) (document).

annual, a. anual.

annuity, n. anualidad (f).

annulata, n.pl. ánulos (m.pl).

annulated, a. anulador.

annulus, n. ánulo (m) (Bot, Zoo); anillo (m) (Mat).

anode, n. ánodo (m).

anodize, v. anodizar.

anomalous, a. anómalo/a.

anomaly, n. anomalía (f); **thermal a.**, anomalía térmica.

anorthite, n. anortita (f).

anorthoclase, n. anortosa (f).

ANOVA, n. ANOVA (m); **analysis of variance**, análisis (m) de varianza.

anoxic, a. anóxico/a.

ANSI, n. **American National Standards Institute**, Instituto (m) Nacional Americano de Normalización.

ant, n. hormiga (f).

antagonism, n. antagonismo (m).

Antarctic, n. Antártida (f).

Antarctic, a. antártico/a.

antechamber, n. antecámara (f).

antenna, n. antena (f); **transmitting a.**, antena emisora.

antennae, n.pl. antenas (f.pl.) (see: antenna).

anther, n. antera (f).

antheridia, n.pl. anteridios (m.pl).

antheridium, n. anteridio (m).

anthozoa, n.pl. antozoos (m.pl).

anthracite, n. antracita (f).

anthrax, n. ántrax (m).

anthro-, pf. antro-.

anthropocentric, a. antropocéntrico/a.

anthropogenic, a. antropógeno/a, antrópico/a.

anthropoid, n. antropoide (m), mono (m) antropomorfo.

anthropoid, a. antropoide, antropoideo/a.

Anthropoidea, n. antropoides (m.pl), el suborden Anthropoidea (Zoo).

anthropology, n. antropología (f).

anti-, pf. anti-.

antibiotic, n. antibiótico (m).

antibiotic, a. antibiótico/a.

antibody, n. anticuerpo (m).

anticline, n. anticlinal (m); **breached a.**, anticlinal aportillado; **closed a.**, anticlinal cerrado.

anticlinorium, n. anticlinorio (m).

anticorrosive, a. anticorrosivo/a.

anticorrosives, n.pl. anticorrosivos (m.pl).

anticyclone, n. anticiclón (m).

anticyclonic, a. anticiclónico/a.

antidote, n. antídoto (m).

anti-foaming, a. antiespumante; **a.-f. agent**, agente (m) antiespumante.

antiform, n. antiforma (f).

antigen, 1. n. antígeno (m); 2. a. antigénico/a.

anti-knocking, a. antidetonante.

antimonate, n. antimoniato (m).

antimonite, n. antimonito (m).

antimony, n. antimonio (m) (Sb).

antioxidant, n. antioxidante (m).

anti-pollution, n. antipolución (f), anticontaminación (f); **a.-p. measures**, medidas (f.pl) de antipolución.

antiquity, n. antigüedad (f).

antiseptic, n. antiséptico (m).

antiseptic, a. antiséptico/a.

antisiphoning, n. antisifonaje (m); **a. pipe**, tubería (f) de antisifonaje.

antler, n. cuerna (f); **antlers (pl)**, cornamenta (f).

anular, a. anular.

Anura, n.pl. anuros (m.pl), el orden Anura (Zoo).

anuran, n. anuro (m).

anus, n. ano (m).

aorta, n. aorta (f).

apartment, n. apartamento (m), piso (m); departamento (m) (LAm), condominio (m) (LAm).

apatite, n. apatito (m).

ape, n. mono (m).

apex, n. ápice (m), ápex (m), vértice (m) (of a curve, polygon); **a. of anticline**, charnela (f) del anticlinal, ápice del anticlinal.

aphid, n. áfido (m).

aphrolithic, a. afrolítico/a; **a. lava**, lava (f) afrolítica.

apical, a. apical.

apices, n.pl. ápices (m.pl), apexes (m.pl) (Biol); vértices (m.pl) (of a triangle).

apiculture, n. apicultura (f).

apogee, n. apogeo (m).

apparatus, n. aparato (m).

apparent, a. aparente, evidente.

appeal, n. apelación (f); petición (f) (petition).

appearance, n. apariencia (f) (look); aparición (f) (act); **physical a.**, apariencia física.

appendices, n.pl. apéndices (m.pl).
appendix, n. apéndice (m).
apple, n. manzana (f) (fruit), manzano (m) (tree); a. orchard, manzanar (m).
appliance, n. aparato (m), dispositivo (m), artefacto (m); electrical a., electrodoméstico (m).
application, n. aplicación (f), uso (m) (use); solicitud (f) (request); a. form, formulario (m), solicitud.
appraisal, n. tasación (f), evaluación (f).
appraiser, n. estimador (m, f), tasador (m, f) (US) (Fin).
approach, n. acercamiento (m) (of something), acceso (m) (access), enfoque (m) (to a problem).
approval, n. consentimiento (m) (consent).
approve, v. aprobar, autorizar.
approximate, 1. v. aproximar; 2. a. aproximado/a.
approximation, n. aproximación (f); Dupuit-Forscheimer a., aproximación de Dupuit-Forscheimer; Jacob's logarithmic a., aproximación logarítmica de Jacob.
apricot, n. albaricoque (m) (Sp), damasco (m) (LAm), chabacano (m) (Mex).
apron, n. antepecho (m) (of a building); delantal (m) (garment).
apse, n. ábside (m).
Aptian, n. Aptiense (m).
aquaculture, n. acuicultura (f), acuacultura (f), cultura (f) acuática, piscicultura (fish).
aquamarine, n. aguamarina (f).
aquarium, n. acuario (m).
aquatic, a. acuático/a.
aqueduct, n. acueducto (m).
aqueous, a. acuoso/a, ácueo/a; aguoso/a (watery).
aquiclude, n. acuicludo (m).
aquiculuture, n. acuacultura (f).
aquifer, n. acuífero (m); alluvial a., acuífero aluvial; anisotropic a., acuífero anisótropo; a. decontamination, descontaminación (f) de acuíferos; a. properties, propiedades (f.pl) del acuífero; a. protection, protección (f) del acuífero; a. recharge, recarga (f) del acuífero; a. vulnerability, vulnerabilidad (f) del acuífero; artesian a., acuífero artesiano; calcareous a., acuífero calcáreo; coastal a., acuífero litoral; confined a., acuífero confinado; consolidated a., acuífero consolidado; deep a., acuífero profundo; dual-porosity a., acuífero de doble porosidad; exploitable a., acuífero explotable; fissured a., acuífero fisurado; fractured a., acuífero fracturado; heterogeneous a., acuífero heterogéneo; isotropic a., acuífero isótropo; karstic a., acuífero kárstico; leaky a., acuifugo (m); multi-layered a., acuífero multicapa; multiple a., acuífero múltiple; over-exploited a., acuífero sobreexplotado; overflowing a., acuífero surgente; phreatic a.,

acuífero freático; porous a., acuífero poroso; recharged a., acuífero recargado; semi-confined a., acuífero semiconfinado; stratified a., acuífero estratificado; unconfined a., acuífero no confinado; volcanic a., acuífero volcánico.
aquiferous, a. acuífero/a.
aquifuge, n. acuifugo (m).
Aquitanian, n. Aquitaniense (m).
aquitard, n. acuitardo (m).
Arab, a. árabe.
Arabic, a. árabe, arábigo/a.
arable, a. cultivable, labrantío/a, arable; a. land, tierra (f) cultivable, tierra de cultivo.
arachnid, n. arácnido (m).
arachnid, a. arácnido/a.
Arachnida, n.pl. arácnidos (m.pl), la clase Arachnida (Zoo).
aragonite, n. aragonita (f).
arbitrary, a. arbitrario/a.
arbor, n. pérgola (f), cenador (m); vine a., parral (m).
arboreal, a. arbóreo/a.
arboretum, n. arboreto (m).
arboriculture, n. arboricultura (f).
arbour, n. pérgola (f), cenador (m); vine a., parral (m).
arc, n. arco (m); electric a., arco eléctrico; island a., arco isla; welding a., arco de soldadura.
arcade, n. arcada (f), pasaje (m) con arcos.
arch, v. abovedar.
arch, n. arco (m); a. beam, viga (f) en arco; a. brace, riostra (f) en arco; a. buttress, arbotante (m); a. dam, presa (f) de arco; blind a., arco ciego; concentric a., arco concéntrico; inverted a., arco invertido; relieving a., arco de aligeramiento; semicircular a., arco de medio punto.
archaeological, a. arqueológico/a.
archaeologist, n. arqueólogo/a (m, f).
archaeology, n. arqueología (f).
Archean, n. Arcaico (m).
arched, a. abovedado/a.
archegonium, n. arquegonio (m).
archeology, n. arqueología (f).
Archimedes, n. Arquímedes (m); A. screw (for pumping), tornillo (m) de Arquímedes.
arching, n. abovedamiento (m), encorvamiento (m).
archipelago, n. archipiélago (m).
architect, n. arquitecto (m); landscape architect, arquitecto paisajista.
architectural, a. arquitectural, arquitectónico/a.
architecture, n. arquitectura (f).
archive, v. archivar.
archive, n. archivo (m).
archway, n. arcada (f).
arctic, a. ártico/a; A. Circle, Círculo (m) Polar; A. Ocean, Océano (m) ártico. arcuate, a. arqueado/a.
area, n. área (f); a. of influence, área de

influencia; **catchment a.**, área de captación; **pumping a.**, área de bombeo; **recharge a.**, área de recarga; **screen open a.**, área abierta de la rejilla; **urban a.**, área urbana; **water protection a.**, área protección de aguas; **wildlife a.**, espacio (m) natural.
arenaceous, a. arenáceo/a, arenoso/a.
Arenigian, n. Arenigiense (m).
arenite, n. arenita (f).
areometer, n. areómetro (m), hidrómetro (m).
arête, n. arete (m).
Argentina, n. Argentina (f).
Argentine, 1. n. argentino/a (m, f); 2. a. argentino/a.
argentite, n. argentita (f).
argillaceous, a. arcilloso/a; **a. sediment**, sedimento (m) arcilloso.
argillite, n. argilita (f).
argon, n. argón (m) (Ar).
arid, a. árido/a.
aridity, n. aridez (f).
arithmetic, n. aritmética (f).
arkose, n. arcosa (f).
arkosic, a. arcósico/a.
arm, n. brazo (m); **a. of sea**, brazo de mar.
armature, n. armadura (f).
aromatic, a. aromático/a; **a. hydrocarbons**, hidrocarburos (m.pl) aromáticos.
arrangement, n. arreglo (m), orden (m).
array, n. orden (m), matriz (f), serie (f) homogénea (Mat); dispositivo (m); **Schlumberger a.**, dispositivo Schlumberger; **Wenner a.**, dispositivo Wenner.
arrow, n. flecha (f).
arrowhead, n. punta (f) de flecha; **flint a. (Arch)**, punta de flecha de sílex.
arroyo, n. arroyo (m).
arsenate, n. arsenato (f).
arseneous, a. arsenioso/a.
arsenic, n. arsénico (m) (As).
arsenide, n. arseniuro (m).
arseniosis, n. arseniosis (m).
arsenite, n. arsenito (m).
arsenolite, n. arsenolita (f).
arsenopyrite, n. arsenopirita (f).
arsenous, a. arsenioso/a.
arsine, n. arsina (f).
artefact, n. artefacto (m).
artery, n. arteria (f); **a. traffic route (principal)**, arteria vial, arteria de comunicación.
artesian, a. artesiano/a; **a. aquifer**, acuífero (m) artesiano; **a. basin**, cuenca (f) artesiana; **a. flow**, flujo (m) artesiano; **a. head**, altura (f) artesiana; **a. pressure**, presión (f) artesiana; **a. spring**, manantial (m) artesiano; **a. well**, pozo (m) artesiano.
arthropod, n. artrópodo (m).
arthropod, a. artrópodo/a.
Arthropoda, n.pl. artrópodos (m.pl), el filum Arthropoda (Zoo).
artichoke, n. alcachofa (f).

article, n. artículo (m) (item, for newspaper), cláusula (f) (clause).
articulate, a. articulado/a.
artifact, n. artefacto (m).
asbestos, n. asbesto (m), asbestos (m); **a. cement**, asbestos-cemento; **a. roofing**, techo (m) de asbesto.
asbestosis, n. asbestosis (m).
ASCII, n. **American Standard Code for Information Interchange (Comp)**, código (m) norteamericano normalizado para el intercambio de información.
aseptic, a. aséptico/a.
asexual, a. asexual; **a. reproduction**, reproducción (f) asexual.
ash, n. ceniza (f) (dust); fresno (m) (tree); **a. shower**, lluvia (f) de cenizas; **fly a.**, cenizas; **volcanic a.**, ceniza volcánica.
Ashgillian, n. Ashgiliense (m).
ashlar, n. sillar (m); **a. masonry**, sillería (f) de fábrica.
Asiatic, a. asiático/a.
asparagus, n. espárrago (m).
aspect, n. aspecto (m); **cultural aspects**, aspectos culturales; **economic a.**, aspecto económico; **health aspects**, aspectos de la salud; **social aspects**, aspectos sociales; **socioeconomic aspects**, aspectos socioeconómicos.
aspen, n. álamo (m) temblón.
asphalt, n. asfalto (m); **a. cement**, cemento (m) asfáltico; **a. roofing**, cubierta (f) de asfalto.
asphyxia, n. asfixia (f).
asphyxiant, a. asfixiante.
asphyxiate, v. asfixiar.
asphyxiated, a. asfixiado/a.
asphyxiating, a. asfixiante, asfixiador/a; **a. fumes**, humos (m.pl) asfixiantes, gases (m.pl) asfixiantes.
aspiration, n. aspiración (f) (intake, breathing, desire); ambición (f) (desire); **pump a.**, aspiración de la bomba.
aspirin, n. aspirina (f).
assay, v. ensayar (Chem).
assay, n. ensayo (m) (test, trial), ensaye (m) (of a metal), aquilatamiento (m) (of gold).
assemblage, n. asociación (f); **faunal a.**, asociación de fauna.
assembly, n. reunión (f) (meeting); parlamento (m) (Reg); montaje (m), ensamblaje (m) (Eng); **a. language (Comp)**, lenguaje (m) de ensamblador; **a. line production**, producción (f) en cadena; **a. plant**, planta (f) de montaje, maquiladora (f) (LAm).
assessment, n. valoración (f), evaluación (f); **environmental impact a. (abbr. EIA)**, evaluación de impacto ambiental (abbr. EIA); **risk a.**, evaluación de riesgo.
assets, n.pl. bienes (m.pl) (Fin); **fixed a.**, inmovilizados (m); **real a.**, bienes raíces.
assignment, n. asignación (f), escritura (f) de cesión (document).
assimilate, v. asimilar.

assimilation, n. asimilación (f).
associate, 1. v. asociar; 2. n. socio (m) (Com, partner), asociado (m); 3. a. asociado/a.
association, n. asociación (f), sociedad (f); **plant a.,** asociación vegetal; **professional a.,** asociación profesional.
assorted, a. variado/a.
astatine, n. astato (m) (At).
asteroid, n. asteroide (m).
Asteroidea, n.pl. asteroideos (m.pl).
asthenosphere, n. astenosfera (f).
asthma, n. asma (f).
asthmatic, a. asmático/a.
astroturf, n. césped (m) artificial.
asymmetry, n. asimetría (f).
asymptote, n. asíntota (f).
asymptotic, a. asintótico/a.
Atlantic, a. atlántico/a; **Central A.,** centroatlántica; **the A. Ocean,** el Océano (m) Atlántico.
atlas, n. atlas (m).
atmosphere, n. atmósfera (f).
atmospheric, a. atmosférico/a.
atoll, n. atolón (m); **coral a.,** atolón coralino.
atom, n. átomo (m); **a. bomb,** bomba (f) atómica.
atomic, a. atómico/a; **a. bomb,** bomba (f) atómica; **a. mass unit (abbr. amu),** unidad (f) de masa atómica (abbr. uma); **a. mass,** masa (f) atómica; **a. number,** número (m) atómico; **a. particle,** partícula (f) atómica.
ATP, n. ATP (m); **adenosine triphosphate,** trifosfato (m) de adenosina.
atrazine, n. atrazina (f).
atrium, n. atrio (m).
atrophy, n. atrofia (f).
attack, n. ataque (m); **chemical a.,** ataque químico; **silicate a.,** ataque de los silicatos.
attainment, n. realización (f), obtención (f).
attenuate, v. atenuar; **a. and disperse policy (for landfill),** política (f) de atenuar y dispersar.
attenuation, n. atenuación (f); **leachate a.,** atenuación de lixiviado; **natural a.,** atenuación natural.
attic, n. desván (m) (loft), ático (m) (top floor).
attorney, n. abogado/a (m, f); apoderado/a (m, f) (representative); **power of a.,** poderes (m.pl); procuración (f).
attrition, n. atrición (f), roce (m).
atypical, a. atípico/a.
aubergine, n. berenjena (f).
auction, n. subasta (f).
audible, a. audible; **a. range,** zona (f) audible.
audiovisual, n. y. a. audiovisual.
audit, n. intervención (f), revisión (f), auditoría (f) (Com, of accounts); **a. trail,** pista (f) de auditoría, registro (m) de auditoría, verificación (f) a posteriori; **environmental a.,** auditoría ambiental, auditoría medioambiental.
auditor, n. auditor (m).
auger, v. taladrar.

auger, n. barrena (f) helicoidal, taladro (m) de suelo; **a. bit,** boca (f) de barrena.
augite, n. augita (f).
augment, v. aumentar.
augmentation, n. aumento (m); **river a. scheme,** proyecto (m) para el aumento del río.
aureole, n. aureola (f); **contact a.,** aureola de contacto.
auriferous, a. aurífero/a.
austral, n. austral (m) (Arg, unit of currency 1985-91).
Australia, n. Australia (f).
Australian, a. australiano/a.
autecology, n. autecología (f).
authoritative, a. autoritario/a (person); autorizado/a (influencial); **authoritatively (ad),** con autoridad, autorizadamente.
authority, n. autoridad (f); **port a.,** autoridad portuaria; **regulatory a.,** autoridad reguladora; **to have a. to (v),** tener autoridad para; **with a. (ad),** con autoridad, autorizadamente.
authorization, n. autorización (f), permiso (m).
authorize, v. autorizar, aprobar.
authorized, a. autorizado/a.
auto-, pf. auto-.
autoabsorption, n. autoabsorción (f).
autochthonous, a. autóctono/a.
autoclast, n. autoclasto (m).
autoclastic, a. autoclástico/a.
autoclave, n. autoclave (m) (Chem).
autocorrelation, n. autocorrelación (f).
autogamy, n. autogamia (f).
automate, v. automatizar.
automatic, a. automático/a.
automation, n. automatización (f).
automization, n. automización (f).
automobile, n. automóvil (m), turismo (m) (private car), carro (m) (LAm), auto (m) (LAm).
autonomous, a. autónomo/a.
autotroph, n. autótrofo (m).
autotrophic, a. autótrofo/a (of the animal); autotrófico (of the property).
autumn, n. otoño (m).
auxiliary, a. auxiliar, subordinado/a.
availability, n. disponibilidad (f).
available, a. disponible; **a. oxygen,** oxígeno (m) disponible.
avalanche, n. alud (m), avalancha (f); **snow a.,** alud de nieve.
average, 1. n. promedio (m), media (f) (Mat); **annual a.,** media anual; **moving a.,** media móvil; 2. a. medio/a; **a. temperature,** temperatura (f) media.
Aves, n.pl. aves (f.pl), la clase Aves (Zoo).
avifauna, n. avifauna (f).
avogadro, n. avogadro (m) (Chem).
avulsion, n. avulsión (f).
awareness, n. conciencia (f) (consciousness); conocimiento (m) (knowledge); **environmental a.,** educación (f) ambiental.

awash, a. de inundado/a, a flor de agua.

awl, n. punzón (m) para madera.

awning, n. toldo (m).

axe, n. hacha (f); **stone a.,** hacha de piedra.

axil, n. axila (f).

axiom, n. axioma (m).

axiomatic, a. axiomático/a.

axis, n. eje (m); **a. of symmetry,** eje de simetría; **a. of tilt,** eje de inclinación; **anticlinal a.,** eje de anticlinal; **Earth's a.,** eje de la Tierra; **fold a.,** eje del pliegue; **vertical a.,** eje vertical; **x-a.,** eje x; **y-a.,** eje y.

axle, n. eje (m).

azeotrope, n. azeótropo (m).

azide, n. azida (f).

azimuth, n. acimut (m), azimut (m).

azurite, n. azurita (f).

B

bacilli, n.pl. bacilos (m.pl).
bacillus, n. bacilo (m).
back, 1. n. lomo (m) (of an animal); espalda (f) (of a person); dorso (m), revés (m), envés (m) (reverse side, e.g. of paper); parte (f) de atrás (e.g. of car); **b. flow**, reflujo (m); **b. pressure**, presión (f) interior; 2. a. trasero/a, posterior; **b. filling**, terraplenado (m).
back-actor, n. excavadora (f) de retrodescarga.
backbone, n. espina (f) dorsal, columna (f) vertebral.
backdigger, n. retrocavador (m), pala (f) retrocavadora.
backfill, n. relleno (m) (Eng), terraplén (m) (earthwork); **clay b.**, relleno arcilloso.
background, n. ambiente (m) medio (environment); antecedentes (m.pl) (history); fondo (m) (in the distance, behind); **b. concentration**, concentración (f) de fondo; **b. values**, valores (m.pl) normales, fondo (m).
backlog, n. trabajo (m) acumulado, trabajo atrasado (of work).
backplate, n. placa (f) de apoyo (Con).
backtrack, v. retroceder.
back-up, a. de reserva; **b. file (Comp)**, fichero (m) de reserva.
backwash, n. retroceso (m) (Eng); **b. curve**, curva (f) de retroceso.
backwater, n. agua (f) estancada (behind an obstruction), remanso (m) hidráulico (still water); lugar (m) atrasado (Fig); bayou (m) (US); **b. flap**, válvula (f) de retorno.
bacteria, n. bacterias (f.pl); **aerobic b.**, bacterias aerobias; **anaerobic b.**, bacterias anaerobias; **chemosynthetic b.**, bacterias quimiosintéticas; **denitrifying b.**, bacterias desnitrificantes; **enteric b.**, bacterias entéricas; **iron b.**, ferrobacterias; **nitrate b.**, bacterias de nitrato; **nitrifying b.**, bacterias nitrificantes; **nitrite b.**, bacterias de nitrito; **nitrogen-fixing b.**, bacterias fijadoras de nitrógeno; **photosynthetic b.**, bacterias fotosintéticas; **purple b.**, bacterias purpúreas; **soil b.**, bacterias del suelo; **sulphur b.**, bacterias de azufre.
bacterial, a. bacteriano/a.
bactericidal, a. bactericida.
bactericide, n. bactericida (m).
bacteriological, a. bacteriológico/a.
bacteriologist, n. bacteriólogo/a (m, f).
bacteriology, n. bacteriología (f).
bacteriophage, n. bacteriófago (m).
bacteriostatic, a. bacteriostático/a.
bacterium, n. bacteria (f).
badlands, n. malas tierras (m); huaiqueria (f) (LAm); tierras estériles.

baffle, n. deflector (m); **b. plate**, deflector, placa (f) de desviación.
bail, 1. v. achicar (from boat); 2. n. libertad (f) bajo fianza.
bait, n. cebo (m).
Bajocian, n. Bajociense (m).
bakelite, n. baquelita (f).
balance, v. equilibrar, pesar (to weigh).
balance, n. balanza (f) (for weighing); balance (m) (financial statement); **analytical b.**, balanza analítica; **b. of accounts**, balanza de pagos; **b. of trade**, balanza comercial; **b. sheet**, balance; **chemical b.**, balance químico; **ecological b.**, balance ecológico; **energy b.**, balance energético, balance de energía; **hydrodynamic b.**, balance hidrodinámico; **ionic b.**, balance iónico; **moisture b.**, balance de humedad; **soil water b.**, balance de agua en el suelo; **spring b.**, balanza de resorte; **water b.**, balance hídrico.
balboa, n. balboa (m) (Pan, unit of currency).
balcony, n. balcón (m).
bale, n. embalar (e.g. waste), achicar (water).
balk, n. madera (f) escuadrada (of timber), viga (f) (beam).
ballast, v. lastrar.
ballast, n. balasto (m), lastre (m).
ballasting, n. lastrado (m) (weigh down).
ballustrade, n. balaustrada (f).
bamboo, n. bambú (m), cañabrava (f) (LAm).
banana, n. plátano (m), banana (f), banano (m) (LAm), cambur (m) (Ven); **b. dealer**, platanero (m); **b. plantation**, platanera (f), platanal (m), platanar (m); **b. tree**, platanero (m).
band, n. cinta (f) (ribbon); tira (f) (strip); anillo (m), sortija (f) (LAm) (ring); grupo (m) (people); banda (f) (radio); capa (f) (layer); **elastic b., rubber b.**, cinta (f) de goma, cinta elástica.
banded, a. bandeado/a.
bank, 1. v. amontonar (for road); 2. n. banco (m) (Geog, Com); orilla (f), ribera (f) (of a river, lake); dique (m) (artificial dyke); loma (f) (raised ground); pendiente (f) (slope); bajo (m), bajío (m) (underwater elevation); **bottle b.**, contenedor (m) de vidrio, iglúe (m) (Fam); **lending b., loans b. (US)**, banco de préstamos; **mortgage b.**, banco hipotecario; **sandbank**, banco de arena; **World B. (Com)**, Banco Mundial.
banker, n. banquero (m).
bankfull, n. cauce (m) lleno; **b. discharge**, caudal (m) de cauce lleno.
bankrupcy, n. bancarrota (f) (Fin).
bankrupt, n. quebrado (m).

bankside, n. cerca (f) de la ribera.

bar, n. banco (m), barra (f) (in a river); resistencia (f) (of electric fire); unit of measurement of pressure (Phys); **b. chart**, cuadro (m) de barras; **bus b. (Elec)**, barra colectora; **cuspate b.**, barra en cúspide; **looped b.**, barra en lazo; **offshore b.**, cordón (m) litoral; **rivermouth b.**, barra de desembocadura de río; **submarine b.**, banco submarino.

barb, n. barba (f) (Bot); barba (f) (of feather); barbilla (f) (of fish).

barbacan, n. barbacana (f) (for defence).

barbed, a. con lengüeta; **b. wire**, alambre (m) de púas, alambre de espino.

barbiturate, n. barbiturato (m), barbitúrico.

barchan, n. barchán (m).

barge, n. barcaza (f), gabarra (f).

baric, a. bárico/a.

barite, n. barita (f).

barium, n. bario (m) (Ba); **b. chloride**, cloruro (m) bárico.

bark, n. corteza (f) (of a tree); ladrido (m) (of a dog).

barley, n. cebada (f); **b. field**, alcacel (m), alcacer (m).

barn, n. granero (m) (for grain), establo (m) (for cows), cuadra (f) (for horses).

barograph, n. barógrafo (m).

barometer, n. barómetro (m); **aneroid b.**, barómetro aneroide; **mercury b.**, barómetro de mercurio.

barrage, n. presa (f) de contención (for water).

barrel, n. tonel (m), barril (m) (of wine, of crude oil) (volume equal to 158.98 litres); bóveda (f) (vault); **b. arch**, bóveda de cañón; **b. vault**, bóveda de medio punto.

Barremian, n. Barremiense (m).

barren, a. árido/a (soil); estéril (sterile).

barrier, n. barrera (f), valla (f), obstáculo (m); **gas b.**, barrera antigás; **impermeable b.**, barrera impermeable; **injection b.**, barrera de inyección; **marine intrusion b.**, barrera de intrusión marina; **physical b.**, barrera física; **pumping b.**, barrera de bombeo.

barrister, n. abogado (m) que puede defender causas en los tribunales superiores (GB).

barrow, n. carretilla (f) (wheelbarrow), colina (f) (hill), túmulo (m) (grave mound); **long b.**, túmulo largo.

Bartonian, n. Bartoniense (m).

barytes, n. barita.

basal, a. basal.

basalt, n. basalto (m).

base, n. base (f); **air b.**, base aérea; **b. of aquifer**, base del acuífero; **b. plate**, placa (f) de asiento; **chemical b.**, base química.

baseflow, n. caudal (m) base; **b. index**, índice (m) de caudal base.

basement, n. basamento (m) (Geol, Con); substrato (m) (Geol); sótano (m) (cellar); **b. rocks**, rocas (f) del basamento.

basic, a. básico/a; **b. wage**, sueldo (m) base, salario (m) base.

basicity, n. basicidad (f).

basidiomycete, n. basidiomiceto (m).

basin, n. cuenca (f) (Geog); balsa (f) (pool); fosa (f) (trough); bacía (f) (vessel); **artesian b.**, cuenca artesiana; **b. catchment**, cuenca colectora; **down-warped b.**, cuenca de alabeada; **evaporation b.**, balsa de evaporación; **infiltration b.**, balsa de infiltración; **ocean b.**, cuenca oceánica; **purification b.**, balsa de depuración; **recharge b.**, balsa de recarga; **river b.**, cuenca del río; **sedimentary b.**, cuenca sedimentaria; **settling b.**, balsa de decantación; **stilling b.**, cuenca amortiguadora; **the Ebro b.**, la cuenca del Ebro; **wash b. (in bathroom)**, lavamano (m), lavabo (m).

basiphilic, a. basifilo/a, basófilo/a.

basis, n. base (f); **b. of calculation**, b. de cálculo; **b. of comparison**, base de comparación.

bass, n. lubina (f), baila (f) (fish).

bat, n. murciélago (m) (mammal).

bathing, n. baño (m), natación (f); **b. water (EurU)**, agua (f) de baño, agua de natación.

Bathonian, n. Bathoniense (m).

bathyal, a. batial.

bathylimnion, n. batilimnio (m).

bathymetric, a. batimétrico/a.

bathymetry, n. batimetría (f).

BATNEEC, n. **Best Available Technology Not Entailing Excessive Cost**, mejor tecnología (f) disponible a precio razonable.

batten, n. tabla (f) para piso de madera, listón (m) (piece of wood); **b. plate**, chapa (f) de refuerzo.

battening, n. colocación (f) de listones.

batter, n. desplome (m) (Con).

battery, n. batería (f) (group), pila (f) (electric); **dry b.**, batería seca; **storage b.**, batería de acumuladores.

battlefield, n. campo (m) de batalla.

battlement, n. muralla (f) almenada, almena (f).

bauxite, n. bauxita (f).

bay, n. bahía (f).

BDAT, n. **Best Demonstrated Available Technology (US)**, tecnología disponible que fue demostrada mejor.

beach, n. playa (f); **raised b.**, playa elevada; **sand b.**, playa de arena; **storm b.**, playa de tormenta.

beak, n. pico (m) (of birds, of insects).

beaker, n. cubeta (f) de precipitación (Chem); cubilete (m), copa (f) (drinking vessel); jarra (f) (tumbler, large glass).

beam, n. balancín (f) (of a machine); viga (f) (Con); rayo (m) (light); **b. engine**, motor (m) de balancín, máquina (f) de balancín; **bearing b.**, viga (f) maestra; **H b.**, viga en H; **laser b.**, rayo laser; **main b.**, viga maestra, viga principal; **reinforced concrete b.**, viga de hormigón armado; **secondary b.**, vigueta (f).

bean, n. alubia (f) (Sp); judía (f); frijol (m) (LAm); poroto (m) (LAm); habichuela (f), haba (f); **broad b.**, haba, habichuela, judión (f); **coffee b.**, grano (m) de café; **french b.**, **green b.**, judía verde; **haricot b.**, judía blanca; **string b.**, judía, habichuela verde; **white b.**, alubia, judía blanca.

bear, n. oso (m); **Brown B.**, Oso pardo.

bearing, n. porte (m), talente (m), maneras (f.pl) (mannerism); producción (f), cosecha (f) (yield); apuntalamiento (m), soporte (m) (Con); azimut (m), orientación (f) (direction); **b. axle**, eje (m) portador; **b. capacity**, carga (f) admisible, capacidad (f) de carga; **b. pile**, pilote (m) de carga; **b. plate**, placa (f) de asiento; **roller b.**, cojinete (m) de bolas, rodamiento (m).

beat, 1. v. golpear (to strike, to blow); batir (e.g. to flap wings); batir (to stir, to mix); batir (Eng, to hammer, to strike metal); latir, pulsar (Med); 2. n. latido (m) (of pulse), pulsación (f).

Beaufort, n. Beaufort (m); **B. Scale**, escala (f) de Beaufort (for wind measurement).

becquerel, n. becquerel (m) (abbr. Bq).

bed, n. cama (f), banco (m), lecho (m); capa (f), estrato (m) (Geol), cauce (m) (of river); **ash b.**, banco de ceniza; **b. load**, carga (f) de fondo; **b. profile**, sección (f) de capa, perfil (m) de lecho; **filter b.** (Treat), lecho bacteriano, lecho fitrante; **fluidized b.**, lecho fluidificado; **incompetent b.** (Geol), banco incompetente; **marker b.**, capa guía; **passage b.** (Geol), capa de transición; **peat b.**, banco de turba.

bedding, n. estratificación (f); **b. plane**, plano (m) de estratificación; **cross-b.**, estratificación cruzada; **graded b.**, estratificación gradada; **hassock b.**, estratificación rizada; **thick-bedded (a)**, estratificación gruesa; **thin-bedded (a)**, estratificación fina.

bedrock, n. basamento (m), substrato (m).

bee, n. abeja (f); **b. keeping**, apicultura (f); **honey b.**, abeja melífera; **queen b.**, abeja reina, abeja maestra; **worker b.**, abeja obrera, abeja neutra.

beech, n. haya (f).

beehive, n. colmena (f).

beeswax, n. cera (f) de abejas.

beet, n. remolacha (f), betarraga (f) (Ch, Perú); **sugar-b.**, remolacha azucarera.

beetle, n. escarabajo (m).

beetroot, n. remolacha (f), betarraga (f) (Ch, Perú); betavel (m) (Méx);.

behavior, n. comportamiento (m).

behaviour, n. comportamiento (m).

being, n. ser (m); **human b.**, ser humano; **living b.**, ser vivo.

belemnite, n. belemnite (m).

bell-shaped, a. acampanado/a.

bellows, n.pl. fuelle (m).

belt, n. cinturón (m), cinta (f); **conveyor b.**, cinta (f) transportadora; **green b.**, zona verde.

bench, n. banco (m) (work table); tribunal (m) (court); desnivel (m) (shelf in ground); **b. mark (reference point)**, punto (m) de referencia.

benchmark, n. cota (f), punto (m) acotado (GB, altitud sobre nivel del mar); punto de referencia (reference point); prueba (f) patrón (pattern).

bend, n. curva (f) (Gen); ángulo (m) (in pipe); comba (f) (warp, sag); recodo (m) (in a river).

bending, n. flexión (f); **b. moment**, momento (m) de flexión.

benefit, n. beneficio (m); **cost b.**, **cost-b.**, costo-beneficio; **direct b.**, beneficio directo; **environmental b.**, beneficio ambiental; **net b.**, beneficio neto; **social b.**, beneficio social.

benthic, a. bentónico/a; **b. flora and fauna**, flora (f) y fauna (f) bentónicas.

benthos, n. bentos (m) (bottom-dwelling organisms of sea, lake, river).

bentonite, n. bentonita (f); **b. seal in a piezometer**, sello (m) de bentonita en un piezómetro.

benzaldehyde, n. benzaldehído (m).

benzene, n. benzol (m), benceno (m).

benzenecarbaldehyde, n. bencenocarbaldehído (m).

benzoyl, n. benzoílo (m).

beriberi, n. beriberi (m).

berm, n. berma (f); albardón (m) (LAm) (naturally-raised river bank), arcén (m) (US); andén (m) (Con).

Berriasian, n. Berriasense (m).

berry, n. baya (f) (fruit); hueva (f) (of fish); grano (m) (of coffee).

berth, n. amarradero (m) (place at wharf).

beryl, n. berilo (m).

beryllium, n. berilio (m) (Be).

beta, n. beta (f).

bevel, n. bisel (m), ángulo (m) oblícuo; **b. edge**, chaflán (f).

bevelled, a. biselado/a.

beverage, n. bebida (f).

bi-, pf. bi-.

biannual, a. semestral.

biaxial, a. biáxico/a.

bicarbonate, n. bicarbonato (m); **calcium b.**, bicarbonato de calcio; **sodium b.**, bicarbonato de sodio.

biconvex, a. biconvexo/a.

bicycle, n. bicicleta (f).

bid, n. licitación (f) (at auction).

biennial, 1. n. planta (f) bienal (Bot); 2. a. bienal.

bifurcated, a. bifurcado/a.

bifurcation, n. bifurcación (f); **b. ratio**, relación (f) de bifurcación.

big, a. grande.
bight, n. rada (f) (Geog).
bilateral, a. bilateral; **b. symmetry,** simetría (f) bilateral.
bile, n. hiel (f).
bilge, n. sentina (f), pantoque (m) (part of hull of a boat); **b. water,** agua (f) de sentina.
bilharzia, n. bilarcia (f), bilarciasis (f).
bilinear, a. bilineal.
bill, n. factura (f) (invoice); cuenta (f) (in shop, restaurant); minuta (f) (Jur); proyecto (m) de ley (Jur, act); pico (m) (of a bird); hoja (f), cartel (m) (advertisement); billete (m) de banco (US, banknote); promontory (Geog); **b. of sale,** contrato (m) de venta; **draft b. (Jur),** anteproyecto (m) de ley.
billing, n. facturación (f) (Fin).
billion, n. billón (m), millardo (m), millón (m) de millones.
bimodal, a. bimodal.
binary, a. binario/a; **b. scale,** zona (f) binaria.
bindweed, n. enredadera (f).
binoculars, n. binoculares (m.pl), gemelos (m.pl), prismáticos (m.pl), largavistas (f.pl) (Arg, Bol, Ch).
binomial, a. binomial; **b. distribution,** distribución (f) binomial, distribución binómica; **b. equation,** ecuación (f) binomial.
bio-, pf. bio-.
bioaccumulation, n. bioacumulación (f).
bio-aeration, n. bioaireación (f).
bioassay, n. bioensayo (m), ensayo (m) biológico.
bioavailability, n. biodisponibilidad (f).
bio-barrier, n. biobarrera (f).
biocenology, n. biocenología (f).
biocenosis, n. biocenosis (f).
biocentric, a. biocéntrico/a.
biochemical, a. bioquímico/a; **b. attenuation,** atenuación (f) bioquímica; **b. oxygen demand (abbr. BOD),** demanda (f) bioquímica de oxígeno (abbr. DBO).
biochemist, n. bioquímico/a (m, f).
biochemistry, n. bioquímica (f).
biocide, n. biocida (m).
bioclast, n. bioclasto (m).
bioclastic, a. bioclástico/a.
biocoenosis, n. biocenosis (f).
biodegradability, n. biodegradabilidad (f).
biodegradable, a. biodegradable.
biodegradation, n. biodegradación (f); **anaerobic b.,** biodegradación anaerobia.
biodigester, n. biodigestor (m).
biodisc, n. biodisco (m), biocilindro (m) (Treat).
biodiversity, n. biodiversidad (f).
bioenergetics, n. bioenergética (f).
biofacies, n.pl. biofacies (f).
biofilm, n. biofilm (m), película (f) biológica.
biofilter, n. biofiltro (m).
bioflocculation, n. biofloculación (f).
biogas, n. biogás (m).

biogenesis, n. biogénesis (f).
biogenetic, a. biogenético/a.
biogenic, a. biogénico/a.
biogeography, n. biogeografía (f).
bioindicator, n. indicador (m) biológico, bioindicador (m).
biological, a. biológico/a; **b. control (Agr),** control (m) biológico; **b. film,** capa (f) biológica; **b. filter,** filtro (m) biológico.
biologist, n. biólogo/a (m, f).
biology, n. biología (f); · **marine b.,** biología marina.
bioluminescence, n. bioluminiscencia (f).
bioluminescent, a. bioluminescente.
biomagnification, n. biomagnificación (f).
biomass, n. biomasa (f).
biome, n. bioma (m).
biometry, n. biometría (f).
biophysics, n. biofísica (f).
bioreactor, n. bioreactor (m).
bioremediation, n. bioremediación (f); **in situ b.,** bioremediación en el sitio; **intrinsic b.,** bioremediación intrínseca; **on site b.,** bioremediación en el sitio.
biorhythm, n. biorritmo (m), ritmo (m) biológico.
biosorption, n. bioabsorción (f).
biosphere, n. biosfera (f); **B. Reserves (pl),** Reservas (f.pl) de la Biosfera.
biostratigraphy, n. biostratigrafía (f).
biostrome, n. biostromo (m).
biosystem, n. biosistema (m).
biota, n. biota (f).
biotechnology, n. biotecnología (f).
biotic, a. biótico/a; **b. factor,** factor (m) biótico; **b. index,** índice (m) biótico.
biotite, n. biotita (f).
biotope, n. biótopo (m).
biotransformation, n. biotransformación (f).
bioturbation, n. bioturbación (f).
biotype, n. biotipo (m).
bioventing, n. bioventilación (f).
biowaste, n. residuos (m.pl) biodegradables.
birch, n. abedul (m).
bird, n. ave (f), pájaro (m); **b. of passage,** ave pasajera, ave de paso; **b. of prey,** ave de rapiña, ave de presa; **flightless b.,** ave corredora (large birds, e.g. ostrich, emu); **migratory b.,** ave pasajera, ave de paso; **sea b.,** ave marina; **water b.,** ave acuática.
birefringence, n. birrefringencia (f).
birth, n. nacimiento (m) ·(med); comienzo (m) (start); origen (m) (origin); **b. control,** control (m) de la natalidad, regulación (f) de nacimientos.
bisector, n. bisectriz (f), bisector (m).
bisexual, a. bisexual.
bismuth, n. bismuto (m) (Bi).
bisulphate, n. bisulfato (m).
bit, n. barrena (f), trépano (m) (for drilling), bit (m), tricono (m), corona (f); bit (m) (Comp).

bitumen, n. bitumen (m), chapopote (m) (Méx).

bituminous, a. bituminoso/a; **b. coal,** hulla (f) bituminosa.

bivalent, a. bivalente.

bivalve, 1. n. bivalvo (m); 2. a. bivalvo/a.

blackberry, n. zarzamora (f), mora (f).

blackcurrant, n. grosella (f) negra.

blackish, a. negruzco/a.

bladder, n. vejiga (f); vejiga urinaria (of urine); blister (of the skin); cámara de aire (Eng. of a tyre); vesícula (f) (Bot); **air b.,** vejiga de aire (of fish); **gall b.,** vejiga de la bilis, vesícula biliar; **gas b.,** vejiga de aire (of fish); **swim b.,** vejiga natatoria.

blast, v. volar (to blow up), explotar (to explode).

blasthole, n. barreno (m) (Min).

blasting, n. golpeteo (m) (Geol), voladura (f), trabajo (m) con explosivos; **b. cap,** detonador (m); **b. charge,** carga (f) explosiva.

blastoid, n. blastoide (m).

bleach, 1. v. blanquear, decolorar (to remove colour); 2. n. lejía (f), decolorante (m).

blight, n. roya (f) (on plants).

blindness, n. ceguera (f); **river-b.,** ceguera de río.

blistered, a. globuloso/a (Geol), ampollado/a.

blistering, n. vesiculación (f).

blizzard, n. ventisca (f), ventisquero (m); **to blow a b. (v),** ventiscar, ventisquear.

block, n. bloque (m) (Gen, Comp); bloque (m), cuadra (f) (LAm), manzana (f) (building); **apartment b.,** bloque de apartamentos; **b. and tackle,** aparejo (m) de poleas; **b.-diagram,** bloque diagrama; **b.-faulted,** bloque fallado; **b. of flats,** bloque de apartamientos; **b.-relief,** relieve (m) de bloques; **en bloc,** en bloque; **fault b.,** bloque de falla; **rift b.,** bloque rift; **Asiatic b.,** bloque asiático.

blockhouse, n. fortín (m).

blocky, a. forma (f) de bloques.

blood, 1. n. sangre (f); **b. group,** grupo (m) sanguíneo; **b. poisoning,** envenenamiento (m) de la sangre; **b. transfusion,** transfusión (f) sanguínea; **red b. cell,** glóbulo (m) rojo sanguíneo; 2. a. sanguíneo/a.

bloodstone, n. heliotropo (m).

bloom, n. flor (f) (flower); floración (f), florecimiento (f) (algae); pelusilla (f) (on fruit); fluorescencia (f); **algal b.,** floración de algas.

blow, n. choque (m), golpe (m), impacto (m).

blowing-up, n. voladura (f) (explosion).

blowlamp, n. soplete (m).

blowtorch, n. soplete (m).

blow up, v. volar, detonar (to detonate), explosionar (to explode).

blue, 1. n. azul (m); **indigo b.,** azul índigo; 2. a. azul.

blueprint, n. proyecto (m) original.

bluff, n. risco (m), farallón (m), peñasco (m) (Geog); fanfarronada (f), farol (m) (act of bluffing).

blunt, a. romo/a.

boa, n. boa (f) (snake).

board, n. tabla (f), tablero (m) (e.g. at school); tablón (m) (sheet, placard); junta (f), consejo (m) (group of officials); comisión (f) (gas, water); **b. of directors,** consejo de administración, junta directiva; **B. of Trade,** Ministerio (m) de Comercio; **b. of trustees,** junta directiva; **b. room,** sala (f) de consejo; **drawing b.,** tablero de dibujo.

boarding, n. entablado (m), entarimado (m) (of floor); embarque (m) (Mar).

boatyard, n. astillero (m).

BOD, n. DBO; **biochemical oxygen demand,** demanda (f) bioquímica de oxígeno.

body, n. cuerpo (m); sólido (m) (Mat); conjunto (m) (group); organismo (m) (organization); **igneous b.,** cuerpo igneo; **ore b.,** cuerpo mineralizado.

bog, n. pantano (m), ciénaga (f), marisma (f), marjal (m).

boggy, a. pantanoso/a.

boil, v. hervir; **to come to the b. (v),** entrar en ebullición, comenzar a hervir.

boiler, n. caldera (f).

boiling, 1. n. ebullición (f); **b. point,** punto (m) de ebullición; 2. a. hirviendo/a, hirviente (Gen); **b. sand,** licuefacción (f) de arena; **b. water,** agua (f) hirviendo.

bole, n. tronco (m) de un árbol.

bolivar, n. bolívar (m) (Ven, unit of currency).

Bolivia, n. Bolivia (f).

Bolivian, 1. n. boliviano/a (m, f); 2. a. boliviano/a.

boliviano, n. boliviano (m) (Bol, unit of currency).

bolt, n. perno (m); cerrojo (m) (heavy bolt); pestillo (m) (small bolt); rayo (m) (of lightning); **U bolt,** perno en U.

bomb, n. bomba (f); **time b.,** bomba de relojería; **volcanic b.,** bomba volcánica.

bond, n. lazo (m), vínculo (m); enlace (m) (Chem); bono (m) (Fin); aparejo (m) (Con); **atomic b.,** enlace atómico; **chemical b.,** enlace químico; **covalent b.,** enlace covalente; **double b.,** enlace doble; **double Flemish b. (brickwork),** doble aparejo flamenco; **Dutch b. (brickwork),** aparejo a la holandesa; **English b. (brickwork),** sardinel; **English cross b. (brickwork),** aparejo cruz inglesa; **garden b. (brickwork),** aparejo tipo jardín; **header b. (brickwork),** aparejo de tijón; **herringbone b. (brickwork),** aparejo en espina; **ionic b.,** enlace iónico; **single b.,** enlace simple, único enlace; **triple b.,** enlace triple.

bone, n. hueso (m); **b. marrow,** médula (f) ósea; **fish b.,** espina (f).

bonemeal, n. harina (f) de huesos.

bonito, n. bonito (m).

bony, a. huesoso/a (with many bones); óseo/a (like bone).

boolean, a. booleano/a (Mat); **b. algebra,** álgebra (f) booleana.

boom, n. barrera (f) (across harbour), aguilón (m) (of crane); auge (m), boom (m) (sudden increase); **construction b.,** boom de la construcción; **oil b. (against pollution),** barrera contra aceite.

boost, v. aumentar; elevar (Elec); promover (to promote).

booster, n. elevador (m) de voltaje (Elec).

boracic, a. bórico/a.

borane, n. borano (m).

borate, n. borato (m).

borax, n. bórax (m).

bordering, a. vecino/a, limítrofe; **b. country,** país (m) limítrofe.

bore, v. barrenar, taladrar, perforar, sondear.

borehole, n. sondeo (m) (Gen), perforación (f), pozo (m) (well); **b. log,** registro (m) del sondeo; **b. logging,** testificación (f) (en sondeos), diagrafía (f); **deep b.,** sondeo profundo; **exploratory b.,** sondeo de exploración; **investigatory b.,** sondeo de reconocimiento; **observation b.,** sondeo de observación; **overflowing b.,** sondeo surgente; **percussion b.,** sondeo a percusión; **rotary b.,** sondeo a rotación; **small-diameter b.,** sondeo de pequeño diámetro; **test b.,** sondeo de ensayo.

borer, n. taladrador/a (m, f).

boric, a. bórico/a.

boring, 1. n. perforación (f), sondeo (m), taladro (m); 2. a. taladrador/a, perforado/a (drilling); aburrido/a (tiresome); **b. machine,** máquina (f) taladradora, taladro (m).

bornite, n. bornita (f).

boron, n. boro (m) (B).

borough, n. municipio (m) (municipality); ciudad (f) (town); distrito (m) (urban district).

borrow, v. pedir prestado, tomar prestado.

borrow-pit, n. zanja (f) de préstamos.

boss, n. jefe (m), gerente (m) (manager); dobladora (f) (Con).

botanical, a. botánico/a.

botanist, n. botanista (m, f).

botany, n. botánica (f).

bottle, 1. v. embotellar; 2. n. botella (f), frasco (m) (small flask); **b. bank,** contenedor (m) de vidrio, iglúe (m) (Fam); **Mariotte b.,** frasco de Mariotte; **Winckler b.,** botella Winckler.

bottled, a. embotellado/a.

bottom, n. fondo (m); **b. of sea, sea floor,** fondo del mar.

botulism, n. botulismo (m).

boudinage, n. formas (f.pl) arriñonadas.

bough, n. rama (f) (branch of tree).

boulder, n. bolo (m), bloque (m), roca (f); **b. clay,** boulder-clay.

boundary, n. límite (m), borde (m); **aquifer b.,**

límite del acuífero; **catchment b.,** límite del área de captación; **constant-head b.,** límite de nivel constante; **drainage b.,** límite de drenaje; **impermeable b.,** límite impermeable; **recharge b.,** límite de recarga; **rectangular b.,** borde rectangular.

bourne, n. arroyo (m) intermitente, rambla (f), riachuelo (m) seco con corriente subterránea, corriente (m) en terrenos calcáreos.

bovine, a. bovino/a, vacuno/a.

bovines, n.pl. bovino (m) (cows, etc.).

box, n. caja (f); arca (f), arcón (m) (large crate); **b. girder,** viga (f) tubular; **junction b.,** caja de conexión; **sand b.,** caja de arena.

brace, n. abrazadera (f) (clasp); berbiquí (m) (of drill); riostra (f), puntal (m) (Con).

brachiopod, n. braquiópodo (m).

bracket, n. brazo (m) (of an object); soporte (m) (support); repisa (f) (on a wall); paréntesis (m), cuadrado (m) (Mat).

brackish, a. salobre.

bract, n. bráctea (f).

braided, a. anastomosado/a (Geog), trensado/a (hair), galoneado/a (decoration); **b. river,** río (m) anastomosado.

brain, n. cerebro (m).

brake, n. freno (m); **b. shoe,** zapata (f) de freno.

bramble, n. zarzal (m).

bran, n. salvado (m).

branch, n. rama (m) (of a tree, of a business).

branchia, n. branquia (f).

branchiae, n.pl. branquias (f.pl) (see: branchia).

branchial, a. branquial.

brash, n. fragmentos (m) de rocas (Min), hielo (m) fragmentado (ice).

brass, n. latón (m), cobre (m) amarillo.

Brazil, n. Brasil (m).

Brazilian, a. brasileño/a, brasilero/a.

breach, 1. v. romper (to break), abrir brecha en (Mil), desportillar (to break open), resquebrajar (to crack open); 2. n. abertura (f) (of law), brecha (f) (gap), infracción (f) (infringement); **b. of law,** violación (f) de la ley; **b. of the regulations,** infracción (f) de los reglamentos.

breached, a. resquebrajado/a.

break, v. quebrar, romper.

breakage, n. destrozo (m), ruina (f).

breakaway, n. escapada (f); **b. curve (Mat),** curva (f) de escape.

breakdown, n. desglose (m) (editing); avería (f) (of a car); **b. of costs,** desglose de los costos.

breakers, n.pl. rompientes (m) (in sea); **line of b.,** línea (f) de rompientes.

break-point, n. punto (m) de rotura.

breakthrough, n. avance (m), adelanto (m) (progress); ruptura (f) (open up).

breakwater, n. escollera (f), rompeolas (f.pl), tajamar (m).

bream, n. **sea b.,** dorado (m), herrera (f), breca (f), besugo (m).

breast, n. pecho (m) (of a person); seno (m) (of a woman); mama (f) (mamma); pechuga (f) (of a bird).

breastfeed, v. dar el pecho.

breath, n. aliento (m); respiración (f) (respiration).

breathe, v. respirar; **to b. in,** aspirar; **to b. out,** exhalar.

breathing, n. respiración (f).

breccia, n. brecha (f); **fault b.,** brecha de falla; **volcanic b.,** brecha volcánica.

breeding, n. cría (f) (of animals); **intensive b.,** cría intensiva.

breeze, n. brisa (f); **fresh b.,** brisa fresca, recencio (m); **light b.,** brisa leve; **sea b.,** brisa del mar; **strong b.,** brisa fuerte.

brewery, n. fábrica (f) de cerveza.

brick, n. ladrillo (m); adobe (m) (sun-dried brick); **b. facing,** pared (f) de ladrillo; **engineer's b.,** ladrillo prensado; **fire b.,** ladrillo de sílice; **glazed b.,** ladrillo vidriado; **refractory b.,** ladrillo refractario; **silica b.,** ladrillo silicocalcáreo; **to b. in (v), to b. up (v),** tapiar con ladrillos.

bricklayer, n. enladrillador (m), albañil (m).

brickworks, n. ladrillera (f), fábrica (f) de ladrillos.

brickyard, n. fábrica (f) de ladrillos, almacén (m) de ladrillos.

bridge, 1. v. tender un puente sobre (a river); **to b. the gap (v),** llenar un vacío; 2. n. puente (m); **railway b.,** puente ferroviario; **suspension b.,** puente colgante; **swing b.,** puente giratorio; **transporter b.,** puente transportador; **Wheatstone b. (Elec),** puente de Wheatstone.

bridleway, n. camino (m) de herradura.

brightener, n. blanqueador (m), abrillantador (m); **optical b.,** blanqueador óptico.

brightness, n. resplandor (m), brillo (m).

brilliance, n. brillantez (f).

brine, n. salmuera (f).

Britain, n. Gran Bretaña; **Great B.,** Gran Bretaña.

British, a. británico/a.

broadcast, n. emisión (f) (TV, radio, etc.).

broadleaved, a. caducifolio/a, de hoja caduca.

broken, a. roto/a, quebrado/a.

broker, n. agente (m, f), intermediario (m) financiero; **stockbroker,** corredor/a (m, f) de bolsa.

bromate, n. bromato (m).

bromide, n. bromuro (m).

bromination, n. bromación (f).

bromine, n. bromo (m) (Br).

bromochloromethane, n. bromoclorometano (m).

bronchii, n.pl. bronquios (m.pl).

bronchiole, n. bronquiolo (m).

bronchitis, n. bronquitis (f).

bronchus, n. bronquio (m).

bronze, n. bronce (m).

brood, n. camada (f), nidada (f), cría (f).

brook, n. riachuelo (m), arroyo (m).

broom, n. retama (f).

brow, n. ceja (f) (Geog, Anat).

brushwood, n. matorral (m).

bryology, n. briología (f).

Bryophyta, n.pl. briofitas (f.pl), la división Bryophyta (Bot).

bryophyte, n. briofita (f).

bryozoa, n.pl. briozoos (m.pl), Bryozoa (f.pl) (Zoo, phylum); **Bryozoa (Zoo, clase),** Bryozoa (f.pl).

bryozoan, n. briozoo (m).

BS, n. **British Standard (GB, Reg),** norma (f) británica.

BSI, n. **British Standards Institute,** Instituto (m) Británico de Normalización.

bubble, n. burbuja (f).

bucket, n. cubo (m), balde (m) (esp. LAm), tacho (m) (LAm); cangilón (m) (of waterwheel), paleta (f) (of turbine), cuchara (f) (of dredger).

buckling, n. pandeamiento (m) (bending).

bud, n. brote (m), yema (f) (shoot); capullo (m) (of flower); **axillary/lateral/terminal b.,** yema axilar/lateral/terminal.

budget, n. presupuesto (m); **b. reform,** reforma (f) presupuestaria; **to balance the b. (v),** equilibrar el presupuesto.

buffalo, n. búfalo (m).

buffer, 1. v. tamponar (Chem), amortiguar; 2. n. amortiguador (m) (deadening device); tope (m) (of railway wagons); buffer (m), tapón (m) (Chem); tampón (m) (Med, Chem); **b. capacity (Chem),** capacidad (f) tampón; **b. solution (Chem),** solución (f) buffer, solución tampón, disolución (f) tampón; **b. zone,** zona (f) tampón, zona intermediaria.

buffering, n. amortiguación (f); **b. capacity (Chem),** capacidad (f) de amortiguación.

bug, n. chinche (f) (Zoo); bicho (m) (Fam, insect); microbio (m) (germ); falto (m), error (m) (Comp).

build, v. construir (to construct), erigir (to erect).

builder, n. constructor/a (m, f).

building, n. edificio (m), edificación (f); construcción (f) (action); **b. materials,** materiales (m.pl) de construcción.

building, 1. n. edificio (m), edificación (f); construcción (f) (action); **b. pathology,** patología (f) de la edificación; **b. stone,** piedra (f) de construcción; **listed b. (GB),** edificio protegido; **sick b. syndrome,** síndrome (m) del edificio enfermo; 2. a. constructor/a; **b. company, b. firm,** empresa (f) constructora; **b. society,** sociedad (f) inmobiliaria, caja (f) de ahorros.

build-up, n. acumulación (f).

built-up, a. urbanizado/a; **b. area,** zona (f) construida, zona urbanizada.

bulb, n. bulbo (m) (Bot); bombilla (f) (of a lamp); cubeta (f) (e.g. of thermometer).

bulkhead, n. mampara (f), cerramiento (m).

bulldozer, n. motoniveladora (f), excavadora (f), bulldozer (m), topadora (f) (Arg, Méx, Ur).

bulletin, n. boletín (m), comunicado (m); **news b.,** boletín informativo.

bulrush, n. espadaña (f), anea (f).

bulwark, n. baluarte (m), bastión (m) (defensive wall).

bumblebee, n. abejorro (m), abejón (m).

bund, n. albardón (f) (earth bank).

bundle, n. bulto (m), haz (f); **b. of rays,** haz de rayos.

bungalow, n. bungalow (m), chalé (m), chalet (m).

buoy, 1. v. balizar, aboyar (to mark with buoys); 2. n. boya (f).

buoyancy, n. flotación (f), flotabilidad (f); empuje (m) (Phys).

Burdigalian, n. Burdigaliense (m).

bureaucracy, n. burocracia (f).

buret, n. bureta (f).

burette, n. bureta (f).

burial, n. enterramiento (m) (grave), soterramiento (m).

burin, n. buril (m) (Arch, borer).

burner, n. quemador (m); **Bunsen b.,** mechero (m) Bunsen; **gas b.,** quemador de gas.

burrow, n. madriguera (f) (of animal), conejera (f) (of rabbit).

bus, n. autobús (m), bus (m) (LAm), ómnibus (m) (LAm), camión (f) (Mex), góndola (f) (Bol, Ch, Peru), guagua (Cu); colectivo (m) (LAm), micro (m) (Arg, Ur), microbús (m), buseta (f), (LAm) (small bus); bus (m) (Comp).

bush, n. arbusto (m), matorral (m); breña (f) (rough land).

business, n. negocio (m), comercio (m) (trade); empresa (f) (company); profesión (f) (profession); **b. deal,** trato (m) comercial; **b. letter,** carta (f) comercial; **b. opportunity,** oportunidad (f) de negocio; **b. park,** parque (m) empresarial, parque comercial; **businessman (pl. businessmen),** hombre (m) de negocios; **businesswoman (pl. businesswomen),** mujer (f) de negocios; **to set up a b. (v),** montar un negocio, poner un negocio.

butadiene, n. butadieno (m).

butaldehyde, n. butaldehído (m).

butanal, n. butanal (m).

butane, n. butano (m).

butanol, n. butanol (m).

butene, n. buteno (m).

butte, n. relicto (m) de erosión.

butterfly, n. mariposa (f), ropalócero (m); **b. nut,** mariposa.

buttress, n. contrafuerte (m), espolón (m) (Con); estribación (f) (Geog, spur); **b. root,** raiz (f) zanco, raiz caulógena.

butyl, n. butilo (m).

butylene, n. butileno (m).

buy, v. comprar; **to b. out,** comprar la parte de.

buyer, n. comprador/a (m, f).

buzzard, n. ratonero (m) (GB, Buteo buteo); **turkey b. (US, Cathartes aura),** buitre (m), gallinazo (m) (LAm), zopilote (m) (CAm, Mex); carancho (m) (CSur).

bypass, n. carretera (f) de circunvalación, bypass (m).

bypassing, n. circunvalación (f).

by-product, n. producto (m) secundario, subproducto (m), derivado (m); **industrial b.,** producto secundario industrial.

byte, n. byte (m).

C

cabbage, n. berza (f).
cabin, n. cabina (f) (of aircraft, of instruments); cabaña (f), choza (small house); **instrument c., recorder hut,** cabina de medición.
cable, n. cable (m) (Elec); cable (unit of length, 185.19 m); **armoured c.,** cable armado; **c. box,** caja (f) de conexiones; **c. sheathing,** cable de entubación; **electrical c.,** cable eléctrico; **high tension c.,** cable de alta tensión; **trunk c.,** cable principal.
cableway, n. transportador (m) aéreo.
cacao, n. cacao (m).
cactus, n. cactus (m), cacto (m).
CAD, n. DAO (m), DAC (LAm); **computer-aided design,** diseño (m) asistido por ordenador/computadora, proyecto (m) con ayuda de calculadoras.
caddisfly, n. tricóptero (m).
cadmium, n. cadmio (m) (Cd).
caecum, n. intestino (m) ciego.
caesium, n. cesio (m) (Cs).
caffeine, n. cafeína (f).
cage, n. jaula (f) (Gen, Min), montacarga (f) (Min); crate (container); lockup (garage).
cairn, n. montón (m) de piedras.
caisson, n. cajón (m), compuerta (f) de dique (used as a gate); **c. pile,** pilote (m) tubular.
calamine, n. calamina (f).
calcarenite, n. calcarenita (f).
calcareous, a. calcáreo/a.
calcdolomite, n. calcodolomita (f).
calcic, a. cálcico/a.
calcicole, n. calcícola (f).
calciferous, a. calcífero/a.
calcifuge, n. calcífuga (f).
calcilutite, n. calcilutita (f).
calcine, v. calcinar.
calcirudite, n. calcirrudita (f).
calcite, n. calcita (f).
calcitic, a. calcítico/a.
calcium, n. calcio (m) (Ca); **c. carbonate,** carbonato (m) cálcico.
calcspar, n. calcita (f).
calculate, v. calcular.
calculating, a. calculador/a, astuto/a; **c. machine,** calculadora (f), máquina de calcular.
calculation, n. cálculo (m); **statistical c.,** cálculo estadístico.
calculator, n. calculador/a (m, f), máquina (f) de calcular.
calculus, n. cálculo (m) integral.
caldera, n. caldera (f).
calendar, n. calendario (m).
calgon, n. calgón (m).
caliber, n. calibre (m), diámetro (m) interior.

calibrate, v. calibrar.
calibration, n. calibración (f); **c. test,** ensayo (m) de calibración.
calibre, n. calibre (m), diámetro (m) interior.
caliche, n. caliche (m).
call, n. llamada (f).
calliper, n. compás (m) de calibrador, calibrador (m); **vernier c.,** calibrador micrométrico.
Callovian, n. Calloviense (m).
callow, n. vega (f).
calm, 1. n. calma (f), tranquilidad (f); **dead c.,** calma chicha; 2. a. tranquilo/a; **c. sea,** mar (m) tranquilo.
calmness, n. tranquilidad (f).
calorie, n. caloría (f) (abbr. cal), caloría gramo, caloría pequeña.
calorific, a. calorífico/a; **c. value,** potencia (f) calorífica, valor (m) calorífico.
calorimetry, n. calorimetría (f).
calyx, n. cáliz (f).
camber, n. combadura (f), comba (f) (in road).
Cambrian, n. Cámbrico (m).
camp, n. campamento (m) (group of tents), campo (m) (LAm), camping (m) (proper camping site); **refugee c.,** campo de refugiados.
campaign, n. campaña (f).
Campanian, n. Campaniense (m).
campground, n. cámping (m).
camphor, n. alcanfor (m).
campsite, n. cámping (m).
campus, n. campus (m).
can, n. lata (f), bote (m); caneca (f) (LAm); tarro (m) (LAm); **tin c.,** lata (f), bote (m).
canal, n. canal (m) (for barge); tubo (m) (Anat); **Panama C.,** Canal de Panamá.
canalization, n. canalización (f) (a river).
canalize, v. canalizar.
cancer, n. cáncer (m).
cancerigenic, a. cancerígeno/a.
cancerogenic, a. cancerígeno/a.
candle, n. vela (f); bujía (f), esperma (f) (LAm), veladora (f) (LAm).
cannery, n. fábrica (f) enlatadora, planta conservera.
canoe, n. canoa (f); **dug-out c.,** bongo (m) (LAm).
canopy, n. bóveda (f) (of a wood); marquesina (f) (roof-like covering).
cantilever, a. viga (f) voladizo, de cantiléver; **c. bridge,** puente (m) voladizo; **c. crane,** grúa (f) voladiza.
canvas, n. lona (f).
canyon, n. cañón (m); **submarine c.,** cañón submarino.

cap, v. capsular (Geol), tapar.

CAP, n. PAC (f); **Common Agricultural Policy (EurU)**, Política (f) Agraria Común.

capacitance, n. capacitancia (f) (Elec).

capacitor, n. condensador (m) (Elec).

capacity, n. capacidad (f); **carrying c. (Ecol)**, capacidad de carga; **drainage c.**, capacidad de drenaje; **field c.**, capacidad de campo; **infiltration c.**, capacidad de infiltración; **recharge c.**, capacidad de recarga; **specific c.**, capacidad específica; **well c.**, capacidad del pozo.

cape, n. cabo (m); **C. Horn**, Cabo de Hornos; **C. of Good Hope**, Cabo de Buena Esperanza.

capillarity, n. capilaridad (f).

capillary, a. capilar; **c. action**, acción (f) capilar; **c. fringe**, franja (f) capilar; **c. pressure**, presión (f) capilar; **c. rise**, ascenso (m) capilar; **c. tube**, tubo (m) capilar; **c. zone**, zona (f) capilar.

capital, 1. n. capital (m) (finance); capital (f), capi (m) (Ch, Méx) (city); capitel (m) (of a building); **initial c.**, capital (m) inicial; **venture c.**, fondos (m.pl) de capital riesgo; 2. a. capital; **c. city**, ciudad (f) capital.

capitalization, n. capitalización (f) (Fin).

capping, n. capsulado (m) (Con), albardilla (f) (Con); **c. machine**, capsuladora (f).

capstan, n. cabrestante (m).

capsule, n. cápsula (f).

captive, a. cautivo/a.

capture, v. capturar.

capture, n. captura (f); **neutron c.**, captura de neutrones; **river c.**, captura de ríos.

car, n. coche (m), automóvil (m), turismo (m) (private car), carro (m) (LAm), auto (m) (LAm); **c. lot (US)**, aparcamiento (m), parking (m).

Caradocian, n. Caradocense (m).

carapace, n. carapazón (m).

carbaryl, n. carbaríl (m).

carbide, n. carburo (m).

carbohydrate, n. carbohidrato (m), hidrato (m) de carbón.

carbolic, a. fénico/a.

carbon, n. carbono (m) (C) (element); carbón (m) (material); **activated c.**, carbón activado, carbón activo; **c. 14**, carbono catorce; **c. dating**, datación (f) por carbono radioactivo; **c.-nitrogen ratio, (abbr. C/N ratio)**, relación (f) carbono-nitrógeno, relación C/N; **organic c.**, carbón orgánico.

carbonaceous, a. carbonoso/a; **c. oxidation**, oxidación (f) carbonosa.

carbonate, n. carbonato (m), carbonático/a (a); **calcium c.**, carbonato de calcio; **c. hardness**, dureza (f) carbonática; **sodium c.**, carbonato de sodio.

carbonic, a. carbónico/a.

carbonization, n. carbonización (f).

carbontetrachloride, n. tetracloruro (m) de carbono.

carborundum, n. carborundo (m); **c. stone**, piedra (f) de carborundo.

carboxylic, a. carboxílico/a.

carcinogen, n. carcinógeno (m).

carcinogenic, a. carcinógeno/a.

carcinoma, n. carcinoma (m).

card, n. tarjeta (f) (visiting, etc.), carnet (m) (membership), ficha (f) (playing); cartulina (f) (thin board); **credit c.**, tarjeta de crédito; **punched c.**, tarjeta perforada.

cardboard, n. cartón (m), papel (m) cartón, papel madera (Arg, Ch); **corrugated c.**, cartón corrugado.

cardiac, a. cardíaco/a.

cardinal, a. cardinal.

care, n. cuidado (m); **health c.**, atención (f) sanitaria; **parental c. (Zoo)**, cuidado parental.

career, n. profesión (f) (occupation), carrera (f) profesional (long-term).

careless, a. descuidado/a; negligente (negligent).

Carib, n. Caribe (m, f).

Caribbean, a. caribe; **C. Sea**, Mar (m) Caribe.

CARICOM, n. CMCC (f); **Caribbean Community and Common Market**, Comunidad (f) y Mercado Común del Caribe.

carnelian, n. carneola (f).

Carnian, n. Carniense (m).

carnivore, n. carnívoro (m).

carnivorous, a. carnívoro/a.

carp, n. carpa (f).

car park, n. coche (m), automóvil (m), turismo (m) (private car), carro (m) (LAm), auto (m) (LAm); **c. park (GB)**, aparcamiento (m), parking (m), turismo (m) (private car), parqueadero (m) (LAm).

carpel, n. carpelo (m).

carpenter, n. carpintero (m).

carpentry, n. carpintería (f).

carriage, n. carroza (f), coche (m); vagón (m), carro (m) (LAm) (railway); transporte (m) (of goods).

carriageway, a. calzada (f), pista (f); **dual c.**, autovía (f), carretera (f) de doble calzada.

carrier, n. transportista (m), portador/a (m, f) (person carrying goods); carrier (m) (Zoo, Med); empresa de transportes, acarreador (m) (haulier).

carrion, n. carroña (f).

carrot, n. zanahoria (f).

carry out, v. realizar (a scheme), llevar a cabo.

Cartesian, a. cartesiano/a; **C. coordinates**, coordenadas (f.pl) cartesianas.

cartographer, n. cartógrafo/a (m, f).

cartographic, a. cartográfico/a.

cartography, n. cartografía (f).

cascade, n. cascada (f), salto (m) de agua.

case, n. caja (f) (box); causa (f), proceso (m) (Jur); **c. history**, historia (f) clínica; **worst-c.**

conditions, condiciones (f.pl) del caso más desfavorable.

case-hardening, n. cementación (f) (Eng).

cash, n. dinero (m) contante, dinero efectivo, metálico (m); **c. payment,** pago (m) al contado; **c. register,** caja (f) registradora; **petty c.,** gastos (m) menores.

casing, n. entubado (m), entubación (f), entubamiento (m); **temporary c.,** entubado provisional; **well c.,** entubado del pozo.

cassiterite, n. casiterita (f).

cast, 1. v. fundir (e.g. iron); 2. n. calco (m), contramolde (m); **c. iron,** hierro (m) colado; **flow c.,** calco corriente; **internal c.,** calco interno.

Castilian, 1. n. castellano/a (m, f); 2. a. castellano/a.

casting, n. fundida (f) (Eng).

cast-iron, a. de hierro fundido.

catabolic, a. catabólico/a.

catabolism, n. catabolismo (m).

Catalan, n. Catalán (m) (language and person).

Catalan, a. catalán/a.

catalog, n. catálogo (m).

catalogue, n. catálogo (m); **library c.,** catálogo de biblioteca.

Catalonia, n. Cataluña (f).

catalysis, n. catálisis (f).

catalyst, n. catalizador (m).

catalytic, a. catalítico/a; **c. converter,** catalizador (m).

catalyze, v. catalizar.

cataract, n. catarata (f) (Geog, Med).

catastrophe, n. catástrofe (m).

catastrophic, a. catastrófico/a; **c. flood,** inundación (f) catastrófica.

catchment, n. captación (f), cuenca (f) (basin); **c. area,** zona (f) de captación; **c. management,** gestión (f) de cuenca; **experimental c.,** cuenca experimental; **mist c.,** captación de neblina.

catchpit, n. cimbra (f) (in river).

categorization, n. clasificación (f).

caterpillar, n. oruga (f) (Zoo); **c. tractor,** tractor (m) oruga.

cathode, n. cátodo (m).

cathodic, a. catódico/a; **c. protection,** protección (f) catódica.

cation, n. catión (m).

cationic, a. catiónico/a.

cattle, n. ganado (m), ganado vacuno, hacienda (f) (Arg); **c. drover,** vaquero (m) (cowboy), tropero (m) (Arg); **c. ranch,** hacienda (f) (LAm, esp. Arg).

cattle-raising, n. agroganadero (m).

catwalk, n. pasarela (f) de servicio, pasadizo (m).

caudal, a. caudal.

cauliflower, n. coliflor (f).

caulking, n. calafateo (m), calafateadura (f).

causal, a. causal.

cause, n. causa (f); **c. and effect,** causa y efecto.

causeway, n. terraplén (m).

cautious, a. precavido/a.

cave, n. cueva (f), caverna (f); **c. painting,** pintura (f) rupestre; **sea c.,** gruta (f) marina.

caved, a. derrumbado/a.

cave-dwelling, a. cavernícola.

caveman, n. troglodita (m, f), hombre cavernícola (m, f).

cavern, n. caverna (f).

cavernous, a. cavernoso/a, oqueroso/a; **c. limestone,** caliza (f) oquerosa, caliza cavernosa.

cavitation, n. formación (f) de cavidades, cavitación (f).

cavity, n. cavidad (f), hueco (m) (hollow); **solution c.,** cavidad de disolución.

CBD, n. **Central Business District (US),** distrito (m) de negocios.

CCTV, n. **closed-circuit television,** televisión (f) por circuito cerrado.

CD, n. CD (m) (abbr. of disco compacto); **CD-Rom (abbr. of Compact Disc, Read-Only-Memory),** CD-Rom.

cedar, n. cedro (m).

ceiling, n. techo (m) (interior), tejado (m) (roof); **c. joist,** vigueta (f) de techo; **c. light,** lámpara (f) de techo.

celery, n. apio (m).

celestine, n. celestina (f).

cell, n. célula (f); celdilla (f) (of bees); celda (f) (Chem); **c. division,** división (f) celular; **c. membrane,** membrana (f) celular; **c. wall,** pared (f) celular; **plant c.,** célula vegetal.

cellar, n. sótano (m); bodega (f) (wine cellar).

cellular, a. celular.

celluloid, n. celuloide (m).

cellulose, n. celulosa (f).

cellulose, a. celulósico/a.

celophane, n. celofán (m).

Celsius, a. celsius, centígrado/a.

Celto-Iberian, a. celtíbero/a.

cement, 1. v. cementar; 2. n. cemento (m); **c. gun,** cañón (m) de cemento; **c. mortar,** mortero (m) de cemento; **c. paste,** pasta (f) de cemento; **c. works,** cementera (f), fábrica (f) de cemento; **Portland c.,** cemento Portland.

cementation, n. cementación (f).

cemetery, n. cementerio (m), panteón (m).

Cenomanian, n. Cenomaniense (m).

Cenozoic, n. Cenozoico (m).

census, n. censo (m), recuento (m), empadronamiento (m); catastro (m) (LAm); **population c.,** censo de población.

cent, n. cent (m) (US), centavo (m), céntimo (m) (unit of currency).

centavo, n. centavo (m) (unit of currency).

center, n. centro (m).

centesimo, n. centésimo (m) (unit of currency).

centi-, pf. centi- (10.E2).

centigrade, a. centígrado/a, celsius.

centimetre, n. centímetro (m).

centimo, n. céntimo (m) (unit of currency).

centipede, n. ciempiés (m).

central, a. central.

centralize, v. centralizar.

centre, n. centro (m); **c. for waste generation,** centro de aportación de residuos; **c. of gravity,** centro de gravedad; **information c.** (e.g. **for tourists),** centro de interpretación.

centrifugal, a. centrífugo/a; **c. pump,** bomba (f) centrífuga.

centrifuge, 1. v. centrifugar; 2. n. centrífuga (f), centrifugadora (f).

centrifuging, n. centrifugación (f).

cephalon, n. cefalón (m).

cephalopod, n. cefalópodo (m).

Cephalopoda, n.pl. cefalópodos (m.pl), la clase Cephalopoda (Zoo).

CEPIS, n. CEPIS; **Pan American Center for Sanitary Engineering and Environmental Sciences (LAm),** Centro (m) Panamericano de Ingeniería Sanitaria y Ciencias de Ambiente.

ceramic, a. cerámico/a; **c. tile,** pieza (f) cerámica.

cereal, n. cereal (m).

cerebrum, n. cerebro (m).

cerium, n. cerio (m) (Ce).

certification, n. certificación (f).

cerussite, n. cerusita (f).

cesspit, n. fosa (f) negra, pozo (m) negro.

cesspool, n. fosa (f) negra, pozo (m) negro; **c. contents,** aguas (f.pl) negras.

Cetacea, n.pl. cetáceos (m.pl), el orden Cetacea (Zoo).

cetacean, n. cetáceo (m).

CFC, n. CFC (m); **chlorofluorocarbon,** clorofluorocarbono (m), halogenuro (m) de carbono.

chaeta, n. queta (f).

chaetae, n.pl. quetas (f.pl).

chain, n. cadena (f) (Gen); cadena (unit of length, 22 yards or 20.117 metres); **c. reaction,** reacción (f) en cadena; **c. tongs,** llave (f) de cadena; **food c.,** cadena alimentaria; **Markov c. (Mat),** cadena de Markov; **mountain c.,** cadena de montañas.

chalcedony, n. calcedonia (f).

chalcopyrite, n. calcopirita (f).

chalcosite, n. calcocita (f).

chalk, n. creta (f), tiza (f), gis (m); **c. puddling (e.g. for stream bed lining),** pudelado (m) con creta; **nodular c.,** creta nodulosa.

chalybite, n. chalibita (f).

chamber, n. cámara (f) (of parliament), despacho (m) (of judge); **c. of commerce,** cámara de comercio; **combustion c.,** cámara de combustión; **septic c.,** fosa (f) séptica, cámara séptica (LAm); **the Upper/Lower C. (GB Parliament),** la Cámara Alta/Baja.

change, n. cambio (m); moneda (f) (coins); **c. of facies,** cambio de facies; **c. of policy,** cambio de política; **climate c.,** cambio climático; **eustatic c.,** cambio eustático; **isotopic c.,** cambio isotópico.

channel, 1. v. canalizar, encauzar; 2. n. canal (m), estrecho (m) (strait), cauce (m) (in river); conducto (m), medio (m) (for cables); **c. capacity,** capacidad (f) canal; **flood relief c.,** canal aliviadero; **small irrigation c.,** reguera (f), regadera (f); **the C. Tunnel,** el túnel (m) del Canal de la Mancha; **the English C.,** el Canal (m) de la Mancha; **tidal c.,** canal de marea.

channelling, n. encauzamiento (m) (a river), canalización (action).

chaos, n. caos (m); **c. theory,** teoría (f) del caos.

chapter, n. capítulo (m).

characteristic, 1. n. característica (f), rasgo (m); **aquifer characteristics,** características del acuífero; **hazardous waste characteristics,** características de los residuos peligrosos; **hydraulic characteristics,** características hidráulicas; **soil characteristics,** características del suelo; **wastewater characteristics,** características de las aguas residuales; 2. a. característico/a.

characterization, n. caracterización (f); **site c.,** caracterización de sitio; **waste c.,** caracterización de residuos.

characterize, v. caracterizar.

characterized, a. caracterizado/a.

charcoal, n. carbón (m) de leña, carbón vegetal; **activated c.,** carbón activado; **c. burning,** carboneo (m).

charge, n. gasto (m) coste (m), costa (f) (Fin); carga (f) (explosive, electric); **c. account,** cuenta (f) abierta, un crédito; **fixed c.,** gastos fijos; **free of c.,** gratis (ad); **freight c.,** gastos de flete; **handling c.,** gastos de tramitación, gastos de manejo; **to bring a c. against (v),** hacer una acusación en contra; **to take c. of (v),** hacerse cargo de.

chargé d'affaires, n. encargado (m) de negocios.

charnokite, n. charnokita (f), charnoquita (f).

chart, n. tabla (f), cuadro (m) (table); gráfico/a (m,f) (graph); carta (f), mapa (f) (Mar); organigrama (m); **Admiralty c.,** carta del Ministerio de Marina, carta de navegación, carta náutica; **flow c.,** organigrama funcional; **hydrographic c.,** mapa hidrográfico; **navigation c.,** carta de navegación, carta marítima; **pie c.,** areograma (m).

chasm, n. sima (Geol); abismo (m) (Fig).

Chattian, n. chattiense (m).

check, 1. v. comprobar, averiguar, verificar; 2. n. comprobación (f), verificación (f), prueba (f).

checkable, a. comprobable, verificable.

checking, n. comprobación (f).

check-list, n. check-list (m), lista (f).

chelating, a. quelante; **c. agent,** agente (m) quelante.

chemical, a. químico/a.

chemist, n. químico/a (m, f).
chemistry, n. química (f).
chemoautotroph, n. quimioautótrofo (m).
chemoautotrophe, n. quimioautótrofo (m).
chemoheterotroph, n. quimioheterótrofo (m).
chemoheterotrophe, n. quimioheterótrofo (m).
chemosynthesis, n. quimiosíntesis (f).
chemotaxis, n. quimiotactismo (m), quimio-taxia (f).
chemotropism, n. quimiotropismo (m).
cherimoya, n. chirimoya (f) (fruit), chirimoyo (m) (tree).
chernozem, n. chernozem (m).
cherry, n. cereza (f) (fruit); cerezo (m) (tree); **c. orchard,** cerezal (m).
chert, n. sílex (m).
chestnut, n. castaño (m) (tree); castaña (f) (nut); **horse c.,** castaño de Indias.
chi, n. ji (m); **C.-squared distribution,** distribución (f) ji-cuadrada.
chicken, n. pollo (m).
chicken-pox, n. varicela (f).
chick-pea, n. garbanzo (m).
childbirth, n. parto (m), alumbramiento (m).
Chile, n. Chile (m).
Chilean, 1. n. chileno/a (m, f); 2. a. chileno/a.
chimney, n. chimenea (f); **c. stack,** cañón (m) de la chimenea, salida (f) de la chimenea.
China, n. China (f).
china, n. china (f), cerámica (f), porcelana (f); **c. clay,** caolín (m), caolinita (f).
chink, n. rendija (f).
chip, 1. v. astillar, desportillar, descantillar; 2. n. esquirla (f), lasca (f), brizna (f) (of stone); chip (m) (Comp).
chipped, a. astillado/a, mellado/a.
chipping, n. astillamiento (m), desportilla-miento (m).
chirimoya, n. chirimoya (f) (fruit), chirimoyo (m) (tree).
chironomid, n. quironómido (m).
Chironomidae, n.pl. quironómidos (m.pl), la clase Quironomidae (Zoo).
Chiroptera, n.pl. quirópteros (m.pl) (Zoo, order).
chiropteran, n. quiróptero (m).
chitin, n. quitina (f).
chives, n.pl. cebolleta (f).
chloramine, n. cloramina (f).
chlorate, n. clorato (m); **sodium c.,** clorato sódico.
chloric, a. clórico/a.
chloride, n. cloruro (m); **calcium c.,** cloruro cálcico.
chlorinate, v. tratar con cloro.
chlorinated, a. clorado/a; **c. hydrocarbon,** hidrocarburo (m) clorado; **c. solvent,** disol-vente (m) clorado.
chlorination, n. cloración (f), clorinización (f), desinfección (f) con cloro; **break-point c.,** cloración rotura; **water c.,** cloración de agua.
chlorinator, n. clorinador (m).

chlorine, n. cloro (m) (Cl); **c. demand,** demanda (f) de cloro; **free c.,** cloro libre; **residual c.,** cloro residual.
chlorite, n. clorito (m).
chloritoid, n. cloritoide (m).
chlorobenzene, n. clorobenzol (m).
chlorobiphenol, n. clorobifenol (m).
chlorodibromomethane, n. clorodibromome-tano (m).
chlorofluorocarbon, n. clorofluorocarbono (m) (abbr. CFC).
chloroform, n. cloroformo (m).
Chlorophyceae, n.pl. cloroficeas (f.pl).
chlorophyl, n. clorofila (f) (also chlorophyll).
Chlorophyta, n.pl. clorofitos (m.pl), la división Chlorophyta (Bot).
chlorophyte, n. clorofito (m).
chloroplast, n. cloroplasto (m).
chlorosis, n. clorosis (f).
cholera, n. cólera (f).
choppy, a. picado/a, agitado/a (the sea).
Chordata, n.pl. cordados (m.pl), el filum Chordata (Zoo).
chordate, 1. n. cordado (m); 2. a. cordado/a.
chromate, n. cromato (m).
chromatic, a. cromático/a.
chromatography, n. cromatografía (f); **gas c.,** cromatografía de gases; **gas-liquid c.,** cromatografía de gases-líquidos.
chromite, n. cromita (f).
chromitite, n. cromitita (f).
chromium, n. cromo (m) (Cr); **c. plating,** cromado (m).
chromosome, n. cromosoma (m).
chronic, a. crónico/a.
chrono-stratigraphy, n. cronostratigrafía (f).
chronological, a. cronológico/a.
chronology, n. cronología (f).
chrysalis, n. crisálida (f), pupa (f).
chrysolite, n. crisolita (f).
churchyard, n. campo (m) santo.
chute, n. conducto (m); chiflón (m) de descarga (Min) (LAm); **rubbish c. (apartments),** vertedero (m) de basuras.
chyme, n. quimo (m).
cidaroid, n. cidaroide (m).
cilia, n.pl. cilios (m.pl).
ciliary, a. ciliar.
Ciliata, n.pl. ciliados (m.pl), la clase Ciliata (Zoo).
ciliate, 1. n. ciliado (m); 2. a. ciliado/a.
cilium, n. cilio (m).
cinder, n. carbonilla (f) (of coal); escoria (f) (from a furnace); **cinders (pl),** cenizas (ash).
cinnabar, n. cinabrio (m).
cipher, n. cifra (f).
circadian, a. circadiario/a; **c. rhythm,** ritmo (m) circadiario.
circle, 1. v. cercar, rodear (to surround); girar alrededor de, dar la vuelta a (move round); 2. n. círculo (m); **Antarctic C.,** Círculo

Antártico; **Arctic C.**, Círculo Polar; **in business circles**, in círculos comerciales.

circuit, n. circuito (m), gira (f); circuito (Elec); **c.-breaker**, cortacircuitos (m); **closed-c. television (abbr. CCTV)**, televisión (f) por circuito cerrado; **integrated c.**, circuito integrado; **printed c. board**, placa (f) de circuito impreso; **short c.**, cruce (m), cortocircuito (m).

circular, a. circular, redondo/a.

circulate, v. circular.

circulation, n. circulación (f); **atmospheric c.**, circulación atmosférica; **reverse c.**, circulación inversa; **traffic c.**, circulación del tráfico.

circumference, n. circunferencia (f).

circumpolar, a. circumpolar.

circumscribed, a. circunscrito/a.

circumstance, n. circunstancia (f).

cirque, n. circo (m) glaciar.

cirrus, n.pl. cirros (m); **cirrostratus**, cirroestratos.

cirtoconical, a. cirtocónico/a.

cistern, n. cisterna (f); aljibe (m) (covered domestic water reservoir); **vaulted c.**, aljibe abovedado.

citadel, n. ciudadela (f).

citation, n. citación (f) (Jur).

citric, a. cítrico/a.

citrine, n. cetrino (m).

citrus, a. cítrico/a; **c. fruits**, cítricos (m.pl).

city, n. ciudad (f); **c. council**, ayuntamiento (m), concejo (m) municipal; **c. dweller**, ciudadano/a (m, f), citadino/a (m, f) (Arg, Mex, Perú, Ven); **c. hall**, ayuntamiento (m) (building), municipalidad (f), municipio (m); **c. planning**, urbanismo (m).

civil, a. civil; **c. law**, derecho (m) civil; **c. servant**, funcionario/a (m, f); **c. service**, administración (f) pública.

claim, n. demanda (f), petición (f) (of something); pretensión (f) (to a title); derecho (m) (right); **to lay c. to (v)**, tener pretensiones de.

clamp, n. grapa (f), zincho (m); laña (f) (dog iron); **adjustable c.**, zincho ajustable.

clarification, n. clarificación (f), aclaración (f).

clarifier, n. clarificador (m); **primary c.**, clarificador primario; **upward-flow c.**, clarificador de flujo ascendente.

clarify, v. aclarar, clarificar; clarificar (a liquid).

clarity, n. claridad (f), nitidez (f) (of liquids).

class, n. clase (m).

classification, n. clasificación (f), grado (m); **chemical c.**, clasificación química; **soil c.**, clasificación de suelos.

classifier, n. clasificador (m) (machine), clasificador/a (m, f) (person).

clast, n. clasto (m).

clastic, a. clástico/a; **c. sediment**, sedimento (m) clástico.

claw, n. garra (f) (Zoo, talon), pinza (f) (Zoo, e.g. of crab); garfio (m) (Eng).

clay, n. arcilla (f), barro (m); **brick c.**, arcilla para ladrillos; **c. lining**, revestimiento (m) de arcilla; **fire c.**, arcilla refractaria; **refractory c.**, arcilla refractaria; **thixotropic c.**, arcilla tixotrópica.

clayey, a. argiláceo/a, arcilloso/a.

claypan, n. capa (f) de arcilla, capa arcillosa compacta; **hard c.**, capa dura, tierra (f) endurecida.

claypit, n. barrero (m); **waterfilled c.**, adobera (f), barrero lleno de agua.

claystone, n. argilita (f).

clean, a. limpio/a.

cleaning, n. limpieza (f), limpia (f); **bed c. (Treat)**, limpieza de banco; **dry c.**, limpieza en seco.

cleanliness, n. limpieza (f).

cleansing, 1. n. purificación (f); limpieza (f) (Med); 2. a. limpiador/a, purificador/a.

clean-up, n. limpieza (f), saneamiento (m) (of pollution, rubbish); **c.-u. costs**, costos (m.pl) de limpieza; **c.-u. criteria**, criterios (m.pl) de limpieza.

clear, v. aclarar, despejar; desbrozar (to remove rubbish or scrub).

clearance, n. desmonte (m), roza (f) (de terreno), claro (m).

clear-cut, a. bien definido/a, claro/a.

clearing, n. desmonte (m) (of trees).

clearness, n. nitidez (f) (e.g. of atmosphere).

cleavage, n. clivaje (m), foliación (f), quebradura (f); **slaty c.**, fisilidad (f), pizarrosidad (f) foliación pizarreña (LAm).

cleft, n. grieta (f), hendidura (f), hendedura (f) (Geol).

cliff, n. acantilado (m), precipicio (m) (precipice); **sea c.**, acantilado marino.

climate, n. clima (m) (Gen); ambiente (m) (Fig); **c. change**, cambio (m) climático; **maritime c.**, clima marítimo; **temperate c.**, clima templado; **the c. of opinion**, la opinión (f) general.

climatic, a. climático/a; **c. change**, cambio (m) climático.

climatology, n. climatología (f).

climax, 1. n. clímax (m); **c. community**, comunidad (f) estable, comunidad clímax, comunidad climácica, comunidad final; 2. a. clímax; **c. forest (Ecol)**, bosque (m) clímax.

climbing, n. montañismo (m).

cline, n. cline (m), clina (f).

clinoform, a. clinoformo/a.

clinograph, n. clinográfico (m).

clinometer, n. clinómetro (m).

clinopyroxenite, n. clinopiroxenita (f).

clippers, n.pl. tijeras (f.pl) (for plants); cizalla (f) (for metal sheet).

clisere, n. cliserie (f).

clod, n. champa (f) (Arg); **c. of earth**, terrón (m).

clogging, n. colmatación (f), atascado (m), tamponamiento (m).

clog up, v. atascar, introducir (to insert).

clone, n. clon (m).

closet, n. retrete (m) (lavatory); armario (m), ropero (m) (for clothes); **earth c.**, retrete seco; **water c.**, sanitario (m).

clostridium, n. clostridium (m).

closure, n. cierre (m); **anticlinal c.**, terminación (f) periclinal; **factory c.**, cierre de fábrica; **structural c.**, cierre estructural.

clot, n. coágulo (m).

cloth, n. tela (f), paño (m).

cloud, 1. v. enturbiar (to make cloudy); 2. n. nube (f); **c. seeding,** estimulación (f) de lluvias.

cloudburst, n. chaparrón (m), chaparrón violento.

cloudiness, n. nubosidad (f) (tiempo); turbidez (m), turbiedad (f) (líquido).

cloud over, v. nublar.

cloudy, a. nuboso/a, nublado/a (weather); turbio/a (liquid).

clover, n. trébol (m).

cluster, n. grupo (m), conglomerado (m); **c. analisis (Mat),** análisis (m) de conglomerados/grupos/agrupamientos.

clutch, n. embrague (m).

clypeasteroid, n. clypeasteroide (m).

CNS, n. SNC (m); **central nervous system,** sistema (m) nervioso central.

coadaptation, n. coadaptación (f).

coagulant, n. coagulante (m).

coagulate, v. coagular.

coagulation, n. coagulación (f); **chemical c.,** coagulación química.

coal, n. carbón (m), hulla (f) (soft coal); **bituminous c.,** carbón bituminoso; **c. basin,** cuenca (f) de carbón; **c. industry,** industria (f) del carbón; **c. measure,** capas (f.pl) de carbón (GB); **c. mine,** mina (f) de carbón, mina de hulla; **c. seam,** capa (f) de carbón; **c. tar,** alquitrán (m) mineral.

coalescence, n. coalescencia (f).

coalescent, a. coalescente.

coalface, n. frente (m) de carbón.

coalfield, n. campo (m) de minas de carbón.

coarse, a. grosero/a, tosco/a, basto/a.

coarseness, n. grosería (f).

coast, n. costa (f) (shore), litoral (m) (coastline); **indented c.,** costa dentada; **low c.,** costa baja.

coastal, a. litoral, costero/a, costeño (LAm).

coaster, n. buque (m) costero, barco (m) de cabotaje.

coastguard, n. guardacosta (m).

coastline, n. línea (f) de costa; litoral (coastal land or sea).

coating, n. recubrimiento (m).

cobalt, n. cobalto (m) (Co).

cobaltic, a. cobáltico/a.

cobble, n. guijo (m), canto (m) rodado; guijón (m) (Arg).

cobbled, a. empedrado/a.

cobblestone, n. guijarro (m).

cocaine, n. cocaína (f).

coccolith, n. cocolito (m).

cockroach, n. cucaracha (f), barata (f) (Ch).

cocoa, n. cacao (m); **c. plantation,** cacaotal (m).

cocoon, n. capullo (m).

cod, n. bacalao (m).

COD, n. DQO; **chemical oxygen demand,** demanda (f) química de oxígeno.

code, n. código (m) (laws); clave (m), cifra (f) (cipher); **bar c.,** código de barras; **civil c.,** código civil; **c. of behaviour, c. of conduct,** código de conducta; **c. of practice,** código de buena práctica, código profesional; **machine c. (Comp),** código de máquina; **penal c.,** código penal; **post c.,** código postal.

co-disposal, n. coeliminación (f), codisposición (f) (LAmer).

coefficient, n. coeficiente (m); **c. of expansion,** coeficiente de expansión; **c. of friction,** coeficiente de fricción; **c. of permeability,** coeficiente de permeabilidad; **c. of sorting,** coeficiente de selección; **Coriolis c.,** coeficiente de Coriolis; **regression c.,** coeficiente de regresión; **temperature c.,** coeficiente térmico.

Coelenterata, n.pl. celentéreos (m.pl), el filum Coelenterata (Zoo).

coelenterate, n. celentéreo (m).

coenzyme, n. coenzima (f).

coevolution, n. coevolución (f).

coffee, n. café (m); **c. plantation,** cafetal (m).

cofferdam, n. ataguía (f), compartimiento (m) estanco.

cog, n. rueda (f) dentada.

cogeneration, n. cogeneración (f).

cognate, a. cognado/a semejante; consanguíneo (by blood).

coherent, a. coherente.

cohesion, n. cohesión (f).

cohesive, a. cohesivo/a.

cohort, n. cohorte (f).

coign, n. esquina (f) (corner), parte (m) saliente (overhanging part) (Con).

coil, n. rollo (m) (of rope); carrete (m), bobina (f) (Elec); espiral (f) (of smoke); serpentín (m) (of pipe); **c. spring,** muelle (m) en espiral, resorte (m).

coin, 1. n. moneda (f); 2. v. acuñar.

coincidence, n. coincidencia (f), conformidad (f).

coke, n. coque (m) (coal); Coca (f) (P) (drink); cocaine.

coking, n. coquificación (f), coquización (f); **c. plant,** planta (f) coquizadora.

col, n. col (Geog); **glacial c.,** col glaciar, col glacial.

cold, 1. n. frío (m) (Gen); resfriado (m) (Med); **common c. (Med),** resfriado común; **to be c.**

(v (Met)), hacer frío; **to catch a c. (v)**, acatarrarse, resfriarse; **to catch c. (v)** coger frío; **to have a c. (v)**, estar resfriado; 2. a. frío/a; **c. front**, frente (m) frío; **c. snap, c. spell**, ola (f) de frío; **c. storage**, conservación (f) en cámara frigorífica; **c.-blooded**, de sangre (f) fría.

cold-blooded, a. de sangre fría; **c. animal**, animal (m) de sangre fría.

coldness, n. frialdad (f).

Coleoptera, n.pl. coleópteros (m.pl), el orden Coleoptera (Zoo).

coleopteran, n. coleóptero (m).

coliform, n. coliforme (m); **c. organism**, organismo (m) coliforme; **faecal c., fecal c. (US)**, coliforme fecal; **total coliforms**, coliformes totales.

collaborate, v. colaborar; cooperar (to cooperate).

collaboration, n. colaboración (f), cooperación (f).

collagen, n. colágeno (m).

collapse, 1. v. hundirse (of building), derrumbarse (to cave in); 2. n. derrumbamiento (m), desplome (m) (falling down); colapso (m) (of building); hundimiento (m) (of government); caída (f) (of plans);.

collar, n. cuello (m) (Gen), collar (m) (Con); **c. beam**, tirante (m) falso.

collate, v. colacionar.

collect, v. recoger, pasar por (LAm) (to pick up, to harvest); reunir, juntar (assemble); coleccionar (valubles).

collection, n. recolección (f), recogida (f) (of post), recaudación (f) (of taxes); **c. of samples, sampling**, recolección de muestras, toma (f) de muestras.

collector, n. colector (m), coleccionista (m) (person), recaudador/a (m, f) (of taxes); recogedor (m) (machine); **grit c. (Treat)**, recogedor de partículas.

college, n. colegio (m), universidad (f).

colliery, n. mina (f) de carbón; **c. spoil**, escombros (m.pl) de minas de carbón.

collimation, n. colimación (f); **c. error**, error (m) de colimación.

colloid, n. coloide (m).

colloidal, a. coloidal, coloideo/a.

colluvial, a. coluvial.

colluvium, n. coluvión (f).

Colombia, n. Colombia (f).

Colombian, 1. n. colombiano/a (m, f); 2. a. colombiano/a.

colon, n. colon (m) (Anat), colón (m) (unit of currency) (CRica, ElS).

colonization, n. colonización (f).

colonnade, n. columnata (f).

colony, n. colonia (f); **c. count (Biol)**, contaje (f) de colonias.

color, n. color (m).

coloration, n. coloración (f).

colored, a. colorado/a.

colorimetric, a. colorimétrico/a.

colorimetry, n. colorimetría (f).

colour, n. color (m).

coloured, a. colorado/a.

column, n. columna (f); **c. anchorage**, anclaje (m) de pilares; **extraction c. (Chem)**, columna de extracción; **geological c.**, columna geológica.

combe, n. valle (m) estrecho entre cerros.

combination, n. combinación (f).

combustible, a. combustible.

combustion, n. combustión (f).

comma, n. coma (f).

commensalism, n. comensalismo (m).

commerce, n. comercio (m); **Chamber of C.**, Cámara (f) de Comercio.

commercial, a. comercial.

commercialization, n. comercialización (f).

commercialize, v. comercializar.

comminutor, n. conminutador (m).

commissariat, n. comisaría (f).

commission, n. comisión (f); **European C.**, Comisión Europea; **European Economic C.**, Comisión Económica Europea.

commissioning, n. puesta (f) en servicio (launching).

commodity, n. mercancía (f); mercadería (f) (US).

common, 1. n. ejido (m), campo (m) comunal; **c. land**, ejido, campo comunal; 2. a. común, general.

Commons, n. Comunes (m.pl) (GB, parlamento); **the House of C. (GB)**, La Cámara (f) de los Comunes (centro político del parlamento británico).

Commonwealth, n. la Mancomunidad (f) (Británica) (abbr. the C.); **C. of Independent States**, Comunidad (f) de Estados Independientes.

communicate, v. comunicar, contagiar (to infect).

communication, n. comunicación (f).

communiqué, n. comunicado (m) oficial.

community, n. comunidad (f), colectividad (f), sociedad (f); **c. property**, bienes (m.pl) municipales, comunidad de bienes; **climax c. (Ecol)**, comunidad clímax, comunidad final, comunidad climácica; **European C. (abbr. EC)**, Comunidad Europea (abbr. CE); **European Economic C. (abbr. EEC)**, Comunidad Económica Europea (abbr. CEE); **pioneer c. (Ecol)**, comunidad pionera.

commutator, n. conmutador (m).

compact, 1. v. condensar, comprimir (to compress), apretar (to squeeze); 2. a. compacto/a.

compaction, n. compactación (f), compresión (f); **c. ratio (in landfilling)**, proporción (f) de compactación.

compactness, n. compacidad (f).

compactor, n. compactador (m).

company, n. compañía (f), empresa (f), sociedad (f); **building c., construction c.**,

empresa constructora; **insurance c.**, compañía de seguros; **joint stock c.**, sociedad anónima (abbr. s.a.), sociedad por acciones; **limited c. (abbr. Ltd)**, sociedad anónima (abbr. s.a.); **limited liability c. (abbr. Ltd)**, sociedad (de responsibilidad) limitada (abbr. s.l.), sociedad en comandita; **public limited c. (abbr. plc)**, sociedad anónima (abbr. s.a.); **stock c.**, sociedad anónima, sociedad de acciones; **trading c.**, sociedad mercantil.
compare, v. comparar, equiparar.
comparison, n. comparación (f); **c. of costs**, comparación de costos; **in c. with**, en comparación con.
compartment, n. compartimiento (m); **watertight c.**, compartimiento estanco.
compass, n. compás (m), brújula (f); **magnetic c.**, compás magnético.
compasses, n.pl. compás (m).
compatibility, n. compatibilidad (f).
compendium, n. compendio (m).
compensation, n. compensación (f), indemnización (f), contrapartida (f) (Fin).
competence, n. competencia (f), habilidad (f).
competent, a. competente; **c. authority**, autoridad (f) competente.
competition, n. competición (f).
competitiveness, n. competitividad (f).
competitor, n. competidor/a (m, f).
compilation, n. recopilación (f).
compile, v. compilar, recopilar.
compiler, n. compilador/a (m, f); recopilador (m).
complete, v. completar, acabar (to finish).
complex, 1. n. complejo (m); **basal c.**, complejo basal; **industrial c.**, complejo industrial; 2. a. complejo/a; **c. number**, número (m) complejo.
complexation, n. complejación (f); **c. reactions**, reacciones (f.pl) de complejación.
complexing, a. complexional; **c. agent**, sustancia (f) capaz de formar iones complejos.
complexity, n. complejidad (f).
complexometric, a. complexométrico/a.
compliance, n. conformidad (f), obediencia (f); **c. with**, de acuerdo con.
comply, v. acatar (with the law); cumplir (with rules); acceder (with wishes).
component, 1. n. componente (m); 2. a. componente.
Compositae, n.pl. compuestas (f.pl) (Bot, family).
composite, a. compuesto/a; **c. liner (for landfill)**, revestimiento (m) compuesto; **c. material**, material (m) compuesto.
composition, n. composición (f); **c. of water**, composición del agua; **chemical c.**, composición química.
compost, 1. v. abonar; convertir en abono; 2. n. compostado (m), abono (m), compost (m).
composter, n. abonadora (f), compostador (m).
composting, n. compostación (f), compostifica-

ción (f), compostaje (m), abonado (m); **c. plant**, planta (f) de abonado, planta de compostificación (LAm); **sludge c.**, compostación de lodos.
compound, 1. v. componer (to make by combining); 2. n. compuesto (m) (Chem); recinto (m), cercado (m) (enclosed land); **chemical c.**, compuesto químico; **volatile organic compounds (abbr. VOCs)**, compuestos orgánicos volátiles; 3. a. compuesto/a.
comprehensive, a. amplio/a (extensive), comprensivo/a (understanding).
compress, v. comprimir, apretar (to squeeze).
compressed, a. comprimido/a; **c. air**, aire (m) comprimido.
compressibility, n. compresibilidad (f); **aquifer c.**, compresibilidad del acuífero.
compression, n. compresión (f); **c. bar (Eng)**, barra (f) de compresión; **c. failure**, rotura (f) por compresión; **c. filter**, filtro (m) de compresión; **c. test**, ensayo (m) de compresión; **differential c.**, compresión diferencial.
compressive, a. compresivo/a; **c. strength**, fuerza (f) de compresiva.
compressiveness, n. compresividad (f).
compressor, n. compresor (machine); **air c.**, compresor de aire.
comprise, v. comprender, englobar.
compromise, v. llegar a un arreglo (to come to an agreement); transigir (to yield); comprometer (to jeopardize).
computation, n. cómputo (m), cálculo (m); computación (f) (esp. LAm).
computer, n. ordenador (m); computador (m) (LAm), computador/a (m, f) (person); **c. graphics**, proceso (m) de información gráfica; **c. science**, informática (f); **personal c. (abbr. PC)**, ordenador personal (abbr. PC); **powerful c.**, ordenador potente.
computerization, n. computerización (f).
computerize, v. computerizar.
concave, a. cóncavo/a.
conceal, v. ocultar, esconder.
concentrate, v. concentrar.
concentration, n. concentración (f); **background c.**, concentración de fondo; **evaporative c.**, concentración por evaporación; **ionic c.**, concentración iónica; **maximum permissible c.**, concentración máxima permisible.
concentric, a. concéntrico/a.
concept, n. concepto (m).
conceptual, a. conceptual; **c. models**, modelos (m.pl) conceptuales.
concession, n. concesión (f); **c. or licence holder**, concesionario (m).
concessionary, a. concesionario/a.
conclusion, n. conclusión (f) (termination, deduction); **to come to the c. that (f)**, llegar a la conclusión de que.
concoidal, a. concoidal, concoideo/a.
concrete, 1. v. cubrir con hormigón; 2. n. hormigón (m), concreto (m) (LAm); **aerated**

c., hormigón aireado; c. **blocks,** bloques (m.pl) de hormigón; c. **mixer,** hormigonera (f); c. **wall,** muro (m) de hormigón (concreto); **mass c.,** hormigón en masa; **post-tensioned c.,** hormigón postensado; **precast c.,** hormigón prefabricado; **pre-stressed c.,** hormigón pretensado; **ready-mixed c.,** hormigón preamasado; **reinforced c.,** hormigón armado.

concretion, n. concreción (f).

condensate, n. líquido (m) condensado.

condensation, n. condensación (f).

condense, v. condensar (vapour); abreviar, resumir (text); comprimir (to compress).

condenser, n. condensador (m) (Elec).

condition, n. condición (f) (stipulation); estado (m) (state); **climatic conditions,** condiciones climáticas; **initial conditions,** condiciones iniciales; **weather conditions,** tiempo (m), condiciones atmosféricas; **working conditions,** condiciones de trabajo.

conditioner, n. acondicionador (m); **air c.,** acondicionador de aire; **soil c.,** acondicionador de suelo.

conditioning, n. acondicionamiento (of place, substance); **sludge c.,** acondicionamiento de lodos; **soil c.,** acondicionamiento del suelo.

condom, n. condón (m), preservativo (m).

condominium, n. condominio (m).

conductance, n. conductancia (f); **electrical c.,** conductancia eléctrica.

conductivimeter, n. conductivímetro (m).

conductivity, n. conductividad (f); **electrical c.,** conductividad eléctrica; **hydraulic c.,** conductividad hidráulica; **thermal c.,** conductividad térmica.

conduit, n. conducto (m), caño (m); **open conduits,** conductos abiertos, conducciones abiertas.

cone, n. cono (m); piña (f) (of pine trees); **alluvial c.,** abanico (m) aluvial; **ash c.,** cono de cenizas volcánicas; **cinder c.,** cono de escorias volcánicas; **pine c.,** piña.

confederation, n. confederación (f).

conference, n. conferencia (f), congreso (m); **press c.,** conferencia de prensa.

confidence, n. confianza (f); **c. limits,** límites (m) de confianza; **in c.,** en confianza.

configuration, n. configuración (f).

confined, a. confinado/a; **c. aquifer,** acuífero (m) confinado.

confinement, n. confinamiento (m).

confirmation, n. confirmación (f); homologación (f) (acknowledgement); ratificación (f) (of agreement).

confluence, n. confluencia (f); **river c.,** confluencia de los ríos.

conformity, n. conformidad (f), concordancia (f) (Geol).

confront, v. encarar, enfrentar, hacer frente.

conglomerate, n. conglomerado (m); **basal c.,**

conglomerado basal; **well-sorted c.,** conglomerado de cantos bien seleccionados.

congress, n. congreso (m) (meeting), asamblea (f) legislativa (US).

congruence, n. congruencia (f).

congruous, a. congruente.

Coniacian, n. Coniacense (m).

conifer, n. conífera (f).

Coniferae, n.pl. coníferas (f.pl), el orden Coniferae (Bot).

Coniferales, n.pl. coníferas (f.pl), el orden Coniferales (Bot).

Coniferophyta, n.pl. coníferas (f.pl), el orden Coniferophyta (Bot).

conjugate, a. conjugado/a.

connection, n. conexión (f), comunicación (f); **crossed c.,** conexión cruzada; **household c.,** comunicación domiciliaria; **illegal c.,** conexión ilícita; **river-aquifer hydraulic c.,** conexión hidráulica entre ríos y acuíferos.

conodont, n. conodonto (m).

consensus, n. consenso (m).

consent, n. permiso (m), aprobación (f), consentimiento (m).

consequence, n. consecuencia (f); **grave c.,** consecuencia grave.

consequential, a. consiguiente.

conservation, n. conservación (f) (e.g. of wildlife); preservación (f) (of natural resources); **c. status,** estado (m) de conservación; **nature c.,** conservación de la naturaleza.

conservationist, n. conservacionista (m, f).

conserve, v. conservar.

consolidate, v. consolidar.

consolidation, n. consolidación (f).

consolidometer, n. edómetro (m).

constant, n. constante (m); **arbitrary c.,** constante arbitraria; **c. of disintegration,** constante de desintegración; **decay c.,** constante de desintegración; **dielectric c.,** constante dieléctrica.

constant-head, a. a descenso constante; **c.-h. tank,** tanque (m) a descenso constante.

constellation, n. constelación (f).

constituent, a. constitutivo/a.

constitution, n. constitución (f) (Jur); estatutos (m.pl) (statutes).

construct, v. construir.

construction, n. construcción (f); **steel-frame c.,** construcción metálica.

constructional, a. constructivo/a.

constructive, a. constructivo/a.

consultancy, n. consultoría (f), asesoría (f); **engineering c.,** consultoría de ingeniería.

consultant, n. consultor (m, f); **biological c., c. biologist,** biólogo (m) consultor.

consultation, n. consulta (f); **public c.,** consulta pública.

consulting, n. consultoría (f); **c. engineer,** ingeniero (m) consultor/asesor/consejero.

consumer, n. consumidor/a (m, f); **c. goods,**

bienes (m.pl) de consumo; **c. society,** sociedad (f) de consumo; **primary c. (Ecol),** consumidor primario; **secondary c. (Ecol),** consumidor secundario.

consumption, n. consumo (m) (of food, resources); tisis (Med); **energy c.,** consumo (m) energético; **industrial water c.,** consumo (m) industrial de agua; **unaccounted-for water c.,** consumo (m) no contabilizado de agua.

consumptive, a. destructivo/a; **c. use,** consumo (m) total.

contact, n. contacto (m).

contactor, n. contactor (m); **rotary c. (Treat),** contactor rotatorio.

container, n. envase (m) (packaging, usually of liquids); recipiente (m), contenedor (m) (e.g. skip); **c. ship,** portacontenedor (m).

containerization, n. contenerización (f), introducción (f) en containers.

containment, n. contención (f); **c. site,** lugar (m) de contención.

contaminant, n. contaminante (m), contaminador (m); **biological c.,** contaminante biológico; **chemical c.,** contaminante químico; **conservative c.,** contaminante conservativo; **c. removal,** eliminación (f) de contaminantes; **c. transport,** transporte (m) de contaminantes; **non-biodegradable c.,** contaminante no biodegradable; **radioactive c.,** contaminante radioactivo.

contaminate, v. contaminar.

contaminated, a. contaminado/a; **c. land,** suelo (m) contaminado; **c. land clean-up,** limpieza (f) de suelos contaminados, saneamiento (m) de suelos contaminados; **c. land recovery,** recuperación (f) de suelo contaminado; **highly c.,** altamente contaminado.

contamination, n. contaminación (f); **atmospheric c.,** contaminación atmosférica; **bacterial c.,** contaminación bacteriológica; **beach c.,** contaminación de playas; **biological c.,** contaminación biológica; **c. by hydrocarbons,** contaminación por hidrocarbonos; **c. by pesticides,** contaminación por pesticidas; **c. plume,** penacho (m) de contaminación; **faecal c.,** contaminación fecal; **groundwater c.,** contaminación de las aguas subterráneas; **indirect c.,** contaminación indirecta; **industrial c.,** contaminación industrial; **radioactive c.,** contaminación radioactiva; **water c.,** contaminación de agua.

contemporaneous, a. contemporáneo/a.

content, n. contenido (m) (amount contained); capacidad (f) (capacity); **carbon c.,** contenido en carbón.

continent, n. continente (m); **the C. of Europe,** el continente europeo.

continental, a. continental; **c. drift,** deriva (f) continental.

contingency, n. contingencia (f).

contingent, a. contingente, aleatorio/a.

continual, a. continuo/a.

continue, v. continuar.

continuity, n. continuidad (f); **equation of c.,** ecuación (f) de la continuidad.

continuous, a. continuo/a.

contorted, a. contorsionado/a.

contortion, n. contorsión (f).

contour, n. contorno (m); **c. line,** curva (f) de nivel; **structural c.,** curva estructural.

contouring, n. delineación (f).

contraception, n. contracepción (f) (contraceptive method).

contraceptive, n. anticonceptivo (m).

contract, 1. v. contratar (action); 2. n. contrato (m); **construction c.,** contrato para construcción.

contracting, a. contratante; **the c. parties,** las partes (f.pl) contratantes.

contraction, n. contracción (f); **differential c.,** contracción diferencial.

contractor, n. contratista (m, f); **building c.,** contratista de obras.

contrast, n. contraste (m); **in c. with,** en contraste con.

contravene, v. contravenir.

contribution, n. contribución (f), aportación (f), aporte (m), (LAm); **average yearly c.,** aportación media anual; **state aid c. (Com),** aportación estatal.

control, n. control (m) (Gen); mando (m), dirección (f) (leadership); **automatic c.,** control automático; **biological c. (Agr, Ecol),** control biológico; **birth c.,** control de la natalidad; **chemical c.,** control químico; **c. gate (Eng),** compuerta (f) de control; **c. of water level,** control del nivel freático; **c. room,** sala (f) de control; **c. rules (pl),** reglas (f.pl) de control; **c. tower,** torre (m) de control; **leakage c.,** control de fugas; **out of c.,** fuera de control; **pH c.,** control de pH; **process c.,** control de proceso; **quality c.,** control de calidad; **telemetry c.,** control (m) remoto, teledirección (f); **under c.,** bajo control.

controlable (US), controllable, a. controlable.

conurbation, n. conurbación (f).

convection, n. convección (f); **c. current,** corriente (f) de convección.

convective, a. convectivo/a.

convention, n. convenio (m) (agreement); congreso (m) (meeting); convención (f) (custom).

conventional, a. convencional; **c. energy,** energía (f) convencional.

converge, v. convergir.

convergence, n. convergencia (f); **optical c.,** convergencia óptica.

conversion, n. conversión (f); **energy c.,** conversión energética.

convex, a. convexo/a.

convexity, n. convexidad (f) curvatura (f).

conveyance, n. traspaso (m) (of a property), transporte (m) (transport).

conveyer, n. transportador (m), conductor (m); **c. belt,** banda (f) transportadora.

convolute, v. enrollar.

convoluted, a. enrollado/a, intraplegado/a, convoluto/a; **c. folding,** plegamiento (m) convolucionado.

convolution, n. circunvolución (f).

cooking, n. cocción (f) (Gen) (boiling, cooking, baking).

cool, a. fresco/a; **c. wind,** viento (m) fresco.

cooling, n. enfriamiento (m); **c. ponds,** lagunas (f.pl) de enfriamiento, balsas (f.pl) de enfriamiento; **c. tower,** torre (m) de refrigeración.

coomb, n. valle (m) estrecho entre cerros.

cooperate, v. cooperar; colaborar (collaborate).

cooperation, n. cooperación (f), colaboración (f); **economic c.,** cooperación económica; **technical c.,** cooperación técnica.

cooperative, 1. n. cooperativa (f); **c. society,** sociedad (f) cooperativa; 2. a. cooperativo/a.

coordinate, 1. v. coordinar; 2. n. coordenada (f); **cylindrical c.,** coordenada cilíndrica; **geographical coordinates,** coordenadas geográficas; **spherical c.,** coordenada esférica.

coordinated, a. coordinado/a.

coordination, n. coordinación (f); **lack of c.,** descoordinación (f).

copepod, n. copépodo (m).

Copepoda, n.pl. copépodos (m.pl), la subclase Copepoda (Zoo).

coping, n. albardilla (f).

copolimerization, n. copolimerización (f).

copper, n. cobre (m) (Cu).

copperplating, n. cobreado (m) galvánico.

coppice, n. soto (m), bosquecillo (m), arboleda (f) (grove).

coprecipitation, n. coprecipitación (f).

copse, n. bosquecillo (m) (small wood).

copulation, n. cópula (f) (Biol).

copyright, n. copyright (m), de propiedad (f) literaria, derechos (m.pl) del autor (royalties).

coral, 1, n. coral; 2, a. coralino/a; **c. limestone,** caliza (f) coralina; **c. reef,** arrecife (m) coralino.

corallitic, a. coraliforme.

cordierite, n. cordierita (f).

cordillera, n. cordillera (f) (mountain range).

cordoba, n. córdoba (m) (Nic, unit of currency).

core, n. testigo (m), centro (m) (of Earth), núcleo (m), corazón (m) (of fruit); **c. barrel,** sacatestigos (m), portatestigos (m); **c. box,** caja (f) de testigos; **drilling c.,** testigo de perforación; **oriented c.,** testigo orientado.

Coriolis, n. Coriolis (m); **C. effect,** efecto (m) de Coriolis.

cork, n. corcho (m) (wood); alcornoque (m) (tree).

corm, n. cormo (m), tallo (m) bulbosa (f).

corn, n. trigo (m) (GB, wheat), maíz (m) (US, maize), choclo (m) (LAm), guate (m) (LAm), elote (m) (LAm, on the cob); **c. exchange,** alhóndiga (f); **c. field (of maize),** milpa (f) (LAm); **c. merchant,** alhondiguero (m).

corned beef, n. carne (f) vacuno, carne en conserva, carne de vaca acecinada.

cornfield, n. trigal (m), campo (m) de trigo (GB, of wheat); maizal (m), campo de maíz (US, of maize).

cornice, n. cornisa (f).

corolla, n. corola (f).

corollary, n. corolario (m).

corona, n. corona (f); **luminous c. (Met),** corona luminosa.

corporation, n. sociedad (f) anónima (compañía); corporación; **municipal c.,** ayuntamiento (m).

corpse, n. muerto/a (m, f); cadáver (m); cuerpo (m) (body).

corpuscle, n. glóbulo (m), corpúsculo (m); **blood c.,** corpúsculo sanguíneo, glóbulo sanguíneo.

correction, n. corrección (f); **c. factor,** factor (m) de corrección; **level c.,** corrección de niveles; **pH c.,** corrección de pH.

corrective, a. correctivo/a, corrector/a, reparador/a (remedial); **corrective measures,** medidas (f.pl) correctoras; **to take c. action (v),** tomar las medidas correctivas.

correlate, v. correlacionar.

correlation, n. correlación (f); **coefficient of c., c. coefficient,** coeficiente (m) de correlación; **degree of c.,** grado (m) de correlación; **multiple c.,** correlación múltiple; **stratigraphical c.,** correlación estratigráfica.

correspondence, n. correspondencia (f).

corridor, n. corredor (m), pasillo (m); **river c. (Ecol),** corredor de ribera, galería (f) (LAm); **wildlife c. (Ecol),** corredor natural.

corrie, n. circo (m) glaciar.

corroborate, v. corroborar.

corroboration, n. corroboración (f).

corrode, v. corroer.

corrosion, n. corrosión (f); **bimetallic c.,** corrosión bimetálica; **electrochemical c.,** corrosión electroquímica; **electrolitic c.,** corrosión electrolítica; **fatigue c.,** corrosión por desgaste.

corrosive, a. corrosivo/a.

corrosivity, n. corrosividad (f).

corrugated, a. corrugado/a; **c. cardboard,** cartón (m) corrugado; **c. iron,** chapa (f) ondulada, placa (f) ondulada, catamina (f) (Ch, Peru).

corrugation, n. corrugación (f).

cortisone, n. cortisona (f).

corundum, n. corindón (m).

cos, n. cos (m) (Mat, abbr. for cosine).
cosec, n. cosec (m) (Mat, abbr. for cosecant).
cosecant, n. cosecante (m) (Mat, abbr. cosec).
cosine, n. coseno (m) (Mat, abbr. cos).
cosmetic, a. cosmético/a.
cosmopolitan, a. cosmopolita.
cosolvent, n. cosolvente (m).
cost, n. coste (m) (expense); costo (m), precio (m) (price); gastos (m.pl) (expenses); costas (f.pl) (Jur); **annual administration costs**, gastos anuales de administración; **c.-benefit**, coste beneficio; **c. less depreciation**, costo menos depreciación; **c. of living**, costo de la vida; **c. recovery**, recuperación (f) de costos; **depreciation costs**, gastos de depreciación; **energy costs**, costes de energía, gastos de energía; **financial costs**, gastos financieros; **indirect costs**, gastos indirectos; **installation c.**, coste de instalación; **maintenance costs**, gastos de mantenimiento; **manufacturing costs**, costos de fabricación; **marginal c.**, coste marginal; **variable costs**, costes variables.
Costa Rica, n. Costa Rica (f).
Costa Rican, 1. n. costarricense (m, f), costarriqueño/a (m, f); 2. a. costarricense, costarriqueño/a.
cost-benefit, n. costos-beneficios (m.pl).
cost-cutting, n. reducción (f) en el coste.
cost-effectiveness, n. costo-eficacia (m), efectividad (f) del coste.
costing, n. cálculo (m) de costos.
cost-plus, n. coste (m) más, costo (m) más; **c.-plus-commission basis**, base (m) del coste más comisión; **c.-plus contract**, contrato (m) al costo más beneficio; **c.-plus-fixed-fee contract**, contrato (m) al coste más honorarios fijos; **c.-plus-maintenance depreciation**, depreciación (f) de costo y mantenimiento; **c.-plus-percentage contract**, contrato (m) por el costo más un porcentaje; **c.-plus-profit basis**, base (m) del coste más beneficio.
cost-reduction, n. reducción (f) de costos.
cotan, n. ctg (m) (Mat, abr. for cotangent).
cotangent, n. cotangente (m) (Mat, abbr. ctg).
cotyledon, n. cotiledón (m).
cough, n. tos (f); **whooping c.**, tos ferina.
coulee, n. barranco (m) (ravine).
coulomb, n. culombio (m) (abbr. C, unit of electrical charge).
coulombmeter, n. culombiómetro (m).
council, n. consejo (m) (committee), junta (f); concejo municipal, delegación (f) (LAm) (local government); reunión (f), sesión (f) (of meeting); **C. of Europe**, Consejo de Europa; **city c.**, **town c.**, ayuntamiento (m), delegación (f) (LAm).
counsel, v. asesorar.
counter, n. contador (m).
counteract, v. contrarrestar; oponer (to oppose).

counterbalance, 1. v. contrapesar, contrabalancear, compensar; 2. n. contrapeso (m).
counterbracing, n. contradiagonal (m).
counterclockwise, a. sinistrorso/a, en sentido opuesto a las agujas del reloj.
countermeasure, n. contramedida (f).
counterpressure, n. contrapresión (f).
counter-radiation, n. contrarradiación (f).
countersink, v. avellanar.
counterweight, n. contrapeso (m).
country, n. país (m) (region, state); campo (m) (out of town); **developing c.**, país en vías de desarrollo.
countryside, n. paisaje (m), campiña (f).
country-wide, a. nacional.
course, n. dirección (f), ruta (f) (route); curso (m) (of river, education); rumbo (m) (of ship); **in the c. of construction**, en vías de construcción; **the c. of a river**, el curso de un río.
court, n. corte (f) (Jur); patio (m); **High C.**, Tribunal (m) Supremo; **Supreme C.**, Corte Suprema.
Couvinian, n. Couviniense (m).
covalence, n. covalencia (f).
covalent, a. covalente.
covariance, n. covarianza (f).
cove, n. caleta (f), cala (f), ensenada (f) (Geog).
cover, 1. v. cubrir (Gen), tapar (to cover over), esconder (to hide); 2. n. cubierta (f) (Gen); tapa (f) (lid); pantalla (f) (screen); abrigo (m), cobijo (m) (shelter); **impermeable c.**, pantalla impermeable; **rock c**, cubierta rocosa; **soil c.**, cubierta de suelo; **tree c.**, cubierta arbórea.
coverage, n. cobertura (f).
cowboy, n. vaquero (m), gaucho (m) (Arg); **c. operator (Com)**, pirata (m), gestor (m) pirata.
CPU, n. UPC (m); **Central Processing Unit (Comp)**, Unidad (m) de Procesamiento Central.
cpue, n. cpue (f); **catch per unit effort (fishing, trapping)**, captura (f) por unidad de esfuerzo.
crack, n. grieta (f), fractura (f), rajadura (f); **sun c.**, grieta de desecación, grieta de retracción.
cracking, n. craqueo (m), cracking (m) (Chem).
craftsman, n. artesano (m).
crag, n. peñasco (m), risco (m); **c. and tail (Geol)**, risco y cola.
crane, n. grulla (f) (Zoo); grúa (f) (hoist); güinche (m) (LAm); **gantry c.**, grúa de pórtico; **tower c.**, grúa de torre.
cranium, n. cráneo (m).
crankshaft, n. cigüeñal (m).
crater, n. cráter (m); **meteorite c.**, cráter meteorítico; **side c.**, cráter lateral.
craton, n. cratón (m).
crazing, n. grietas (f.pl) cruciformes.

credit, n. crédito (m) (Fin), prestigio (m) (prestige), honor (m) (honour); **c. account,** cuenta (f) de crédito.

creek, n. cala (f), ensenada (f), riachuelo (m) (US), caleta (f); arroyuelo (small creek).

creep, 1. v. reptar (e.g. soils), deslizar; 2. n. reptación (f), deslizamiento (m); **soil c.,** reptación del suelo.

creeper, n. enredadera (f) (plant).

crematorium, n. horno (m) crematorio.

crenulate, a. crenulado/a.

crepuscular, a. crepuscular.

crescent, n. creciente (m); **Fertile C. (Geog),** Creciente Fértil.

cress, n. berro (m); **cressbed,** prado (m) de berros.

crest, n. cresta (f), cima (f); copa (f) (of a tree); charnela (f) (hinge); **anticlinal c.,** charnela del anticlinal.

Cretaceous, n. Cretácico (m).

crevasse, n. grieta (f).

cricket, n. grillo (m) (insect); criquet (m) (sport).

crinoid, n. crinoide (m).

Crinoidea, n.pl. crinoideos (m.pl), la clase Crinoidea (Zoo).

criosphere, n. criosfera (f).

crisis, n. crisis (f); **c. management,** control (m) de crisis.

cristobalite, n. cristobalita (f).

criterion, n. criterio (m).

critical, a. crítico/a; **c. path method,** método (m) del camino crítico.

criticism, n. crítica (f).

crockery, n. vajilla (f).

crop, n. cosecha (f); **c. protection,** protección (f) de cubiertas vegetales; **c. spraying,** fumigación (f) con plaguicidas; **cover c.,** cosecha actual; **standing c.,** cosecha estable, cosecha permanente.

cropland, n. tierra (f) de cultivo, tierra agrícola.

cropper, n. cosechadora (f) (machine).

cropping, n. cosecha (f); **continuous c.,** cosecha continua.

cross, 1. v. cruzar; atravesar (to pass through); 2. n. cruz (f) (in form of X); cruce (m) (crossing); cruzamiento (m) (Bot, Zoo, hybridization); cruce (m), cruzado/a (m, f) (hybrid); **back-c.,** retrocruzamiento (m), cruzamiento prueba; **c.-fertilization,** fertilización (f) cruzada; **c.-pollination,** polinización (f) cruzada; **Red C.,** Cruz Roja.

crossbeam, n. viga (f) transversal.

cross-bed, n. laminación (f) cruzada; entrecruzamiento (m) (LAm).

cross-bedded, a. con laminación cruzada; entrecruzado/a (LAm).

crossbreed, 1. v. cruzar (to hybridize); 2. n. híbrido (m), cruzado/a (m, f).

cross-country, 1. a. campo a través, a campo traviesa, a campo través; 2. n. **c.-c. skiing** esquí (m) de fendo; **c.-c. race** cross (m).

cross-current, n. contracorriente (f).

cross-examination, n. interrogatorio (m) hecho para comprobar lo declarado anteriormente; **c.-e. questions,** repreguntas (f.pl).

cross-flow, n. corriente (f) cruzada, flujo (m) transversal.

cross-grained, a. de fibras irregulares, de fibras cruzadas (for wood).

crossing, n. travesía (f) (by sea), cruce (m) (intersection), paso (m) (passage); **pedestrian c.,** cruce (m) de peatones.

cross-pollination, n. polinización (f) cruzada.

cross-reference, n. referencia (f), remisión (f).

crossroads, n. cruce (f).

cross-section, n. corte (m) transversal.

cross-vaulting, n. bóveda (f) de arista.

crosstie, n. traviesa (f) (US).

crosswalk, n. cruce (m) de peatones (US).

crowbar, n. palanca (f).

crowd, n. muchedumbre (m), multitud (f); choclón (m) (Ch).

crown, n. corona (f) (Gen); copa (de un árbol); copete, cresta (of a bird); coronamiento (Con).

crucible, n. crisol (m).

crumbly, a. fragmentado/a, desmoronable.

crush, v. aplastar; moler, triturar (to grind).

crust, n. costra (f), corteza (f); **ice c.,** costra de hielo; **the Earth's c.,** la corteza terrestre.

Crustacea, n.pl. crustáceos (m.pl), Crustácea (f.pl) (Zoo, class).

crustacean, 1. n. crustáceo (m); 2. a. crustáceo/a.

crustal, a. cortical.

cryolite, n. criolita (f).

cryoscopy, n. crioscopia (f).

cryptocrystalline, a. criptocristalino/a.

crystal, n. cristal (m); **c. lattice,** enrejado (m) cristalino; **liquid c.,** cristal líquido.

crystalline, a. cristalino/a.

crystallization, n. cristalización (f); **fractional c.,** cristalización fraccionada.

crystallography, n. cristalografía (f).

Cuba, n. Cuba (f).

Cuban, 1. n. cubano/a (m, f); 2. a. cubano/a.

cube, 1. v. elevar al cubo (Mat); 2. n. cubo (m); **c. root (Mat),** raíz (f) cúbica.

cubic, a. cúbico/a.

cucumber, n. pepino (m).

cul-de-sac, n. callejón (m) sin salida, calle (f) ciega (LAm).

cullet, n. chatarra (f) de vidrio, desperdicios (m.pl) de vidrio.

culmination, n. culminación (f).

cultivable, a. cultivable.

cultivar, n. variedad (f) (Agr, Bot, variety).

cultivate, v. cultivar.

cultivation, n. cultivo (m).

culture, n. cultura (f) (heritage); cultivo (m) (cultivation); cría (f) (e.g. of shellfish); **c. medium (Biol),** caldo (m) de cultivo; **c. of microorganisms,** cultivo de microorganismos; **tissue c.,** cultivo celular, cultivo in vitro.

culvert, n. alcantarilla (f); encajamiento (m) (boxed in); **box c.,** alcantarilla cuadrada.

cumec, n. medida (f) de flujo igual a metro cúbico por segundo.

cumulative, a. cumulativo/a, acumulativo/a; **c. distribution,** distribución (f) acumulativa.

cumulonimbus, n. cumulonimbos (m).

cumulous, a. cúmulo/a.

cumulus, n. cúmulos (m); **altocumumlus,** altocúmulos.

cupola, n. cúpula (f).

cupric, a. cúprico/a.

cuprite, n. cuprita (f).

cupronickel, n. cuproníquel (m).

cuprous, a. cuproso/a.

curdling, n. cuajada (f).

curie, n. curie (m) (abbr. Ci, unit of measurement of radioactivity).

currant, n. grosella (f); **c. bush,** grosellera (f).

current, 1. n. corriente (f) (water, electricity, etc.); **alternating c.,** corriente alterna; **coastal c.,** corriente litoral; **c. meter (for rivers),** cuadalímetro (m), molinete (m), micromolinete; **direct c.,** corriente directa; **Humboldt c.,** corriente de Humboldt; **longshore c.,** corriente marginal; **rip c.,** corriente de resaca; **suspension c.,** corriente de suspensión; **three-phase c.,** corriente trifásica; **tidal c.,** corriente de marea; **turbidity c.,** corriente de turbidez; 2. a. corriente, actual; **c. assets,** cuenta (f) corriente; **c. year,** año (m) en curso.

curriculum, n. currículum (m); **c. vitae,** currículum vitae.

curtain, n. cortina (f) (for windows), telón (m) (screen); **grout c.,** gonitado (m), telón de lechada.

curvature, n. curvatura (f).

curve, n. curva (f); **backwater c.,** curva de remanso; **breakaway c.,** curva de escape; **cumulative c.,** curva acumulativa; **c.-fitting,** coincidencia (f) de curvas; **double-mass c.,** curva de doble masa; **drying c.,** curva de secado; **envelope c.,** curva envolvente; **frequency c.,** curva de frecuencia; **gaussian c.,** curva gausiana; **hypsometric c.,** curva hipsométrica; **logarithmic c. (abbr. log curve),** curva logarítmica; **population growth c.,** curva de crecimiento de población; **pump characteristic c.,** curva característica de una bomba; **rainfall intensity duration c.,** curva de intensidad de lluvia-duración; **recession c.,** curva de recesión; **regression c.,** curva de regresión; **reservoir storage c.,** curva de volumen embalsado; **time-drawdown c.,** curva de descenso-tiempo; **type c.,** curva tipo; **well c.,** curva de pozo.

curved, a. encorvado/a.

cusec, n. medida (f) de flujo igual a pie cúbico por segundo.

cushioning, a. amortiguador/a; **c. effect,** efecto (m) amortiguador.

cusp, n. cúspide (f).

cuspate, a. apuntado/a (pointed); **c. foreland,** antepaís (m) apuntado.

custard-apple, n. chirimoya (f) (fruit).

custodian, n. custodio (m), guardián/guardiana (m, f), conservador/a (m, f) (of a museum).

cut, v. cortar.

cut-and-fill, n. desmonte (m) y terraplén, desmonte (m) y relleno (m), corte (m) y relleno.

cutaneous, a. cutáneo/a.

cuticle, n. cutícula (f).

cut-off, n. corte (m), cese (m); brazo (m) muerte (US, of a river).

cut-out, n. interruptor; **automatic-time c.-out/ switch,** interruptor (m) eléctrico automático; **low-level c.-out,** interruptor de nivel mínimo.

cutter, n. cortadora (f) (cutting machine).

cutting, n. esqueje (m) (plant); desmonte (m) (road, railway).

cuttings, n.pl. recortes (m.pl); detritos (m.pl) (Min); **drill c.,** detritos de perforación; **newspaper c.,** recortes (m) de periódicos.

cwm, n. circo (m) glaciar (País de Gales).

cyanamide, n. cianamida (f).

cyanate, n. cianato (m).

cyanide, n. cianuro (m).

cyanite, n. cianita (f).

cyanobacteria, n.pl. cianobacteria (f.pl).

cyanogen, n. cianógeno (m).

Cyanophyta, n.pl. cianófitas (f.pl), la división Cyanophyta (Bot).

cyanophyte, n. cianófita (f).

cybernetics, n. cibernética (f).

cycle, n. ciclo (m); bicicleta (f) (vehicle); **carbon c.,** ciclo de carbono; **cell c.,** ciclo celular; **cycles per second (abbr. c/s),** ciclos por segundo; **economic c.,** ciclo económico; **food c.,** ciclo alimentario; **geochemical c.,** ciclo geoquímico; **hydrological c.,** ciclo hidrológico; **lunar c.,** ciclo lunar; **menstrual c.,** ciclo menstrual; **nitrogen c.,** ciclo de nitrógeno; **oxygen c.,** ciclo de oxígeno; **sedimentary c.,** ciclo sedimentario; **solar c.,** ciclo solar; **water c.,** ciclo de agua.

cyclic, a. cíclico/a.

cyclical, a. cíclico/a.

cycloalkane, n. cicloalcano (m).

cycloalkene, n. cicloalquino (m).

cyclohexane, n. ciclohexano (m).

cyclohexanol, n. ciclohexanol (m).

cyclone, n. ciclón (m).

cyclonic, a. ciclónico/a.

cycloparaffins, n. cicloparafinas (f.pl).

cyclotherm, n. cicloterma (f).

cylinder, n. cilindro (m).

cylindrical, a. cilíndrico/a.

cytochemistry, n. citoquímica (f).

cytology, n. citología (f).

cytolysis, n. citólisis (f).

cytoplasm, n. citoplasma (m).

D

dacite, n. dacita (f).

dacron, n. dacrón (m).

dairy, n. vaquería (f) (part of a farm); lechería (f) (creamery, shop); tambo (m) (Arg, Par, Ur, small farm); **d. products,** productos (m.pl) lácteos.

dam, 1. v. represar, embalsar; 2. n. presa (f), embalse (m); **arch d.,** presa de arco; **buttress d.,** presa con contrafuertes; **coffer d.,** presa encofrada; **earth d.,** presa de tierra; **gravity d.,** presa de gravedad; **impounding d.,** presa de retención; **masonary d.,** presa de mampostería; **rockfill d.,** presa de relleno.

damage, n. daño (m); **d. evaluation,** evaluación (f) de daños.

damming, n. represamiento (m) (action).

dampen, v. mojar.

dampening, n. amortiguación (f).

damper, n. regulador (m) de tiro (for chimney), comprobador (m) de vibraciones (for vibrations), amortiguador (m) (shock absorber).

damp-proofing, n. hidrofugación (f).

damselfly, n. libélula (f).

damson, n. ciruela (f) damascena.

dandelion, n. diente (m) de león.

danger, n. peligrosidad (f).

dangerous, a. peligroso/a, arriesgado/a (hazardous).

Danian, n. Daniense (m).

data, n.pl. datos (m.pl); **available d.,** datos disponibles; **climatic d.,** datos climáticos; **d. analysis,** análisis (m) de datos; **d. collection,** toma (f) de datos; **d. logger,** registrador (m) automático de datos; **d. processing,** procesamiento (m) de datos; **d. set,** conjunto (m) de datos; **experimental d.,** datos experimentales; **ranked d.,** datos categorizados, datos ordenados por sus rangos; **raw d.,** datos a tratar, datos no evaluados, datos no analizados, datos sin procesar, datos en bruto; **statistical d.,** datos estadísticos.

database, n. base (f) de datos; **computer d., computerized d.,** base de datos informática.

date, 1. v. datar; 2. n. fecha (f) (time); dátil (m) (fruit); **d. tree,** palmera (f), palmera datilera; **expiry d.,** fecha de caducidad; **out-of-d.,** pasado de moda; **sell-by d.,** fecha de caducidad; **up-to-d.,** moderno, al dia.

dating, n. datación (f); **geological d.,** datación geológica; **radiocarbon d.,** datación con radiocarbono; **tritium d.,** datación con tritio.

datum, n. dato (m).

daughter, n. hija (f).

dawn, n. amanecer (m), primera luz (f); **at d.,** al amanecer.

daybreak, n. amanecer (m), primera luz (f); **at d.,** al amanecer.

daylight, n. luz (f) del día, a plena luz.

DC, n. CC; **direct current (Elec),** corriente (f) continua.

DDT, n. DDT; **dichlorodiphenyltrichloroethane,** diclorodifenil-tricloroetano (m).

dead, 1. v. estar muerto (to be dead); 2. a. muerto/a; **d. centre,** punto (m) muerto.

deadline, n. fecha (f) límite, fecha tope.

deadman, n. macizo (m) de anclaje (Con, to resist force).

deaeration, n. desaireación (f).

dealer, n. negociante (m, f), comerciante (m, f) (merchant).

dealing, n. compraventa (f) (buying and selling).

death, n. muerte (f); fallecimiento (m) (de personas); **the Black D.,** muerte negra, peste (f) negra, peste bubónica.

debark, v. descortezar (a tree).

debate, n. debate (m).

debilitation, n. debilitamiento (m), debilitación (f).

debris, n. despojo (m), detrito (m), detritus (m).

debug, v. depurar.

debugging, n. depuración (f).

decade, n. decenio (m).

decant, v. decantar, transvasar.

decantation, n. decantación (f).

decanter, n. decantador (m) (Treat).

decanting, n. decantación (f).

decapod, n. decápodo (m).

Decapoda, n.pl. decápodos (m.pl), Decapoda (f.pl) (Zoo, order).

decarbonation, n. decarbonatación (f).

decay, 1. v. descomponer, pudrirse (e.g. of plants); decaer (e.g. of building); 2. n. descomposición (f) (e.g. of plants).

decentralization, n. descentralización (f).

decentralize, v. descentralizar.

dechlorination, n. desclorinización (f), descloración (f).

decibel, n. decibelio (m), decibel (m).

deciduous, a. decíduo/a, caduco/a, de hoja caduca.

decimal, 1. n. decimal (m); 2. a. decimal.

decipher, v. descifrar, solucionar (solve).

decision, n. decisión (f); **d. making process,** proceso (m) de toma de decisiones.

declare, v. manifestar.

declination, n. declinación (f); **magnetic d.,** declinación magnética.

decommissioning, n. cierre (m) definitivo.

decompose, v. descomponer.

decomposition, n. descomposición (f); **aerobic**

d., descomposición aerobia; **anaerobic d.,** descomposición anaerobia.

decompression, n. descompresión (f).

decongestion, n. descongestión (f).

decontaminate, v. descontaminar.

decontamination, n. descontaminación (f).

decoration, n. decoración (f), ornamentación (f).

decrease, n. disminución (f), reducción (f); rebaja (f) (in wages).

decree, n. decreto (m) (order), estatuto (m) (law).

deduction, n. deducción (f), inferencia (f); descuento (m) (discount).

deem, v. juzgar, considerar.

deep, a. profundo/a, hondo/a; **d. well,** pozo (m) profundo.

deepen, v. profundizar.

deep-freeze, n. congelador (m), congeladora (f) (LAm).

deep-seated, a. emplazamiento (m) profundo (Geol).

defect, n. defecto (m).

defective, a. defectivo/a, imperfecto/a, defectuoso/a.

defence, n. defensa (f); **fire d.,** defensa contraincendio.

defensive, a. defensivo/a.

deficiency, n. deficiencia (f) (Gen); falta (f) (lack); defecto (m) (defect); carencia (f) (scarcity); **d. disease,** enfermedad (f) carencial.

deficit, n. déficit (m); **saturation d.,** déficit de saturación; **soil moisture d.,** déficit de humedad en el suelo.

defile, n. desfiladero (m), cañada (f) (ravine).

define, v. definir.

defined, a. definido/a; **well-d.,** bien definido.

definite, a. definido/a.

definition, n. definición (f).

deflation, n. deflación (f).

deflection, n. desviación (of a stream); flexión (f) (flexure).

deflocculent, n. anticoagulante (m).

defluoridation, n. defluorización (f), defluoruración (f).

defoamant, a. antiespumante.

deforest, v. deforestar, despoblar de árboles.

deforestation, n. deforestación (f).

deformation, n. deformación (f); **elastic d.,** deformación elástica.

defray, v. sufragar; costear, pagar (to pay); **to d. the costs,** sufragar los gastos.

degas, v. desgasificar.

degassed, a. desgasificado/a.

degassing, n. desgasificación (f).

degeneration, n. degeneración (f).

degradable, a. degradable.

degradation, n. degradación (f).

degrade, v. degradar, disminuir.

degrease, v. desengrasar.

degreaser, n. desengrasador (m) (Trat, machine).

degreasing, n. desengrasado (m); desgrasado (m) (LAm).

degree, n. grado (m) (on a scale); punto (m) (extent); título (m) (university); **d. absolute,** grado absoluto; **d. Celsius,** grado Celsius; **d. centigrade,** grado centígrado; **d. Fahrenheit,** grado Fahrenheit; **d. of freedom,** grado de libertad; **d. of saturation,** grado de saturación; **d. of uncertainty,** grado de incertidumbre; **first d.,** licenciatura (f) (de una universidad); **second d.,** posgrado (m).

dehalogenation, n. deshalogenación (f).

dehiscent, a. dehiscente.

dehumidifier, n. deshumidificador (m).

dehumidify, v. deshumidificar.

dehydrate, v. deshidratar.

dehydrated, a. deshidratado/a.

dehydration, n. deshidratación (f).

deincrustation, n. desincrustación (f); **chemical d.,** desincrustación química.

deionization, n. desionización (f).

deionize, v. desionizar.

delay, 1. v. retrasar; 2. n. retraso (m), tardanza (f), lag (m).

delayed, a. retardado/a, retrasado/a; diferido/a (deferred); **d. storage, d. yield,** drenaje (m) retardado.

delegate, v. delegar.

delegation, n. delegación (f).

delimit, v. delimitar.

delimiting, n. delimitación (f).

delineate, v. delinear, deslindar; trazar (to trace).

delineating, n. deslinde (m).

deliquesce, v. licuarse.

deliquescence, n. delicuescencia (f).

deliquescent, a. delicuescente.

delta, n. delta (m); **arcuate d.,** delta encorvado; **cuspate d.,** delta triangular, delta cuspidada.

deltaic, a. deltáico/a.

deltoid, n. deltoide.

deluge, n. diluvio (m) (of rain).

demand, n. demanda (f); **biochemical oxygen d. (abbr. BOD),** demanda bioquímica de oxígeno (abbr. DBO); **chemical oxygen d. (abbr. COD),** demanda química de oxígeno; **fluctuating d.,** demanda variable/oscilante/elástica; **industrial d.,** demanda industrial; **oxygen d.,** demanda de oxígeno.

demineralization, n. desmineralización (f).

demineralize, v. desmineralizar.

demographic, a. demográfico/a.

demography, n. demografía (f).

demolish, v. demoler (a building).

demolition, n. demolición (f), derribo (m) (of a building).

demonstrate, v. demostrar, probar.

demonstration, n. demostración (f) (proof, display); manifestación (f) (mass meeting); **d. plant,** planta (f) de demostración.

denature, v. desnaturalizar.
dendritic, a. dendrítico/a.
dendrochronologist, n. dendrocronólogo/a (m, f).
dendrochronology, n. dendrocronología (f).
dendroid, n. dendroide (m).
denitrification, n. desnitrificación (f).
denitrify, v. denitrificar.
denitrifying, a. desnitrificante.
dense, a. denso/a; espeso/a (thick).
denseness, n. espesura (f).
densimeter, n. densímetro (m).
density, n. densidad (f); **absolute d.,** densidad absoluta; **apparent d.,** densidad aparente; **bulk d.,** densidad de masa, densidad aparente; **d. of saline water,** densidad del agua salina; **population d.,** densidad de población; **probability d.,** densidad de probabilidad; **relative d.,** densidad relativa; **spectral d.,** densidad espectral.
dental, a. dental.
denudation, n. denudación (f).
denunciation, n. denuncia (f) (accusation); denunciación (f) (action).
deodorization, n. desodorización (f), deodorización (f).
deodourization, n. desodorización (f), deodorización (f).
deoxidant, n. desoxidante (m).
deoxidation, n. desoxidación (f).
deoxidising, a. desoxidante.
deoxidization, n. desoxidación (f).
deoxidize, v. desoxidar.
deoxy-, pf. desoxi-.
deoxygenate, v. desoxigenar.
deoxygenation, n. desoxigenación (f).
department, n. departamento (m), sección (f) (administrative section); ramo (m) (branch); **D. of Justice (US),** Departamento de Justicia; **D. of Trade,** Departamento de Comercio, Ministerio (m) de Comercio; **d. store,** gran almacén (m), tienda (f) de departamentos.
depleted, a. agotado/a, empobrecido/a; vacío/a (empty).
depletion, n. disminución (f) (lessening of stocks); agotamiento (m) (exhaustion of stocks); depleción (f); **d. of resources,** agotamiento de recursos.
deployment, n. despliegue (m).
depopulate, v. despoblar.
depopulation, n. despoblación (f), despoblamiento (m).
deposit, 1. v. depositar, sedimentar; 2. n. depósito (m), yacimiento (m); **alluvial d.,** depósito aluvial; **channel-filled d.,** depósito de relleno de cauce; **estuarine d.,** depósito de estuario; **shallow-water d.,** depósito de aguas someras.
deposition, n. deposición (f); **sedimentary d.,** sedimentación (f).
depositional, a. deposicional.

depreciate, v. depreciar.
depreciation, n. depreciación (f).
depression, n. depresión (f) (Gen, Met); hoya (f) (hollow); **pumping d.,** depresión de bombeo.
depth, n. profundidad (f), hondura (f); **ten metres in d.,** diez metros de profundidad.
deregulate, v. desregular, liberalizar.
deregulation, n. desregulación (f), liberalización (f).
derelict, a. abandonado/a, negligente (in duty); **d. land,** tierras (f.pl) abandonadas.
derivation, n. derivación (f).
derivative, n. derivada (f) (Mat); derivativo (m) (Chem); **partial d. (Mat),** derivada parcial; **petroleum d.,** derivado del petróleo.
derive, v. derivar.
dermal, a. dérmico/a.
dermatitis, n. dermatitis (f).
dermatology, n. dermatología (f).
dermis, n. dermis (f).
derogate, v. derogar, ir en contra de.
derogation, n. derogación (f) (de una ley), menosprecio (m) (contempt).
derrick, n. grúa (f) (crane in port); torre (f) de perforación (oil rig); **oil d.,** torre de extracción.
DES, n. **Department of Education and Science (GB),** departamento (m) de educación y ciencia.
desalinate, v. desalinizar, desalinar, desalar.
desalinating, a. desalinizador/a; **d. capacity,** capacidad (f) desalinizadora; **d. plant,** planta (f) desalinizadora.
desalination, n. desalinización (f), desalación (f); **d. by hydration,** desalinización por hidratación; **d. by ion exchange,** desalinización por intercambio iónica; **d. by reverse osmosis,** desalinización por ósmosis inversa; **d. installed capacity,** capacidad (f) instalada de desalinización; **d. systems,** sistemas (m.pl) de desalinización; **nuclear d.,** desalinización nuclear.
desander, n. desarenador (m) (machine) (Treat).
desert, 1. n. desierto (m); **low-latitude d.,** desierto de baja latitud; **sand d.,** desierto de arena; 2. a. desértico/a.
deserted, a. despoblado/a, abandonado/a.
desertification, n. desertificación (f) (through Man's activities), desertización (natural process).
desiccate, v. desecar.
desiccation, n. desecación (f).
design, n. diseño (m); **d. fault,** fallo (m) en el diseño.
designer, n. diseñador/a (m, f), trazador/a (m, f).
desilicification, n. desilicificación (f).
desilification, n. desilificación (f).
desilting, n. descolmatación (f).
desludging, n. desenlodamiento (m).

desorption, n. desorción (f).
dessicated, a. desecado/a.
destabilize, v. desestabilizar, inestabilizar.
destructive, a. destructivo/a, destructor/a.
desulfurization, n. desulfurización (f).
desulphurization, n. desulfurización (f).
detachment, n. separación (f) (separation), objetividad (objectivity).
detail, 1. v. detallar; pormenorizar (to itemize); 2. n. detalle (m); pormenor (m) (item, particular).
detailed, a. detallado/a; pormenorizado/a (itemized).
detection, n. detección (f); **d. limit,** límite (f) de detección.
detector, n. detector (m).
detention, n. detención (f), arresto (m); **d. time (Treat),** período (m) de detención.
detergent, n. detergente (m); **anionic d.,** detergente aniónico; **synthetic d.,** detergente sintético.
deterioration, n. deterioro (m), empeoramiento; daño (m) (damage); **d. in water quality,** deterioro en la calidad de las aguas.
determinand, n. determinante (m).
determination, n. determinación (f); **field d.,** determinación de campo; **laboratory d.,** determinación de laboratorio.
deterministic, a. determinista, determinístico/a; **d. model,** modelo (m) determinista.
detonate, v. detonar, explosionar (to explode), volar (to blow up).
detonator, n. detonador (m).
detoxification, n. desintoxicación (f).
detoxify, v. desintoxicar.
detriment, n. detrimento (m), perjuicio (m) (prejudicial); **to the d. of.,** en detrimento de, en mengua de, con menoscabo de.
detrimental, a. perjudicial.
detrital, a. detrítico/a.
detritivore, n. detritívoro (m).
detritus, n. detrito (m).
deuterium, n. deuterio (m).
devastate, v. asolar, devastar.
devastation, n. asolamiento (m), devastación (f).
developer, n. desarrollador (m) (urban developer), especulador/a (m, f) urbanizador.
developing, a. revelado (m) (photography); **d. country,** país (m) en vías de desarrollo.
development, n. desarrollo (m) (Gen); urbanización (f) (building); explotación (f) (resources); novedad (f), cambio (m) (change in situation); **aquifer d.,** desarrollo del acuífero; **building d.,** urbanización (f), conjunto (m) urbanístico; **d. agency,** agencia (f) de desarrollo; **d. area,** zona (f) de desarrollo, polo (m) de desarrollo; **d. bank,** banco (m) de desarrollo; **d. company,** companía (f) de explotación; **d. corporation,** corporación (f) de desarrollo, corporación de promoción; **d. plan,** plan (m)

de desarrollo; **economic d.,** explotación (f) económica; **effective d.,** desarrollo efectivo; **housing d.,** conjunto (m) urbanístico; **rural d.,** desarrollo rural; **sustainable d.,** desarrollo sostenible, explotación sostenible; **well d. by overpumping,** desarrollo del pozo por sobrebombeo.
deviation, n. desviación (f); **standard d.,** desviación estándar, desviación uniforme.
device, n. aparato (m), mecanismo (m) (gadget, etc.); dispositivo (m); **output d. (Comp),** dispositivo de salida.
devitrification, n. devitrificación (f).
Devonian, n. Devónico (m).
dew, n. rocío (m), sereno (m) (night dew); **d. point,** punto (m) de rocío, temperatura (f) saturación; **d. pond,** lodar (m), charca (f) de condensación, laguna (f).
dewater, v. desaguar.
dewatering, n. deshidratación (f), agotamiento (m) (of mines); **sewage sludge d.,** deshidratación de los fangos cloacales.
dextrin, n. dextrina (f).
diabase, n. diabasa (f).
diachronous, a. diacrónico/a.
diagenesis, n. diagénesis (m).
diagenetic, a. diagenético/a.
diagnose, v. diagnosticar.
diagnosis, a. diagnóstico (m).
diagnostic, a. diagnóstico/a.
diagonal, 1. n. diagonal (f); 2. a. diagonal.
diagonalize, n. trazar la diagonal, diagonalizar.
diagram, n. diagrama (m); **circular d.,** diagrama circular; **contour d.,** diagrama de frecuencia; **flow d.,** diagrama de flujo; **logarithmic d.,** diagrama logarítmico; **scatter d.,** diagrama de dispersión; **triangular d.,** diagrama triangular; **tridimensional d.,** diagrama tridimensional; **two-dimensional d.,** diagrama bidimensional.
dial, n. dial (m), esfera (f).
dialysis, n. diálisis (f).
dialyze, v. dializar.
diameter, n. diámetro (m); **effective d.,** diámetro eficaz; **pump d.,** diámetro de la bomba; **well d.,** diámetro del pozo.
diamond, n. diamante (m).
diapause, n. diapausa (f).
diaphragm, n. diafragma (m); **d. wall,** diafragma de contención.
diapir, n. diapiro (m) (Geol).
diapositive, n. diapositiva (f).
diarrhea, n. diarrea (f).
diarrhoea, n. diarrea (f).
diatom, n. diatomea (f), diatomita (f).
diatomaceous, a. diatomáceo/a; **d. earth,** tierra (f) de diatomeas, diatomita (f).
Diatomeae, n.pl. diatomeas (f.pl), la clase Diatomaea (Zoo).
dichlorobenzene, n. diclorobenzol (m).
dichloroethane, n. dicloretano (m).
dichograptid, n. dichograpto (m).

dichotomy, n. dicotomía (f).
dichromate, n. bicromato (m), dicromato (m).
dicotyledon, n. dicotiledónea (f).
Dicotyledonae, n.pl. dicotiledóneas (f.pl), Dicotyledonae (f.pl) (Bot, class).
dicotyledonous, a. dicotiledóneo/a.
dictionary, n. diccionario (m).
dicyclical, a. dicíclico/a.
didactic, a. didáctico/a.
diductor, n. diductor (m).
didymograptid, n. didymograptido (m).
dieback, v. secarse.
dieldrin, n. dieldrín (m).
diet, n. dieta (f), alimentación (f); régimen (m) (restricted intake).
die-off, n. decaimiento (m); **bacterial d.**, decaimiento bacteriano.
dietary, a. alimenticio/a; **d. regime**, alimentación (f), régimen (m) alimenticio.
difference, n. diferencia (f), discrepancia (f); **finite d.**, diferencia finita; **head d. (Eng)**, diferencia de nivel.
differential, a. diferencial; **d. equation**, ecuación (f) diferencial; **d. pressure**, presión (f) diferencial; **d. settlement**, hundimiento (m) diferencial.
differentiate, v. diferenciar.
differentiation, n. diferenciación (f).
diffraction, n. difracción (f).
diffuse, v. difundir, extender.
diffuser, n. difusor (m); **dome d.**, domo (m) difusor.
diffusion, n. difusión (f); **coefficient of d.**, coeficiente (m) de difusión; **fluid d.**, difusión de fluido; **molecular d.**, difusión molecular.
diffusive, a. difusor/a.
diffusivity, n. difusividad (f); **coefficient of d.**, coeficiente (m) de difusividad; **hydraulic d.**, difusividad hidráulica; **molecular d.**, difusividad molecular.
diffusor, n. difusor (m) (Treat).
dig, v. excavar (Arch, Eng), profundizar (to make a hole), buscar (to search for), escarbar (animal); **d.-and-haul cleanup (US)**, recuperación (f) de suelos contaminados a base de excavación y transporte.
digest, v. digerir.
digested, a. digestivo/a; **d. sludge**, fango (m) de digestión.
digester, n. digestor (m); **sewage sludge d.**, digestor (m) de fangos de alcantarilla, asimilador de fangos de alcantarilla.
digestion, n. digestión (f); **acid d.**, digestión por ácido; **aerobic d.**, digestión aerobia; **anaerobic d.**, digestión anaerobia; **bacterial d.**, digestión bacteriana; **d. tank**, tanque (m) de digestión; **mesophilic d.**, digestión mesofílica; **sludge d.**, digestión de fango.
digestive, a. digestivo/a; **d. system**, aparato (m) digestivo.
digger, n. excavador/a (m, f) (e.g. for Arch); waste tip scavenger (Ven).

digit, n. dígito (m) (Mat); dedo (m) (Anat).
digital, a. digital; **d. mapping**, cartografía (f) digital, mapeo (m) (LAm).
digitalization, n. digitalización (f).
digitation, n. digitación (f).
digitize, v. digitalizar.
dike, n. dique (m) (protective wall), zanja (f) (ditch).
dilate, v. dilatar.
dilation, n. dilatación (f).
dilemma, n. dilema (m), disyunctiva (f).
dilute, v. diluir; **d. and disperse policy (for landfilling)**, política (f) de diluir y dispersar.
dilution, n. dilución (f).
dimension, n. dimensión (f).
dimensional, n. dimensional; **three-d.**, tridimensional; **two-d.**, bidimensional.
dimethylamine, n. dimetilamina (f).
dimethylbenzene, n. dimetilbenzol (m).
diminish, v. disminuir.
diminished, a. disminuido/a, menguado/a (decreased).
diminution, n. disminución (f), mengua (f) (detriment).
diminutive, a. diminutivo/a.
dinosaur, n. dinosaurio (m).
diode, n. diodo (m).
dioecious, a. dioico/a.
diopside, n. diópsido (m).
diorite, n. diorita (f).
dioxane, n. dioxano (m).
dioxide, n. dióxido (m); **carbon d.**, dióxido de carbono; **sulphur d.**, dióxido de azufre; **titanium d.**, dióxido de titanio.
dioxin, n. dioxina (f).
dip, 1. v. inclinar, buzar; 2. n. buzamiento (m), inclinación (f) (Geol); **apparent d.**, buzamiento aparente; **down-d.**, buzamiento hacia abajo; **regional d.**, buzamiento regional; **true d.**, buzamiento verdadero; **up-d.**, buzamiento hacia arriba.
diphenyl, n. difenol (m).
diphteria, n. difteria (f).
diploid, a. diploide.
diplopod, n. diplópodo (m).
Diplopoda, n.pl. diplopódos (m.pl), la clase Diplopoda (Zoo).
dipole, n. dipolo (m).
Diptera, n.pl. dípteros (m.pl), el orden Diptera (Zoo).
dipteran, 1. n. díptero (m); 2. a. díptero/a.
direct, a. directo/a.
direction, n. dirección (f), rumbo (m), sentido (m); **preferred d.**, dirección preferente.
directive, n. directriz (f), directiva (f), orden (f); **EC D. (Jur), European D.**, directiva europea, directiva comunitaria.
director, n. director (m); **managing d.**, director gerente.
directorate, n. directiva (f) (asamblea de directores).
directory, n. directorio (m) (book of instruc-

tions); **telephone d.**, guía (f) telefónica, anuario (m) telefónico.

dirt, n. suciedad (f) (dirtiness); tierra (f) (earth); excremento (m) (faeces); **d. farmer (US, Fam)**, cultivador (m), pequeño granjero (m); **d. road, d. track**, camino (m) de tierra, camino sin firme, pista (f) de ceniza.

dirty, a. sucio/a.

disagreement, n. desacuerdo (m).

disappear, v. desaparecer.

disappearance, n. desaparición (f).

disapproval, n. desaprobación (f).

disaster, n. desastre (m); **man-made d.**, desastre provocado por el hombre; **natural d.**, desastre natural, catástrofe (f) natural.

disc, n. disco (m); **compact d. (abbr. CD)**, disco compacto (abbr. CD); **d. drive**, disquetera (f).

discharge, n. descarga (f) (unloading); desecho (m), vertido (m) (of waste); **d. pipe**, tubería (f) emisora, cañón (m) de vertido, tubo (m) de descarga; **direct d.**, vertido directo; **groundwater d.**, descarga de agua subterránea; **indirect d.**, vertido indirecto; **induced d.**, descarga inducida; **industrial d.**, desechos industriales, agua (f) industrial; **natural d.**, descarga natural; **spring d.**, descarga de manantiales.

discharging, n. descarga (f).

disclosure, n. divulgación (f), revelación (f).

discoidal, a. discoidal.

discolour, v. decolorar.

discolouration, n. decoloración (f).

disconnection, n. desconexión (f).

discontinuity, n. discontinuidad (f).

discount, n. descuento (m); **d. rate (Fin)**, tipo (m) de descuento.

discover, v. descubrir.

discovery, n. descubrimiento (m), divulgación (f); hallazgo (m) (finding).

discrepancy, n. discrepancia (f).

discriminant, a. discriminante.

discrimination, n. discriminación (f).

discritization, n. discretización (f); **spatial d.**, discretización espacial.

discritize, v. discretizar.

disease, n. enfermedad (f); **Chagas' d.**, mal (m) de Chagas; **communicable d.**, enfermedad transmisible; **foot and mouth d.**, fiebre (f) aftosa; **gastrointestinal d.**, enfermedad gastrointestinal; **parasitic d.**, enfermedad parasitaria; **respiratory d.**, enfermedad respiratoria; **waterborne d.**, enfermedad de origen hídrico.

diseased, a. enfermo/a.

disentailment, n. desamortización (f) (Jur) (freeing of land).

dish, n. plato (m), placa (f); **Petri d. (Biol)**, placa de Petri.

disinfect, v. desinfectar.

disinfectant, n. desinfectante (m).

disinfection, n. desinfección (f); **UV d. (Treat)**, desinfección por luz UV.

disinformation, n. desinformación (f).

disintegrate, v. desintegrarse, disgregarse.

disintegration, n. desintegración (f).

disintegrator, n. desintegrador (m).

disk, n. disco (m); **d. filter**, filtro (m) de disco; **floppy d.**, disquete (f), diskette (m).

diskette, n. diskette (m), disquete (f) (Comp).

dislocated, a. dislocado/a.

dislocation, n. dislocación (f), falla (f); tiro (m) (throw).

dismantling, n. desmantelamiento (m).

dismissal, n. despido (m) (from a job), licenciamiento (of employees).

disparate, a. dispar.

dispel, v. ahuyentar.

dispersal, n. dispersión (f).

dispersant, n. dispersante (m).

disperse, v. dispersar; **attenuate and d. policy (for landfilling)**, política (f) de atenuar y dispersar.

dispersed, a. disperso/a.

dispersion, n. dispersión (f); **hydrodynamic d.**, dispersión hidrodinámica; **isotropic d.**, dispersión isotrópica; **lateral d.**, dispersión lateral; **longitudinal d.**, dispersión longitudinal; **molecular d.**, dispersión molecular.

dispersive, a. dispersivo/a; **d. model**, modelo (m) dispersivo.

dispersivity, n. dispersividad (f); **coefficient of d.**, coeficiente (m) de dispersividad.

displace, v. desplazar.

displacement, n. desplazamiento (m), deslizamiento (m); **mass d.**, deslizamiento de masas; **piston d.**, desplazamiento del pistón.

display, v. desplegar, exhibir.

disposal, n. descarga (f), vertido (m) (discharge); destrucción (f), eliminación (f) (elimination); arreglo (m), colocación (f), disposición (f) (arrangement); **d. of hazardous waste to sea**, descarga/vertidos de residuos peligrosos al mar; **sewage d.**, evacuación (f) de las aguas residuales.

disposing, n. descarga (f) (dumping).

disposition, n. disposición (f), trazado (m) (layout).

disrepair, n. desarreglo, mal estado (bad state).

disrupt, v. interrumpir (meeting); alterar (plans).

dissect, v. disecar.

dissection, n. disección (f), disecación (f).

disseminate, v. diseminar, propagar (propaganda).

dissemination, n. diseminación (f).

dissimilate, v. desasimilar.

dissimilation, n. desasimilación (f).

dissipate, v. disipar (e.g. uncertainty); derrochar (e.g. wealth).

dissociate, v. disociar.

dissociation, n. disociación (f).

dissoluble, a. disoluble, soluble.

dissolution, n. disolución (f).
dissolvable, a. soluble, disoluble.
dissolve, v. disolver.
dissolved, a. disuelto/a; **d.-air flotation,** flotación (f) por aire disuelto; **d. oxygen,** oxígeno (m) disuelto; **d. salts,** sales (f.pl) disueltas; **d. solids,** sólidos (m.pl) disueltos.
dissolvent, n. disolvente, solvente.
dissuasion, n. disuación (f), desincentivización (f).
distance, n. distancia (f), lejanía (f).
distancing, n. distanciamiento (m), espaciamiento (m).
distant, a. alejado/a.
distil, v. destilar.
distillate, n. destilado (m).
distillation, n. destilación (f); **d. by thermocompression,** destilación por termocompresión; **flash d.,** destilación instantánea; **fractional d.,** destilación fraccional; **multistage d.,** destilación multiefecto; **solar d.,** destilación solar; **steam d.,** destilación de vapor; **vacuum d.,** destilación al vacío.
distilled, a. destilado/a.
distort, v. deformar, torcer; distorsionar (Geol).
distortion, n. torcimiento (m), deformación (f), distorsión (f); **geometric d.,** deformación geométrica.
distributed, a. distribuido/a.
distribution, n. distribución (f); reparación (f) (division); **binomial d.,** distribución binomial, distribución binómica; **d. panel,** panel (m) de distribución; **d. system,** partidor (m), sistema (m) de distribución; **frequency d.,** distribución de frecuencias; **gamma d.,** distribución gamma; **Gaussian d.,** distribución de Gaus; **log-normal d.,** distribución logarítmico-normal; **Poisson d.,** distribución de Poisson; **statistical d.,** distribución estadística; **Student's t d.,** distribución de t de Student.
distributor, n. distribuidor (m) (machine), distribuidor/a (person), partidor (m) (network).
district, n. región (f), comarca (f) (of country); distrito (m), barrio (m) (of city); **residential d.,** barrio residencial.
disturbance, n. perturbación (f), disturbio (m).
disturbed, a. perturbado/a.
disuasive, a. disuasorio/a.
disymmetry, n. disimetría (f).
ditch, 1. v. zanjar (to dig a ditch or trench); 2. n. zanja (f) (general), cuneta (f) (at roadside), acequia (f) (for irrigation), foso (m) (for defence); **drainage d.,** zanja de drenaje, zanja para avenamiento; **open d.,** zanja a cielo abierto; **oxidation d.,** zanja de oxidación; **recharge d.,** zanja de recarga.
diuretic, 1. n. diurético (m); 2. a. diurético/a.
diurnal, a. circadiano/a, diurno/a; **d. rhythm,** ritmo (m) diurno, ritmo circadiano; **d.**

variation, variación (f) circadiana, variación diurnal.
dive, v. bucear (to swim under water).
divergence, n. divergencia (f); **river d.,** divergencia de ríos.
divergent, a. divergente.
diversification, n. diversificación (f).
diversify, v. diversificar.
diversion, n. diversión (f) (pastime); desviación (f); desvío (m) (in road); **d. channel,** canal (m) de desviación, canal desviador; **d. dam,** presa (f) de desviación; **river d.,** desvío del río.
diversity, n. diversidad (f); **biological d.,** diversidad biológica; **d. index,** índice (m) de diversidad; **d. spatial d.,** diversidad espacial; **species d.,** diversidad de especies.
divide, 1. n. divisoria (f); 2. v. separar, dividir, desunir.
divided, a. dividido/a, separado/a, partido/a; **d. by,** dividido por.
dividend, n. dividendo (m) (Fin).
dividing, a. divisorio/a; **d. line,** línea (f) divisoria.
diving, n. buceo (m) (skindiving); **d. team,** equipo (m) de buceo.
division, n. división (f).
DNA, n. ADN (m); **deoxyribonucleic acid,** ácido (m) desoxirribonucleico.
dock, n. dársena (f), muelle (m), dique (m), puerto (m) (port); acedera (f), ramazo (m) (Bot); **docks,** muelles; **dry d.,** dique seco, dique en dársena, varadero (m); **floating d.,** dique flotante.
docker, n. descargador (m).
dockyard, n. astillero (m).
documentation, n. documentación (f).
dodecahedron, n. dodecaedro (m).
DOE, n. **Department of the Environment (GB),** Departamento (m) del Medioambiente.
dog rose, n. escaramujo (m), rosal (m) silvestre.
doldrums, n.pl. zona (f) de calmas ecuatoriales.
dolerite, n. dolerita (f).
doline, n. dolina (f) (karstic depression), bassa (f).
dollar, n. dólar (m).
dolmen, n. dolmen (m) (Arch).
dolomite, n. dolomita (f).
dolomitic, a. dolomítico/a; **d. limestone,** caliza (f) dolomítica.
dolomitization, n. dolomitización (f).
dolphin, n. delfín (m).
domain, n. dominio (m); **public d.,** dominio público.
dome, n. domo (m), cúpula (f); **exfoliation d.,** domo de exfoliación; **injection d.,** domo de inyección; **periclinal d.,** domo periclinal; **salt d.,** domo de sal; **volcanic d.,** domo volcánico.
domestic, a. doméstico/a; **d. rubbish, d. waste,** basura (f) doméstica.
dominant, a. dominante.

doming, n. formación (f) de domos (in landfilling).

Dominican, 1. n. dominicano/a (m, f); 2. a. dominicano/a.

Dominican Republic, n. República (f) Dominicana.

dominion, n. dominio (m).

donate, v. donar.

donation, n. donativo (m).

donor, n. donante (m, f).

doorway, n. portal (m).

dormancy, n. dormancia (f), letargo (m) (Zoo, Bot).

dormant, a. inactivo/a, latente.

dormin, n. dormina (f).

dorsal, a. dorsal.

dory, n. gallo (m), pez (f) de San Pedro; **john d.,** gallo (m) (pez).

dosage, n. dosificación (f).

dose, 1. v. dosificar; 2. n. dosis (f); **permissible d.,** dosis permisible; **radiation d.,** dosis de radiación.

dosimeter, n. dosímetro (m), dosificador (m).

dosing, n. dosificación (f); **d. chamber,** cámara (f) de dosificación, vasija (f) de dosificación; **d. siphon,** sifón (m) de dosificación.

DOT, n. **Department of Transport (GB), Department of Transportation (US),** departamento (m) de transporte.

double, a. doble; **d.-action pump,** bombeo (m) con acción doble.

doubling, n. doblamiento (m).

down-gradient, n. descenso (m).

downcut, v. profundizar, excavar.

downgrade, v. degradar.

downhill, n. cuesta (f) abajo.

down-hole, a. fondo del sondeo, fondo de la perforación; **d. log,** registro (m) del sondeo, registro de perforación; **d. logging,** testificación (f) del sondeo, registrado del sondeo; **d. treatment,** tratamiento (m) a fondo de pozo.

downpour, n. chaparrón (m), aguacero (m) fuerte, tromba (f) de agua; **heavy d.,** fuerte aguacero, palo (m) de agua.

downslope, n. pendiente (f) descendente.

downstream, n. aguas (f.pl) abajo, corriente (m) abajo.

downthrow, n. tirado (m) hacia abajo.

downward, p. hacia abajo.

downwarp, n. alabeo (m) hacia abajo.

dowser, n. rabdomante (m), hidroscopista (m, f); zahorí (m) (also soothsayer).

dowsing, n. rabdomancia (f), hidroscopia (f); **d. rod,** varita (f) de avellano para la rabdomancia.

dozer, n. bulldozer (m), excavadora (f).

dracunculosis, n. dracunculosis (m).

draft, 1. v. redactar (text), esbozar (drawing); 2. n. borrador (m), redacción (f); **first d.,** primera redacción.

draftsman, n. delineante (m).

drag, 1. v. arrastrar (a cable); dragar, rastrear (a river); 2. n. resistencia (f) (aerodynamic); **d. coefficient,** coeficiente (m) de resistencia.

dragline, n. excavadora (f) de cuchara de arrastre, funidraga (f), dragalina (f).

dragonfly, n. caballito (m) de diablo, libélula (f).

drain, 1. v. agotar (to empty); desaguar (e.g. a mine, a river); avenar (Agr); secar (ropas); desecar (a marsh); vaciar (a boiler); 2. n. desagüe (f), dren (m), drenaje (m) (e.g. from house), boca (f) de alcantarilla, sumidero (m) (in street), alcantarillado (m) (main sewer), desaguadero (m) (for water); **d. hole,** desaguadero (m), agujero (m) de desagüe; **d. pipes,** tubos (m) de drenaje; **french d.,** dren francés, drenaje francés, zanja (f) de avenamiento rellena de grava; **house d.,** desagüe interior, desagüe dentro del edificio; **mole d.,** dren de topo, drenaje de topo; **reverse d.,** dren inverso, drenaje inverso; **storm d.,** drenaje para tormenta; **tile d.,** zanja (f) para drenaje con tubos de barro cocido.

drainable, a. drenable, desaguable.

drainage, n. desagüe (m) (e.g. from house); avenamiento (m) (by rivers); saneamiento (m) (of land); drenaje (m) (artificial); desecación (f) (of lake); alcantarillado (m) (sewage system); **agricultural d.,** drenaje agrícola; **braided d.,** drenaje anastomosado; **coastal d.,** drenaje litoral; **delayed d.,** drenaje diferido; **dendritic d.,** avenamiento dendrítico; **d. area (river catchment),** área (f) de drenaje, cuenca (f) hidrográfica; **d. basin,** cuenca (f) hidrográfica, cuenca vertiente, área (f) de recepción, hoya (f) de avenamiento; **d. by a river,** drenaje por un río; **d. canal,** canal (m) de drenaje, canal de desagüe, canal de alcantarillado; **d. ditch,** zanja (f) para avenamiento, zanja para salida del agua sobrante del riego; **d. level (Min),** galería (f) de desagüe, socavón (m); **d. network density,** densidad (f) de la red de drenaje; **d. pattern,** diseño (m) de avenamiento; **d. pipe,** tubo (m) de drenaje; **d. pump,** bomba (f) de agotamiento, bomba de desagüe (Min); **d. shaft,** pozo (m) de desagüe; **d. sump,** poceta (f) de desagüe; **d. system,** sistema (m) de drenaje; **d. tube,** tubo (m) de drenaje; **d. tunnel,** túnel (m) de drenaje; **gravitational d.,** drenaje gravitacional; **internal d.,** drenaje interno; **land d.,** saneamiento del terreno, drenaje del terreno; **main d. (for sewage),** colector (m) de drenaje; **road d.,** drenaje de carreteras; **surface d.,** drenaje superficial.

drained, a. drenado/a, desecado/a, purgado/a; **tile d.,** drenado con tubos de barro cocido.

drainpipe, n. colector (m) de drenaje.

draught, n. corriente (f) de aire.

draughtsman, n. delineante (m).

drawbridge, n. puente (m) levadizo.

drawdown, n. descenso (m) (Geol); **distance-d.,**

descenso-distancia; **groundwater d.**, descenso de nivel de agua subterránea; **observed d.**, descenso observado; **piezometric d.**, descenso piezométrico; **residual d.**, descenso residual; **specific d.**, descenso específico; **theoretical d.**, descenso teórico; **time-d.**, descenso-tiempo.

drawing, n. dibujo (m); **scale d.**, dibujo a escala.

draw up, v. trazar (to prepare); **to d. up a programme**, trazar un programa.

dredge, 1. v. dragar (p.e. a harbour); 2. n. draga (f).

dredger, n. draga (f).

dredging, n. dragado (m) (action), fango (m) de dragado (material); **d. machine**, draga (f).

drift, n. deriva (f) (deviation); cambio (m) (change of direction); derrubio (m) (Geog, deposit); **continental d.**, deriva continental, deriva de los continentes; **d. net**, red (f) de deriva; **genetic d.**, deriva genética; **glacial d.**, derrubio glaciar, derrubio glacial; **longshore d.**, deriva (f) litoral.

drill, 1. v. perforar, sondear, barrenar; horradar (to bore through); 2. n. barreno (m) (machine); taladradora (f); **d. bit**, barrena (f), trépano (m); **pneumatic d.**, taladradora neumática.

driller, n. perforador (m), taladrador (m).

drillhole, n. perforación (f), sondeo (m).

drilling, 1. n. perforación (f), sondeo (m); **air d.**, perforación con aire comprimido; **diamond d.**, perforación con corona de diamantes; **direct circulation d.**, perforación con circulación directa; **percussion d.**, perforación a percusión; **reverse circulation d.**, perforación con circulación inversa; **rotary d.**, perforación rotativa; **rotary percussion d.**, sondeo de rotopercusión; **small-diameter d.**, perforación de pequeño diámetro; 2. a. perforado/a.

drink, n. bebida (f); **bottled drinks**, bebidas envasadas; **soft d.**, bebida refrescante, refresco (m).

drive in, v. hincar (to insert); **to d. in a piezometer**, hincar un piezómetro.

drizzle, 1. v. lloviznar; 2. n. llovizna (f), lluvia (f) ligera; garúa (f) (LAm); **freezing d.**, llovizna helada.

drogue, n. ancla (f) flotante.

drone, n. zángano (m) (insect); zumbido (m) (noise).

drop, n. gota (f) (of liquid); caída (f), disminución (f).

droplet, n. gotita (f).

dropsy, n. hidropesía (f).

drought, n. sequía (f); **d. sequence**, secuencia (f) de sequías, serie (m) de sequías.

drown, v. anegar (to flood), ahogar (to suffocate).

drowned, a. anegado/a (Gen), ahogado/a (e.g. animals), inundado/a (land), mojado/a (soaked); **d. weir**, vertedero (m) anegado/a.

drug, n. droga (f).

drum, n. tambor (m) (barrel, drum of winch), bidón (m) (can); **oil d.**, tambor de aceite.

drumlin, n. drumlin (m) (glacial hillock).

drupe, n. drupa (f).

dry, 1. v. secar; 2. a. seco/a.

dryer, n. secador (m) (device), desecador (m); **band d.**, banda (f) secador.

drying, 1. n. desecación (f); 2. a. secante.

dryness, n. sequedad (f).

DTI, n. **Department of Trade and Industry (GB)**, departamento (m) de comercio y industria.

dual-carriageway, n. autovía (f), carretera (f) de doble calzada.

duality, n. dualidad (f).

duck, n. pato (m).

duct, n. conducto (for water, gas, electricity); canal (m) (Anat).

ductile, a. dúctil.

dullness, n. deslustre (m).

dump, 1. v. descargar (to unload), verter (to dispose of); 2. n. vertedero (m) (tip); cancha (f) (Min), escombrera (f) (also slagheap); **rubbish d.**, vertedero; **d. truck**, volquete (m).

dumping, n. basculamiento (m) (tipping), descarga (f) (discharging); **d. ground**, basculamiento (m), vertedero (waste tip).

dumptruck, n. dumper (m) (US).

dune, n. duna (f), médano (m); **coastal d.**, duna costera; **longitudinal d.**, duna longitudinal; **sand d.**, duna de arena.

dung, n. estiércol (m), excrementos (m.pl).

dungheap, n. muladar (m).

durabilty, n. durabilidad (f).

duration, n. duración (f); **flow d. curve**, curva (f) de duración-caudal; **short-lived d.**, poca duración, efímero (m).

dust, n. polvo (m), cenizas (f.pl); **coal d.**, polvo de carbón; **d. cloud**, polvareda (f), tolvanera (f); **volcanic d.**, ceniza volcánica.

dustbin, n. cubo (m) de basura, balde (m) (esp. LAm), pipote (m) (Ven).

dustcloud, n. tolvanera, terral (LAm).

Dutch, a. holandés/holandesa.

duty-free, a. exento de aduanas.

dye, n. tinte (m) (colourant), colorante (m), color (m) (colour); **d. tracer**, trazador (m) de color; **fast d.**, color sólido.

dyke, n. dique (m) (barrier), canal (m), acequia (f) (channel), calzada (f) (causeway, embankment); **d. swarm**, enjambre (m) de diques, haz (m) de diques; **Dutch d.**, dique holandés; **radial d.**, dique radial; **ring d.**, dique anular.

dynamic, a. dinámico/a.

dynamics, n. dinámica (f); **fluid d.**, dinámica de fluidos; **population d.**, dinámica poblacional.

dyphenyl, n. difenilo (m).

dysentery, n. disentería (f).

dysprosium, n. disprosio (m) (Dy).

dystrophic, a. distrófico/a.

E

ear, n. oreja (f) (Gen); oído (m) (Zoo); espiga (f) (e.g. of corn); mazorca (f) (of maize); **inner e.,** oído interno; **middle e.,** oído medio; **outer e.,** oído externo.

eardrum, n. tímpano (m).

earth, n. tierra (f) (ground), suelo (m) (soil), La Tierra (f) (planet Earth); **black e.,** tierra negra; **brown e.,** tierra parda; **diatomaceous e.,** tierra de diatomeas, tierra diatomácea; **e. closet,** retrete (m) seco; **electrical e.,** toma (f) de tierra; **loose e.,** tierra floja, tierra suelta.

earthmoving, n. movimiento (m) de tierras; **e. equipment,** equipo (m) para el movimiento de tierras.

earthquake, n. terremoto (m), sismo (m), seísmo (m).

earthwork, n. terraplén (m); relleno (m) (backfill), movimiento (m) de tierras.

earthworm, n. lombriz (f).

earthy, a. terroso/a.

easement, n. servidumbre (f), de paso, derecho (m) de paso.

East, 1. n. oriente (m); **Far E.,** Extremo Oriente, Lejano Oriente; **Middle E.,** Oriente Medio; 2. a. este, del este.

eastern, a. oriental.

ebb, 1. v. menguar, bajar (the tide); 2. n. reflujo (m) (of tide); **e. and flow,** el flujo (m) y reflujo; **e. tide,** marea (f) menguante, marea baja.

ebonite, n. ebonita (f).

ebulloscope, n. ebuloscopio (m).

EC, n. CE (f); **European Community,** Comunidad (f) Europea.

eccentric, a. excéntrico/a.

eccentricity, n. excentricidad (f).

ecdysis, n. ecdisis (f).

echinoderm, n. equinodermo (m).

Echinodermata, n.pl. equinodermos (m.pl), el filum Echinodermata (Zoo).

Echinoidea, n. equinoideos (f.pl).

echninoid, n. equinoide (m).

echo, n. eco (m); **e. sounder,** ecosondador (m), ecosonda (f); sonda (f) acústica.

ECLA, n. CEPAL (f); **Economic Commission for Latin America,** Comisión (f) Económica para América Latina.

eclogite, n. eclogita (f).

ECM, n. MCE (m); **European Common Market,** Mercado (m) Común Europeo.

eco-, pf. eco-.

ecobalance, n. ecoequilibrio (m).

ecocentric, a. ecocéntrico/a.

ecodevelopment, n. ecodesarrollo (m).

ecohydrologic, a. ecohidrológico/a.

ecolabelling, n. eco-etiquetado (m), etiquetado (m) ecologista.

eco-labelling, n. eco-etiquetado (m), etiquetado (m) ecologista.

ecological, a. ecológico/a; **e. succession,** sucesión (f) ecológica.

ecologist, n. ecólogo/a (m, f); ecologista (m,f) (environmentalist).

ecology, n. ecología (f); **aquatic e.,** ecología acuática.

economic, a. económico/a; **e. prosperity,** prosperidad (f) económica.

economically, ad. económicamente; **e. viable,** económicamente viable.

economist, n. economista (m, f).

economize, v. economizar.

economy, n. economía (f); **market e.,** economía de mercado.

ecosensitive, a. ecosensible.

ecosphere, n. ecosfera (f).

ecosystem, n. ecosistema (m), biosistema (m); **Amazon e.,** ecosistema (m) amazónico; **Andean e.,** ecosistema (m) andino.

ecotone, n. ecotono (m), interfase (f).

ecotourism, n. ecoturismo (m).

ecotoxicological, a. ecotoxicológico/a.

ecotoxicology, n. ecotoxicología (f).

ecotype, n. ecotipo (m).

ECSC, n. CECA (f); **European Coal and Steel Community,** Comunidad (f) Europea del Carbón y del Acero.

ectoparasite, n. ectoparásito (m).

ECU, n. ECU (f); **European Currency Unit,** Unidad (f) Monetaria Europea.

Ecuador, n. El Ecuador (m).

Ecuadoran, 1. n. ecuatoriano/a (m, f); 2. a. ecuatoriano/a.

Ecuadorian, 1. n. ecuatoriano/a (m, f); 2. a. ecuatoriano/a.

eczema, n. eczema (m).

edaphic, a. edáfico/a; **e. factor,** factor (m) edáfico.

edaphology, n. edafología (f), pedología (f).

eddy, n. contracorriente (m) (countercurrent), remolino (m) (whirling movement).

EDF, n. FED (m); **European Development Fund,** Fondo (m) Europeo de Desarrollo.

edge, n. arista (f) (Geog), borde (m) (of town), afueras (f.pl) (of lake), orilla (f) (of object), margen (m) (of paper); **on e.,** de canto.

edging, n. cenefa (f).

edifice, n. edificio (m).

editor, n. director/a (m, f) (of newspaper), redactor/a (m, f) (of publisher), montador/a (of TV, cinema, radio).

EEC, n. CEE (f); **European Economic Community,** Comunidad (f) Económica Europea.

eel, n. anguila (f); **conger e.,** congrio (m); **moray e.,** morena (f).

eelworm, n. anguílula (f).

effect, n. efecto (m); **barometric e.,** efecto barométrico; **diurnal temperature e.,** efecto de las temperaturas diurnas; **greenhouse e.,** efecto invernadero; **osmotic e.,** efecto osmótico; **skin e. (flow of water),** efecto pelicular, efecto Kelvin.

effective, a. efectivo/a; **e. porosity,** porosidad (f) efectiva; **e. rainfall,** lluvia (f) efectiva, precipitación (f) pluvial.

effectiveness, n. efectividad (f); **e. of treatment,** efectividad de tratamiento.

effervescence, n. efervescencia (f).

efficacy, n. eficacia (f).

efficiency, n. eficiencia (f); rendimiento (m) (of machine); **barometric e.,** eficiencia barométrica; **energy e.,** eficiencia energética; **relative e.,** eficiencia relativa; **well e.,** eficiencia de pozo.

efflorescence, n. eflorescencia (f).

efflorescent, a. eflorescente.

effluent, n. efluente (m), emisión (f); **domestic e.,** efluente doméstico; **e. discharge,** descarga (f) de efluente; **e. quality,** calidad (f) de emisión, calidad de efluente; **e. standard,** norma (f) de emisión; **final e.,** agua (f) depurada; **raw e.,** efluente crudo; **treated e.,** efluente tratado.

effort, n. esfuerzo (m) (work); intento (m) (attempt); **fishing e.,** esfuerzo de pesca.

effusion, n. efusión (f).

EFTA, n. AELC (f); **European Free Trade Association,** Asociación (f) Europea de Libre Comercio.

egg, n. huevo (m).

eggplant, n. berenjena (f); **e. field,** berenjenal.

egyptology, n. egiptología (f) (Arch).

Eh, n. redox potencial.

EIA, n. EIA (f); **environmental impact assessment,** evaluación (f) de impacto ambiental.

Eifelian, n. Eifeliense (m).

eigenvalue, n. valor (m) propio.

eigenvector, n. vector (m) propio.

EIS, n. DIA (f); **Environmental Impact Statement (GB),** Declaración (f) de Impacto Ambiental (Sp).

eject, v. eyectar, echar, arrojar, expeller.

ejected, a. arrojado/a.

ejector, n. eyector (m); **pneumatic e.,** eyector neumático.

elaborate, v. elaborar.

elastic, a. elástico/a.

elasticity, n. elasticidad (f); **coefficient of e.,** coeficiente (m) de elasticidad.

elastomer, n. elastómero (m).

elder, n. saúco (m).

eldrin, n. eldrín (m).

electric, a. eléctrico/a; **e. current,** corriente (m) eléctrica; **e. heater,** estufa (f) eléctrica.

electrical, a. eléctrico/a; **e. appliance,** electrodoméstico (m); **e. engineer,** ingeniero/a (m, f) eléctrico/a; **e. engineering,** ingeniería (f) eléctrica.

electrician, n. electricista (m, f).

electricity, n. electricidad (f).

electrification, n. electrificación (f).

electrify, v. electrificar (e.g. railway system).

electro-plating, n. electroplastia (f).

electroanalysis, n. electroanálisis (m).

electrocardiogram, n. electrocardiograma (m) (Med).

electrochemical, a. electroquímico/a.

electrochemistry, n. electroquímica (f).

electrochromatography, n. electrocromatografía (f).

electrode, n. electrodo (m); **conductivity meter e.,** electrodo conductivimétrico; **hydrogen e.,** electrodo de hidrógeno; **negative e.,** electrodo negativo; **positive e.,** electrodo positivo.

electrodialysis, n. electrodiálisis (m).

electrodynamic, a. electrodinámico/a.

electroencephalogram, n. electroencefalograma (m) (Med).

electrofishing, n. electropesca (f).

electrogravimetry, n. electrogravimetría (f).

electrolysis, n. electrólisis (f).

electrolyte, n. electrolito (m).

electrolytic, a. electrolítico/a; **e. process,** proceso (m) electrolítico.

electrolyze, v. electrolizar.

electromagnet, n. electroimán (m).

electromagnetic, a. electromagnético/a; **e. field,** campo (m) electromagnético.

electron, n. electrón (m); **e. microscope,** microscopio (m) electrónico.

electron-volt, n. electrón-voltio (m) (abbr. eV).

electronegative, a. electronegativo/a.

electronic, a. electrónico/a; **e. mail,** correo (m) electrónico, correo informático.

electronics, n. electrónica (f).

electroosmosis, n. electroósmosis (f).

electroosmotic, a. electroósmótico/a.

electrophoresis, n. electroforesis (f).

electroplating, n. galvanostegia (f), galvanotecnia (f), galvanoplastia (f); **e. industry,** industria (f) galvanoplástica.

electropositive, a. electropositivo/a.

electrostatic, a. electrostático/a; **e. precipitator,** precipitador (m) electrostático.

element, n. elemento (m); **toxic e.,** elemento tóxico; **trace e.,** elemento traza, microelemento (m); oligoelemento (m).

elephant, n. elefante (m).

elevation, n. elevación (f) (of hill), altitud (f) (above sea level); alzado (m) (of a building); **front e. (Con),** alzado (m) de fachada.

elevator, n. ascensor (m).

elimination, n. eliminación (f).

elision, n. elisión (f).

ellipse, n. elipse (f).

ellipsoid, n. elipsoide (m); **strain e.,** elipsoide de deformación; **stress e.,** elipsoide de esfuerzo.

elliptical, a. elíptico/a.

elm, n. olmo (m).

El Niño, n. El Niño (m) (Mar).

elongate, v. alargar, estirar.

elongated, a. alargado/a, estirado/a.

elongation, n. elongación (f), alargamiento (m) (extension).

El Salvador, n. El Salvador (m).

eluent, n. eluyente (m) (Chem).

elute, v. eluir (Chem), lixiviar (to leach), lavar (to wash).

elution, n. elución (f).

elutriation, n. elutriación (f), lavado (m) por decantación.

eluvial, a. eluvial.

eluviation, n. eluviación (f).

eluvium, n. eluvial (m), eluvión (f).

elvan, n. pórfido (m) feldespático; elvan (m) (LAm).

elytra, n.pl. élitros (m.pl) (see: elytron).

elytron, n. élitro (m).

elytrum, n. élitro (m).

emanate, v. emanar.

emanation, n. emanación (f).

embank, v. encauzar, poner diques (river); terraplenar (road).

embanking, n. encauzamiento (m), encauzar (the channelling of).

embankment, n. terraplén (m) (railway, road); dique (m), malecón (m) (by water).

embargo, n. embargo (m).

embayment, n. bahía (f); engolfamiento (m) (LAm).

embed, v. engastar, empotrar.

embrasure, n. barbacana (f), alféizar (m) (Arch); tronera (f) (Mil).

embryo, n. embrión (m).

embryonic, a. embrionario/a.

emerald, n. esmeralda (f).

emergency, n. emergencia (f); **in case of e.,** en caso de emergencia.

emergent, a. emergente.

emery, n. esmeril (m); **e. cloth,** tela (f) de esmeril.

emetic, n. emético (m), vomitivo (m).

emetic, a. emético/a, vomitivo/a.

emission, n. efluente (m), emisión (f), inmisión (f); **e. standard,** criterio (m) de emisión; **fixed e. limit,** límite (m) de emisión fija; **fugitive e.,** emisión fugitiva; **gaseous emissions,** emisiones gaseosas; **stack emissions,** emisiones de chimeneas.

emit, v. emitir.

emitter, n. emisor (m); **alpha/beta/gamma e.,** emisor alpha/beta/gamma.

emitting, n. fuga (f) (vapour, gas), reboso (m) (spillage).

emphasis, n. énfasis (m); **special e.,** énfasis especial.

emphasize, v. enfatizar, recalcar.

emphysema, n. enfisema (f).

empirical, a. empírico/a.

empiricism, n. empirismo (m).

emplace, v. ubicar (to put in place).

emplacement, n. emplazamiento (m), instalación (f).

employee, n. empleado/a (m, f).

employment, n. empleo (m), colocación (f); **conditions of e.,** condiciones (f.pl) de empleo; **e. agency,** agencia (f) de empleo; **full e.,** pleno empleo.

empty, a. vacío/a.

emptying, n. vaciamiento (m); descarga (f) (de vagones), agotamiento (m) (pozos).

EMU, n. UEM (f); **European Monetary Union,** Unión (f) Económica y Monetaria.

emulsifier, n. emulsionador (m).

emulsify, v. emulsionar.

emulsifying, a. emulsionante; **e. agent,** agente (m) emulsionante.

emulsion, n. emulsión (f).

enact, v. promulgar, dar fuerza de ley a (a bill).

enamel, n. esmalte (m), barniz (m) (varnish).

enamelling, n. esmaltado (m); **protective e.,** esmaltado protector.

encapsulation, n. encapsulación (f).

enclose, v. encerrar, rodear (to surround), incluir (to contain).

enclosed, a. encerrado/a, cercado/a.

enclosure, n. clausura (f), cercado (m).

encompass, v. envolver (to envelope), abarcar (to include), rodear (to surround).

encrinite, n. encrinita (f).

encroach, v. invadir (Gen), abusar (rights).

encroachment, n. usurpación (f).

encrustation, n. incrustación (f).

end, n. término (m), tope (m), fin (m), extremo (m), cabo (m).

end-point, n. punto (m) de equivalencia (Chem).

endanger, v. poner en peligro.

endangered, a. en peligro, comprometido/a; expuesto/a (exposed); **e. species,** especie (f) comprometida, especie (f) en peligro.

endeavour, n. intento (m), tentativa (f) (attempt); esfuerzo (m) (effort).

endemic, 1. n. endémico (m); 2. a. endémico/a.

endive, n. endibia (f).

endo-, pf. endo-.

endogenetic, a. endógeno/a; **e. processes,** procesos (m.pl) endógenos.

endogenous, a. endógeno/a.

endoparasite, n. endoparásito (m).

endoskeleton, n. endoesqueleto (m).

endosperm, n. endospermo (m).

endothermic, a. endotérmico/a; **e. reaction,** reacción (f) endotérmica.

endowment, n. dotación (f) (act), donación (f) (amount), dote (f).

energy, n. energía (f); **alternative e.**, energía alternativa, energía blanda; **chemical e.**, energía química; **conventional e.**, energía convencional, energía dura; **electrical e.**, energía eléctrica; **e. efficiency**, eficiencia (f) energética; **e. supply**, abastecimiento (m) de energía, suministro (m) de energía; **e. use**, utilización (f) de energía; **fossil e.**, energía fósil; **geothermal e.**, energía geotérmica; **heat e.**, energía calorífica; **hydraulic e.**, energía hidráulica; **hydroelectric e.**, energía hidroeléctrica; **hydrothermal e.**, energía hidrotérmica; **kinetic e.**, energía cinética; **mechanical e.**, energía mecánica; **nuclear e.**, energía nuclear; **radiant e.**, energía radiante; **renewable e.**, energía renovable; **solar e.**, energía solar; **thermal e.**, energía térmica; **tidal e.**, energía mareal; **waste-to-e. plant**, planta (f) de recuperación energética; **wave e.**, energía de olas; **wind e.**, energía eólica.

energy-saving, a. que ahorra energía.

enforce, v. hacer cumplir (e.g. law); imponer (impose).

enforceable, a. ejecutorio/a (contract), aplicable (law).

enforcement, n. aplicación (f) (of the law); coacción (f) (coercion); entrada (f) en vigor (putting into effect).

engine, n. motor (m), máquina (f); **beam e.**, motor de balancín; **diesel e.**, motor diesel; **internal combustion e.**, motor de combustión interna, motor de explosión; **jet e.**, motor de reacción; **petrol e.**, motor de gasolina; **steam e.**, motor de vapor, máquina (f) de vapor.

engineer, n. ingeniero (m); **agricultural e.**, ingeniero agrónomo; **civil e.**, ingeniero civil, ingeniero de caminos, canales y puertos; **consulting e.**, ingeniero consultor; **electrical e.**, ingeniero eléctrico; **environmental e.**, ingeniero medioambiental; **highway e.**, ingeniero de caminos, ingeniero de carreteras; **mechanical e.**, ingeniero mecánico; **mining e.**, ingeniero de minas; **municipal e.**, ingeniero municipal; **project e.**, ingeniero de proyecto; **sanitary e.**, ingeniero sanitario; **water e.**, ingeniero hidráulico.

engineering, n. ingeniería (f); **agricultural e.**, ingeniería agrícola; **civil e.**, ingeniería civil; **coastal e.**, ingeniería litoral; **electrical e.**, ingeniería eléctrica; **environmental e.**, ingeniería ambiental, ingeniería medioambiental.

England, n. Inglaterra (f).

English, n. inglés (language).

English, a. inglés, inglesa.

enhancement, n. aumento (m), incremento (m) (increase); realce (m) (of splendour).

enlarge, v. agrandar.

enlargement, n. aumento (m) (increase size); ensanchamiento (m), ensanche (m) (of a town); ampliación (f) (of an organization).

en masse, a. en masa.

enquiry, n. pregunta (f) (question); investigación (f) (investigation).

enrich, v. enriquecer.

enrichment, n. enriquecimiento (m); **iron e.**, enriquecimiento (m) en hierro.

entail, v. suponer (require), acarrear (hardship).

enterobacter, n. enterobácter (m).

enterobacteria, n.pl. enterobacterias (f.pl).

enterococci, n. enterococos (m.pl).

enterovirus, n. enterovirus (m).

enterprise, n. empresa (f), ventura (f); **private e.**, empresa privada.

enthalpy, n. entalpía (f).

entomologist, n. entomólogo/a (m, f).

entomology, n. entomología (f).

entomophilous, a. entomófilo/a.

entrainment, n. arrastre (m); **e. separator**, separador (m) de arrastre.

entrepreneur, n. empresario/a (m, f) (Com); capitalista (m, f) (Fin).

entropy, n. entropía (f).

entry, n. contabilizado (m) (in a record); acceso (m) (access).

enumerate, v. enumerar.

enumeration, n. enumeración (f).

envelope, v. envolver.

enveloping, a. envolvente; **e. line**, línea (f) envolvente.

environment, n. ambiente (m) (Gen); medioambiente (m), medio ambiente (m), entorno (m); **aquatic e.**, ambiente acuático; **chemical e.**, ambiente químico; **controlled e.**, medio ambiente controlado; **E. Agency (abbr. EA) (England and Wales)**, Agencia (f) de Medio Ambiente (abr. AMA) (Sp); **human e.**, ambiente humano; **marine e.**, ambiente marino; **reducing e.**, ambiente reductor; **sedimentary e.**, ambiente sedimentario; **socio-economic e.**, entorno socioeconómico.

environmental, a. ambiental, del medio ambiente; **e. awareness**, sensibilidad (f) ambiental, educación (f) ambiental; **e. health**, salud (f) ambiental; **e. impact**, impacto (m) ambiental; **e. impact assessment (abbr. EIA)**, evaluación (f) de impacto ambiental (abbr. EIA); **e. law**, ley (f) de medio ambiente; **e. lobby**, ecologistas (m.pl) (environmentalists); **e. policy**, política ambiental, política (f) respetuosa con el medio ambiente; **e. pressure group**, grupo (m) ecologista; **e. protection**, protección (f) ambiental; **E. Protection Agency (abbr. EPA) (US)**, Agencia (f) de protección del medio ambiente; **e. science**, ciencia (f) del medio ambiente; **e. scientist**, científico/a (m,f) ambiental, científico/a medioambiental.

environmentalism, n. medioambientalismo (m).

environmentalist, n. ecologista (m, f), medioambientalista (m, f).

environs, n. alrededores (m.pl).

enzyme, n. enzima (f).
eolian, a. eólico/a.
eolinite, n. eolinita (f).
eon, n. eón (m).
eosin, n. eosina (f).
EPA, n. **Environmental Protection Agency (US),** agencia (f) para la protección del medio ambiente.
ephemeral, a. efímero/a; **e. stream,** cauce (m) efímero.
epi-, pf. epi-.
epicentre, n. epicentro (m).
epidemic, n. epidemia (f).
epidemiology, n. epidemiología (f).
epidermis, n. epidermis (f).
epidiorite, n. epidiorita (f).
epidote, n. epidota (f).
epigenesis, n. epigénesis (m).
epigenic, a. epigino/a.
epilimnion, n. epilimnio (m).
epinephrin, n. adrenalina (f).
epiphyte, n. epífito (m).
epithermal, a. epitermal.
epoch, n. época (f), era (f).
epsilon, n. épsilon (f).
EQO, n. **Environmental Quality Objective,** objectivo (m) de calidad ambiental.
equation, n. ecuación (f); **continuity e.,** ecuación de continuidad; **differential e.,** ecuación diferencial; **e. of conservation of mass,** ecuación de conservación de masa; **Laplace e.,** ecuación de Laplace; **mass transport e.,** ecuación de transporte de masa; **non-linear e.,** ecuación no lineal; **parabolic e.,** ecuación parabólica; **partial differential e.,** ecuación diferencial parcial; **second order e.,** ecuación de segundo grado; **simple e.,** ecuación de primer grado.
equator, n. ecuador (m).
equi-, pf. equi-.
equidistance, n. equidistancia (f).
equidistant, a. equidistante.
equilateral, a. equilátero/a.
equilibrium, n. equilibrio (m); **chemical e.,** equilibrio químico; **dynamic e.,** equilibrio dinámico; **e. conditions,** condiciones (f.pl) de equilibrio; **hydraulic e.,** equilibrio hidráulico; **rock-water e.,** equilibrio agua-roca; **stable e.,** equilibrio estable, estado (m) estacionario; **unstable e.,** equilibrio inestable.
equinox, n. equinoccio (m); **autumn e.,** equinoccio de otoño; **vernal e.,** equinoccio de primavera.
equipment, n. equipo (m); herramientas (f.pl) (tools), utillaje (m); **electrical e.,** equipo eléctrico; **pumping e.,** equipo de bombeo.
equipotential, a. equipotencial; **e. line,** línea (f) equipotencial.
equipping, n. equipamiento (m).
equity, n. equidad (f) (fairness), beneficio (m), valor (m) líquido (Fin); **e. capital,** capital (m) propio.

equivalence, n. equivalencia (f).
equivalent, a. equivalente; **e. weight,** peso (m) equivalente.
era, n. era (f), época (f).
eradicate, v. erradicar.
eradication, n. erradicación (f).
erbium, n. erbio (m) (Er).
ERDF, n. FEDER (m); **European Regional Development Fund,** Fondo (m) Europeo de Desarrollo Regional.
erg, n. erg (m), ergio (m) (Phys); sandy desert (Geog).
ergonomics, n. ergonomía (f).
ergot, n. ergot (m).
erode, v. erosionar.
eroded, a. erosionado/a.
erosion, n. erosión (f); **glacial e.,** erosión glaciar, erosión glacial; **sheet e.,** erosión en mantos.
erosive, a. erosivo/a; **e. agent,** agente (m) erosivo.
erroneous, a. erróneo/a, equivocado/a (mistaken).
error, n. error (m), equivocación (a mistake); **analytical e.,** error analítico; **e. handling (Comp),** tratamiento (m) de errores, manipulación (f) de errores; **human e.,** error humano; **residual e.,** error residual; **systematic e.,** error sistemático; **truncation e.,** error de truncación.
erupt, v. entrar en erupción (volcano).
eruption, n. erupción (f); **volcanic e.,** erupción volcánica.
eruptive, a. eruptivo/a.
escalator, n. escalera (f) mecánica.
escape, n. fuga (f) (Gen); huida (f) (flight); evasión (f) (evasion); **garden e. (Ecol),** planta (f) no autóctona en el campo silvestre que se dispersa de jardines.
escaphopod, n. escafópodo (m).
escaping, n. fuga (f) (of vapour, gas, air), desprendimiento (of gases).
escarpment, n. escarpe (m), escarpadura (f); escarpa (f) (LAm); **chalk e.,** escarpe de creta.
esker, n. esker (m).
esparto, n. esparto (m) (Bot); **e. grass fibre,** fibra (f) de esparto.
essence, n. esencia (f), extracto (m).
essential, a. esencial (quality); fundamental (important).
establish, v. establecer, fundar.
established, a. establecido/a; **e. practice,** práctica (f) establecida.
establishment, n. establecimiento (m), institución (f), instituto (m) (body); instauración (f) (setting up); **e. of a management regime,** instauración de un régimen de gestión; **the E. (Fig),** clase (f) dirigente (ruling class).
estate, n. propiedad (f) (property); finca (f), hacienda (f) (LAm), estancia (f) (LAm, land); polígono (m) (industrial); **e. agency,** agente (m) inmobilario; **housing e.,** núcleo

(m) residencial, urbanización (f), residencial (f) (LAm); **industrial e.**, polígono (m), zona (f) industrial; **real e.**, inmuebles (m.pl), bienes (m.pl) raíces.

Estefanian, n. Estefaniense (m).

ester, n. éster (m).

esthetic, a. estético/a.

estimate, n. estimación (f), apreciación (f), cálculo (m) aproximado; **population e.**, estimación de la población; **rough e.**, tanteo (m).

estimation, n. estimación (f), cálculo (m), apreciación (f); **point e. (Mat),** estimación puntal.

estivation, n. estivación (f).

estuarine, a. estuarino/a.

estuary, n. estuario (m).

ethanamide, n. etanamido (m).

ethane, n. etano (m).

ethanol, n. etanol (m).

ethene, n. eteno (m).

ether, n. éter (m); **ethyl e.**, éter etílico.

ethnic, a. étnico/a.

ethnographic, a. etnográfico/a.

ethology, n. etología (f).

ethoxy-, pf. etoxi-.

ethyl, n. etilo.

ethylbenzene, n. etilobenzol (m).

ethyne, n. etino (m).

eucalpyti, n.pl. eucaliptos (m.pl) (see eucalyptus).

eucalyptus, n. eucalipto (m).

Eucaryota, n.pl. eucarioatas (f.pl), el grupo Eucaryota (Zoo, Bot).

eucaryote, n. eucariota (f).

eukaryote, n. eucariota (f).

euphotic, a. eufótico/a; **e. zone (Biol),** zona (f) eufótica.

Eurasian, a. euroasiático/a.

European, 1. n. europeo/a (m, f) (person); 2. a. europeo/a; **E. Commission,** Comisión (f) Europea; **E. Community (abbr. EC),** Comunidad (f) Europea (abbr. CE); **E. Economic Community (abbr. EEC),** Comunidad (f) Económica Europea (abbr. CEE); **E. Union (abbr. EU),** Unión (f) Europea (abbr. UE).

europium, n. europio (m) (Eu).

eustacy, n. eustasia (f).

eustatic, a. eustático/a.

eutectic, a. eutéctico/a.

eutrophic, a. eutrófico/a.

eutrophicate, v. eutrofizar.

eutrophication, n. eutrofización (f).

eV, n. eV (electron-volt).

evaluate, v. evaluar, valorar; interpretar.

evaluation, n. evaluación (f), valoración (f); interpretación (f); tasación (f) (appraisal).

evapograph, n. evaporígrafo (m).

evaporate, v. evaporar.

evaporation, n. evaporación (f); **actual e.**, evaporación real; **e. pan,** cubeta (f) de evaporación; **e. tank,** tanque (m) de evapora-

ción; **potential e.**, evaporación potencial; **solar e.**, evaporación solar.

evaporator, n. evaporador (m).

evaporimeter, n. evaporímetro (m); **balancing e.**, evaporímétro de balanza.

evaporite, n. evaporita (f).

evapotranspiration, n. evapotranspiración (f); **actual e.**, evapotranspiración real; **potential e.**, evapotranspiración potencial.

evapotranspirometer, n. evapotranspirómetro (f).

even-grained, a. equigranular.

everglade, n. tierra (f) baja pantanosa cubierta de altas hierbas (US).

evergreen, a. perenne, de hoja perenne.

evidence, n. evidencia (f) (Jur); hechos (m.pl), datos (m.pl) (facts).

evolution, n. evolución (f); **theory of e.**, teoría (f) de la evolución.

evolutionary, a. evolutivo/a.

evolve, v. evolucionar (Bot, Zoo); desarrollarse.

exactness, n. exactitud (f).

example, n. ejemplo (m).

excavation, n. excavación (f).

exceed, v. exceder (quantity); sobrepasar, repasar (a limit); **to e. the quality standards,** sobrepasar los límites de calidad.

exchange, n. cambio (m), intercambio (m); **base e.**, intercambio de bases; **corn e.**, bolsa (f) de cereales; **e. rate,** tipo (m) de cambio; **ionic e.**, intercambio iónico; **stock e.**, bolsa (f).

exchanger, n. intercambiador (m); **heat e.**, intercambiador de calor.

excitation, n. excitación (f).

excretion, n. excreción (f).

excursion, n. excursión (f).

execute, v. ejecutar, cumplir (an order); ejecutar (a criminal); **to e. an action plan,** ejecutar un plan de acciones.

executive, 1. n. ejecutivo/a (m, f) (businessman/ businesswoman); **e. board,** consejo (m) de dirección; 2. a. ejecutivo/a.

exempt, a. exento/a, libre (free from).

exemption, n. exención (f); **legal e.**, exención jurídica.

exfiltration, n. exfiltración (f).

exfoliate, v. exfoliar.

exfoliation, n. exfoliación (f).

exhale, v. exhalar.

exhaust, 1. v. agotar; 2. n. escape (m); **e. emissions, e. fumes, e. gases,** gases (m.pl) de escape; **e. pipe,** tubo (m) de escape.

exhaustible, a. agotable; **e. resources,** recursos (m.pl) agotables.

exhume, v. exhumar.

exo-, pf. exo-.

exodus, n. éxodo (m).

exogenous, a. exógeno/a.

exoskeleton, n. exoesqueleto (m).

exothermic, a. exotérmico/a; **e. reaction,** reacción (f) exotérmica.

expand, v. expandir; extender (to extend); hinchar (to swell), ampliar (to enlarge).

expansion, n. expansión (f) (e.g. of gas), desarrollo (m) (of activity, trade); ensanche (m) (urban); dilatación (f) (gas, metal); desarrollo (m) (Mat); **e. joint,** junta (f) de expansión; **e. tank,** tanque (m) de expansión; **thermal e.,** dilatación termal.

expectancy, n. esperanza (f); **life e.,** esperanza de vida.

expectation, n. esperanza (f).

expel, v. echar, expeler, expulsar.

expenditure, n. gasto (m), desembolso (m) (of money, etc.); gastos (of money); **income and e.,** ingresos y gastos.

experience, n. experiencia (f).

experiment, 1. v. experimentar, hacer experimentos; examinar; probar (to prove); 2. n. experimento (m).

experimental, a. experimental.

experimentation, n. experimentación (f).

expert, n. experto/a (m, f), especialista (m, f); **e. opinion,** juicio (m) de peritos.

expertise, n. pericia (f).

expertness, n. pericia (f), habilidad (f), peritación (f).

expiry, n. caducidad (f); **e. date,** fecha (f) de caducidad; **e. of licence, e. of permit,** caducidad del permiso.

explain, v. explicar.

explanation, n. explanación (f).

explanatory, a. explicativo/a.

explicit, a. explícito/a; **e. function,** función (f) explícita.

explode, v. explotar, volar (to blow up).

exploitability, n. explotabilidad (f).

exploitable, a. explotable.

exploitation, n. explotación (f); **e. of resources,** explotación de recursos; **mineral e.,** explotación minera; **optimum e.,** explotación óptima.

exploited, a. explotado/a.

exploration, n. exploración (f), aprovechamiento (m).

exploratory, a. exploratorio/a, aventurero/a (adventurous); **e. adit,** galería (f) de exploración, socavón (m) aventurero.

explore, v. explorar.

explosimeter, n. explosímetro (m).

explosion, n. explosión (f); **nuclear e.,** explosión nuclear; **population e.,** explosión demográfica; **underground e.,** explosión subterránea.

explosive, 1. n. explosivo (m); **plastic e.,** goma (f) explosiva; 2. a. explosivo/a; **e. wastes,** residuos (m.pl) explosivos.

exponential, a. exponencial; **e. function,** función (f) exponencial.

export, 1. v. exportar; 2. n. exportación (f); **e. goods,** artículos (m.pl) de exportación, bienes (m.pl) de exportación; **e. trade,** comercio (m) de exportación.

exportation, n. exportación (f).

expose, v. aflorar (Geol), exponer (to leave uncovered), descubrir (to discover).

exposure, n. asomo (m) (Geol), exposición (f) (to light, etc.), revelación (f) (e.g. of a secret).

express, v. expresar.

expressed, a. expresado/a.

expression, n. expresión (f); **mathematical e.,** expresión matemática.

expressly, ad. expresamente, específicamente.

expropriate, v. expropiar.

expropriation, n. expropiación (f).

extension, n. extensión (f), propagación (f); ampliación (f); **e. of capital,** ampliación (f) de capital.

extensive, a. extenso/a (vast); extensivo/a (Agr); **e. cultivation,** cultivo (m) extensivo.

extensometer, n. extensímetro (m).

extinct, a. extinguido/a, extinto/a; **e. species (Zoo, Bot),** especie (f) extinguida.

extinction, n. extinción (f).

extinguish, v. extinguir.

extinguished, a. extinguido/a.

extinguisher, n. extintor/a (m, f); **fire e.,** extintor de fuego.

extract, 1. v. extraer, sacar (to remove); 2. n. extracto (m).

extraction, n. extracción (f); **soil vapour e.,** extracción de vapor del suelo; **solvent e.,** extracción de disolvente; **vacuum e.,** extracción por vacío.

extractive, a. extractivo/a; **e. industry,** actividades (f.pl) extractivas.

extrapolate, v. extrapolar.

extrapolation, n. extrapolación (f).

extremity, n. extremidad (f).

extrusive, a. extrusivo/a; **e. rock,** roca (f) extrusiva.

exudation, n. exudación (f).

exude, v. exudar.

eye, n. ojo (m) (Anat); yema (f), botón (m) (Bot); **compound e.,** ojo compuesto; **simple e.,** ojo simple; **tiger's e. (Geol),** ojo de tigre.

eyeball, n. globo (m) del ojo, globo ocular.

eyelid, n. párpado (m).

eyewitness, n. testigo (m) ocular.

F

fabric, n. fábrica (f) (structure); tela (f) tejido (m) (cloth); **deformation f. (Geol),** fábrica de deformación, fábrica deformacional.
fabricate, v. fabricar (to construct); falsificar (to falsify).
fabrication, n. fabricación (f) (manufacture); falsificación (f) (evidence).
facade, n. fachada (f).
face, n. cara (f), rostro (m) (Anat), frente (m); **coal f.,** frente de carbón; **shore f.,** frente de ribera; **slip f.,** frente de deslizamiento; **working f. (Min),** frente de corte.
facet, n. faceta (f).
face up to, v. encarar, hacer frente.
facies, n. & n.pl. facies (f) (pl. facies); **lacustrine f.,** facies lacustres; **marine f.,** facies marinas; **shelly f.,** facies conchíferas.
facility, n. facilidad (f), instalación (f); **fuelstorage f.,** facilidad para el almacenamiento de combustible; **recovery f., (of wastes),** instalación/facilidad de recuperación; **waste management f. (US),** vertedero (m) controlado.
fact, n. hecho (m); dato (m) (piece of information); realidad (f) (not fiction); **facts and figures,** datos (m.pl).
factor, n. factor (m); **biotic f.,** factor biótico; **climatic f.,** factor climático; **edaphic f.,** factor edáfico; **growth f.,** factor de crecimiento; **limiting f.,** factor limitador, factor limitante; **political f.,** factor político; **social f.,** factor social; **socioeconomic f.,** factor socioeconómico.
factorial, a. factorial.
factorization, n. factorización (f).
factory, n. fábrica (f).
facultative, a. facultativo/a; **f. parasite,** parasitario (m) facultativo.
faecal, a. fecal; **f. streptococci,** estreptococos (m.pl) fecales.
faeces, n.pl. heces (f.pl), excreta (f), excrementos (m.pl).
Fahrenheit, n. Fahrenheit (m).
failure, n. fracaso (m) (Gen); fallo (m) (fault); rotura (f) (break); corte (f) (cut); interrupción (f); **shear f.,** rotura de cizalla; **tension f.,** rotura por tracción.
fairground, n. parque (m) de atracciones.
fall, n. caída (f), desprendimiento (m) (of rocks); corrimiento (m) (of earth); derrumbamiento (m) (of building); aguacero (m) (of rain); diminución (f) (decrease); otoño (m) (US autumn); **ice f.,** caída de hielo; **rock f.,** caída de roca.
fallout, n. residuos (m.pl) atmosféricos, lluvia (f) radioactiva, polvillo (m) radioactivo;

radioactive f., residuos (m.pl) atmosféricos radioactivos.
fallow, n. barbecho (m); **to leave f. (v),** dejar en barbecho.
falls, n.pl. saltos (m), salto (m) de agua (Geog).
false, a. falso/a; **f. bedding,** falsa estratificación (f), estratificación entrecruzada.
family, n. familia (f).
famine, n. hambre (f), hambruna (f); escasez (f) (LAm), hambrusia (f) (LAm).
Fammenian, n. Fammeniense (m).
fan, n. abanico (m), cono (m), bajada (f) (Geol); ventilador (m) (de aire); **f.-shaped (a),** en forma de abanico; **rock f.,** abanico de roca.
fang, n. colmillo (m); diente (m) (of snake).
FAO, n. OAA (f); **Food and Agriculture Organization,** Organización (f) de las Naciones Unidas para la Alimentación y la Agricultura.
farad, n. faradio (m) (abbr. F, unit of electrical capacitance).
farm, 1. v. cultivar; labrar (to till); ser agricultor (to be a farmer); 2. n. granja (f), estancia (f), quinta (f) (LAm); hacienda (f), rancho (m) (Mex); **collective f.,** granja colectiva; **f. implements,** instrumentos (m) de labranza, aperos (m) de labranza; **f. labourer (laborer, US),** jornalero/a (m, f), peón (m), obrero/a (m, f) agrícola; **f. produce,** productos (m.pl) agrícolas; **fish f.,** piscifactoría (f); **poultry f.,** granja avícola.
farmer, n. agricultor/a (m, f), granjero (m), campesino (m); haciendo/a (m, f) (LAm); **peasant f.,** labrador/a (m, f); **small f.,** agricultor pequeño, campesino.
farmhand, n. labriego/a (m, f), peón (m), trabajador (m) agrícola, trabajador del campo.
farmhouse, n. cortijo (m); alquería (f) (LAm), casa (f) de hacienda (LAm).
farming, 1. n. labranza (f), cultivo (m) (tillage of land); agricultura; cría (f) (animal-rearing); 2. a. agrícola.
farmland, n. labranza (f), labrantío (m) (arable land), suelos (m.pl) agrícolas.
farmstead, n. alquería (f), granja (f), finca (f).
farmyard, n. corral (m).
fasten, v. sujetar (to attach), fijar (to fix), zunchar (to fasten with metal band).
fat, n. grasa (f).
fatal, a. fatal, mortal.
fate, n. destino (m), suerte (f); **f. of contaminants,** destino de contaminantes.
fathom, n. braza (f) (English fathom is 1.83m, Spanish fathom is 1.671m).

fatigue, n. fatiga (f), cansancio (m); **metal f.**, fatiga del metal.

fattened, a. cebado/a (livestock).

faucet, n. grifo (m), llave (f), canilla (f) (LAm).

fault, n. falla (f), defecto (m), tara (f) (in production); **boundary f.**, falla de borde; **complex f.**, complejo (m) de falla; **drag f.**, falla de arrastre; **f. block**, bloque (m) de falla; **f. breccia**, brecha (f) de falla; **f. line**, línea (f) de falla; **f. scarp**, escarpe (m) de falla; **f. zone**, zona (f) de falla; **high-angle f.**, falla de alto ángulo; **limb of f.**, flanco (m) de falla; **normal f.**, falla normal; **step f.**, falla en escalón, falla de gradería; **tear f.**, falla de cizalla, falla de rasgadura; **throw of f.**, salto (m) vertical de una falla; **thrust f.**, cabalgamiento (m), falla de empuje.

faulted, a. fallado/a.

fauna, n. fauna (f); **aquatic f.**, fauna acuática; **shelly f.**, fauna conchífera.

faunal, a. faunal.

favourable, a. favorable, propicio/a.

fear, n. miedo (m), temor (m).

feasibility, n. factibilidad (f), viabilidad (f); **f. study**, análisis (m) de factibilidad, estudio (m) de viabilidad.

feasible, a. factible (plan), plausible (theory).

feather, n. pluma (f); pestaña (f), reborde (m) (Eng, flange); **down f.**, pluma de fondo; **primary/secondary f.**, pluma primaria/secundaria.

feather-edge, n. borde (m) delgado.

feature, n. rasgo (m), característica (f); **geographical f.**, rasgo geográfico; **structural f.**, rasgo estructural.

fecal, a. fecal.

feces, n.pl. heces (f.pl), excreta (f), excrementos (m.pl).

fecundity, n. fecundidad (f).

federal, a. federal; **f. body**, entidad (f) federal.

federation, n. federación (f); **European F. of Geologists (abbr. EFG)**, Federación Europea de Geólogos (abbr. FEG).

feed, 1. v. alimentar; 2. n. cebo (m), pienso (m), forraje (m) (for livestock).

feedback, n. retroalimentación (f) (within systems); realimentación (f) (Elec); reaprovechamiento (m), feedback (m) (of information); retroacción (f) (Med); **negative f.**, feedback negativo; **positive f.**, feedback positivo.

feedstock, n. materias (f.pl) primas (industrial raw material).

feeler, n. antena (f) (e.g. of insects).

feet, n.pl. pies (m) (Anat, Mat); **it is six f. (ft) long**, tiene seis pies de largo.

feldspar, n. feldespato (m).

feldspathic, a. feldespático/a.

fell, v. talar (to cut).

fell, n. collado (m) (Geog).

felled, a. talado (of trees).

felling, n. tala (f), corte (m) (of trees).

felspar, n. feldespato (m).

female, 1. n. mujer (f) (of a woman); hembra (f) (Zoo); 2. a. hembra, femenino/a, de mujer.

fen, n. pantano (m), marjal (m); **the Fens (GB)**, las tierras (f.pl) bajas de Lincolnshire y Cambridgeshire.

feral, a. feral, silvestre.

ferment, 1. v. fermentar; 2. n. fermento (m).

fermentation, n. fermentación (f); **anaerobic f.**, fermentación anaerobia.

fern, n. helecho (m).

ferrate, n. ferrato (m).

ferric, a. férrico/a.

ferrite, n. ferrita (f).

ferrocement, n. ferrocemento (m), hormigón (m) armado (reinforced concrete).

ferrocyanide, n. ferrocianuro (m).

ferromagnesian, a. ferromagnésiano/a.

ferromagnetic, a. ferromagnético/a.

ferromanganese, n. ferromanganeso (m).

ferronickel, n. ferroníquel (m).

ferrous, a. ferroso/a.

ferruginous, a. ferruginoso/a.

fertile, a. fértil (land); fecundo/a (Zoo); feraz.

fertility, n. fertilidad (f); feracidad (f) (of soil); fecundidad (f) (fecundity).

fertilization, n. fertilización (f), fecundación (f); **cross-f.**, fertilización cruzada, sobrecruzamiento (m).

fertilize, v. abonar (of soil); fecundar (of reproduction); fertilizar.

fertilizer, n. abono (m), fertilizante (m); **chemical f.**, fertilizante químico; **compound f.**, abono compuesto; **fast-acting f.**, abono de acción rápida; **organic f.**, abono orgánico, fertilizante orgánico; **phosphate f.**, abono fosfatado; **slow-acting f.**, abono de acción lenta; **synthetic f.**, fertilizante sintético.

fertilizing, a. fertilizante.

fetal, a. fetal.

fetch, n. alcance (m) (waves); **f. of waves**, alcance de las olas.

fetus, n. feto (m).

fever, n. fiebre (f), calentura (f); **typhoid f.**, fiebre tifoidea; **yellow f.**, fiebre amarilla.

fevered, a. febril.

fiber, n. fibra (f), hilo (m) (thread); **dietary f.**, fibra dietética, fibra alimenticia.

fiberglass, n. fibra (f) de vidrio, fiberglass.

fibre, n. fibra (f), hilo (m) (thread); **dietary f.**, fibra dietética, fibra alimentícia; **fibreglass**, fibra de vidrio; **man-made f.**, fibra artificial; **optical f.**, fibra óptica; **synthetic f.**, fibra sintética; **textile f.**, fibra textil.

fibreboard, n. cartón-fibra (f).

fibreglass, n. fibra (f) de vidrio, fiberglass.

fibrocement, n. fibrocemento (m) (asbestos cement).

fibrolite, n. fibrolita (f).

fibrous, a. fibroso/a.

field, n. campo (m), terreno (m), banco (m);

electromagnetic f., campo electromagnético; **f. biology**, biología (f) de campo; **f. capacity**, capacidad (f) del campo; **f. of activity**, campo de actividad; **f. of wheat**, campo de trigo; **goldfield**, campo aurífero; **ice f.**, campo de hielo; **oil f.**, campo petrolífero; **playing f.**, campo de juego; **pumping f.**, campo de bombeo; **sown f.**, sembrado (m), sembrío (m) (LAm); **well-f.**, campo de pozos.

fieldbook, n. libreta (f) de campo.

fieldwork, n. trabajo (m) de campo.

fig, n. higo (m), breva (f); **f. tree**, higuera (f).

figure, n. cifra (f), número (m) (number); precio (m) (price); suma (f), cantidad (f) (sum); **geometric f.**, cifra geométrica; **production f.**, cifra de producción; **significant figures (Mat)**, cifras significativas.

filament, n. filamento (m); **gill f.**, filamento branquial.

filamentous, a. filamentoso/a; **f. organism**, organismo (m) filamentoso.

file, n. fichero (m), archivo (m) (for documents); expediente (m) (dossier); lima (f) (tool); **back-up f. (Comp)**, fichero de reserva.

fill, 1. v. rellenar (to fill a hole); cargar (to load); ocupar (a vacancy); 2. n. relleno (m) (Eng), ripio (m) (rubble), terraplén (embankment); **clay f.**, relleno arcilloso.

filler, n. material (m) de relleno.

filling, n. relleno (m); terraplenado (m), colmatación (f) (of a ditch, pond).

film, n. película (f); capa (f) (thin layer); cine (m) (cinema); **f. industry**, industria (f) cinematográfica; **f. library**, filmoteca (f), cinemateca (f); **infrared f.**, película infrarroja; **water f.**, película de agua.

filter, 1. v. filtrar; 2. n. filtro (m); **bag f.**, filtro de manga; **biological f.**, filtro biológico; **biologically-aerated f.**, filtro biológico aireado; **coarse-mesh f.**, filtro de mallas anchas; **compression f.**, filtro de compresión; **f. bed (Treat)**, lecho (m) bacteriano, lecho filtrante, lecho percolador; **f. blanket (Treat)**, colchón (m) filtrador; **f. cake (Treat)**, torta (f) de filtro; **f. feeder (Zoo)**, alimentador (f) por filtración; **f. medium**, medio (m) filtrante; **f. press**, prensa (f) de filtro; **fine-mesh f.**, filtro de mallas pequeñas; **gravel f.**, filtro de grava; **grease f.**, filtro para grasa; **membrane f. press**, prensa (f) filtradora de membrana; **membrane f.**, filtro de membrana; **nylon f.**, filtro de nilón; **percolating f.**, filtro de goteo, filtro percolador; **press f.**, filtro prensa; **pressure f.**, filtro a presión; **rapid gravity sand f.**, filtro rápido de arena por gravedad; **rotating biological f.**, filtro biológico rotativo; **trickling f.**, **percolating f.**, filtro percolador; **upward-flow sand f.**, filtro arenoso de flujo ascendente; **vacuum f.**, filtro por vacío.

filth, n. suciedad (f), porquería (f), inmundicia (f) (dirt).

filthy, a. asqueroso/a (very dirty, obscene).

filtrability, n. filtrabilidad (f).

filtrate, n. filtrado (m).

filtration, n. filtración (f); **extended-f. process**, proceso (m) de filtración extendida; **f. press**, filtración en prensa; **gravity f.**, filtración por gravedad; **mechanical f.**, filtración mecánica; **rapid sand f.**, filtración rápida por arena.

fin, n. aleta (f); **paired fins**, aletas pares.

finalization, n. finalización (f).

finalize, v. finalizar, terminar (to end).

finance, n. finanzas (f.pl), fondos (m.pl), financiamiento (m); **f. director**, director/a (m, f) de finanzas; **Minister of F.**, Ministro (m) de Hacienda.

financial, a. financiero/a; **f. analysis**, análisis (m) financiero; **f. backing**, respaldo (m) financiero; **f. return**, retorno (m) financiero; **f. statement**, estado (m) financiero; **f. status**, capacidad (f) financiera; **F. Times Index**, índice (m) bursátil del Financial Times; **f. year**, ejercicio (m) financiero.

financier, n. financiero (m).

financing, n. financiación (f), financiamiento (m).

find, v. encontrar, hallar; **to f. a solution**, encontrar una solución; **to f. water**, encontrar agua.

finding, n. hallazgo (m), descubrimiento (m) (discovery).

find out, v. averiguar.

fineness, n. finura (f).

fines, n.pl. finos (m.pl) (Min).

finger, n. dedo (m).

fingerprint, n. impresión (f) digital.

finite, a. finito/a; **f.-difference model**, modelo (m) por diferencias finitas; **f.-element model**, modelo (m) por elementos finitos; **implicit f.-difference approximation**, aproximación (f) de diferencias finitas implícitas.

fir, n. abeto (m).

fire, n. fuego (m) (Gen); incendio (m) (del campo); estufa (f) (stove); **f. brick**, ladrillo (m) refractario; **f. brigade (GB)**, cuerpo (m) de bomberos; **f. clay**, arcilla (f) refractaria; **f. engine**, coche (m) de bomberos; **f. fighting measures**, defensas (f.pl) contra los incendios, defensas contraincendios; **f. gap**, cortafuego (m); **f. hazard**, riesgo (m) de incendio; **f. hydrant**, hidrante (m); **f. precautions**, defensas (f.pl) contra los incendios, defensas contraincendios; **f. resistance**, ignifugación (f); **f. station**, parque (m) de bomberos; **forest f.**, incendio forestal, siniestro (m) forestal; **to catch f. (v)**, encenderse, incendiarse.

firebreak, n. cortafuego (m).

firedamp, n. grisú (m).

firefighter, n. bombero (m).

fire-proof, a. incombustible, ignífugo/a.

fire-proofing, n. ignifugación (f).
fire-resistant, a. ignífugo/a.
firewood, n. leña (f).
firm, 1. n. compañía (f), empresa (f); 2. a. firme, sólido/a; estable (stable).
firmness, a. firmeza.
firth, n. estrecho de mar (Geog); **F. of Forth**, Estrecho de Forth.
fiscal, a. fiscal.
fish, 1. v. faenar (to work fishing grounds), pescar (to go fishing); 2. n. & n.pl. pez (m) (alive); pescado (m) (food, catch of fish); pesca (f) (quantity of catch); **coarse f.**, peces de aguas dulces no salmón ni trucha (GB); **f. farm**, piscifactoría (f), criadero (m) de peces, vivero (m) de peces; **f. farm**, piscifactoría (f), granja (f) piscícola; **f. farming**, piscicultura (f); **f. hatchery**, piscifactoría (f), granja (f) piscícola; **f. ladder**, escalera (f) para peces, rampa (f) para peces; **f. stock**, población (f) de peces; **freshwater f.**, pez/peces de agua dulce; **game f.**, peces deportivos de aguas dulces (salmón y trucha); **marine f.**, pez marino, (pl. peces marinos); **saltwater f.**, pez/peces de agua salada; **shellfish**, mariscos (m.pl), crustáceos (m.pl).
fisherman, n. pescador (m).
fishery, n. pesquería (f) (fishing grounds), pesca (f) (industry); **coarse f.**, pesca de aguas dulces no salmón ni trucha; **f. protection vessel**, buque (m) guardapesca; **game f.**, pesca deportiva (de salmón y trucha en aguas dulces).
fishing, n. pesca (f); **coastal f.**, pesca de bajura, pesca de litoral, pesca costera (f); **common f. policy (EurU)**, Política (f) Pesquera Común; **deep sea f.**, pesca de altura; **f. boat**, pesquero (m), barco (m) de pesca; **f. gear**, artes (m.pl) de pesca; **f. grounds (pl)**, pesquería, zona (f) de pesca; **f. industry**, industria (f) pesquera; **f. licence**, licencia (f) para pescar; **f. net**, red (f) de pesca; **f. port**, puerto (m) pesquero; **f. rod**, caña (f) de pescar; **f. tackle**, aparejo (m) de pesca; **f. techniques**, técnicas (f.pl) de pesca; **inshore f.**, pesca de bajura, pesca de litoral, costera (f); **underwater f.**, pesca submarina.
fishmeal, n. harina (f) de pescado.
fissibilty, n. fisibilidad (f).
fissile, a. fisil.
fission, n. fisión (f); **nuclear f.**, fisión nuclear.
fissure, n. fisura (f), vena (f).
fissured, a. fisurado/a; **f. aquifer**, acuífero (m) fisurado.
fissuring, n. agrietamiento (m).
fit, n. ajuste (m) (de dimensión); ataque (m) (Med); **goodness-of-fit test**, test (m) de buen ajuste, prueba (f) de bondad del ajuste.
fit out, v. acondicionar (to arrange).
fitting, n. adecuación (f).
fix, v. fijar, asegurar (Gen); arreglar (Gen, to mend); fijar (by bacteria).

fixation, n. fijación (f); **chemical f.**, fijación química; **nitrogen f.**, fijación de nitrógeno.
fixed, a. fijo/a.
fixer, n. fijador (m); **nitrogen f.**, fijador de nitrógeno.
fjord, n. fiordo (m).
flaggy, a. lajoso/a (Geol).
flagstone, n. laja (f), roca (f) lajosa.
flake, 1. v. descascarar (to peel off); 2. n. copo (m) (fragment), escama (f), lámina (f), hojuela (f), hoja (f); **m. flake**, lámina de mica; **stone f.**, hojuela de roca.
flaking, n. descascaramiento (m).
flammable, a. inflamable, easily set on fire.
flange, n. pestaña (f); reborde (m) (projecting rim); ceja (f) (on wheel), collarín (m) (on pipe); **f. pipe**, tubería (f) con brida.
flank, n. ala (f) (wing); flanco (m) (of hill); ijar (m), ijada (f) (of animal); **f. of a fold**, flanco de un pliegue; **f. of anticline**, flanco de anticlinal.
flap, n. solapa (f) (envelope), aletazo (m) (wing), válvula (f) (valve); **tidal f.**, válvula de marea.
flash, n. luz (m) relámpago (light), instante (m) (short time), destello (m) (of light), flash (m) (photography); **f. distillation**, destilación (f) flash; **f. of lightning**, relámpago (m), centella (f), refucilo (m) (LAm); **f. point**, temperatura (f) de inflamabilidad.
flask, n. matraz (m) (Chem); termo (m) (thermos); caja de moldear (in foundry); **conical f.**, matraz cónico; **filter f.**, matraz filtro; **graduated f.**, matraz graduado.
flat, n. apartamento (m), piso (m); condominio (m) (LAm); llanura (f) (Geog); **mud f.**, llanura de fangos; **tidal f.**, llanura mareal.
flatness, n. llanura (f) (of land); planitud (f); monotonía (f) (Fig, monotony).
flattening, n. nivelación (f), aplanamiento (m) (levelling), alisado (m).
flatworm, n. gusano (m) plano, platelminto (m).
flavor, n. sabor (m), gusto (m).
flavour, n. sabor (m), gusto (m).
flax, n. lino (m).
fleet, n. flota (f); **deep sea fishing f.**, flota pesquera de altura; **factory fishing f.**, flota congeladora; **inshore fishing f.**, flota de bajura, flota de fresco, costera (f); **ocean-going fishing f.**, flota pesquera de altura.
fleeting, a. pasajero/a.
flexibility, n. flexibilidad (f), adaptabilidad (f).
flexture, n. flexión (f); pliegue (m) (fold).
flexure, n. flexión (f); pliegue (m) (fold).
flight, n. fuga (f) (escape); vuelo (m) (airline).
flint, n. sílex, pedernal; **f. implements**, utensilios (m.pl) de pedernal.
float, 1. v. boyar, flotar; lanzar (launch); 2. n. flotador (m) (Gen); **f.-operated switch**, interruptor (m) operado por un flotador; **f.-operated valve**, válvula (f) operada por un flotador.

floating, a. flotante; **f. bridge,** pontón (m) flotante; **f. dock,** dique (m) flotante.

floc, n. flóculo (m), coágulo (m).

flocculant, n. floculante (m).

flocculate, v. flocular.

flocculating, a. floculante; **f. agent,** agente (m) floculante, agente de floculación.

flocculation, n. floculación (f); **m. flocculation,** floculación (f) mecánica.

flocculator, n. floculador (m); **mechanical f.,** floculador mecánico.

flock, n. bandada (f) (of birds); rebaño (m) (of sheep, goats).

floe, n. témpano (m); **ice f.,** témpano de hielo.

flood, 1. v. inundar, anegar (to drown); **to f. a field,** anegar un campo; 2. n. avenida (f), crecida (f) (by a river); inundación (f) (of land, property); **annual f. (highest flow in a year),** crecida anual; **f. banks,** margen (m) de crecidas; **f. channel,** canal (m) de desagüe de crecidas; **f. control,** control (m) de inundaciones; **f. deposit,** sedimentos (m.pl) aluviales; **f. flow,** caudal (m) de avenida, caudal de crecimiento, inundación (f) fluvial; **f. forecast,** pronóstico (m) de inundaciones, previsión (f) de crecidas; **f. plain,** llanura (f) de inundación, planicie (f) de inundación, bajial (m) (LAm); **f. protection,** obras (f.pl) de protección contra las crecidas; **f. relief,** protección (f) contra crecidas; **f. risk,** riesgo (m) de crecidas (de ríos); **f. routing,** encauzamiento (m) de la inundación, regulación (f) de crecidas; **f. warning,** aviso (m) de inundaciones.

floodable, a. anegacido/a, inundable.

floodgate, n. compuerta (f), esclusa (f).

flooding, n. anegamiento (m), inundación (f).

floodlighting, n. iluminación (f) (of a stadium).

floodplain, n. llanura (f) de inundación, llanura aluvial.

floor, n. piso (m), suelo (m); fondo (m) (bottom); **f. drain,** desagüe (m) de piso; **ocean f.,** fondo oceánico; **valley f.,** fondo de valle.

floppy, a. flojo/a; **f. disc, f. disk (Comp),** disquete (m), diskette (m), disco (m) flexible.

flora, n. flora (f); **aquatic f.,** flora acuática; **endemic f.,** flora endémica.

flotation, n. flotación (f); lanzamiento (m) (of a company); **dissolved-air f.,** flotación por aire disuelto.

flotsam, n. pecios (m, pl) (ship wreckage).

flourmill, n. fábrica de harina (f).

flow, n. flujo (m), caudal (m) (volume); corriente (f) (current); escorrentía (f) (runoff); curso (m) (course); circulación (f) (e.g. of blood); **basalt f.,** flujo de basalto; **base f.,** flujo basal, caudal de base; **critical f.,** caudal crítico; **dry weather f. (Treat),** caudal de estación seca; **exploitable f.,** caudal explotable; **fissure f.,** flujo por grietas; **flood f.,** caudal de avenida, caudal de crecida; **f. characteristics,** caracter-ísticas (f.pl) del caudal; **f. hydrograph,** hidrograma (m) de flujo; **f. line,** línea (f) de flujo; **f. measuring tank,** recipiente (m) de aforo; **f. structure,** estructura (f) de flujo; **groundwater f.,** flujo de agua subterránea; **heat f.,** flujo calorífico; **instantaneous f.,** caudal instantáneo; **karstic f.,** flujo kárstico; **laminar f.,** flujo laminar; **multiphase f.,** flujo multifase; **open-channel f.,** flujo en canales abiertos; **overland f.,** escorrentía superficial; **perennial f.,** caudal perenne; **regional f.,** flujo regional; **river f.,** caudal fluvial, caudal del río; **sheet f.,** flujo laminar, flujo en manto; **specific f.,** flujo específico; **storm weather f. (Treat),** caudal de aguas de lluvia; **subcritical f.,** flujo subcrítico; **turbulent f.,** corriente turbulenta, flujo turbulento; **two-dimensional f.,** flujo bidimensional.

flower, n. flor (f); **composite f.,** flor compuesta.

flowing, a. fluente, fluyente; surgente (over-flowing).

flow into, v. desembocar.

flowmeter, n. caudalímetro (m).

flu, n. gripe (f); gripa (f) (LAm), influenza (f).

fluctuate, v. fluctuar, oscilar.

fluctuation, n. fluctuación (f), variación (f); **diurnal f.,** fluctuación diurna; **statistical f.,** fluctuación estadística.

flue-gas, n. gas (m) de la combustión; **f.-gas desulphurization,** desulfurización (f) del gas de combustión; **f.-gas removal plant,** instalación (f) extractora de gas de combustión; **f.-gas scrubber,** lavador (m) del gas de combustión; **f.-gas washing plant,** instalación (f) lavadora del gas de combustión.

fluid, 1. n. fluido (m); 2. a. fluido/a.

fluidity, n. fluidez (f).

fluidization, n. fluidificación (f).

fluidized, a. fluido/a, fluidificado/a; **f.-bed incineration,** incinerador (m) de lecho fluido.

fluke, n. uña (f) (of anchor, harpoon, etc.); aleta (f) (of whale); trematodo (m) (worm); chiripa (f) (stroke of luck).

flume, n. canaleta (f), saetín (m); aforador (m) (gauging device); **Parshall f.,** aforador Parshall; **trapezoidal f.,** aforador trapezoidal.

fluorate, n. fluorar.

fluoresce, v. ser fluorescente.

fluorescein, n. fluorescencia (f).

fluorescene, n. fluoresceina (f).

fluorescent, a. fluorescente; **f. tracer,** trazador (m) fluorescente.

fluoridation, n. fluorización (f).

fluoride, n. fluoruro (m).

fluoridization, n. fluorinazión (f).

fluorine, n. flúor (m) (F).

fluorite, n. fluorita (f).

fluorometer, n. fluorómetro (m).

fluorspar, n. fluorita (f).

flush, 1. v. enrasar (to make level); 2. n. cisterna (f), descarga (f) de agua (lavatory);

brote (m) (vegetation); gran limpieza (f) con agua (cleaning); enrase (m) (level).

flushing, n. limpieza (f) (with water), baldeo (m); descarga (f) de agua (jetting); **steam f.**, descarga de vapor.

flute, n. acanaladura (f) (vertical groove).

fluvial, a. fluvial.

fluvio-deltaic, a. fluviodeltaico/a.

fluvio-lacustrine, a. fluviolacustre.

fluvio-marine, a. fluviomarino/a.

flux, n. flujo (m) (flow); fundente (m) (of metals), liga (f) (alloy); **zero f. plane (Geog)**, plano (m) de flujo cero.

fly, 1. v. volar; 2. n. mosca (f) (Biol); **f. ash (waste dust)**, cenizas (f.pl), cenizas volantes, hollín (m) (soot).

fly-ash, n. hollín (m) (soot).

flyover, n. paso (m) elevado.

flysch, n. flysch (m).

flywheel, n. volante (m).

foal, n. potro (m).

foam, 1. v. espumar, hacer espuma; 2. n. espuma (f); **f. fractionation**, fraccionamiento (m) por espuma.

foaming, a. espumoso/a.

focus, n. foco (m), punto (m) de origen.

focusing, n. enfoque (m); **ecological f.**, enfoque ecológico.

fodder, n. forraje (m), pienso (m).

FOE, n. **Friends of the Earth (GB)**, organización (f) ecologista.

foetal, a. fetal.

foetus, n. feto (m).

FOG, n. **fats, oils and greases (Treat)**, grasas (f.pl) y aceites (m.pl).

fog, n. niebla (f); camanchacas (f.pl) (LAm); **f. collector**, atrapaniebla (f); **freezing f.**, niebla helada.

foggy, a. nebuloso/a, brumoso/a; **it is f. (v)**, hay niebla.

fold, 1. v. plegar; 2. n. pliegue (m) (Geol); dobladura (f), flexión (f); **cross f.**, pliegue cruzado; **drag f.**, pliegue de arrastre; **f. axis**, eje (m) del pliegue; **isoclinal f.**, pliegue isoclinal; **minor f.**, pliegue secundario; **overthrust f.**, pliegue por cabalgamiento; **plunging f.**, pliegue tumbado, pliegue buzante (LAm); **recumbent f.**, pliegue recumbente.

folded, a. plegado/a; **down-f.**, plegado hacia abajo; **up-f.**, plegado hacia arriba.

folding, n. plegamiento (m); **flexure-slip f.**, plegamiento por flexo-deslizamiento.

foliaceous, a. foliado/a (Geol).

foliage, n. follaje (m).

foliation, n. foliación (f), clivaje (m) (Geol).

foliole, n. folíolo (m).

follow-up, n. reiteración (f) (repeat), secuencia (f) (sequence).

food, n. alimento (m); **f. chain**, cadena (f) alimentaria, cadena trófica; **f cycle**, ciclo (m) alimentario; **f. industry**, industria (f) alimenticia; **f. processing**, procesamiento (m) de alimentos; **f. web**, ciclo (m) alimentario; **irradiated f.**, alimento irradiado.

foodstuffs, n.pl. productos (m.pl) alimenticios.

foot, n. pie (m) (Anat, Mat) (unit of length equal to 0.3048m); **at the f. of**, al pie de; **f. of mountain**, pie de la montaña; **on f.**, a pie; **square f.**, pie cuadrado.

footbridge, n. pasarela (f); **highway f.**, pasarela de vía urbana.

foothill, n. falda (f), estribaciones (f.pl) (foothills).

footing, n. zapata (f) (of building); **f. beam**, viga (f) principal.

footprint, n. pisada (f), rastro (m).

forage, n. forraje (m).

Foraminifera, n.pl. foraminíferos (m.pl) (Zoo, group).

force, n. fuerza (f); **by f.**, por fuerza; **centrifugal f.**, fuerza centrífuga; **in f. (Jur)**, vigente.

forced, a. forzado/a, duro/a.

forceful, a. energético/a, fuerte.

ford, 1. v. vadear; 2. n. vado (m) (across a river).

forecast, n. previsión (f); **flood f.**, previsión de avenidas; **water demand f.**, previsión de demanda de agua; **weather f.**, previsión del tiempo.

foreground, n. primer plano (m), primer término (m).

foreland, n. promontorio (m), cabo (m); antepaís (m) (LAm).

foreman, n. jefe (m) de taller, capataz (m).

foreseeable, a. previsible.

foreshore, n. ribazo (m), playa (f) entre pleamar y bajamar, cercanía (f) de costa.

forest, 1. n. selva (f), bosque (m); **temperate f.**, bosque de clima templado; **tropical rain f.**, selva tropical; **virgin f.**, bosque virgen, selva virgen; 2. a. forestal, selvático/a, de la selva; **f. fire**, incendio (m) forestal.

forestation, n. forestación (f), aforestación (f).

forestry, n. explotación (f) forestal, silvicultura (f).

foreword, n. prólogo (m) (of a book, report, etc.).

forge, v. forjar.

forged, a. forjado/a; **f. steel**, acero (m) forjado.

forklift, n. carretilla (f) elevadora, montecarga (f) de horquilla.

form, n. forma (f) (Gen); clase (f), tipo (m) (kind); manera (f), forma (f) (means); **application f.**, solicitud (f); **land f.**, forma del terreno.

formaldehyde, n. formaldehído (m).

formalin, n. formalina (f).

format, n. formato (m); **disk f. (Comp)**, formato de disco.

formation, n. formación (f); **calcareous f.**, formación calcárea; **fissured f.**, formación fisurada; **karstic f.**, formación kárstica; **soft f.**, formación blanda; **stratified f.**, formación estratificada; **tight f. (Geol)**, formación compacta; **unconsolidated f.**, formación no consolidada; **water-bearing f.**, formación acuífero.

formic, a. fórmico/a.

formula, n. fórmula (f); **chemical f.**, fórmula química; **empirical f.**, fórmula empírica; **mathematical f.**, fórmula matemática; **molecular f.**, fórmula molecular.

formulate, v. formular.

formulation, n. formulación (f).

formwork, n. encofrado (m).

fort, n. fortaleza (f); **small f.**, fortín (m).

fortification, n. fortificación (f).

fortuitous, a. fortuito/a.

fossa, n. fosa (f) (ditch).

fossil, n. fósil (m); **f. energy**, energía (f) fósil.

fossiliferous, a. fosilífero/a.

fossilization, n. fosilización (f).

foster, v. fomentar, promover (to encourage); patrocinar (a project).

foul, a. asqueroso/a (revolting, filthy); sucio/a (dirty); fétido/a (smell); contrario/a (wind, tide); obstruido/a (obstructed); **f. air**, aire (m) viciado; **f. sewer**, cloaca (f) de aguas negras.

foundation, n. cimientos (m.pl) (of building), firme (m) (roadbed); fundamento (m) (base, basis); fundación (f) (establishment); **piled f.**, pilotaje (m), cimientos en los pilotes; **raft f.**, cimientos de una balsa; **to lay the foundations of a building (v)**, echar los cimientos de un edificio.

foundry, n. fundición (f), fundidora (f) (LAm); **iron f.**, fundición de hierro.

fountain, n. fuente (f); **drinking f.**, fuente de beber.

four-wheel, a. de cuatro ruedas, todoterreno; **f.-wheel drive vehicle**, vehículo (m) todoterreno (m), campero (m).

foxglove, n. dedalera (f).

fractal, a. fractal.

fraction, n. fracción (f), quebrado (m); **representative f.**, fracción (f) representativa.

fractional, a. fraccionado/a; **f. crystalization**, cristalización (f) fraccionada.

fractionating, n. fraccionamiento (m); **f. column**, columna (f) de fraccionamiento; **f. tower**, torre (m) de fraccionamiento.

fractionation, n. fraccionamiento (m); **foam f.**, fraccionamiento (m) por espuma.

fractioning, n. fraccionamiento (m).

fracture, n. fractura (f); rotura (f) (rupture).

fractured, a. fracturado/a.

fracturing, n. dislocación (f), fracturación (f); **hydraulic f., hydro-f.**, fracturación hidráulica.

fragile, a. frágil, quebradizo/a, deleznable.

fragment, n. fragmento (m), trozo (m).

fragmentary, a. fragmentario/a.

framework, n. estructura (structure); armazón (f), entramado (m) (Con).

France, n. Francia (f).

franchise, n. franquicia (f).

francium, n. francio (m) (Fr).

Frasnian, n. Frasniaense (m).

free, a. libre; desocupado (vacant); **f. radical (Chem)**, radical (m) libre; **f. trade**, libre comercio (m), libre cambio (m).

freeboard, n. borde (m) libre, francobordo (boats), cota (f) de seguridad.

freedom, n. libertad (f); **f. of information**, libertad de información (falta de censura); **F. of Information Act (US)**, libertad de información para estudiar casi todos los registros referentes a un departamento de la administración pública.

free-lance, a. independiente (de trabajar).

free-phase, n. fase (m) libre.

freeway, n. autopista (f) sin peaje (US).

freewheel, n. rueda (f) libre.

freeze, v. congelar.

freeze-dried, a. liofilizado/a.

freeze-drying, n. liofilización (f).

freezing, n. congelación (f); **5 degrees below f.**, 5 grados bajo cero; **f. point**, punto (m) de congelación, temperatura (f) de congelación.

freight, n. flete (m), cargo (m), mercancías (f.pl); **f. car (US)**, vagon (m) de mercancías; **f. collect (US)**, flete por cobrar.

French, a. francés/francesa.

freon, n. freón (m).

frequency, n. frecuencia (f); **f. distribution**, distribución (f) de frecuencia; **high-f. waves**, ondas (f.pl) de alta frecuencia; **low-f. waves**, ondas (f.pl) de baja frecuencia; **mains f.**, frecuencia de red; **radio f. (abbr. RF)**, radiofrecuencia (f) (abbr. FR); **size f. (Geol)**, frecuencia granulométrica; **stream f.**, frecuencia de los ríos.

fresco, n. pintura (f) al fresco.

freshet, n. avenida (f).

freshwater, n. agua (f) dulce.

friability, n. friabilidad (f).

friable, a. friable.

friction, n. fricción (f); **f. loss**, pérdida (f) por fricción.

fridge, n. refrigerador (m), frigorífico (m).

fringe, n. franja (f); linde (m), lindero (m) (of wood); periferia (f) (of city), orla (f) (of vegetation); faja (f) (strip); **capillary f.**, franja capilar; **f. vegetation (Ecol)**, orla (f); **interference f.**, franja de interferencia; **subtropical f.**, franja subtropical; **urban f.**, franja urbana.

frond, n. fronda (f).

front, n. frente (m); fachada (f) (of building); **atmospheric f.**, frente atmosférico; **cold f.**, frente frío; **displacement f.**, frente de desplazamiento; **occluded f.**, frente ocluido; **recharge f.**, frente de agua recargada; **warm f.**, frente cálido.

frontage, n. fachada (f).

frontal, a. frontal.

frontier, 1. n. frontera (f) (border), línea (f)

divisoria (dividing line); 2. a. fronterizo/a; **f. zone,** zona (f) fronteriza.

frost, n. helada (f) (weather), escarcha (f) (ground frost); **f. action,** acción (f) de la helada; **hoar f.,** helada blanca.

frosted, a. mate, deslustrado/a, satinado/a.

froth, n. espuma (f).

frozen, a. congelado/a.

frucose, n. frucosa (f).

fructose, n. fructosa (f).

fruit, n. fruta (f), fruto (m); **f. farming,** fruticultura (f); **f. grower,** fruticultor/a (m, f); **f. tree,** árbol (m) frutal, frutal (m); **seasonal f.,** fruta del tiempo.

fruit-bearing, a. frutal.

ft, n. pie (m) (Mat) (abbr. de foot, unidad de longitud igual a 0,3048m).

fucose, n. fucosa (f).

fuel, n. combustible (m), energéticos (m.pl); gasolina (f) (US: petrol, gas); **f.-oil,** fuel (m), fuel-oil (m), combustóleo (m), aceito (m) pesado; **fossil f.,** combustible fósil; **solid f.,** combustible sólido.

fuel-oil, n. fuel-oil (m); combustóleo (m) (LAm).

fugitive, a. fugitivo/a; **f. emission,** emisión (f) fugitiva.

fulfilment, n. realización (f).

fulminant, a. fulminante.

fumarole, n. fumarola (f).

fume, 1. v. humear, echar humo (Chem); 2. n. vapor (m), gas (m), humo (m); **factory fumes,** humo de las fábricas; **f. cupboard,** campana (f) de humo.

fumes, n.pl. vapores (m.pl), gases (m.pl), humo (m); **noxious f.,** gases nocivos, humos nocivos.

fumigant, n. fumigante (m).

fumigate, v. fumigar.

fumigation, n. fumigación (f).

function, n. función (f) (Mat); acto (m) (ceremony); **Bessel functions,** funciones de Bessel; **error f.,** función de error; **mathematical f.,** función matemática; **probability f.,** función de probabilidad; **transfer f.,** función de transferencia; **well f.,** función de pozo.

functional, a. funcional.

functioning, n. funcionamiento (m), el trabajar (m) (the working of).

fund, n. fondo (m) (Gen), reserva (f) (reserve).

fundamental, a. fundamental.

fungi, n.pl. hongos (m.pl).

fungicide, n. fungicida (m).

fungus, n. hongo (m); **slime f.,** moho (m.pl) mucoso.

funnel, n. embudo (m) (for pouring liquids); chimenea (for boat, etc.); conducto (m) (ventilation, etc); **filter f.,** embudo de filtro; **separation f.,** embudo de separación; **thistle f. (Chem),** embudo cardo.

furan, n. furano (m).

furlong, n. medida de longitud de 201.17 metros.

furnace, n. horno (m); boliche (m) (small furnace); estufa (f) (domestic heating); **blast f.,** horno alto; **mouth of blast f.,** tragante (m); **reverberatory f.,** horno de reverbero.

furniture, n. mobilario (f); **s. furniture (e.g. street lights),** mobilario urbano.

furrow, n. surco (m), ranura (f).

furrowed, a. surcado/a, acanalado/a.

fuse, 1. v. fundir (to melt); fusionar (to join); **to blow a f. (Elec),** fundirse un fusible, fundirse un plomo; 2. n. plomo (m), fusible (m) (Elec); fulminante (m) (detonator); **f. box,** caja (f) de fusibles.

fusiform, a. fusiforme.

fusion, n. fusión (f); **f. point,** punto (m) de fusión; **nuclear f.,** fusión nuclear.

G

gabbro, n. gabro (m).
gabion, n. cestón (m), gabión (m) metálico.
gable, n. gablete (m); **g. end,** hastial (m), muro (m) piñón.
gadolinium, n. gadolinio (m) (Gd).
galaxy, n. galaxia (f).
gale, n. vendaval (m), ventarrón (m), ventolada (f) (LAm); **force 10 g.,** vendaval de fuerza 10; **g. warning,** aviso (m) tormenta.
galena, n. galena (f).
galenite, n. galenita (f).
galium, n. galio (m) (Ga).
gallery, n. galería (f) (Gen, Min, Eng); qanat (m) (Arab); **infiltration g.,** galería de infiltración; **interceptor g.,** galería de captación; **well g.,** galería de pozo.
gallon, n. galón (m) (measure of volume, 4.55 litres in GB, 3.79 litres in US).
galvanitic, a. galvánico/a; **g. cell,** pila (f) galvánica.
galvanize, n. galvanizar.
galvanized, a. galvanizado/a; **g. sheet,** lámina (f) galvanizada.
galvanometer, n. galvanómetro (m).
game, n. juego (m) (play or sport); caza (f) (hunting); **g. bird,** ave (m) de caza; **g. fishing,** pesca (f) deportiva; **g. warden,** guardabosque (m), guarda (f) de caza; guardamonte(s) (m).
gamete, n. gameto (m).
gamma, n. gamma (f); **g. rays,** rayos (m.pl) gamma.
ganat, n. ganat (m), qanat (m), khanat (m) (Arab); galería (f) (Ch).
gangrenous, a. gangrenoso/a.
gangue, n. ganga (f), estéril (m).
gangway, n. pasarela (f).
gannister, n. ganister (m).
gantry, n. caballete (m) (Gen), pórtico (m) (of crane).
gap, n. hueco (m), vacío (m) (Gen); brecha (f), quebrada (f), desfiladero (m) portillo (m), abra (f) (Geog, narrow pass); **fire g.,** abra de incendio; **water g.,** abra de agua; **wind g.,** abra de viento.
garage, n. garaje (m) (in house); taller (m) mecánico (for servicing cars).
garbage, n. basura (f); **city g.,** residuos municipales; **g. can (US),** cubo (m) de basura, balde (m) (LAm), pipote (m) (Ven); **g. heap,** vertedero (m), basurero (m).
garden, n. jardín (m); **g. centre, g. center (US),** vivero (m), centro (m) de jardinería; **public gardens,** parque (m); **vegetable g.,** huerto (m).
gardener, n. jardinero (m); **landscape g.,** jardinero paisajista.

gardening, n. jardinería (f).
garigue, n. garriga (f).
garnet, n. granate (m).
gas, n. gas (m); gasolina (f), nafta (f) (CSur), bencina (f) (Ch) (US, gasoline); **biogas,** biogás (m), gas de origen bioquímico; **dissolved g.,** gas disuelto; **g. chromatography-mass spectrometry,** cromatografía (f) de gases-espectrometría de masas; **g. detector,** detector (m) de gases; **g. extraction,** extracción (f) de gases; **g. liquor,** licor (m) de gases; **g. meter,** contador (m) de gas; **g. saturation,** saturación (f) con gas; **g. scrubbing,** lavado (m) de gases; **g. station,** gasolinera (f), estación (f) de servicio, grifo (m) (Peru), bencinera (f) (Ch); **g. tanker,** gasolinero (m); **g. ventilation,** ventilación (f) de gases; **greenhouse gases,** gases (m.pl) de efecto invernadero; **inert g.,** gas inerte; **landfill g.,** gas de vertido controlado; **marsh g.,** gas de los pantanos; **natural g.,** gas natural; **noble gases,** gases nobles; **rare gases,** gases raros, gases nobles; **sewer g.,** gas de cloaca.
gaseous, a. gaseoso/a; **g. emissions,** emisiones (f.pl) gaseosas.
gasholder, n. gasómetro (m).
gasification, n. gasificación (f).
gasify, v. gasificar.
gasohol, n. gasohol (m).
gasoline, n. gasolina (f), nafta (f) (CSur), bencina (f) (Ch); **g. station,** gasolinera (f), estación (f) de servicio, grifo (m) (Peru), bencinera (f) (Ch).
gasometer, n. gasómetro (m) (gas holder).
gastric, a. gástrico/a.
gastroenteritis, n. affección (f) de tipo gastrointestinal.
gastropod, n. gasterópodo (m), gastrópodo (m).
Gastropoda, n.pl. gasterópodos (m.pl), la clase Gastropoda (Zoo).
gate, n. verja (f) (in fence), puerta (f) (of city); **g. valve,** válvula (f) de compuerta; **sluice g.,** compuerta (f).
GATT, n. GATT (m); **General Agreement on Tariffs and Trade,** acuerdo (m) general sobre aranceles aduaneros.
gauge, v. aforar.
gauge, n. aparato (m) de medida, aforo (m); gálibo (m) (Phys), galga (f) (calliper); ancho (m) de vía (railway); **g. board (for river level),** escalilla (f); **raingauge,** pluviómetro (m); **stream g.,** estación (f) de medición de caudales.
gauging, n. medición (f), aforo (m), aforamiento (m); **chemical g.,** aforo químico;

float g., aforo con flotadores; **flow g.**, aforo de caudal; **g. device**, aforador (m); **g. station**, estación (f) de medición de caudales, estación de aforo; **g. with bailer**, aforo por cuchareo; **g. with radioactive tracers**, aforo con trazadores radioactivos; **stream g.**, aforo de caudales.

gaussian, a. gaussiano/a; **g. distribution**, distribución (f) gaussiana.

gazelle, n. gacela (f).

GB, n. **Great Britain**, Gran Bretaña (f).

GDP, n. PIB (m) (Sp), PGB (m) (Ch), PTB (m) (Peru); **Gross Domestic Product**, Producto (m) Interior Bruto (Sp), Producto Geográfico Bruto (Ch), Producto Territorial Bruto (Peru).

gear, n. equipo (m) (equipment); herramientas (f.pl) (tools); velocidad (f), marcha (f) (for a car); **winding g.**, manubrio (m), cabrestante (m).

gearbox, n. caja (f) de cambios, caja de velocidades.

gecko, n. geco (m).

Gedinnian, n. Gedinnense (m).

GEF, n. GEF (m); **Global Environment Facility**, Fondo (m) para el Medio Ambiente Mundial.

Geiger, n. Geiger; **G. counter**, contador (m) Geiger.

gel, n. gel (m) (pl. gel, geles).

gelatine, n. gelatina (f).

gelatinous, a. gelatinoso/a.

gelignite, n. gelignita (f).

gene, n. gene (m), gen (m); **g. pool**, pool (m) génico.

genera, n.pl. géneros (m.pl).

general, a. general.

generality, n. generalidad (f).

generalized, a. generalizado/a; **g. linear model**, modelo (m) lineal generalizado.

generation, n. generación (f); **data g.**, generación de datos.

generator, n. generador (m) (Elec); **electrical g.**, generador eléctrico; **g. set**, grupo (m) electrógeno.

generatrix, n. generatriz (f) (Mat).

generic, a. genérico/a; **g. name (Biol)**, nombre (m) genérico; **g. exchange**, intercambio (m) genético; **g. variation**, variación (f) genética.

genetics, n. genética (f).

genome, n. genoma (m).

genotype, n. genotipo (m).

genus, n. género (m).

geo-, pf. geo-.

geoarchaeology, n. geoarqueología (f).

geochemical, a. geoquímico/a.

geochemist, n. geoquímico/a (m, f).

geochemistry, n. geoquímica (f).

geode, n. geoda (f).

geodesic, a. geodésico/a.

geodesy, n. geodesia (f).

geodetic, a. geodésico/a.

geodynamic, a. geodinámico/a.

geoelectrical, a. geoeléctrico/a.

geogenic, a. geogénico/a.

geographer, n. geógrafo/a (m, f).

geographical, a. geográfico/a.

geography, n. geografía (f).

geohazard, n. georiesgo (m).

geohydrochemical, a. geohidroquímico/a.

geohydrochemistry, n. geohidroquímica (f).

geohydrological, a. geohidrológico/a.

geohydrologist, n. geohidrólogo/a (m, f).

geohydrology, n. geohidrología (f).

geoid, n. geoide (m).

geological, a. geológico/a; **British G. Survey (GB)**, Instituto (m) Geológico y Minero de España (Sp).

geologist, n. geólogo/a (m, f).

geology, n. geología (f).

geomembrane, n. geomembrana (f).

geometric, a. geométrico/a; **g. progression**, progresión (f) geométrica.

geometry, n. geometría (f).

geomorphological, a. geomorfológico/a.

geomorphologist, n. geomorfólogo/a (m, f).

geomorphology, n. geomorfología (f).

geophone, n. geófono (m).

geophysical, a. geofísico/ca.

geophysicist, n. geofísico/a (m, f).

geophysics, n. geofísica (f).

geopolitics, n. geopolítica (f).

georeference, n. georreferenciado (m).

Georgian, n. Georgiense (m).

geosphere, n. geosfera (f).

geostatistics, n. geoestadística (f).

geosynclinal, n. geosinclinal.

geosyncline, n. geosinclinal (m).

geosynthetic, a. geosintético/a; **g. lining (for landfill)**, revestimiento (m) geosintético.

geotechnics, n. geotecnia (f), geomecánica (f).

geotextile, 1. n. geotextil (m); 2. a. geotextil; **g. lining (for landfill)**, revestimiento (m) geotextil.

geothermal, a. geotérmico/a; **g. reaction**, reacción (f) geotérmica.

geotropism, n. geotropismo (m), haptotropismo (m).

germ, n. germen (m) (Bot, brote); germen, microbio (m) (microorganism); **g. cell**, célula (f) germinal.

German, a. alemán/a.

germanium, n. germanio (m) (Ge).

Germany, n. Alemania (f).

germinate, v. germinar.

germination, n. germinación (f).

geyser, n. geyser (m).

giardia, n. giardia (f).

giga-, pf. giga- (10.E9).

gigabyte, n. gigabyte (m).

gill, n. agalla (f), branquia (f) (of fish); lámina (f) (of fungi); laminilla (f) (of plants); (GB, medida de líquidos de un cuarto de una pinta).

giraffe, n. jirafa (f).

girder, n. girder (m), viga (f); **box g.**, viga tubular.

GIS, n. SIG (m); **Geographical Information System**, Sistema (m) de Información Geográfica.

Givetian, n. Givetiense (m).

glacial, 1. n. glacial (m); 2. a. glaciar, glacial.

glaciated, a. que ha sufrido glaciación, afectado por glaciación.

glaciation, n. glaciación (f).

glacier, n. glaciar (m); **piedmont g.**, glaciar de piedemonte, glaciar pedemontano (LAm); **valley g.**, glaciar de valle.

glaciofluvial, a. glaciofluvial.

glade, n. claro (in a wood); ciénaga (f), calvero (m) (wet ground).

glades, n.pl. praderas (f) encharcadizas (US); **Everglades (Florida)**, Everglades (m.pl).

gland, n. glándula (f).

glandular, a. glandular.

glass, n. vidrio (m); vaso (m), copa (f) (for drinking); lente (f) (lens); cristal (m) (window pane); gemelos (m.pl) (binoculars); **container g.**, contenedor (m) de vidrio; **g. fibre, g. fiber (US)**, fibra (f) de vidrio; **glasses**, gafas (f.pl), anteojos (m.pl) (LAm); **ground g.**, vidrio esmerilado; **magnifying g.**, lupa (f).

glassfibre, n. fibra (f) de vidrio.

glauconite, n. glauconita (f).

glaucophane, n. glaucófana (f).

glaze, 1. v. vidriar (ceramics); barnizar (to varnish); poner cristales a (windows); 2. n. vidriado (m) (ceramics); barniz (varnish); aguanieve (f) (US, sleet).

glazed, a. vidriado/a (surface); con cristal (GB, of door, window).

glazing, n. cristal (m), vidrio (m); encristalado (m) (action).

glen, n. hocino (m), cañada (f) (Scottish valley).

gley, n. suelo (m) arcilloso; gley (m) (LAm).

glitch, n. fallo (m), espúreo (m).

GLM, n. **generalized linear model** (Mat), modelo (m) lineal generalizado.

global, a. global; mundial (worldwide); esférico/a (spherical); **g. warming**, calentamiento (m) global.

globe, n. globo (m), esfera (f) (sphere); mundo (m) (the world); **the Globe (the Earth)**, globo terráqueo, el mundo.

globule, n. glóbulo (m).

glossary, n. glosario (m); **g. of terms**, glosario de términos.

glow, n. brillo (m) (of lamp), luminosidad (f), incandescencia (f).

glucosamine, n. glucosamina (f).

glucose, n. glucosa (f).

glume, n. gluma (f).

glutamine, n. glutamina (f).

gluten, n. gluten (m).

glyceraldehyde, n. gliceraldehído (m).

glyceride, n. glicérido (m).

glycerine, n. glicerina (f).

glycerol, n. glicerol (m).

glycogen, n. glucógeno (m).

glycol, n. glicol (m).

glycolysis, n. glucólisis (f).

GMT, n. TMG (m); **Greenwich Mean Time**, Tiempo (m) Medio de Greenwich.

gnat, n. mosquito (m), jején (m) (LAm).

gneiss, n. gneis (m), neis (m).

GNP, n. PNB (m); **Gross National Product**, Producto (m) Nacional Bruto.

goal, n. gol (m) (deportes); objetivo (m), meta (f), fin (m) (objective); **water quality g. (US)**, estándar (m) de calidad de las aguas.

goat, n. cabra (f) (female); chivo (m), macho (m) cabrío (male); **mountain g.**, cabra montés.

goethite, n. goethita (f).

goiter, n. papera (f), bocio (m) (Med).

goitre, n. papera (f), bocio (m) (Med).

gold, n. oro (m) (Au); **alluvial g.**, oro de aluvión; **fool's g. (Fig)**, oro de los tontos, pirita (f); **g. field**, placer (m), campo (m) de oro.

golf, n. golf (m); **g. course, g. links**, campo (m) de golf.

gonad, n. gónada (f).

goniatite, n. goniatita (f).

goods, n.pl. bienes (m.pl); mercancías (f.pl) (merchandise); géneros (m.pl), artículos (m.pl) (Com); **consumer g.**, bienes de consumo.

gooseberry, n. grosella (f) espinosa, uva (f) espina.

gorge, n. garganta (f), gorja (f), desfiladero (m), pongo (m) (LAm).

gorse, n. aulaga (f), tojo (m).

gouging, n. encajamiento (m), exhondamiento (m) (Geog).

gourd, n. calabaza (f) (legumbre), auyama (f) (LAm), guacal (m) (LAm), jícara (f) (LAm).

governing, a. gobernador/a; **g. board**, junta (f) de gobierno.

government, n. gobierno (m); oficialismo (m) (LAm, of a state); dirección (f), administración (f) (of a company).

governmental, a. gubernamental, gubernativo/a; gubernista (LAm).

governor, n. gobernador/a (m, f) (of state), intendente (m) (CSur); director/a (m, f) (of school), jefe (m), (boss); **civil g.**, gobernador/a civil; **provincial g.**, gobenador/a provincial.

GPS, n. global positioning system.

grab, n. cuchara (f); **hydraulic g.**, cuchara hidráulica.

graben, n. graben (m).

grade, 1. v. clasificar, graduar; 2. n. grado (m), categoría (f), clase (m); **high/low g.**, grado alto/bajo; **metamorphic g.**, grado metamórfico.

grader, n. máquina (f) niveladora, máquina para igualar terreno.

gradient, n. pendiente (f), cuesta (f), gradiente (m); **concentration g.**, gradiente de concentración; **geothermal g.**, gradiente geotérmico; **hydraulic g.**, gradiente hidráulico; **piezometric g.**, gradiente piezométrico; **threshold g.**, gradiente umbral.

grading, n. clasificación (f), graduación (f).

gradual, a. paulatino/a.

graduation, n. graduación (f) (Mat); entrega (f) de título (e.g. at university).

graft, 1. v. injertar (Bot); 2. n. injerto (m) (Bot).

grain, n. grano (m), cereales (m.pl), trigo (m), fibra (f) (wood), veta (f) (stone); **coarse-grained (a)**, grano grueso; **fine-grained (a)**, grano fino; **g.-size**, tamaño (m) de grano.

grainy, a. granudo/a.

gram, n. gramo (m).

Gram, a. Gram; **G.'s stain**, tinción (f) Gram; **G.-negative**, Gram negativo/a; **G.-positive**, Gram positivo/a.

gram-atom, n. átomo-gramo (m).

Graminaceae, n.pl. gramíneas (f.pl), la familia Graminaceae (Bot).

graminaceous, a. gramíneo/a.

Graminae, n.pl. gramíneas (f.pl), la familia Graminae (Bot).

gramineous, a. gramíneo/a.

gram-ion, n. ión-gramo (m).

gramme, n. gramo (m).

gram-molecule, n. molécula-gramo (m).

granite, n. granito (m).

granitization, n. granitización (f).

granodiorite, n. granodiorita (f).

granophyre, n. granófiro (m).

granular, a. granular.

granule, n. gránulo (m).

granulite, n. granulita (f).

granulometry, n. granulometría (f).

grape, n. uva (f); **bunch of grapes**, racimo (m) de uvas; **g. harvest**, vendimia (f); **g. must**, mosto (m).

grapefruit, n. pomelo (m), toronja (f).

grapevine, n. vid (f), parra (f); sarmiento (m) (shoot or branch of a vine).

graph, n. gráfico (m), gráfica (f); **g. paper**, papel (m) cuadriculado; **time-drawdown g.**, gráfico de descenso-tiempo; **water level-time g.**, gráfico de niveles-tiempos; **yield-drawdown g.**, gráfico de caudal-descenso.

graphic, a. gráfico/a.

graphical, a. gráfico/a.

graphite, n. grafito (m).

grapple, n. garfio (m).

graptolite, n. graptolita (f).

grass, n. hierba (f), herbácea (f); césped (m), pasto (m), grama (f) (LAm) (lawn); **permanent g.**, pradera (f) natural.

grasshopper, n. saltamontes (m), chapulín (m) (CAm).

grassland, n. pradera (f), pampa (f) (LAm), prado (m), hierbal (f) (LAm); pasto (m) (pasture); **fertilized g.**, pasto fertilizado; **natural g.**, pradera natural.

grating, n. verja (f), enrejado (m), reja (f), parrilla (f).

grave, n. sepultura (f), tumba (f) (tomb).

gravel, n. grava (f), pedregullo (m) (Arg, Ch, Uru); **beach g.**, grava de playa; **flood g.**, grava de inundación, grava de crecida; **g. pack**, empaque (m) de gravas.

gravelly, a. guijarroso/a.

gravestone, n. lápida (f) sepulcral, lápida mortuoria.

graveyard, n. cementario (m).

gravimetry, n. gravimetría (f).

gravitate, v. gravitar.

gravitation, n. gravitación (f).

gravitational, a. gravitacional.

gravity, n. gravedad (f); **centre of g.**, centro (m) de gravedad; **g. drainage**, drenaje (m) por gravedad; **g. filter**, filtro (m) por gravedad; **g. system (e.g. for water supply)**, sistema (m) por gravedad.

graywacke, n. grauvaca (f).

graze, v. pacer.

grazing, n. pastoreo (m) (type of farming); pasto (m) (pasture); **g. land**, pasto (m).

grease, 1. v. engrasar, lubricar; 2. n. grasa (f); **fats, oils and greases (Treat)**, grasas y aceites; **g. trap**, interceptor (m) de grasas.

green, 1. v. verdear (plants turning green); 2. a. verde (colour); verde (unripe); nuevo/a (inexperienced); crédulo/a (gullible); **g. belt**, zona (f) verde; **g. issues**, temas (m.pl) ecológicos; **G. Party**, Los Verdes, Partido Verde (politics); **g. politics**, política (f) ecologista; **g. pound (sterling)**, libra (f) verde; **The Greens**, Los Verdes, Partido Verde (politics); **village g.**, césped (m) de uso común (LAm).

greengage, n. ciruela (f) claudia.

greenhouse, n. invernadero (m); **g. effect**, efecto (m) invernadero; **g. gases**, gases (m.pl) de efecto invernadero.

greensand, n. arena (f) glauconítica.

greenstone, n. rocas (f.pl) verdes.

greisen, n. greisen (m).

greywacke, n. grauvaca (f).

grid, n. cuadrícula (f), reja (f), verja (f).

gridline, n. línea (f) de reticulado.

grille, n. rejilla (f); reja (f) (of window, door), verja (f) (screen).

grime, n. mugre (m), suciedad (f).

grind, v. moler, afilar, esmerilar (polish).

grinder, n. triturador (m).

grinding, n. molienda (f) (milling), esmerilado (m) (e.g. of valves).

grindstone, n. muela (f), piedra de afilar (f).

grit, n. arenisca (f), asperón (f); **g. removal**, extracción (f) de arenisca.

groin, n. espolón (m), muro (m) de defensa (Con).

groove, n. ranura (f), acanaladura (f).

grooved, a. acanalado/a, surcado/a.

gross, a. bruto/a (weight, profit); grueso/a, grande (fat, large); ordinario (vulgar); total (overall); **g. national product (abbr. GNP)**, producto (m) nacional bruto (abbr. PNB); **g. negligence**, gran negligencia (f); **g. ton**, tonelada (f) larga; **g. weight**, peso (m) en bruto.

grossularite, n. grosularita (f).

grotto, n. gruta (f) (cave).

ground, 1. n. terreno (m), suelo (m); **boggy g.**, bodonal (m), tolla (f), arribanzo (m); **rough g.**, terreno abrupto, terreno desigual; **stony g.**, terreno pedregoso; 2. a. molido/a, esmerilado/a (polished).

groundnut, n. cacahuete (m), cacahuate (m) (LAm), maní (m) (LAm).

groundswell, n. mar de fondo (m), marejada (f).

groundwater, n. agua (f) subterránea; **artesian g.**, agua subterránea artesiana; **free g.**, agua subterránea libre; **g. flow**, flujo (m) de agua subterránea; **g. level**, nivel (m) de agua subterránea; **g. rebound**, rebote (m) de agua subterránea; **g. recession**, retroceso (m) de agua subterránea; **g. recharge**, recarga (f) de agua subterránea; **g. runoff**, escorrentía (f) de agua subterránea; **g. storage**, almacenamiento (m) de agua subterránea; **perched g.**, agua subterránea colgada.

group, n. grupo (m), juego (m) group; **blood g.**, grupo sanguíneo; **pressure g.**, grupo de presión.

grouper, n. mero (m) (fish).

grout, 1. v. gunitar; 2. n. gunita (f); **g. curtain**, gunitado (m).

grouted, a. gunitado/a.

grouting, n. gunitado (m), lechada de cemento; **pressure g.**, inyección (f) de mortero de cemento a presión.

grove, n. arboleda (f); **almond g.**, almendral (m); **chestnut g.**, castañar (m), castañeda (f); **lemon g.**, limonar (m); **orange g.**, naranjal (m).

grow, v. crecer (Gen), cultivar (to farm).

growth, n. crecimiento (m); **bacterial g.**, crecimiento bacterial; **g. curve**, curva (f) de crecimiento; **g. ring (Bot)**, anillo (m) de crecimiento; **irruptive g.**, crecimiento irruptivo; **population g.**, crecimiento de la población, crecimiento demográfico; **rate of g.**, rapidez (f) de crecimiento; **secondary g.**, crecimiento secundario; **sustained economic g.**, crecimiento económico sostenido.

groyne, n. espolón (m), muro (m) de defensa.

grub, n. larva (f), gusano (m) (Zoo).

guano, n. guano (m).

guarani, n. guaraní (m) (Par, language, unit of currency).

guarantee, 1. v. garantizar; 2. n. garantía (f) (surety), caución (f) (guarantor); **g. certificate**, certificado (m) de garantía; **g. of supply**, garantía de suministro.

Guatemala, n. Guatemala (f).

Guatemalan, 1. n. guatemalteco/a (m, f); 2. a. guatemalteco/a.

guesstimate, n. estimación (f) aproximada, tanteo (m).

Guiana, n. Guayana (f).

guide, n. guía (f).

guideline, n. directriz (f), pauta (f); **to lay down guidelines (v)**, marcar la pauta.

gulch, n. quebrada (f).

gulf, n. golfo (m); **the G. of Mexico**, el golfo de México/Méjico.

gull, n. gaviota (f).

gullet, n. abertura (f) entre estratos.

gulley, n. cañada (f), cárcava (f), barranco (m), quebrada (f); alcantarilla (f) (sewer).

gully, n. cañada (f), cárcava (f), barranco (m), quebrada (f); alcantarilla (f) (sewer).

gum, n. goma (f); **g. arabic**, goma arábica; **g. tree**, gomero (m).

gunite, n. gunita (f).

gush, v. brotar (water flow).

gust, 1. v. soplar racheado; 2. n. ráfaga (f) (of wind); racha (f), aguacero (m), chaparrón (m) (of rain); ventolina (f) (LAm).

gutter, n. arroyo (m) cuneta (f) (in street); canal (m), canalón (m) (on roof).

guttering, n. canales (m.pl) (Con).

Guyana, n. Guayana (f).

Gymnophyta, n.pl. gimnofitas (f.pl), Gymnophyta (f.pl) (Bot, division).

gymnosperm, n. gimnosperma (f).

Gymnospermae, n.pl. gimnospermas (f.pl), la división Gymnospermae (Bot).

gypsite, n. gipsita (f).

gypsum, n. yeso (m); **g. deposit**, yacimiento de yeso, yesera (f).

gypsy, n. gitano/a (m, f) (W. Europe), zíngaro/a (m, f) (C. Europe).

gyrocone, n. girocono (m).

H

habitat, n. habitat (m), hábitat (m).
hachis, n. marihuana (f), mariguana (f), marijuana (f).
hacker, n. pirata (f) (Comp).
haddock, n. eglefino (m).
hade, n. inclinación (f) de fallas.
haematite, n. hematita (f), hematites (m).
haematology, n. hematología (f).
haemocyanin, n. hemocianina (f).
haemoglobin, n. hemoglobina (f).
haemophilia, n. hemofilia (f).
haemophiliac, n. hemofílico/a (m, f).
haemophilic, a. hemofílico/a.
haemorrhage, n. hemorragia (f).
hafnium, n. hafnio (m) (Hf).
hail, 1. v. granizar; 2. n. granizo (m).
hailstone, n. pedrisco (m) (large hailstone).
hailstorm, n. granizada (f).
hair, n. pelo (m) (of people, of animals), cabello (m) (of people); pelo (m), pelusa (f) (Bot); **root h. (Bot),** pelo radical.
Haiti, n. Haití (m).
Haitian, 1. n. haitiano/a (m, f); 2. a. haitiano/a.
hake, n. merluza (f).
half, n. mitad (f), medio (m); **at half price,** a mitad de precio.
half-life, n. vida (f) media; periódo (m) de semidesintegración (Phys).
half-open, a. entreabierto/a.
halide, n. haluro (m); **volatile h.,** haluro volátil.
halite, n. halita (f), sal de roca (f).
halo, n. halo (m) (Met).
halogen, n. halógeno (m), halogenuro (m).
halogenated, a. halogenado/a; **h. hydrocarbon,** hidrocarburo (m) halogenado; **h. solvent,** disolvente (m) halogenado.
halogenation, n. halogenación (f).
halogenesis, n. halogénesis (m).
halophilic, a. halófilo/a.
halophyte, n. halófito (m).
halteres, n.pl. alterio (m), balancín (m).
hammer, n. martillo (m) (tool), percusor (m) (of firearm); **air h.,** martillo de aire comprimido; **drop h.,** martillo, martillo pilón; **pneumatic h.,** martillo neumático.
hammermill, n. molino (m) de martillo (for pulverising waste) (Eng).
handaxe, n. hacha (f) de mano.
handbook, n. guía (f), manual (m).
handling, n. manejo (m) (e.g. a machine), gestión (f) (management), tratamiento (m) (treatment).
haphazard, a. casual, fortuito/a.
haploid, a. haploide.
hapteron, n. zarcillo (m).
harbour, n. puerto (m), refugio (m).

hard, a. duro/a, tenaz.
hardboard, n. madera (f) prensada.
hardcore, n. ripio (m), pedreplén (m) (rough stone), mampostería (f) (rubblework).
hardness, n. dureza (f); **bicarbonate h.,** dureza bicarbonatada; **carbonate h.,** dureza del carbonato; **h. of water,** dureza del agua; **permanent h.,** dureza permanente; **temporary h.,** dureza temporal; **total h.,** dureza total.
hardpan, n. tosca (f), costra (f) continua (Geol).
hardstanding, n. instalación (f) provista de zonas afirmadas.
hardware, n. ferretería (f), quincallería (f) (de metal); hardware (m), equipo físico (Comp).
hardwood, n. madera (f) dura.
harm, n. daño (m) (damage), perjuicio (m) (against one's interests).
harmful, a. perjudicial (reputación), dañino/a, nocivo/a (substance).
harmonic, a. armónico/a; **h. series,** serie (f) armónica.
harmonize, v. armonizar.
harness, v. captar (water power).
harpoon, n. arpón (m).
harvest, 1. v. cosechar, recoger; 2. n. cosecha (f), siega (f) (of cereals), recolección (f) (of vegetables), vendimia (f) (of grapes).
harvester, n. segador/a (person), cosechadora (f) (machine); **combine h.,** segadoratrilladora (f).
hashish, n. hachís (m).
hassock, n. estratificación (f) rizada (Geol); cojín ((m) (cushion).
hatchery, n. criadero (m) (de peces).
haulier, n. transportista (m), acarreador (m).
Haunterivian, n. Hauteriviense (m).
hawthorn, n. espino (m), tejocote (m) (LAm).
hay, n. heno (m), hierba (f) seca; **h. fever,** fiebre (f) del heno; **h. field,** henar (m).
haystack, n. almiar (m).
hazard, n. azar (m), riesgo (m) (risk), peligro (m) (danger).
hazardous, a. arriesgado/a, peligroso/a (dangerous); **h. waste,** residuos (m.pl) peligrosos.
haze, n. calina (f); niebla (f) (mist).
hazel, n. avellano (m) (tree), avellana (f) (tree).
H-bomb, n. bomba (f) H.
HCB, n. HCB (m); **hexachlorobenzene,** hexaclorobenzol.
HCBD, n. HCBD (m); **hexachlorobutadiene,** hexaclorobutadieno.
HCH, n. HCH (m); **hexachlorocyclohexane,** hexaclorociclohexano.
head, n. cabeza (f) (of body, organization); cabeza, res (f) (of livestock); cabezuela (f)

(of flowers); culata (f) (Eng, of cylinder); espiga (f) (of corn); **h., headland,** punta (f), promontorio (m), cabo (m); **h. of pressure, h. of steam,** presión (f); **h. of water,** altura (f) de caída; **h., headwater,** nacimiento (m), cabecera (f) (of a river); **pressure h.,** presión (f); **recording h.,** tape **h.,** cabeza sonora.

headache, n. dolor (m) de cabeza.

header, n. tijón (m) (Con).

headframe, n. torre (f) (Min).

heading, n. galería (f); qanat (m) (Arab), socavón (m).

headland, n. punta (f), cabo (m), promontorio (m).

headquarters, n. sede (f) (of an organization); oficina (f) central (main office).

headwaters, n.pl. cabecera (f), chortal (m) (in southern Spain); **h. of river,** cabecera de río.

headwind, n. viento (m) contrario, viento en contra.

headworks, n. obras (f.pl) de cabecera (Con).

health, n. salud (f), sanidad (f); **h. risk.,** riesgo (m) para la salud; **occupational h.,** salud ocupacional, medicina (f) de trabajo; **primary h. care,** atención (f) primaria de la salud; **public h.,** sanidad pública.

heap, n. pila (f), montón (m).

hearing, n. oído (m), audición (f) (sense); vista (f), audiencia (f) (Jur).

heart, n. corazón (m).

hearth, n. hogar (m), chimenea (f) (fireplace).

heartwood, n. duramen (m).

heat, 1. n. calor (m); **h. energy,** energía (f) calorífica; **h. of combustion,** calor de combustión; **h. of reaction,** calor de reacción; **latent h. of evaporation,** calor latente de evaporación; **radiant h.,** calor radiante; **solar h.,** calor solar; **specific h.,** calor específico; 2. a. calorífico/a.

heater, n. calentador (m).

heath, n. brezal (m); páramo (m) (esp. LAm) (wilderness).

heather, n. brezo (m).

heating, n. calefacción (f), calentamiento (m); **central h.,** calefacción central; **geothermal h.,** calentamiento geotérmico; **solar h.,** calefacción solar.

heat-resistent, a. calorífugo/a, refractario/a.

heat up, v. calentar.

heatwave, n. ola (f) de calor.

heave, n. levantamiento (m) (lifting), hinchamiento (m) (swelling); **ground h.,** levantamiento del terreno.

hectare, n. hectárea (m) (10,000 sq. metres; equiv. to 2.471 acres).

hecto-, pf. hecto-.

hectogram, n. hectogramo (m).

hectogramme, n. hectogramo (m).

hectoliter, n. hectolitro (m).

hectolitre, n. hectolitro (m).

hectometer, n. hectómetro (m) (see hectometre).

hectometre, n. hectómetro (m) (unit of volume equal to 10.E4 cubic metres).

hedge, n. seto (m), seto vivo.

hedgerow, n. seto (m), seto vivo.

height, n. altura (f) (measurement); estatura (f) (of person); altitud (f), cota (f) (above sea level); **capillary h.,** altura capilar; **flight h.,** altura de vuelo; **manometer h.,** altura manométrica; **piezometric h.,** altura piezométrica; **topographic h.,** altura topográfica.

heliograph, n. heliógrafo (m); **Campbell-Stokes h. (sunshine recorder),** heliógrafo de Campbell-Stokes.

heliophytic, a. heliofítico/a.

heliotrope, n. heliotropo (m).

helium, n. helio (m) (He); **h. nucleus,** helión (m).

helix, n. hélice (f).

help, 1. v. ayudar; 2. n. ayuda (f) (assistance), auxilio (m) (from danger).

hematite, n. hematita (f), hematites (m).

hematology, n. hematología (f).

Hemiptera, n.pl. hemípteros (m.pl), el orden Hemiptera (Zoo).

hemipteran, 1. n. hemíptero (m); 2. a. hemíptero/a.

hemipteron, n. hemíptero (m).

hemipterous, a. hemíptero/a.

hemisphere, n. hemisferio (m); **northern h.,** hemisferio norte; **southern h.,** hemisferio sur.

hemocyanin, n. hemocianina (f).

hemoglobin, n. hemoglobina (f).

hemophilia, n. hemofilia (f).

hemophiliac, n. hemofílico/a (m, f).

hemophilic, a. hemofílico/a.

hemorrhage, n. hemorragia (f).

henequen, n. agave (m), henequén (m).

hepatic, a. hepático/a.

Hepaticae, n.pl. hepáticas (f.pl), la clase Hepaticae (Bot).

hepatitis, n. hepatitis (f).

heptachlor, n. heptocloro (m).

heptagon, n. heptágono (m).

heptaldehyde, n. heptaldehído (m).

heptane, n. heptano (m).

heptanol, n. heptanol (m).

heptene, n. hepteno (m), heptileno (m).

heptylene, n. heptileno, hepteno.

heptyne, n. heptino (m).

herb, n. hierba (f), plant (f) herbacea; **h. layer (Ecol),** capa (f) herbácea.

herbaceous, a. herbáceo/a.

herbicide, 1. n. herbicida (m); 2. a. herbicida.

herbivore, n. herbívoro/a (m, f).

herbivorous, a. herbívoro/a.

herd, 1. v. llevar en manada; tropear (Arg); 2. n. rebaño (m), manada (f); piara (f) (of pigs).

hereditary, a. hereditario/a.

heritage, n. herencia (f), patrimonio (m); **national h.,** patrimonio nacional.

hermaphrodite, 1. n. hermafrodita (m, f); 2. a. hermafrodita.

heroin, n. heroína (f) (drug).

herringbone, a. cola (f) de pez, de espinapez; **h. system (for drainage),** sistema (m) de espinapez.

hetero-, pf. hetero-.

heterocyclic, a. heterocíclico/a.

heterogeneity, n. heterogeneidad (f).

heterogeneous, a. heterogéneo/a.

heterolysis, n. heterólisis (m).

heterosexual, a. heterosexual.

heterotroph, n. heterótrofo (m).

heterotrophe, n. heterótrofo (m).

heterotrophic, a. heterotrófico/a.

heterozygote, n. heterocigoto (m).

heterozygous, a. heterocigótico/a.

Hettangian, n. Hettangiense (m).

heuristic, a. heurístico/a.

hexa-, pf. hexa-.

hexachlorobenzene, n. hexaclorobenzol (m), hexacloruro (m) de benceno (abbr. HCB).

hexachlorobutadiene, n. hexaclorobutadieno (m) (abbr. HCBD).

hexachlorocylcohexane, n. hexaclorociclohexano (m) (abbr. HCH); **alpha/beta/gamma/epsilon h.,** alfa/beta/gamma/épsilon hexaclorociclohexano.

hexachloroethane, n. hexacloroetano (m).

hexagon, n. hexágono (m).

hexagonal, a. hexagonal.

hexaldehyde, n. hexaldehído (m).

hexamethylenetetramine, n. hexametilenotetramina (f), hexamina (f).

hexamine, n. hexamina (f).

hexanal, n. hexanal (m).

hexane, n. hexano (m).

hexanol, n. hexanol (m).

Hexapoda, n.pl. hexápodos (m.pl) (Zoo, grupo).

hexavalent, a. hexavalente.

hexene, n. hexeno (m).

hexilene, n. hexileno (m).

hiatus, n. hiato (m).

hibernate, v. hibernar (of animals), invernar (of people).

hibernation, n. hibernación (f).

hierarchial, a. jerárquico/a; **h. order,** orden (m) jerárquico.

hierarchical, a. jerarquizado/a.

hierarchy, n. jerarquía (f).

hieroglyph, n. jeroglífico (m) (Arch).

high, a. alto/a.

highland, a. serrano/a.

highlands, n. tierras (f.pl) altas.

highlight, v. destacar, hacer resaltar.

highlights, n.pl. parte (f) destacable.

high-risk, a. de alto riesgo.

highway, n. carretera; **h. code,** código (m) de la circulación.

hiker, n. excursionista (m, f); **hitch-h.,** autoestopista (m, f).

hiking, n. senderismo (m) (walking).

hill, n. colina (f), cerro (m), loma (f) (esp. LAm), cuesta (f) (slope); **h. farming,** agricultura (f) de montaña; **h. fort (Arch),** castro (m).

hillock, n. montecillo (m), montículo (m); altozano (m), loma (f) (knoll), cuesta (f) (rise).

hillside, n. ladera (f), falda (f).

hilltop, n. cumbre (f), cima (f) (summit).

hilly, a. serrano/a, montañoso/a.

hindrance, n. obstáculo (m); impedimento (m) (prevention).

hinge, n. gozne (m), bisagra (f) (of door); charnela (f).

hinged, a. articulado/a.

hinterland, n. hinterland (m), región (f) interior.

hiring, n. contratación (f); **h. of a consultant,** contratación de un consultor.

Hispano-America, n. Hispanoamérica (the Spanish-speaking countries of North, Central and South America).

histamine, n. histamina (f).

histidine, n. histidina (f).

histogram, n. histograma (m).

histological, a. histológico/a.

histology, n. histología (f).

historian, n. historiador/a (m, f).

hitch-hike, v. hacer autostop.

hitch-hiker, n. autostopista (m, f).

hoe, n. azada (f); **large h.,** azadón (m).

hogsback, n. espinazo (m) (ridge) (Geog).

hoist, n. levantamiento (m) (action), montacargas (m) (lift, elevator), cabria (f) (mechanism), grúa (f) (crane).

holasteroid, n. holasteroide (m).

holdfast, n. zarcillo (m) (Bot); grapa (f) (Eng).

hole, n. hoya (f), agujero (m) (Gen); bache (m) (in road); boquete (m), hueco (m) (opening); brecha (f) (in wall); hoyo (m) (pit, depression); **kettle h. (Geog),** hoya glaciar.

holistic, a. holístico/a.

Holland, n. Holanda (f), Países Bajos (m.pl).

hollow, n. hoyo (m), hoyada (f) (in ground); hondonada (f) (small valley); hueco (m) (hole).

hollow, a. hueco/a.

holly, n. acebo (m); **h. oak,** encina (f).

holmium, n. holmio (m) (Ho).

holm-oak, n. encina (f).

holo-, pf. holo-.

Holocene, n. Holoceno (m).

holocrystalline, a. holocristalino/a.

holophytic, a. holofítico/a.

holozoic, a. holozoico/a.

home, n. hogar (m), casa (f) (house), domicilio (m) (domicile).

homeopathic, a. homeopático/a (of medicine); homeópata (of a doctor).

homeopathy, n. homeopatía (f).

homeostasis, n. homeostasis (f).

homeothermic, a. homeotermo/a.

homestead, n. caserío (m); casa (f) (US), granja (f) (farm).

homoeopathic, a. homeopático/a (of medicine); homeopáta (of a doctor).

homoeopathy, n. homeopatía (f).

homogeneity, n. homogeneidad (f).

homogeneous, a. homogéneo/a.

homogenization, n. homogeneización (f).

homography, n. homografía (f).

homoiothermic, a. homeotermo/a.

homologation, n. homologación (f) (confirmation).

homologous, a. homólogo/a.

homology, n. homología (f).

homolysis, n. homólisis (m).

homosexual, a. homosexual.

homotaxis, n. homotaxis (m).

homozygous, a. homocigótico/a.

Honduran, 1. n. hondureño/a (m, f); 2. a. hondureño/a.

Honduras, n. Honduras (f).

honey, n. miel (f).

honeycombed, a. alveolado/a.

honeysuckle, n. madreselva (f).

hook, n. banco (m) encorvado (Geog); gancho (m), garfio (m) (for holding); anzuelo (m) (for fishing); aldabilla (f) (on door).

hopper, n. tolva (f) (in factory); tragante (m) (of furnace); insecto (m) saltador (Zoo).

horizon, n. horizonte (m); **soil h.,** horizonte de suelo; **stratigraphical h.,** horizonte estratigráfico.

horizontal, a. horizontal.

hormonal, a. hormonal.

hormone, n. hormona (f); **growth h.,** hormona de crecimiento; **plant h.,** hormona vegetal; **sex h.,** hormona sexual.

horn, n. cuerno (m); asta (f) (of deer); bocina (f) (claxon).

hornblendite, n. hornblendita (f).

horneblende, n. hornblenda (f).

hornfel, n. cornubianita (f), corneana (f).

hornstone, n. piedra (f) córnea.

horse, n. caballo (m).

horsepower, n. caballo (m) inglés, caballo (m) de vapor.

horst, n. horst (m), bloque (m) tectónico elevado.

horticultural, a. hortícola.

horticulturalist, n. horticultor/a (m, f).

horticulture, n. horticultura (f).

hose, n. manguera (f), manga (f) (flexible pipe); **garden h.,** manga de riego, manguera.

hosepipe, n. manguera (f), manga (f) de riego; **garden h.,** manga de riego, manguera.

hospital, n. hospital (m), nosocomio (m) (LAm); **h. incinerator,** incinerador (m) de hospital; **h. wastes,** residuos (m.pl) hospitalarios.

host, n. huésped (m. & f) (Zoo, Bot e.g. of parasite).

hot, a. caliente.

hotel, n. hotel (m); **h. sector (Com),** sector (m) hostelero; **the h. trade,** la hostelería (f).

hotspot, n. sitio (m) de intenso calor, zona (f) de mayor temperatura, punto (m) caliente; **radioactive h.,** zona (f) de gran radioactividad.

hour, n. hora (f); **fifty miles an h., fifty miles per h.,** cincuenta millas por hora; **rush h.,** hora punta, rush-hour (m).

house, 1. v. albergar (e.g. a machine); 2. n. casa (f) (Gen); cámara (f) (legislature); **H. of Commons (GB),** Cámara de los Comunes; **H. of Representatives (US),** Cámara de Representantes; **Houses of Parliament (e.g. GB),** Parlamento (m); **Lower H.,** Cámara Baja.

household, 1. n. familia (f), hogar (m) (home); 2. a. casero/a.

housing, n. vivienda (f); alojamiento (m) (accommodation); casas (f.pl); **h. development,** urbanización (f); **h. estate,** urbanización (f), fraccionamiento (m) (Mex).

hovercraft, n. aerodeslizador (m).

human, a. humano/a; **h. being,** ser (m) humano; **h. environment,** ambiente (m) humano.

humic, a. húmico/a; **h. acid,** ácido (m) húmico.

humid, a. húmedo/a.

humidification, n. humectación (f).

humidifier, n. humedecedor (m), humectador (m).

humidifying, n. humedecimiento (m).

humidity, n. humedad (f); **absolute h.,** humedad absoluta; **degree of h.,** grado (m) de humedad; **relative h.,** humedad relativa; **specific h.,** humedad específica.

hummock, n. morón (m) , montecillo (m).

hump, n. giba (f) (Geol), joroba (f) (animal).

humus, n. humus (m), mantillo (m).

hundredweight, n. quintal (m) (Castilla) (unit of weight).quintal (Castilla) (46 kg), quintal métrico (100 kg).

hunger, n. hambre (m).

hunter-gatherer, n. cazador recolector (m).

hunting, n. cacería (f), caza (f); **h. licence, h. license (US),** licencia (f) de caza, permiso (m) de caza.

Huronian, n. Huronense (m).

hurricane, n. huracán (m).

husk, n. cáscara (f) (of cereals).

hyalite, n. hialita (f), hialófana (f).

hyalophane, n. hialófana (f).

hybrid, 1. n. híbrido (m), cruzado/a (m, f); **h. vigour, h. vigor (US),** vigor (m) híbrido; 2. a. híbrido/a.

hybridization, n. hibridización (f), hibridación (f).

hydra, n. hidra (f).

hydrant, n. boca (f) de riego; **fire h.,** boca de incendio.

hydrate, n. hidrato (m).

hydrated, a. hidratado/a.

hydration, n. hidratación (f).

hydraulic, a. hidráulico/a; **h. energy,** energía (f)

hidráulica; **h. gradient,** gradiente (m) hidráulico; **h. jack,** gato (m) hidráulico; **h. joint,** junta (f) hidráulica; **h. jump,** salto (m) hidráulico; **h. loading,** carga (f) hidráulica; **h. pressure,** presión (f) hidráulica; **h. radius,** radio (m) hidráulico; **h. ram,** ariete (m) hidráulico; **h. structure,** construcción (f) hidráulica; **h. valve,** válvula (f) hidráulica; **h. works,** obras (f.pl) hidráulicas.
hydraulics, n. hidráulica (f); **open-channel h.,** hidráulica de canal a cielo abierto; **well h.,** hidráulica de pozos.
hydrazine, n. hidracina (f).
hydric, a. hídrico/a.
hydride, n. hidruro (m).
hydro-, pf. hidro-.
hydrobiology, n. hidrobiología (f).
hydrocarbon, n. hidrocarburo (m); **chlorinated h.,** hidrocarburo clorado; **halogenated hydrocarbons,** hidrocarburos halogenados; **volatile h.,** hidrocarburo volátil.
hydrochloric, a. clorhídrico/a.
hydroclastic, a. hidroclástico/a.
hydrocracking, n. hidrocraqueo (m).
hydrodialysis, n. hidrodiálisis (m).
hydrodynamic, a. hidrodinámico/a; **h. model,** modelo (m) hidrodinámico.
hydrodynamics, n. hidrodinámica (f).
hydroelectric, a. hidroeléctrico/a; **h. scheme,** proyecto (m) hidroeléctrico.
hydrofracturing, n. fracturación (f) hidráulica.
hydrogen, n. hidrógeno (m) (H).
hydrogenate, v. hidrogenar.
hydrogenation, n. hidrogenación (f).
hydrogencarbonate, n. hidrogenocarbonato (m); **calcium h.,** hidrogenocarbonato cálcico.
hydrogeochemistry, n. hidrogeoquímica (f).
hydrogeological, a. hidrogeológico/a.
hydrogeologist, n. hidrogeólogo/a (m, f).
hydrogeology, n. hidrogeología (f).
hydrogeomorphology, n. hidrogeomorfología (f).
hydrogeophysics, n. hidrogeofísica (f).
hydrograph, n. hidrograma (m), hidrografía (f); **complex h.,** hidrograma complejo; **compound h.,** hidrograma compuesto; **flow h.,** hidrograma de caudales; **h. analysis,** análisis (m) de hidrogramas; **h. separation,** separación (f) de hidrogramas; **unit h.,** hidrograma unitario.
hydrographic, a. hidrográfico/a.
hydrography, n. hidrografía (f).
hydrological, a. hidrológico/a.
hydrologist, n. hidrólogo/a (m, f).
hydrology, n. hidrología (f); **groundwater h.,** hidrología subterránea; **karstic h.,** hidrología kárstica; **surface h.,** hidrología superficial, hidrología de superficie; **urban h.,** hidrología urbana.
hydrolysis, n. hidrólisis (m).
hydrolyze, n. hidrolizar.
hydromechanics, n.pl. hidromecánica (f).

hydrometeorology, n. hidrometeorología (f).
hydrometer, n. hidrómetro (m), areómetro (m).
hydrometric, a. hidrométrico/a; **h. scheme,** esquema (m) hidrométrico.
hydrometry, n. hidrometría (f).
hydrophilic, a. hidrofílico/a.
hydrophobic, a. hidrófobo/a, hidrofóbico/a.
hydrophylic, a. hidrófilo/a.
hydrophyte, n. hidrofita (m, f).
hydropneumatics, n. hidroneumática.
hydroponic, a. hidropónico/a.
hydroponics, n. hidroponia (f), cultivo (m) hidropónico.
hydropower, n. energía (f) hidráulica, hidroenergía (f).
hydrosere, n. hidroserie (f).
hydrosphere, n. hidrosfera (f).
hydrostatic, a. hidrostático/a; **h. head,** altura (f) hidrostática; **h. level,** nivel (m) hidrostático.
hydrostatics, n. hidrostática (f).
hydrothermal, a. hidrotermal, hidrotérmico/a; **h. energy,** energía (f) hidrotérmica; **h. spring,** manantial (m) hidrotermal.
hydrous, a. hidroso/a.
hydroxide, n. hidróxido (m); **h. ions,** hidroxilones (m.pl); **sodium h.,** hidróxido sódico.
hydroxyl, n. hidróxilo (m), oxhídrilo (m).
hydrozoa, n.pl. hidrozoos (m.pl).
hyena, n. hiena (f).
hygiene, n. higiene (f); **environmental h.,** higiene ambiental.
hygienic, a. higiénico/a.
hygro-, pf. higro-.
hygrometer, n. higrómetro (m).
hygrometry, n. higrometría (f).
hygroscope, n. higroscopio (m).
hygroscopic, a. higroscópico/a.
hygroscopicity, n. higroscopicidad (f).
Hymenoptera, n.pl. himenópteros (m.pl), el orden Hymenoptera (Zoo).
hymenopteran, n. himenóptero (m).
hyper-, pf. hiper-.
hyperbola, n. hipérbola (f).
hyperbolic, a. hiperbólico/a.
hypersensitivity, n. hipersensibilidad (f).
hypertonic, a. hipertónico/a.
hypha, n. hifa (f).
hyphae, n.pl. hifas (f.pl).
hypo-, pf. hipo-.
hypocaust, n. hipocausto (m) (Arch).
hypochlorite, n. hipoclorito (m).
hypolimnion, n. hipolimnio (m).
hypophosphite, n. hipofosfito (m).
hypothermal, a. hipotermal.
hypothermia, n. hipotermia (f).
hypothesis, n. hipótesis (f); **null h.,** hipótesis nula.
hypothetical, a. hipotético/a.
hypotonic, a. hipotónico/a.
hypsometric, a. hipsométrico/a.
hysteresis, n. histéresis (m).

I

ice, n. hielo (m); **broken i.**, hielo desgajado; **dry i.**, hielo seco; **i. age**, periódo (m) glacial; **i. breccia**, brecha (f) de hielo; **i. cap**, casquete (m) polar; **i. field**, campo (m) de hielo; **i. floe**, bandejón (m) (LAm); **i. sheet**, manto (m) de hielo; **pack i.**, banco (m) de hielo, banquisa (f), hielo empacado, hielo de pack (LAm); **polar i.**, hielo polar; **rotten i.**, hielo podrido.
iceberg, n. iceberg (m).
icebound, a. bloqueado por el hielo (road).
icebreaker, n. rompehielos (m.pl).
icecap, n. casquete (m) glaciar, calota (m) de hielo (LAm).
Iceland, n. Islandia (f).
icepack, n. paquete (m) de hielo.
ice up, v. helarse, congelarse.
icicle, n. carámbano (m).
ICJ, n. CIJ (f); **International Court of Justice**, Corte (f) Internacional de Justicia.
icy, a. glacial (weather).
IDB, n. BID; **Inter-American Development Bank**, Banco (m) Interamericano de Desarrollo.
identification, n. identificación (f).
idol, n. ídolo (m).
IEB, n. BEI (m); **European Investment Bank**, Banco (m) Europeo de Inversiones.
IFAD, n. FIDA (m); **International Fund for Agricultural Development**, Fondo (m) Internacional de Desarrollo Agrícola.
igneous, a. ígneo/a; **i. body**, cuerpo (m) ígneo, masa (f) ígnea; **i. intrusion**, intrusión (f) ígnea.
ignimbrite, n. ignimbrita (f).
ignitable, a. ignitable, inflamable.
ignition, n. ignición (f), combustión (f); **i. point**, punto (m) de combustión; **i. temperature**, temperatura (f) de ignición.
iguana, n. iguana (f).
ileum, n. íleon (m).
ilex, n. acebo (m) (holly tree).
iliac, a. ilíaco/a.
ill, a. enfermo/a.
illegal, a. ilegal.
illite, n. illita (f).
illness, n. enfermedad (f).
illuminate, v. iluminar.
illuminated, a. alumbrado/a, iluminado/a.
illumination, n. iluminación (f), alumbrado (m) (system); alumbramiento (m) (action); **artificial i.**, iluminación indirecta.
illustrate, v. ilustrar.
illustration, n. ilustración (f); ejemplo (m) (example).
ilmenite, n. ilmenita (f).

image, n. imagen (f); **stereoscopic i.**, imagen estereoscópica.
imago, n. imago (m).
imbed, v. engastar.
imbibation, n. imbibición (f).
imbibe, v. embeber, absorber.
immersion, n. inmersión (f), submersión (f).
immobilization, n. inmovilización (f).
immune, a. inmune, inmunitorio/a; **i. system**, sistema (m) inmune, sistema inmunitorio.
immunity, n. inmunidad (f) (Zoo, Bot).
immunization, n. inmunización (f).
immunize, v. inmunizar.
immunoassay, n. inmunoensayo (m).
immunological, a. inmunológico/a.
immunology, n. inmunología (f).
IMO, n. OMI (f); **International Maritime Organization**, Organización (f) Marítima Internacional.
impact, n. impacto (m), choque (m) (blow); **environmental i. assessment (abbr. EIA)**, evaluación (f) de impacto ambiental (abbr. EIA); **environmental i.**, impacto ambiental; **meteorite i.**, impacto de meteorito; **visual i.**, impacto visual.
impairment, n. deterioro (m), daño (m) (damage).
impedance, n. impedancia (f).
impeller, n. impulsor (m); **pump i.**, impulsor de una bomba.
impenetrable, a. impenetrable.
imperfect, a. imperfecto/a, defectivo/a, defectuoso/a.
impermeability, n. impermeabilidad (f).
impermeabilization, n. impermeabilización (f).
impermeable, a. impermeable; **i. boundary**, límite (m) impermeable.
impermissible, a. inadmisible.
impervious, a. impermeable.
impingement, n. usurpación (f) (encroachment).
implantation, n. implantación (f).
implement, 1. v. realizar, implementar; ejecutar (to undertake); aplicar (the law); 2. n. implemento (m).
implementability, n. implementabilidad (f).
implementation, n. realización (f) (fulfilment); aplicación (f) (of regulations); puesta (f) en ejecución (measures).
implication, n. implicación (f), inferencia (f); implicancia (f) (LAm).
implicit, a. implícito/a; **i. function**, función (f) implícita.
import, 1. v. importar; 2. n. importación (f); **i. goods**, bienes (m.pl) de importación, artículos (m.pl) de importación.

impound, v. embalsar (water), confiscar (goods).

impounding, a. embalsamiento (m), represamiento (m); **i. reservoir**, depósito (m) de captación.

impoverish, v. empobrecer.

impoverishment, n. depauperación (f), empobrecimiento (m), esquilmo (m).

imprecision, n. imprecisión (f).

impregnate, v. impregnar.

impregnation, n. impregnación (f).

impression, n. impresión (f).

improve, v. mejorar; acondicionar (e.g. the roads); progresar (to progress); perfeccionar (to perfect).

improvement, n. mejora (f), acondicionamiento (m); **channel i.**, acondicionamiento de cauces.

impulse, n. impulso (m); **nerve i.**, impulso nervioso.

impulsion, n. impulsión (f).

inadequacy, n. insuficiencia (f).

inarticulate, a. inarticulado/a.

inauguration, n. inauguración (f).

inbreeding, n. cruzamiento (m) consanguíneo, inbreeding.

Inc., a. s.a.; **Incorporated (US) (abbr. Inc)**, sociedad (f) anónima (abbr. s.a.).

incandescence, n. incandescencia (f).

incandescent, a. incandescente.

incentive, n. incentivo (m).

inch, n. pulgada (f) (unit of length equal to 2.54 cm).

incidence, n. incidencia (Phys); frecuencia (number); extensión (of a disease); **angle of i.**, ángulo (m) de incidencia.

incinerate, v. quemar (rubbish); incinerar (cremation).

incineration, n. incineración (f), combustión (f); **i. plant**, instalación (f) de combustión, planta (f) incineradora, planta de incineración.

incinerator, n. incinerador (m), incineradora (f); **sewage sludge i.**, incinerador de fangos de alcantarilla; **waste i.**, incinerador de basura.

incised, a. excavado/a, profundizado/a; **i. meander**, meandro (m) excavado.

inclination, n. inclinación (f); **angle of i.**, ángulo (m) de inclinación; **axis of i.**, eje (m) de inclinación.

incline, n. chiflón (m) (LAm).

inclined, a. inclinado/a, en pendiente; propenso/a (favoured).

include, v. incluir, englobar (to lump together).

included, a. incluido/a.

inclusion, n. inclusión (f).

incoherent, a. incoherente.

incombustible, a. incombustible, ignífugo/a.

income, n. ingresos (m.pl) (revenue), renta (f) (private income), rédito (m) (interest); **gross**

i., renta bruta; **i. and expenditure**, ingresos y gastos.

incompetent, a. incompetente.

incomprehensible, a. incomprensible.

incompressibility, n. incompresibilidad (f).

inconsistent, a. inconsistente.

incorporated, a. incorporado/a; **Incorporated (US, Com) (abbr. Inc)**, sociedad (f) anónima (abbr. s.a.).

increment, n. incremento (m) (Mat), aumento (m) (increase).

incrustation, n. incrustación (f); **carbonate i.**, incrustación carbonática; **iron i.**, incrustación de hierro; **silica i.**, incrustación de sílice.

incubation, n. incubación (f).

incubator, n. incubadora (f); **i. test**, ensayo (m) de incubadora.

indemnification, n. indemnización (f).

indemnify, v. indemnizar.

indentation, n. abolladura (f), muesca (f); sangría (f) (texto).

independent, a. independiente.

index, n. índice (m); **base-exchange i.**, índice de cambio de bases; **b. index**, índice biótico; **card i.**, fichero (m), clasificador (m); **cost-of-living i.**, índice del coste de la vida; **diversity i.**, índice de diversidad; **hazard i.**, índice de riesgo; **hydrochemical i.**, índice hidroquímico; **quality i.**, índice de calidad; **refractive i.**, índice de refracción; **saturation i.**, índice de saturación; **sorting i. (Geol)**, índice granulométrico.

indication, n. indicación (f).

indicator, 1. n. indicador (m); **i. organisms**, organismos (m.pl) indicadores; **oil pressure i.**, indicador de presión del aceite; **pollution i.**, indicador de contaminación, indicador de polución; **temperature i.**, indicador de temperatura; **water level i.**, indicador de nivel del agua; 2, a. indicador/a; **i. species**, especie (f) indicadora.

indictment, n. auto (m) de procesamiento.

indigenous, a. autóctono/a, indígena, nativo/a.

indirect, a. indirecto/a.

indispensable, a. imprescindible, esencial.

indissoluble, a. indisoluble.

indium, n. indio (m) (In).

Indoarabian, a. indo-arábigo/a.

induction, n. inducción (f).

indurated, a. endurecido/a.

industrial, a. industrial; **i. action**, huelga (f); **i. discharge**, desecho (m) industrial, agua (f) industrial.

industrialization, n. industrialización (f).

industry, n. industria (f); **agrifood i.**, (industria) agroalimentaria; **car i.**, industria automovilística; **dairy i.**, industria láctea; **extractive i.**, actividades (f.pl) extractivas; **food processing i.**, industria de procesamiento de alimentos; **heavy i.**, industria pesada; **key i.**, industria clave; **oil i.**, industria petrolera; **olive i.**, industria olivarera; **pulp and paper i.**, industria de pulpa y papel; **steel i.**, industria siderúrgica.

inefficiency, n. ineficiencia (f).

inelastic, a. inelástico/a, rígido/a.

inelasticity, n. inelasticidad (f).

inequality, n. desigualdad (f).

inert, a. inerte; **chemically i.**, inerte químicamente.

inertia, n. inercia (f).

infect, v. infeccionar, infectar, contagiar.

infected, a. infecto/a, infectado/a.

infection, n. infección (f).

infectious, a. infeccioso/a.

inference, n. inferencia (f), implicación (f).

inferior, a. inferior.

infilling, n. colmatación (f) (e.g. of lake, with sediments or/& vegetation), relleno (m).

infiltrate, v. infiltrarse.

infiltration, n. infiltración (f); **annual i.**, infiltración anual; **induced i.**, infiltración inducida; **i. basin,** balsa (f) de infiltración; **i. capacity,** capacidad (f) de infiltración; **i. efficiency,** eficacia (f) de infiltración, infiltración eficaz; **i. gallery,** galería (f) de infiltración; **i. rate,** velocidad (f) de infiltración.

infiltrometer, n. infiltrómetro (m).

infinite, a. infinito/a.

infinitesimal, a. infinitesimal.

infinity, n. infinidad (f).

inflammable, a. inflamable (easily set on fire).

inflammation, n. inflamación (f).

inflate, v. hinchar, inflar; **to i. with a pump,** hinchar con una bomba.

inflexion, n. inflexión (f); **i. point,** punto (m) de inflexión.

inflow, n. afluencia (f), afluente (m).

influence, n. influencia (f); **cone of i.,** cono (m) de influencia; **pumping i.,** influencia del bombeo; **to have an i. upon (v),** ejercer una influencia sobre.

influent, a. influente; **i. stream,** río (m) influente.

influenza, n. gripe (f), influenza (f) (esp. LAm).

information, n. información (f); **i. highway,** autopista (f) de la información; **i. technology,** tecnología (f) de la información.

infraction, n. infracción (f).

infrared, a. infrarrojo/a.

infrastructure, n. infraestructura (f); **traffic i.,** infraestructura viaria.

infringement, n. infracción (f).

infuse, v. infundir.

infusion, n. infusión (f).

ingest, v. ingerir.

ingestion, n. ingestión (f).

ingot, n. lingote (m).

ingress, n. ingreso (m) (going in), acceso (access); **i. of water,** ingreso de agua.

inhabitant, n. habitante (m).

inhale, v. inhalar, aspirar.

inherent, a. inherente; **i. error,** error (m) inherente.

inhibition, n. inhibición (f).

inhibitor, n. inhibidor (m); **corrosion i.,** inhibidor de corrosión; **growth i.** (Bot, Zoo), inhibidor de crecimiento.

inhomogeneity, n. inhomogeneidad (f).

initiative, n. iniciativa (f); **to use one's i. (v),** obrar por propia iniciativa.

inject, v. inyectar.

injected, a. inyectado/a.

injection, n. inyección (f); **borehole i.,** inyección en sondeo; **cement i.,** inyección de cemento; **chemical i.,** inyección química; **deep well i.,** inyección de pozo en profundidad; **hazardous waste deep well i.,** inyección subterránea de residuos peligrosos; **i. of compressed air,** inyección de aire comprimido; **i. of tracer,** inyección del trazador; **pressure i.,** inyección a presión; **steam i.,** inyección de vapor; **wastewater subsurface i.,** inyección subterránea de aguas residuales.

injunction, n. requerimiento (m), entredicho (m) (Jur); mandato (m), orden (f) (command).

injured, a. herido/a.

inland, n. tierra (f) adentro.

inlet, n. ensenada (f), cala (f), caleta (f) (small bay), bocana (f) (of river); **tidal i.,** bocana de marea.

inlier, n. relicto (m) interior (Geol).

innovation, n. innovación (f), novedad (f); **technical i.,** innovación técnica; **technological i.,** innovación tecnológica.

innovative, a. innovador/a.

inoculate, v. inocular.

inoculation, n. inoculación (f).

inorganic, a. inorgánico/a.

input, n. entrada (f) (power terminal to machine); input (m), entrada (f) (Comp); insumo (Com, materials for production); **computer i./output (abbr. I/O),** entrada (f)/salida (f).

inquiry, n. pregunta (f) (question); investigación (f) (investigation).

insalubrious, a. insalubre.

inscription, n. inscripción (f).

insect, n. insecto (m).

Insecta, n.pl. insectos (m.pl), la clase Insecta (Zoo).

insecticidal, a. insecticida.

insecticide, n. insecticida (m); **contact i.,** insecticida de contacto; **systemic i.,** insecticida endoterápico.

Insectivora, n.pl. insectívoros (m.pl), el orden Insectivora (Zoo).

insectivore, n. insectívoro (m).

insectivorous, a. insectívoro/a.

inselberg, n. inselberg (m).

inseminate, v. inseminar.

insemination, n. inseminación (f); **artificial i.,** inseminación artificial.

insertion, n. inserción (f); encarte (m) (in a document).

inshore, a. costero/a, cercano a la orilla; **i. fishing fleet**, flota (f) costera.
in situ, ad. in situ.
insolation, n. insolación (f).
insoluble, a. insoluble (Chem); irresoluble (problema).
insolvency, n. insolvencia (f) (bankruptcy).
inspection, n. inspección (f), registro (m), examen (m); **i. chamber**, cámara (f) de inspección; **sanitary i.**, inspección sanitaria.
inspiration, n. inspiración (f); inhalación (f) (inhalation).
instability, n. inestabilidad (f).
instalation (US) installation, n. emplazamiento (m), instalación (f), introducción (f); **electrical i.**, instalación eléctrica; **heating i.**, instalación de calefacción; **plumbing i.**, instalación de fontanería.
instantaneous, a. instantáneo/a.
instar, n. instar (m).
instigate, v. fomentar (e.g. new ideas), instigar (e.g. riot).
instigation, n. instigación (f).
instigator, n. instigador/a (m, f).
instinct, n. instinto (m).
instinctive, a. instintivo/a.
institute, n. instituto (m), establecimiento (m), institución (f); **Geological I. of Spain**, Instituto Geológico de España (IGE); **Spanish Hydrological I.**, Instituto de Hidrología de España.
instruction, n. instrucción (f).
instrument, n. instrumento (m); **i. panel**, tablero (m) de instrumentos.
instrumentation, n. instrumentación (f); **downhole i.**, instrumentación del sondeo, equipo de sondeo.
insulate, v. aislar.
insulating, a. aislante; **i. tape**, cinta (f) aislante.
insulation, n. aislamiento (m); **i. material**, material (m) aislante; **thermal i.**, aislamiento térmico.
insulin, n. insulina (f).
insurance, n. seguro (m); **fully comprehensive i.**, seguro a todo riesgo; **i. company**, compañía (f) de seguros; **i. policy**, póliza (f) de seguro; **i. premium**, prima (f) de seguros; **National I. (GB)**, **social i.**, seguro social.
intake, n. válvula (f) de admisión (for fuel, water, etc.).
integer, n. entero (m) (Mat).
integral, a. integral, integrante; **i. calculus**, cálculo (m) integral.
integrated, a. integrado/a; **i. circuit**, circuito (m) integrado.
integration, n. integración (f).
integrator, n. integrador (m).
integument, n. tegumento (m), integumento (m).
intelligence, n. inteligencia (f); noticia (f) (information); **artificial i.**, inteligencia artificial.

intensity, n. intensidad (f); **light i.**, intensidad de la luz; **maximum rainfall i.**, intensidad máxima pluvial.
intensive, a. intensivo/a.
intention, n. intención (f); intencionalidad (f) (intentionality).
inter-, pf. inter-.
interact, v. interaccionar.
interaction, n. interacción (f).
interbedded, a. interstratificado/a.
interbedding, n. intercalación (f).
interbreed, v. cruzar (of hybrids).
intercalation, n. intercalación (f).
intercellular, a. intercelular.
interception, n. intercepción (f).
interceptograph, n. interceptógrafo (m).
interceptor, n. interceptor (m); **i. sewer**, interceptor.
interchange, n. intercambio (m).
intercommunicate, v. intercomunicarse.
interconnect, v. interconectar.
interconnection, n. interconexión (f).
intercontinental, a. intercontinental.
interdependence, n. interdependencia (f).
interdisciplinary, a. interdisciplinario/a; **i. study**, estudio (m) interdisciplinario.
interest, n. interés (m), beneficio (m), provecho (m); **actualized i.**, interés actualizado; **compound i.**, interés compuesto; **public i.**, interés público; **simple i.**, interés simple; **ten per cent i.**, un interés de un diez por ciento; **to take an i. in (v)**, tener interés por.
interesting, a. interesante.
interface, n. interfase (f); **air-water i.**, interfase aire-agua; **depth of i.**, profundidad (f) de la interfase; **dynamic i.**, interfase dinámica; **equilibrium i.**, interfase de equilibrio; **freshwater-seawater i.**, interfase agua dulce-agua marina.
interference, n. interferencia (f); **well i.**, interferencia de pozos.
interfingering, n. interdigitación (f).
interflow, n. interflujo (m).
interfluve, n. interfluvio (m).
interglacial, a. interglaciar, interglacial.
intergrowth, n. intercrecimiento (m) (Geol).
interior, a. interior, interno/a.
interlock, v. entrelazarse, enclavar.
intermittent, a. intermitente, pasajero/a; **i. discharge**, descarga (f) intermitente; **i. spring**, manantial (m) intermitente.
intermix, v. entremezclar.
internal, a. internal, interno/a.
INTERNET, n. INTERNET (m).
interpolation, n. interpolación (f).
interpret, v. interpretar.
interpretation, n. interpretación (f); **geophysical i.**, interpretación geofísica; **photographic i.**, interpretación fotográfica.
interquartile, a. intercuartil; **i. range**, alcance (m) intercuartil.
inter-relation, n. interrelación (f).

interruption, n. interrupción (f), corte (f) (break).

intersection, n. intersección (f), cruce (m) (crossroads).

interspecific, a. interespecífico/a.

intersperse, v. esparcir, entremezclar.

interspersed, a. esparcido/a.

interstice, n. intersticio (m).

interstitial, a. intersticial; **i. fluid,** fluido (m) intersticial.

interstratification, n. interstratificación (f).

interstratified, a. interstratificado/a.

intertidal, a. intermareal, mareal, intercotidal (LAm), intertidal; **i. zone,** zona (f) de oscilación de las mareas, zona intertidal.

interval, n. intervalo (m); **adjustment i.,** intervalo de ajuste; **time i.,** intervalo de tiempo.

intervention, n. intervención (f).

intestinal, a. intestino/a.

intestine, n. intestino (m); **large i.,** intestino grueso; **small i.,** intestino delgado.

intracellular, a. intracelular.

intraspecific, a. intraespecífico/a.

intrinsic, a. intrínseco/a; **i. permeability,** permeabilidad (f) intrínseca.

introduce into, v. involucrar.

introduction, n. introducción (f), presentación (f) (one person to another); prólogo (m) (of a book).

intrude, v. intruir.

intrusion, n. intrusión (f); **igneous i.,** intrusión ígnea; **marine i.,** intrusión marina; **saline i.,** intrusión salina; **visual i.,** intrusión visual.

inundate, v. inundar, encharcar.

inundation, n. inundación (f).

invariant, a. invariante.

inventory, n. inventario (m) (itemized list), existencias (f.pl) (stock).

inverse, a. inverso/a; **i. function,** función (f) inversa.

inversion, n. inversión (f); **salinity i.,** inversión de salinidad; **temperature i.,** inversión de temperatura.

invert, n. invertido (m).

Invertebrata, n.pl. invertebrados (m.pl) (Zoo, group).

invertebrate, n. invertebrado (m).

inverted, a. invertido/a.

invest, v. invertir (Fin).

investigation, n. investigación (f); **field i.,** investigación de campo.

investigator, n. investigador (m, f).

investor, n. inversor/a (m, f), inversionista (m).

invoice, 1. v. facturar; 2. n. factura (f).

invoicing, n. facturación (f).

involuted, a. involuto/a.

iodide, n. ioduro (m).

iodine, n. iodo (m) (I); **i. disinfection,** yoduración (f).

ion, n. ión (m); **calcium i.,** ión calcio; **complex i.,** ión complejo; **i. exchange,** cambio (m) iónico; **metallic i.,** ión metálico; **minor i.,** ión menor.

ionic, a. iónico/a; **i. balance,** balanza (f) iónica.

ionization, n. ionización (f); **i. energy,** energía (f) de ionización.

ionize, v. ionizar.

ionizing, a. ionizante; **i. radiation,** radiación (f) ionizante.

ionosphere, n. ionosfera (f).

IPCC, n. IPCC; **International Panel on Climatic Change,** Panel (m) Internacional sobre el Cambio Climático.

Ireland, n. Irlanda (f); **Northern I.,** Irlanda del Norte; **Republic of I.,** República (f) de Irlanda.

iridium, n. iridio (m) (Ir).

Irish, n. irlandés/irlandesa; **the I. Sea,** el Mar irlandés.

iron, n. hierro (m) (Fe), fierro (m) (LAm); plancha (f) (for clothes); **cast i.,** hierro fundido; **dissolved i.,** hierro disuelto; **i. reduction,** reducción (f) de hierro; **i. solubilization,** solubilización (f) de hierro; **pig i.,** hierro en lingotes; **soft i.,** hierro dulce; **wrought i.,** hierro forjado.

iron-like, a. férreo/a (of iron).

ironstone, n. roca (f) ferruginosa.

ironwork, n. carpintería (f) de hierro, taller (m) de hierro.

irradiate, v. irradiar.

irradiation, n. irradiación (f).

irreducible, a. irreductible.

irregular, a. irregular.

irregularity, n. irregularidad (f).

irrigate, v. regar, irrigar.

irrigated, a. de regadío/a; **i. land,** regadío (m), tierra (f) de regadío.

irrigation, n. riego (m), regadío (m), irrigación (f); **agricultural i.,** riego agrícola; **basin i.,** riego por compartimientos; **channel i.,** riego por surcos; **crop i.,** riego agrícola; **dripfeed i.,** riego por goteo, riego gota a gota; **i. farming,** cultivo (m) de regadío; **overhead i., spray i., sprinkler i.,** riego por aspersión; **wastewater i.,** riego con aguas residuales.

irritation, n. irritación (f).

irruptive, a. irruptivo/a; **i. growth,** crecimiento (m) irruptivo.

islet, n. isleta (f), islote (m).

ISO, n. ISO; **International Standards Organization,** Organización (f) Internacional de Normalización.

iso-, pf. iso-.

isobar, n. isobara (f).

isobaric, a. isobárico/a.

isobutane, n. isobutano (m).

isobutylene, n. isobutileno (m).

isochrone, n. línea (f) isócrona.

isochronous, a. isócrono/a.

isoclinal, a. isoclinal.

isocyclic, a. isocíclico/a.

isodrin, n. isodrín (m).

isohyet, n. isoyeta (f).
isohyetal, a. isoyetal; **i. map,** mapa (m) de isoyetas.
isolate, v. aislar.
isolated, a. aislado/a.
isolation, n. aislamiento (m).
isoline, n. isolínea (f).
isomer, n. isómero (m).
isomerism, n. isomería (f).
isomerization, n. isomerización (f).
isomorphic, a. isomórfico/a.
isomorphism, n. isomorfismo (m).
isopach, n. isopaca (f).
isopachyte, n. isopaca (f).
isopleth, n. isopleta (f).
isoprene, n. isopreno (m).
isopropyl, n. isopropilo (m).
Isoptera, n.pl. isópteros (m.pl), el orden Isoptera (Zoo).
isopteran, n. isóptero (m).
isosceles, a. isósceles.
isostacy, n. isostasia (f).
isostatic, a. isostático/a.
isotherm, n. isoterma (f).
isothermic, a. isotérmico/a.

isotonic, a. isotónico/a.
isotope, n. isótopo (m); **radioactive i.,** isótopo radioactivo.
isotopic, a. isotópico/a; **i. change,** cambio (m) isotópico.
isotopy, n. isotopía (f).
isotropic, a. isotrópico/a.
isotropy, n. isotropía (f).
isthmus, n. istmo (m).
Italian, a. italiano/a.
Italy, n. Italia (f).
iterate, v. iterar.
iteration, n. iteración (f).
iterative, a. iterativo/a.
itinerary, n. itinerario (m).
IUCN, n. UICN; **International Union for the Conservation of Nature and Natural Resources,** Unión (f) Internacional para la Conservación de la Naturaleza y Recursos Naturales.
ivy, n. hiedra (f), yedra (f).
IWSD, n. DIAS (f); **International Water and Sanitation Decade,** Década (f) Internacional del Agua y Saneamiento.

J

jack, n. gato (m) (for raising loads); **hydraulic j.,** gato hidráulico.
jackal, n. chacal (m).
jackhammer, n. martillo (m) neumático (pneumatic drill).
jade, n. jade (m).
jadeite, n. jadeíta (f).
jagged, a. dentado/a, melado/a.
jam, n. mermelada (conserve); atasco (traffic congestion).
Jamaica, n. Jamaica (f).
jamb, n. jamba (f) (of door).
jar, n. tarro (m); **jam j.,** tarro de mermelada.
jarosite, n. jarosita (f).
jasper, n. jaspe (m), jaspilita (f).
jaspilite, n. jaspilita (f), jaspe (m).
jaundice, n. ictericia (f).
jaundiced, a. ictérico/a.
jaw, n. mandíbula (f); quijada (f) (of animals); mordaza (f), mandíbula (Eng).
jeep, n. jeep (m), todoterreno (m), coche (m) todoterreno, campero (m) (Col).
jellyfish, n. medusa (f); aguamala (f) (LAm).
jetting, n. limpieza (f) con chorro de agua a presión; **j. machine,** máquina (f) de limpieza con chorro de agua a presión.
jetty, n. malecón (m) (pier), muelle (m) (wharf), embarcadero (m).
jig, n. plantilla (f) para montaje, plantilla para taladrar.
job, n. trabajo (m), puesto (m), empleo (m) (employment); tarea (f), trabajo (m), chamba (f) (Mex), laburo (m) (Arg) (task, work).
join, v. unir.
joined, a. conjunto/a, combinado/a, unido/a.

joiner, n. ebanista (m), carpintero (m).
joinery, n. ebanistería (f), carpintería (f).
joining, n. unión (f).
joint, 1. n. diaclasa (f) (Geol), unión (f) (place), cuarto (m) (of meat), articulación (f) (Med); **cross j.,** diaclasa transversal; **dovetail j.,** cola (f) de pato, cola (f) de milano; **j. surface,** superficie (f) de diaclasas; **j. system,** sistema (m) de diaclasas; **major j.,** diaclasa principal; 2. a. diaclasa (f) (Geol), unión (f) (place), cuarto (m) (of meat), articulación (f) (Med); **j. account,** cuenta (f) común; **j. interest,** coparticipación (f); **j. liability,** responsibilidad (f) solidaria; **j. ownership,** copropiedad (f), propiedad (f) común; **j. partner,** copartícipe (f); **j. venture,** empresa (f) conjunta, sociedad (f) conjunta (compañía); joint-venture (m) (acto).
jointing, n. diaclasamiento (m), fisuración (f), disjunción (f); **columnar j.,** disjunción (f) columnar.
joist, n. vigueta (f); **j. anchor,** anclaje (m) de vigueta.
joule, n. julio (m).
journey, n. viaje (m), recorrido (m).
judicial, a. jurídico/a, judicial.
judiciary, n. judicatura (f), judiciario (m).
juice, n. jugo (m); zumo (m) (fruit); **digestive juices,** jugos digestivos; **gastric juices,** jugos gástricos.
jungle, n. jungla (f).
juniper, n. enebro (m).
Jupiter, n. Júpiter (m).
jurisdiction, n. jurisdicción (f).
jurisprudence, n. jurisprudencia (f).
justification, n. justificación (f).

K

kame, n. depósito (m) glaciar, delta (f) glaciar.
kaolin, n. caolín (m).
kaolinite, n. caolinita (f).
kaolinization, n. caolinización (f).
karst, n. karst (m), carst (m); **k. aquifer**, acuífero (m) kárstico; **k. development**, desarrollo (m) de karst; **k. system**, sistema (m) kárstico.
karstic, a. kárstico/a, cárstico/a.
karstification, n. karstificación (f), carstificación (f).
katharometer, n. catarómetro (m).
keep, n. torre (m) del homenaje (of a castle).
kelp, n. ocle (f), golfe (m).
Kelvin, n. Kelvin (Chem, degrees absolute, abbr. K).
keratin, n. queratina (f).
kerosene, n. keroseno (m), kerosene (m) (LAm), querosén (m) (LAm).
ketone, n. cetona (f).
kettlehole, n. depresión (f) glaciar.
keyboard, n. teclado (m) (Comp).
keystone, n. clave (m).
khanat, n. khanat (m), ganat (m), qanat (m) (m) (Arab); galería (f).
khanate, n. qanat (m), kanat (m), kanato (m) (Arab), galería (f).
kidney, n. riñón (m).
kiln, n. horno (m) de calcinación.
kiln-dried, n. secado (f) en estufa.
kilo-, pf. kilo-.
kilobyte, n. kilobyte (m).
kilocalorie, n. kilocaloría (f) (abbr. kCal).
kilocycle, n. kilociclo (m).
kilogramme, n. kilogramo (m) (abbr. kg).

kilometre, n. kilómetro (m) (abbr. km).
kilovolt, n. kilovoltio (m).
kilowatt, n. kilovatio (m).
kilowatt-hour, n. kilovatio-hora (f) (abbr. KWh, unit of electrical consumption).
Kimmeridgian, n. Kimmeridgense (m).
kinematics, n.. cinemática (f); **fluid k.**, cinemática de fluidos.
kinetic, a. cinético/a; **k. energy**, energía (f) cinética.
kingdom, n. reino (m); **animal k.**, reino animal; **plant k.**, reino vegetal; **United K. (abbr. UK)**, Reino Unido.
kink, n. empuje (m), retorcimiento (m), kink (m).
Kjeldahl, n. Kjeldahl (m); **K. nitrogen**, nitrógeno (m) Kjeldahl.
knap, v. picar piedra (Arch).
knapping, n. acción (f) de picar piedra (Arch).
knee, n. rodilla (f).
knickpoint, n. resalto (m) de pendiente (Geog).
knock down, v. derrumbar, derrocar (to demolish, to tear down).
knoll, n. altozano (m), montecillo (m), montículo (m).
knot, n. nudo (m) (one nautical mile per hour).
knotted, a. nudoso/a.
knotty, a. nudoso/a.
know, v. conocer (to be acquainted with), saber (to have knowledge of).
knowledge, n. conocimiento (m).
known, a. conocido/a.
krypton, n. kriptón (m) (Kr).
kyanite, n. cianita (f).

L

LAATI, n. ALADI (f); **Latin American Association for Trade Integration**, Asociación (f) Latinoamericana de Integración.

label, n. etiqueta (f), rótulo (m).

labelling, n. etiquetado (m).

labia, n.pl. labios (m.pl).

labium, n. labio (m).

laboratory, n. laboratorio (m); **l. analysis**, análisis (m) de laboratorio.

labour, n. trabajo (m) (work); tarea (f), faena (f) (task); mano de obra (work force); parto (m) (Zoo, childbirth); laborismo (m) (politics); **L. Party**, partido (m) laborista; **manual l.**, trabajo (m) manual; **skilled l.**, mano (f) de obra especializada.

labourer, n. peón (m), obrero (m).

labra, n.pl. labros (m.pl).

labrum, n. labro (m) (de insectos).

laccolith, n. lacolito (m).

lacquer, n. laca (f).

lactate, v. lactar.

lactation, n. lactación (f), lactancia (f).

lactose, n. lactosa (f).

lacustrine, a. lacustre.

ladder, n. escalera (f) de mano; **fish l.**, escala (f) para peces, rampa (f) para peces.

ladle, n. cucharón (m) (large spoon); caldera (f) (for furnace).

LAFTA, n. ALALC; **Latin American Free Trade Association**, Asociación (f) Latinoamericana de Libre Comercio.

lag, n. lag (m), retardo (m), retraso (m).

lagging, n. revestimiento (m).

lagomorph, n. lagomorfo (m).

Lagomorpha, n.pl. lagomorfos (m.pl), el orden Lagomorpha (Zoo).

lagoon, n. laguna (f), albufera (f) (esp. in Valencia); **tidal l.**, laguna mareal, albufera laguna (f), albufera (f) (small reservoir) (esp. in Valencia); **aerated l. (Treat)**, laguna aireada; **oxidation l. (Treat)**, laguna de oxidación.

lahar, n. lahar (m) (mudflow).

laid out, a. trazado (e.g. a road).

lake, n. lago (m), laguna (f) (small lake); **cirque l.**, laguna glaciar; **L. District (GB)**, País (m) de los Lagos; **ox-bow l.**, laguna en forma de semiluna; **saltwater l.**, lago de agua salada.

lamella, n. lámina (f).

lamellibranch, n. lamelibranquio (m).

Lamellibranchia, n.pl. lamelibranquios (m.pl), Lamellibranchia (f.pl) (Zoo, class).

laminar, a. laminar; **l. flow**, flujo (m) laminar; **l. velocidad**, velocidad (f) laminar.

laminated, a. laminado/a.

lamination, n. laminación (f); **cross-l.**, laminación cruzada.

lamprey, n. lamprea (f).

land, n. tierra (f) (country); terreno (m), suelo (m) (ground); país (m) (nation); **arable l.**, tierra de labranza, tierra de cultivo, tierra cultivable; **buildable l.**, suelo urbanizable (Sp); **common l.**, campo (m) comunal, acotada (f), abertal (m); **contaminated l.**, suelo contaminado; **crown l.**, patrimonio (m) real; **cultivable l.**, tierra cultivable; **derelict l.**, terreno abandonado; **dry l.**, tierra firme; **fen l.**, terreno pantanoso; **fertile l.**, tierra fértil; **flat l.**, terreno plano; **high l.**, tierra alta; **irrigated l.**, tierra de regadío; **l. conservation**, conservación (f) del suelo; **l. drainage**, saneamiento (m) de terreno, drenaje (m) del terreno; **l. improvement**, mejoramiento (m) de tierras; **l. irrigation**, regadío (m) del terreno; **l. reclamation (por expulsión de agua del mar)**, terrenos ganados al mar; **l. reform**, reforma (f) agraria; **l. reclamation**, **l. restoration**, restauración (f) del terreno, corrección (f) de terrenos; **l. subject to frequent flooding**, aguachinal (m), aguadizo (m); **l. subsidence**, hundimiento (m) del suelo; **l. use**, uso (m) del suelo, ocupación (f) de suelo, ocupación de terreno; **level l.**, terreno horizontal; **low l.**, terreno bajo; **lowland**, tierra baja; **rough l.**, tierra desigual; **sterile l.**, tierra estéril; **uncultivated l.**, erial (m); **unirrigated l.**, tierra de secano.

landfarming, n. tratamiento (m) biológico para residuos.

landfill, n. vertedero (m), vertido (m) en terraplén, relleno (m) de tierra; **controlled l.**, vertedero controlado; **l. gas**, gas (m) de vertedero controlado; **municipal l.**, vertedero público, vertedero municipal; **secure l.**, relleno de seguridad.

landfilling, n. práctica (f) de verter basuras sólidas urbanas en una depresión del terreno.

landform, n. forma (f) fisiográfica, forma de relieve.

Landinian, n. Landiniense (m).

landlocked, a. cercado de tierra.

landmark, n. mojón (m), marca (f) terrestre, señal (f), punto (m) de referencia.

Landsat, n. Landsat, satélite (m) artificial que describe 14 órbitas diarias alrededor de la tierra.

landscape, 1. n. paisaje (m); **l. work**, paisajismo (m); **relic l. feature**, rasgo (m) residual del paisaje; 2. a. paisajístico/a; **l. value**, valor (m) paisajístico.

landscaping, n. paisajismo (m).
landslide, n. alud (m), deslizamiento (m) de tierra, desprendimiento (m) de tierras, corrimiento (m) de tierras; huayco (m) (Peru).
landslip, n. deslizamiento (m) de tierra.
land-use, n. uso (m) del suelo; **l. planning,** ordenación (f) de suelo, ordenación de territorio, ordenación territorial.
Langhian, n. Languiense (m).
language, n. lenguaje (m), lengua (f), idioma (f); **machine l.** (Comp), lenguaje de máquina; **scientific l.,** lenguaje científico.
lanolin, n. lanolina (f).
lanoline, n. lanolina (f).
lanthanide, n. lantánido (m).
lanthanum, n. lantano (m) (La).
lapilli, n. lapilli (m).
lapis lazuli, n. lapislázuli (m).
laptop, n. ordenador (m) portátil plegable (Comp).
larch, n. alerce (m).
large, a. grande (Gen), extenso/a (extensive), amplio/a, extensivo/a.
larva, n. larva (f); cresa (f) (maggot).
larvae, n.pl. larvas (f.pl) (see: larva).
larval, a. larval (Zoo).
laryngeal, a. laríngeo/a.
larynges, n.pl. laringes (f.pl).
laryngitis, n. laringitis (f).
larynx, n. laringe (f).
laser, n. láser (m).
latent, a. latente; **l. heat,** calor (m) latente.
lateral, a. lateral; **l. line (of fish),** línea (f) lateral.
laterite, n. laterita (f).
latest, a. novísimo/a, último/a; **the l. development,** el último avance.
latex, n. látex (m).
lathe, n. torno (m), máquina (f) herramienta (machine tool).
latifundia, n.pl. latifundios (m.pl); **l. system,** latifundismo (m).
latifundium, n. latifundio (m); **l. system,** latifundismo (m).
Latin America, n. Latinoamérica (f), Hispanoamérica (f) (the Spanish speaking countries of North, Central and South America), Iberoamérica (f) (excluding Brazil), América (f) latina, América (f).
Latin-American, a. latinoamericano/a, hispanoamericano/a, iberoamericano/a.
latite, n. latita (f).
latitude, n. latitud (f); **degrees of l.,** grados (m) de latitud; **horse latitudes,** calmas (f.pl) tropicales.
latitudinal, a. latitudinal.
latosol, n. suelo (m) laterítico.
latrine, n. letrina (f), retrete (m); **dry l.,** letrina seca; **pit l.,** letrina de hoyo seco.
launch, 1. v. botar (a ship), lanzar (a missile, a new product); sacar al mar (a lifeboat); 2. n. lancha (f); **m. launch,** lancha motora.

lava, n. lava (f); **l. cone,** cono (m) de lava; **l. field,** campo (m) de lava; **l. flow,** colada (f) de lava; **l. plateau,** meseta (f) de lava; **pahoehoe l.,** lava pahoehoe; **pillow l.,** lava almohadilla.
lavant, n. arroyo (m) intermitente (GB), rambla (f).
lavender, n. espliego (m), lavanda (f), alhucema (f).
law, n. ley (f), derecho (m); **against the l.,** contra la ley, ilegal; **civil l.,** derecho civil; **common l.** (GB), derecho consuetudinario; **constitutional l.,** ley orgánica; **current l.,** vigente; **customary l.,** derecho consuetudinario; **Darcy's L.,** ley de Darcy; **environmental l.,** ley del medio ambiente; **Federal Water L.** (US), Ley Federal de Aguas; **l. and order,** orden (m) público; **l. court,** tribunal (m) de justicia; **l. in force,** ley vigente; **L. Lords** (GB), jueces (m.pl) que son miembros de la Cámara de los Lores; **l. of diminishing returns,** ley de los decrecientes; **l. of mass action** (Phys), ley de acción de masas; **l. of nature,** ley de la naturaleza; **l. of superposition** (Geol), ley de superposición; **l. of supply and demand,** ley de la oferta y la demanda; **laws of physics,** leyes de la física; **maritime l.,** derecho marítimo; **outside the l.,** al margen de la ley, fuera de la ley; **physical l. of conservation of mass and energy,** ley física de la conservación de la materia y la energía; **physical laws,** leyes de la física; **water abstraction l.,** ley de extracción de agua.
lawful, a. legal (in accordance with the law), lícito/a (authorized by law), legítimo/a (recognized by law), válido/a (by contract).
lawn, n. césped (m).
lawsuit, n. pleito (m) (case), juicio (m) (justice), litigio (m) (litigation).
lawyer, n. abogado (m); jurista (m, f) (legal expert).
layer, n. camada (f), lecho (m), capa (f); **boundary l.,** capa límite; **flow layering,** camada de flujo; **herb l.** (Ecol), capa herbácea, estrato (m) herbáceo.
layout, n. replanteo (m), (of a building); trazado (m), disposición (f) (disposition).
lay out, v. replantear (a new building).
lazurite, n. lazurita (f).
LCD, n. PCD (f); **liquid crystal display,** pantalla (f) de cristal líquido.
leach, v. lixiviar, lavar.
leachate, n. lixiviado (m), filtrado (m), rezumo (m) percolado; **l. collection,** recolección (f) de lixiviado; **l. treatment,** tratamiento (m) de lixiviado.
leaching, n. lixiviación (f), filtración (f), infiltración (f).
lead, n. plomo (m) (Pb); mina (f) (of a pencil); sonda (f), escandallo (m) (Mar); cable (m) (for electricity); acequia (f) (irrigation

canal); **l.-free petrol**, sin plomo; **l. poisoning**, envenenamiento (m) por plomo.

leader, n. guía (f), jefe (m).

leadership, n. jefatura (f), dirección (f).

leaf, n. hoja (f); **l. blade**, limbo (m); **l. litter**, broza (f).

leaf-fall, n. abscisión (f).

leafiness, n. frondosidad (f).

leaflet, n. hojuela (f), folíolo (m) (Bot); folleto (m) (pamphlet).

leafy, a. frondoso/a.

leak, n. fuga (f), gotera (f) (e.g. in roof), salida (f) (escape), pérdida (f) (loss); **distribution l.**, fuga de distribución; **reservoir l.**, fuga de embalse.

leaky, a. agujereado/a.

learning, n. conocimientos (m.pl) (knowledge); aprendizaje (m) (process); **l. curve**, curva (f) de apredizaje.

leat, n. seatín (m), caz (m) (of water mill).

lecithin, n. lecitina (f).

lecturer, n. conferenciante (m, f) (speaker).

ledge, n. resalte (m), proyección (f).

lee, n. de sotavento (m) (Mar), abrigo (m) (Fig); **on the leeside**, sotavento (m).

leech, n. sanguijuela (f).

leek, n. puerro (m).

leeward, n. sotavento (m).

leg, n. pata (f) (of an animal); pierna (f) (of a person); trayecto (m) (Mar, distance covered).

legal, a. jurídico/a, legal (related to law), lícito/a (permitted by law), legítimo/a (recognised by law); **l. advisor**, asesor (m) jurídico; **l. contract**, contrato (m) legal; **l. procedure**, procedimiento (m) jurídico; **to take l. action against someone (v)**, entablar un pleito contra alguien.

legend, n. leyenda (f) (on maps).

legislate, v. legislar, establacer por ley.

legislation, n. legislación (f); **draft l.**, anteproyecto (m) de ley; **EC l.**, legislación comunitaria europea; **environmental l.**, legislación ambiental, legislación medioambiental; **European Community l.**, legislación comunitaria europea; **existing l.**, normativa (f) vigente; legislación existente; **national l.**, legislación estatal, legislación nacional; **pollution l.**, legislación sobre contaminación; **private l.**, legislación privada; **proposed l.**, legislación propuesta; **state l.**, legislación estatal; **water l.**, ley (f) de aguas.

legislative, a. legislativo/a.

legislator, n. legislador/a (m, f) (person).

legume, n. legumbre (f), leguminosa (f).

Leguminosae, n.pl. leguminosas (f.pl), la familia Leguminosae (Bot).

leguminous, a. leguminoso/a.

leisure, n. ocio (m); **l. industry**, industria (f) que produce lo que pide la gente para ocupar su tiempo libre; **l. time**, ratos (m.pl) de ocio.

lemon, n. limón (m) (fruta); limonero (m) (tree).

lempira, n. lempira (m) (Hond, unit of currency).

lend, v. prestar.

lending, n. acción (f) de prestar.

length, n. longitud (f), largo (m) (size); duración (f) (time); extensión (f) (extent); **wave l.**, longitud de onda.

lens, n. lente (m); capa (f) lenticular (Geol); **magnifying l.**, lupa (f), lente de aumento.

lentic, a. lenítico/a.

lenticular, a. lenticular.

lentil, n. lenteja (f).

leper, n. leproso/a (m, f).

lepidolite, n. lepidolita (f).

Lepidoptera, n.pl. lepidópteros (m.pl), el orden Lepidoptera (Zoo).

lepidopteran, n. lepidóptero (m).

lepidopteran, a. lepidóptero/a.

lepidopteron, n. lepidóptero (m).

leprosy, n. lepra (f).

leptospirosis, n. leptospirosis (m).

lethal, a. letal, mortal; mortífero/a; **l. dose**, dosis (f) mortal; **l. weapon**, arma (f) mortífera.

letter, n. letra (f) (alphabet); carta (f) (correspondence); **business l.**, carta comercial; **registered l.**, carta certificada; **special delivery l.**, carta urgente.

leucite, n. leucita (f).

leucitophyre, n. leucitófiro (m).

leucocyte, n. leucocito (m).

leukocyte, n. leucocito (m).

levée, n. terraplén (m), ribero (m), dique (m); albardón (m) (LAm).

level, 1. v. nivelar; 2. n. nivel (m) (altitude), llanura (f) (plain), ras (m) con ras (horizontal); **base l.**, nivel de base; **datum l.**, nivel cero, nivel de datum; **flood l**, nivel de inundación; **ground l.**, nivel del suelo; **groundwater l.**, nivel de agua subterránea; **l. with**, a ras de; **mean sea l.**, nivel medio del mar; **piezometric l.**, nivel piezométrico; **sea l.**, nivel del mar; **spirit l.**, nivel de burbuja de aire; **tide l.**, nivel de la marea; **water l.**, nivel del agua.

levelling, n. nivelación (f), nivelamiento (m), aplanamiento (m); **l. of the ground**, aplanamiento del suelo.

lever, n. palanca (f); cigüeñal (m) (Arab) (for raising water); apoyo (m) (Fig).

levy, n. tributación (f).

liability, n. responsabilidad (f) (legal), sujeción (f) (to duty); **public l.**, responsabilidad civil.

liana, n. liana (f).

liane, n. liana (f).

Lias, n. Liásico (m).

liberalization, n. liberalización (f).

lice, n.pl. piojos (m.pl).

licence, n. licencia (f), permiso (m).

licensee, n. persona (f) que tiene un permiso o una licencia.

licensing, n. licencia (f), autorización (f), concesión (f).

lichen, n. líquen (m).

life, n. vida (f); **half-l.,** vida media; **l. cycle assessment,** valoración (f) del ciclo de vida, análisis (m) ciclo vida; **span,** duración (f) de la vida; **plant l.,** vida vegetal (Bot); **useful l.,** vida útil.

life-sized, a. de tamaño natural.

life-span, n. duración (f) de la vida.

lift, n. ascensor (m), elevador (LAm) (elevator).

ligament, n. ligamento (m).

light, n. luz (f); día (m) (daylight); semáforo (m) (traffic light); faro (m) (headlight); **electric l.,** luz eléctrica; **flashing l.,** luz intermitente; **half-l.,** media luz; **in the l. of day,** a la luz del día; **l. and shade,** luz y sombra; **natural l.,** luz natural; **pilot l.,** piloto (m); **polarized l.,** luz polarizada; **to come to l. (v),** salir a la luz; **warning l.,** piloto (m).

lighter, n. lanchón (m) (boat); encendedor (m) (para cigarrillo).

lighthouse, n. faro (m).

lighting, n. alumbrado (m), iluminación (f) (system); alumbramiento (m); **street l.,** alumbrado público.

lighting up, n. alumbramiento (m).

lightning, n. relámpago (m); **flash of l.,** centella (f), refusilo (m) (LAm); **l. conductor,** pararrayos (m.pl); **sheet l.,** relámpago difuso.

lightship, n. buque (m) faro.

light-year, n. año (m) luz, año de luz.

ligneous, a. leñoso/a.

lignin, n. lignina (f).

lignite, n. lignito (m), carbón pardo (m) (brown coal), carbón lignítico.

likelihood, n. verosimilitud (f).

limb, n. miembro (m).

lime, n. cal (f) (Chem); lima (f) (fruit); limero (m), limo (m) (LAm) (tree); **l. kiln,** horno (m) de cal; **quick l.,** cal viva; **slaked l.,** cal apagada, cal muerta.

limekiln, n. calera (f).

limestone, n. caliza (f); **coral l.,** caliza coralina; **shelly l.,** caliza bioclástica.

limit, n. límite (m); máximo (m) (maximum restriction); **fixed emission l.,** límite de emisión; **l. of detection,** límite de detección; **liquid l.,** límite líquido; **plastic l.,** límite plástico; **tolerance l.,** límite de tolerancia.

limitation, n. limitación (f), restricción (f).

limited, a. limitada; **l. company, l. liability company (abbr. ltd),** sociedad (f) (de responsibilidad) limitada (abbr. s.l.).

limnigram, n. limnigrama (m).

limnigraph, n. limnígrafo (m).

limnimeter, n. limnímetro (m).

limnology, n. limnología (f).

limonite, n. limonita (f).

lindane, n. lindano (m).

line, n. línea (f) (Gen); trazo (m) (trace); raya (f) (drawn); límite (m) (boundary); **base l.,** línea de base; **centre l. (center l., US),** línea central; **coastline,** línea de costa; **dividing l.,** línea divisoria; **dotted l.,** línea de puntos; **drainage l.,** línea de drenaje; **equipotential l.,** línea equipotencial; **fault l.,** línea de falla; **flight l.,** línea de vuelo; **flow l.,** línea de flujo; **hinge l.,** charnela (f); **ice l.,** límite del hielo; **on-line (Comp),** en línea, en directo, conectado; **Plimsoll l.,** línea de carga máxima; **plumb l.,** línea de plomada, línea vertical; **railway l.,** línea férrea; **snow l.,** límite de la nieve perenne; **straight l.,** línea recta; **tree l.,** límite de los árboles, límite del bosque; **waterline,** línea de agua, línea de flotación.

linear, a. lineal (design), de longitud (measure); **l. interpolation,** interpolación (f) linear; **non-l. equation,** ecuación (f) no linear.

linearization, n. linearización (f).

lineation, n. lineación (f).

liner, n. revestimiento (m), forro (m) (lining); camisa (f) (of a cylinder); barco (m) de pasajeros (ship); **clay l. (for landfill),** revestimiento arcilloso; **composite l. (for landfill),** revestimiento compuesto; **landfill l.,** revestimiento de vertedero; **membrane l. (for landfill),** revestimiento de membrana.

lining, n. entubado (m), entubación (f); **telescopic l.,** entubado telescópico; **temporary l.,** entubado provisional.

link, n. nexo (m).

links, n.pl. terreno (m) arenoso; campo (m) de golf.

lintel, n. dintel (m) (of door).

lion, n. león (m).

lip, n. labio (m).

lipid, n. lípido (m).

lipolysis, n. lipólisis (f).

liquefaction, n. licuación (f), licuefacción (f).

liquefy, v. licuar, licuefacer.

liquid, 1. n. líquido (m); 2. a. líquido/a.

liquidity, n. liquidez (m); **l. index (Eng),** índice (m) de fluidez.

liquify, v. licuar, licuefacer.

liquor, n. licor (m) (Chem); **gas l.,** licor de gases; **mixed l. (Treat),** licor mixto; **process liquors,** aguas (f.pl) de tratamiento.

list, n. lista (f).

listing, n. listado (m) (Comp), registro (m).

literature, n. literatura (f); obras (f.pl) (works), publicitarios (m.pl) (printed material), documentación (of a subject); **l. search,** búsqueda (f) sistemática de la literatura.

lithic, a. lítico/a.

lithification, n. litificación (f).

lithium, n. litio (m) (Li).

lithoclase, n. litoclasa (f).

lithofacies, n. litofacies (f).

lithogenesis, n. litogénesis (m).

lithological, a. litológico/a.

lithology, n. litología (f).

lithophilous, a. litófilo/a.

lithosere, n. litoserie (f).

lithosphere, n. litosfera (f).

lithospheric, a. litosférico/a; **l. plates**, placas (f) litosféricas.

lithostatic, a. litostático/a.

lithostratigraphic, a. litostratigráfico/a.

litigation, n. litigio (m), pleito (m).

litmus, n. tornasol (m); **l. paper**, papel (m) de tornasol; **l. test**, prueba (f) de tornasol.

litre, n. litro (m).

litter, n. basura (f) (rubbish); mantillo (m) (humus); camada (f), cría (f) (new-born animals); hojarasca (f) (fallen leaves); **l. bin**, basurero (m).

littoral, 1. n. litoral (m); , 2. a. litoral, costero/a.

live, a. vivo/a; con corriente, cargado/a (Elec); en directo, en vivo (broadcast).

liver, n. hígado (m).

liverwort, n. hepática (f).

livestock, n. ganado (m).

living, 1. n. vida (f); **standard of l.**, nivel (m) de vida; 2. a. vivo/a, viviente; **l. organism**, organismo (m) vivo.

lixiviated, a. lixiviado/a.

lixiviation, n. lixiviación (f).

lizard, n. lagarto (m) (larger species), lagartija (f) (smaller species).

Llandeilian, n. Llandeilo (m).

Llandoverian, n. Llandoveriense (m).

Llanvirnian, n. Llanvirniense (m).

load, n. carga (f); **electrical l.**, carga en el circuito, carga eléctrica; **hydraulic l.**, carga hidráulica; **vehicle l.**, carga de un vehículo.

loader, n. cargadora (f) (machine); cargador (m) (person); **front-end l.**, pala (f) cargadora de ataque frontal.

loadstone, n. magnetita (f).

loam, n. suelo (m) arcilloso; loam (m) (LAm).

loan, n. préstamo (m).

lobby, n. lobby (m), grupo (m) de presión (activists); vestíbulo (m) (anteroom).

lobbying, n. presiones (f.pl), cabildeo (m); **environmental l.**, presiones ambientales.

lobe, n. lóbulo (m); **solifluction l.**, lóbulo de solifluxión.

lobster, n. langosta (f), cigarra (f).

local, a. local.

locality, n. localidad (f); **type l.**, localidad tipo.

location, n. situación (f), sitio (m).

lock, n. esclusa (f) (on a canal); cerradura (f) (on door); cerrojo (m) (bolt).

lockage, n. sistema (m) de esclusas (use of water between locks).

lockjaw, n. trismo (m), tétano (m).

locomotive, n. locomotora (f).

locus, n. lugar (m) geométrico.

lode, n. filón (m) mineralizado.

loess, n. loess (m).

log, 1. v. anotar, apuntar (to record); perfilar (Geol, to profile); **to l. in (Comp)**, entrar en el sistema, acceder; **to l. off (Comp)**, salir del sistema, terminar de operar; **to log on (Comp)**, entrar en el sistema, acceder; **to log out (Comp)**, salir del sistema, terminar de operar; 2. n. registro (m) (graphical record); tronco (m) (of tree); diario (m) (diary); log (m) (Mat, abbr. of logarithm); diagrafía (f) (Geol); **conductivity l.**, registro de conductividad; **differential temperature l**, registro de temperatura diferencial; **driller's l.**, informe (m) diario del sondista; **flight l.**, diario de vuelo; **gamma-gamma l.**, registro de gamma-gamma; **geophysical l.**, registro geofísico; **l. phase**, fase (f) logarítmica, fase exponencial; **natural gamma l.**, registro de gamma natural; **resistivity l.**, registro de resistividad; **salinity l.**, registro de salinidad; **seismic l.**, registro sísmico; **ship's l.**, diario de barco; **spontaneous potential l.**, registro de potencial espontáneo.

logarithm, n. logaritmo (m) (abbr. log).

logarithmic, a. logarítmico/a; **l. phase**, fase (f) logarítmica, fase exponencial.

logging, n. realización (f) de registro, diagrafía (f), explotación (f); perfilaje (m) (LAm); **downhole l.**, registro (m) simultáneo con la perforación.

long-distance, n. de fondo (m).

longitude, n. longitud (f); **degrees of l.**, grados (m) de longitud.

longitudinal, a. longitudinal.

longshore, a. marginal, litoral, costero, longitudinal; **l. current**, corriente (f) marginal; **l. drift**, deriva (f) litoral, desplazamiento (m) costero.

longsightedness, n. visión (f) de larga, hipermetropía (f); presbicia (f).

look up, v. buscar, consultar (to search for information); **look-up table (n)**, tabla (f) de consulta.

loophole, n. aspillera (f) (Mil).

loose, a. suelto/a.

lopolith, n. lopolito (m).

lorry, n. camión (f); **l. driver**, camionero (m).

loss, n. pérdida (f); **distribution l.**, pérdida en la red de distribución; **friction l.**, pérdida por fricción; **head l.**, pérdida de carga; **head losses**, pérdidas en cabeza; **pump losses**, pérdidas en las bombas; **water l.**, pérdida de agua; **water main losses**, pérdidas en tuberías.

lost, a. perdido/a.

lot, n. parcela (f) (plot); lote (m); **building l.**, solar (m); **parking l.**, aparcamiento (m), parking (m).

louse, n. piojo (m).

low, a. bajo/a; **l. grade**, baja calidad; **l. relief**, bajo relieve; **l. tide**, marea (f) baja; **l. water**, bajamar (f); **on l. ground**, a nivel del mar; **the L. Countries**, Los Países (m.pl) Bajos.

low-cost, a. económico/a.

lower, a. inferior.

lowermost, a. lo más bajo.

lowland, n. tierra (f) baja, bajial (m) (LAm), abajeño (m) (LAm), abajino (m) (LAm).
low-lying, a. bajo/a.
low-profile, a. discreto/a.
low-risk, a. de bajo riesgo.
Ltd, a. s.l.; **limited company,** sociedad (f) limitada.
lubricant, n. lubricante (m).
lubricate, v. lubricar.
lucerne, n. mielga (f).
Ludlovian, n. Ludloviense (m).
lumber, n. madera (f); **l. industry,** industria (f) de la madera; **l. yard,** aserradero (m), maderería (f) (LAm).
luminescence, n. luminiscencia (f).
luminosity, n. luminosidad (f).
lunar, a. lunar.
lung, n. pulmón (m).
luster, n. brillo (m), lustre (m).
lustre, n. brillo (m), lustre (m); **metallic l.,** brillo metálico; **vitreous l.,** brillo vitreo.

lustrous, a. lustroso/a.
Lutetian, n. Luteciense (m).
lutetium, n. lutecio (m) (Lu).
lutite, n. lutita (f), esquisto (m) arcilloso.
luxullianite, n. luxullianita (f).
luxuriance, n. frondosidad (f) (Bot, of vegetation).
luxuriant, a. frondoso/a (of vegetation).
lymph, n. linfa (f).
lymphatic, a. linfático/a; **l. system,** sistema (m) linfático.
lymphocyte, n. linfocito (m).
lynx, n. lince (m).
lyophilic, a. liofílico/a; **l. colloid,** coloide (m) liofílico.
lyophobic, a. liofóbico/a; **l. colloid,** coloide (m) liofóbico.
lysimeter, n. lisímetro (m); **irrigated l.,** lisímetro humectado.

M

Maastrichtian, n. Maastrichtense (m).
macerator, n. macerador (m) (machine).
machine, n. máquina (f); **calculating m.**, calculadora (f), máquina de calcular; **m. tool**, máquina herramienta; **trenching m.**, zanjadora (f); **vending m.**, máquina automática.
machinery, n. maquinaria (f).
mackerel, n. caballa (f), bonito (m); **horse m.**, **jack m.**, jurel (m).
macro-, pf. macro-.
macroartifact, n. macroartefacto (m).
macroeconomics, n. macroeconomía (f).
macroinvertebrate, n. macroinvertebrado (m); **benthic m.**, macroinvertebrado bentónico.
macronutrient, n. macronutriente (m).
macrophyte, n. macrofito (m).
macroscopic, a. macroscópico/a.
MACT, n. **Maximum Achievable Control Technology (US)**, máxima tecnología (f) de control alcanzable.
madreporic, a. madrepórica; **m. plate**, placa (f) madrepórica.
maelstrom, n. torbellino (m).
MAFF, n. MAPA (m); **Ministry of Agriculture, Fisheries and Food (GB)**, Ministerio (m) de Agricultura, Pesca y Alimentación (Sp).
mafic, a. máfico/a.
maggot, n. cresa (f); gusano (m) (worm).
magistrate, n. magistrado (m), juez (m).
magma, n. magma (f).
magmatic, a. magmático/a.
magnesic, a. magnésico/a.
magnesite, n. magnesita (f).
magnesium, n. magnesio (m) (Mg).
magnet, n. imán (m).
magnetic, a. magnético/a; **m. anomaly**, anomalía (f) magnética; **m. disc**, disco (m) magnético; **m. field**, campo (m) magnético; **m. north**, norte (m) magnético; **m. tape**, cinta (f) magnética.
magnetism, n. magnetismo (m).
magnetite, n. magnetita (f).
magnetometer, n. magnetómetro (m).
magnitude, n. magnitud (f).
mahogany, n. caoba (f).
main, n. tubería (f), cañería (f), colector (m); cable (m) (Elec); **m. cable**, cable (m) principal; **rising m.**, tubo (m) de impulsión; **trunk m. (for water)**, cañería (f) principal; **water m. leaks**, fugas (f.pl) en cañerías; **water m. losses**, pérdidas (f.pl) en tuberías.
mainland, n. tierra (f) firme, continente (m).
maintenance, n. mantenimiento (m) (of a machine), manutención (f) (of a building); **m. costs**, gastos (m.pl) de mantenimiento; **machine m.**, mantenimiento de máquinas;

plant m., mantenimiento de planta; **preventive m.**, mantenimiento preventivo; **pump m.**, mantenimiento de bombas.
maize, n. maíz (m), choclo (m) (Arg, Bol, Ch, Ec), guate (m) (CAm), elote (m) (on the cob); jojoto (m) (LAm, ear of maize); **m. field**, maizal (m), milpa (f) (CAm); **m. grower**, milpero (m) (LAm).
majority, n. mayoría (f).
malachite, n. malaquita (f).
malaria, n. paludismo (m), malaria (f).
malarial, a. palúdico/a.
malathion, n. malatión (m).
male, 1. n. macho (m), varón (m); 2. a. macho, varón, masculino/a.
malignant, a. maligno/a.
malleability, n. maleabilidad (f).
malleable, a. maleable.
malnutrition, n. malnutrición (f), desnutrición (f).
malodour, n. malos olores.
malt, v. maltear.
malt, n. malta (f) (grain) (Chem); **m. sugar**, maltosa (f).
maltase, n. maltasa (f).
malted, a. malteado/a; **m. milk**, leche (f) malteada.
malthouse, n. maltería (f), fábrica (f) de malta.
malting, n. maltaje (m).
maltose, n. maltosa (f).
mamma, n. mama (f).
mammal, n. mamífero (m).
Mammalia, n.pl. mamíferos (m.pl), la clase Mammalia (Zoo).
mammalian, a. mamífero/a.
mammary, a. mamario/a; **m. gland**, mama (f).
mammiferous, a. mamífero/a.
management, n. dirección (f), administración (f), gestión (f), gerencia (f); **environmental m.**, gestión ambiental, gerencia ambiental (LAm); **integrated m.**, gestión integrada; **sustainable m.**, gestión sostenible.
manager, n. gerente (m), director (m), jefe (m); **business m.**, director comercial.
manageress, n. jefa (f), directora (f), administradora (f).
mandarin, n. mandarina (f) (fruit), mandarino (m) (tree).
mandate, 1. v. poner bajo el mando (de); 2. n. mandato (m).
mandatory, a. obligatorio/a.
mandible, n. mandíbula (f).
manganese, n. maganeso (m) (Mn).
mangrove, n. mangle (m) (tree); **m. swamp**, manglar (m).
manhole, n. caja (f) de registro (for meter),

buzón (m) (box); **inspection m.**, buzón de visita, pozo (m) de revisión.

manifest, 1. n. manifiesto (m) (list); 2. a. manifiesto/a, evidente (evident).

manifestion, n. manifestación (f), demostración (f); **hydrothermal m.**, manifestación hidrotérmica.

manipulate, v. manipular.

manipulation, n. manipulación (f).

man-made, a. sintético/a, artificial.

manner, n. manera (f).

manometer, n. manómetro (m).

manpower, n. mano (m) de obra; hombres (m.pl) (men).

mantle, n. manto (m) (Geol), capa (f) (layer); **rock m.**, cobertera (f) rocosa, manto rocoso; **soil m.**, manto de suelo; **the Earth's m.**, el manto de la Tierra.

manual, n. manual (m) (book).

manufacture, 1. v. fabricar, manufacturar; 2. n. fabricación (f).

manufacturer, n. fabricante (m), industrial (m).

manufacturing, a. manufacturero/a, industrial, fabril; **m. costs**, costos (m.pl) de fabricación.

manure, 1. v. abonar, estercolar; 2. n. estiércol (m), abono (m).

many-sided, a. polifacético/a.

map, 1. v. levantar un mapa de; trazar, dibujar (a plan); **to m. out**, proyectar, planear; 2. n. mapa (m); **aerial m.**, mapa aéreo; **base m.**, mapa base; **contour m.**, mapa de curvas de nivel; **isopachyte m.**, mapa de isopacas; **m. overlay**, mapa superpuesto; **piezometric m.**, mapa piezométrico; **structural m.**, mapa estructural; **subsurface m.**, mapa de subsuelo; **topographic m.**, mapa topográfico.

maple, n. arce.

mapping, n. cartografía (f), levantamiento (m) de planos; **digital m.**, cartografía (f) digital.

maquis, n. maquis (m).

marble, n. mármol (m).

marbling, a. marmolado/a; **m. effect**, efecto (m) marmolado.

marcasite, n. marcasita (f).

margin, n. margen (m), borde (m); **chilled m. (Geol)**, borde enfriado; **continental m.**, margen continental; **m. of error**, margen de error; **safety m.**, margen de seguridad.

marginal, a. marginal.

marijuana, n. marihuana (f), mariguana (f), marijuana (f).

marine, a. marino/a.

maritime, a. marítimo/a, de marina.

mark, 1. v. señalar, indicar (to show); marcar (trace); 2. n. huella (f) (trace); signo (m) (sign); señal (f), mancha (f) (stain); marca (f) (brand, trademark); nivel (m) (standard); **landmark**, marca terrestre; **Plimsoll m.**, línea (f) de carga máxima; **reference m.**, punto (m) de referencia; **scour m.**, huella de desgaste; **trademark**, marca registrada, marca paten-

tada, marca de fábrica; **watermark**, nivel de agua.

marker, n. referencia (f); trazador (m) (tracer).

market, n. mercado (m); **domestic m.**, mercado nacional, mercado interior; **free m.**, libre mercado; **home m.**, mercado nacional, mercado interior; **m. analysis**, análisis (m) de mercado(s); **m. demand**, demanda (f) del mercado; **m. equilibrium**, equilibrio (m) del mercado; **m. forces**, fuerzas (f.pl) del mercado; **m. garden**, huerto (m) (large), huerta (f) (small); **m. leader**, líder (m) del mercado; **m. research**, estudios (m.pl) de mercado; **overseas m.**, mercado exterior; **single m.**, mercado único; **stock m.**, bolsa (f).

market-garden, n. cultivo (m) para la venta, hortaliza (f), huerta (f) de legumbres.

marketing, n. mercadeo (m), mercadotecnia (f) (market research), comercialización (f) (of goods).

marketplace, n. mercado (m).

marl, n. marga (f).

marlite, n. marga (f); marlita (f) (LAm).

marlstone, n. marga (f); margalita (f) (LAm).

marly, a. margoso/a.

marrow, n. calabacín (m), zapallo (m) (LAm) (vegetable); médula (f) (Anat); **bone m.**, médula ósea.

Mars, n. Marte (m).

marsh, n. pantano (m), ciénaga (f); marjal (m), estero (m) (LAm); **m. gas**, gas (m) de los pantanos; **salt m.**, marisma (f); **tidal m.**, marisma (f) de marea.

marshland, n. pantano (m); bañado (m) (Arg).

marshy, a. pantanoso/a.

marsupial, 1. n. marsupial (m), didelfo (m); 2. a. marsupial.

mason, n. albañil (m); cantero (m) (of stone).

masonry, n. mampostería (f) (stonework), albañilería (f) (trade); **dry m.**, mampostería seca.

mass, n. masa (f) (Phys); multitud (f), muchedumbre (f) (of people); **atomic m. unit (abbr. a.m.u.)**, unidad (f) de masa atómica (abbr. u.m.a.); **land m.**, masa terrestre; **m. of concrete**, macizo (m) de hormigón; **volcanic m.**, macizo (m) volcánico.

massif, n. macizo (m).

mass-produce, v. producir en serie.

mast, n. mástil (m), palo (m) (Mar); poste (m) (radio); torre (f) (Eng).

material, 1. n. material (m); **building m.**, material de construcción; **granular m.**, material granular; **molten m.**, material en fusión; **parent m.**, material de partida; **raw m.**, materia prima; 2. a. material; **m. assets**, bienes (m.pl) materiales.

materialization, n. materialización (f).

mathematical, a. matemático/a; **m. expression**, expresión (f) matemática; **m. model**, modelo (m) matemático.

mathematician, n. matemático/a (m, f).
mathematics, n. matemáticas (f, pl).
maths, n. mates (f.pl) (abbr. of matemática).
mating, n. apareamiento (m), acoplamiento (m).
matrix, n. matriz (f); **rock m.,** matriz de la roca.
matter, n. materia (f); asunto (m) (subject); material (m); **carbonaceous m.,** materia carbonosa; **dissolved m.,** materia disuelta; **insoluble m.,** materia insoluble; **living m.,** materia viviente; **m. in suspension,** materia en suspensión; **organic m.,** materia orgánica; **particulate m.,** material granular, material particulado (LAm); **suspended m.,** materia en suspensión.
mattock, n. azadón (m), zapapico (m) (pick).
maturation, n. maduración (f); **m. pond (Treat),** laguna (f) de maduración.
mature, a. maduro/a.
maturity, n. madurez (f); **river m.,** madurez de un río.
maturity, n. madurez (f); **sexual m.,** madurez sexual.
mausoleum, n. mausoleo (m).
maxilla, n. maxilar (m), maxilar superior.
maxillae, n.pl. maxilas (f.pl) (see: maxilla).
maximum, a. máximo/a.
meadow, n. prado (m), pradera (f); **water m.,** almarcha (f), almarge (m).
meal, n. comida (f); harina (flour); **bonemeal,** harina de hueso; **fishmeal,** harina de pescado.
mean, 1. n. media (f), promedio (m) (Mat); **arithmetic m.,** media aritmética; **geometric m.,** media geométrica; 2. a. medio/a.
meander, 1. v. serpentear (a river); 2. n. meandro (m); **abandoned m. (ox-bow lake),** meandro abandonado, banco (m) (US); **incised m.,** meandro excavado; **m. belt,** zona (f) de meandros; **m. loop,** rizo (m) de meandro; **river m.,** meandro del río.
meandering, 1. n. con meandros, serpenteo (m) de ríos, divagación (f); **stream m.,** serpenteo de ríos; 2. a. que forma meandros, con meandros.
measles, n.pl. sarampión (m); **German m.,** rubéola (f), sarampión alemán.
measure, 1. v. medir; 2. n. medida (f); metro (m) (rule, yardstick); **coal measures (GB),** yacimientos (m.pl) de carbón, capas (f.pl) de carbón; **corrective m.,** medida correctiva; **cubic m.,** medida de volumen; **preventative m.,** medidas preventivas; **square m.,** medida de superficie; **tape m.,** cinta (f) métrica; **volumetric m.,** medida de volumen.
measurement, n. medida (f), medición (f); **air temperature m.,** medida de la temperatura; **insolation m.,** medida de la insolación; **pressure m.,** medida de la presión; **rainfall m.,** medida de la lluvia; **soil moisture m.,**

medida de la humedad del suelo; **water level m.,** medida de niveles de agua.
measurer, n. medidor (m, f).
measuring, n. medición (f), medida (f); **m. cylinder,** cilindro (m) de medir; **m. cylinder, m. tape,** cinta (f) métrica, huincha (f) (Bol, Ch, Peru).
mechanic, n. mecánico (m).
mechanical, a. mecánico/a.
mechanics, n. mecánica (f); **rock m.,** mecánica de rocas; **soil m.,** mecánica de suelos, geomecánica, geotecnia; **wave m.,** mecánica ondulatoria.
mechanist, n. mecánico (m).
mechanization, n. mecanización (f).
mechanize, v. mecanizar.
mechanized, a. mecanizado/a; **m. farming,** motocultivo (m).
media, n.pl. medios (m.pl) (see: medium); medios de comunicación (news).
median, 1. n. mediana (f) (Mat, middle term); 2. a. mediano/a; **m. strip (in road),** franja (f) central, bandeja (f) (Ch), camellón (m) (Col, Mex).
mediation, n. mediación (f).
medicinal, a. medicinal, medicamentoso/a; **m. waters,** aguas (f.pl) medicinales.
Mediterranean, 1. n. Mediterráneo (m); 2. a. mediterráneo/a.
medium, 1. n. medio (m) (means, Phys, Biol); instrumento (m); **anaerobic m.,** medio anaerobio; **anisotropic m.,** medio anisótropo; **culture m.,** medio de cultivo, caldo (m) de cultivo; **filter m.,** medio filtrante; **happy m.,** un justo (m) medio; **heterogeneous m.,** medio heterogéneo; **isotropic m.,** medio isotrópico; **porous m.,** medio poroso; **semipermeable m.,** medio semipermeable; 2. a. mediano/a, medio/a.
medusa, n. medusa (f).
meeting, n. encuentro (encounter); reunión (f) (assembly); cita (f) (appointment); mitin (m) (rally); **annual general m.,** junta (f) general (anual).
mega-, pf. mega- (10.E6).
megabyte, n. megabyte (m).
megalith, n. megalito (m) (Arch).
megalithic, a. megalítico/a (Arch).
megalitre, n. unidad de volumen igual a 10.E3 metros cúbicos, abbr. Ml (GB).
megalopolis, n. megalópolis (f) (US).
megaton, n. megatón (m).
megavolt, n. megavoltio (m).
megawatt, n. megavatio (m).
melanin, n. melanina (f).
melanism, n. melanismo (m); **industrial m.,** melanismo industrial.
melon, n. melón (m); **m. dealer, m. grower,** melonero/a (m, f); **m. patch,** melonar (m); **watermelon,** sandía (f), melón de agua.
melted, a. fundido/a, en fusión.

melting, a. fundente (m), punto de fusión (m); **m. pot,** crisol (m).

meltwater, n. aguanieve (f); **glacial m.,** aguanieve de deshielo glaciar.

member, n. miembro (m); **m. countries,** los países (m.pl) miembros; **M. of Parliament (GB),** diputado (m), miembro de Parlamento; **m. state,** Estado (m) miembro.

membrane, n. membrana (f); **cell m.,** membrana celular; **damp-proof m.,** membrana hidrófuga; **filter m.,** membrana filtrante; **mucous m.,** membrana mucosa; **semi-permeable m.,** membrana semipermeable.

memory, n. memoria (f), recuerdo (m); **computer m.,** memoria del ordenador; **direct-access m. (Comp),** memoria de acceso directo; **random-access m. (abbr. RAM) (Comp),** memoria de acceso aleatorio; **read-only m. (abbr. ROM) (Comp),** memoria de sólo lectura.

meniscus, n. menisco (m).

mensuration, n. medición (f), mensuración (f), medida (f).

menthol, n. mentol (m).

mercantile, a. mercantil.

mercaptan, n. mercaptano (m).

merchandise, n. mercancías (f.pl), mercaderías (f.pl).

merchant, n. comerciante (m, f), negociante (m, f), comercializador (m); **wholesale m.,** negociante al por mayor, mayorista (m).

mercuric, a. mercúrico/a.

mercurous, a. mercurioso/a.

mercury, n. mercurio (m) (Hg).

Mercury, n. Mercurio (m) (planet).

mere, n. lago (m) superficial.

meridian, n. meridiano (m).

meridional, a. meridional.

meristem, n. meristemo (m); **apical m.,** meristemo apical.

mesh, 1. v. engranar; 2. n. malla (f), tela (f) (web, fabric); **capacitor-resistor m.,** malla de capacidades y resistencias; **grid m.,** malla cuadriculada; **polygonal m.,** malla poligonal; **rectangular m.,** malla rectangular; **square m.,** malla cuadrada; **wire m.,** tela metálica.

meshed, a. engranado/a.

mesophilic, a. mesófilo/a; **m. digestion (Treat),** digestión (f) mesófila.

mesophylic, a. mesofílico/a.

mesophyte, n. mesofito/a (m, f).

mesosphere, n. mesosfera (f).

Mesozoic, n. Mesozoico (m).

message, n. mensaje (m), nota (f).

Messinian, n. Messinense (m).

meta-, pf. meta-.

metabolic, a. metabólico/a; **basal m. rate,** metabolismo (m) basal; **m. activity,** actividad (f) metabólica.

metabolism, n. metabolismo (m); **base m.,** metabolismo basal.

metabolite, n. metabolito (m).

metal, n. metal (m); **heavy m.,** metal pesado; **non-ferrous m.,** metal no ferroso.

metallic, a. metálico/a.

metalloid, n. metaloide (m).

metallurgy, n. metalurgía (f); **ferrous m.,** siderometalurgía (f).

metalwork, n. metalistería (f).

metamorphic, a. metamórfico/a.

metamorphism, n. metamorfismo (m).

metamorphose, v. metamorfosear.

metamorphosis, n. metamorfosis (f), metamórfosis (m).

metaphosphate, n. metafosfato (m).

metaquartzite, n. metacuarcita (f).

metasediment, n. metasedimento (m).

metasomatism, n. metasomatismo (m).

Metazoa, n.pl. metazoos (m.pl), el subreino Metazoa (Zoo).

metazoan, n. metazoo (m), metazoario (m).

meteor, n. meteoro (m).

meteorite, n. meteorito (m), aerolito (m).

meteorological, a. meteorológico/a.

meteorologist, n. meteorólogo/a (m, f).

meteorology, n. meteorología (f).

meter, n. metro (m) (length); contador (m) (gauge); medidor (m) (gauge) (LAm); **current m.,** correntómetro (m), molinete (m), micromolinete; **domestic flow m.,** medidor de carga de agua; **Venturi m.,** medidor de Venturi; **water m.,** contador de agua.

methaemoglobinaemia, n. metahemoglobinamia.

methaldehyde, n. metaldehído (m).

methane, n. metano (m).

methanogenesis, n. metanogénesis (f).

methanol, n. metanol (m).

methanotrophic, a. metanotrófico/a; **m. bacteria (pl),** bacterias (f.pl) metanotróficas.

method, n. método (m); manera (f); modo (m) (of payment); procedimiento (m) (means); **analogue m.,** método analógico; **bailer m.,** método de cuchara; **critical path m.,** método de vía crítica, método de ruta crítica (LAm); **dilution m.,** método de dilución; **explicit m.,** método explícito; **finite-difference m.,** método de diferencias finitas; **graphical m.,** método gráfico; **gravimetric m.,** método gravimétrico; **implicit m.,** método implícito; **iterative m.,** método iterativo; **logarithmic m.,** método logarítmico; **m. of images,** método de las imágenes; **m. of matrix inversion,** método de inversión de matrices; **m. of separation of variables,** método de separación de variables; **m. of superposition,** método de superposición; **optimization m.,** método de optimización; **recovery m.,** método de recuperación; **relaxation m.,** método de relajación; **traditional methods (e.g. fishing),** artes (m.pl) tradicionales.

methodical, a. metódico/a.

methodology, n. metodología (f).

methoxychlor, n. metoxicloro (m).

methyl, n. metilo (m); **m. orange,** naranja (f) de metilo.

methylamine, n. metilamina (f).

methylbenzene, n. metilbenzol (m).

methylene, n. metileno (m); **m. blue,** azul (m) de metileno.

methylmercury, n. metilmercurio (m).

methylparathion, n. metilparathión (m).

metre, n. metro (m); **cubic m.,** metro cúbico.

metric, a. métrico/a; **m. system,** sistema (m) métrico.

metro, n. metro (m), ferrocarril (m) subterráneo.

metropolis, n. metrópoli (f).

metropolitan, a. metropolitano/a.

Mexican, 1. n. mejicano/a (m, f), mexicano/a (m, f); 2. a. mexicano/a, mejicano/a.

Mexico, n. México, Méjico; **M. City,** ciudad (f) de México, ciudad de Méjico.

mezzanine, n. mezzanine (m), piso (m) entresuelo.

mho, n. mho (former unit of electrical conductance).

mica, n. mica (f); **m. schist,** micaesquisto (m), micacita (f) (LAm).

micrinite, n. micrinita (f).

micro-, pf. micro- (Mat, 10.E-6).

microanalysis, n. microanálisis (m).

microartifact, n. microartefacto (m) (Arch).

microbe, n. microbio (m).

microbial, a. microbiano/a; **m. contamination,** contaminación (f) microbiana.

microbic, a. microbiano/a.

microbiology, n. microbiología (f).

microblast, n. microblasto (m).

microchip, n. microchip (m), microplaqueta (f) (Comp).

microclimate, n. microclima (m).

microclimatology, n. microclimatología (f).

microcline, n. microclina (f).

microcrystalline, a. microcristalino/a.

microfarad, n. microfaradio (m) (unit of electrical capacitance).

microfiltration, n. microfiltración (f).

microfissuring, n. microfisuración (f).

microfossil, n. microfósil (m).

micrograph, n. micrografía (f); **electron m.,** micrografía electrónica.

microhm, n. microhmio (m), microhm (m) (unit of electical resistence).

microlog, n. microlog (m), microregistro (m).

micrometeorology, n. micrometeorología (f).

micrometer, n. micrómetro (m).

micrometre, n. micrómetro (m).

micrometric, a. micrométrico/a.

micron, n. micra (f), micrón (m), micrómetro (m).

micronutrient, n. micronutriente (m).

microorganism, n. microorganismo (m).

microprocessor, n. microprocesador (m).

microquake, n. microterremoto (m).

microscope, n. microscopio (m); **electron m.,** microscopio electrónico; **polarizing m.,** microscopio de luz polarizada; **simple m.,** microscopio simple.

microscopic, a. microscópico/a; **m. examination,** examen (m) microscópico.

microscopy, n. microscopia (f).

microscreen, n. microtamiz (m), microfiltro (m).

microscreening, n. microtamizado (m).

microsiemen, n. microsiemen (m) (unit of electrical conductance).

microsonic, a. microsónico/a.

microstraining, n. microcribado (m).

microwave, n. microonda (f).

mid-Atlantic, a. mid-atlántico/a.

midday, n. mediodía (f); **at m.,** a mediodía.

midden, n. muladar (m), estercolero (m) (Arch, rubbish dump).

middle-distance, n. de medio fondo (m).

middleman, n. intermediario (m).

midnight, n. medianoche (f).

migmatite, n. migmatita (f).

migmatization, n. migmatización (f).

migraine, n. migraña (f), jaqueca (f).

migrant, a. migratorio/a; **m. worker,** trabajador (m) migratorio.

migrate, v. emigrar.

migration, n. migración (f); **lateral gas m.,** migración lateral de gases; **leachate m.,** migración de lixiviado; **upward gas m.,** migración hacia arriba de gases.

migratory, a. migratorio/a.

mildew, n. moho (m); roya (f), añublo (m) (on plants); mildiu, mildiú, mildeu (m) (on vines).

mile, n. milla (f) (1609 metres); **nautical m. (1852 metres),** milla marina.

miliampere, n. miliamperio (m) (unit of electrical current).

milk, n. leche (f); **full-cream m.,** leche completa, leche entera; **m. diet,** dieta (f) láctea; **m. powder,** leche en polvo; **m. sugar,** lactosa (f); **powdered m.,** leche en polvo; **skimmed m.,** leche desnatada.

Milky Way, n. Vía (f) Láctea.

mill, n. molino (m) (for grain); **sugar m.,** azucarera (f), trapiche (m) (Mex); **watermill,** molino de agua.

miller, n. molinero (m).

millet, n. mijo (m).

milli-, pf. mili- (10.E-3).

millibar, n. milibar (m) (unit of atmospsheric pressure).

milligram, n. miligramo (m).

milligramme, n. miligramo (m).

milliliter, n. mililitro (m).

millilitre, n. mililitro (m).

millimeter, n. milímetro (m); **m. of mercury,** milímetro de mercurio.

millimetre, n. milímetro (m); **m. of mercury,** milímetro de mercurio.

millipede, n. milpiés (m).

millrace, n. saetín (m), caz (m).
millstone, n. rueda (f) de molino, volandera (f).
mimic, n. mímico (m).
mimicry, n. mimetismo (m).
mine, n. mina (f); **abandoned m.,** mina abandonada; **coal m.,** mina de carbón.
miner, n. minero (m).
mineral, 1. n. mineral (m); **heavy minerals,** minerales pesados; **m. deposit,** yacimiento (m) mineral; **m. oil,** aceite (m) mineral; **m. salts,** sales (m.pl) minerales; **m. water,** agua (f) mineral; 2. a. mineral.
mineralization, n. mineralización (f).
mineralize, v. mineralizar.
mineralogy, n. mineralogía (f).
mineralometry, n. mineralometría (f).
Minimata, n. Minimata (f); **M. disease,** enfermedad (f) de Minimata.
minimization, n. minimización (f).
minimize, v. minimizar.
minimum, a. mínimo/a.
mining, n. minería (f), explotación (f) de minas; **adit m.,** explotación (f) por socavón; **cut-and-fill m.,** explotación (f) por fajas aisladas; **drift m.,** explotación por galerías; **m. area,** zona (f) minera, campo (m) de minas (LAm).
minister, n. ministro (m); **Prime M.,** Primer Ministro.
ministry, n. ministerio (m); sacerdocio (m) (religion); **M. for the Environment,** Ministerio del Medio Ambiente; **M. of Education,** Ministerio de Educación; **M. of Housing,** Ministerio de la Vivienda; **M. of Industry and Commerce,** Ministerio de Industria y Comercio; **M. of Public Works,** Ministerio de Obras Públicas.
minus, p. menos, sin.
Miocene, n. Mioceno (m).
mirage, n. espejismo (m).
mire, n. pantano (m).
mirror, n. espejo (m).
MIS, n. **management information system,** sistema (m) de información para la gestión.
miscibility, n. miscibilidad (f).
misfit, a. desproporcionado/a; **m. river,** río (m) desproporcionado.
misinformation, n. mala información (f).
mist, n. neblina (f) (Gen), calina (f), garúa (f) (LAm); **m. catchment,** captación (f) de neblina; **m. collector,** captador (m) de neblina, colector (m) de neblina (LAm).
misty, a. brumoso/a, nebuloso/a.
mite, n. garrapata (f).
mitigate, v. mitigar; aliviar (alleviate).
mitigation, n. mitigación (f); alivio (m) (alleviation).
mix, v. mezclar (combine).
mixed, a. mixto/a, variado/a.
mixer, n. mezclador (m) (machine).
mixture, n. mezcla (f), mixtura (f).
mobility, n. movilidad (f).

mobilization, n. movilización (f).
mobilize, v. movilizar.
modal, a. modal.
modality, n. modalidad (f).
mode, n. modo (m), manera (f), forma (f) (manner); moda (f) (Mat, most common value); **m. of occurrence,** manera de presentarse.
model, 1. v. modelar; 2. n. modelo (m); **advection m.,** modelo de advección; **analogue m.,** modelo analógico; **analytical m.,** modelo analítico; **autoregressive m.,** modelo autoregresivo; **conceptual m.,** modelo conceptual; **deterministic m.,** modelo determinístico; **diffusion m.,** modelo de difusión; **digital m.,** modelo digital; **finite-difference m.,** modelo por diferencias finitas; **global climate m. (abbr. GCM),** modelo de cambio climático; **hybrid m.,** modelo híbrido; **hydrodynamic m.,** modelo hidrodinámico; **implicit finite-difference approximation m.,** modelo de aproximación de diferencias finitas implícitas; **lumped-parameter m.,** modelo de parámetro concentrado; **mathematical m.,** modelo matemático; **m. validation,** validación (f) de un modelo; **one-dimensional m.,** modelo unidimensional; **physical m.,** modelo físico; **probabilistic m.,** modelo probabilístico; **stochastic m.,** modelo estocástico; **time-dependent m.,** modelo dependiente del tiempo; **transient m.,** modelo transitorio; **transport m.,** modelo de transporte; **two-dimensional m.,** modelo bidimensional.
modeler, n. modelista (m, f).
modeling, 1. n. modelización (f), modelado (m); 2. a. modelador/a.
modeller, n. modelista (m, f).
modelling, 1. n. modelización (f), modelado (m); 2. a. modelador/a.
modem, n. módem (m).
modern, a. moderno/a.
modernization, n. modernización (f), actualización (f).
modification, n. modificación (f).
modular, a. modular; **m. flow,** flujo (m) modular.
module, n. módulo (m).
modulus, n. módulo (m); **m. of elasticity,** módulo de elasticidad.
Moho, n. Moho (m) (Geol); **M. layer (abbr. Mohorovicic discontinuity),** capa (f) de Moho (abbr, discontinuidad Mohorovicic).
moistening, n. humedecimiento (m), humectación (f).
moisture, n. humedad (f); **soil m. deficit,** déficit (m) de humedad del suelo.
moisturize, v. humedecer.
mol, n. mol (m).
molal, a. molal.
molality, n. molalidad (f).
molar, 1. n. muela (f) (Zoo); 2. a. molar; **m. solution,** solución (f) molar.

molarity, n. molaridad (f).
molasses, n. melaza (f).
molding, n. moldura (f).
mole, n. topo (m) (Zoo); malecón (m) (jetty), muelle (m) (harbour); mol (Chem); **m. drain,** dren (m) de topo.
molecular, a. molecular.
molecule, n. molécula (f).
mollusc, n. molusco (m).
Mollusca, n.pl. moluscos (m.pl), el fílum Mollusca (Zoo).
molluscicide, n. molusquicida (m).
mollusk, n. molusco (m).
molt, 1. v. mudar; 2. n. muda (f).
molten, a. en estado fundido, en fusión; **m. rock,** roca (f) fundida.
molting, n. muda (f).
molybdate, n. molibdato (m).
molybdenum, n. molibdeno (m) (Mo).
momentum, n. ímpetu (m), velocidad (f) (speed); momento (m) (Phys).
monazite, n. monacita (f).
monel, n. monel (m) (metal).
monitor, n. monitor (m) (Comp).
monitoring, n. monitorización (f), monitoreo (m), vigilancia (f), control (m); **environmental m.,** control ambiental, monitoreo (m) del ambiente; **m. network,** red (f) de vigilancia; **m. point, m. station,** estación (f) de vigilancia, punto (m) de control.
mono-, pf. mono-.
monoclinal, a. monoclinal; **m. block,** bloque (m) monoclinal.
monocline, n. monoclinal (m).
monocotyledon, n. monocotiledónea (f).
Monocotyledonae, n.pl. monocotiledóneas (f.pl), la clase Monocotyledonae (Bot).
monoculture, n. monocultivo (m) (single-crop farming).
monocyclic, a. monocíclico/a.
monocyclical, a. monocíclico/a.
monoecious, a. monoico/a.
monofilament, n. monofilamento (m); **nylon m. (fishing),** monofilamento de nilón.
monograph, n. monografía (f).
monograptid, n. monograptido (m).
monohydrate, n. monohidrato (m).
monolith, n. monolito (m).
monomer, n. monómero (m).
monorail, n. monocarril (m).
monosaccharide, n. monosacárido (m).
monospecies, n. monoespecie (f).
monotrophy, n. monotrofía (f).
monovalent, a. monovalente.
monoxide, n. monóxido (m); **carbon m.,** monóxido de carbono.
monsoon, n. monzón (m).
Montenian, n. Monteniense (m).
montmorillonite, n. montmorillonita (f).
monument, n. monumento (m); **natural m. (US),** sitios (m.pl)/parajes (m.pl) de interés nacional (Sp).

monzonite, n. monzonita (f).
monzonite, n. monzonita (f).
moon, n. luna (f); **full m.,** luna llena; **half m.,** media luna; **new m.,** luna nueva.
moonlight, n. luz (f) de luna.
moonlit, a. iluminado/a por la luna.
moonstone, n. piedra luna (f), albita (f).
moor, n. páramo (m), brezal (m).
mooring, n. fondeadero (m), amarre (m).
moorland, n. páramo (m).
moraine, n. morrena (f); **basal m.,** morrena basal; **central m.,** morrena central; **terminal m.,** morrena terminal.
morass, n. cenegal (m).
moratorium, n. moratoria (f).
morbidity, n. morbosidad (f).
morphine, n. morfina (f).
morphological, a. morfológico/a.
morphology, n. morfología (f); **leaf m. (Bot),** morfología foliar.
mortality, n. mortalidad (f); **infant m.,** mortalidad infantil.
mortar, n. mortero (m) (cement).
mortgage, n. hipoteca (f) (Fin).
Moslem, n. musulmán, musulmana (m, f).
mosquito, n. mosquito (m), zancudo (m) (LAm).
mosquitoes, n.pl. mosquitos (m.pl).
moss, n. musgo (m).
mossy, a. musgoso/a, cubierto de musgo.
motel, n. motel (m).
moth, n. mariposa (f) nocturna; polilla (f) (in clothes).
mother-of-pearl, n. nácar (m), madreperla (f).
motile, a. motil, móvil.
motor, n. motor (m); coche (m) (car); **electric m.,** motor eléctrico; **m.-driven,** impulsado por un motor; **outboard m.,** motor fuera borda; **starter m.,** motor de arranque.
motorboat, n. motora (f), lancha (f) motora.
motorcar, n. coche (m), automóvil (m), turismo (m) (private car), carro (m) (LAm), auto (m) (LAm).
motorway, n. autopista (f); **toll-paying m.,** autopista de peaje.
motte, n. montículo (m) (Arch); **m. and bailey,** montículo y banqueta.
mottle, n. mota (f).
mottled, a. abigarrado/a, moteado/a.
mould, n. molde (m) (container); forma (f) (shape); mantillo (m) (leaf litter); moho (m) (fungus); **leaf m., vegetable m.,** mantillo.
moulding, n. moldura (f).
mouldy, a. mohoso/a, enmohecido/a.
moult, 1. v. mudar; 2. n. muda (f).
moulting, n. muda (f).
mountain, n. montaña (f); **m. chain,** cordillera (f), cadena (f) de montañas; **m. range,** sierra (f).
mountaineering, n. alpinismo (m).
mountainous, a. montañoso/a, gigantesco/a (gigantic); **m. range,** serranía (f).

mountainside, n. ladera (f) de montaña.

mouse, n. ratón (m) (Zoo, Comp).

mouth, n. boca (f) (Anat); embocadura (f), desembocadura (f) (of river); abertura (f) (opening); entrada (f) (of cave); **m. of river, river m.**, ría (f), desembocadura de río, boca de río.

move, n. paso (m), marcha (f), movimiento (m).

movement, n. movimiento (m); **earth m.**, movimiento de la tierra; **m. of contaminants**, movimiento de contaminantes; **m. of underground water**, movimiento del agua subterránea; **rotational m.**, movimiento rotacional; **strike-slip m.**, movimiento de desplazamiento en dirección; **wave m.**, movimiento de las olas.

mow, v. segar, cortar.

mucilage, n. mucílago (m).

muck, n. fango (m) (sludge), estiércol (m) (manure), suciedad (f) (dirt).

mucosa, n. mucosa (f).

mucosae, n.pl. mucosas (f.pl) (see: mucosa).

mucous, a. mucoso/a; **m. membrane**, membrana (f) mucosa.

mucus, n. moco (m), mucus (m), mucús (m).

mud, n. fango (m), lodo (m), barro (m), légamo (m); **bentonite m.**, lodo bentonítico; **biodegradable m.**, lodo biodegradable; **drilling m.**, lodo de perforación; **organic m.**, lodo orgánico.

muddy, a. túrbido/a (turbid), fangoso/a (mucky), cenagoso/a (river), turbio/a (liquid).

mudflat, n. marisma (f), barreal (m) (LAm).

mudflow, n. corriente (f) de barro, lahar (m).

mudstone, n. fangolita (f), lodolita (f).

muffled, a. amortiguado/a, sordo/a, apagado/a.

muffler, n. silenciador (m), mofle (m) (LAm), mofler (m) (Mex, PRico, RDom).

mulberry, n. mora (f); **m. tree**, moral (m), morera (f).

mulch, n. pajote (m) (Agr).

mulching, n. mulching (m).

mulitphase, n. multifase (f).

mullet, n. salmonete (m).

multi-, pf. multi-.

multicellular, a. pluricelular, multicelular.

multi-coloured, a. abigarrado/a.

multidisciplinary, a. multidisciplinario/a.

multi-disciplinary, a. pluridisciplinario/a.

multimedia, a. multimedia.

multinational, a. multinacional.

multiparametric, a. multiparamétrico/a.

multiple, 1. n. múltiplo (m); **lowest common m.**, mínimo común múltiplo; 2. a. múltiplo/a; **m. regression**, regresión (f) múltiple.

multiplication, n. multiplicación (f).

multispectral, a. multiespectral.

multivariate, a. multivariante; **m. analysis**, análisis (m) de múltiples variables.

mummy, n. momia (f) (Arch).

mumps, n. paperas (f.pl).

municipal, a. municipal; **m. district**, distrito (m) municipal, delegación (f) (LAm).

municipality, n. municipalidad (f).

muscle, n. músculo (m); **smooth m.**, músculo liso.

muscovite, n. moscovita (f).

mush, n. hielo (m) troceado (US) (Met).

mushroom, n. seta (f), hongo (m), champiñón (m).

must, n. mosto (m).

mutagen, n. mutágeno (m), agente (m) mutágeno.

mutagenic, a. mutagénico/a.

mutagenicity, n. mutagenicidad (f).

mutant, a. mutante.

mutate, v. mudar.

mutation, n. mutación (f); **gene m., genetic m.**, mutación genética.

mutualism, n. mutualismo (m).

mycelia, n.pl. micelios (m.pl).

mycelium, n. micelio (m).

mycology, n. micología (f).

mycophyte, n. micófito (m).

mycorrhiza, n. micorriza (f).

mylonite, n. milonita (f).

mylonitization, n. milonitización (f).

myo-, pf. mio-.

myopia, n. miopía (f).

myopic, a. miope.

N

nail, 1. v. clavar; 2. n. clavo (m) (of metal); uña (f) (Zoo).

name, n. nombre (m); **Christian n.**, nombre, nombre de pila; **family n.**, apellido (m); **first n.**, nombre, nombre de pila; **full n.**, nombre y apellidos; **generic n.** (Biol), nombre genérico; **place n.**, topónimo (m); **proper n.**, nombre propio; **specific n.** (Biol), nombre específico; **surname,** apellido (m); **systematic n.**, nombre sistemático; **trivial n.**, nombre corriente.

Namurian, n. Namuriense (m).

nano-, pf. nano (m) (10.E-9).

nanoplankton, n. nanoplancton (m).

naphthalene, n. naftalina (f).

nappe, n. manto (m) de corrimiento (Geol); cobijadura (f) (LAm).

narcotic, a. narcótico/a, estupefaciente.

nationalization, n. nacionalización (f).

native, 1. n. nativo/a (m, f) (inhabitant), indígena (m, f), natural (m, f); 2. a. nativo/a, indígena; **n. gold,** oro (m) nativo; **n. soil,** suelo (m) nativo.

NATO, n. OTAN (f); **North Atlantic Treaty Organization,** Organización (f) del Tratado de Atlántico Norte.

natural, a. natural; **n. resources,** recursos (m.pl) naturales; **n. selection,** selección (f) natural.

nature, n. naturaleza (f) (wildlife); esencia (f) (character or property); índole (f) (manner); **laws of n.,** leyes (f.pl) de la naturaleza; **Mother N.,** la madre (f) naturaleza; **n. park, n. reserve,** espacio (m) natural protegido.

nautical, a. de marina, marítimo/a; **n. chart,** carta (f) náutica; **n. mile (1852 metres),** milla (f) marina.

nautilocone, n. nautilocono (m).

nautiloid, n. nautiloide (m).

navel, n. ombligo (m).

navigable, a. navegable.

navigation, n. navegación (f); **coastal n.,** navegación costera, navegación de cabotaje; **n. chart,** carta (f) náutica; **river n.,** navegación fluvial.

navy, n. marina (f) (people, organization); armada (f), flota (f) (ships); **merchant n.,** marina mercante.

neap, n. marea (f) muerta; **n. tide,** marea muerta.

nearctic, a. neártico/a.

necessitate, v. necesitar; exigir (to demand, to exact).

necessity, n. necesidad (f).

necropolis, n. necrópolis (m).

necrosis, n. necrosis (f).

nectar, n. néctar (m).

need, 1. v. necesitar, exigir, requerir; 2. n. necesidad (f).

negative, a. negativo/a; **n. charge,** carga (f) negativa; **n. electrode,** electrodo (m) negativo; **n. ion,** ión (m) negativo; **n. pole,** polo (m) negativo; **n. taxis,** taxia (f) negativa; **n. tropism,** tropismo (m) negativo.

negligence, n. negligencia (f); **gross n.,** gran negligencia.

negligent, a. negligente, descuidado/a.

negligible, a. insignificante, despreciable.

neighboring, a. vecino/a, limítrofe; **n. country,** país (m) vecino, país limítrofe.

neighbourhood, n. barrio (m), vecindad (f), inmediaciones (f.pl) (area); vecindario (m), vecinos (m.pl) (people).

neighbouring, a. vecino/a, limítrofe; **n. country,** país (m) vecino, país limítrofe.

nekton, n. necton (m).

Nematoda, n.pl. nemátodos (m.pl), el filum Nematoda (Zoo).

nematode, n. nemátodo (m).

neodynium, n. neodimio (m) (Nd).

Neogene, n. Neógeno.

Neolithic, n. neolítico (m).

neon, n. neón (m) (Ne).

neoprene, n. nepreno (m).

nepheline, n. nefelina (f).

nephelite, n. nefelinita (f).

nephrite, n. nefrita (f).

neptunism, n. neptunismo (m).

neptunite, n. neptunita (f).

neptunium, n. neptunio (m) (Np).

neritic, a. nerítico/a.

nerve, n. nervio (m); **n. centre,** centro (m) nervioso; **n. ending,** terminación (f) nerviosa; **n. gas,** gas (f) neurotóxico; **n. impulse,** impulso (m) nervioso; **optic n.,** nervio óptico.

nervous, a. nervioso/a; **central n. system (abbr. CNS),** sistema (m) nervioso central (abbr. SNC).

NESHAP, n. **National Emission Standard for Hazardous Air Pollutants (US),** valor (m) límite nacional para contaminantes peligrosos del aire.

net, 1. n. red (f); **drag n.,** red de arrastre; **drift n. (fishing),** arte (m) de deriva; **flow n. (groundwater),** red de flujo, red de percolación; **gill n. (fishing),** red de enmallado; **purse seine n. (fishing),** cerco (m) de jareta; 2. a. red (f); **n. amount,** cantidad (f) neta; **n. weight,** peso (m) neto.

Netherlands, n. Holanda (f), Países Bajos (m.pl).

netting, n. malla (f) (wire mesh), tela (f)

(screen); red (f) (for fishing); **wire n.**, tela metálica.
network, n. red (f); **channel n.**, red de canales; **distribution n.**, red de distribución; **drainage n.**, red de drenaje; **flow gauging n.**, red de estaciones de aforos; **monitoring n.**, red de vigilancia; **rail n.**, red de ferrocarril; **road n.**, red de carreteras; **seismic n.**, red sísmica.
neural, a. neural, de los nervios.
neurology, n. neurología (f).
neurone, n. neurona (f); **motor n.**, neurona motora.
Neuroptera, n.pl. neurópteros (m.pl), el orden Neuroptera (Zoo).
neuropteran, n. neuróptero (m).
neurotoxic, a. neurotóxico/a.
neurotoxicity, n. neurotoxicidad (f).
neuter, a. neutro/a.
neutral, a. neutral; **n. centre,** punto (m) muerto; **n. point,** punto (m) muerto.
neutralization, n. neutralización (f).
neutralize, v. neutralizar.
neutron, n. neutrón (m).
neutron-gamma, n. neutrón-gamma (m); **n.-g. sonde,** sonda (f) de neutrón-gamma.
neutron-neutron, n. neutrón-neutrón (m); **n.-n. sonde,** sonda (f) de neutrón-neutrón.
newsprint, n. papel (m) de periódico.
newt, n. tritón (m).
newton, n. newton (m) (unit of force).
New Zealand, n. Nueva Zelanda (f), Nueva Zelandia (f) (LAm).
NGO, n. ONG; **non-governmental organization (US),** organización (f) no gubernamental.
niacin, n. niacina (f).
Nicaragua, n. Nicaragua (f).
Nicaraguan, n. nicaragüense (m, f).
Nicaraguan, a. nicaragüense.
niche, n. nicho (m); **ecological n.**, nicho ecológico.
nickel, n. níquel (m) (Ni).
nickel-iron, n. níquel-hierro (m).
nickpoint, n. resalte (m) de pendiente (Geog).
nicotine, n. nicotina (f).
nimbostratus, n. nimboestrato (m) (Met).
NIMBY, n. expresión que refiere a la actitud de gente que aprueba el concepto de una nueva planta/un vertedero etc, pero que no lo quiere en su propio barrio; **"not in my back yard",** "en el patio trasero de mi casa, ¡no!".
niobium, n. niobio (m) (Nb).
nipple, n. pezón, tetilla (f) (of mammals); protruberancia (f) (knob); pezón, boquilla de unión (Eng, connection).
nitrate, n. nitrato (m); **Chilean n.**, nitrato de Chile, salitre (m) (Ch); **n. fertilizer,** abono (m) nitrogenado.
nitric, a. nítrico/a.
nitride, n. nitruro (m).
nitrification, n. nitrificación (f).
nitrify, v. nitrificar.

nitrifying, a. nitrificante; **n. bacteria,** bacterias (f.pl) nitrificantes.
nitrite, n. nitrito (m).
nitro-cellulose, n. nitrocelulosa (f).
nitroaniline, n. nitroanilina (f).
nitrobenzene, n. nitrobenzol (m).
nitrocellulose, n. nitrocelulosa (f).
nitrogen, n. nitrógeno (m) (N); **albuminoid n.**, nitrógeno albuminoideo; **n. cycle,** ciclo (m) de nitrógeno; **n. fertilizer,** abono (m) nitrogenado; **n. fixation,** fijación (f) de nitrógeno; **n. removal,** eliminación (f) de nitrógeno.
nitrogenation, n. nitrogenación (f).
nitrogenous, a. nitrogenoso/a.
nitroglycerine, n. nitroglicerina (f).
nitrosyl, n. nitrosilo (m).
nitrous, n. nitroso (m).
nitryl, n. nitrilo (m).
node, n. nodo (m); nudo (m) (Bot).
nodular, a. nodular.
nodule, n. nódulo (m); **root n. (Bot),** nódulo radical
noise, n. ruido (m); sonido (m) (sound); **background n.**, ruido de fondo; **level of n.**, nivel (m) de ruidos, nivel sonoro, nivel acústico; **n. and vibration,** ruidos y vibraciones; **n. control,** control (m) de ruido; **n. level,** nivel (m) de ruidos, nivel sonoro, nivel acústico; **n. nuisance,** ruido (m) irritante; **n. pollution,** agresión (f) acústica, ruido irritante; **n. reduction,** reducción (f) de ruido; **road n.**, ruido callejero.
noiseproofing, n. aislamiento (m) acústico, insonorización (f).
nomenclature, n. nomenclatura (f).
nomogram, n. nomograma (m).
non-aqueous, a. no acuoso/a, anhidro/a; **n.-aqueous phase,** fase (f) anhidro.
non-compliance, n. incumplimiento (m).
nonconformity, n. inconformidad (f).
non-ferrous, a. no ferroso/a.
non-linear, a. no lineal.
noon, n. mediodía (m).
non-productive, a. improductivo/a.
Norian, n. Noriense (m).
norite, n. norita (f).
norm, n. norma (f) (standard).
normal, a. normal, perpendicular (Mat).
normality, n. normalidad (f) (Chem).
normalization, n. normalización (f).
north, 1. n. norte (m); **grid n.**, norte de la malla; **magnetic n.**, norte magnético; **true n.**, norte verdadero; 2. a. norte, del norte; **N. America,** América (f) del Norte, Norteamérica (f).
North American, a. norteamericano/a, estadounidense.
northerly, a. norte, norteado/a; **in a n. direction,** en dirección (f) norte.
northern, a. norteño/a, del norte; **in the n. part,** en la parte del norte.

nose, n. nariz (f); **anticlinal n.**, punta (f) anticlinal, nariz del anticlinal.

nostril, n. ventanilla (f) de la nariz, ventana (f) de la nariz;; **nostrils,** narices (f.pl).

notary, n. notario (m); **public n.**, notario público.

notation, n. notación (f).

notch, n. muesca (f), corte (m) (dent).

notifiable, a. de declaración obligatoria; **n. waste,** desecho (m) de declaración obligatoria.

notifier, n. avisador (m) (person), notificador (m).

notify, v. notificar, avisar (advise).

nourish, v. nutrir, alimentar, sustentar.

nourishing, a. nutritivo/a.

nourishment, n. alimento (m), sustento (m).

NOX, n. NOX (abbr. for the nitrogen oxides).

noxious, a. nocivo/a.

noxiousness, n. nocividad (f).

NPK, n. NPK (abbr. for nitrogen, phosphorous and potassium); **NPK fertilizer,** abono (m) NPK.

nth, a. enésimo/a (Mat); **to raise to the power n (v),** elevar a la enésima potencia.

nubosity, n. nubosidad (f).

nuclear, a. nuclear; **n. energy,** energía (f) nuclear; **n. fission,** fisión (f) nuclear; **n. fusion,** fusión (f) nuclear; **n. waste,** residuos (m.pl) nucleares.

nucleation, n. nucleación (f).

nucleon, n. nucleón (m).

nucleotide, n. nucleótido (m).

nucleus, n. núcleo (m); **atomic n.,** núcleo atómico.

nuclide, n. nuclídico (m).

nuée ardent, n. nube (f) ardiente.

nuisance, n. molestia (f); perjuicio (m) (Jur).

nullify, n. anular, invalidar.

number, n. número (m); **atomic n.,** número atómico; **cardinal n.,** número cardinal; **complex n.,** número complejo; **ordinal n.,** número ordinal; **prime n.,** número primo; **reference n.,** número de referencia; **round n.,** número redondo.

numeration, n. numeración (f).

numerator, n. numerador (m).

numerical, a. numérico/a; **n. model,** modelo (m) numérico.

nunatak, n. nunatak (m).

nursery, n. semillero (m) (Agr).

nut, n. nuez (f) (fruit); tuerca (f) (Eng).

nutmeg, n. nuez (f) moscada.

nutrient, n. nutriente (m), nutrimiento (m); **available nutrients,** nutrientes disponibles.

nutrition, n. nutrición (f), alimentación (f).

nylon, n. nilón (m), nylon (m).

nymph, n. ninfa (f).

O

oak, n. roble (m); **evergreen o.,** encina (f); **holly o., holm-o.,** encina (f); **o. apple,** agalla (f) de roble; **o. grove,** robledal (m).

OAS, n. OEA; **Organization of American States,** Organización (f) de Estados Americanos.

oasis, n. oasis (m).

oats, n.pl. avena (f).

obesity, n. obesidad (f).

object, n. objetivo (m) (article); motivo (m), tema (m) (subject-matter); propósito (m) (aim).

objective, 1. n. objetivo (m), propósito (m); **primary o.,** objetivo principal, objetivo primordial; 2. a. objetivo/a.

oblate, a. oblato/a; **o. spheroid,** esferoide (m) oblato.

obligate, a. obligado/a; **o. parasite,** parásito (m) obligado.

obligation, n. obligación (f); **free of o.,** exento (m) de toda obligación.

obligatory, a. obligatorio/a, preceptivo/a.

oblige, v. obligar (to compel).

oblique, a. oblicuo/a.

obscene, a. obsceno/a, grosero/a.

observe, v. observar.

observed, a. observado/a.

obsidian, n. obsidiana (f).

obsolescence, n. obsolescencia (f), caída (f) en desuso.

obsolescent, a. obsolescente.

obstacle, n. obstáculo (m); traba (f) (Fig).

obstruct, v. obstaculizar.

obtain, v. obtener, conseguir; adquirir (acquire).

obtuse, a. obtuso/a.

occluded, a. ocluido/a; **o. front (Met),** frente (m) ocluido.

occlusion, n. oclusión (f) (Met).

occupancy, n. ocupación (f), posesión (f) (possession), tenencia (f) (tenancy).

occurrence, n. suceso (m), caso (m) (happening); **a common o.,** un caso frecuente.

ocean, n. océano (m); **Atlantic/Indian/Pacific O.,** Océano Atlántico/Índico/Pacífico.

ocean-going, a. transatlántico/a.

oceanic, a. oceánico/a; **o. plate (Geol),** placa (f) oceánica.

oceanographer, n. oceanógrafo/a (m, f).

oceanography, n. oceanografía (f).

oceanology, n. oceanología (f).

ocher, n. ocre (m).

ochre, n. ocre (m).

octahedrite, n. octaedrita (f), anatasa (f).

octahedron, n. octaedro (m).

octanal, n. octanal (m).

octane, n. octano (m); **o. rating,** octanaje (m).

octanol, n. octanol (m).

octene, n. octeno (m).

octocorals, n.pl. octocorales (m.pl).

octyne, n. octino (m).

OD, n. **Ordnance Datum (GB),** nivel (m) medio del mar usado en topografía.

odd, a. impar (number); unos, unas (fifty-odd pounds); suelto/a (isolated); **o. number,** número (m) impar; **o.-numbered day,** día (m) impar.

Odonata, n.pl. odonatos (m.pl), el orden Odonata (Zoo).

odor, n. olor (m).

odour, n. olor (m); **elimination of o.,** eliminación (f) de olor.

OECD, n. OCDE (f); **Organization for Economic Cooperation and Development,** Organización (f) para la Cooperación y el Desarrollo Económico.

oedema, n. edema (m).

oedometer, n. oedómetro (m).

OEEC, n. OECE (f); **Organization for European Economic Cooperation,** Organización (f) Europea para la Cooperación Económica.

oestrogen, n. estrógeno (m).

oestrus, n. estro (m).

offence, n. ofensa (f) (acción), delito (m) (Jur), escándalo (m) (scandal).

offer, n. oferta (f); **to make an o. (v),** hacer una oferta.

office, n. oficina (f); **head o.,** oficina principal, oficina central; **Meteorological O. (GB),** Instituto (m) Nacional de Meteorología (Sp); **o. block,** edificio (m) de oficinas; **o. hours,** horas (f.pl) de oficina; **site o.,** oficina al pie de la obra.

official, n. funcionario/a (m, f), empleado/a (m, f) público/a; personero (m) (LAm, representante gubernamental).

offlap, n. traslape (m) regresivo (Geol).

offset, n. renuevo (m), vástago (m) (Bot); retranqueo (m), retallo (m) (Con); estribación (f) (Geol).

offshore, a. cercano/a a la costa; **o. wind** un viento (m) de la costa.

offspring, n. progenie (f) (Zoo).

ogee, n. cimacio (m), gola (f) (Con); **o. arch,** arco (m) cimacio.

ohm, n. ohmio (m), ohm (m) (unit of electrical current).

ohm-metre, n. ohmio-metro (m).

oil, n. aceite (m); petróleo (m) (Min); **crude o.,** crudo (m), petróleo crudo; **diesel o.,** aceite diesel; **fuel o.,** aceite combustible; **heavy o.,** aceite pesado, combustóleo; **light o.,** aceite ligero; **linseed o.,** aceite de linaza; **mineral o.,** aceite mineral; **o. industry,** industria (f) del petróleo; **o. pollution,** contaminación (f) petrolífera; **o. reserves,** reservas (f.pl) de petróleo; **o. slick,** marea (f) negra; **o. tanker,** petrolero (m); **o. terminal,** terminal (f) petrolífera; **o. well,** pozo (m) de petróleo; **paraffin o.,** aceite de parafina; **vegetable o.,** aceite vegetal.

oilfield, n. yacimiento (m) petrolífero, campo (m) petrolífero.

oilshale, n. pizarra (f) bituminosa.

oilstone, n. piedra (f) de afilar (for sharpening tools).

oleander, n. adelfa (f).

olefine, n. olefina (f).

olfaction, n. olfacción (f).

Oligocene, n. Oligoceno (m).

Oligochaetae, n. oligoquetos (m.pl), la clase Oligochaetae (Zoo).

oligochaete, n. oligoqueto (m).

oligoclase, n. oligoclasa (f).

oligotrophic, a. oligotrófico/a.

olive, n. olivo (m) (tree); oliva (f), aceituna (f) (fruit).

olive-grove, n. olivar (m).

olivine, n. olivina (f).

olivinite, n. olivinita (f).

ombudsman, n. Defensor (m) del Pueblo (Jur).

omnivore, n. omnívoro (m).

omnivorous, a. omnívoro/a.

onboard, a. a bordo; **o. computer (Mar),** ordenador (m) de a bordo.

one-dimensional, a. monodimensional, unidimensional.

onion, n. cebolla (f).

on-line, a. on-line, en línea (Comp).

onshore, a. hacia la tierra; **o. wind,** un viento (m) en tierra.

on-site, a. en el sitio, in situ, a pie (m) de obra; **o.-site remediation,** remediación (f) en el sitio; **o.-site treatment,** tratamiento (m) en el sitio.

Ontarian, n. Ontariense (m).

onyx, n. ónice (m), onix (m).

oolite, n. oolito (m).

oolith, n. oolito (m).

oolitic, a. oolítico/a.

ooze, 1. v. rezumar; 2. n. fango (m); **calcareous o.,** fango calcáreo.

opacimeter, n. opacímetro (m).

opacity, n. opacidad (f).

opal, n. ópalo (m).

opaque, a. opaco/a.

open, 1. v. abrir; **to o. a trench,** abrir un surco; 2. a. abierto/a.

opencast, 1. n. explotación (f) a cielo abierto (Min); 2. a. a cielo abierto; **o. mining,** explotación (f) minera a cielo abierto.

opening, n. abertura (f).

operand, n. operando (m).

operative, n. operario/a (m, f).

operator, n. operario/a (m, f), maquinista (f).

opercula, n.pl. opérculos (m.pl).

operculum, n. opérculo (m).

ophiolite, n. ofiolita (f).

ophitic, a. ofítico/a.

opinion, n. opinión (f), parecer (m) (belief); dictamen (m) (legal opinion); **expert o.,** juicio (m) de peritos, peritaje; **public o.,** opinión pública; **public o. poll,** sondeo (m) de la opinión pública; **to be of the o. that (v),** ser de la opinión que; **to express an o. (v),** emitir un juicio.

opium, n. opio (m).

opportunist, a. oportunista.

opportunity, n. oportunidad (f).

opposition, n. oposición (f); **strong o.,** oposición fuerte.

optical, a. óptico/a; **o. disc (Comp),** disco (m) óptico.

optimal, a. óptimo/a.

optimization, n. optimización (f).

optimize, v. optimizar, mejorar (to improve); perfeccionar (to make perfect).

optimum, a. óptimo/a, optimal; **under o. conditions,** en las condiciones más favorables.

option, n. opción (f), escogimiento (m), alternativa (f); **economic o.,** opción económica.

optional, a. facultativo/a, optativo/a.

oral, a. oral.

orange, n. naranja (f) (fruit); naranjo (m) (tree); **methyl o. (Chem),** naranja de metilo.

orbit, n. órbita (f).

orbital, a. orbital.

orchard, n. huerto (m); **apple o.,** manzanar (m); **cherry o.,** cerezal (m); **pear o.,** peraleda (f).

orchid, n. orquídea (f).

order, 1. v. encargar, pedir, hacer un pedido (to place an order); mandar, ordenar (to command); poner en orden (to put in order, to sort); 2. n. orden (m) (serial place, class); orden (f) (command); pedido (m), encargo (m) (Com); grado (m) (Mat); mandato (m), fallo (m) (Jur); **alphabetical o.,** orden alfabético; **chronological o.,** por orden cronológico; **in o. (a),** en orden, ordenado (of a room); **o. of magnitude,** orden de magnitud; **o. of reaction,** orden de reacción; **second o. equation,** ecuación (f) de segundo grado; **stream o.,** orden fluvial.

ordinal, a. ordinal.

ordinance, n. ordenanza (f) (Jur).

ordinate, n. ordenada (f) (Mat).

Ordovician, n. Ordovícico (m).

ore, n. mineral (m), mena (f); **iron o.,** mineral de hierro; **o. body,** masa (f) de mineral, cuerpo (m) de mineral (LAm).

organ, n. órgano (m); **reproductive o.,** órgano reproductivo; **sense o.,** órgano de sentido.

organic, a. orgánico/a; **o. growth,** crecimiento

(m) orgánico; **o. material,** materia (f) orgánica.

organism, n. organismo (m); **living o.,** organismo vivo.

organization, n. organización (f); **nongovernment o. (US, abbr. NGO),** organización no gubernamental (abbr. ONG).

organize, v. organizar.

organochlorine, n. organocloro; **o. pesticides,** pesticidas (m.pl) organoclorados.

organoleptic, a. organoléptico/a.

orientate, v. orientar.

orientation, n. orientación (f), rumbo (m) (course, direction).

orifice, n. orificio (m); **o. plate (for measuring flow),** vertedero (m) de orificio.

origin, n. origen (m).

ornamentation, n. ornamentación (f), decoración (f).

ornithological, a. ornitológico/a.

ornithologist, n. ornitólogo/a (m, f).

ornithology, n. ornitología (f).

orogenesis, n. orogénesis (m).

orogenic, a. orogénico/a; **o. belt,** cinturón (m) orogénico.

orogeny, n. orogenia (f).

orographic, a. orográfico/a; **o. rain,** lluvia (f) orográfica.

orography, n. orografía (f).

ortho-, pf. orto-.

orthoclase, n. ortoclasa (f), ortosa (f).

orthoconical, a. ortocónico/a.

orthogonal, a. ortogonal.

orthophosphate, n. ortofosfato (m).

Orthoptera, n.pl. ortópteros (m.pl), el orden Orthoptera (Zoo).

orthopteran, n. ortóptero (m).

orthopyroxenite, n. ortopiroxenita (f).

oscillate, v. oscilar, fluctuar.

oscillation, n. oscilación (f), fluctuación (f); balanceo (m) (swinging); **climatic o.,** oscilación climática; **long-period o.,** oscilación de periódo largo; **sinusoidal o.,** oscilación sinusoidal.

osmium, n. osmio (m) (Os).

osmometer, n. osmómetro (m).

osmoregulation, n. osmoregulación (f).

osmosis, n. osmosis (f), ósmosis (f); **reverse o.,** osmosis inversa.

osmotic, a. osmótico/a; **o. potential,** potencial (m) osmótico; **o. pressure,** presión (f) osmótica.

osmotrophic, a. osmotrófico/a.

ossicle, n. osículo (m).

ossification, n. osificación (f).

ossify, v. osificar.

ossuaria, n. osario (m) (for bones).

ostracod, n. ostracodo (m).

ounce, n. onza (f) (unit of weight, 28.35 grams).

outcrop, 1. v. aflorar (Geol); 2. n. afloramiento (m) (Geol).

outcry, n. protesta (f); **public o.,** protesta popular, rechazo (m) popular.

outfall, n. desembocadura (f), desaguadero (m), emisor (m); **long sea o.,** desembocadura en el mar; **marine o.,** desembocadura marina; **sewage o.,** desembocadura de aguas residuales.

outflow, n. derrame (m), rebose (m); **flood-plain o.,** derrame de llanura aluvial.

outflowing, a. surgente.

outings, n.pl. excursionismo (m).

outlay, n. gastos (m.pl), desembolso (m); **initial o.,** pago (m) inicial.

outlet, n. salida (f) (for water), desagüe (m) (drain), desembocadura (f) (of river); tienda (f) (shop); orificio (m) (opening).

outlier, n. valor (m) atípico; relicto (m) exterior (Geol).

outline, n. esquema (m); contorno (m), silueta (f) (de forma).

outlined, a. dibujado/a, perfilado/a.

outpouring, n. derramamiento (m).

output, 1. v. imprimir (from computer); 2. n. producción (f) (of factory); productividad (f) (of person); rendimiento (m) (of machine); emisión (f) (emission); output (m), salida (f) (Comp); **o. device (Comp),** dispositivo (m) de salida; **pumped o.,** caudal (m) de bombeo (of a well).

outskirts, n.pl. cercanías (f), afueras (f), alrededores (m).

outwash, n. detrito (m) glaciar, cono (m) de transición.

outweigh, v. pesar más que; valer más que (to be more valuable).

ovary, n. ovario (m).

oven, n. horno (m), estufa (f).

overbreak, n. sobre excavación (Min).

overburden, n. cubierta (f), manto (m) (rock cover), sobrecarga (f) (excess weight).

overcast, a. encapotado/a, cubierto/a (Met).

overconsolidation, n. sobreconsolidación (f) (of clays).

overconsumption, n. sobreconsumo (m).

overcrop, v. esquilmar (Agr).

over-crowding, n. masificación (f) (e.g of beaches).

overdeveloped, a. superdesarrollado/a.

overdevelopment, n. superdesarrollo (m).

overdraft, n. giro (m) en descubierto (Fin).

overexploitation, n. sobreexplotación (f); **o. of aquifers,** sobreexplotación de acuíferos.

overexploited, a. sobreexplotado/a; **o. aquifer,** acuífero (m) sobreexplotado.

overfishing, n. sobrepesca (f).

overflow, 1. v. desbordar; 2. n. caudal (m), surgencia (f), surgente (m) caudal, desbordamiento (m) (from a river); **o. channel,** aliviadero (m); **stormwater o.,** aliviadero (m) de crecidas.

overflowing, 1. n. desbordamiento (m); rebasamiento (m) (from reservoir); 2. a. surgente, desbordante.

overgrazing, n. sobrepastoreo (m).

overgrowth, n. crecimiento (m) excesivo.

overhang, 1. v. sobresalir; 2. n. sobrecolgante (m).

overhaul, n. revisión (f), examen (m) detenido.

overheads, n.pl. gastos (m) generales.

overland, a. terrestre, por tierra, por vía terrestre.

overlap, 1. v. superponer, imbricar, traslapar; 2. n. traslape (m), solape (m), superposición (f); recubrimiento (m) (overlie, Geol).

overlapping, n. imbricación (f), superposición (f).

overlay, n. hoja (f) superpuesta (on a diagram).

overlie, v. suprayacer, superponer.

overload, 1. v. sobrecargar; 2. n. sobrecarga (f) (excess weight); **electrical o.**, sobrecarga eléctrica.

overlying, a. suprayacente.

overpopulated, a. superpoblado/a.

overpopulation, n. superpoblación (f).

overproduction, n. superproducción (f).

over-pumping, n. sobrebombeo (m).

overseas, a. exterior (trade); extranjero (foreign, abroad); **o. trade,** comercio (m) exterior.

oversight, n. descuido (m).

over-simplification, n. sobresimplificación (f).

oversizing, n. sobredimensionamiento (m); **o. of works,** sobredimensionamiento de obras.

overspend, v. gastar más de la cuenta.

overstock, v. abarrotar.

overthrust, n. cabalgamiento (m), sobrecorrimiento (m) (LAm).

overturn, n. vuelco (m) (of a car), zozobra (f) (of a ship), derrocamiento (m) (of government); la mezcla de las capas de aguas frías y calientes (in lakes, etc.).

overview, n. resumen (m) (report), visión (f) de conjunto.

overwatering, n. exceso (m) de riego, sobrerriego (m).

oviparous, a. ovíparo/a.

ovipositor, n. ovipositor (m).

ovule, n. óvulo (m).

owl, n. lechuza (f), búho (m); **Barn O.**, lechuza común; **Eagle O.**, búho real; **Little O.**, mochuelo (m); **Long-eared O.**, búho chico; **Pygmy O.**, mochuelo (m); **Scops' O.**, autillo (m); **Short-eared O.**, lechuza campestre; **Tawny O.**, cárabo (m).

owner, n. dueño/a (m, f); propietario/a (m, f) (of land or buildings); **joint o.**, copropietario/a (m, f).

oxalate, n. oxalato (m).

oxalic, a. oxálico/a.

ox-bow, n. meandro (m) abandonado, madre (f) vieja, tipischa (f) (Peru); **o. lake,** lago (m) de meandro abandonado, madre (f) vieja.

Oxfordian, n. Oxfordiense (m).

oxic, a. óxico/a.

oxidant, n. oxidante (m).

oxidation, n. oxidación (f); **o. ditch (Treat)**, zanja (f) para oxidación; **o.-reduction potential (abbr. redox potential)**, potencial (m) de oxidorreducción (abbr. potencial redox).

oxide, n. óxido (m).

oxidize, v. oxidar.

oxidizing, a. oxidante; **o. environment,** medio (m) oxidante.

oxy-, pf. oxi-.

oxyacetylene, a. oxiacetilénico/a; **o. welding,** soldadura (f) oxiacetilénica.

oxygen, n. oxígeno (m) (O); **o. sag,** hundimiento (m) de oxígeno.

oxygenate, v. oxigenar.

oxygenated, a. oxigenado/a.

oxygenation, n. oxigenación (f); **o. capacity,** capacidad (f) de oxigenación.

oxyhydroxide, n. oxihidróxido (m).

ozone, n. ozono (m); **o. hole,** agujero (m) de ozono; **o. layer,** capa (f) de ozono.

ozonide, n. ozónido (m).

ozonization, n. ozonización (f).

ozonize, v. ozonizar, ozonar.

ozonizer, n. ozonizador (m).

ozonosphere, n. ozonosfera (f).

P

Pacific, a. Pacífico/a; **the P. Ocean** el Océano Pacífico.

pack, v. empaquetar, rellenar (to fill).

package, n. paquete (m); **computer p.**, lote (m) de programa, paquete.

packaging, n. embalaje (m), envases (m.pl) y embalajes (m.pl).

packed, a. empaquetado/a.

packer, n. envasadora (f), empaquetador (m) (for wrapping); sellador (m), obturador (m) (Geol) (for boreholes); **p. test (in boreholes),** ensayo (m) de sellador.

packing, n. empaquetadura (f), empaque (m), empaquetado (m); envase (m) (action); **gravel p.**, empaque de gravas.

paddy, n. arrozal (m); **rice p.**, arrozal.

paediatrics, n. pediatría (f).

pagoda, n. pagoda (f).

PAHO, n. PAHO (f); **Pan American Health Organization**, Organización (f) Panamericana de la Salud.

pain, n. dolor (m).

paint, n. pintura (f); **anticorrosive p.**, pintura anticorrosiva; **spray p.**, pintura con pistola.

painting, n. pintura (f); cuadro (m) (picture); **cave p.**, pintura rupestre.

pair, n. par (m); pareja (f) (people, animals); **base p.**, par de bases; **breeding pairs**, parejas reproductoras; **stereo p.**, par estereoscópico.

palaeo-, pf. paleo-.

palaeobotanist, n. paleobotánico/a (m, f).

palaeobotany, n. paleobotanía (f).

Palaeolithic, n. Paleolítico (m); **Lower P.**, Paleolítico inferior; **Middle P.**, Paleolítico medio; **Upper P.**, Paleolítico superior.

palaeontology, n. paleontología (f).

Palaeozoic, n. Paleozoico (m).

paleo-, pf. paleo-.

paleobotany, n. paleobotánica (f).

Paleocene, n. Paleoceno (m).

paleoclimate, n. paleoclima (f).

paleoclimatology, n. paleoclimatología (f).

Paleogene, n. Paleógeno (m).

paleogeography, n. paleogeografía (f).

paleographer, n. paleógrafo/a (m, f).

paleography, n. paleografía (f).

paleolimnology, n. paleolimnología (f).

paleontologist, n. paleontológico/a (m, f).

paleontology, n. paleontología (f).

Paleozoic, n. Paleozoico (m).

palinogenesis, n. palinogénesis (m).

palladium, n. paladio (m) (Pd).

palliate, v. paliar, aliviar (alleviate).

palm, n. palmera (f).

palynology, n. palinología (f).

pampas, n. pampa (f) (LAm).

pan, n. cacerola (f), cazo (m) (metal container); cazuela (f) (earthenware); sartén (m) (frying pan); batea (f) (for gold panning); mortero (m) (for crushing); capa (f) (Geol); **iron p.** (Geol), capa ferruginosa.

Panama, n. Panamá (m).

Panamanian, 1. n. panameño/a (m, f); 2. a. panameño/a.

pan-american, a. panamericano/a.

pancreas, n. páncreas (m).

pandemic, 1. n. pandemia (f); 2. a. pandémico/a.

panel, n. panel (m), tablero (m) (of plywood); lienzo (m) (of a wall), artesón (m) (of a ceiling); grupo (m) (of experts); **p. of experts,** panel de expertos, grupo (m) de expertos; **solar p.**, panel solar.

panelling, n. revestimiento (m) (of walls), artesonado (m) (of a ceiling).

panhandle, n. enclave (m), entrada (f) angosta de un territorio en otro (Geog).

panning, n. bateado (m) (Geol); **gold p.**, bateado para oro.

panorama, n. panorama (m).

panoramic, a. panorámico/a; **p. view (viewing point)**, mirador (m).

pantile, n. teja (f) flamenca.

paper, n. papel (m) (material); **abrasive p.**, papel de lija; **double-logarithmic p.**, papel doble logarítmico; **filter p.**, papel de filtro; **graph p.**, papel cuadriculado; **litmus p.**, papel tornasol; **logarithmic-probability p.**, papel probalístico-logarítmico; **recycled p.**, papel reciclado; **semi-logarithmic p.**, papel semilogarítmico; **waste p.**, papeles viejos, papel usado.

papers, n.pl. papeles (m.pl) (writings); documentación (f) (official documents); periódicos (m.pl), prensa (f) (press).

papilla, n. papila (f).

papillae, n.pl. papilas (f.pl).

papyrus, n. papiro (m).

para-, pf. para-.

parabola, n. parábola (f) (Mat).

parabolic, a. parabólico/a.

paradox, n. paradoja (f).

paraffin, n. keroseno (m), petróleo lampante (m).

paraffin, n. parafina (f) (Chem).

paragenesis, n. paragénesis (m).

Paraguay, n. Paraguay (m).

Paraguayan, 1. n. paraguayo/a (m, f); 2. a. paraguayo/a.

paraldehyde, n. paraldehído (m).

parallax, n. paralaje (f).

parallel, n. y a. paralelo (m) (e.g. latitude),

paralelo/a; **p. processing**, proceso (m) en paralelo.
parallelogram, n. paralelogramo (m).
parameter, n. parámetro (m).
parametric, a. paramétrico/a; **non-p. techniques (Mat)**, técnicas (f.pl) no paramétricas.
parapet, n. pretil (m), antepecho (m) (of bridge, balcony, etc.), parapeto (m) (Mil, of trench), brocal (m) (curb).
parasite, n. parásito (m); **facultative p.**, parásito facultativo; **obligate p.**, parásito obligado.
parasitic, a. parásito/a, parasitario/a.
parasitical, a. parásito/a, parasitario/a.
parasitism, n. parasitismo (m).
parasitology, n. parasitología (f).
parathion, n. paratión (m).
parent, n. padre (m) (parent or father), madre (f) (mother); **parents**, padres (m.pl).
parental, a. parental; **p. care**, cuidado (m) parental.
parity, n. paridad (f).
park, n. parque (m); **amusement p.**, parque de atracciones; **leisure p.**, campo (m) recreativo; **national p.**, parque nacional; **p. and ride (v)**, aparcar y conducir; **zoological p. (abbr. zoo)**, parque zoológico.
parliament, n. parlamento (m); **European P.**, Parlamento Europeo; **Houses of P. (GB)**, Parlamento; **Member of P. (GB)**, miembro (m) del Parlamento, diputado (m).
parsnip, n. chirivía (f), pastinaca (f).
parthenogenesis, n. partenogénesis (f).
partial, a. parcial.
participation, n. participación (f).
particle, n. partícula (f); **airborne p.**, partícula viable en el aire; **alpha/beta/gamma p.**, partícula alfa/beta/gamma; **suspended p.**, partícula suspendida.
particular, a. particular, especial.
particularity, n. particularidad (f).
particulate, a. particulado/a; **p. material**, material (m) particulado.
partition, n. partición (f), división (f); tabique (m) (thin wall); **p. coefficient**, coeficiente (m) de partición.
partitioning, n. particionamiento (m), reparto (allocating).
partly, ad. parcialmente.
partnership, n. asociación (f); sociedad (f) (company); **limited p.**, sociedad limitada, sociedad comanditaria (LAm).
partridge, n. perdiz (f).
parturition, n. parto (m).
pass, n. puerto (m); paso (m) (esp. LAm); desfiladero (m), garganta (f); boquete (m) portillo (m) (narrow pass); portezuela (m).
passage, n. pasaje (m) (between buildings, underground); pasillo (m) (corridor); travesía (f) (voyage); paso (m), tránsito (m) (travel through).
paste, n. pasta (f).

pasteurization, n. pasterización (f), pasteurización (f).
pasteurize, v. pasterizar, pasteurizar.
pasteurized, a. pasterizado/a, pasteurizado/a.
pastoral, a. pastoral; **p. farming**, pastoreo (m) (shepherding).
pasture, n. pasto (m), grama (f) (Arg, Ur); **fertilized p.**, pasto fertilizado; **groundwater-fed p.**, baén (m), frescal (m).
patch test, n. prueba (f) de emplasto (Med).
patella, n. patela (f), rótula (f).
patellae, n.pl. patelas (f.pl).
path, n. camino (m), sendero (m), senda (f); **critical p.**, camino crítico.
pathogen, n. patógeno (m).
pathogenic, a. patógeno/a; **p. agent**, agente (m) patógeno.
pathology, n. patología (f).
pathway, n. vía (f) de acceso (for pollutants); camino (m) (road), senda (f) (path); **exposure p. (for contaminants)**, vía de acceso de exposición.
pattern, n. diseño (m), modelo (m); **drainage p.**, diseño de avenamiento.
PAU, n. UPA (f); **Panamerican Union**, Unión (f) Panamericana.
pavement, n. pavimento (m) (paving); acera (f), vereda (f) (LAm), andén (m) (CAm, Col), banqueta (f) (Mex) (roadside walkway); **brick p.**, enladrillado (m); **desert p.**, pavimiento del desierto; **flagstone p.**, enlosado (m); **limestone p.**, pavimento de caliza.
paw, n. pata (f); zarpa (f) (with claws).
pay, v. pagar; costear (to pay for).
PC, n. PC; **personal computer**, ordenador (m) personal.
PCB, n. PCB (m); **polychlorbiphenyl**, policlorobifenilo.
pea, n. guisante (m), chícharo (m) (LAm), arveja (f) (LAm).
peach, n. melocotón (m); durazno (m) (LAm); **p. tree**, melocotón (m), melocotonero (m), duraznero (m) (LAm).
peak, n. punta (f), cumbre (m) (summit); máximo (m) (maximum); cúspide (m) (apex); **flood p.**, punta de avenida.
peanut, n. cacahuete (m); cacahuate (m) (LAm), maní (m) (LAm).
pear, n. pera (f) (fruit), peral (m) (tree); **p. orchard**, peraleda (f); **prickly p.**, chumba (f), higo (m) chumbo (fruit); chumbera (f) (plant).
pearl, n. perla (f); **mother-of-p.**, madreperla (f), nácar (m); **p. oyster**, madreperla (f), ostra (f) perlífera.
peasant, n. campesino/a (m, f), paisano/a (m, f) (LAm), huaso (m) (Bol, Ch), ranchero/a (m, f) (Mex), chagra (m, f) (Ec).
peat, n. turba (f); **p. bed**, capa (f) de turba; **p. bog**, pantano (m) de turba, turbera (f); **p. formation**, formación (f) de turba, turbalización (f) (LAm).

pebble, n. guijarro (m), canto (m) rodado, guija (f); **flat p.,** guijarro aplanado; **river p.,** canto rodado de río.

pectin, n. pectina (f).

pectoral, a. pectoral.

pedestal, n. pedestal (m).

pedestrian, 1. n. peatón/peatona (m, f); **p. crossing,** paso (m) de peatones; 2. a. peatonal; **p. area,** zona (f) peatonal.

pedestrianization, n. peatonalización (f).

pediatrics, n. pediatría (f).

pediment, n. pedimento (m).

pediplain, n. llanura (f) de pedimentos.

pediplane, n. planos (m) de pedimentos.

pedology, n. pedología (f), edafología (f).

pedometer, n. pedómetro (m).

pedostratigraphic, a. pedostratigráfico/a.

peduncular, a. peduncular.

peeling, n. descamación (f) (Geol).

pegmatite, n. pegmatita (f).

pelagic, a. pelágico/a.

pelite, n. pelita (f).

pelitic, a. pelítico/a.

pellet, n. bolita (f), pelotilla (f), pellet (m) (aglomerado); **faecal p.,** bolita fecal.

pelletization, n. pelletización (f); **p. plant,** planta (f) de pelletización.

pellicle, n. película (f).

pelvic, a. pélvico/a, pelviano/a.

pelvis, n. pelvis (f).

penalty, n. pena (f), castigo (m) (punishment); multa (f) (fine).

pendulum, n. péndulo (m).

penecontemporaneous, a. penecontemporáneo/a.

peneplain, n. peneplanicie (f), penillanura (f); **dissected p.,** peneplanicie disectada.

peneplanation, n. peneplanización (f).

penetrability, n. penetrabilidad (f).

penetrable, a. penetrable.

penetration, n. penetración (f); **marine p.,** penetración marina; **saline p.,** penetración salina; **well p.,** penetración del pozo.

penetrometer, n. penetrómetro (m).

penicillin, n. penicilina (f).

peninsula, n. península (f).

penis, n. pene (m).

penstock, n. compuerta (f) (with sluice gate); caz (m) (millrace).

penta-, pf. penta-.

pentachlorobiphenol, n. pentaclorobifenol (m).

pentachlorophenol, n. pentaclorofenol (m).

pentane, n. pentano (m).

pentanol, n. pentanol (m).

pentavalent, a. pentavalente.

pentene, n. penteno (m).

penthouse, n. ático (m), cámara (f) (apartment); cobertizo (m) (shed).

pentoxide, n. pentóxido (m).

pentyne, n. pentino (m).

pepper, n. pimiento (m) (vegetable); pimienta (f) (spice).

pepsin, n. pepsina (f).

peptide, n. péptido (m).

peptone, n. peptona (f).

per-, pf. per-.

per capita, a. per cápita.

per cent, n. por ciento (m); **a twenty p. c. increase,** un aumento (m) del veinte por ciento.

percentage, 1. n. porcentaje (m); **p. dry weight,** porcentaje de peso seco; 2. a. porcentual.

percentile, n. percentil (m).

perceptible, a. perceptible, sensible; **p. difference,** diferencia (f) sensible.

perch, n. percha (f), espolón (m) (for birds); perca (f) (fish).

perched, a. colgado/a; **p. groundwater,** freático (m) colgado.

perchlorate, n. perclorato (m).

percolate, 1. v. percolar; 2. n. percolado (m), filtrado (m).

percolation, n. percolación (f); **p. gauge,** percolímetro (m).

percussion, n. percusión (f); **p. drilling,** perforación (f) por percusión, sondeo (m) por percusión; **rotary p.,** rotopercusión (f).

perennial, 1. n. planta (f) perenne (Bot) 2. a. perenne; **p. spring,** manantial (m) perenne.

perfect, v. perfeccionar.

performance, n. realización (f), ejecución (f) (execution), verificación (f) (confirmation).

perfumery, n. perfumería (f).

pergola, n. pérgola (f).

perianth, n. periantio (m).

pericline, n. periclinal (m), domo (m).

peridotite, n. peridotita (f).

perigee, n. perigeo (m).

periglacial, a. periglaciar, periglacial.

perimeter, n. perímetro (m); **wetted p.,** perímetro mojado.

period, n. período (m), plazo (m); período (Med, menstruation); **gestation p.,** período de gestación; **glacial p.,** período glaciar, período glacial; **half-life p.,** período de semidesintegración; **latent p.,** período de latencia; **menstrual p.,** período menstrual; **pluvial p.,** período de lluvias, período pluvial; **recovery p.,** período de recuperación; **refractory p. (Zoo),** período refractario; **return p.,** período de retorno.

periodic, a. periódico/a.

periodicity, n. periodicidad (f).

peripheral, a. periférico/a; **p. drainage,** drenaje (m) periférico.

peristalsis, n. peristaltismo (m).

peristaltic, a. peristáltico/a; **p. pump,** bomba (f) peristáltica.

perlite, n. perlita (f).

permafrost, n. permafrost (m), pergelisol (m) (LAm).

permanganate, n. permanganato (m); **p. index,** índice (m) de permanganato; **potassium p.,** permanganato potásico.

permeability, n. permeabilidad (f); **Darcy p.**, permeabilidad de Darcy; **horizontal p.**, permeabilidad horizontal; **intrinsic p.**, permeabilidad intrínseca; **primary p.**, permeabilidad primaria; **secondary p.**, permeabilidad secundaria; **unsaturated p.**, permeabilidad no saturada.

permeable, a. permeable; **p. formation**, formación (f) permeable.

permeameter, n. permeómetro (m); **differential p.**, permeómetro diferencial; **falling-head p.**, permeómetro de carga variable; **fixed-head p.**, permeómetro de carga fija.

Permian, n. Pérmico (m).

permissible, a. permisible, lícito/a.

permission, n. permiso (m), autorización (f).

permit, n. permiso (m), licencia (f).

permutation, n. permutación (f) (Mat).

peroxidation, n. peroxidación (f).

peroxide, n. peróxido (m); **hydrogen p.**, peróxido de hidrógeno, agua (f) oxigenada.

perpendicular, a. perpendicular.

persistence, n. persistencia (f).

persistent, a. persistente.

personnel, n. personal (m).

perspective, n. perspectiva (f).

perspex, n. perspex (m).

perspiration, n. sudor (m), transpiración (f).

persuade, v. persuadir, convencer.

perthite, n. pertita (f).

perthosite, n. perthosita (f).

perturbation, n. perturbación (f), disturbio (m).

Peru, n. Perú (m).

Peruvian, n. peruano/a (m, f).

pervious, a. permeable.

peso, n. peso (m) (unit of currency).

pest, n. plaga (f); organismo (m) nocivo; **p. control**, control (m) de plagas.

pesticide, n. pesticida (m), plaguicida (m); **organochlorine p.**, pesticida organoclorado; **p. residues**, residuos (m.pl) de pesticidas; **p. spraying**, pulverización (f) con pesticidas.

pestilence, n. pestilencia (f), peste (f).

petal, n. pétalo (m).

petering out, n. adelgazamiento (m), acuñamiento (m) (Geol).

petiole, n. peciolo (m), pecíolo (m).

petition, n. petición (f).

petrification, n. petrificación (f).

petrochemical, a. petroquímico/a; **p. industry**, industria (f) petroquímica.

petrogenesis, n. petrogénesis (m).

petrographic, a. petrográfico/a.

petrography, n. petrografía (f).

petrol, n. gasolina (f), nafta (f) (CSur), bencina (f) (Ch); **p. station**, gasolinera (f), estación (f) de servicio, grifo (m) (Peru), bencinera (f) (Ch); **p. tanker**, gasolinero (m).

petroleum, n. petróleo (m) (Min); **p. reserves**, reservas (f.pl) de petróleo.

petroliferous, a. petrolífero/a.

PFA, n. **pulverized fuel ash**, ceniza (f) de combustible pulverizado.

pH, n. pH (hydrogen ion concentration); **field pH**, pH de campo; **pH meter**, pHímetro (m).

Phaeophyta, n.pl. feofíceas (f.pl), feofitos (m.pl) algas (f.pl) pardas.

phage, n. fago (m), bacteriófago (m).

phanerite, n. fanerita (f).

pharmaceutical, a. farmacéutico/a; **p. industry**, industria (f) farmacéutica.

pharmacology, n. farmacología (f).

pharmacy, n. farmacia (f).

phase, 1. v. poner en fase; escalonar (to plan in stages); **to p. out**, desfasar; 2. n. fase (f); **advanced p.**, fase avanzada; **immiscible p.**, fase inmiscible; **in p. (a)**, estar en fase; **out of p. (a)**, desfasado/a; **p. difference**, desfasaje (m), desfase (m); **p. shift**, desfase (m), desfasaje (m); **three-p. current**, corriente (f) trifásica.

phenocryst, n. fenocristal (m).

phenol, n. fenol (m).

phenolphthalein, n. fenolftaleína (f).

phenomena, n.pl. fenómenos (m.pl).

phenomenon, n. fenómeno (m).

phenotype, n. fenotipo (m).

phenyl, n. fenilo (m).

phenylamine, n. fenilamina (f).

phenylene, n. fenileno (m).

pheromone, n. feromona (f).

phi, a. phi; **p.-scale**, escala (f) de phi.

philosophy, n. filosofía (f).

phloem, n. floema (m).

Phoenician, n. fenicio/a (m, f).

phosgene, n. fosgeno (n).

phosphate, n. fosfato (m); **p. fertilizer**, fosfato abono; **p. removal**, eliminación (f) de fosfato.

phosphatic, a. fosfático/a.

phosphatize, v. fosfatar.

phosphine, n. fosfina (f).

phosphite, n. fosfito (m).

phospholipid, n. fosfolípido (m).

phosphoresce, v. fosforescer.

phosphorescence, n. fosforescencia (f).

phosphorite, n. fosforita (f).

phosphorous, n. fósforo (m) (P).

photic, a. fótico/a; **p. zone (Geog)**, zona (f) fótica.

photo-, pf. foto-.

photo-index, n. fotoíndice (m).

photochemical, a. fotoquímico/a.

photocolorimeter, n. fotocolorímetro (m).

photoelectric, a. fotoeléctrico/a; **p. cell**, célula (f) fotovoltaica, célula fotoeléctrica.

photogeology, n. fotogeología (f).

photogrammetry, n. fotogrametría (f).

photograph, n. fotografía (f), foto (f).

photographic, a. fotográfico/a.

photography, n. fotografía (f); **aerial p.**, fotografía aerea, aerofotogrametría (f) (LAm).

photoheterotroph, n. fotoheterótrofo (m).

photoheterotrophe, n. fotoheterótrofo (m).
photointerpretation, n. fotointerpretación (f).
photoluminescence, n. fotoluminescencia (f).
photolysis, n. fotólisis (m).
photometry, n. fotometría (f).
photomontage, n. fotomontaje (m).
photomosaic, n. fotomosaico (m).
photomultiplier, n. fotomultiplicador (m).
photon, n. fotón (m).
photoperiodism, n. fotoperiodicidad (f), foto-periodismo (m).
photosynthesis, n. fotosíntesis (f).
photosynthesize, v. fotosintetizar.
photosynthetic, a. fotosintético/a.
phototaxis, n. fototactismo (m), fototaxia (f).
phototropism, n. fototropismo (m).
photovoltaic, a. fotovoltaico/a; **p. cell,** célula (f) fotovoltaica.
phragmocone, n. fragmocono (m).
phreatic, a. freático/a.
phreatophyte, n. freatofita (f).
phtalein, n. ftaleína (f).
phycomycete, n. ficomiceto (m); **Phycomycetes (Zoo, Bot, class),** ficomicetos (pl).
phyllite, n. filita (f).
phyllonite, n. filonita (f).
phylloxera, n. filoxera (f).
physical, a. físico/a; **p. model,** modelo (m) físico.
physicist, n. físico/a (m, f).
physico-chemical, a. físico-químico/a.
physics, n. física (f).
physiographic, a. fisiográfico/a.
physiology, n. fisiología (f).
phyto-, pf. fito-.
phytogenetics, n. fitogenética (f).
phytogeography, n. fitogeografía (f).
phytometer, n. fitómetro (m).
phytophagous, a. fitófago/a.
phytoplankton, n. fitoplancton (m); **p. bloom,** emergencia (f) de fitoplancton, floración (f).
phytosanitary, a. fitosanitario/a.
phytosociology, n. fitosociología (f).
phytotoxic, a. fitotóxico/a.
phytotoxicity, n. fitotoxicidad (f).
pi, n. pi (f) (Mat).
Piacenzian, n. Piacenciense (m).
pick, n. pico (m), piqueta (f) (tool); selección (f), elección (f) (choice).
pickaxe, n. piqueta (f), zapapico (m).
pickle, v. decapar (to descale metal).
pickling, n. tratamiento (m) desoxidante, piclaje (m), decapado (m); **p. plant,** instalación (f) para decapar; **p. tank,** cuba (f) de decapado.
pickup, n. fonocaptor (m) (of sound).
pico-, pf. pico- (10.E-12).
picrate, n. picrato (m).
picrite, n. picrita (f).
piecework, n. trabajo (m) a destajo.
piedmont, n. piedemonte (m), pedemontano (m).

pier, n. malecón (m), embarcadero (m) (landing stage), estribo (m), pila (f) (of bridge).
pierce, v. horadar, perforar (to drill, to perforate), sondear (to drill borehole).
piezometer, n. piezómetro (m); **coastal p.,** piezómetro costero, piezómetro litoral.
piezometric, a. piezométrico/a; **p. surface,** nivel (m) piezométrico.
piezometry, n. piezometría (f).
pig, n. cerdo (m), puerco (m), chancho (m) (LAm); guarro (m) (LAm); **p. iron,** arrabio (m), hierro (m) colado.
piggery, n. pocilga (f) (pigsty).
piglet, n. cerdito (m), conchinillo (m), lechon (m).
pigment, n. pigmento (m).
pigsty, n. porqueriza (f).
pilchard, n. sardina (f).
pile, n. pilote (m) (support), estaca (f) (stake); pila (f) (Phys); pila (f), montón (m) (e.g. of sand); **atomic p.,** pila atómica; **p. driver,** martinete (m); **sheet p.,** tablestaca (f).
piledriver, n. martillo (m) pilón.
piling, n. pilotaje (m); **sheet p.,** tablestacado (m).
pill, n. píldora (f); **the p. (Med, contraceptive),** la píldora.
pillar, n. pillar (m) (Min), columna (f); **p. and stall (Min),** cámaras (f.pl) y pilares.
pilot, a. piloto/a, modelo/a; **p. plant,** planta (f) pilota; instalación (f) de ensayo; **p. scheme,** proyecto (m) piloto.
pinacoid, n. pinacoide.
pincers, n.pl. pinzas (f.pl) (Zoo, e.g. of lobsters); tenaza (f), tenazas (f.pl) (Eng).
pinching out, n. acuñamiento (m), adelgazamiento (m) (Geol).
pine, n. pino (m).
pineapple, n. piña (f); piña de América; ananás (m) (LAm).
pinnacle, n. pináculo (m), cúspide (m), apogeo (m), cima (f) (peak).
pinnate, a. pinado/a, pinnado/a.
pint, n. pinta (f) (unit of volume equal to 0.568 litres in GB and 0.473 litres in US).
pioneer, 1. n. pionera (f); **p. community (Ecol),** comunidad (f) pionera; 2. a. pionero/a.
pip, n. pepita (f) (of fruit).
pipe, 1. v. entubar, encañar, conducir por tubería (water, gas, etc.); 2. n. tubo (m), tubería (f), conducto (m), cañería (f), cañón (m); **asbestos cement p.,** tubería de fibro-cemento; **bamboo pipes,** caña (f) de bambú, tubería de bambú; **copper p.,** tubería de cobre; **discharge p.,** cañón de vertido; **drain p.,** tubo de desagüe; **drainage p.,** tubo de drenaje; **ductile iron p.,** tubería de acero estirado; **exhaust p.,** tubo de escape; **galvanized p.,** tubo galvanizado; **gas p.,** tubo de gas; **perforated p.,** tubería perforada; **p. duct,** conducto para tuberías; **p. fittings,**

accesorios (m.pl) para tubos; **p. flange,** brida (f) de tubo; **p. laying,** tendido (m) de redes; **p. networks,** redes (f.pl) de canalización; **p. relining,** revestimiento (m) de tubos; **p. wrapping,** envuelta (f) de tubos; **PVC p.,** tubería de PVC; **reinforced concrete p.,** tubería de hormigón armado; **rising main p.,** tubería de impulsión; **service p.,** tubería de servicio; **steel p.,** tubería de acero; **vent p.,** purga (f) de aire; **waste p.,** tubo de desagüe, desagüe (m); **water p.,** tubo de agua; **water supply p.,** tubería de conducción de agua.

pipe-laying, n. tendido (m) de tuberías.

pipeline, n. tubería (f) (Gen); **gas p.,** gaseoducto (m); **oil p.,** oleoducto (m); **water p.,** cañería (f).

pipette, n. pipeta (f).

piping, n. cañerías (f.pl), entubación (f), tubería (f); conducción (f) (of liquids); sofosión (f) (action of upward water pressure); **telescopic p.,** entubación telescópica; **temporary p.,** tubería provisional.

Pisces, n.pl. peces (m.pl), el grupo Pisces (Zoo, group).

piscivore, n. piscívoro/a (m, f).

pisolite, n. pisolita (f).

pisolitic, a. pisolítico/a.

piston, n. pistón (m).

pit, n. hoyo (m), hoya (f), hoyuelo (m) (hole in ground); mina (f) (mine); cantera (f) (quarry); trampa (f) (trap); **borrow p. (Eng),** zona (f) de préstamo; **coal p.,** mina (f) de carbón.

pitch, n. alquitrán (m), brea (f), chapopote (m) (Mex) (substance); lanzamiento (m) (throw); cabeceo (m), buzamiento (m) (Geol); pendiente (m) (of roof); **p. of roof,** pendiente del tejado.

pitchblende, n. pechblenda (f), uraninita (f).

pith, n. médula (f) (medulla); médula, esencia (f) (Fig, essential theme).

pithead, n. bocamina (f).

pitot, n. pitot (m); **p. tube,** tubo (m) pitot; tubo de pitot.

pitted, a. picado de viruelas (pockmarked), deshuesado/a (US, fruit); alveolado/a (honeycombed), hoyuelo/a (dimpled).

pivot, n. pivote (m).

place, 1. v. ubicar (to emplace, to put in position); 2. n. lugar (m), sitio (m) (spot, position); puesto (m), empleo (m) (post); plaza (f) (square); casa (f) (house), edificio (m) (building); **p.-name,** topónimo (m).

placebo, n. placebo (f).

placenta, n. placenta (f).

placentae, n.pl. placentas (f.pl).

placer, n. placer (m) (Min); **p. deposit,** depósito (m) de placer, placer aluvial.

plagioclase, n. plagioclasa (f).

plagioclimax, n. plagioclímax (m).

plagionite, n. plagionita (f).

plague, n. peste (f), plaga (f); **bubonic p.,** peste bubónica; **pneumonic p.,** peste neumónica.

plain, n. llanura (f), llanada (f) (Geog); planicie (f) (flat ground); **coastal p.,** llanura costera, llanura litoral; **flood p.,** llanura de inundación, planicie de inundación, bajial (m) (LAm); **treeless p.,** raso (m).

plan, 1. v. hacer el plan de (to design); planificar (production), planear (e.g. a journey); 2. n. plano (m) (design, map); plan (m) (scheme); planta (f) (floor plan); **according to p.,** como estar previsto; **business p.,** plan comercial; **development p.,** plan de desarrollo; **five-year p.,** plan quinquenal; **master p.,** plan maestro; **protection p.,** plan de protección; **three-year p.,** plan trienial; **town p.,** plan urbanístico.

Planaria, n.pl. planarios (m.pl), la clase Planaria (Zoo).

planarian, n. planario (m).

planation, n. rebajamiento (m) (Geog).

plane, 1. v. cepillo (m) (tool); 2. n. plano (m); **axial p.,** plano axial; **bedding p.,** plano de estratificación; **fault p.,** plano de falla; **joint p.,** plano de diaclasa; **pediplane,** plano de pedimentos.

planimeter, n. planímetro (m).

plank, 1. v. entablar, entarimar (to cover with boards); 2. n. tablón (m), tabla (f) (board); punto (m) (Fig. of policy).

plankton, n. plancton (m).

planner, n. planificador/a (m, f); **city p. (US),** **town p.,** urbanista (m, f).

planning, n. planificación (f), planeamiento (m), ordenamiento (m), ordenación (f); **environmental p.,** ordenamiento (m) ambiental; **family p.,** planificación familiar; **rural p.,** planificación rural; **top-down p.,** planificación de arriba abajo; **water resources p.,** ordenamiento (m) de los recursos hídricos.

plant, 1. n. planta (f) (vegetation); taller (m), fábrica (f) (factory); equipo (m), maquinaria (f) (machinery); estación (f) (works); **annual p.,** planta anual; **biennial p.,** planta bienal; **demonstration p.,** planta de demostración; **desalination p.,** planta de desalinización, planta desalinizadora, planta desaladora (LAm); **hydroelectric p.,** planta hidroeléctrica; **incineration p.,** planta incineradora, instalación (f) de combustión; **industrial p.,** planta industrial; **oxidation p. (Treat),** planta de oxidación; **perennial p.,** planta perenne; **pilot p.,** planta de pruebas, planta piloto; **p. life,** vida (f) vegetal; **processing p.,** planta de tratamiento; **treatment p.,** planta de tratamiento, estación (f) depuradora; **vascular plants (Bot),** plantas vasculares; **water treatment p.,** planta potabilizadora, estación (f) de tratamiento de agua potable; 2. a. vegetal.

plantation, n. plantación (f).

plaque, n. placa (f).

plasma, n. plasma (m).
plasmolysis, n. plasmólisis (f).
plaster, n. yeso (m) (for walls, ceilings, etc.), mortero (m) (mortar); emplasto (m) (Med).
plasterboard, n. cartón-yeso (m).
plastered, a. enlucido/a.
plasterwork, n. enlucido (m) (coat of plaster).
plastic, 1. n. plástico (m); 2. a. plástico/a.
plasticity, n. plasticidad (f).
plate, n. plato (m), lámina (f), placa (f), plancha (f); **continental p.**, placa continental; **oceanic p.**, placa oceánica; **p. boundary,** borde (m) de placa; **p. count,** recuenta (f) en placa; **p. tectonics,** tectónica (f) de placas.
plateau, n. meseta (f), altiplano (m) (LAm); **high p.**, altiplanicie (m).
platelet, n. plaquita (f).
platform, n. plataforma (f); **drilling p.**, plataforma de perforación.
plating, n. recubrimiento (m) electrolítico; **zinc p.**, zincado (m) electrolítico.
platinoid, n. platinoide (m).
platinum, n. platino (m) (Pt).
platyhelminth, n. platelminto (m).
Platyhelminthes, n.pl. platelmintos (m.pl), el filum Platyhelminthes (Zoo).
plc, n. s.a. (f); **public limited company (GB),** sociedad (f) anónima.
Plecoptera, n.pl. plecópteros (m.pl), la clase Plecoptera (Zoo).
plecopteran, n. plecóptero (m).
Pleistocene, n. Pleistoceno (m).
plexiglass, n. plexiglás (m).
Pliensbachian, n. Pliensbaquiense (m).
plinth, n. zócalo (m) (of a wall), plinto (m) (of a column).
Pliocene, n. Plioceno (m).
plot, n. parcela (f), terreno (m), haza (f) (of land); solar (m) (building site); gráfico (m) (Mat); **logarithmic p.**, diagrama (m) logarítmico, gráfico (m) logarítmico.
plotter, n. trazador (m) (de gráficos).
plough, 1. v. arar, roturar; 2. n. arado (m), avenador (m), zanjador (m), plantador (m).
ploughing, n. arado (m); **contour p.**, arado por curvas de nivel, surcado (m) en contorno.
plow, 1. v. arar, roturar; 2. n. arado (m), avenador (m), zanjador (m), plantador (m).
plowing, n. arado (m); **contour p.**, arado por curvas de nivel, surcado (m) en contorno.
plug, n. tapón (m); **bentonite p. in a piezometer,** tapón de bentonita en un piezométro.
plugging, n. taponamiento (m).
plum, n. ciruela (f) (fruit), ciruelo (m) (tree).
plumbing, n. fontanería (f); gasfitería (f), plomería (f) (LAm).
plumbocalcite, n. plumbocalcita (f).
plume, n. pluma (f) (of feather, of smoke); penacho (m) (crest of bird, of smoke); **contamination p.**, pluma de contaminación; **leachate p.**, pluma de lixiviado; **p. of smoke,** penacho de humo; **sewage p.**,

pluma de desagües; **thermal p.**, pluma térmica.
plunge, 1. v. sumergir, hundir (to immerse); caer, inclinar (Geol); 2. n. buzamiento (m), inclinación (f) (Geol).
plunging, a. inclinado/a, buzando/a (Geol).
plus, n. signo (m) más (Mat); **2 plus 2 is 4,** 2 más 2 son 4.
Pluto, n. Plutón (m).
pluton, n. plutón (m).
plutonic, a. plutónico/a.
plutonium, n. plutonio (m) (Pu).
pluvial, a. pluvial.
pluviograph, n. pluviógrafo (m).
pluviometry, n. pluviometría (f).
pluviosity, n. pluviosidad (f).
plywood, n. madera (f) contrachapada.
pneumatic, a. neumático/a; **p. tyre,** neumático (m).
pneumatolysis, n. neumatolisis (m).
pneumoconiosis, n. neumoconiosis (m).
pneumonia, n. neumonía (f), pulmonía (f).
PO, n. **Post Office,** correos (m.pl), oficina (f) de correos.
poaching, n. caza (f) furtiva (hunting).
pocket, n. bolsillo (m), bolsa (f); **p. of mineral (Min),** bolsada (f).
pod, n. vaina (f) (Bot); manada (f) (Zoo); ranura (f) (Eng, groove).
podsol, n. podsol (m), suelo (m) podsol.
poikilotherm, n. poiquilotermo (m).
poikilothermic, a. poiquilotermo/a.
point, n. punto (m), punta (f); **control p.**, punto de comprobación, punto de control; **dew p.**, punto de rocío, temperatura (f) de saturación; **flash p.**, temperatura (f) de inflamabilidad, punto de inflamación; **focal p.**, punto focal; **melting p.**, punto de fusión; **nodal p.**, nodo (m); **observation p.**, punto de observación; **permanent wilting p.**, punto de marchitez permanente; **p. of inflexion,** punto de inflexión; **stagnation p. (Mat),** punto de estancamiento.
pointing, n. rejuntado (m) (of brickwork).
poise, n. poise (m) (unit of viscosity).
poison, 1. v. envenenar; 2. n. veneno (m), tóxico (m).
poisoning, n. envenenamiento (m); **blood-p.**, envenenamiento de la sangre.
poisonous, a. venenoso/a (e.g. of a snake); tóxico/a (e.g. of a chemical), intoxicado/a (e.g. of gas).
polar, a. polar; **p. cap,** casquete (m) polar; **p. circle,** círculo (m) polar; **p. lights,** aurora (f) boreal; **p. wandering (Geol),** deriva (f) de los polos.
polarimetry, n. polarimetría (f).
polarization, n. polarización (f).
polarizing, a. polarizador/a; **p. microscope,** microscopio (m) polarizador.
polder, n. polder (m).
pole, n. palo (m), poste (m) (stick); polo (m)

(extreme, focus); **like poles,** polos iguales; **North/South P.,** Polo Norte/Sur; **opposite poles,** polos contrarios; **telegraph p.,** poste telegráfico (m).

policy, n. política (f); **agricultural p.,** política agraria.

polimerization, n. polimerización (f).

polio, n. polio (f), poliomielitis (f).

poliomyelitis, n. poliomielitis (f).

polish, 1. v. pulir (Gen); lustrar, pulimentar; 2. n. pulimento (m), lustre (m).

politics, n. política (f); **green p.,** política verde, política ecologista.

pollen, n. polen (m); **p. analysis,** palinología (f); **p. count,** recuento (m) polínico; **p. grain,** grano (m) de polen.

pollinate, v. polinizar.

pollination, n. polinización (f); **cross-p.,** polinización cruzada.

pollutant, n. contaminador (m), contaminante (m); **aquatic p.,** contaminante acuático.

pollute, a. contaminar.

polluted, a. contaminado/a, poluto/a.

polluter, n. contaminador (m); **"p. pays" principle,** principio (m) de "quien contamina paga".

polluting, a. contaminante.

pollution, n. contaminación (f), polución (f); **diffuse p.,** contaminación difusa (of watercourses); **point source p.,** contaminación directa; **p. control,** control (m) de la contaminación; **p. load,** carga (f) de contaminación; **river p.,** contaminación de los ríos.

polonium, n. polonio (m) (Po).

poly-, pf. poli-.

polyamide, n. poliamida (f).

Polychaetae, n. poliquetos (m.pl), la clase Polychaetae (Zoo).

polychaete, n. poliqueto (m).

polychlorinated, a. policlorado/a.

polycyclic, a. policíclico/a; **p. aromatic hydrocarbon (abbr. PAH),** hidrocarburo (m) policíclico aromático (abbr. HPA).

polyelectrolyte, n. polielectrolito (m).

polyester, n. poliéster (m).

polyethylene, n. polietileno (m), polieteno (m).

polygamous, a. polígamo/a.

polygon, n. polígono (m); **frequency p.,** poligono de frecuencia; **Thiessen p.,** poligono de Thiessen.

polygonal, a. poligonal.

polyhalite, n. polihalita (f).

polyhedron, n. poliedro (m).

polymer, n. polímero (m).

polynomial, a. polinómico/a, polinomial.

polyp, n. pólipo (m).

polypeptide, n. polipéptido (m).

polyphosphate, n. polifosfato (m).

polyploid, a. poliploide.

polypropene, n. polipropileno (m), polipropileno (m).

polypropylene, n. polipropileno (m), polipropeno (m).

polysaccharide, n. polisacárido (m).

polysaprobic, a. polisapróbico/a.

polystyrene, n. poliestireno (m).

polythene, n. polieteno (m).

polythene, n. polieteno (m), polietileno (m).

polyurethane, n. poliuretano (m).

polyvinyl, n. polivinilo (m).

pomegranate, n. granada (f) (fruit), granado (m) (tree).

pond, n. charca (f) (natural), estanque (m) (artificial), charcón (m), laguna (f), poza (f); **maturation p. (Treat),** laguna de maduración; **stagnant p.,** charca de agua estancada, pecinal (m).

pontoon, n. pontón (m) (floating bridge); **p. bridge,** puente (m) de pontones.

pony, n. poney (m) (horse).

pool, n. charca (f) (natural), estanque (m) (artificial); piscina (f) (swimming pool); pozo (m) (in river); charco (m) (puddle); reserva (f) (common fund); **stagnant p.,** pecinal (m); **swimming p.,** piscina (f).

poplar, n. chopo (m), álamo (m).

poppy, n. amapola (f).

populated, a. poblado/a; **densely p.,** densamente poblado.

population, n. población (f); **centre of p.,** núcleo (m) de población; **increase in p.,** aumento (m) de la población; **p. centre,** núcleo (m) de población; **p. explosion,** explosión (f) demográfica; **p. increase,** crecimiento (m) demográfico; **rural p.,** población rural; **working p.,** población activa.

porcelain, n. porcelana (f).

porcellanite, n. porcelanita (f).

porch, n. porche (m) (of a house).

porcupine, n. puerco (m) espín.

pore, n. poro (m); **p. pressure,** presión (f) de poro; **p. size,** tamaño (m) de poros; **p. water pressure,** presión (f) de poro intersticial.

porifer, n. porífero (m).

Porifera, n.pl. poríferos (m.pl), el filum Porifera (Zoo).

porosimetry, n. porosimetría (f).

porosity, n. porosidad (f); **effective p.,** porosidad efectiva, porosidad eficaz; **primary/secondary p.,** porosidad primaria/secundaria; **theoretical p.,** porosidad teórica.

porous, a. poroso/a.

porphyrite, n. pórfido (m).

porphyritic, a. porfídico/a.

porphyroblast, n. porfidoblasto (m).

porphyroclast, n. porfidoclasto (m).

porphyroid, n. porfiroide.

porphyry, n. pórfido (m); **diorite p.,** pórfido diorítico; **granite p.,** pórfido granítico.

portable, a. portátil.

portal, n. pórtico (m), portal (m) (Con); porta (f) (Anat).

Portlandian, n. Portlandiense (m).

Portugal, n. Portugal (m).

Portugese, a. portugués/potuguesa.

Portuguese, n. portugués, portuguesa.

position, n. posición (f), sitio (m), ubicación (f); **to put in p. (v),** ubicar, colocar.

positive, a. positivo/a; **p. charge,** carga (f) positiva; **p. electrode,** electrodo (m) positivo; **p. ion,** ión (m) positivo; **p. pole,** polo (m) positivo; **p. taxis,** taxia (f) positiva; **p. tropism,** tropismo (m) positivo.

positron, n. positrón (m).

possibility, n. posibilidad (f).

post, n. poste (m) (pole); estaca (f), palo (m) (stake); correos (m.pl) (postal service); cartas (f.pl) (letters); empleo (m) (job); **door p.,** jamba (f) de la puerta.

Postdamian, n. Postdamiense (m).

poster, n. cartel (m), letrero (m).

posterior, a. posterior.

post-glacial, a. postglaciar, postglacial.

posthole, n. hoyo (m) de poste (Arch).

post-mortem, n. autopsia (f), postmortem (m); **p. examination,** autopsia (f).

postpone, v. aplazar.

pot, n. olla (f), marmita (f), puchero (m) (for cooking); vasija (f) (vessel); sombrerete (m) (of chimney); **earthenware p.,** vasija de barro; **porous p.** olla porosa.

potability, n. potabilidad (f); **bacterial p.,** potabilidad bacteriológica; **chemical p.,** potabilidad química; **p. analysis,** análisis (m) de potabilidad.

potable, a. potable; **p. water,** agua (f) potable.

potamology, n. potamología (f).

potash, n. potasa (f).

potassic, a. potásico/a.

potassium, n. potasio (m) (K); **p. chloride,** cloruro (m) potásico.

potato, n. patata (f), papa (f) (LAm); **p. field,** patatal (m) (LAm), patatán (LAm); **sweet p.,** batata (f), boniato (m), camote (m) (LAm).

potential, 1. n. potencial (m); voltaje (m) (Elec); **redox p., reduction-oxidation p.,** potencial redox, potencial de reducción-oxidación; 2. a. potencial; **p. energy,** energía (f) potencial; **p. evaporation,** evaporación (f) potencial.

potentiometer, n. potenciómetro (m).

potentiometric, a. potenciométrico/a; **p. surface,** superficie (f) potenciométrica.

potentiometry, n. potenciometría (f).

pothole, n. marmita (f) de gigante (Geol), gruta (f) (cavern), cadozo (m) (in river or lake), bache (m) (in road).

potholing, n. espeleología (f).

potsherd, n. tiesto (m), casco (m) (Arch).

poulterer, n. pollero/a (m, f).

poultry, n. avícola (f), aves (f.pl) de corral (alive); aves (as food).

pound, n. libra (f) (money) (unit of weight, 16 ounces or 453.6 grams); redil (m) (for sheep); depósito (m) (for animals, cars); ruido (m) (noise); **p. sterling,** libra esterlina.

pouring, n. derrame (m) (outflow, spillage).

power, n. fuerza (f), potencia (f) (strength); poder (m) (authority); facultad (f) (ability); **executive p.,** poder ejecutivo; **full powers,** plenos poderes; **hydro p.,** fuerza hidráulica; **p. station,** central (f), central eléctrica; **purchasing p.,** poder adquisitivo; **to raise to the p. of three (v),** elevar a la potencia tres.

powerful, a. potente.

ppm, n. ppm; **parts per million,** partes (m.pl) por millón.

practicability, n. factibilidad (f).

practical, a. práctico/a.

practice, n. bufete (m) (e.g. lawyers office); práctica (f) (exercise); ejercicio (m) (of profession); **code of good p.,** código (m) de buena práctica; **code of p.,** código (m) de práctica; **good p.,** buena práctica.

practise, v. practicar (US, to practice); ejercer (profession).

pragmatic, a. pragmático/a.

pragmatism, n. pragmatismo (m).

prairie, n. pradera (f); llanura (f) (US); pampa (f) (LAm).

praseodynium, n. praseodimio (m) (Pr).

prawn, n. gamba (f), camarón (m).

pre-, pf. pre-.

preaeration, n. preaireación (f).

Pre-Cambrian, n. Precámbrico (m).

precaution, n. precaución (f); **fire precautions,** precauciones contraincendios; **to take precautions against (v),** tomar precauciones, precaver.

precautionary, a. preventivo/a, con precaución; **p. measures,** medidas (f.pl) preventivas.

precipice, n. precipicio (m), despeñadero (m).

precipitate, v. causar, producir (to cause); arrojar (to throw down); precipitar (Chem).

precipitate, n. precipitado (m); **chemical p.,** precipitado químico.

precipitation, n. precipitación (f); **chemical p.,** precipitación química; **convective p.,** precipitación convectiva; **cyclonic p.,** precipitación ciclónica; **effective p.,** precipitación eficaz; **frontal p.,** precipitación frontal; **orographic p.,** precipitación orográfica.

precipitator, n. precipitador (m); **electrostatic p.,** precipitador electrostático.

precision, n. precisión (f), exactitud (f).

precycling, n. preciclaje (m).

predation, n. depredación (f).

predator, n. depredador (m), animal (m) de rapiña.

predatory, a. depredador/a.

prediction, n. predicción (f).

pre-exisiting, a. preexistente.

prefabricate, v. prefabricar.

prefabricated, a. prefabricado/a.

preference, n. preferencia (f).

preferred, a. preferente; **p. option,** opción (f) preferente.

prefilter, n. prefiltro (m).

prefiltration, n. prefiltración (f), filtración previa.

prefinished, a. preacabado/a.

prefix, n. prefijo (m).

pregnancy, n. preñez (f) (Zoo); embarazo (m) (of women).

pregnant, a. preñado/a (Zoo); embarazada (of women).

preheating, n. precalentamiento (m).

prehensile, a. prensil.

prehistoric, a. prehistórico/a.

prejudicial, a. perjudicial.

preliminary, a. preliminar.

preloading, n. precargo (m).

premises, n.pl. locales (m.pl), oficinas (f.pl), instalaciones (f.pl); **industrial p.,** instalación (f) industrial; **on the p.,** en los locales, en las oficinas.

preparedness, n. preparación (f), estado (m) de preparación.

prerequisite, 1. n. requisito (m) previo; 2. a. previamente necesario.

prescreening, n. prefiltrado (m).

prescribed, a. reglamentario/a (by regulation); **p. flow,** flujo (m) reglamentario.

prescriptive, a. prescriptivo/a, de prescripción, establecido/a (established by usage); **p. right (established by usage),** derecho (m) de prescripción.

presence, n. presencia (f) (Gen); asistencia (f) (attendance); **p. of mind,** presencia de ánimo.

present, 1. n. actualidad (f); **p.-day methods,** métodos (m.pl) actuales; 2. a. actual (time), presente (in attendance); **at p.,** en el momento, en el presente.

presentation, n. presentación (f) (lecture); introducción (to someone).

preservation, n. preservación (f) (protection), conservación (f) (e.g. of foodstuffs).

preservative, n. preservativo (m), agente (m) de conservación.

preserves, n.pl. conservas (f.pl); **fish p.,** conservas de pescados; **meat p.,** conservas de carnes; **tinned (US, canned) p.,** conservas alimenticias.

president, n. presidente (m).

pressure, n. presión (f) (Phys), peso (m) (weight); **absolute p.,** presión absoluta; **at full p.,** a toda presión; **atmospheric p.,** presión atmosférica; **blood p.,** presión arterial, presión sanguínea; **capillary p.,** presión capilar; **confining p.,** presión de confinamiento; **critical p.,** presión crítica; **high p.,** presión alta; **hydrostatic p.,** presión hidrostática; **injection p.,** presión por inyección; **interstitial p.,** presión intersticial; **low p.,** presión baja; **osmotic p.,** presión osmótica; **partial vapour p.,** presión parcial del vapor; **p. filter,** filtro (m) de presión; **p.**

gauge, manométrica (f); **p. group,** grupo (m) de presión; **p. head of water,** columna (f) de agua, presión de agua; **p. head,** presión de columna estática; **p. ridge,** cresta (f) de presión; **p. transducer,** transductor (m) de presión; **p.-reducing valve,** válvula (f) piezorreductora; **p.-relief valve,** válvula (f) de seguridad; **reduced p. head,** presión reducida del agua; **water p.,** presión del agua.

pressurize, v. presurizar.

prestressed, a. pretensado/a; **p. concrete,** hormigón (m) pretensado.

presuppose, v. presuponer.

presupposition, n. presupuesto (m).

pretreatment, n. pretratamiento (m), tratamiento preliminar; **p. of waste,** pretratamiento de residuos; **wastewater p.,** pretratamiento de aguas residuales.

prevailing, a. predominante; **p. wind,** viento (m) predominante.

prevent, v. impedir.

prevention, n. prevención (f); **flood p.,** prevención de inundaciones.

preventive, a. preventivo/a; **p. measures,** medidas (f) preventivas.

prey, n. presa (f); **bird of p.,** ave (f) de presa.

prey on, v. alimentarse de.

Priabonian, n. Priaboniense (m).

price, n. precio (m); **cost p.,** precio de coste; **net p.,** precio neto; **p. rise,** subida (f) de precio; **unit p.,** precio por unidad, precio unitario.

prickly, a. espinoso/a; **p. heat (Med),** sudamina (f).

primary, a. primario/a, principal; **p. cell (Elec),** pila (f); **p. clarifier,** clarificador (m) primario; **p. colour,** color (m) primario; **p. education,** enseñanza (f) primaria; **p. feathers,** plumas (f) primarias, primarias (f.pl); **p. school,** escuela (f) primaria; **p. sedimentation tank,** tanque (m) para sedimentación primaria; **p. treatment,** tratamiento (m) primario.

primate, 1. n. primate (m); **Primates (Zoo, order),** primates (m.pl); 2. a. primate.

prime, a. primo/a (Mat); primero/a (first); principal, fundamental (main); original, primitivo/a (primary); **P. Minister,** primer ministro; **p. number,** número (m) primo.

principle, n. principio (m); **Archimedes p.,** principio de Arquímedes; **"polluter pays" p.,** principio de "quien contamina paga"; **p. of superposition,** principio de superposición.

printed, a. impreso/a.

printer, n. impresor/a (m,f); **laser p.,** impresora láser.

prioritize, v. priorizar.

priority, n. prioridad (f); **high p.,** alta prioridad.

prism, n. prisma (f); **nicol p.,** prisma de nicol; **quartz p.,** prisma de cuarzo.

private, a. privado/a; **p. waters**, aguas (f.pl) privadas.
privatization, n. privatización (f).
privilege, n. privilegio (m); prerogativa (prerrogative); honor (m) (bestowed).
probability, n. probabilidad (f); **p. distribution**, distribución (f) de probabilidad(es).
probe, n. sonda (f) (instrument), sondeo (m) (act).
probing, n. sondeo (m), investigación (f), exploración (f).
problem, n. problema (m).
problematical, a. problemático/a.
proboscides, n.pl. probóscides (f.pl).
proboscis, n. probóscide (f); trompa (f).
Procaryota, n.pl. procariotas (f.pl).
procaryote, n. procariota (f).
procedure, n. procedimiento (m) (Gen); gestión (f) (management); trámite (m) (step); **legal p.**, trámite legal.
process, n. proceso (m); **anaerobic p.**, proceso anaerobio; **Bessemer p. (for steel making)**, proceso Bessemer; **biochemical p.**, proceso bioquímico; **diagenetic p.**, proceso de diagénesis; **endogenic p.**, proceso endógeno; **extended-aeration p.**, proceso de aireación extendida; **iterative p.**, proceso iterativo; **karstification p.**, proceso de karstificación; **life p.**, proceso de la vida; **mechanical p.**, proceso mecánico; **modelling p. (Mat)**, proceso de modelado; **over-relaxation p.**, proceso de sobre-relajación; **p. liquors (Treat)**, aguas (f.pl) de tratamiento; **thermal p.**, proceso térmico, proceso termal; **treatment p.**, proceso de depuración.
processing, n. procesamiento (m), tratamiento (m); **data p.**, procesamiento de datos; **food p.**, procesamiento de alimentos; **parallel p.**, proceso (m) en paralelo; **p. of raw materials**, tratamiento de materias primas; **serial p.**, proceso (m) en serie.
processor, n. procesador (m); **word p.**, procesador de palabras.
produce, v. producir; fabricar (to manufacture); **to mass-p.**, producir en serie.
producer, n. productor/a (m, f); fabricante (m, f) (manufacturer).
producing, a. productor/a.
product, n. producto (m); **canned products**, conservas (f.pl), enlatados (m.pl); **dissolved p.**, producto disuelto; **end p.**, producto final; **free p. (Chem)**, producto libre; **gross domestic p. (abbr. GDP)**, producto interno bruto (Sp, abbr. PIB), producto geográfico bruto (Ch, abbr. PGB), producto territorial bruto (Peru, abbr. PTB); **gross national p. (abbr. GNP)**, producto nacional bruto (abbr. PNB); **manufactured products**, productos manufacturados; **primary products**, productos primarios.
production, n. producción (f); **primary p.**, producción primaria.

productivity, n. productividad (f) (of person), rendimiento (m), capacidad (f) (of machine).
proenzyme, n. proenzima (f).
profession, n. profesión (f).
professional, a. profesional, de profesión.
profile, n. perfil (m); **geophysical p.**, perfil geofísico; **in p.**, de perfil; **longitudinal p.**, perfil longitudinal; **p. section**, perfil transversal; **salinity p.**, perfil de salinidad; **shore p.**, perfil de ribera; **soil moisture p.**, perfil de humedad del suelo; **to keep a low p. (v)**, tratar de pasar desapercibido.
profiled, a. perfilado/a.
profit, n. provecho (m), beneficio (m), ganancia (f); **gross p.**, beneficio bruto; **p. margin**, margen (m) de beneficio; **p.-sharing**, participación (f) en los beneficios; **yearly p.**, beneficio anual.
profitability, n. rentabilidad (f).
profitable, a. rentable, provechoso/a; **to make p. (v)**, hacer rentable.
progeny, n. progenie (f).
progesterone, n. progesterona (f).
program, n. programa (m); **applications p. (Comp)**, programa de aplicación(es).
programmable, a. programable.
programme, n. programa (m); **development p.**, programa de desarrollo.
programmed, a. programable.
programmer, n. programador/a (m, f) (e.g. of computers).
programming, n. programación (f); **computer p.**, programación en ordenador, p. informática; **dynamic p.**, programación dinámica; **linear p.**, programación lineal.
progression, n. progresión; **geometric p.**, progresión geométrica.
prohibit, v. prohibir.
project, n. proyecto (m).
projection, n. proyección (f); **map p.**, proyección de mapas; **stereographic p.**, proyección estereográfica.
Prokaryota, n.pl. procariotas (f.pl).
prokaryote, n. procariota (f).
proleg, n. propata (f), protoextremidad (f), protomiembro (m).
proliferate, v. proliferar.
proliferation, n. proliferación (f).
prolong, v. prolongar, extender.
prolongation, n. prolongación (f), extensión (f).
prolonged, a. alargado/a.
promenade, n. paseo (m); **seaside p.**, paseo marítimo.
promethium, n. prometio (m) (Pr).
promontory, n. promontorio (m); cabo (m) (cape).
promoter, n. promotor/a (m, f).
promotion, n. fomento (m) (fostering); promoción (f) (rank); promoción (sales).
promptness, n. prontitud (f).
promulgate, v. promulgar.
promulgating, a. promulgador/a.

promulgation, n. promulgación (f).
pronota, n.pl. pronotos (m.pl).
pronotum, n. pronoto (m).
prop, v. entibar (to install in mine), apuntalar (support a wall).
propaganda, n. propaganda (f).
propagate, v. propagar.
propagation, n. propagación (f), difusión (f); **vegetative p.,** propagación vegetativa.
propagator, n. propagador/a (m, f).
propagule, n. propágulo (m).
propanal, n. propanal (m).
propane, n. propano (m).
propanol, n. propanol (m).
propel, v. impeler, impulsar, propulsar.
propeller, n. propulsor (m) (means of being propelled), hélice (f) (of ship or aircraft).
propene, n. propileno (m).
property, n. propiedad (f) (possession); dominio (m) (dominion); característica (f) (characteristic); bienes (m.pl) (goods); **intellectual p. right,** derechos (m) de propiedad intelectual; **physical p.,** propiedad física; **p. developer,** promotor (m) inmobiliario; **p. market,** mercado (m) inmobiliario; **public p.,** dominio público; **soil properties,** propiedades del suelo.
prophylactic, 1. n. profiláctico (m); 2. a. profiláctico/a.
propitious, a. propicio/a, favorable.
proportion, 1. v. proporcionar, adecuar; 2. n. proporción (f); razón (f) (Mat).
proportional, a. proporcional; **inversely p.,** inversamente proporcional.
proportionate, a. proporcionado/a.
proposal, n. propuesta (f), proposición (f); **to make a p. (v),** hacer una propuesta.
propose, v. proponer; **to p. a plan (v),** proponer un proyecto.
proposition, n. proposición (f), propuesta (proposal).
proprietor, n. propietario/a (m, f).
propyl, n. propilo (m).
propylene, n. propileno (m).
propyne, n. propino (m).
pros and cons, n.pl. los pros y los contras (m.pl).
prosecute, v. procesar, enjuiciar (Jur).
prosecution, n. enjuiciamiento (m), procesamiento (m) (action).
prospect, 1. v. prospectar (Min); catear (Min) (LAm); investigar (investigate); buscar (search for); 2. n. (LAm) cateo (m) (Min).
prospecting, n. investigando (m), prospectando (m); cateando (m) (Min) (LAm).
prospection, n. prospección (f).
prospector, n. prospector (m); cateador (m) (Min) (LAm).
protactinium, n. protactinio (m) (Pa).
protection, n. protección (f); **aquifer p.,** protección de acuíferos; **cathodic p.,** protección catódica; **environmental p.,** protección

ambiental; **groundwater p.,** protección de aguas subterráneas; **p. against corrosion,** protección contra la corrosión; **p. measures,** medidas (f) de protección; **sanitary well p.,** protección sanitaria del pozo.
protective, a. protector/a, de protección, preventivo/a; **p. coating,** revestimiento (m) protector.
protein, n. proteína (f); **soluble p.,** proteína soluble.
proteolysis, n. proteolisis (f).
Proterozoic, n. Proterozoico (m).
protest, n. protesta (f), queja (f) (complaint); objeción (f) (objection).
protium, n. protio (m).
proto-, pf. proto-.
protocol, n. protocolo (m).
proton, n. protón (m).
prototype, n. prototipo (m).
protozoa, n.pl. protozoos (m.pl), el filum (o subreino) Protozoa (Zoo); protozoarios (m.pl).
protozoan, n. protozoo (m).
protractor, n. transportador (Math, instrument).
protuberence, n. protuberancia (f).
prove, v. probar, demostrar.
provenance, n. procedencia (f).
provide, v. proveer, proporcionar; suministrar (with electricity, water, etc.).
provincial, a. provincial.
proximity, n. proximidad (f), cercanía (f).
prudent, a. prudente.
prune, n. ciruela (f) pasa.
pruning, n. poda (f); **p. hook,** podadera (f).
psammite, n. samita (f); psammita (f) (LAm).
psammitic, a. samítico/a; psammítico/a (LAm).
pseudo-, pf. pseudo-, seudo-.
pseudobreccia, n. pseudobrecha (f).
pseudomonas, n. pseudomonas (f).
pseudomorph, n. pseudomorfo (m).
pseudomorphic, a. pseudomórfico/a.
psychiatry, n. psiquiatría (f).
psychology, n. psicología (f).
psychrometer, n. psicrómetro (m).
psychrophylic, a. psicrofílico/a.
Pteridophyta, n.pl. pteridofitas (f.pl), la división Pteridophyta (Bot).
pteridophyte, n. pteridofita (f).
puberty, n. pubertad (f).
public, a. público/a; **p. works,** obras (f.pl) públicas.
publication, n. publicación (f).
puddle, n. charca (f).
puddling, n. pudelado (m); **chalk p.,** pudelado con creta.
Puerto Rican, a. portorriqueño/a.
Puerto Rico, n. Puerto (m) Rico.
pulley, n. polea (f); motón (m) (Mar); **cable rig p. (for drilling),** polea de cable de perforación.
pulmonary, a. pulmonar.

pulp, n. pulpa (f), pasta (f); **paper/wood p.**, pasta de papel/madera.

pulsation, n. pulsación (f) (beat), vibración (f) (vibration).

pulse, 1. v. pulsar, latir; vibrar; 2. n. leguminosa (f), legumbre (f) (vegetable); pulso (m) (Anat), pulsación (f) (of heart).

pulverize, v. pulverizar.

pulverizer, n. pulverizador (m) (machine).

pumice, n. pómez (f), pumita (f); **p. stone**, piedra (f) pómez.

pump, n. bomba (f); surtidor (m) (in petrol station); **air-lift p.**, bomba de aire comprimido; **Archimedes screw p.**, bomba en forma de tornillo de Arquímedes; **axial flow p.**, bomba de flujo axial; **booster p.**, bomba de refuerzo; **centrifugal p.**, bomba centrífuga; **diaphragm p.**, bomba de diafragma; **double-action p.**, bomba de doble efecto; **drainage p.**, bomba de desagüe (from mine), bomba de agotamiento (m) (dewatering); **duplex p.**, bomba dúplex; **electrical submersible p.**, electrobomba (f) sumergida; **fire p.**, bomba contra incendios; **hand p.**, bomba de mano; **peristaltic p.**, bomba peristáltica; **positive-displacement p.**, bomba de desplazamiento positivo; **ram p.**, bomba por fuerza; **reciprocating p.**, bomba oscilante; **rotary p.**, bomba rotatoria; **scavenger p.**, bomba de barrido; **submersible p.**, bomba sumergible; **suction p.**, bomba aspirante; **surface p.**, bomba de superficie; **triple-action p.**, bomba de triple efecto; **wind-pump**, bomba de aeromotor.

pumpage, n. volumen (m) bombeado.

pumped-storage, n. almacenamiento (m) bombeado.

pumphouse, n. casa (f) de bombas.

pumping, n. bombeo (m), el bombear (m); **constant-rate p.**, bombeo a caudal constante; **interference p.**, bombeo por interferencia; **intermittent p.**, bombeo intermitente; **p. station**, estación (f) de bombeo; **p. steps**, etapas (f.pl) de bombeo, niveles (m.pl) de bombeo; **p. test**, ensayo (m) de bombeo; **p. with compressed air**, bombeo por aire comprimido.

pumpkin, n. calabaza (f).

pungent, a. acre, muy picante.

pupa, n. pupa (f), crisálida (f).

pupae, n.pl. pupas (f.pl).

pupil, n. alumno/a (m, f) (student); pupila (f) (of eye).

Purbeckian, n. Purbeckiense (m).

purchase, 1. v. comprar (to buy), conseguir (to obtain); 2. n. compra (f), adquisición (f); palanca (Eng, lever).

purge, v. purgar, limpiar (to cleanse).

purging, n. expiación (f), purgación (f), limpieza (f).

purification, n. purificación (f); depuración (f) (waste treatment); **biological p.**, depuración biológica; **preliminary water p.**, depuración previa; **primary water p.**, depuración primaria; **secondary water p.**, depuración secundaria; **tertiary water p.**, depuración terciaria; **water p.**, depuración de agua.

purifier, n. purificador/a (m,f), depurador (m).

purify, v. purificar; depurar (to treat).

purifying, a. purificador/a; depurador/a (treating); **p. plant**, estación (f) depuradora, planta (f) purificadora.

purity, n. pureza (f); **atmospheric p.**, pureza atmosférica; **water p.**, pureza del agua.

purple, 1. n. púrpura (f); 2. a. purpúreo/a.

purpose-made, a. hecho de encargo.

pus, n. pus (m).

putrefaction, n. putrefacción (f).

putrefied, a. putrefacto/a.

putrefy, v. pudrir; gangrenarse (Med).

putrescent, a. putrescente.

putrescible, a. putrescible; **p. waste**, desechos (m.pl) putrescibles.

putrid, a. pútrido/a, putrefacto/a, podrido/a; gangrenoso/a (Med).

putty, n. masilla (f).

PVC, n. PVC (m); **polyvinylchloride**, polivinilcloruro (m), policloroeteno (m), cloruro (m) de polvinilo.

PWR, n. PWR; **pressurized water reactor (power generation)**, reactor (m) de agua a presión.

pylon, n. poste (m), torre (f) metálica (for cables).

pyramid, n. pirámide (f).

pyramidal, a. piramidal.

pyranometer, n. piranómetro (m).

pyrethrum, n. piretro (m).

pyrex, n. pyrex (m).

pyrgeometer, n. pirgeómetro (m).

pyrheliograph, n. pirheliógrafo (m).

pyrheliometer, n. pirheliómetro (m).

pyridine, n. piridina (f).

pyridoxine, n. piridoxina (f) (vitamin B6).

pyrite, n. pirita (f); **iron p.**, pirita de hierro.

pyrites, n. pirita (f); **copper p.**, pirita de cobre; **iron p.**, pirita de hierro; **tin p.**, pirita de estaño; **white p.**, pirita blanca de hierro.

pyritization, n. piritización (f).

pyro-, pf. piro-.

pyroclast, n. piroclasto (m).

pyroclastic, a. piroclástico/a.

pyrolusite, n. pirolusita (f).

pyrolysis, n. pirólisis (f).

pyrometer, n. pirómetro (m).

pyrophosphate, n. pirofosfato (m).

pyroxene, n. piroxeno (m).

pyroxenite, n. piroxenita (f).

Q

qanat, n. qanat (m), ganat (m), khanat (m) (Arab); galería (f).
quadrangle, n. patio (m) rectangular (Con), cuadrilátero (m) (figure with four sides).
quadrant, n. cuadrante (m).
quadrat, n. cuadrado (m) (Bot).
quadratic, a. cuadrático/a; **quadratic e.**, ecuación (f) cuadrática.
quadripole, n. cuadripolo (m).
quadrivalent, a. cuadrivalente, tetravalente.
quagmire, n. cenegal (m), lodazal (m).
quake, n. terremoto (m), temblor (m) (LAm); **sea q.**, maremoto (m).
qualification, n. reserva (f), restricción (f) (restriction), calificación (f) (evaluation); competencia (aptitude).
qualitative, a. cualitativo/a.
quality, n. calidad (f); **bacterial q.**, calidad bacteriológica; **industrial q.**, calidad industrial; **physical-chemical q.**, calidad físicoquímica; **q. control,** control (m) de calidad; **q. objective**, norma (f) de calidad; **q. of life,** calidad de vida; **q. standard,** norma (f) de calidad.
quango, n. organización (f) no gubernamental (abbr. ONG) (Fig).
quantifiable, a. cuantificable.
quantification, n. cuantificación (f).
quantile, n. cuantil (m).
quantitative, a. cuantitativo/a.
quantity, n. cantidad (f), (amount), cuantía (f) (distinction); **q. surveyor,** aparejador (m).
quantum, n. quantum (m) (pl. quanta), cuanto (m) (Phys); parte (f) (portion); **q. mechanics,** mecánica (f) cuántica; **q. theory,** teoría (f) cuántica.

quark, n. quark (m).
quarry, n. cantera (f); **limestone q.**, cantera de caliza.
quarrying, n. explotación (f) de canteras.
quarryman, n. cantero (m).
quarterly, a. trimestral.
quartile, n. cuartil (m); **lower q.**, cuartil más bajo, cuartil 25%, cuartil inferior; **upper q.**, cuartil superior.
quartz, n. cuarzo (m); **q.-bearing (a),** que contiene cuarzo; **rose q.**, cuarzo rosado; **smoky q.**, cuarzo ahumado.
quartziferous, a. cuarcífero/a.
quartzite, n. cuarcita (f).
quartzose, a. cuarzoso/a.
Quaternary, n. Cuaternario (m).
quay, n. muelle (m), espigón (m).
quechuan, a. quechua (language).
queen, n. reina (f); **q.-bee**, abeja (f) reina, abeja maestra.
questionaire, n. cuestionario (m).
quetzal, n. quetzal (m) (Guat, unit of currency).
quicklime, n. cal (f) viva.
quicksand, n. arenas (f.pl) movedizas, ñocle (m) (LAm).
quiescence, n. quietud (f).
quill, n. púa (f) (e.g. of hedgehog); cañón (m) (of a feather).
quinine, n. quinina (f), quina (f).
quota, n. cupo (m), cuota (f).
quotation, n. cotización (f), oferta (f); **q. price,** precio (m) de cotización.
quotient, n. cociente (m); **respiratory q. (abbr RQ),** cociente respiratorio.

R

rabid, a. rábico/a.
rabies, n. rabia (f).
race, n. raza (f) (people); carerra (f) (contest).
raceme, n. racimo (m).
rachis, n. raquis (m).
rack, n. ronza (f), detritos (m, pl) del río (Geog); estante (m), anaquel (m) (shelf); soporte (m) (for bicycles); redecilla (f) (in a train).
R & D, n. I & D; **research and development,** investigación (f) y desarrollo (m).
radar, n. radar (m); **r. screen,** pantalla (f) de radar.
radial, a. radial.
radiant, a. radiante.
radiation, n. radiación (f); **alpha/beta/gamma r.,** radiación alfa/beta/gamma; **ionizing r.,** radiación ionizante; **short-wave r.,** radiación de onda corta; **solar r.,** radiación solar; **ultraviolet r.,** radiación ultravioleta.
radical, 1. n. radical (m) (Chem); **acid r.,** radical ácido; **free r.,** radical libre; 2. a. radical; **r. measures,** medidas (f) radicales.
radicle, n. radícula (f) (Bot).
radio, n. radio (m); **r. beacon,** radiofaro (m); **r. frequency,** radiofrecuencia (f), frecuencia de radio; **r. receiver,** radiorreceptor (m); **r. telephone,** radioteléfono (m).
radioactive, a. radioactivo/a; **low level r. waste,** residuos (m.pl) radioactivos de baja actividad; **r. emissions,** emisiones (f.pl) radioactivas.
radioactivity, n. radiactividad (f), radioactividad (f).
radiocarbon, n. radiocarbono (m).
radiochemistry, n. radioquímica (f).
radiocommunication, n. radiocomunicación (f).
radiodiagnostic, a. radiodiagnóstico/a.
radioisotope, n. radioisótopo (m).
radiola, n. radiola (f).
radiolaria, n.pl. radiolarios (m.pl).
radiolarite, n. radiolarita (f).
radiometer, n. radiómetro (m); **r. of net radiation,** radiómetro de radiación neta.
radionucleus, n. radionúclido (m).
radionuclide, n. radionucleido (m).
radiosonde, n. radiosonda (f).
radish, n. rábano (m), rabanito (m).
radium, n. radio (m) (Ra).
radius, n. radio (m); **hydraulic r.,** radio hidráulico; **r. of curvature,** radio de curvatura; **r. of influence,** radio de influencia.
radon, n. radón (m) (Ra).
radula, n. rádula (f).
radulae, n.pl. rádulas (f.pl).
raft, n. balsa (f) (floating platform), masa (f)

flotante (of ice); **r. foundation,** balsa de fundición.
rafter, n. viga (f).
railings, n.pl. enrejado (m).
railway, n. ferrocarril (m); camino (m) de hierro (LAm) línea (f) férrea; vía (f) férrea.
rain, v. llover.
rain, n. lluvia (f); **acid r.,** lluvia ácida; **artificial r. seeding,** siembra (f) de lluvia artificial; **heavy rains,** lluvias abundantes, lluvias copiosas; **r. cloud,** nubarrón (m); **r. imprint,** impresión (f) de lluvia; **r. shadow,** zona (f) de abrigo de la lluvia, sombra (f) de lluvia.
rainbow, n. arco (m) iris.
raindrop, n. gota (f) de lluvia.
rainfall, n. lluvia (f), lluvias caídas, precipitación (f); **annual r.,** lluvia anual; **automatic siphoning r. recorder,** pluviógrafo (m) de sifonación automática; **average r.,** lluvia media; **daily r.,** lluvia diaria; **effective r.,** lluvia efectiva; **r. distribution,** distribución (f) de la lluvia; **r. measurement,** medida (f) de la lluvia, medición (f) de la lluvia; **r. measurement network,** red (f) de lluvia (para la medición de la lluvia); **r. recharge,** recarga (f) por lluvia; **r.-runoff relationship,** relación (f) precipitación-escorrentía; **tilting bucket r. recorder,** pluviógrafo (m) de cangilones.
rainforest, n. selva (f); **the Amazon r.,** la selva amazónica; **tropical r.,** selva tropical.
raingauge, n. pluviómetro (m).
rainshed, n. divisoria (f) de aguas (US), divisoria (f) fluvial (US).
rainstorm, n. tempestad (f) de lluvia.
rainwash, n. erosión (f) pluvial; levigación (f) pluvial (LAm), lavaje (m) pluvial.
rainwater, n. agua (f) de lluvia; **r. catchment,** captación (f) de precipitaciones.
rainy, a. lluvioso/a, pluvioso/a; **r. season,** temporada (f) lluviosa.
raise, v. subir, levantar (to raise); elevar (Mat), erguir, (Con); **to r. to the power n (Mat),** elevar a la enésima potencia.
ram, n. ariete (m) (battering ram); carnero (m) (Zoo); **hydraulic r.,** ariete hidráulico.
RAM, n. RAM (f); **random access memory (Comp),** memoria (f) de acceso aleatorio.
ramp, n. rampa (f) (slope), elevador (m); **hydraulic r.,** elevador hidráulico.
rampart, n. muralla (f), defensa (f), camino (m) de ronda.
Ramsar, n. Ramsar; **R. sites,** Humedales (m.pl) Ramsar.
ranch, n. rancho (m), hacienda (f), estancia (Arg, Ch, Col, Ur, Ven), hacendado (m)

Arg), fundo (Ch, Peru); **cattle r.**, hacienda (f) (LAm, esp. Arg), hato (m) (Bol, Col, Cuba, Ven).

random, a. aleatorio/a (Mat); caprichoso/a (indiscriminate); fortuito/a (fortuitous); **at r.**, al azar; **r. noise**, ruido (m) aleatorio; **r. sampling**, muestreo (m) aleatorio; **r. variable**, variable (f) aleatoria.

randomization, n. randomización (f).

randomizing, n. randomización (f).

randomness, n. aleatoridad (f).

range, n. distancia (f) (distance); radio (m) (radius); cadena (f) (mountains); campo (m) (Fig, field); alcance (m) (scope); escala (f), gama (f) (of a series); **interquartile r.**, alcance intercuartil; **r. of variation (Mat),** rango (m) de variación.

rangeland, n. pastizal (m).

rank, n. grado (m), categoría (f), clase (f).

rap, n. ripio (m) (broken stone).

rapidity, n. rapidez (f).

rapids, n.pl. rápidos (m), rabiones (m); correntera (f) (LAm); correntada (f) (Ch, Arg, Ur).

rare, a. raro/a.

rarefaction, n. rarefacción (f), enrarecimiento (m).

rarity, n. rareza (f).

raspberry, n. frambuesa (f) (fruit), frambueso (m) (bush); **r. cane**, frambueso (m).

rate, n. velocidad (f), ritmo (m) (speed); tasa (f), coeficiente (m) (coefficient); precio (m) (price); tipo (m) (interest on money); índice (m) (index); proporción (f) (ratio); razón (f) (Mat, ratio); frecuencia (f) (Med); **birth r.**, tasa de natalidad; **death r.**, índice (m) de mortalidad, porcentaje (m) de defunciones; **exchange r.**, tipo de cambio; **fertility r.**, índice (m) de fertilidad; **flow r.**, caudal (m) medio (of water); régimen (m) (Elec); **free market r.**, precio de libre mercado; **interest r.**, tasa de interés; **pay r.**, sueldo (m), tarifa (f); **r. of exchange**, tipo de cambio; **r. of flow**, caudal (m) medio (of water); régimen (m) (Elec); **r. of increase**, coeficiente de incremento; **r. of interest**, tasa de interés; **r. of production**, ritmo de producción; **r. of work**, ritmo de trabajo; **rates (now Council Tax, GB)**, impuestos (m.pl) municipales, contribución (f) municipal; **second r.**, de segunda categoría (f).

ratification, n. ratificación (f), confirmación (f).

rating, n. valoración (f), estimación (f) (estimate), rango (m) (rank); clasificación (US) (of a pupil); **r. curve**, curva (f) de aforo (gauging), curva de caudales en función del nivel (stage-discharge curve).

ratio, n. proporción (f), relación (f) (Mat); razón (f); **bifurcation r.**, relación de bifurcación; **carbon 14 r.**, proporción del carbono 14; **carbon-nitrogen r., (abbr. C/N r.),** relación carbono-nitrógeno (abbr. r. C/

N); **cost-benefit r.**, relación costo-beneficio; **deuterium-oxygen 18 r.**, relación de deuterio-oxígeno 18; **freshwater-saltwater r.**, relación de agua dulce-agua salada; **helium-argon r.**, relación de helio-argón; **hydraulic r.**, relación hidráulica; **isotope r.**, relación isotópica; **storativity/transmissivity r.**, relación de storatividad/transmisividad (abbr. relación de S/T); **submergence r.**, relación de sumersión; **void r.**, proporción de vacío.

rationalization, n. racionalización (f).

rationalize, v. racionalizar.

ravine, n. barranco (m), cañada (f), quebrada (f) (esp. LAm).

raw, a. crudo/a (meat), bruto/a (oil), puro/a (alcohol), primo/a (materials); **c. sewage**, desagües (m.pl) crudos.

ray, n. rayo (m) (e.g. of light); raya (fish); **cosmic rays**, rayos cósmicos; **infrared rays**, rayos infrarrojos; **ultraviolet rays**, rayos ultravioletas; **X-rays**, rayos X.

rayon, n. rayón (m).

reach, n. tramo (m) (of road); tramo abierto, tramo recto (of a river) alcance (range).

react, v. reaccionar.

reactant, n. reactante (m) (Chem).

reaction, n. reacción (f); **chemical r.**, reacción química; **exothermic r.**, reacción exotérmica; **nuclear r.**, reacción nuclear; **oxidation-reduction r.**, reacción de oxidación-reducción; **redox r.**, reacción redox; **thermonuclear r.**, reacción termonuclear.

reactive, a. reactivo/a.

reactor, n. reactor (m); **nuclear r.**, reactor nuclear.

reader, n. lector/a (m, f) (person who reads); profesor adjunto (senior lecturer); **punched-tape r.**, lectora de cinta perforada.

reaeration, n. reaireación (f).

reagent, n. reactivo (m).

real, n. real (m) (Bra, unit of currency).

realgar, n. rejalgar (m).

realignment, n. realineación (f), realineamiento (m).

real-time, a. tiempo (m) real.

ream, v. escariar (to enlarge).

reamer, n. escariador (m).

reaming, n. escariado (m) (widening).

reap, v. segar.

rearing, n. cría; **cattle r.**, cría de ganado.

reasoning, n. razonamiento (m); cálculos (m.pl) (calculation).

rebound, n. rebote (m); **groundwater r.**, rebote de agua subterránea.

receiver, n. recibidor/a, receptor/a (m, f) (person who receives); auricular (m) (earpiece); receptor (m) (television or radio set); recipiente (receptacle) (Chem).

receiving, a. receptor/a; **r. station**, estación (f) receptora; **r. water (Treat),** agua (f) del medio receptor, cauce (m) receptor.

recent, a. reciente.

receptacle, n. recipiente (m), receptáculo (m).
receptor, n. receptor (m).
recession, n. recesión (f); **r. curves,** curvas (f) de recesión.
recessive, a. recesivo/a.
recharge, 1. v. recargar; 2. n. recarga (f); **aquifer r.,** recarga del acuífero; **artificial r.,** recarga artificial; **coastal r.,** recarga litoral; **groundwater r.,** recarga de aguas subterráneas; **induced r.,** recarga inducida; **natural r.,** recarga natural; **net r.,** recarga neta; **rainfall r.,** recarga por lluvia; **r. capacity,** capacidad (f) de recarga; **river r.,** recarga del río; **wastewater r.,** recarga con aguas residuales.
reciprocal, n. recíproco (m).
reciprocating, a. oscilante.
recirculate, v. recircular.
recirculation, n. recirculación (f).
reclamation, n. reclamación (f), restauración (f) (restoration); roturación (f) (Agr, ploughing); aprovechamiento (m) (of land); saneamiento (m) (draining of marshland); **land r.,** saneamiento (m), terrestrificación (f) (del mar), restauración de terreno.
recolonize, v. recolonizar.
recommend, v. recomendar.
recommendation, n. recomendación (f).
reconnaissance, n. reconocimiento (m), exploración (f) preliminar (survey).
reconnoiter, v. reconocer.
reconnoitre, v. reconocer.
reconvert, n. reconvertir.
record, 1. v. grabar (sound), tomar nota de, marcar (a scale); 2. n. registro (m), anotación (f), documento (m) (document); récord (m) (the best, the most remarkable event); **analogue r.,** registro analógico; **digital r.,** registro digital; **driller's r.,** parte (m) del sondista; **geological r.,** registro geológico; **r. card,** ficha (a); **r. holder,** actual poseedor/a (m, f).
recorder, n. registrador (m); medidor (m) (instrument), grabador (m) (tape recorder); **acoustic r.,** registrador acústico; **automatic r.,** registrador automático; **float r.,** medidor de flotador; **pneumatic r.,** medidor neumático; **punched tape r.,** medidor de cinta perforada.
recording, n. grabación (f) (of sound), registro (m); **r. head,** cabeza (f) de grabación.
recover, v. recobrar, recuperar (to salvage).
recovery, n. recuperación (f), recobro (m); **r. time,** tiempo (m) de recuperación; **solvent r.,** recuperación de disolventes.
recrystallization, n. recristalización (f).
rectangle, n. rectángulo (m).
rectangular, a. rectangular; **r. weir (for measuring flow),** vertedero (m) rectangular.
rectification, n. rectificación (f).
rectify, n. rectificar.
rectifying, a. rectificado/a.
rectilinear, a. rectilíneo/a.

recuperator, n. recuperador (m).
recurrence, n. recurrencia (f), reproducción (f) (Med); **r. interval,** intervalo (m) de recurrencia.
recurrent, a. recurrente.
recursive, a. recursivo/a.
recyclable, a. reciclable.
recycle, v. reciclar.
recycler, n. reciclador (machine).
recycling, n. reciclaje (m), reciclado (m); **r. plant,** planta (f) de reciclaje.
red, a. rojo/a; **R. Book (Ecol),** Libro (m) Rojo.
redeposit, v. redepositar.
redeposited, a. redepositado/a.
redeposition, n. redeposición (f).
redevelopment, n. redesarrollo (m).
redox, n. redox (m), oxi-reducción (f); **r. conditions,** condiciones (f.pl) de oxi-reducción; **r. potential,** potencial (m) redox (abbr. Eh); **r. reaction,** reacción (f) redox.
redrilling, n. reperforación (f).
reducer, n. reductor (m) (Chem).
reducible, a. reductible.
reducing, a. reductor/a; **r. environment,** medio (m) reductor.
reduction, n. reducción (f); rebaja (f), descuento (m) (in price); **bacterial r.,** reducción bacteriana; **iron r.,** reducción de hierro; **nitrate r.,** reducción de nitrógeno; **sulphate r.,** reducción de sulfatos.
reed, n. caña (f), junco (m), carrizo (m), bejuco (m) (LAm) (rattan), otake (m) (Mex); **r.-bed,** cañaveral (m), carrizal (m), gandara (f), totoral (m) (Arg, Bol, Ec, Peru).
reef, n. arrecife (m); **artificial r.,** arrecife artificial; **coral f.,** arrecife de coral, arrecife coralino; **fringing r.,** arrecife costero en orla; **limestone r.,** arrecife calcáreo.
reference, n. referencia (f); **r. book,** libro (m) de consulta; **r. datum,** nivel (m) de referencia; **r. plane,** plano (m) de referencia.
refinery, n. refinería (f); **oil r.,** refinería de petróleo.
refining, n. refinamiento (m), refinación (f), refinado (m); **oil r.,** refinamiento de petróleo; **r. by smelting,** refinamiento por fundición.
reflection, n. reflexión (f).
reflex, 1. n. reflejo (m); **conditioned/unconditioned r. (Biol),** reflejo condicionado/incondicionado; **passive r. (Biol),** reflejo pasivo; 2. a. reflejo/a; **r. action,** acción (f) refleja.
reflux, n. reflujo (m); **r. valve,** válvula (f) de reflujo, válvula de retención.
refolding, n. replegamiento (m).
reforestation, n. reforestación (f), repoblación (f) forestal.
reform, n. reforma (f); **agrarian r.,** reforma agraria; **land r.,** reforma agraria; **legislative r.,** reforma legislativa.
refraction, n. refracción (f); **wave r.,** refracción de ondas.

refractive, a. refractivo/a; **r. index,** índice (m) de refracción.

refractory, a. refractario/a.

refrigerant, 1. n. refrigerante; 2. a. refrigerante.

refrigeration, n. refrigeración (f).

refrigerator, n. refrigerador (m), frigorífico (m).

refuse, n. basura (f), desechos (m.pl); **domestic r.,** basura doméstica; **r. collection,** recogida (f) de basura; **r. dump,** basural (m), basurero (m), tiradero (m).

regenerate, v. regenerar.

regeneration, n. regeneración (f); **industrial r.,** regeneración industrial.

regenerative, a. regenerativo/a, regenerador/a; **r. capacity,** capacidad (f) regenerativa.

regime, n. régimen (m) (pl. regímenes), regla (f); **flow r.,** régimen de flujo.

region, n. región (f), área (f) (Gen); comarca (f); **abyssal r.,** región abisal; **disturbed r.,** región perturbada.

regional, a. regional, comarcal; **r. development,** desarrollo (m) regional.

register, n. registro (m); **land r.,** catastro (m), registro de la propiedad.

registered, a. registrado/a; certificado (correo); **r. letter, r. post,** carta (f) certificada.

regolith, n. regolito (m).

regression, n. regresión (f); **linear r.,** regresión lineal; **marine r.,** regresión marina; **multiple r.,** regresión múltiple; **polynomial r.,** regresión polinómica.

regularity, n. regularidad (f).

regulation, n. reglamento (m), reglamentación (f); normativa (f), reglas (f.pl) (rule), regulación (f) (control); **basin r.,** regulación de cuenca; **EU Regulation,** Reglamento de la Unión Europea; **government r.,** reglamentos del gobierno; **river r.,** regulación fluvial.

regulations, n.pl. regulación (f) (action); reglamentación (f) (setting rules); reglamento (m), reglas (f) (set of rules); leyes (f.pl) de un concurso (statutory regulations).

regulator, n. regulador (m).

regulatory, a. regulador, de control; **r. agency,** organismo (m) de control.

rehabilitate, v. restaurar (to good condition), rehabilitar (rank).

rehabilitation, n. rehabilitación (f), recuperación (f); **aquifer r.,** recuperación del acuífero.

reinforce, v. reforzar, fortalecer; armar (concrete).

reinforced, a. armado/a (concrete).

reinforcement, n. refuerzo (m), fortalecimiento (m); armazón (f) (of concrete); **r. bars,** armadura (f).

reinstate, v. reintegrar, rehabilitar.

reintroduction, n. reintroducción (f).

reinvest, v. reinvertir.

reinvestment, n. reinversión (f).

reject, v. desechar (to throw out, to scrap).

rejuvenation, n. rejuvenecimiento (m), reactivación (f).

relative, a. relativo/a; **r. density,** densidad (f) relativa.

relaxation, n. descanso (m) (after activity); relajamiento (f) (of rules); disminución (f) (of effort); relajación (f) (Mat); **over-r. (Mat),** sobrerelajación (f).

relay, n. relé (m) (Elec).

release, v. liberar (to free), descargar (from duty), arrojar (e.g. wastes), emitir (e.g. smoke), soltar (to let go), desbloquear (to unblock).

reliability, n. fiabilidad (f) (Gen); seguridad (f) (soundness).

reliable, a. fiable (seguro); **r. yield (of a reservoir),** caudal (m) de seguridad.

relic, n. reliquia (f).

relict, n. relicto (m), vestigio (m) (Geol).

relief, n. relieve (m) (Geog); alivio (m), descanso (Med); socorro (m), ayuda (f) (aid); **low r. (Geog),** bajo relieve; **r. channel,** canal (m) suplementario; **r. map,** mapa (m) de relieve.

relining, n. revestimiento (m); **borehole r.,** revestimiento del pozo.

remain, v. permanecer, quedarse.

remainder, n. resto (m) (Gen, Mat).

remains, n.pl. despojos (m.pl); restos (m.pl) (human); **mortal r.,** restos mortales.

remedial, a. remediador/a, reparador/a; **r. measures,** medidas (f.pl) reparadoras.

remediate, v. remediar.

remediation, n. remediación (f), recuperación (f); **contaminated land r.,** recuperación de suelo contaminado; **groundwater r.,** recuperación de agua subterránea; **in situ biological r.,** recuperación biológica in situ; **landfill r.,** recuperación de confinamiento; **on-site r.,** recuperación en el sitio, recuperación in situ; **r. technologies,** tecnologías (f.pl) de recuperación.

remedy, n. remedio (m).

remobilization, n. removilización (f).

remote-controlled, a. teledirigido/a.

removal, n. despido (m) (discharge); traslado (m) (transfer); eliminación (f) (elimination); **phosphate r.,** eliminación (f) de fosfato.

remove, v. apartar, remover, quitar.

rendering, n. revoque (m), enlucido (m) (Con).

rendzina, n. rendzina (f).

renegotiate, v. renegociar, negociar de nuevo.

renegotiation, n. renegociación (f), nueva negociación (f).

renew, v. renovar.

renewable, a. renovable; **r. energy,** energía (f) renovable; **r. resources,** recursos (m.pl) renovables.

renewal, n. renovación (f), reanudación (f) (continuation after a break); **urban r.,** renovación urbana.

rennet, n. renina (f) (rennin); cuajo (m) (curds).

rennin, n. renina (f).

renovation, n. renovación (f).
reorganization, n. reorganización (f).
reoxygenation, n. reoxigenación (f).
repeal, v. revocar, anular, abrogar.
repercussion, n. repercusión (f).
replace, v. reemplazar; suplantar, desbancar (supplant).
replacement, n. reemplazo (m).
replenishment, n. relleno (m) (refill), reabastecimiento (m) (resupply); **beach r.**, regeneración (f) de playa.
replica, n. réplica (f) (Bot, Zoo); copia (f), reproducción (f).
replicate, v. replicar.
replication, n. replicación (f) (Bot, Zoo).
repopulation, n. repoblación (f).
report, 1. v. informar; denunciar (report accident or accuse wrongdoer); 2. n. informe (m) (account); reportaje (m) (by the media); noticia (f) (piece of news), parte (m); **annual r.**, informe anual; **weather r.**, parte meteorológico.
repository, n. depósito (m); **safe r. (for waste),** depósito de seguridad.
representation, n. representación (f); descripción (f) (description); **double-log r.**, representación doble logarítmica; **graphical r.**, representación gráfica; **semi-logarithmic r.**, representación semilogarítmica.
representative, a. representativo/a; **r. sample,** muestra (f) representativa.
reprocessing, n. reprocesamiento (m); **r. plant,** planta (f) de reprocesamiento.
reproduce, v. reproducir.
reproduction, n. reproducción (f); **asexual/ sexual r.**, reproducción asexual/sexual; **vegetative r.**, reproducción vegetativa.
reproductivity, n. reproductividad (f).
reptile, n. reptil (m).
Reptilia, n.pl. reptiles (m.pl), la clase Reptilia (Zoo).
reptilian, a. reptil.
request, v. pedir, solicitar (to ask), rogar (to beg).
requirement, n. exigencia (f).
research, n. investigación (f); **r. and development (abbr. R&D),** investigación y desarrollo (abbr. I&D); **scientific r.**, investigación científica.
resemblance, n. semejanza (f).
reservation, n. reserva (f); **Indian R. (US),** territorio (m) reservado para los Indios (US).
reserve, n. reserva (f); **Biosphere Reserves (pl),** Reservas de la Biosfera; **game r.**, coto (m) de caza; **mineral r.**, reservas minerales; **nature r.**, reserva de la naturaleza; **oil reserves, petroleum reserves,** reservas de petróleo.
reservoir, n. embalse (m), represa (f); presa (f) (dam); reservorio (m) (LAm); depósito, tanque (m) (e.g. of fuel); **balancing r., compensation r.**, embalse de compensación;

covered **r.**, embalse cubierto; **flood-control r.**, embalse para control de crecidas; **impounding r.**, embalse de contención; **pumped-storage r.**, embalse de almacenamiento bombeado; **regulating r.**, embalse de regulación; **r. lining**, revestimiento (m) de embalse; **r. management**, gestión (f) de embalse; **r. operation**, operación (f) de embalse; **r. storage**, embalse de almacenamiento; **service r.**, embalse de servicio.
residential, a. residencial.
residual, a. residual.
residue, n. residuo (m); **chemical r.**, residuo químico; **dry r.**, residuo seco; **radioactive r.**, residuo radioactivo; **toxic r.**, residuo tóxico.
resilience, n. elasticidad (f) (of an object); resistencia (f) (of human body); fuerza (f) moral (temperament); rebote (m) (rebound).
resin, n. resina (f); **epoxy r.**, resina epoxi.
resistance, n. resistencia (f).
resistivimeter, n. resistivímetro (m).
resistivity, n. resistividad (f); **apparent r.**, resistividad aparente; **electrical r.**, resistividad eléctrica; **ground r.**, resistividad de terreno.
resite, v. ubicar (to reposition).
resiting, n. reubicación (f).
resolution, n. resolución (f).
resolve, v. resolver; solucionar (solve).
resonance, n. resonancia (f).
resorption, n. resorción (f), reabsorción (f).
resource, n. recurso (m).
resources, n.pl. recursos (m.pl); **available r.**, recursos disponibles; **financial r.**, recursos financieros; **genetic r.**, recursos genéticos; **groundwater r.**, recursos de agua subterránea; **human r.**, recursos humanos; **natural r.**, recursos naturales; **non-renewable r.**, recursos no renovables; **renewable r.**, recursos renovables; **water r.**, recursos hídricos.
respiration, n. respiración (f); **aerobic r.**, respiración aerobia; **anaerobic r.**, respiración anaerobia; **artificial r.**, respiración artificial; **tissue r.**, respiración tisular.
respirator, n. respirador (m).
respirometer, n. respirómetro (m).
response, n. respuesta (f); reacción (f) (Med, Zoo, Bot); **r. time**, tiempo (m) de respuesta.
responsibility, n. responsabilidad (f); **joint r.**, responsabilidad conjunta; **placing r.**, responsabilización (f).
rest, n. descanso (m) (relaxation); resto (m) (remainder).
rest on, v. yacer (Geol), descansar.
restoration, n. restauración (f), restablecimiento (m); **land r.**, restauración de terrenos; **site r.**, restauración de lugares contaminados.
restore, v. restaurar.
restriction, n. restricción (f), limitación (f); **water r.**, restricción del consumo de agua.
restructure, n. restructuración (f).

résumé, n. resumen (m) (summary); currículum vitae (US).

resuscitate, v. reanimar.

resuscitation, n. reanimación (f); **r. apparatus,** aparato (m) de reanimación.

retail, a. al por menor, al detall.

retailer, n. minorista (m, f), comerciante (m, f) al por menor.

retardation, n. retardo (m), retraso (m), atraso (m); **r. coefficient,** coeficiente (m) de retardo.

retention, n. retención (f); **flood r.,** retención de avenidas; **r. time (Treat),** período (m) de retención.

reticular, a. reticular; **r. network,** red (f) reticular.

retina, n. retina (f).

retinae, n.pl. retinas (f.pl).

retort, n. retorta (f) (Chem); réplica (f) (reply).

retouching, n. retoque (m).

retraction, n. retracción (f).

retrieval, n. recuperación (f) (Comp); reparación (f) (of a mistake); **information r. (Comp),** recuperación informativa.

return, n. retorno (m), regreso (m); vuelta (f) (journey); **diminishing r.,** rendimiento (m) decreciente; **financial r.,** retorno financiero; **r. period,** período (m) de retorno; **wastewater r.,** retorno de aguas residuales.

returnable, a. retornable.

reuse, n. reutilización (f); **wastewater r.,** reutilización de aguas residuales.

revegetation, n. revegetación (f).

revenue, n. ingresos (m.pl), entrada (f) (income); rentas (f.pl) (from taxes).

reversal, n. inversión (f).

reverse, a. inverso/a (inverse); opuesto/a, contrario/a (opposite).

revetment, n. revestimiento (m).

review, n. revisión (f), revista (f); examen (m) (examination); **literature r.,** examen de documentación; **price r.,** revisión de precios.

revitalize, v. revitalizar.

revoke, v. revocar, anular.

revolution, n. revolución (f) (Fig); vuelta (f) (turn); revolución (f), gira (f) (rotation); **green r.,** revolución verde; **industrial r.,** revolución industrial; **revolutions per minute,** revoluciones por minuto (abbr. rpm).

reward, n. galardón (m), premio (m).

reworked, a. retrabajado/a.

RF, n. FR (f); **radio frequency,** radiofrecuencia (f).

Rh, a. Rh; **rhesus,** rhesus (m).

rhabdosome, n. rhabdosoma (f).

Rheatian, n. Retiense (m).

rhenium, n. renio (m) (Re).

rheology, n. reología (f).

Rhesus, a. Rhesus; **R. factor,** factor (m) Rhesus; **r. monkey,** macaco (m) de la India.

rheumatism, n. reumatismo (m), reuma (f), reúma (f).

rhinoceros, n. rinoceronte (m).

rhizoid, n. rizoide (m).

rhizome, n. rizoma (m).

rhizopod, n. rhizópodo (m).

rhizosphere, n. rizoesfera (f).

rhodamine, n. rhodamina (f), rodamina (f).

rhodium, n. rodio (m) (Rh).

rhodochrosite, n. rodocrosita (f).

Rhodophyta, n.pl. rodofíceas (f.pl), Rhodophyta (f.pl) (Bot, class).

rhomboid, n. romboide (m) (Mat).

rhynchonellid, n. rhynconélido (m).

rhyolite, n. riolita (f).

rhythm, n. ritmo (m); **diurnal r.,** ritmo diurno, ritmo circadiano.

ria, n. ría (f) (estuary).

ribbing, n. varillaje (m).

riboflavin, n. riboflavina (f) (vitamin B2).

rice, n. arroz (m); **r. farmer,** arrocero/a (m, f); **r. field,** arrozal (m); **r. grower,** arrocero/a (m, f); **r. paddy,** arrozal (m).

rickets, n. raquitismo (m).

riddlings, n.pl. cerniduras (f.pl), granzas (f.pl) (screenings).

ridge, n. cadena (f) (Geog); caballete (m) (of roof), caballón (m) (between field furrows); **hilly r.,** lomada; (f); **low r.,** loma (f); **whaleback r.,** loma (f) en forme de lomo de ballena.

riebeckite, n. riebeckita (f).

riffle, n. rabión (m) (rapids), rápido (m) de poca altura.

rift, n. grieta (f), fisura (f), rift (m) (fissure); claro (m) (in clouds); **r. block,** bloque (m) de rift.

rig, n. equipo (m), aparejo (m); torre (m) (mast); **drilling r.,** torre de perforación, equipo perforador.

rigging, n. montaje (m) (of a machine), equipo (m) (equipment), cableado (m) (cabling); aparejo (m), jarcia (f) (Mar).

right, n. derecho (m), justicia (f); bien (m); **by r.,** por derecho; **civil rights,** derechos, derechos civiles; **r. of, r. to,** derecho a.

right-of-way, n. servidumbre (f) de paso, derecho de paso (land); preferencia de paso (roads), prioridad (f) (roads).

rigid, a. rígido/a, inelástico/a.

rigidity, n. rigidez (f).

rigor, n. rigor (m).

rigorous, a. riguroso/a.

rigour, n. rigor (m).

rill, n. cava (f) (Geog).

rimstone, n. roca (f) de aureola de contacto.

ring, n. sonido (m) (sound), círculo (m) (circle), anillo (m) (of tree, on finger), anilla (f) (for birds), red (f) (Chem); **annual r. (Bot),** anillo anual; **r. main (Elec),** canalización (f) circular, red (f) electrica circular; **r. pull,** anilla (f); **r. road (esp. GB),** vía (f) de circunvalación.

ringworm, n. tiña (f).

riparian, a. ribereño/a, ripario/a; ripícola (river-dwelling); **r. rights,** derechos (m.pl) ribereños.

ripening, n. maduración (f).

ripple, n. onda (f), rizo (m) (small wave); murmullo (f) (of sound); **r. mark,** ondulita (f), ondula (f), marca (f) de corriente, ripple mark.

ripply, a. rizado/a.

riprap, n. ripio (m) (broken stone); pedraplén (m) (rough stone base).

rise, 1. v. subir, levantarse; 2. n. ascenso (m), ascensión (f), elevación (f), subida (f) (Gen); cuesta (f) (of slope); nacimiento (m) (of river); surgencia (f) (upward movement); **capillary r.,** ascenso capilar; **continental r.,** ascensión continental; **saline r.,** ascenso salino, elevación salina.

risk, n. riesgo (m), peligro (m); **health r.,** riesgo para la salud; **r. analysis,** análisis (m) de riesgos; **r. assessment,** evaluación (f) de riesgos, valoración (f) de riesgos; **r. management,** gestión (f) de riesgos.

river, n. río (m); **braided r.,** río anastomosado; **graded r.,** río en equilibrio; **intermittent r.,** río intermitente, río estacional; **misfit r.,** río desproporcionado; **reach of a r.,** tramo (m) recto de un río, parte (f) recta de un río; **r. bank,** margen (m) de río; **r. basin,** cuenca (f) del río; **r. bed,** lecho (m) del río, madre (m) del río; **r. capture,** captura (f) fluvial; **r. catchment,** captación (f) en ríos; **r. channel,** cauce (m), cauce de un río; **r. meandering,** serpenteo (m) de río, meandros (m.pl) fluviales; **r. mouth,** boca (f) de río, desembocadura (f) de río; **r. pollution,** contaminación (f) de los ríos, contaminación fluvial; **r. regulation,** regulación (f) fluvial; **r. valley,** valle (m) fluvial; **source of r.,** nacimiento (m) de un río; **truncated r.,** río truncado.

riverbank, n. ribera (f), orilla (f), banda (f).

riverbed, n. lecho (m) de río.

riverside, n. orilla (f), ribera (f), margen (m) de río.

rivet, 1. v. remachar; 2. n. remache (m) (also riveting).

rivulet, n. arroyo (m).

RNA, n. ARN; **ribonucleic acid,** ácido (m) ribonucleico.

road, n. camino (m) (route), carretera (f) (main road); **approach r.,** carretera de acceso, camino de acceso; **class A r. (GB),** carretera general, carretera nacional; **class B r. (GB),** carretera secundaria, carretera comarcal; **country r.,** carretera vecinal; **main r.,** carretera principal; **ring r. (bypass),** carretera de circunvalación; **r. network,** red (f) de carreteras; **r. safety,** seguridad (f) vial; **r. works,** obras (f.pl); **secondary r.,** carretera secundaria, carretera comarcal; **unmade r.,** **unmetalled r., unsurfaced r.,** carretera sin firme.

roadside, 1. n. borde (m) de la carretera; 2. a. al borde de la carretera; **r. plants (Bot),** plantas (f.pl) viarias.

robotics, n. robótica (f).

roche mountonnée, n. roca (f) aborregada.

rock, n. roca (f), peña (f); **altered r.,** roca alterada; **bedrock,** roca firme, sustrato (m) rocoso; **country r.,** roca madre; **igneous r.,** roca ígnea; **metamorphic r.,** roca metamórfica; **reservoir r.,** roca almacén, roca productiva; **r. cover,** cubierta (f) rocosa; **r. fall,** caída (f) de roca; **r. mantle,** cubierta (f) de roca, regolito (m); **r. salt,** sal (f) de roca, halita (f); **r. waste,** ganga (f), estéril (m); **sedimentary r.,** roca sedimentaria.

rocket, n. cohete (m).

rockfall, n. desprendimientos (m.pl) de roca.

rocky, a. rocoso/a; **r. ground,** peñascal (m); **r. place,** roquedal (m); **r. spur,** contrafuerte (m).

rod, n. barra (f) (bar, pole); vástago (m) (of piston); caña (f) (for fishing); medida de longitud equivalente a 5.029 metros; **drilling r.,** vástago de perforación; **ranging r.,** jalón (m); **r. and line (fishing),** palillo.

rodding, n. formación (f) de estrías (Geol).

rodent, n. roedor (m).

rodenticide, n. rodenticida (m).

roe, n. corzo/a (m, f) (deer); hueva (f) (fish eggs), freza (f) (spawn).

roentgen, n. roentgen (m), roentgenio (m); **r. equivalent man (abbr. REM),** roentgen equivalente para el hombre.

roll, n. rollo (m) (of paper); carrete (m) (of film); nómina (f) (list of names); registro (m) (register); balanceo (m) (swaying movement).

roller, n. apisonadora (for road), rodillo (m) (cylinder); ola (f) grande (sea wave); **road r.,** apisonadora; **sheepsfoot r.,** rodillo con patas de cabra; **vibrating r.,** apisonadora vibratoria, rana.

ROM, n. ROM (f); **read-only memory (Comp),** memoria (f) de sólo lectura.

Roman, a. romano/a.

roof, n. tejado (m); techo (m) (ceiling); **r. pitch,** inclinación (f) de tejado; **r. tie,** tirante (m) de tejado; **r. truss,** armadura (f) de cubierta; **sliding r.,** techo corredizo.

roofing, n. techumbre (m), techo (m).

root, n. raíz (f); **adventitious roots,** raíces adventicias; **buttress roots,** raíces caulógenas, raíces zancos; **fibrous roots,** raíces fibrosas; **r. constant,** constante (m) de raíz; **r. hair (Bot),** pelo (m) radical, pelo absorbente; **r. nodule,** radulo (m) radicular; **r. pressure,** presión (f) radical; **r. zone,** zona (f) de raíz; **square r. (Mat),** raíz cuadrada; **to r. up (v) (Agr),** desarraigar, arrancar de raíz; **to take r. (v) (Agr),** echar raíces.

rose, n. rosal (m) (bush); rosa (f) (flower);

alcachofa (f) (on watering can); **wind r.** (Met), rosa de los vientos.

rosemary, n. romero (f).

rostrum, n. rostro (m).

rot, 1. v. pudrirse, descomponerse; 2. n. putrefacción (f).

rotary, a. rotatorio/a, giratorio/a, rotativo/a; **r. drilling**, perforación (f) por rotación.

rotation, n. rotación (f); giro (m) (turn); **crop r.**, rotación de la cosecha, rotación de cultivos.

rotational, a. rotacional; **r. slide**, deslizamiento (m) rotacional.

rotavation, n. roturación (f).

rotavirus, n. rotavirus (m).

rotor, n. rotor (m).

rotten, a. podrido/a, putrefacto/a.

roughcast, n. revestimiento (m) tosco.

roughness, n. aspereza (f), rugosidad (f) (of surface); desigualdad (f) (of road); agitación (f) (of sea); **r. coefficient**, coeficiente (m) de rugosidad.

roundabout, n. cruce (m) giratorio; glorieta (f) (on roads).

roundness, n. redondez (f).

roundworm, n. ascáride (m).

routine, a. rutinario/a; **r. check**, prueba (f) rutinaria.

rpm, n. rpm; **revolutions per minute**, revoluciones (f.pl) por minuto.

rubber, n. goma (f), caucho (m), hule (m) (LAm) (substance); gomero (m), árbol (m) de caucho (tree); goma de borrar (eraser); condón (m) (US); **foam r.**, goma-espuma, hule-espuma (LAm); **r. industry**, industria (f) de caucho, industria de cauchera; **r. plantation**, cauchal (m).

rubbish, n. basura (f) (refuse, trash); escombros (m.pl) (rubble); **r. bin**, basurero (m); **r. collection**, recogida (f) de basura; **r. dump**, vertedero (m), tiradero (m) (Mex); **r. sack**, bolsa (f) de basura; **r. tip**, vertedero (m), tiradero (m) (Mex).

rubble, n. ripio (m) (riprap), escombros (m) (ruins), detrito (m) (detritus) (m), cascajo (m) (screenings, scrap), roca (f) fragmentada (broken rock).

rubblework, n. mampostería (f).

rubella, n. rubéola (f).

rubidium, n. rubidio (m) (Rb).

ruby, n. rubi (m).

rudaceous, a. rudáceo/a.

ruderal, a. ruderal.

rudite, n. rudita (f).

rugged, a. abrupto/a.

rugosity, n. rugosidad (f); **coefficient of r.**, coeficiente (m) de rugosidad.

ruin, 1. v. asolar (to destroy); arruinar (to destroy financially); 2. n. ruina (f), escombros (m.pl) (of a building).

rules, n. regla (f), reglamento (m).

rumen, n. rumen (m), herbario (m).

ruminant, n. rumiante (m).

runnel, n. riatillo (m).

runner, n. rastrero (m), tallo (m) rastrero (Bot).

runoff, n. escorrentía (f), agua (f) de escorrentía, flujo (m) de escorrentía; **r. coefficient**, coeficiente (m) de escorrentía; **seasonal r.**, escorrentía estacional.

runway, n. pista (f) de aterrizaje.

Rupelian, n. Rupeliense (m).

rupestrian, a. rupestre.

rupture, n. ruptura (f), rotura (f) (fracture), hernia (f) (Med).

rural, a. rural; **r. development**, desarrollo (m) rural; **r. planning**, planificación (f) rural; **r. population**, población (f) rural; **r. water supply**, abastecimiento (m) rural de agua.

rush, n. ímpetu (m) (impetus); prisa (f), apuro (m) (LAm); junco (m) (Bot); **r. hour**, hora (f) punta; **r. job**, trabajo (m) urgente; **r. matting**, estera (f) de juncos.

rush-bed, n. juncal (m).

rust, n. orín (m), herrumbre (f), moho (m) (of metal); corrosión (f), oxidación (f) (acción); roya (f), tizon (m) (hongos).

rustication, n. labrado (m) tosco (old looking).

rusty, a. oxidado/a, mohoso/a, herrumbroso/a.

ruthenium, n. rutenio (m) (Ru).

rutile, n. rutilo (m).

rye, n. centeno (m).

S

sabin, n. sabinio (m) (unit of sound absorption).

sac, n. saco (m), bolsa (f); **air s.,** saco aéreo; **pollen s.,** saco polínico; **yolk s.,** saco vitelino.

saccharide, a. sacárido/a.

saccharine, n. sacarina (f).

saccharomycete, n. sacaromiceto (m).

saccharose, n. sacarosa (f).

saddle, n. estructura (f) en silla de montar (Geog); depresión (f) estructural (Geol); silla (f) (of horse).

safe, a. seguro/a.

safeguard, 1. v. salvaguardar, proteger (to protect); 2. n. salvaguardia (f), salvaguarda (f), protección (f).

safekeeping, n. custodia (f); **to put into s. (v),** poner a buen recaudo.

safety, n. seguridad (f), protección (f); **s. device,** dispositivo (m) de seguridad; **s. equipment,** equipo (m) de seguridad; **s. glass,** vidrio (m) de seguridad; **s. lamp,** lámpara (f) de seguridad; **s. margin,** margen (m) de seguridad; **s. measures, s. precautions,** medidas (f.pl) de seguridad; **s. valve,** válvula (f) de seguridad.

saffron, n. azafrán (m).

sag, 1. v. combar; 2. n. hundimiento (m); **oxygen s.,** hundimiento de oxígeno.

sailboarding, n. windsurf (m), surf (m) a vela.

sailing, n. navegación (f); **s. boat,** barco (m) de vela, velero (m).

sailor, n. marinero (m), marino (m).

salamander, n. salamanquesa (f), salamandra (f).

sale, n. venta (f) (transaction); liquidación (f), saldo (m), rebajas (m) (of goods at reduced price); subasta (f) (auction); **"for sale",** "se vende"; **s. price,** precio (m) de venta; **sales check (US),** factura (f).

salesman, n. vendedor (m) (seller); representante (m) (representative); **travelling s.,** viajante (m); viajante de comercio.

saleswoman, n. vendedor (f) (seller); representante (f) (representative).

salicylate, n. salicilato (m).

saline, a. salino/a; **s. wedge (Geol),** cuña (f) salina.

salinity, n. salinidad (f).

salinization, n. salinización (f); **soil s.,** salinización de suelo.

saliva, n. saliva (f).

salmon, n. salmón (m).

salt, 1. v. salar; 2. n. sal (f); **common s.,** sal común; **dissolved salts,** sales disueltas; **Epsom Salts,** sal de Epsom; **mineral salts,** sales minerales; **rock s.,** sal gema; **s. lake bed,**

playa (f) (Mex, US); **s. lake,** salar (m); **s. marsh,** marisma (f); **s. mine,** salina (m); **s. pan,** costa (f) salina, salita (f); **s. works,** salina (m).

saltation, n. erosión (f) por saltación.

salting, n. saladura (f); salazón (m) (LAm).

saltmarsh, n. marisma (f), salador (m).

saltpeter, n. salitre (m), nitrato (m) potásico.

salty, a. salado/a.

Salvadoran, 1. n. salvadoreño/a (m, f); 2. a. salvadoreño/a.

Salvadorian, 1. n. salvadoreño/a (m, f); 2. a. salvadoreño/a.

samarium, n. samario (m) (Sm).

sample, 1. v. muestrar; 2. n. muestra (f); **grab s.,** muestra tomada por cubeta (by sampler), muestra tomada por draga (by dredger), muestra tomada al azar (at random); **invertebrate s.,** muestra de invertebrados; **random s.,** muestra aleatoria; **representative s.,** muestra representativa; **river s.,** muestra de río; **unaltered s.,** muestra inalterada; **water s.,** muestra de agua.

sampler, n. muestreador (m) (equipment); muestreador/a (m, f) (person); **automatic s.,** muestreador automático.

sampling, n. muestreo (m), recolección de muestra (f); **s. point,** punto (m) de muestreo.

sanction, n. penalización (f), sanción (f).

sanctuary, n. santuario (m) (sacred place), refugio (m) (e.g. for birds).

sand, 1. v. enarenar; 2. n. arena (f); **quartz s.,** arena de cuarzo; **s. bar,** barra (f) arenosa, barra costera; **s. drift,** acumulación (f) de arena, depósitos (m) de arena; **s. pit,** arenero (m); **sands (pl) (beach),** playa (f); **sharp s.,** arena angulosa; **silica s.,** arena de sílice; **volcanic s.,** arena volcánica.

sandbank, n. bajío (m), banco (m) de arena.

sandblast, 1. v. golpetear con arena; 2. n. chorro (m) de arena.

sandblasting, n. golpeteo (m) de arena, chorreo (m) con arena.

sandpit, n. cantera (f) de arena.

sandstone, n. arenisca (f); **coarse s.,** arenisca de grano grueso.

sandstorm, n. tormenta (f) de arena.

sandy, a. arenoso/a.

sanitary, a. sanitario/a; **s. engineering,** ingeniería (f) sanitaria; **s. installation,** instalación (f) sanitaria.

sanitation, n. saneamiento (m), higiene (f) del medio; **emergency s.,** saneamiento de emergencia; **rural s.,** saneamiento rural; **s. policy,** política (f) de saneamiento.

sanitize, v. esterilizar, higienizar, depurar.

Santonian, n. Santoniense (m).

sap, n. savia (f); **cell s.,** sustancias (f.pl) de reserva.

saponification, n. saponificación (f).

sapphire, n. zafiro (m).

saprobiotic, a. saprobiótico/a; **s. zone,** zona (f) saprobiótica.

saprolite, n. saprolito (m).

saprolith, n. saprolito (m).

saprophyte, n. saprófito (m).

saprophytic, a. saprófito/a.

saprozoic, a. saprozoico/a.

sapwood, n. albura (f).

sarcoma, n. sarcoma (m).

sardine, n. sardina (f).

sarsen, n. bloque (m) errático de arenisca silicea (sarsen stone).

satellite, n. satélite (m); **s. remote sensing,** teledetección (f) por satélite; **s. dish,** antena (f) parabólica; **s. town,** ciudad (f) satélite; **space s.,** satélite espacial; **telecommunications s.,** satélite de telecomunicaciones.

saturated, a. saturado/a; **s. zone,** zona (f) saturada.

saturation, n. saturación (f); **critical s.,** saturación crítica; **degree of s.,** grado (m) de saturacion; **s. index,** índice (m) de saturación; **s. point,** temperatura (f) de saturacion.

Saturn, n. Saturno (m).

savanna, n. sabana (f); pampa (f) (CSur), llano (m) (Ven) (llanura).

savannah, n. sabana (f); pampa (f) (CSur), llano (m) (Ven) (llanura).

saving, n. ahorro (m), economía (f); **energy s.,** ahorro energético; **water s.,** ahorro del agua.

saw, n. sierra (f) (tool); **band s.,** sierra de cinta; **pit s.,** sierra abrazadera; **power s.,** aserradora (f).

sawdust, n. serrín (m).

sawmill, n. aserradero (m), serrería (f).

sawyer, n. aserrador (m).

scabies, n. sarna (f).

scaffolding, n. andamio (m), andamiaje (m).

scalar, a. escalar.

scale, 1. v. escalar; dibujar a escala (map, drawing); **to s. down,** reducir proporcionalmente, reducir a escala (map, drawing); **to s. up,** aumentar escala (map, plan); aumentar proporcionalmente; 2. n. escala (f); amplitud (f) (of a project); extensión (f) (of an event); escama (f) (of a fish); incrustaciones (f.pl) (in a boiler); óxido (m) (rust); **Beaufort s. (for wind measurement),** escala de Beaufort; **centigrade s.,** escala centígrada; **drawn to s. (a),** hecho/a a escalar; **Fahrenheit s.,** escala de Fahrenheit; **large-s. (a),** a gran escala; **pH s.,** escala de pH; **Richter S.,** escala de Richter; **s. drawing,** dibujo (m) hecho a escala; **s. model,** modelo (m) a escala; **s. of magnitude,** escala de magnitud; **s. of measurement,** escala graduada; **small-s. (a),** a pequeña escala.

scaling, n. descascarillado (m) (removing scale).

scallop, n. venera (f), vieira (f) (shellfish); festón (m), onda (f) (shape).

scaly, a. escamoso/a.

scandium, n. escandio (m) (Sc).

scanner, n. detector (m); **direction s.,** detector direccional.

scar, n. cicatriz (f).

scarcity, n. escasez (f), falta (f); rareza (f) (rarity).

scarifier, n. escarificador (m).

scarify, v. escarificar.

scarp, n. escarpe (m); **fault s.,** escarpe de falla; **s. slope,** pendiente (f) del escarpe.

scattered, a. disperso/a, esparcido/a.

scavenger, n. (on waste tips) ciruja (f) (Arg, Ur), cartonero (m) (Arg, Ch), segregador (Bol, Par, Perú), catador (m) (Br), cartonero (m) (Ch), cachuero (m) (Ch), pepenador (m) (C.Rica, ElS, Méx, Nic, Pan, rebuscador (m) (Ven); **s. beetle,** necróforo (m).

scenario, n. escenario (m).

schedule, n. lista (f), inventario (m) (list); programa (m), plan (m) (programme); **work s.,** plan (m) de trabajo.

schematic, a. esquemático/a.

schematized, a. esquematizado/a.

scheme, n. esquema (m), proyecto (m) (project), programa (m) (programme); **hydroelectric s.,** proyecto hidroeléctrico.

schist, n. esquisto (m).

schistose, a. esquistoso/a.

schistosity, n. esquistosidad (f).

schlieren, n. schlieren (m).

school, n. escuela (f).

science, n. ciencia (f); **computer s.,** informática (f); **environmental s.,** ciencia ambiental, ciencia del medio.

scientific, a. científico/a.

scientist, n. científico/a (m, f); **environmental s.,** científico/a ambiental.

scintillation, n. centelleo (m); **s. counter,** contador (m) de centelleo.

scintillator, n. centelleante (m).

scion, n. retoño (m), renuevo (m).

sclerophyllous, a. esclerófilo/a; **s. forest,** bosque (m) esclerófilo.

scoop, 1. v. cavar (to excavate); achicar (to bail water from a boat); sacar (to profit); 2. n. cuchara (f) (of an excavator); achicador (m) (for bailing out); recogedor (m) (for rubbish).

scope, n. ámbito (m) (ambit), competencia (f) (ability).

scoriaceous, a. escoriáceo/a; **s. lava,** lava (f) escoriácea.

scour, v. restregar, fregar (to rub); derrubiar, erosionar (Geol).

scouring, n. desgaste (m), escariación (f), estropajado (m); **stream s.,** escariación fluvial.

scrap, n. pedacito (m), fragmento (m) (piece); restos (m.pl), sobras (f.pl) (leftovers); chatarra (f) (metal, etc.); **s. iron,** chatarra de hierro, desecho (m) de hierro; **s. metal,** chatarra de metal.

scraper, n. raspador (m), escarbadora (f), raedora (f) (machine); rascador (m) (Treat); **earth s.,** máquina (f) escarbadora.

scrapyard, n. parque (m) de chatarra.

scratch, n. raspadura (f).

scree, n. derrubio (m), pedrero (m), escombros (m) de talud, canchal (m); **s. slope,** pendiente (m) de derrubios.

screen, n. reja (f) (on window), rejilla (f) (gridwork), criba (f) (sieve), biombo (m) (moveable partition), pantalla (f) (computer), malla (f) (mesh); **band s.,** criba de cinta; **bar s.,** cinta (f) de barras; **bridge-slotted well s.,** filtro (m) de puentecillos, rejilla de pozo de puentecillos; **drum s.,** criba de tambor, tromel; **slotted well s.,** filtro (m) en rejilla, rejilla de pozo de persiana.

screening, n. examen (m) (Chem, characterization); chequeo (m) (Med); protección (f) (protection); cribado (m) (sieving); procedimiento (m) de selección (selection of candidates).

screenings, n.pl. cerniduras (f.pl); granzas (f.pl) (siftings, riddlings), cribaduras (f.pl) (e.g. of crushed stone).

screw, n. tornillo (m); tuerca (f) (female screw); **Archimedes s.,** tornillo de Arquímedes; **s. jack,** gato (m) de husillo.

scrub, 1. v. fregar (Gen); depurar (Eng, e.g. gas); cancelar (Fam); 2. n. matorral (m), maleza (f) (bushes); monte (m) bajo (undergrowth); manigua (swampy scrubland).

scrubbing, n. fregado (m) (e.g. clothes); depuración (f), lavado (m) (gases); **dry s. (Treat),** captación (f) seca de gases; **gas s.,** lavado (m) de gases; **wet s. (Treat),** captación (f) húmeda de gases.

scrubby, a. achaparrado/a (Bot, stunted).

scrubland, n. monte (m) bajo, pajonal (m) (LAm).

scum, n. espuma (f) (dirty foam).

Scythian, n. Escitiense (m).

sea, n. mar (m. y f.); **beside the s.,** a orillas (f.pl) del mar; **choppy s.,** mar rizada; **heavy s.,** mar gruesa; **rough s.,** mar agitado; **s. lane,** ruta (f) marítima.

seaboard, n. costa (f) marina.

seal, 1. v. sellar; cerrar, pegar (put on document); 2. n. sello (m) (official stamp); sellador (m) (in borehole); pegamento (m) (of package); junta (f) (e.g. of door); foca (f) (Zoo); **monk s.,** foca monje.

sealant, n. sellante (m).

sealed, a. sellado/a, cerrado/a (closed).

seam, n. veta (f), filón (m), manto (m), criadero (m), capa (f) (Min); **coal s.,** capa de carbón.

seaman, n. marino (m), marinero (m); **merchant s.,** marino mercante.

seamount, n. monte (m) submarino.

seaplane, n. hidroavión (f) (flying boat).

search, 1. v. buscar (to look for); registrar (e.g. in a drawer); 2. n. búsqueda (f) (look for), registro (m) (e.g. of a house).

searching, n. rebuscador (m).

seashell, n. caracol/a (m, f).

seashore, n. orilla (f) del mar, borde (m) del mar.

seaside, n. borde (m) del mar.

season, n. temporada (f), estación (f); **high/low s.,** alta/baja temporada; **mating s.,** época (f) de reproducción.

seasonal, a. estacional (of the seasons); temporal (temporary, e.g. of work).

seated, a. sentado/a; **deep-s.,** emplazamiento (m) profundo.

seawall, n. rompeolas (f, pl), muelle (m), malecón (m), dique costero para preservar de la erosión del mar.

seaweed, n. alga (f), alga marina.

seaworthy, a. marinero/a, en buen estado para navegar.

sebaceous, a. sebáceo/a.

sec, n. sec (m) (Mat, abbr. of secant).

secant, n. secante (m) (Mat, abbr. sec).

secondary, 1. n. secundario (m); 2. a. secundario/a; **s. cell (Elec),** acumulador (m); **s. school,** escuela (f) secundaria; **s. settling tank,** sedimentador (m) secundario; **s. treatment,** tratamiento (m) secundario.

secrete, v. secretar.

secretion, n. secreción (f).

section, n. sección (f) (Gen); sector (m) (of community), barrio (m) (of town); página (f) (of newspaper); corte (m) (cut); **cross-s.,** corte transversal; **profile,** a. perfil (m) transversal; **thin-s. slide (for microscope),** lámina (f) delgada; **type s.,** sección típica.

sector, n. sector (m); **energy s.,** sector energético; **industrial s.,** sector industrial; **private s.,** sector privado; **public s.,** sector público, sector estatal.

security, n. seguridad (f); garantía (f) (guarantee); **social s.,** seguridad social.

sedative, n. sedante (m), sedativo (m).

sedentary, a. sedentario/a.

sediment, n. sedimento (m); **s. load,** carga (f) de sedimento; **s. transport,** transporte (m) de sedimento; **unconsolidated s.,** sedimento no consolidado.

sedimentary, a. sedimentario/a.

sedimentation, n. sedimentación (f).

sedimentology, n. sedimentología (f).

seed, n. semilla (f); pepita (f) (of a fruit); simiente (f) (Agr); freza (f) (spawn); **s. drill,** sembradora (f).

seed-coat, n. tegumento (m).

seeding, n. el sembrar (n), siembra (f); **artificial rain s.,** siembra (f) para lluvia artificial.

seedling, n. planta (f) de semillero, plantón (m).

seep, v. filtrarse, rezumar.

seepage, n. rezume (m) (leak), filtración (f) (through), infiltración (f) (into), secreción (f) (secretion), absorbencia (f) (absorption).

segment, n. segmento (m).

segregation, n. segregación (f).

seine, n. red (f) de arrastre; jábega (f) (sweep net, dragnet); rapeta (f) (Galicia, Sp); **purse s. net (fishing) (Galicia, Sp)**, cerco (m) de jareta; **s. fishing**, pesca (f) con jábega (f).

seismic, a. sísmico/a; **s. network**, red (f) sísmica; **s. phenomena**, fenómeno (m) sísmico; **s. resistant**, sismorresistente.

seismicity, n. sismicidad (f).

seismograph, n. sismograma (m).

seismology, n. sismología (f).

seismometer, n. sismómetro (m).

selection, n. selección (f); **artificial s. (Biol)**, selección artificial; **natural s. (Biol)**, selección natural.

selectivity, n. selectividad (f).

selenite, n. selenita (f).

selenium, n. selenio (m) (Se).

self-assessment, n. autoevaluación (f).

self-cleaning, a. autolimpiable.

self-fertilization, n. autofecundación (f).

self-financing, a. autofinanciado/a.

self-generation, n. autogeneración (f).

self-governing, a. autónomo/a.

self-pollination, n. autopolinización (f).

self-protection, n. autoprotección (f).

self-purification, n. autodepuración (f) (e.g. of rivers); **biological s.-p. (e.g. in rivers)**, autodepuración biológica.

self-regulating, a. autorregulador/a.

self-regulation, n. autorregulación (f).

self-sufficiency, n. autosuficiencia (f); auto-abastecimiento (m) (self-supply).

sell, v. vender.

seller, n. vendedor/a (m, f) (who sells); comerciante (m,f) (dealer).

selva, n. selva (f).

semen, n. semen (m).

semiarid, a. semiárido/a.

semichord, n. semicuerda (f).

semiconductor, n. semiconductor (m).

semi-confined, a. semiconfinado/a.

semi-confinement, n. semiconfinamiento (m).

semi-consolidated, a. semiconsolidado/a.

semi-empirical, a. semiempírico/a.

semi-infinite, a. semiinfinito/a.

seminar, n. seminario (m).

seminatural, a. seminatural; **s. ecosystem**, ecosistema (m) seminatural.

senate, n. senado (m).

senescence, n. senescencia (f).

senility, n. senilidad (f).

sense, n. sentido (m) (Gen); sensación (f) (feeling); juicio (m) (sanity); **common s.**, sentido común; **good s.**, buen sentido; **s. of**

direction, sentido de la dirección; **s. of smell**, olfato (m); **s. organ**, órgano (m) del sentido.

sensing, n. detección (f); **remote s.**, teledetección.

sensitive, a. sensible; **less s. waters (EU)**, zona (f) menos sensible.

sensitivity, n. sensibilidad (f); **s. analysis (Mat)**, análisis (m) de sensibilidad.

sensor, n. sensor (m); **remote s.**, sensor remoto.

sepal, n. sépalo (m).

separate, v. separar, dividir.

separated, a. separado/a, dividido/a.

separation, n. separación (f), espaciamiento (m).

separator, n. separador (m); **cyclone s.**, separador ciclónico; **inclined plate s. (Treat)**, separador con platos inclinados; **lamella s.**, separador lamella; **vortex s.**, separador de vórtice.

sepiolite, n. sepiolita (f).

septic, a. séptico/a; **s. tank**, fosa (f) séptica, tanque (m) séptico.

septicaemia, n. septicemia (f), envenenamiento (m) de la sangre.

septicemia, n. septicemia (f).

septicity, n. septicidad (f).

septum, n. septo (m).

sequence, n. secuencia (f), orden (f).

sequestration, n. confiscación (f) (of property), secuestro (m) (of a person).

sera, n.pl. sueros (m.pl) (see: serum).

serac, n. serac (m).

sere, n. serie (f).

sericite, n. sericita (f).

series, n.pl. serie (f); **calc-alkali s.**, serie alcalina cálcica; **chronological s.**, serie cronológica; **continuous s.**, serie continua; **deterministic s.**, serie determinista; **harmonic s.**, serie armónica; **soil s.**, serie de suelos; **stochastic s.**, serie estocástica; **synthetic time s.**, serie sintética de tiempo.

serpentine, n. serpentina (f).

serpentinite, n. serpentinita (f).

serrate, a. serrado/a.

serrated, a. serrado/a.

Serravallian, n. Serravaliense (m).

serum, n. suero (m); **blood s.**, suero sanguíneo.

service, n. servicio (m); **24-hour s.**, servicio permanente, servicio de 24 horas; **after-sales s.**, servicio postventa; **public s.**, servicio público; **repair s.**, servicio de reparaciones; **social s.**, servicio social.

servomechanism, n. servomecanismo (m).

servosystem, n. servosistema (m).

sesame, n. sésamo (m).

sesquioxide, n. sesquióxido (m).

sessile, a. sésil.

set, n. juego (m), aparato (m) (apparatus), grupo (m), familia (f); **joint s. (Geol)**, familia (f) de diaclasas; **s. of laws**, conjunto (m) de leyes; **s. of measures**, serie (f) de medidas; **s. of sieves**, juego de tamices.

setback, n. contratiempo (m).

setting, n. colocación (f) (putting down); fraguado (m) (cement); puesta (f) (of the sun); escenario (m) (of a film).

setting up, n. montaje (m); **s. u. an instrument**, montaje de un instrumento.

settle, v. asentar, sedimentarse.

settlement, n. hundimiento (m), asiento (m) (of a building); colonización (f), población (f) (of land, country), asentamiento (m) (emplacement); **human s.**, asentamiento humano; **rural s.**, asentamiento rural; **squatter s.** (shanty town), asentamiento precario, barrio (m) marginal, barriada (f), conventillo (m), favela (f), tugurio (m), villa (f) miseria, zona (f) marginal.

settling, n. sedimentación (f), asentamiento (m), hundimiento (m); **s. basin**, balsa (f) de asentamiento; **s. tank**, cubo (m) de clarificación.

severity, n. severidad (f).

sewage, n. aguas (f.pl) residuales, aguas cloacales, aguas negras, aguas servidas, líquidos (m.pl) cloacales; **settled s.**, aguas cloacales decantadas; **s. digester**, digestor (m) de los fangos cloacales; **s. fungus (in rivers)**, hongo (m) de aguas cloacales; **s. lagoon**, laguna (f) para verter las aguas cloacales; **s. purification**, depuración (f) de aguas residuales; **s. sludge**, fangos (m.pl) cloacales, lodos (m.pl) de aguas cloacales; **s. sludge dewatering**, deshidratación (f) de los fangos cloacales; **s. sludge utilization on land**, utilización (f) agrícola de los fangos de alcantarilla; **s. treatment**, depuración (f) de las aguas residuales; **s. works**, planta (f) de depuración de aguas residuales.

sewer, 1. v. dotar de alcantarillado; 2. n. alcantarilla (f), cloaca (f), albañal (m); **foul s.**, cloaca de aguas negras; **intercepting s.**, alcantarilla interceptadora; **main s.**, colector (m), sistema (m) colector; **relief s.**, aliviadero (m); **s. gas**, gases (m.pl) cloacales; **s. system**, red (f) cloacal, red de alcantarillado; **storm s.**, drenaje (m) para tormentas; **stormwater s.**, alcantarillado (m) para recoger solamente las aguas pluviales.

sewerage, n. alcantarillado (m), sistema (m) de alcantarillado; redes (f.pl) de desagüe (LAm); **combined s. overflow**, rebose (m) de alcantarillado combinado; **combined s. systems**, sistemas de alcantarillado combinado.

sewered, a. con alcantarillado; **s. inlet**, boca (f) de alcantarilla.

sex, n. sexo (m); **s. hormone**, hormona (f) sexual.

sexual, a. sexual.

shade, 1. v. dar sombra; 2. n. sombra (f), tinte (m) (colour).

shadow, n. sombra (f), oscuridad (f) (darkness); **rain s.**, zona (f) al abrigo de las lluvias.

shady, a. umbrío/a.

shaft, n. pozo (m) (of mine, of well); mango (m) (handle); árbol (m), eje (m) (transmission axis); aguja (f) (of tower); manga (f) (e.g. air shaft); **air s.**, pozo aireado, pozo con respiración; **cam s.**, eje de levas; **ventilation s.**, manga (f) de ventilación.

shake, v. agitar, sacudir.

shaker, n. vibrador (m), criba (f).

shaking, n. sacudida (f) (e.g. of earthquake).

shale, n. lutita (f), esquisto (m) arcilloso; **oil s.**, esquisto bituminoso, lutita bituminosa.

shallow, a. poco profundo/a, playo/a (Arg, Mex), somero/a.

shallowness, n. poca profundidad (f).

shaly, a. lutítico/a.

shanty town, n. chabola (f), favela (f); jacal (m) (Mex), bohío (m) (CAm), callampa (f) (Ch); villa (f) miseria (Arg, Ur), ciudad perdida (Mex), pueblo (m) joven (Peru), barriadas (f.pl) (Peru), cantegriles (m.pl) (Ur), ranchos (m.pl) (Ven).

shape, n. forma (f), estado (m), tipo (m).

shapeless, a. informe, deforme, sin forma.

shard, n. triza (f), fragmento (m) encorvado.

share, n. parte (f) (portion); acción (f) (Com); contribución (f) (contribution); aportación (f) (capital).

shareholder, n. accionista (m, f).

shareholding, n. accionariado (m).

sharing-out, n. reparto (m), distribución (f); **s.-o. of costs**, reparto de costos.

shark, n. tiburón (m).

shear, 1. v. esquilar (e.g. sheep); cizallar (to cut); 2. n. desviación (f), tijera (f), cizalla (f); deslizamiento (m) (Geol); **s. diagram (Phys)**, diagrama (m) de esfuerzos cortantes; **s. legs (for lifting)**, cabria (f).

shearing, n. cizallamiento (m), cizalladura (f); **zone of s.**, zona (f) de cizallamiento.

shears, n.pl. cizalla (f), cizallas (f.pl), tijeras (f.pl).

sheathing, n. revestimiento (m), envuelta (f).

sheep, n. oveja (f) (Zoo); ganado (m) lanar, ovinos (m.pl) (Agr); **s. dog**, perro (m) pastor; **s. farming**, cría (f) de ovejas, ganadería (f) lanar.

sheep-dip, n. desinfectante (m) para ovejas.

sheepherder, n. pastor (m) (US).

sheepshearer, n. esquilador/a (m, f) (person); esquiladora (f) (machine).

sheepshearing, n. esquileo (m).

sheet, n. chapa (f) (of metal, etc.), capa (f) (of water, snow, ice), cortina (f) (of rain); **balance s.**, balance (m. or f); **corrugated s.**, lamina (f) corrugada; **s. ice**, capa de hielo; **s. piles**, tablestacas (f.pl); **steel s.**, chapa de acero.

sheeting, n. revestimiento (m).

shell, n. concha (f) (on beach); cáscara (f) (of egg, of nut); armazón (m) (Con).

shellfish, n. & n.pl. mariscos (m.pl), crustáceos (m.pl).

shellfisherman, n. mariscador/a (m, f), marisquero/a (m, f).

shelter, n. abrigo (m).

shepherd, n. pastor (m).

shepherdess, n. pastora (f).

shepherding, n. pastoreo (m).

shield, n. escudo (m); **the Laurentian S.,** el Escudo Laurentiano.

shift, n. turno (m); **day s.,** turno diurno; **night s.,** turno nocturno.

shingle, n. grava (f) de playa (on beach), guijarros (m.pl) (pebbles), playa (f) de guijarros (shingle beach); teltejamaní (m), tejamanil (m) (US) (on roof).

ship, n. buque (m), barco (m); **cargo s.,** barco de carga, barco mercante; **passenger s.,** barco de pasajeros; **sailing s.,** barco de vela.

shipbuilding, n. construcción (f) marina.

shipment, n. carga (f) (load); cargamento (m) (goods shipped).

shipping, 1. n. naviera (f); 2. a. naviero/a.

shipyard, n. astillero (m).

shoal, n. bajío (m), banco (m) de arena sumergido.

shock, n. sacudida (f) (strike, blow); sacudida (f) (shaking, tremor); shock (m) (Med); **electric s.,** calambre (m), sacudida de electricidad, descarga (f); **state of s. (Med),** estado (m) de shock.

shoot, n. brote (m), retoño (m), renuevo (m) (of a plant); rápido (m) (river); coto (m) de caza (hunting party).

shooting, n. caza (f) (of animal); disparo (m), tiro (m) (e.g. a gun shot); salida (f) (of twigs); **s. party,** cacería (f).

shop, n. tienda (f); almacén (m) (large store); **closed s.,** coto (m) cerrado (de sindicatos).

shopping, n. compra (f); **s. centre, s. center (US), s. precinct,** centro (m) comercial.

shore, n. orilla (f) (of sea, of lake); playa (f) (beach); costa (f) (coast); **longshore drift,** desplazamiento (m) costero.

shoreline, n. línea (f) de costa (coastline); línea de ribera (of river, estuary).

shoreward, n. hacia la costa (f).

shoring, n. entibación (f).

shortage, n. falta (f) (lack of), insuficiencia (f), desabastecimiento (m) (lack of supply).

short-circuit, n. cortocircuito (m).

shortcoming, n. punto (m) flaco (defect), defecto (m).

shorten, v. acortar.

shortening, n. acortamiento (m).

shortsightedness, n. visión (f) corta, miopía (f).

shotblast, n. chorro (m) de granalla.

shotcreting, n. hormigón (m) bombeado.

shovel, 1. v. mover con un pala; 2. n. pala (f);

mechanical s., pala mecánica, pala cargadora, excavadora (f).

shovelful, n. paletada (f).

shower, n. chubasco (m), chaparrón (m) (Met); ducha (f) (baño); **heavy s.,** tronada (f), tormenta (f).

showery, a. lluvioso/a.

shredder, n. desfibradora (f), desmenuzador (m); **waste s.,** desfibradora de desecho.

shrew, n. musaraña (f).

shrimp, n. camarón (m), quisquilla (f); gamba (f) (US).

shrub, n. arbusto (m), matorral (m).

shutdown, n. cierre (m), paro (m); **industrial s.,** cierre industrial.

shutter, n. contraventana (f), postigo (m) (on window); obturador (m) (in camera).

shuttering, n. entibación (f), maderaje (m), maderamen (m) (Con).

SIAL, n. SIAL (f) (Geol).

sickly, a. malsano/a.

sicula, n. sicula (f).

side, n. costado (m), lado (m), labio (m); **downthrow s.,** labio hundido; **upthrow s.,** labio levantado; **valley s.,** costado del valle.

siderite, n. siderita (f).

sidewalk, n. acera (f); vereda (f) (LAm), andén (m) (CAm, Col), banqueta (f) (Mex).

Siegenian, n. Siegeniense (m).

sieve, 1. v. cribar; 2. n. tamiz (m), cedazo (m); **set of s.,** juego (m) de tamices.

sieving, n. cribado (m).

siftings, n.pl. cerniduras (f.pl) (residue).

sight, n. vista (f) (seeing); visión (f) (vision); espectáculo (m) (spectacle); **long s.,** presbicia (f), visión de largo; **short s.,** visión corta, miopía.

sightseeing, n. turismo (m).

sigmoid, a. sigmoideo/a.

signal, n. señal (f), tono (m) (Elec); sintonía (f) (TV, radio); **distress s. (abbr. SOS),** llamada (f) de socorro (abbr. S.O.S.); **electromagnetic s.,** señal electromagnética; **mechnical s.,** señal mecánica; **traffic signals,** semáforos (m.pl).

significant, a. significativo/a, importante; **s. figures,** cifras (f.pl) significativas.

signpost, n. poste (m) telegráfico.

silage, n. ensilaje (m); **s. clamp,** pinza (f) de ensilaje.

silencer, n. silenciador (m).

silex, n. sílex (m).

silexite, n. silexita (f).

silhouette, n. silueta (f).

silica, n. sílice (f); **s. sand,** arena (f) de sílice.

silicate, n. silicato (m).

siliceous, a. silíceo/a.

silicification, n. silicificación (f).

silicon, n. silicio (m) (Si); **s. carbide,** carburo (m) de silicio.

silicosis, n. silicosis (f).

silk, n. seda (f); **s. culture,** sericultura (f).

silkworm, n. gusano (m) de seda.
sill, n. zócalo (m) (de una casa); filón (m) estrato (Geol).
sillimanite, n. sillimanita (f).
silo, n. silo (m), hoyo (m); **grain s.,** silo.
siloxane, n. siloxano (m).
silt, n. limo (m).
silting, n. colmatación (f), enlame (m).
siltstone, n. limolita (f).
silty, a. limoso/a.
Silurian, n. Silúrico (m).
silver, n. plata (f) (Ag).
silver-plated, a. plateado/a.
silvered, a. plateado/a.
silviculture, n. silvicultura (f).
SIMA, n. SIMA (f) (Geol).
simazine, n. simazina (f).
simplification, n. simplificación (f); **over-s.,** sobresimplificación.
simulate, v. simular.
simulation, n. simulación (f); **Monte Carlo s.,** técnica (f) de Monte Carlo; **numerical s.,** simulación numérica.
simulid, n. simúlido (m).
Simulidae, n.pl. simúlidos (m.pl), la clase Simulidae (Zoo).
simultaneity, n. simultaneidad (f).
simultaneous, a. simultáneo/a.
sin, n. sen (m) (Mat, abbr. for sine).
sine, n. seno (m) (Mat, abbr. sen).
Sinemurian, n. Sinemuriense (m).
sink, 1. v. hundirse; 2. n. depresión (f) kárstica (Geol); fregadero (m) (in kitchen); lavabo (m) (in bathroom), sumidero (m) (drain), sima (f) (Fig, depths); **energy s.,** sumidero de energía.
sinkhole, n. sumidero (m), sima (f); **collapse s.,** sumidero por desplome.
sinking, n. hundimiento (m).
sinter, n. sinter (m).
sinuate, a. sinuado/a.
sinuosity, n. sinuosidad (f).
sinuous, a. sinuoso/a.
sinus, n. seno (m).
sinusoidal, a. sinusoidal.
siphon, n. sifón (m) (of liquids); **inverted s.,** sifón invertido.
siphoning, n. sifonación (f), sifonamiento (m).
sirocco, n. siroco (m).
sisal, n. pita (f), sisal (m); henequén (m) (LAm).
site, n. solar (m) (building site); lugar (m), sitio (m) (place); situación (f) (situation); **building s.,** solar (m), sitio de construcción; **development s.,** lugar de desarrollo; **industrial s.,** zona (f) industrial; **landfill s.,** vertedero (m), vertido (m) controlado; **s. characterization,** caracterización (f) de sitio.
site-specific, a. sitio específico.
siting, n. localización (f), emplazamiento (m), situación (f); **s. of reservoirs,** localización de embalses.

situation, n. situación (f), localización (f), sitio (m).
size, n. tamaño (m), talla (f) (of garments), magnitud (f); **grain s.,** tamaño de grano; **life s.,** tamaño natural.
sizing, n. dimensionamiento (m) (of an installation).
skarn, n. skarn (m).
skeleton, n. esqueleto (m).
skewed, a. sesgado/a; **s. distribution,** distribución (f) sesgada; **s. statistics,** estadísticas (f) sesgadas.
skewness, n. asimetría (f) (Mat); **coefficient of s.,** coeficiente (m) de asimetría.
skim, v. espumar.
skimmed, a. desnatado/a (of milk).
skimming, n. espumado (m); **s. tank,** tanque (m) de espumado.
skin, n. piel (f) (Gen), cutis (m) (Anat).
skin-diver, n. buceador (m), submarinista (m, f).
skip, n. eskip (m), balde (m) (container); vasija (f) (container for rubbish); cestón (m) (large bin).
sky, n. cielo (m); **overcast s.,** cielo encapotado.
skylight, n. tragaluz (m), claraboya (f).
skyline, n. línea (f) del horizonte; contorno (m) (of city).
skyscraper, n. rascacielos (m.pl).
slab, n. losa (f), laja (f).
slabby, a. lajoso/a.
slack, n. depresión (f) pantanosa junto a una colina (Geog); **s. water,** agua (f) estancada.
slag, n. escoria (f); **blastfurnace s.,** escoria de alto horno; **s. heap,** escombrera (f) de escoria, escorial (m).
slagheap, n. escombrera (f).
slate, n. pizarra (f).
slaty, a. pizarroso/a; pizarreño/a (LAm); **s. cleavage,** foliación (f) pizarrosa, pizarrosidad (f).
slaughterhouse, n. matadero (m), rastro (m); **s. wastes,** desechos (m.pl) de matadero/de rastro.
sleet, n. aguanieve (f), cellisca (f).
sleeve, n. manguito (m) (joint); manga (f) (garment); **threaded s. (Eng),** manguito rosgado.
slice, n. rebanada (f).
slick, n. arena (f) negra (Geol), mancha (f), película (f); **oil s.,** mancha (f) de petróleo, marea (f) negra.
slickenside, n. espejo (m) de falla, slickenside (m).
slide, n. deslizamiento (m), derrumbe (m) (movement); superficie (f) (smooth surface); portaobjeto (m) (of a microscope); diapositiva (f), transparencia (f) (photograph).
slime, n. cieno (m), fango (m) (mud); baba (f) (of snail); moco (m) (mucus); **s. fungi,** mohos (m.pl) mucosos, mixomicetos (m.pl), mixomicofitos (m.pl).

slimy, a. limoso/a.

slip, n. resbalamiento (m), desplazamiento (m), deslizamiento (m) (Geol); **s. circle,** círculo (m) de desplazamiento; **s. surface,** superficie (f) de deslizamiento; **soil s.,** resbalamiento de suelo, deslizamiento de suelo.

slipping, n. deslizamiento (m); arrastramiento (m) (dragging).

slipway, n. grada (f).

slope, n. cuesta (f), pendiente (f) (up); declive (m), bajada (f) (down); **on the north s.,** en la vertiente (m) norte; **steep s.,** escarpadura (f), escarpe (m); **windward s.,** vertiente (m) del lado del viento.

sloping, a. en pendiente, inclinado/a, en cuesta.

slough, n. depresión (f) inundada.

slow-cooling, a. enfriado lento, enfriado paulatino.

sludge, n. fango (m), lodo (m), cieno (m); **activated s.,** fango activado, cieno activado, lodo activado; **digested s.,** fango digestivo, lodo digestivo; **extended aeration activated s.,** lodos activados por aireación extendida; **mixed s.,** fango mixto, lodos mixtos; **raw s.,** cieno (m) sin tratar; **sewage s.,** lodo de depuración; **s. dewatering,** deshidratación (f) de fangos, deshidratación de lodos; **s. digestion,** digestión (f) de lodos; **s. disposal to sea,** descarga (f) de lodos al mar; **s. drying,** secado (m) de fangos, secado de lodos; **s. incineration,** incineración (f) de lodos; **s. incinerator,** incinerador (m) de los fangos de alcantarilla; **s. liquor,** licor (m) de fango; **s. treatment,** tratamiento (m) de fango, tratamiento de lodos.

sluice, n. compuerta (f), esclusa (f); **s. gate,** compuerta de esclusa.

slum, n.pl. barrio (m) marginal (Gen, Ec, Nic), villa (f) miseria (Arg, Par), favela (f) (Br), tugurio (m) (Ch, C.Rica, ElS), campamento (m) (Ch), barriada (Pan, Perú), pueblo (m) joven (Perú), ranchería (m) (Ur), suburbio (m) (Ven).

slumping, n. desmoronamiento (m), deslizamiento (m), asentamiento (m), slumping (m); **marine s.,** desmoronamiento submarino.

slurry, n. lechada (f), pasta (f) aguada.

slush, n. aguanieve (m), hielo (m) grasiento.

small, a. pequeño/a.

smallholding, n. parcela (f), minifundio (m); chacra (f) (CSur), chácara (f) (CSur), conuco (m) (Ven).

smear, n. frotis (m).

smectite, n. esmectita (f).

smell, 1. v. oler, olfatear; 2. n. olor (m); **sense of s.,** olfato (m).

smelly, a. maloliente.

smelt, v. fundir; **to s. iron,** fundir hierro.

smelted, a. fundido/a.

smelter, n. fundidor (m).

smelting, n. fundición (f); **s. furnace,** horno (m) de fundición.

smog, n. smog (m), esmog (m) (Met).

smoke, n. humo (m).

smoothing, n. smoothing (m) (Mat).

smoulder, v. arder sin llama.

snail, n. caracol (m).

snake, n. serpiente (f), culebra (f).

snout, n. morro (m), hocico (m).

snow, 1. v. nevar; **to s. lightly,** neviscar; 2. n. nieve (f); **s. gauge,** medidor (m) de nieves; **s. line,** límite (m) de las nieves perpetuas.

snow-covered, a. nevado/a.

snowdrift, n. nieve (f) acumulada, nevero (m).

snowfall, n. nevada (f); **light s.,** nevisca (f).

snowfield, n. nevero (m).

snowflake, n. copo (m) de nieve.

snowplough, n. quitanieves (m).

snowplow, n. quitanieves (m).

snowstorm, n. nevasca (f); nevazón (f) (LAm).

soak, v. empapar.

soakage, n. empapamiento (m).

soakaway, n. sumidero (m) ciego, excavación (f) de infiltration (Treat).

soaking-up, n. imbibición (f), absorción (f).

soap, n. jabón (m); **s. hole (Geog),** laguna (f) en la pradera con freatofitas (US); **s. powder,** jabón en polvo.

soapstone, n. esteatita (f), talco (m), roca (f) jabón.

social, a. social; **s. worker,** asistente (m, f) social.

society, n. sociedad (f); **consumer s.,** sociedad de consumo.

socio-economic, a. socioeconómico/a; **s. planning,** planificación (f) socioeconómica.

sociologist, n. sociólogo/a (m, f).

socket, n. glena (f) (of joint); cuenca (f) (of eye); enchufe (m) (elec); casquillo (m) (for bulb); **ball and s.,** rótula (f).

soda, n. sosa (f) (Chem); **caustic s.,** sosa cáustica.

sodalite, n. sodalita (f).

sodic, a. sódico/a.

sodium, n. sodio (m) (Na); **s. chloride,** cloruro (m) sódico.

soffit, n. sofito (m), cimbra (f) (centre of arch).

soft, a. blando/a; **s. water,** agua (f) blanda.

softener, n. suavizante (m).

softening, n. ablandamiento (m); **s. by ion exchange,** ablandamiento por intercambio iónico; **water s.,** ablandamiento del agua.

software, n. software (m); **application s.,** software de aplicación; **s. support,** soporte (m) software.

softwood, n. madera (f) blanda.

soil, n. suelo (m), tierra (f); **boggy s.,** suelo turboso; **laterite s.,** suelo laterítico; **loamy s.,** suelo franco; **prairie s.,** suelo de pradera; **red desert s.,** suelo rojo desértico; **residual s.,** suelo residual; **saline s.,** suelo salino; **saturated s.,** suelo saturado; **s. acidity,** acidez (f) del suelo; **s. contamination,** contaminación (f) del suelo; **s. creep,** reptación (f) del suelo; **s. erosion,** erosión (f) del suelo; **s. mechanics,** mecánica (f) de suelos; **s. moisture deficit,** déficit (m) de humedad del suelo; **s. profile,** perfil (m) del suelo; **s. stripping,** desmonte (m); **s. structure,** estructura (f) del suelo; **tundra s.,** suelo de tundra.

soilslip, n. deslizamiento (m) de suelo, deslizamiento de tierra.

sol, n. sol (m); **unit of currency (Peru),** nuevo sol.

solar, a. solar; **s. power,** energía (f) solar; **s.-power technology,** tecnología (f) fotovoltáica.

solarimeter, n. solarímetro (m).

solar-powered, a. de energía solar.

solder, v. estañar, soldar.

soldering, n. estañado (m), estañadera (f).

sole, n. lenguado (m) (fish); suela (f) (of shoe).

solicitor, n. procurador/a (m, f) (lawyer who prepares a case); notario (m) (for wills, deeds); abogado (m) (lower courts).

solid, 1. n. sólido (m); **dry solids content,** contenido (m) de sólidos secos; **suspended solids,** sólidos (m.pl) suspendidos; 2. a. macizo/a, sólido/a; **s. case,** asunto (m) macizo; **s. concrete,** hormigón (m) macizo; **s. rock,** roca (f) maciza.

solidification, n. solidificación (f).

solid-state, n. estado (m) sólido (Elec).

solifluction, n. solifluxión (f); **s. lobe,** lóbulo (m) de solifluxión.

solstice, n. solsticio (m); **summer s.,** solsticio de verano; **winter s.,** solsticio de invierno.

solubility, n. solubilidad (f); **gas s.,** solubilidad de gases.

solubilization, n. solubilización (f).

solubilize, v. solubilizar.

soluble, a. soluble; **water-s.,** soluble en agua.

solute, n. soluto (m).

solution, n. solución (f); **normal s.,** solución normal; **oversaturated s.,** solución sobresaturada; **saturated s.,** solución saturada.

solve, v. solucionar.

solvency, n. solvencia (f).

solvent, n. disolvente (m), solvente (m); **chlorinated s.,** disolvente clorado; **s. recovery,** recuperación (f) de disolventes.

sonde, n. globo (m) sonda, globo piloto.

soot, n. tizne (m), hollín (m).

sorghum, n. sorgo (m).

sori, n.pl. soros (m.pl).

sorption, n. absorción (f).

sorptive, a. absorbible; **s. capacity,** capacidad (f) de absorción.

sort, v. seleccionar (to select), clasificar (to classify), separar (to sort out).

sorted, a. seleccionado/a; **size-s.,** seleccionado granulométricamente.

sorting, n. selección (f), clasificación (f).

sorus, n. soro (m).

SOS, n. SOS, llamada (f) de socorro; **to pick up an SOS (v),** captar un SOS.

sound, n. sonido (m); ruido (m) (noise); estrecho (m) (Mar, channel); brazo (m) (Mar, inlet).

sounding, n. sonda (f) (probe), sondeo (m) (Mar); **echo s.,** sondeo acústico; **s. line,** cable (m) de remolque, sonda (f).

soundproofed, a. insonorizado/a; **s. chamber,** cabina (f) de insonorización.

soundproofing, n. insonorización (f).

source, n. fuente (f), nacimiento (m) (of river); origen (m), fuente (f) (of information); **diffuse s. (of pollution),** fuente difusa (de contaminación); **from a reliable s.,** de fuente fidedigna; **point s.,** fuente fija, fuente puntual; **radioactive s.,** fuente radioactiva; **s. language (Comp),** lenguaje (f) de partida original, lenguaje (f) fuente; **s. of energy,** fuente de energía; **s. of food,** fuente de alimentos; **s. of supply,** fuente del suministro.

South, n. sur (m); **S. America,** América (f), América (f) del Sur, Suramérica (f).

south, a. sur, del sur.

southern, a. del sur, meridional, hacia el sur.

sovereignty, n. soberanía (f).

sow, v. sembrar.

sowing, n. siembra (f) (of plants).

spa, 1. n. balneario (m); 2. a. balneario/a; **s. town,** pueblo (m) balneario.

space, 1. v. esparcir, separar; 2. n. espacio (m), sitio (m); **green s.,** zona (f) verde.

spacecraft, n. aeronave (m) espacial, nave (m) espacial.

spacing, n. espaciamiento (m), distanciamiento (m).

spacious, a. espacioso/a.

Spain, n. España (f).

Spaniard, n. español/a (m, f), hispano/a (m, f).

Spanish, a. español/a, castellano, hispano; **S. American,** hispanoamericano/a.

Spanish-American, a. hispanoamericano/a.

spar, n. palo (m) (pole); espato (m) (Geol); **Iceland s.,** espato de Islandia; **Satin S.,** espato satinado.

sparger, n. tubo (m) burbujeador.

sparging, n. burbujeo (m); **air s.,** burbujeo con aire.

spatangoid, n. spatangoide (m).

spate, n. avenida (f) (in river), crecida (f) (flood flow).

spatial, a. espacial.

spawn, 1. v. depositar (of eggs); 2. n. freza (f), desove (m), huevos (m.pl) (of amphibians); hueva (f) (of fish); **s. season,** freza (f).

spearhead, n. azagaya (f), lanza (f).

specialist, n. especialista (m, f).

specialization, n. especialidad (f), especialización (f).

specialize, v. especializar.

speciation, n. especiación; **aqueous s.,** especiación acuosa.

species, n. & n.pl. especie (f); **accidental s.,** especie accidental; **characteristic s.,** especie característica; **endangered s.,** especie en peligro de extinción; **extinct s.,** especie extinguida; **protected s.,** especie protegida; **s. richness,** riqueza (f) de especies; **threatened s.,** especie amenazada; **type s.,** especie tipo, especie nominal; **vulnerable s.,** especie vulnerable.

specific, a. específico/a, determinado/a; **s. capacity,** capacidad (f) específica; **s. name,** nombre (m) específico; **s. retention,** retención (f) específica; **s. yield,** rendimiento (m) específico.

specifically, ad. específicamente, expresamente.

specification, n. especificación (f); prescripción (f); **building s.,** especificación de construcción; **specifications,** presupuesto (m) (project budget); plan (m) detallado (project plans).

specify, v. especificar.

specimen, n. espécimen (m) (pl. especímenes); muestra (f) (sample); **hand s.,** muestra (f) de mano, muestra manual.

speckled, a. abigarrado/a, moteado/a.

spectograph, n. espectrográfo (m).

spectral, a. espectral; **s. analysis,** análisis (m) espectral.

spectrographic, a. espectrográfico/a.

spectrography, n. espectrografía (f).

spectrometer, n. espectrómetro (m); **mass s.,** espectrómetro de masas.

spectrometry, n. espectrometría (f); **mass s.,** espectrometría de masas.

spectrophotocolorimeter, n. espectrofotocolorímetro (m).

spectrophotometer, n. espectrofotómetro (m); **flame s.,** espectrofotómetro de flama.

spectrophotometry, n. espectrofotometría (f).

spectroscope, n. espectroscopio (m).

spectrum, n. espectro (m); **ultraviolet s.,** espectro ultravioleta; **visible s.,** espectro visible.

speculation, n. especulación (f); especulación bursátil (on the stock exchange); **property s.,** especulación inmobiliaria.

speed, 1. v. ir corriendo, correr, apresurarse; acelerar (of an engine); 2. n. velocidad (f), rapidez (f); **s. of light,** velocidad de la luz; **s. of response,** rapidez de respuesta; **s. of sound,** velocidad del sonido.

speleology, n. espeleología (f).

sperm, n. esperma (f).

Spermatophyta, n.pl. espermatofitas (f.pl), la división Spermatophyta (Bot).

spermatophyte, n. espermatofita (f).

spessartite, n. espesartita (f).

sphagnum, n. esfagno (m).

sphalerite, n. esfalerita (f), blenda (f).

sphene, n. esfena (m), titanita (f).

sphere, n. esfera (f); **s. of activity,** esfera de actividad; **s. of influence,** esfera de influencia.

spherical, a. esférico/a.

sphericity, n. esfericidad (f).

spheroid, n. esferoide (m); **oblate s.,** esferoide oblato.

spheryte, n. esferita (f).

spice, n. especia (f).

spicule, n. espícula (f); **sponge s.,** espícula de una esponja.

spider, n. araña (f); **sea s.,** araña de mar, centollo/a (m, f); **s.'s web,** telaraña (f), tela (f) de araña.

spigot, n. empalme (m) de enchufe; **s. and socket joint,** empalme de enchufe y cordón.

spilite, n. espilita (f), spilita (f).

spill, 1. v. derramar, verter, desbordarse; 2. n. vertido (m), descarga (f); derrame (m) (of liquid); **accidental oil s.,** vertido accidental de petróleo; **oil s.,** vertido de aceite, la marea (f) negra, derrame de petróleo; **s. from olive waste,** vertido de alpachín.

spillage, n. vertido (m), rebosamiento (m); derrames (m.pl) (LAm).

spilling, n. derrame (m); reboso (m) (escape).

spillway, n. rebosadero (m), aliviadero (m), vertedero (m).

spinach, n. espinaca (f).

spinal, a. espinal; **s. cord (Anat),** médula (f) espinal.

spine, n. espina (f).

spinel, n. espinela (f).

spinney, n. arboleda (f).

spiral, 1. v. moverse en espiral, dar vueltas; 2. n. espiral (f), hélice (f); **the inflationary s.,** la espiral inflacionaria; 3. a. espiral.

spire, n. aguja (f) (Con); brizna (f) (Bot); espiral (f) (Mat).

spirifid, n. espirífido (f).

spit, n. punto (m) (Geog); banco (m) encorvado (m) (Geog, of sand); salivazo (m) (Med); rocío (m) (de lluvia); **longshore s.,** banco costero.

splash, n. salpicadura (f), rociada (f) (spray); mancha (f) (mark).

splinter, 1. v. astillar; 2. n. astilla (f).

split, 1. v. agrietar; 2. n. hendidura (f).

splitting, n. hendimiento (m).

spoil, 1. v. estropear, arruinar (ruin); dañar (harm); 2. n. escombros (m.pl) (waste); ganga (f), estéril (m) (mine waste); **s. area (Mar),** zona (f) de vertido de productos de dragado; **s. removal,** desmonte (m) de tierra.

spokesman, n. portavoz (m).

spokesperson, n. portavoz (m, f).

sponge, n. esponja (f).

sponsor, n. patrocinador/a (m, f).

sponsorship, n. patrocinio (m).

spontaneity, n. espontaneidad (f).

spoon, n. cuchara (f).

spoonful, n. cucharada (f).

sporadic, a. esporádico/a.

spore, n. espora (f).

sport, n. deporte (m); **water s.,** deportes acuáticos.

spray, 1. v. rociar, regar; pulverizar (to atomize); 2. n. rociada (f) (liquid), ramita (f) (of flowers), chorro (m) (of sea), pulverización (f) (atomizer); **aerosol s.,** atomizador (m).

spraying, n. aspersión (f), atomización (f); fumigación (f), pulverización (f) (with pesticides); rociada (f) (sprinkling); **crop s.,** fumigación de los cultivos; **pesticide s.,** pulverización con pesticidas.

spread, 1. v. extender, difundir; propagar (to propagate); 2. n. propagación (f), difusión (f) (of disease or ideas), extensión (f) (of land).

spreading, n. extensión (f), propagación (f), expansión (f); **sea floor s.,** expansión del fondo oceánico.

spread out, 1. v. tender, extender (e.g. a map); espaciar (to space out), separar (to separate); 2. a. tendido/a.

spreadsheet, n. hoja (f) de cálculo, hoja electrónica de cálculo.

spring, n. manantial (m), vertiente (f) (LAm); fuente (f) (fountain); primavera (f) (season); salto (m) (jump); muelle (m) (mattress); **boiling s.,** manantial en ebullición; **gaseous s.,** manantial gaseoso; **gravity s.,** manantial gravitacional; **hot s.,** manantial de agua termal, terma (f); **intermittent s.,** manantial intermitente; **karst s.,** manantial kárstico; **mineral s.,** manantial mineral; **perennial s.,** manantial perenne; **s. board,** trampolín (m); **s. cleaning,** limpieza (f) general; **s. discharge,** surgencia (f) de un manantial; **s. time,** primavera (f); **thermal s.,** manantial termal.

springhead, n. punto (m) de manantial, nacimiento (m) de río (river source).

sprinkler, n. regadera (f) (Agr); extintor (to extinguish a fire); **s. system,** sistema (m) por aspersión.

sprinkling, n. aspersión (f), rociada (f), salpicadura (f).

sprocket, n. diente (m) de engranaje.

sprout, 1. v. echar, echar brotes, brotar; 2. n. brote (m), retoño (m); **Brussels s.,** col (f) de Bruselas.

spur, n. espolón (m) (Zoo); cornezuelo (m) (Bot); vía (f) muerta, vía apartadero (of railway); espolón, estribación (f) (of a mountain); contrafuerte (m) (buttress).

spurt, v. surgir, chorrear, salir a chorros (to gush forth).

squall, n. borrasca (f), ráfaga (f), racha (f); turbonada (f) (LAm).

squalor, n. inmundicia (f), suciedad (f).

square, 1. v. elevar a la potencia de dos (Mat); 2. n. cuadrado (m), cuadro (m); casilla (f) (on graph paper); plaza (f) (in town); **least**

squares **(Mat),** mínimos (m.pl) cuadrados; 3. a. cuadrado/a; **s. root,** raíz (f) cuadrada.

squatter, n. ocupante (m, f) ilegal; **s. settlement (LAm),** asentamiento (m) precario, barriada (f), barrio (m) marginal, conventillo (m), favela (f), tugurio (m), villa (f) miseria, zona (f) marginal.

squeeze, v. apretar, comprimir (to compress).

squid, n. & n.pl. calamar (m).

stability, n. estabilidad (f); **numerical s. (Mat),** estabilidad numérica; **slope s.,** estabilidad de la pendiente.

stabilization, n. estabilización (f); **s. of levels,** estabilización de niveles; **s. of pressures,** estabilización de presiones; **s. pond,** laguna (f) de estabilización; **s. tank,** estanque (m) de estabilización.

stabilization, n. estabilización (f); **waste s.,** estabilización de residuos.

stabilize, v. estabilizar.

stabilizer, n. estabilizador (m); **formation s.,** estabilizador de formación.

stable, a. estable.

stack, n. pila (f) (pile), montón (m) (heap), chimenea (f) (chimney); **s. emission,** emisión (f) de chimenea.

stage, n. etapa (f) (Gen); estado (m) (Geol); piso (m) (Chronology); nivel (m) (level); altura (f) (height); fase (f) (phase); **in stages,** por etapas; **multi-s.,** etapas múltiples; **old age s.,** estado senil, "tercera edad"; **s.-discharge curve,** curva (f) de caudales-alturas; **s.-discharge relationship,** relación (f) entre caudales-alturas; **two-s.,** dos etapas; **youth s.,** estado juvenil.

staging, n. andamiaje (m) (scaffolding).

stagnant, a. estancado/a.

stagnation, n. estancamiento (m).

stain, n. mancha (f) (mark); tinción (f) (dye); tintura (f) (colour); colorante (m) (for microscope work); **Gram's s.,** tinción Gram.

stainless, a. inoxidable; **s. steel,** acero (m) inoxidable.

staircase, n. escalera (f); **spiral s.,** escalera de caracol.

stake, n. estaca (f) (post).

stalactite, n. estalactita (f).

stalagmite, n. estalagmita (f).

stall, n. caseta (f) (in exhibition), establo (m) (stable).

stamen, n. estambre (m).

stand-alone, a. autónomo/a, independiente.

standard, 1. n. estándar (m), norma (f) (established norm); criterio (m) (criterion); nivel (m), grado (m) (level, degree); **British Standards,** Estándares británicos; **Bureau of Standards (US),** Oficina (f) de pesas y medidas; **environmental quality s.,** norma de calidad ambiental; **s. of living,** nivel de vida; **to be below s. (v),** ser de baja calidad; **to be up to s. (v),** satisfacer los requisitos; **water quality s.,** norma de calidad del agua;

2. a. estándar, estándar; normal (common); **s. practice,** norma (f); **s. solution (Chem),** dosificación (f) (titration).

standardization, n. estandardización (f). estandarización (f), normalización (f).

standardize, v. estandizar, estandardizar, normalizar.

standards, n. reglas (f.pl), normas (f.pl) (rules); estándares (m.pl) (quality); modelos (m.pl), tipos (m.pl) (models); **US Bureau of Standards,** Oficina (f) de pesas y medidas.

standby, 1. n. estado (m) inactividad; espera (f); 2. a. de reserva; **s. plant,** planta (f) de reserva.

standpipe, n. columna (f) de alimentación.

standpost, n. fuente (f) de agua, puesto (m) de agua; **public s.,** fuente pública de agua, puesto público de agua.

stannate, n. estanato (m).

stannite, n. estanina (f), pirita de estaño (f).

starch, n. almidón (m).

starfish, n. estrella (f) de mar.

starting, n. arranque (m).

state, 1. n. estado (m), condición (f); **initial s.,** estado inicial; **physical s.,** estado físico; **solid s.,** estado sólido; **s. of emergency,** estado de emergencia; **s. of the art,** estado actual de la tecnología; **steady s.,** estado equilibrio, estado estacionario; **the States, the United States,** los Estados Unidos; **transitory s.,** estado transitorio; 2. a. estatal; **s. sector,** sector (m) estatal.

statement, n. declaración (f); **Environmental Impact S. (GB, abbr. EIS),** Declaración de Impacto Ambiental (Sp, abbr. DIA).

statesman, n. estadista (m).

state-subsidized, a. subvencionado/a por el estado.

static, a. estático/a; **s. head,** presión (f) estática, altura (f) piezométrica, carga (f) de agua.

station, n. estación (f) (location); puesto (m) (post); central (f) (power plant); **climate s.,** estación climatológica; **gauging s.,** estación de aforo; **monitoring s.,** estación de vigilancia; **nuclear power s.,** central nuclear; **power s.,** central térmica, central eléctrica; **railway s.,** estación de ferrocarril; **rainfall s.,** estación pluviométrica; **research s.,** estación de investigación; **weather s.,** estación meteorológica.

stationary, a. fijo/a, estacionario/a; **s. phase (Mat),** estado (m) estacionario.

statistic, n. estadística (f).

statistical, a. estadístico/a.

statistically, ad. según las estadísticas.

statistician, n. estadístico/a (m, f), estadista (m).

statistics, n. estadística (f) (ciencia); estadísticas (f.pl) (data); **parametric/non-parametric s.,** estadística paramétrica/no paramétrica.

status, n. estado (m); **conservation s.,** estado de conservación; **national s.,** estado nacional.

status quo, n. status quo (m).

statute, n. estatuto (m), decreto (m) (established law); estatutos (m.pl) (of a chartered body); **s. book,** código (m); **s. law,** derecho (m) escrito; **s. mile (GB),** milla (f) terrestre (igual a 1609 m).

statutory, a. estatutario/a (established by statute), reglamentario/a (conforming to statute).

staurolite, n. estaurolita (f).

steady, a. fijo/a (firm); constante (unvarying); **s.-state,** estado (m) equilibrio, estado estacionario.

steam, 1. v. echar vapor (to give off); **to s. up,** empañarse; 2. n. vapor (m) (of water); **s. boiler,** caldera (f) de vapor; **s. engine,** máquina (f) de vapor; 3. a. de vapor.

steamroller, n. apisonadora (f), cilindradora (road roller).

steamship, n. buque (m) de vapor.

stearate, n. estearato (m); **sodium s.,** estearato de sodio.

steatite, n. esteatita (f).

steel, n. acero (m); **forged s.,** acero forjado; **iron and s. industry,** siderurgia (f); industria (f) siderúrgica; **mild s.,** acero dulce; **stainless s.,** acero inoxidable; **s. alloy,** acero de aleación; **s. casting,** fundición (f) de acero; **s. works,** fábrica (f) siderúrgica.

steelmaking, n. aceración (f), fábrica (f) del acero.

steep, a. escarpado/a, abrupto/a, alcantilado/a; **s. slope,** escarpadura (f), escarpe (m).

stem, n. tallo (m) (of plant, flower); tronco (m) (of tree); rabo (m) (of leaf, fruit); vástago (m) (rod).

step, n. escalón (m) (stairs); trámite (m) (procedure); paso (m), medida (f) (measure); **take steps to do something (v),** hacer trámites para hacer alguna cosa.

stepladder, n. escalera (f) de tijera.

steppe, n. estepa (f).

steps, n.pl. escalones (m.pl); **pumping s.,** escalones de bombeo.

stereochemistry, n. estereoquímica (f).

stereogram, n. estereograma (m).

stereoscope, n. estereoscopio (m).

stereoscopic, a. estereoscópico/a; **s. pair,** par (m) estereoscópico.

sterilant, n. esterizante (m).

sterile, a. estéril.

sterilization, n. esterilización (f).

sterilize, v. esterilizar.

sterling, a. esterlina; excelente, de buena calidad; **the pound s.,** la libra (f) esterlina.

steroid, n. esteroide (m).

stewardship, n. gestoría (f), gestión (f); **environmental s.,** gestión medio; gestoría ambiental.

sticky, a. pegajoso/a; **s. plastic clay,** légamo (m).

stiffener, n. rigidizador (m), endurecedor (m), refuerzo (m) (strengthening).

stiffness, n. rigidez (f) (rigidity), dureza (f) (toughness).

stigma, n. estigma (m).

stilling, n. amortiguación (f); **s. basin,** estanque (m) de amortiguación.

stimulant, n. estimulante (m).

stimulate, v. estimular.

stimulation, n. estimulación (f) (action), estímulo (m) (stimulus).

stimuli, n.pl. estímulo (m).

stimulus, n. estímulo (m).

stochastic, a. estocástico/a.

stock, n. reservas (f.pl), existencias (f.pl) (Com); acciones (f.pl), valores (m.pl) (in a stock exchange); capital (m) social (of a company); ganado (m) (livestock); cepa (f) (Bot, trunk); raza (f) (race); estirpe (m) (lineage); **fish s.,** recursos (m.pl) pesqueros; **joint-s. company,** sociedad (f) anónima; **rolling s.,** material (m) rodante (railway); **s. exchange,** bolsa (f); **to take s. (v),** hacer balance.

stockade, n. empalizada (f).

stockbreeding, n. ganadería (f), agroganadero (m).

stockbroker, n. agente (m) de cambio y bolsa, corredor/a (m, f) de bolsa.

stockpile, n. reservas (f.pl).

stoichiometry, n. estequiometría (f).

stolon, n. estolón (m).

stoma, n. estoma (m).

stomach, n. estómago (m).

stomata, n.pl. estomas (m.pl).

stone, n. piedra (f), lápida (f); peso (m) (unit of weight, 14 pounds or 6.350 kilos); **pudding s.,** pudinga (f); **s. slab,** losa (f).

stonecutter, n. cantero (m).

stonework, n. cantería (f).

stony, a. pedregoso/a, pétreo/a.

stopcock, n. llave (f) de paso.

stoping, n. socovación (f) (Min); stoping (m) (LAm); **magmatic s.,** socovación magmática.

stoppage, n. paro (m).

stopper, n. taqué (m), tapón (m), obturador (for pipe).

storage, n. almacenamiento (m); **aquifer s.,** almacenamiento de acuífero; **bankside s.,** almacenamiento en la ribera; **coefficient of s.,** coeficiente (m) de almacenamiento; **specific s.,** almacenamiento específico; **s. capacity,** capacidad (f) de almacenamiento; **s. reservoir,** embalse (m) por almacenamiento; **s. tank,** tanque (m) de almacenamiento; **underground s.,** almacenamiento subterráneo; **well s.,** almacenamiento en el pozo.

storativity, n. storatividad (f).

store, 1. v. almacenar; 2. n. almacenamiento (m), acopio (m).

stored, a. almacenado/a, registrado/a (e.g. on computer).

storm, n. tormenta (f); temporal (m) (severe storm), tempestad (f); escándalo (m) (Fig. uproar); **dust s.,** tormenta de polvo, tolanera (f) (CAm, Mex); **electric s.,** tormenta eléctrica; **sandstorm,** tormenta de arena; **severe s.,** temporal; **short sharp s. (downpour),** tromba (f) de agua; **s. overflow,** aliviadero (m) de crecidas; **s. warning,** aviso (m) de tormenta; **s. water,** aguas (f.pl) pluviales; **to weather the s. (v),** capear el temporal.

stormwater, n. agua (f) de lluvia, agua meteórica, agua pluvial; **s. overflow,** rebose; **s. tank,** tanque (m) de agua de lluvia.

straight, a. recto/a, derecho/a; **s. line,** recta (f).

strain, n. tensión (f) (Phys); raza (f), variedad (f) (Zoo, Bot).

strainer, n. colador (m).

straits, n.pl. estrecho (m) (Geog); **S. of Gibraltar,** Estrecho de Gibraltar.

strata, n.pl. estratos (m.pl); stratos (m.pl) (LAm).

strategic, a. estratégico/a; **s. development,** desarrollo (m) estratégico; **s. planning,** planificación (f) estratégica.

strategy, n. estrategia (f).

stratification, n. estratificación (f); **thermal s.,** estratificación (f) térmica, estratificación termal.

stratified, a. estratificado/a.

stratigraphical, a. estratigráfico/a.

stratigraphy, n. estratigrafía (f).

stratocumulus, n. estratocúmulo (m) (Met).

stratosphere, n. estratosfera (f).

stratospheric, a. estratosférico/a.

stratum, n. estrato (m).

straw, n. paja (f).

strawberry, n. fresa (f); frutilla (f) (LAm) (wild fruit); fresón (m) (cultivated fruit).

streak, n. raya (f).

streaky, a. fajeado/a.

stream, n. arroyo (m), río (m), corriente (f), quebrada (f) (LAm, brook); **underground s.,** corriente subterránea, estavella (f).

streamline, 1. v. modernizar (to modernize), racionalizar (to rationalize); 2. n. línea (f) de corriente.

street, n. calle (f); **dead-end s. (cul-de-sac),** callejón (m) sin salida, calle ciega (LAm); **high s., main s.,** calle mayor; **one-way s.,** calle de sentido único, una vía (LAm); **streetcar,** tranvía (f); **s. cleaner,** barrendero (m); **s. cleaning,** limpieza (f) de vías; **s. lamp, streetlight,** poste (m) del alumbrado, foco (m) (LAm); **s. lighting, streetlights,** alumbrado (m) público; **s. sweeper,** barrendero (m).

strength, n. fuerza (f); resistencia (f) (e.g. of a construction); graduación (f), proporción (f) (Chem, of a solution); **acid s.,** fuerza de un ácido; **compressive s.,** resistencia al aplastamiento, resistencia a la compresión; **rupture s.,** resistencia a la ruptura; **shear s.,** fuerza cizallante, fuerza de corte.

strengthening, n. fortalecimiento (m); refuerzo (m) (reinforcing).

streptococci, n.pl. estreptococos (m.pl).

streptococcus, n. estreptococo (m).

streptomycin, n. estreptomicina (f).

stress, n. esfuerzo (m) (Phys); tensión (f), estrés (m) (Med); hincapié (f) (emphasis); **shear s.**, esfuerzo cizallante; **tensile s.**, esfuerzo de tensión.

stretch, n. tramo (m) (of road); elasticidad (f) (elasticity); trecho (m) (distance); extensión (f) (expanse); período (m) (of time); manga (f) (de un río); **s. of road**, tramo de carretera.

striated, a. estriado/a.

striation, n. estría (f).

stridulate, v. chirriar.

stridulation, n. chirrido (m).

strike, n. dirección (f) horizontal (Geol); golpe (m) (blow); descubrimiento (m) (of oil, etc.); ataque (m) (Mil).

string, n. cuerda (f) (cord, rope); fina (f) (e.g. of hotels); sarta (f) (series); **drilling s.**, sarta de varillas de perforación.

stringbog, n. cuerdas (f) de turbera separades por depresiones (US) (Geog).

stringer, n. zanca (f) (Con).

strip, 1. v. desmontar, descabezar (Min), despojar; 2. n. faja (f), franja (f) (of land); tira (f) (of paper); fleje (m) (of metal); **the Gaza S.**, la Franja de Gaza.

striped, a. listado/a, rayado/a, a rayas.

strip-farming, n. agricultura (f) en franjas.

stripping, n. desmonterado (m) (de vegetación); arranque (m) (Min); arrasamiento (m) (devastation), stripping (Ind); **sulphur s.**, **sulfur s. (US)**, extracción (f) de azufre, arranque del azufre.

stromatolite, n. estromatolito (m).

strontianite, n. estroncianita (f).

strontium, n. estroncio (m) (Sr).

structural, a. estructural.

structure, n. estructura (f), construcción (f); **anticlinal s.**, estructura anticlinal; **columnar s.**, estructura columnar; **management s.**, estructura de dirección; **molecular s.**, estructura molecular; **organizational s.**, estructura organizativa.

strut, n. codal (m), puntal (m) (Con).

strychnine, n. estricnina (f).

stubble, n. rastrojo (m) (plant); **s. burning**, quema (f) de rastrojo; **s. field**, rastrojero (m).

stucco, n. estuco (m).

stud, n. semental (m) (animal); caballeriza (f),cuadra (stable); clavo (m), espiga (f) (pin).

study, 1. v. estudiar; 2. n. estudio (m); **economic s.**, estudio económico; **feasibility s.**, estudio de factibilidad; **geotechnical s.**, estudio geotécnico; **planning s.**, estudio de planificación; **preliminary s.**, estudio preliminar, estudio previo; **water resources s.**, estudio de recursos hidráulicos.

STW, n. EDAR; **sewage treatment works**, estación (f) depuradora de aguas residuales.

style, n. estilo (m).

styrene, n. estireno (m).

styrofoam, n. espuma de poliestireno (m).

sub-system, n. subsistema (m).

subaerial, a. subaéreo/a.

subaqueous, a. subacuático/a.

subatomic, a. subatómico/a; **s. particles**, partículas (f.pl) subatómicas.

subcommittee, n. subcomité (m), subcomisión (f).

subcontracted, a. subcontratado/a.

subcrustal, a. subcortical.

subcutaneous, a. subcutáneo/a.

subdesert, a. subdesértico/a.

subduction, n. subducción (f).

subglacial, a. subglacial, subglaciar.

subgrade, n. subgrado (m) (Treat).

subhumid, a. subhúmido/a.

subject, n. tema (m), concepto (m); asunto (m) (subject matter).

subjective, a. subjetivo/a.

subjectivity, n. subjetividad (f).

subkingdom, n. subreino (m) (Biol).

sublimation, n. sublimación (f).

submarine, 1. n. submarino (m); 2. a. submarino/a.

submariner, n. submarinista (m, f).

submerge, v. sumergir.

submerged, a. sumergido/a.

submergence, n. sumersión (f); **s. ratio**, relación (f) de sumersión.

submersible, a. sumergible, sumergido/a; **s. pump**, bomba (f) sumergida.

submersion, n. sumersión (f), inmersión (m).

suboxide, n. subóxido (m).

subsample, n. submuestra (f).

subsequent, a. subsiguente.

subside, v. hundirse (land); bajar, descender (flood); amainar (wind).

subsidence, n. hundimiento (m).

subsidiary, a. subsidiario/a, secundario/a; **s. company**, sucursal (f), filial (f).

subsidize, n. subvencionar.

subsidy, n. subvención (f) (for a country), subsidio (for a family).

subsistence, n. subsistencia (f).

subsoil, n. subsuelo (m).

substage, n. subpiso (m) (chronology).

substance, n. materia (f) (matter); substancia (f) (compound); esencia (f) (essence); **amber-list s. (EU, pollutants)**, substancia en la lista ámbar; **black/grey/red-list s. (EU, pollutants)**, substancia en la lista negra/gris/roja; **colloidal s.**, substancia coloidal; **hazardous s.**, substancia peligrosa; **humic s.**, sustancia húmica; **noxious s.**, substancia dañina (LAm); **organic s.**, substancia orgánica; **radioactive s.**, substancia radioactiva; **toxic s.**, substancia tóxica.

substation, n. subestación (f).

substitute, v. reemplazar.
substitution, n. substitución (f).
substrata, n.pl. sustratos (m.pl) (see substratum), sustratos (m.pl).
substrate, n. sustrato (m).
substratum, n. sustrato (m); subsuelo (m) (subsoil).
substructure, n. subestructura (f); basamento (m) (basement).
subsurface, a. subterráneo/a.
subterranean, a. subterráneo/a.
subtidal, a. submareal.
subtract, v. restar, sustraer.
subtropical, a. subtrópico/a, subtropical.
suburb, n. suburbio (m); **suburbs,** las afueras (f) de la ciudad.
subway, n. paso (m) subterráneo, subte (m) (metro) (LAm); metro (m) (US, de trenes).
succession, n. sucesión (f); **ecological s.,** sucesión ecológica.
succulent, 1. n. suculento (m) (Bot); 2. a. suculento/a.
sucker, n. chupón (m, f) (of plants); trompa (f) (of insects); ventosa (f) (Zoo, e.g. leech).
sucking, a. chupón/chupona.
sucre, n. sucre (m) (Ec, unit of currency).
suction, 1. n. succión (f) (of liquid), aspiración (f) (of air); 2. a. de succión (liquid), aspirante (air); **s. pump,** bomba (f) aspiradora, bomba de succión.
sudden, a. repentino/a, súbito/a; brusco/a (sharp); inesperado/a (unexpected).
suffocate, v. asfixiar, ahogar, sofocar.
suffocated, a. asfixiado/a.
suffocating, a. sofocador/a, sofocante.
suffocation, n. sofocación (f), sofoco (m), asfixia (f).
suffusion, n. difusión (f).
sugar, n. azúcar (m, f); **cane s.,** azúcar de caña; **fruit s.,** azúcar de la fruta; **s. beet,** remolacha (f) azucarera; **s. mill,** azucarera.
suitability, n. idoneidad (f), conveniencia (f) (convenience), aptitud (f) (aptitude).
suitable, a. oportuno/a, idóneo/a (convenient); apropiado/a, adecuado/a (appropriate).
sulfamide, n. sulfamida (f).
sulfanate, n. sulfanato (m).
sulfate, n. sulfato (m).
sulfide, n. sulfuro (m).
sulfite, n. sulfito (m).
sulfonamide, n. sulfonamida (f).
sulfonated, a. azufrado/a.
sulfonation, n. sulfonación (f).
sulforous, a. azufroso/a.
sulfurate, v. sulfurar.
sulfuric, a. sulfúrico/a (US).
sulphamide, n. sulfamida (m).
sulphane, n. sulfano (m).
sulphate, n. sulfato (m); **s. reduction,** sulfato reducción.
sulphide, n. sulfuro (m); **hydrogen s.,** sulfuro de hidrógeno.

sulphite, n. sulfito (m).
sulphochloride, n. sulfocloruro (m).
sulphonamide, n. sulfonamida (f).
sulphonate, n. sulfanato (m).
sulphonated, a. azufrado/a.
sulphonation, n. sulfonación (f).
sulphorous, a. azufroso/a.
sulphur, n. azufre (m) (S).
sulphurate, v. azufrar, sulfurar.
sulphuric, a. sulfúrico/a.
sum, n. suma (f), cantidad (f), adición (f).
summarize, v. extractar (to abridge).
summary, n. resumen (m); recopilación (f) (of data).
summation, n. sumatorio (m).
summit, n. cima (f), cumbre (f); cúspide (m) (apex); **s. conference,** conferencia (f) cumbre.
sump, n. sumidero (m); pileta (f) (Min); fosa (f) (Geog).
sunbeam, n. rayo (m) de sol, rayo solar.
sundown, n. anochecer (m).
sunflower, n. girasol (m).
sunken, a. hundido/a.
sunlight, n. luz (f) del sol, luz solar.
sunset, n. puesta (f) del sol.
sunshine, n. brillo (m) del sol, resplandor (m) solar, sol (m); **hours of s., s. hours,** horas (f.pl) de sol.
sunspot, n. mancha (f) solar.
superficial, a. superficial, somero/a; **s. waters,** aguas (f.pl) superficiales.
Superfund, n. fondo (m) gubernamental para recuperación y limpieza ambiental (US).
superimpose, v. sobreimponer.
superior, a. superior.
supermarket, n. supermercado (m); automercado (m) (Ven).
supernatent, a. sobrenadante; **s. liquor,** licor (m) sobrenadante.
superoxide, n. superóxido (m).
superphosphate, n. superfosfato (m).
superposition, n. superposición (f).
superstructure, n. superestructura (f).
supervision, n. supervisión (f).
supoena, n. citación (f).
supplant, v. suplantar, desbancar.
supplier, n. proveedor/a (m, f), suministrador/a (m, f).
supply, 1. v. suministrar (with water, electricity), abastecer (to provide, to supply), proveer (to provide); 2. n. abastecimiento (m), suministro (m) (action); provisión (f); **s. and demand,** la oferta y la demanda; **s. network,** red (f) de abastecimiento; **water s.,** abastecimiento de aguas.
support, n. apoyo (m) (aid); sostén (m) (Eng, Con, structure).
supracrustal, a. supracortical.
supression, n. supresión (f) represión (f).
surcharge, n. sobrecarga (f) (overload), rebasamiento (m) (overflow).

surety, n. fianza (f) (sum), caución (f) (person), garantía (f) (guarantee).

surf, n. oleaje (m) (waves), resaca (f) (current).

surface, n. superficie (f) (Gen), firme (m) (of road); **Earth's s.,** superficie terrestre; **free s.,** superficie libre; **level s.,** superficie a nivel, superficie nivelada; **phreatic s.,** superficie freática; **piezometric s.,** superficie piezométrica; **rock s.,** superficie rocosa; **s. area,** área (f) de la superficie; **s. of erosion,** superficie de erosión; **s. route,** ruta (f) de superficie; **topographic s.,** superficie topográfica.

surfactant, n. surfactante (m).

surge, n. oleaje (m) (Mar); sobretensión (f) (Elec); intumescencia (f) (swelling), elevación (f); **s. tank (water supply),** cámara (f) de compensación, tanque (m) de ruptura de carga, tanque de intumescencia.

surgence, n. surgencia (f), manantial (m).

surging, n. contracorriente (f).

surname, n. apellido (m).

surpass, v. sobrepasar, exceder.

surround, v. rodear.

surveillance, n. vigilancia (f).

survey, 1. v. inspeccionar, reconocer (of ground), examinar, medir (to measure); 2. n. inspección (f), reconocimiento (m) (inspect); levantamiento (m) (make a map); examen (m), informe (m) (examine); **aerial s.,** reconocimiento aéreo, levantamiento aéreo; **baseline s.,** anteproyecto (m) de referencia; **land s.,** levantamiento topográfico; **market s.,** estudio (m) de mercado; **Ordnance S. (GB),** equiv. to Instituto (m) Geográfico Nacional (Sp); **quantity s.,** inventario (m) de material; **sampling s.,** encuesta (f) por muestreo.

surveying, a. topografía (f), agrimensura (f), levantamiento (m) de planos (of maps).

surveyor, n. topógrafo/a (m, f), agrimensor (m); **quantity s.,** medidor (m) de cantidades de obra.

survival, n. supervivencia (f); **s. of the fittest,** supervivencia de los más aptos, supervivencia de los mejor dotados; **the fight for s.,** lucha (f) por la supervivencia.

surviving, a. superviviente.

survivor, n. superviviente (m, f).

susceptibility, n. susceptibilidad (f).

susceptible, a. susceptible.

suspended, a. suspendido/a; **s. solids,** sólidos (m.pl) en suspensión.

suspension, n. suspensión (f); **colloidal s.,** suspensión coloidal; **s. bridge,** puente (m) colgante.

sustain, v. sostener, sustentar.

sustainability, n. sostenimiento (m), sostenibilidad (f).

sustainable, a. sostenible; **s. development,** desarrollo (m) sostenible.

sustained, a. sostenido/a.

suture, n. sutura (f).

swabbing, n. estropajado (m) (scouring).

swallow, n. golondrina (f) (Zoo); trago (m) (drink); bocado (m) (food); **s. hole,** sima (f), pozo (m) kárstico; cenote (m) (natural well) (LAm).

swamp, 1. v. inundar; 2. n. pantano (m), bañado (m), ciénaga (f), cenagal (m); guadal (m), estero (m) (LAm).

swampy, a. cenagoso/a.

swarm, n. enjambre (m).

swath, n. ringlera (f) (strip).

Swazian, n. Swaziano (m).

sweat, 1. v. sudar; calentar (Eng); 2. n. sudor (m).

sweatshop, n. fábrica (f) donde se explota al obrero.

sweeper, n. barredora (f) (machine); barrendero/a (m, f) (cleaner).

sweeping, n. barrido (m) (brushing); **mechanical s.,** barrido mecánico; barrido (m) (i.e. cleaning); **street s. and cleaning,** barrido y limpieza de vías.

sweepings, n.pl. barreduras (f) (Min).

swell, 1. v. engrosar, hinchar; 2. n. oleaje (m) (in sea); elevación (f) suave (rise in land); **ocean s.,** oleaje oceánico.

swelling, n. hinchazón (f), hinchamiento (m) (Gen); tumefacción (f) (Med).

swimming, n. natación (f); **s. pool,** piscina (f), natatorio (m) (LAm), pileta (f) de natación (LAm).

swine, n. porcino (m), ganado (m) porcino, cerdo (m); chanco (LAm).

swinging, n. balanceo (m).

swirl, n. remolino (m), torbellino (m) (movement).

switch, n. interruptor (m) (Elec); cambio (m) (change); **trip s.,** interruptor de disparo.

switchgear, n. interruptor (m), conmutador (m), aparato (m) de conexión.

swollen, a. crecido/a (of river); hinchado/a (e.g. gland).

sycamore, n. sicomoro (m).

syenite, n. sienita (f).

sylane, n. silano (m).

sylvanite, n. silvanita (f).

symbiosis, n. simbiosis (f).

symbiotic, a. simbiótico/a.

symbol, n. símbolo (m), signo (m); **equal/minus/plus s.,** símbolo igual/menos/más; **map s.,** símbolo en mapa; **topographic s.,** símbolo topográfico.

symmetrical, a. simétrico/a.

symmetry, n. simetría (f); **bilateral s.,** simetría bilateral; **radial s.,** simetría radial.

symposium, n. simposio (m).

synapse, n. sinapsis (f).

synclinal, a. sinclinal.

syncline, n. sinclinal (f).

synclinorium, n. sinclinorio (m).

syndrome, n. síndrome (m); **acquired immune**

deficiency s. (abbr. AIDS), síndrome de inmunodeficiencia adquirida (abbr. SIDA); **sick building s.**, síndrome del edificio enfermo.
synergism, n. sinergismo (m).
synergistic, a. sinérgico/a.
synergy, n. sinergia (f), sinergismo (m).
syngenetic, a. singenético/a.
synoptic, a. sinóptico/a; **s. forecast**, previsión (f) sinóptica.
syntax, n. sintaxis (f).
synthesis, n. síntesis (m).
synthesize, v. sintetizar.
synthetic, a. sintético/a; **s. data generation**, generación (f) de datos sintéticos; **s. detergent**, detergente (m) sintético; **s. time series**, serie (f) sintética de tiempo.
synthetize, v. sintetizar.
syphilis, n. sífilis (f).
syphon, n. sifón (m) (of liquids); **inverted s.**, sifón invertido.
syringe, n. jeringa (f); jeringuilla (f) (small syringe).

system, n. sistema (m), régimen (m) (pl. regímenes); **administrative s.**, sistema administrativo; **autonomic nervous s.** (abbr. ANS), sistema nervioso autónomo (abbr. SNA); **circulatory s.**, sistema circulatorio; **conjugated s.**, sistema conjugado; **decimal s.**, sistema decimal; **digestive s.**, sistema digestivo, aparato (m) digestivo; **hydraulic s.**, régimen hidráulico; **joint s.** (Geol), sistema de diaclasas; **metric s.**, sistema métrico; **mountain s.**, sistema montañoso; **multiaquifer s.**, sistema multiacuífero; **nervous s.**, sistema nervioso; **non steady-state s.**, régimen no permanente; **river s.**, sistema fluvial; **solar s.**, sistema solar; **steady-state s.**, régimen estacionario; **turbulent s.**, régimen turbulento; **vascular s.**, sistema vascular; **water-vapour s.**, sistema agua-vapor.
systematic, a. sistemático/a.
systematics, n. sistemática (f), taxonomía (f).
systemic, a. sistémico/a.

T

table, n. mesa (f) (furniture); tabla (f), lista (f) (list); cuadro (m), gráfica (f) (Mat); capa (f) (Geol); **logarithm tables,** tabla de logarítmos; **look-up t.,** tabla de consulta; **Periodic T. of Elements,** Tabla Periódica de los Elementos; **rating t.,** tabla de calibración.

tabulate, v. tabular.

tabulation, n. tabulación (f).

tachymetry, n. taquímetro (m).

tackle, n. aparejo (m) (rigging), jarcias (f.pl) (ropes), polipasto (m) (pulley, hoist block); **block and t.,** aparejo de poleas.

tactile, a. táctil.

tadpole, n. renacuajo (m).

tag, n. etiqueta (f), rótulo (m) (label); trazador (m) (tracer).

taiga, n. taiga (f).

tail, n. cola (f) (Gen); rabo (m) (e.g. cattle).

tail-pipe, n. tubo (m) de escape.

tailback, n. cola (f) de tráfico.

tailgate, n. puerta (f) trasera.

tailings, n. colas (f.pl), relaves (m.pl), estériles (m.pl) (Min).

tailpipe, n. tubo (m) de escape (US).

tailwind, n. viento (m) de cola, viento trasero.

take steps, v. tramitar, hacer gestiones.

talc, n. talco (m).

tallow, n. sebo (m).

talus, n. talud (m); **t. stabilization,** estabilización (f) de taludes.

tan, n. tg (m) (Mat, abbr. of tangent).

tangent, n. tangente (m) (Mat, abbr. tg).

tangerine, n. mandarina (f) (fruit); mandarino (m) (tree).

tank, n. tanque (m), alberca (f) (small irrigation tank), estanque (m), fosa (f); **aeration t.,** tanque de aireación; **baffle t.,** tanque con desviadores; **balancing t.,** estanque de estabilización; **constant-head t.,** tanque de descenso constante; **digester t.,** tanque de tratamiento; **equalizing t.,** tanque de compensación; **evaporation t.,** estanque de evaporación; **gauge t.,** tanque de aforo; **septic t.,** fosa séptica; **settling t.,** estanque de sedimentación, sedimentador (m); **storage t.,** tanque de almacenamiento; **surge t.,** tanque de ruptura de carga.

tanker, n. petrolero (m), buque (m) aljibe (Mar); camión (m) cisterna (lorry).

tannery, n. curtiduría (f), tenería (f); **t. wastewater,** aguas (f.pl) residuales de planta de curtidos.

tannin, n. tanino (m).

tanning, 1. n. curtido (m); 2. a. curtido/a; **t. industry,** industria (f) de curtiembres, industria curtidora.

tantalum, n. tántalo (m) (Ta).

tap, 1. v. dar un toque a, dar un golpecito en (to knock gently); hacer una toma de (of water, oil); resinar (to bleed a tree for resin); explotar (resources); 2. n. grifo (m) (of water); canilla (f) (LAm); **water t.,** grifo del agua.

tape, n. cinta (f); **adhesive t.,** cinta adhesiva; **insulating t.,** cinta aislante; **on t. (a),** grabado/a, en cinta; **punched t.,** cinta perforada; **recording t.,** cinta magnetofónica; **red t. (Fig),** papeleo (m), trámites (m.pl); **t. measure,** cinta métrica.

tapering, a. afilado/a.

tapeworm, n. tenia (f).

tar, n. alquitrán (m), brea (f); **coal t.,** alquitrán de hulla.

target, n. blanco (m) (Gen); objetivo (m), meta (f) (objective); **non-t. species,** especie (m) no objectivo; **t. population,** población (f) objectiva.

tariff, n. tarifa (f) tariff.

tarn, n. laguna (f) glaciar, lago (m) de circo glaciar.

tarnish, v. deslustrar, empañarse.

tarnished, a. deslustrado/a.

tarpaulin, n. lona (f) alquitranada, alquitranado (m).

tarragon, n. estragón (m).

tarsal, 1. n. tarso (m); 2. a. tarsiano/a; **t. bones,** huesos (m.pl) tarsianos.

tarsi, n.pl. tarsos (m.pl) (see: tarsus).

tarsus, n. tarso (m) (of insects); tarso parpebral (Med, ankle).

tartar, n. tártaro (m) (Chem).

tartrate, n. tartrato (m).

taste, 1. v. degustar, probar (to test), saborear (flavour); 2. n. gusto (m); sabor (m) (flavour).

taster, n. catador (m) (person).

tasting, n. cata (f), catadura (f); **taste t.,** cata de ensayo.

taxa, n.pl. taxones (m.pl) (see: taxon).

taxis, n. tactismo (m), taxia (f); **negative t.,** taxia negativa; **positive t.,** taxia positiva.

taxon, n. taxón (m).

taxonomic, a. taxonómico/a; **t. unit,** unidad (f) taxonómica.

taxonomy, n. taxonomia (f); **numerical t.,** taxonomia numérica.

TDS, n. SDT (m); **total dissolved solids,** sólidos (m.pl) disueltos totales.

tea, n. té (m).

teak, n. teca (f).

teamwork, n. trabajo (m) de equipo.

teamwork, n. trabajo (m) en equipo.

tear, 1. v. desgarrar; 2. n. lágrima (f) (of eyes); rasgón (m) (rip); **t. duct,** conducto (m) lacrimal.

teat, n. tetilla (of mammals), tetina (f) (of bottles).

technetium, n. tecnecio (m) (Tc).

technical, a. técnico/a; **t. assessment,** evaluación (f) técnica; **t. terminology,** terminología (f) técnica.

technicality, n. tecnicismo (m), tecnicidad (f).

technician, n. técnico/a (m, f); especialista (m, f) (specialist); ayudante (m) (in a laboratory).

technique, n. técnica (f), método (m).

technology, n. tecnología (f); **aerospace t.,** técnica (f) aeroespacial; **alternative t.,** tecnología alternativa; **appropriate t.,** tecnología apropiada; **clean technologies (Ind),** tecnologías limpias; **intermediate t.,** tecnología intermedia; **low-cost t.,** tecnología de bajo costo; **low impact t.,** tecnología de bajo impacto; **soft t.,** tecnología suave/blanda.

tectonic, a. tectónico/a.

tectonics, n. tectónica (f); **plate t.,** tectónica de placas.

teflon, n. teflón (m).

telecommunication, n. telecomunicación (f).

telegraph, n. telégrafo (m).

telemetric, a. telemétrico/a.

telemetry, n. telemetría (f), teledirección (f), telegestión (f).

teleost, 1. n. teleósteo (m); 2. a. teleósteo/a.

Teleostei, n.pl. teleósteos (m.pl).

telephone, n. teléfono (m); **t. exchange,** central (f) telefónica.

television, n. televisión (f), televisor (m) (set); **closed-circuit t. (abbr. CCTV),** televisión (f) por circuito cerrado; **down-hole t.,** televisor en sondeos.

tellurium, n. teluro (m) (Te).

temperate, a. templado/a; **t. zone,** zona (f) templada.

temperature, n. temperatura (f); **absolute t.,** temperatura absoluta; **ambient t.,** temperatura ambiente; **annual mean t.,** temperatura media anual; **room t.,** temperatura ambiente; **t. and rainfall measurement,** termopluviometría (f).

tempest, n. tempestad (f).

template, n. plantilla (f) (metal or wood).

temple, n. templo (m).

temporal, a. temporal.

temporary, a. transitorio/a (measure); temporal, provisional (arrangement).

tench, n. tenca (f).

tendency, n. tendencia (f); **measure of central t. (Mat),** medida (f) de la tendencia central.

tender, n. ténder (m) (of a locomotive); curso (m) legal (document).

tendon, n. tendón (m); **Achilles t.,** tendón de Aquiles, talón (m) de Aquiles.

tendril, n. zarcillo (m).

tenon, n. espiga (f), macho (m), barbilla (f).

tensile, a. tensor/a (exerting tension), extensible (extendable).

tension, n. tensión (f); tracción (f) (traction, pulling); **high-t. cable,** cable (m) de alta tensión; **surface t.,** tensión superficial; **vapour t.,** tensión de vapor.

tensionometer, n. tensiómetro (m).

tensoactive, a. tensoactivo/a.

tensor, n. tensor (m).

tentacle, n. tentáculo (m).

tentaculite, n. tentaculito (m).

tera-, pf. tera- (10.E12).

terbium, n. terbio (m) (Tb).

term, n. término (m) (word); plazo (m), período (m) (time); trimestre (m) (of schools, etc.); condition (of a contract); **glossary of terms,** glosario (m) de términos; **in terms of,** en términos de; **technical t.,** término técnico.

terminal, 1. n. terminal (f) (Comp, Comm); terminal (m) (Elec); **air t.,** terminal aérea; **computer t.,** terminal de un ordenador; 2. a. terminal.

termination, n. terminación (f), finalización (f) conclusión (f).

terminology, n. terminología (f); **technical t.,** terminología técnica.

termite, n. termita (f), comején (m) (LAm).

terpene, n. terpeno (m).

terrace, 1. v. abancalar, aterrazar; 2. n. banco (m) prominente (Geog); terraza (f) (patio); **alluvial t.,** terraza aluvial; **raised beach t.,** terraza elevada; **river t.,** terraza fluvial; **shore t.,** terraza ribereña.

terracing, n. abancalamiento (m), aterrazamiento (m).

terracotta, n. terracota (f).

terrain, n. terreno (m); **broken t.,** terreno incoherente; **fissured t.,** terreno fisurado; **karstic t.,** terreno kárstico; **sandy t.,** terreno arenoso.

terrapin, n. terrapene (m), tortuga (f) acuática.

terrestrial, a. terrestre.

territoriality, n. territorialidad (f).

territory, n. territorio (m).

Tertiary, n. Terciario (m).

terylene, n. terileno (m).

test, 1. v. probar, ensayar (Chem); poner a prueba (e.g. a prototype); analizar (to analyse); 2. n. ensayo (m), prueba (f), experimento (m); análisis (m); test (m); **Ames t. (Biol, Chem),** test de Ames; **aquifer t.,** ensayo de acuífero; **bailer t.,** ensayo de cuchareo; **bench t.,** banco (m) de pruebas; **brown ring t. (Biol, Chem),** test del anillo marrón; **Chi-squared t. (Mat),** prueba jicuadrada; **compression t.,** ensayo de compresión, prueba de compresión; **constant-level t.,** ensayo a nivel constante; **efficiency t.,** ensayo de eficiencia; **environmental t.,** ensayo ambiental, prueba ambiental; **field t.,** ensayo in situ; **fitness t.**, prueba de idoneidad; **flow t.**, ensayo de medida del caudal; **formation t.**, ensayo de formación; **goodness-of-fit t.** (Mat), prueba de bondad de ajuste, test de buen ajuste; **hardness t.**, ensayo de dureza; **injection t.**, ensayo de inyección; **interference t.**, ensayo de interferencia; **jar t.**, prueba de jarras; **laboratory t.**, prueba del laboratorio; **Lassaigne t.** (Biol, Chem), test de Lassaigne; **long-term t.**, ensayo a largo plazo; **Mann-Whitney U t.** (Mat), prueba U de Mann Whitney; **mechanical t.**, ensayo mecánico; **nuclear t.**, prueba nuclear; **packer t.** (Geol), ensayo de sellador; **patch t.** (Med), prueba de emplasto; **performance t.**, prueba de rendimiento, desempeño; **permeability t.**, ensayo de permeabilidad; **pumping t.**, ensayo de bombeo; **recovery t.** (Geol), ensayo de recuperación; **reliability t.**, ensayo de funcionamiento seguro; **screen t.** (Geol), ensayo de criba, ensayo de rejilla; **step t.** (Geol), ensayo escalonado; **Student's t test**, ensayo de la t de Student; **t. data**, datos (m.pl) de los ensayos; datos de las pruebas; **t. kit**, aparato (m) de ensayo; **t. panel**, equipo (m) de pruebas; **t. run**, ensayo; **t. site**, centro (m) de experiencias, campo (m) de experiencias; **t. to destruction**, ensayo destructivo, prueba destructiva; **thermonuclear t.**, ensayo termonuclear; **tracer t.**, ensayo con trazadores; **triaxial t.**, ensayo triaxial; **vane t.**, ensayo de molinete.

testa, n. testa (f).
testes, n.pl. testes (m.pl) (see: testis).
testicle, n. testículo (m).
testing, 1. n. prueba (f); **well t.**, prueba de pozos; 2. a. de pruebas.
testis, n. teste (m).
testosterone, n. testosterona (f).
tetanus, n. tétano (m), tétanos (m).
tetra-, pf. tetra-.
tetrabratulid, n. tetrabratúlido (m).
tetrachloride, n. tetracloruro (m); **carbon t.**, tetracloruro de carbono.
tetrachlorobiphenol, n. tetraclorobifenol (m).
tetrachloroethylene, n. tetracloroetileno (m).
tetragraptid, n. tetragraptido (m).
tetrahedron, n. tetraedro (m).
tetrahydrate, n. tetrahidrato (m).
tetramycin, n. tetramicina (f).
tetranitromethane, n. tetranitrometano (m).
tetrapod, 1. n. tetrápodo (m); 2. a. tetrápodo/a.
tetravalent, a. tetravalente, cuadrivalente;.
textile, a. textil; **t. industry**, industria (f) textil.
texture, n. textura (f); **drainage t.**, textura de avenamiento; **even t.**, textura homogénea.
thalli, n.pl. talos (m.pl).
thallium, n. talio (m) (Tl).
Thallophyta, n.pl. talofitas (f.pl), la división Thallophyta (Bot).
thallus, n. talo (m).

Thanetian, n. Thanetiense (m).
thatch, 1. v. cubrir con paja; 2. n. paja (f); **t. roof**, techo (m) de paja.
thaw, 1. v. descongelar, derretir, deshelar; 2. n. deshielo (m).
thawing, n. deshielo (m), derretimiento (m); **t. of a river**, deshielo de un río.
theca, n. teca (f).
theodolite, n. teodolito (m).
theorem, n. teorema (m).
theoretical, a. teórico/a.
theoretically, ad. teóricamente, en teoría.
theory, n. teoría (f); **atomic t.**, teoría atómica; **quantum t.**, teoría cuántica, teoría de los cuanta.
thermal, a. térmico/a, termal; **t. imaging**, termografía (f) infrarroja, formación (f) de imágenes térmicas; **t. stratification**, estratificación (f) térmica.
thermistor, n. termistor (m).
thermite, n. termita (f).
thermo-, pf. termo-.
thermochemical, a. termoquímico/a.
thermocline, n. termoclina (f).
thermocouple, n. termopar (m), par (m) térmico.
thermodynamic, a. termodinámico/a.
thermodynamics, n. termodinámica (f); **first law of t.**, primera ley (f) de la termodinámica.
thermograph, n. termógrafo (m).
thermography, n. termografía (f); **infrared t.**, termografía infrarroja.
thermometer, n. termómetro (m); **maximum and minimum t.**, termómetro de las máximas y mínimas; **mercury t.**, termómetro de mercurio.
thermonuclear, a. termonuclear; **t. test**, prueba (f) termonuclear.
thermophile, n. termófilo (m).
thermophilic, a. termófilo/a; **t. anaerobic digestion**, digestión (f) anaerobia termófila.
thermoplastic, 1. n. termoplástico (m); 2. a. termoplástico/a.
thermosphere, n. termoesfera (f).
thermostat, n. termostato (m).
thesaurus, n. tesauro (m), diccionario (m) de sinónimos.
thesis, n. tesis (f).
thiamine, n. tiamina (f) (vitamin B1).
thick, a. grueso/a, espeso/a, gordo/a.
thicken, v. espesar; aumentar en espesor (Geol).
thickener, n. espesador (m) (Treat); **sludge t.**, espesador de fangos, espesador de lodos.
thickening, n. espesamiento (m); **t. tank**, tanque de espesamiento.
thicket, n. soto (m), matorral (m), matas (f.pl).
thickness, n. espesor (m); **aquifer t.**, espesor del acuífero; **saturated t.**, espesor saturado; **unsaturated t.**, espesor no saturado.
thigmotropism, n. tigmotropismo (m).

thin, a. delgado/a, flaco/a (person); fino/a (e.g. cloth, steel sheet).

thinning, n. adelgazamiento (m).

thin out, v. atenuarse (Geol).

thio-, pf. tio-.

thiocyanate, n. tiocianato (m).

thiosulphate, n. tiosulfato (m).

thistle, n. cardo (m); **cotton t.**, cardo borriquero.

thixotropic, a. tixotrópico/a.

thixotropy, n. tixotropía (f).

tholeiite, n. toleita (f).

thoracic, a. torácico/a.

thorax, n. tórax (m).

thorium, n. torio (m) (Th).

thorn, n. espina (f).

thorny, a. espinoso/a.

threaded, a. roscado/a (of screw, etc.).

threat, n. amenaza (f); **t. to health**, amenaza para la salud.

threatened, a. amenazado/a; **t. species**, especie(s) (f.pl) amenazada(s).

three-dimensional, a. tridimensional.

thresh, v. trillar.

threshing, n. trilla (f), trilladura (f); **t. ground**, era (f); **t. machine**, trilladora (f).

threshold, n. umbral (m); **tolerance t.**, umbral de tolerancia.

throw, n. rechazo (m) (Geol); tiro (m), lanzamiento (m) (to launch); dislocación (f) (to dislocate); **t. of fault**, salto (m) vertical de una falla, tiro de falla.

thrust, n. cabalgamiento (m) (Geol); **high-angle t.**, cabalgamiento de alto ángulo.

thulium, n. tulio (m) (Tm).

thunder, 1. v. tronar; 2. n. trueno (m).

thunderbolt, n. rayo (m), centella (f).

thunderclap, n. trueno (m).

thunderous, a. estruendoso/a.

thundery, a. tormentoso/a.

thyme, n. tomillo (m).

tibia, n. tibia (f).

tibiae, n.pl. tibias (f.pl).

tick, n. garrapata (f) (Biol); tictac (m) (of clock); palomita (f) (mark).

tidal, a. mareal, de la marea; **t. constituent (Mat)**, constituyente (m) armónico de la marea; **t. datum (Mar)**, plano (m) de referencia; **t. energy**, energía (f) de la marea; **t. estuary**, estuario (m) con mareas; **t. harmonic (Mat)**, constituyente (m) armónico de la marea; **t. range**, amplitud (f) de la marea, alcance (m) de la marea; **t. volume (Anat, of lungs)**, volumen (m) de ventilación pulmonar; **t. wave**, maremoto (m).

tide, n. marea (f); **ebb t.**, marea menguante; **equinoctial t.**, marea equinoccial; **high t.**, marea alta, pleamar (m); **incoming t.**, marea entrante; **low t.**, marea baja, bajamar (m), estiaje (m); **neap t.**, marea muerta; **outgoing t.**, marea descendente, marea saliente; **rising t.**, marea creciente; **spring t.**, marea viva; **t.**

gate, compuerta (f) de marea; **t. tables**, tablas (f.pl) de mareas.

tidemark, n. línea (f) de la marea alta.

tie, n. tirante (m) (Con); traviesa (f) (US, railway sleeper); **t. beam**, tirante de cercha.

tile, n. teja (f) (on roof); baldosa (f) (on floor); azulejo (m) (glazed ceramic); **field t. drain**, drenaje (m) poroso; **ridge t.**, teja de caballete.

till, 1. v. labrar; 2. n. till (m) (Geol); **glacial t.**, till glaciar, till glacial.

tillite, n. tillita (f).

tilt, v. ladear.

tilted, a. inclinado/a, ladeado/a; en pendiente (sloping).

tilting, n. basculamiento (m), inclinación (f).

timber, 1. v. enmaderar (e.g. wall); encofrar (to plank), entibar (a mine); 2. n. madera (f) (for construction); viga (f) (beam); bosque (m) (US); **t. merchant**, negociante (m) de madera.

timbered, a. enmaderado/a (e.g. ceiling), entibado/a (in a mine), arbolado/a (forested land).

timbering, n. entibación (f), entibado (m) (e.g. in a mine); maderaje (m), maderamen (m) (for construction).

timber-line, n. límite (m) de la vegetación arbórea.

timberwork, n. maderaje (m), maderamen (m).

time, n. tiempo (m); época (f) (season, epoch); hora (f) (moment); **arrival t.**, tiempo de llegada; **contact t. (Treat)**, tiempo de contacto; **geological t.**, tiempo geológico; **Greenwich Mean T. (abbr. GMT)**, Tiempo Medio de Greenwich (abbr. TMG); **harvest t.**, cosecha (f), época (f) de la cosecha; **reaction t.**, tiempo de reacción; **recovery t.**, tiempo de renovación; **residence t.**, tiempo de residencia, tiempo de estancia; **response t.**, tiempo de respuesta; **retention t.**, tiempo de retención; **run t. (Comp)**, tiempo corrido; **start t. (máquinas)**, tiempo de arranque; **t. lag**, retardo (m) de tiempo; **t. limit**, limitación (f) de tiempo; **t. of concentration (Geog)**, tiempo de concentración; **t. series**, series (f.pl) temporales; **t. zone**, huso (m) horario; **transit t.**, tiempo de tránsito; **turnaround t.**, tiempo de respuesta.

time-lag, n. retraso (m) de tiempo.

times, v. por; más (Mat); **2 t. 3 is 6**, 2 por 3 son 6; **4 t. as big**, 4 veces más grande.

timescale, n. escala (f) de tiempo.

timetable, n. horario (m) (e.g. of public transport); itinerario (m), programa (m) (of events).

tin, n. estaño (m) (Sn); lata (f) (can).

tinder, n. yesca (f).

tine, n. púa (f), punta (f) (prong); diente (m) (of a fork).

tin-plate, v. estañar.

tin-plating, n. estañado (m), estañadura (f).

tip, n. vertedero (m) (rubbish dump); ápice (m) (point); **licensed t.**, vertedero controlado; **open t.**, vertedero de superficie; **waste t. operative**, basurero (m) (Arg, Ch, Méx, Nic, Par, Perú, Ur), recolector (m) (Bol, Méx), empleado (m) de limpieza, operario (m) de limpieza (LAm).

tipping, n. derrumbamiento (m); **fly t., illegal t.**, derrumbamiento ilegal.

tissue, n. tejido (m); **adipose t.**, tejido adiposo; **connective t.**, tejido conectivo; **t. culture**, cultivo (m) celular, cultivo de tejido; **vascular t.**, tejido vascular.

titanite, n. titanita (f), esfena (f).

titanium, n. titanio (m) (Ti); **t. dioxide**, dióxido (m) de titanio.

titer, n. valoración (f).

titration, n. titración (f), valoración (f), titulación (f); **end-point t.**, valoración a punto final; **t. curve**, curva de titulación.

titre, n. valoración (f) (Chem).

toadstool, n. hongo (m).

Toarcian, n. Toarciense (m).

tobacco, n. tabaco (m).

TOC, n. COT (m); **total organic carbon**, carbono (m) orgánico total.

toe, n. dedo (m) del pie (Anat); puntera (f) (of shoe); pestaña (f) (Con, of bank); **t. drain (of a dam)**, desaguadero (m) de pie; **t. filter (e.g. of a dam)**, filtro (m) de pie.

toilet, n. lavabo (m), retrete (m), water (m) (latrine), servicios (m.pl) (e.g. restaurant), baño (m) (US); arreglo (m), aseo (m) (dressing, shaving, etc.).

tolerance, n. tolerancia (f); **t. limit**, límite (m) de tolerancia.

tollway, n. autopista (f) de peaje; autopista de cuota (LAm).

toluene, n. tolueno (m).

tomato, n. tomate (fruit and plant); **t. dealer, t. grower**, tomatero/a (m, f).

tomatoes, n.pl. tomates (m.pl) (see: tomato).

tomb, n. tumba (f), sepultura (f).

ton, n. tonelada (f); **long t.**, tonelada larga (1,016 kg); **metric t. (tonne)**, tonelada métrica (1,000 kg); **t. (US)**, tonelada corta (907.18 kg); tonelada (f) larga (1,016 kg).

tonalite, n. tonalita (f).

tongs, n.pl. tijeras (f.pl), pinzas (f.pl).

tongue, n. lengua (f); **mother t., native t.**, lengua materna, lengua nativa.

tonne, n. tonelada métrica (f) (1,000 kg).

tonsillitis, n. tonsilitis (f), amigdalitis (f).

tool, n. herramienta (f); **cutting t.**, herramienta de corte; **drilling t.**, herramienta de perforación; **t. box, t. kit, tools**, caja (f) de herramientas, herramental (m).

tooth, n. diente (m).

top, n. cima (f), cumbre (f) (of mountain); tapa (f), tapón (m) (lid); parte (f) superior (upper part).

topaz, n. topacio (m).

topic, n. tema (m).

topographical, a. topográfico/a.

topography, n. topografía (f); **block-faulted t.**, topografía de bloques fallados; **submarine t.**, topografía submarina.

topology, n. topología (f).

toppling, n. vuelco (m).

topsoil, n. capa (f) arable, capa (f) superficial, suelo (m) vegetal.

tornado, n. tornado (m).

torque, n. momento (m) de una fuerza, momento de torsión.

torr, n. torr (m) (unit of measurement of pressure).

torrent, n. torrente (m) (rushing stream), rambla (f) (intermittent stream in southern Spain); **in torrents**, a torrentes.

torrential, a. torrencial; torrentoso/a (LAm).

torsion, n. torsión (f).

tortoise, n. tortuga (f).

Tortonian, n. Tortoniense (m).

tortuosity, n. tortuosidad (f).

touch, n. tacto (m) (sense); toque (m) (tap).

tough, a. duro/a, resistente, tenaz.

tourism, n. turismo (m).

tourmaline, n. turmalina (f).

Tournaisian, n. Tournasiense (m).

tower, n. torre (f); **control t.**, torre de control; **cooling t.**, torre de refrigeración; **t. block**, torre de pisos.

towline, n. cable (m) de remolque.

town, n. ciudad (f); **new t.**, ciudad nueva; **t. council**, Ayuntamiento (m), concejo (m) municipal; **t. councillor**, concejal (m); **t. hall**, Ayuntamiento (m) (building); **t. planner**, urbanista (m, f); **t. planning**, urbanismo (m), planeación (f) urbana.

town-planning, a. urbanístico/a.

township, n. municipio (m); municipalidad (f).

townspeople, n.pl. ciudadanos (m.pl).

towpath, n. camino (m) de sirga (on canal).

toxic, a. tóxico/a; **t. substance**, substancia (f) tóxica; **t. waste**, desecho (m) tóxico.

toxicant, n. tóxico (m).

toxicity, n. toxicidad (f).

toxicology, n. toxicología (f).

toxin, n. toxina (f).

trace, 1. v. trazar, delinear (to delineate); 2. n. rastro (m), traza (f), pisada (f); **t. amount**, trazas; **t. element**, elemento (m) traza, oligoelemento (m).

tracer, n. trazador (m), traza (f) (to trace the path of); **artificial t.**, trazador artificial; **chemical t.**, trazador químico; **dilution t.**, trazador de dilución; **dye t.**, trazador de color, trazador colorante; **inorganic t.**, traza inorgánica; **isotopic t.**, trazador isotópico; **metallic t.**, traza metálica; **organic t.**, traza orgánica; **radioactive t.**, trazador radioactivo; 3. a. trazador/a.

trachea, n. tráquea (f).

tracheae, n.pl. tracheas (f.pl).

trachyandesite, n. traquiandesita (f).
trachybasalt, n. traquibasalto (m).
trachyte, n. traquita (f).
tracing, n. calco (m) (action of drawing), trazado (m) (drawing a line).
track, n. huella (f) (Gen); rastro (m) (of animal); pista (f) (of person); rodado (m) (of vehicle); camino (m) (way); **forest t.**, camino forestal; **the beaten t.**, camino trillado.
traction, n. tracción (f).
tractor, n. tractor (m) (Agr), camión (m) tractor (lorry); **caterpillar t.**, tractor oruga.
trade, n. comercio (m); industria (f) (manufacture); **Department of T.**, Ministerio (m) de Comercio; **domestic t.**, comercio interior; **export t.**, exportación (f); **foreign t.**, comercio exterior; **free t.**, libre comercio; **overseas t.**, comercio exterior; **retail t.**, comercio al por menor; **wholesale t.**, comercio al por mayor.
trademark, n. marca (f) registrada, marca patentada, marca de fábrica.
traffic, n. tráfico (m) (Gen); circulación (f) (e.g. cars); **maritime t.**, tráfico marítimo; **principal t. routes**, arterias (f.pl) de circulación vial; **road t.**, tráfico rodado; **t. congestion**, atasco (m) de tráfico, embotellamiento (m), atorón (m) (Mex); **t. control centre**, centro (m) de control de tráfico; **t. flow**, circulación del tráfico; **t. jam**, atasco (m) de tráfico, embotellamiento (m), atorón (m) (Mex); **t. lights**, semáforos (m.pl); **t. sign**, señal (f) de tráfico; **vehicular t.**, tráfico rodado.
trail, n. camino (m), senda (f) (path); rastro (m), pista (f) (track); estela (of dust); **nature t.**, senda ecológica, itinerario (m) de la naturaleza; **town t.**, itinerario (m) urbano.
trailer, n. remolque (m) (behind a vehicle).
training, n. formación (f), capacitación (f) (job); adiestramiento (m) (Mil); entrenamiento (m) (sport).
trait, n. rasgo (m), característica (f).
trajectory, n. trayectoria (f).
transaction, n. tramitación (f); **business t.**, tramitación de un asunto.
transactions, n.pl. movimiento (m) de caja (Fin).
transboundary, 1. n. transfrontera (f); 2. a. transfronterizo/a.
transducer, n. transductor (m).
transect, n. transecto (m); **linear t.**, transecto lineal.
transfer, n. transferencia (f), traslado (m) (of place); **inter-basin t.**, transferencia de agua entre cuencas; **technology t.**, transferencia de tecnología; **t. function (Mat)**, función (f) de transferencia; **t. station (Treat)**, estación (f) de transferencia.
transformation, n. transformación (f); **Laplace t.**, transformación de Laplace.
transformer, n. transformador (m); **step-up t. (Elec)**, elevador (m).

transfusion, n. transfusión (f); **blood t.**, transfusión sanguínea.
transgression, n. transgresión (f); **marine t.**, transgresión marina.
transient, a. transitorio/a, transeúnte; **t. model**, modelo (m) transitorio.
transistor, n. transistor (m).
transitory, a. transitorio/a.
translocation, n. translocación (f).
transmissibility, n. transmisibilidad (f).
transmission, n. transmisión (f).
transmissivity, n. transmisividad (f).
transmitter, n. emisora (f); **walkie-talkie t.**, emisor receptor.
transom, n. travesaño (m), montante (m) (of windows and doors), dintel (m) (lintel).
transparency, n. transparencia (f).
transparent, a. transparente.
transpiration, n. transpiración (f).
transplant, n. trasplante (m).
transport, 1. v. transportar (goods); deportar (to deport); **to t. goods**, transportar mercancías; 2. n. transporte (m) (Gen); arrastre (m) (haulage, dragging); **means of t.**, medio (m) de transporte; **Ministry of T.**, Ministerio (m) de transporte; **overland t.**, transporte terrestre; **public t.**, transporte colectivo, transporte público; **rail t.**, transporte por ferrocarril; **t. costs**, gastos (m.pl) de transporte; **t. network**, red (f) viaria, red de transporte; **t. of fine particles**, arrastre de partículas finas; **t. of goods**, transporte de mercancías; **t. ship**, buque (m) de transporte; **urban public t.**, transporte público urbano.
transportable, a. transportable.
transportation, n. el transporte de (m); **t. of goods**, el transporte de mercancías.
transporter, n. transportista (m) (conveyor).
transverse, a. transverso/a, transversal.
trap, n. trampa (f) (for hunting); sifón (m) (in a pipe); **grease t., oil t. (Treat)**, interceptor (m) de grasas; **speed t.**, control (m) de la velocidad.
trapezium, n. trapecio (m).
trapezoid, a. trapezoide (m), trapecio (m) (US, Mat).
trapezoidal, a. trapezoidal; **t. weir**, vertedero (m) trapezoidal.
trash, n. basura (f), desecho (m); **t. can**, cubo (m) de basura, balde (m) (esp. LAm), pipote (m) (Ven); **t. heap**, vertedero (m), basurero (m); **t. screen**, rejilla (f) para la basura.
travel, v. viajar; **to t. through**, recorrer.
travertine, n. travertino (m).
trawl, 1. v. arrastrar; 2. n. arrastre (m); **t. net**, red (f) de arrastre, red barredora.
trawler, n. barco (m) rastreador, buque (m) pasquero de rastreo, trainera (f); **t. fleet**, flota (f) de arrastre.
trawling, n. arrastre (m) (fishing); **bottom t.**, arrastre de fondo.
treasure, n. tesoro (m).

treat, n. tratar.

treatability, n. facultad (f) de tratamiento.

treated, a. tratado/a.

treatise, n. tratado (m); **mathematical t.,** tratado de matemáticas.

treatment, n. trato (m); tratamiento (m) (Eng, Chem); **aerobic t.,** tratamiento aerobio; **anaerobic t.,** tratamiento anaerobio; **biological t.,** tratamiento biológico, biotratamiento (m); **grass-plot t.,** tratamiento de pastizal; **ground t.,** tratamiento terrestre; **on-site t.,** tratamiento en origen, tratamiento en la fuente; **preliminary t.,** tratamiento preliminar; **primary t.,** tratamiento primario; **reed-bed t.,** tratamiento de cañaveral; **secondary t.,** tratamiento secundario; **stage t.,** tratamiento por etapas; **tertiary t.,** tratamiento terciario; **t. at source,** tratamiento en origen, tratamiento en la fuente; **or t. of raw materials,** or tratamiento de materias primas; **wastewater t.,** tratamiento de aguas residuales; **water t.,** tratamiento del agua.

tree, n. árbol (m); **t. cover,** cubierta (f) arbórea.

treeless, n. despoblado (m) de árboles.

tree-line, n. límite (m) de la vegetación arbórea.

Tremadocian, n. Tremadociense (m).

Trematoda, n.pl. tremátodos (m.pl), Trematoda (f.pl) (Zoo, class).

trematode, n. tremátodo (m).

tremie, n. tubo (m) bajo el agua para hormigonado.

tremolite, n. tremolita (f).

tremor, n. temblor (m); **earth t.,** temblor de tierra.

trench, n. zanja (f) (Gen), trinchera (f) (Mil), fosa (f); **cut-off t.,** zanja de impermeabilización.

trenching, n. zanjero (m).

trend, n. tendencia (f); dirección (f) (direction); **rising t.,** la tendencia al alza, la tendencia al aumento.

tri-, pf. tri-.

trial, n. prueba (f), ensayo (m) (test); juicio (m) (Jur).

triangle, n. triángulo (m); **equilateral t.,** triángulo equilátero; **isosceles t.,** triángulo isósceles.

triangular, a. triangular; **t. weir,** vertedero (m) triangular.

triangulate, v. triangular.

triangulation, n. triangulación (f).

Triassic, n. Triásico (m).

triaxial, a. triaxial; **t. test,** ensayo (m) triaxial.

triazine, n. triazina (f).

tribromomethane, n. tribromometano (m).

tributary, n. tributario (m), afluente (m) (inflow).

triceps, n. tríceps (m).

trichloroacetaldehyde, n. tricloroacetaldehido (m).

trichloroethene, n. tricloroetileno (m).

trichloroethylene, n. tricloroetileno (m).

trichloromethane, n. triclorometano (m).

Trichoptera, n.pl. tricópteros (m.pl), Trichoptera (f.pl) (Zoo, clase).

trichopteran, n. tricóptero (m), canutillo (m).

trickle, n. chorrito (m) (small flow); gota (f) a gota (drip-feed); **t. irrigation,** riego (m) gota a gota.

tricone, a. tricono/a; **t. bit,** trépano (m) tricono.

trigonometry, n. trigonometría (f).

trihalomethane, n. trihalometano (m).

trilobite, n. trilobites (m.pl).

trimethylamine, n. trimetilamina (f).

trinitroglycerine, n. trinitroglicerina (f).

trinitrophenol, n. trinitrofenol (m).

trinitrotoluene, n. trinitrotolueno (m).

trioxide, n. trióxido (m).

triphenylmethane, n. trifenilmetano (m).

tripod, n. trípode (m).

tritium, n. tritio (m); **t. correlation,** correlación (f) por tritio; **t. dating,** datación (f) por tritio; **t. distribution,** distribución (f) por tritio.

trivalent, a. trivalente.

trivial, a. trivial, baladí.

trochiform, a. troquiforme.

troctolite, n. troctolita (f).

troglodyte, n. troglodita (m, f).

trophic, a. trófico/a; **t. level,** nivel (m) trófico.

tropic, a. trópico/a.

tropical, a. tropical; **t. rainforest,** selva (f) tropical.

Tropics, n.pl. los Trópicos (m.pl) (Geog).

tropism, n. tropismo (m); **negative/positive t.,** tropismo negativo/positivo.

troposphere, n. troposfera (f).

troubleshooting, n. localización de averías, reparación de averías.

trough, n. pesebre (m) (for animal food), seno (m) (between waves), depressión (f) (depression); **t. of low pressure,** zona (f) de bajas presiones.

trout, n. trucha (f).

truck, n. camión (m), vagón (m) de mercancías; **t.-garden (US),** hortaliza (f), de cultivo (m) para la venta.

truncated, a. truncado/a.

truncation, n. truncación (f); truncamiento (m) (Mat, rounding of).

trunk, n. tronco (m) (e.g. of tree).

trunking, n. cable (m) principal (Elec).

tsetse, a. tse-tse, tsé-tsé; **t. fly,** mosca (f) tsé-tsé.

tsunami, n. tsunami (m).

tube, n. tubo (m); trompa (f) (Anat); metro (m) (underground railway); **capillary t.,** tubo capilar; **cathode ray t.,** tubo de rayos catódicos; **inner t. (of tyre),** cámara (f) de aire; **piezometric t.,** tubo piezométrico; **Pitot t.,** tubo de Pitot; **test t.,** tubo de ensayo; **X-ray t.,** tubo de rayos-X.

tuber, n. tubérculo (m); **root t.,** tubérculo radical.

tuberculosis, n. tuberculosis (f).

tubing, n. entubamiento (m), tubo (m); conducción (f) (of liquids); **Visking t. (Biol)**, tubo de Visking.

tubule, n. túbulo (m).

tufa, n. toba (f).

tuff, n. toba (f); **welded tuffs**, tobas soldadas.

tuffaceous, a. tobáceo/a.

tumour, n. tumor (m); **benign t.**, tumor benigno; **malignant t.**, tumor maligno.

tumultous, a. tumultoso/a.

tumulus, n. túmulo (m) (Arch).

tuna, n. atún (m); **yellow fin t.**, rabil (m).

tundra, n. tundra (f).

tungsten, n. tungsteno (m) (W); **t. carbide**, carburos (m.pl) de tungsteno.

tungstite, n. tungstita (f).

tunnel, n. túnel (m); **wind t.**, túnel aerodinámico.

tunnelling, n. tunelización (f).

Turbellaria, n.pl. turbelarios (m.pl), la clase Turbellaria (Zoo).

turbid, a. turbio/a.

turbidimeter, n. turbidímetro (m).

turbidite, n. turbidita (f).

turbidity, n. turbidez (m), turbiedad (f); **t. current (Geog)**, corriente (f) de turbidez.

turbidness, n. turbiedad (f), turbidez (m).

turbine, n. turbina (f); **condensation t.**, turbina de condensación; **counterpressure t.**, turbina de contrapresión.

turbiniform, a. turbiniforme.

turbulence, n. turbulencia (f); **wind t.**, turbulencia del viento.

turbulent, a. turbulento/a; **t. flow**, flujo (m) turbulento.

turf, n. turba (f), césped (m) (lawn).

turgid, a. turgente, túrgido/a.

turgor, n. turgor (m), turgencia (f).

turlough, n. small karstic lake (Ir).

turn, n. vuelta (f), revolución (f) (of a wheel); giro (m), rotación (f) (rotation); curva (f) (of a river); cambio (m) (of time).

turnaround, n. cambio (m) de rumbo; **t. time**, tiempo (m) de vuelta, tiempo de respuesta (Comp).

turnip, n. nabo (m); **t. tops**, grelos (m.pl).

turnkey, a. llave en mano; **t. installation, t. instalation (US)**, planta (f) llave en mano.

turnover, n. movimiento (m) de mercancías (Fin), volumen (m) de negocios (Fin), volumen de ventas (money from sales).

turnpike, n. autopista (f) de peaje.

Turonian, n. Turoniense (m).

turpentine, n. trementina (f); **oil of t. (thinners)**, aguarrás (f).

turquoise, n. turquesa (f).

turtle, n. tortuga (f) marina.

tusk, n. colmillo (m).

twin, n. gemelo (m).

twinned, a. gemelo/a; **t. crystal**, cristal (m) gemelo.

two-dimensional, a. bidimensional.

type, n. tipo (m), clase (f), modelo (m); tipo (tipografía); **t. of cultivation**, tipo de cultivo; **t. species (Bot, Zoo)**, especie (f) nominal; **wild t. (Ecol)**, tipo salvaje.

typescript, n. texto (m) mecanografiado.

typewriter, n. máquina (f) de escribir.

typewritten, a. mecanografiado/a.

typhoid, a. tifoideo/a; **t. fever**, fiebre (f) tifoidea, tiphoidea (f).

typhoon, n. tifón (m).

typhus, n. tifus (m).

typological, a. tipológico/a.

tyre, n. neumático (m).

U

ubiquitous, a. ubicuo/a.
UK, n. United Kingdom, Reino (m) Unido.
ulcer, n. úlcera (f).
ulceration, n. ulceración (f).
ultrabasic, a. ultrabásico/a.
ultracentrifuge, n. ultracentrífuga (f).
ultrafiltration, n. ultrafiltración (f).
ultramaphic, a. ultramáfico/a.
ultrasonic, a. ultrasónico/a, supersónico/a.
ultrasound, n. ultrasonido (m).
ultraviolet, a. ultravioleta.
umbel, n. umbela (f).
umbellifer, n. umbelífera (f).
Umbelliferae, n.pl. umbelíferas (f.pl), la familia Umbelliferae (Bot).
umbo, n. umbo (m).
UN, n. ONU (f); United Nations Organization, Naciones (f.pl) Unidas, Organización (f) de las Naciones Unidas.
unaerated, a. no aireado/a.
unanimity, n. unanimidad (f).
unauthorized, a. no autorizado/a, desautorizado/a.
unblock, v. desbloquear.
unbroken, a. intacto/a.
uncabling, n. descableado (m).
uncertain, a. incierto/a.
uncertainly, n. incertidumbre (f), duda (f) (doubt); scientific u., incertidumbre científica.
unchanged, a. sin variación.
unconfined, a. no confinado/a.
unconformity, n. discordancia (f), inconformidad (f); angular u., discordancia angular.
uncontaminated, a. incontaminado/a, impoluto/a.
underconsumption, n. subconsumación (f).
undercurrent, n. resaca (f).
undercut, a. socavado/a.
underdevelopment, n. subdesarrollo (m).
underdrain, n. subdren (m).
underemployment, n. trabajo (m) encubierto.
underestimation, n. infravaloración (f), estimación (f) demasiado baja.
underflow, n. corriente (f) profunda, flujo (m) inferior.
underground, 1. n. metro (m), subterráneo (m) (LAm) (railway, subway); 2. a. subterráneo/a; u. drainage, drenaje (m) subterráneo.
undergrowth, n. maleza (f), monte (m) bajo, sotobosque (m).
underlie, v. subyacer, infrayacer (Geol).
underline, v. subrayar; to u. the importance of something, subrayar la importancia de algo.
underlying, a. infrayacente (Geol), subyacente.
undermining, n. socavación (f).

underpin, v. apuntalar.
underpinning, n. apuntalamiento (m) (action).
underprivileged, a. desfavorecido/a.
underside, n. superficie (f) inferior, parte (m) de abajo.
undersizing, n. subdimensionamiento (m); u. of works, subdimensionamiento de obras.
understand, v. comprender, entender.
understorey, n. sotobosque (m) (Biol, undergrowth).
undertake, v. emprender (a task), prometer (a promise), realizar (to carry out).
undertaking, n. tarea (f), empresa (f) (task); empresa (f) (business); promesa (f) (promise); ejecución (f) (carrying out); u. of a scheme, ejecución de un esquema.
underwater, a. submarino/a.
undeveloped, a. no desarrollado/a.
undisturbed, a. imperturbado/a, no perturbado/a.
undulating, a. ondulado/a (land).
undulation, n. ondulación (f).
unearth, v. desenterrar; descubrir (to discover).
unemployed, 1. n.pl. desempleados (m.pl), pardos (m.pl) (of people); 2. a. parado/a, desempleado/a, sin trabajo.
unemployment, n. desempleo (m), sin empleo (m); level of u., índice (m) de desempleo; seasonal u., desempleo estacional.
UNEP, n. UNEP (f); UN Environment Programme, programa (m) ambiental de la ONU.
unevenness, n. desnivel (m), desigualdad (f).
unexploited, a. no explotado/a.
unfit, a. no apto (unsuitable); incapaz (incompetent), inepto/a (inapt); u. for human consumption, no apto para el consumo humano.
unforeseen, a. imprevisto/a; u. hazards, riesgos (m.pl) imprevistos.
unhealthiness, n. insalubridad (f).
unhealthy, a. insalubre.
uniform, a. uniforme.
uniformitarianism, n. uniformitarismo (m); theory of u. (Geol), teoría (f) del uniformitarismo.
uniformity, n. uniformidad (f); coefficient of u., coeficiente (m) de uniformidad.
unimpaired, a. sin menoscabo.
uninhabited, a. deshabitado/a, desplobado/a.
union, n. unión (f), acción (f) (joining); trade u., sindicato (m); gremio (m); U. Jack, bandera (f) del Reino Unido.
unique, a. único/a.
unisexual, a. unisexual.
unit, n. unidad (f); atomic mass u. (abbr.

a.m.u.), unidad de masa atómica (abr. u.m.a.); **central processing u. (abbr. CPU) (Comp),** unidad central de proceso (abbr. UPC).

unitary, a. unitario/a.

univalve, n. univalvo (m).

universe, n. universo (m); **the U.,** el Universo.

university, n. universidad (f).

unknown, n. incógnita (f) (Mat); desconocido/a (m, f) (person).

unload, v. descargar.

unloading, n. descarga (f).

UNO, n. ONU (f); **United Nations Organization,** Organización (f) de las Naciones Unidas.

unpolluted, a. impoluto/a, incontaminado/a.

unproductive, a. improductivo/a.

unprofitable, a. poco rentable (Fin).

unprotected, a. desprotegido/a, no protegido/a.

unsafe, a. peligroso/a (machine), arriesgado/a (method).

unsaturated, a. no saturado/a; **u. zone,** zona (f) no saturada.

unscientific, a. poco científico/a.

unstable, a. inestable.

unstratified, a. no estratificado/a.

unsurfaced, a. no cubierta con grava (road).

untreated, a. no tratado/a.

unweathered, a. inalterado/a, no meteorizado/ a.

update, v. actualizar.

upgrade, v. mejorar la calidad de (to improve the quality of).

up-gradient, n. pendiente (m).

uphill, n. cuesta (f) arriba.

upland, n. tierras (f.pl) altas.

uplift, n. levantamiento (m) (Geol), elevación (f).

upper, a. superior.

uppermost, a. más alto.

uproot, v. desarraigar, arrancar de raíz; **to u. a tree,** desarraigar un árbol.

uprooting, n. desarraigo (m); **social u.,** desarraigo social.

uprush, n. embestida (f), avance (m) físico.

upslope, n. pendiente (f) ascendente.

upstream, n. aguas (f.pl) hacia arriba, corriente (m) hacia arriba.

upthrow, n. cabalgamiento (m) vertical, empuje (m) ascendente.

upthrust, n. solevantamiento (m), corrimiento (m) hacia arriba, empuje (m) ascendente.

upward, p. hacia arriba.

upwelling, 1. n. surgencia (f), llamada (f) de fondo, ojo (m) de agua subterránea; afloramiento (Mar); 2. a. surgente.

uralite, n. uralita (f).

uraninite, n. uraninita (f), pechblenda (f).

uranium, n. uranio (m) (U).

Uranus, n. Urano (m).

urban, a. urbano/a, urbanístico/a; citadino/a (LAm); **u. population,** población (f) urbana; **u. renewal,** renovación (f) urbana; **u. sprawl,** extensión (f) urbana.

urbanization, n. urbanización (f).

urea, n. urea (f).

ureter, n. uréter (m).

urethra, n. uretra (f).

urge, v. requerir (to request), exhortar (to exhort), recomendar (to recommend).

urinal, n. urinario (m).

urinary, a. urinario/a.

urinate, v. orinar.

urination, n. urinación (f), micción (f).

urine, n. orina (f), orines (m.pl).

Uruguay, n. Uruguay (m), la Banda Oriental (LAm, esp. CSur).

Uruguayan, 1. n. uruguayo/a (m, f); 2. a. uruguayo/a.

US, n.pl. EEUU; **United States,** Estados (m.pl) Unidos.

USA, n.pl. EEUU (m.pl), EUA (m.pl); **United States of America,** Estados (m.pl) Unidos; Estados Unidos de América.

useful, a. útil.

usefulness, n. utilidad (f).

useless, a. inútil (Gen); inservible (unusable).

user-friendly, a. fácil de utilizar, de utilización sencilla (Comp).

utensil, n. utensilio (m), herramienta (f) (tool).

uteri, n.pl. úteros (m.pl).

uterus, n. útero (m).

utilization, n. utilización (f).

V

V-shaped, a. en forma de V.
vaccinate, v. vacunar.
vaccination, n. vacunación (f).
vaccine, n. vacuna (f).
vacuole, n. vacuola (f).
vacuum, n. vacío (m); **v. extraction**, extracción (f) por vacío; **v. filter**, filtro (m) vacío.
Valangian, n. Valanginense (m).
valence, n. valencia (f).
valency, n. valencia (f).
valid, a. válido/a (Gen); vigente (law); valedero/a (ticket).
validate, v. validar.
validation, n. validación (f); **model v.**, validación del modelo.
validity, n. validez (f).
valley, n. valle (m), vaguada (f) (lowermost part); **alluvial v.**, valle aluvial; **anticlinal v.**, valle anticlinal; **blind v.**, valle ciego; **buried v.**, valle enterrado, valle sepultado; **drainage v.**, valle de drenaje; **drowned v.**, valle ahogado, valle anegado; **dry v.**, valle seco; **fault-line v.**, valle de línea de falla; **hanging v.**, valle colgante; **old age v.**, valle antiguo, valle senil; **rift-block v.**, valle de bloques de rift; **river v.**, valle fluvial; **v. bottom**, fondo (m) de valle; **youthful v.**, valle juvenil.
valuation, n. valoración (f), tasación (f) (valuation); estimación (of person ability).
value, n. valor (m) (Gen); méritos (m.pl) (merit); utilidad (f) (usefulness); **average v.**, valor medio; **commercial v.**, valor comercial; **ecological v.**, valor ecológico; **extreme values**, los valores extremos máximos y mínimos; **increase in v.**, plusvalía (f); **mandatory v. (Reg, of EU Directive)**, valor límite imperativo; **replacement v.**, valor de reemplazamiento; **residual v.**, valor residual; **threshold v.**, valor umbral; **v. added tax (abbr. VAT)**, impuesto (m) al valor añadido/agregado (abbr. IVA).
valuer, n. estimador (m, f), tasador (m, f) (Fin).
valve, n. válvula (f); **air blow-off v.**, válvula de purga de aire; **blow-off v.**, válvula de purga; **butterfly v.**, válvula mariposa; **clack v.**, válvula de clapeta; **control v.**, válvula de control; **discharge v.**, válvula de descarga; **flow-reducing v.**, válvula de control de flujo; **foot v.**, válvula de pie; **gate v.**, válvula de compuerta; **inlet v.**, válvula de admisión; **non-return v.**, válvula de clapeta; **reflux v.**, válvula de reflujo; **regulator v.**, válvula de regulación; **retention v.**, válvula de retención; **safety v.**, válvula de seguridad; **three-way v.**, válvula de tres vías.

valving, n. válvulas (f.pl), valvulería (f).
van, n. camioneta (f), furgoneta (f).
vanadium, n. vanadio (m) (V).
vane, n. veleta (f) (weather vane); paleta (f), pala (f) (of a propeller); vano (m), barbas (f, pl) (of a feather); álabe (m) (of a waterwheel); **v. test (Eng)**, prueba de pala.
vapor, n. vapor (m), vaho (m) (condensation on windows).
vaporizer, n. vaporizador (m); vaporador (m) (LAm).
vapour, n. vapor (m), vaho (m) (condensation on windows); **soil v. extraction**, extracción (f) de vapores en suelos; **v. pressure**, presión (f) de vapor; **v. tension**, tensión (f) de vapor; **water v.**, vapor de agua.
variability, n. variabilidad (f).
variable, n. variable (f); **continuous v.**, variable continua; **discrete v.**, variable discreta; **random v.**, variable aleatoria.
variance, n. varianza (f) (Mat); **analysis of v. (abbr. ANOVA)**, análisis (m) de varianza; **explicit v.**, varianza explícita; **sample v.**, varianza de la muestra; **spectral v.**, varianza espectral; **total v.**, varianza total.
variant, 1. n. variante (f); 2. a. variante, diferente, variable.
variation, n. variación (f); **climatic v.**, variación climática; **genetic v.**, variación genética; **magnetic v.**, variación magnética; **seasonal v.**, variación estacional; **v. in pumped output**, variación de caudal de bombeo; **v. in saturated thickness**, variación de espesor saturado.
variegated, a. abigarrado/a.
variety, n. variedad (f).
variolitic, a. variolítico/a.
various, a. diverso/a, vario/a.
varnish, v. barnizar.
varnish, n. barniz (m).
varve, n. barba (f); varva (f) (LAm); **varved clay**, arcilla (f) barbada, arcilla de varvas (LAm).
vascular, a. vascular; **v. plant**, planta (f) vascular.
vasectomy, n. vasectomía (f).
vaseline, n. vaselina (f) (R).
vasoconstriction, n. vasoconstricción (f).
vasodilation, n. vasodilación (f), vasodilatación (f).
VAT, n. IVA; **value added tax (GB)**, impuesto (m) sobre el valor añadido/agregado (Sp).
vector, 1. n. vector (m); **v. analysis**, análisis (m) vectorial; 2. a. vectorial.
vegetable, 1. n. legumbre (f), hortaliza (f),

verdura (f); **v. market,** mercado (m) de verduras; 2. a. vegetal.

vegetation, n. vegetación (f); **antropic v.,** vegetación antrópica; **natural/semi-natural v.,** vegetación natural/semi-natural.

vegetative, a. vegetativo/a; **v. reproduction,** reproducción (f) vegetativa.

vehicle, n. vehículo (m); **heavy goods v. (GB) (abbr. HGV),** vehículo pesado; **light v.,** vehículo ligero.

vein, n. veta (f), vena (f), filón (m); **quartz v.,** filón de cuarzo.

veinlet, n. filoncillo (m).

velocity, n. velocidad (f); **angular v.,** velocidad angular; **channel v.,** velocidad en el canal; **critical v.,** velocidad crítica; **flow v.,** velocidad de flujo; **terminal v.,** velocidad terminal; **v. head,** altura (f) cinética; **v. of seismic waves,** velocidad de ondas sísmicas; **wind v.,** velocidad del viento.

venation, n. venación (f); nervadura (f); **parallel v.,** venación paralela.

Venezuela, n. Venezuela (m).

Venezuelan, 1. n. venezolano/a (m, f); 2. a. venezolano/a.

venom, n. veneno (m).

venomous, a. venenoso/a.

venous, a. venoso/a.

vent, 1. v. descargar, emitir (to discharge); hacer un agujero en (to make a hole in); 2. n. abertura (f), aguja (f) (aperture); agujero (m) de respiración, cañón (m) de chimenea (chimney); válvula (f) (valve), chimenea (f) (Geol); **central v. (of volcano),** chimenea central.

ventilation, n. ventilación (f), aireación (f); respiración (f) (Med); **v. shaft (e.g. in a mine),** respiradero (m).

ventilator, n. ventilador (m), respiradero (m).

ventral, a. ventral.

ventricle, n. ventrículo (m).

venture, n. empresa (f), negocio (m).

Venturi, n. Venturi (m) (Eng); **V. flume,** canal (m) Venturi; **V. meter,** contador (m) Venturi.

Venus, n. Venus (f).

verifiable, a. verificable, comprobable.

verification, n. verificación (f), comprobación (f).

verify, v. verificar.

verifying, n. comprobación (f).

vermiculite, n. vermiculita (f).

vermin, n. bichos (m.pl) (rats, mice, etc), sabandijas (f.pl) (fleas, lice, etc.).

vernalization, n. vernalización (f).

vertebra, n. vértebra (f).

vertebrae, n.pl. vértebras (f.pl).

vertebral, a. vertebral; **v. column,** columna (f) vertebral.

Vertebrata, n.pl. vertebrados (m.pl), el subfilum Vertebrata (Zoo).

vertebrate, 1. n. vertebrado (m); 2. a. vertebrado/a.

vesicle, n. vesícula (f), hueco (m) (hollow).

vessel, n. vaso (m), recipiente (m) (receptacle), vasija (f) (pot); barco (m), nave (f), navio (m), buque (m) (Mar); **blood v.,** vaso sanguíneo; **earthenware v.,** vasija de barro.

vestigial, a. vestigial.

vesuvianite, n. vesubiana (f).

vet, n. veterinario/a (m, f).

veterinary, 1. n. veterinario/a (m, f); **v. doctor, v. surgeon,** veterinario/a (m, f); 2. a. veterinario/a; **v. medicine, v. science,** veterinaria (f).

viability, n. viabilidad (f); **v. study,** estudio (m) de viabilidad.

viable, a. viable.

viaduct, n. viaducto (m).

vibrate, v. vibrar.

vibration, n. vibración (f).

vibroflotation, n. vibroflotación (f).

view, n. vista (f), panorama (m) (sight); paisaje (m) (landscape); **aerial v.,** vista aérea; **bird's eye v.,** vista de pájaro; **panoramic v.,** vista panorámica.

viewpoint, n. mirador (m) (Geog); punto (m) de vista (Fig).

vigor, n. vigor (m).

vigour, n. vigor (m); **hybrid v.,** vigor híbrido.

vine, n. vid (f), parra (f); sarmiento (m) (shoot or branch of a vine); enredadera (f) (climbing plant); **v.-grower,** viticultor (m), viñatero (m), viñador (m).

vineyard, n. viña (f), viñedo (m).

vinyl, n. vinilo (m).

violation, n. violación (f).

viral, a. viral.

virology, n. virología (f).

virulence, n. virulencia (f).

virulent, a. virulento/a.

virus, n. virus (m) (pl. virus); **computer v.,** virus informático; **mosaic v.,** virus del mosaico.

viscera, n.pl. vísceras (f.pl).

viscosimeter, n. viscosímetro (m).

viscosity, n. viscosidad (f); **coefficient of v.,** coeficiente (m) de viscocidad; **dynamic v.,** viscosidad dinámica; **kinematic v.,** viscosidad cinemática.

Visean, n. Viseense (m).

visibility, n. visibilidad (f).

visible, a. visible; **v. reserves (Com),** reservas (f.pl) visibles.

vision, n. visión (f); vista (f) (view); **range of v.,** alcance (m) de la vista.

visual, a. visual.

visualization, n. visualización (f).

visualize, v. visualizar.

vitamin, n. vitamina (f); **v. C,** vitamina C, ácido (m) ascórbico; **v. deficiency,** avitaminosis (f).

vitelline, a. vitelino/a.

viticulturist, n. viñero (m), viñatero (m), viñador (m), viticultor (m).

vitreous, a. vítreo/a.

vitrification, n. vitrificación (f).
vitrinite, n. vitrinita (f).
vitriol, n. vitriolo (m).
vitroclast, n. vitroclasto (m).
viviparous, a. vivíparo/a.
vivisection, n. vivisección (f).
VOC, n. COV; **volatile organic compound,** compuesto (m) orgánico volátil.
vocabulary, n. vocabulario (m) (i.e. dictionary).
void, n. vacío (m); **v. ratio,** proporción (f) de vacío; **v. space,** espacio (m) de vacío.
volatile, a. volátil (Chem); voluble (situation).
volatility, n. volatilidad (f).
volatilize, v. volatilizar.
volcanic, a. volcánico/a; **v. neck,** chihuido (m); **v. plug,** tápon (m) volcánico.
volcano, n. volcán (m); **active v.,** volcán en actividad; **extinct v.,** volcán extinto; **shield v.,** volcán en escudo.
vole, n. topillo (m).
volt, n. voltio (m) (unit of electrical capacity).
volt-ampere, n. voltamperio (m).

voltage, n. voltaje (m) (electrical potential); **v. rating,** tensión (f) de régimen.
voltaic, a. voltaico/a.
voltameter, n. (US), voltammeter, n. volt-amperímetro (m).
voltmeter, n. voltímetro (m).
volume, n. volumen (m); **critical v.,** volumen crítico; **gram molecular v.,** volumen molecular gramo; **molar v.,** volumen molar.
volumetric, a. volumétrico/a.
vomit, 1. v. vomitar; 2. n. vómito (m).
vomiting, n. vómitos (m.pl).
vortex, n. vórtice (m), torbellino (m).
vulcanism, n. vulcanismo (m), volcanismo (m).
vulcanization, n. vulcanización (f).
vulcanology, n. vulcanología (f).
vulnerability, n. vulnerabilidad (f).
vulnerable, a. vulnerable.
vulture, n. buitre (m); **turkey v.,** buitre (m), gallinazo (m) (LAm), zopilote (m) (CAm, Mex); jote (m) (Ch); urubú (m) (Peru, Ur), zamuro (m) (Ven), aura (f) (Carib).

W

wader, n. zancuda (f), ave (f) zancuda.
wadi, n. wadi (m), uadi (m), guad (m).
wage, n. salario (m).
wagon, n. vagón (m); **covered w.,** vagón cerrado.
wagonette, n. vagoneta (f) (small wagon, skip).
wait, n. espera (f).
waive, v. renunciar (a right), desistir (a claim).
walkie-talkie, n. emisor (m) receptor.
walking, n. andadura (f) (action).
wall, n. muro (m) (outside), pared (f) (inside), muralla (f) (of a city); **cell w. (Biol),** pared celular; **curtain w.,** muro cortina; **drystone w.,** pirca (f) (Arg, Bol); **foot w. (Geog),** pared basal; **hanging w. (Geog),** pared colgante; **load-bearing w.,** muro de contención; **mud w.,** tapia (f); **retaining w.,** muro de contención, barrera (f) de contención; **toe w.,** muro de pie; **valley w.,** pared de valle; **w. anchor,** anclaje (m) para muro; **w. socket,** enchufe (m) de pared.
walnut, n. nuez (f) (nut); nogal (m) (tree).
warehouse, n. almacén (m); **w. dues,** almacenaje (m).
warfarin, n. warfarina (f).
warm, a. caliente; cálido/a (climate); **to have w. weather (v),** hacer calor.
warm-blooded, a. de sangre caliente; **w. animal,** animal (m) de sangre caliente.
warming, n. calentamiento (m); **global w.,** calentamiento global.
warning, n. previo (m), aviso (m) (notice); advertencia (f), aviso (of danger); **flood w.,** aviso de inundación; **w. system,** sistema (m) de aviso; **without w.,** sin previo aviso.
warp, v. alabear.
warped, a. alabeado/a.
warping, n. pandeamiento (m).
warship, n. buque (m) de guerra.
washing, n. lavado (m); depuración (f) (of gas); colgado (m) (clothes); **w. machine,** lavadora (f); **w. powder,** jabón (m) en polvo.
washings, n.pl. mineral (m) tras el lavado (Min); lavazas (f) (Treat, dirty water); colas (f.pl) (Min, tailings).
washland, n. llanura (f) de inundación.
washwater, n. aguas (f.pl) de aclarado.
wasp, n. avispa (f).
waste, n. desecho(s) (m. or m.pl) (material), residuos (m.pl); basura (f) (rubbish); desperdicio (m) (Gen); desgaste (m) (misuse); derroche (m), despilfarro (m) (wastefulness); pérdida (f) (loss); **biowaste,** residuos biodegradables; **colliery w.,** residuos de minas de carbón; **commercial w.,** desecho comercial; **controlled w.,** vertidos controla-

dos; **dangerous waste(s),** desechos peligrosos, residuos peligrosos; **demolition w.,** desecho de demolición; **domestic w.,** basura doméstica; **hazardous w. deep well injection,** inyección (f) subterránea de residuos peligrosos; **hazardous w. disposal,** disposición (f) de residuos peligrosos; **hazardous w. processing,** procesamiento (m) de residuos peligrosos; **hazardous w. storage,** almacenamiento (m) de residuos peligrosos; **hazardous w. transport,** transporte (m) de residuos peligrosos; **hazardous w. treatment,** tratamiento (m) de residuos peligrosos; **hospital w.,** residuos hospitalarios; **household w.,** residuos domésticos; **inert w.,** desecho inerte; **municipal w.,** basura municipal, residuos municipales; **nuclear w.,** residuos nucleares; **poisonous w.,** residuos venenosos; **radioactive w.,** desechos radioactivos, residuos radioactivos; **scrapyard w.,** residuos de parque de chatarra; **solid w. collection,** recolección (f) de residuos sólidos; **solid w. disposal,** disposición (f) de residuos sólidos; **solid w. recovery,** recuperación (f) de residuos sólidos; **solid w. reuse,** reutilización (f) de residuos sólidos; **solid w. segregation,** separación (f) de residuos sólidos; **solid w. segregator,** segregador (m) de r. sólidos (LAm); **special category w.,** residuos de alta peligrosidad; **special w.,** desecho especial, residuos especiales; **toxic w.,** desecho tóxico; **trade w.,** desecho comercial; **w. carrier,** transportista (m) de desechos; **w. collection and disposal,** recogida (f) y tratamiento (m) de residuos; **w. disposal,** destrucción (f) de residuos; **w. heap,** escombrera (f); **w. land,** tierras (f.pl) baldías; **w. liquor,** licores (m.pl) residuales; **w. management,** gestión (f) de residuos; **w. plastic,** residuos plásticos; **w. producer,** productor (m) de desecho; **w. product,** producto (m) de desecho, desperdicios (m.pl); **w. recovery,** recuperación (f) de desechos, aprovechamiento (m) de desechos; **w.-to-energy,** recuperación (f) energética de residuos, aprovechamiento (m) energético de residuos; **w. transport,** transporte (m) de desechos.
wasteland, n. yermo (m) (barren land); erial (m), tierras (f.pl) baldías (uncultivated land).
wastewater, n. aguas (f.pl) residuales, desagües (m.pl); **abattoir w.,** aguas residuales de mataderos, aguas residuales de rastro; **industrial w.,** aguas residuales industriales; **municipal w.,** desagües domésticos; **papermill**

w., aguas residuales de papeleras; **slaughter-house w.**, aguas residuales de mataderos, aguas residuales de rastro; **tannery w.**, aguas residuales de fábricas de curtidos, aguas residuales de curtiembres; **w. discharge**, evacuación (f) de aguas residuales; **w. disposal**, disposición (f) de aguas residuales; **w. reuse**, reutilización (f) de aguas residuales, aprovechamiento (m) de aguas residuales; **w. treatment**, tratamiento (m) de aguas residuales, depuración (f) de aguas residuales.

wasting, n. despilfarro (m) (squandering).

water, 1. v. regar (irrigate, water garden); abrevar, dar de beber a (water livestock); aguar (wine); 2. n. agua (f); **acid w.**, agua ácida; **alkaline w.**, agua alcalina; **artesian w.**, agua artesiana; **bathing w.**, agua de natación, agua de baño; **bottled w.**, agua embotellada, agua envasada; **bottom w.**, agua de fondo; **brackish w.**, agua salobre; **capillary w.**, agua capilar; **clean w.**, agua limpia; **clear w.**, agua clara; **combined w.**, agua combinada; **community w. taps (LAm)**, caños (m.pl) comunales; **compensation w.**, agua de compensación; **connate w.**, agua singénica, agua de formación; **contaminated w.**, agua contaminada; **cooling w.**, agua de enfriamiento; **corrosive w.**, agua corrosiva; **deionized w.**, agua desionizada; **distilled w.**, agua destilada; **drinking w.**, agua potable, agua de bebida; **dual w. supply network**, red (f) dual de agua; **ferruginous w.**, agua ferruginosa; **fluoridated w.**, agua fluorada; **fossil w.**, agua fósil; **fresh w.**, agua dulce; **hard w.**, agua dura; **head of w.**, altura (f) de impulsión; **heavy w.**, agua pesada; **highly-mineralized w.**, aguas duras, agua altamente mineralizada; **high w.**, agua de arriba; **infiltration w.**, agua de infiltración; **interstitial w.**, agua intersticial; **irrigation w.**, agua de regadío; **karst w.**, agua kárstica, agua cárstica; **large-scale w. transfer**, trasvase (m), grandes conducciones de agua (f.pl); **leachate w.**, agua de lixiviación; **lime w.**, agua de cal; **low w. (Mar)**, marea baja, estiaje (m); **magmatic w.**, agua magmática; **melt w**, agua de deshielo, agua de derretimiento; **meteoric w.**, agua meteórica; **mine w.**, agua de mina; **neutral w.**, agua neutra; **oil-field w.**, agua de campo petrolífero; **peaty w.**, agua turbia; **perched w.**, agua colgada; **percolation w.**, agua de percolación; **phreatic w.**, agua freática; **polluted w.**, agua contaminada; **pore w.**, agua intersticial; **pure w.**, agua pura; **receiving w.**, agua del medio receptor, agua receptora; **recirculated w.**, agua recirculada; **reclaimed w.**, agua reciclada, agua recirculada (LAm); **recycled w.**, agua reciclada; **rough w. (Mar)**, marejada; **running w.**, agua corriente; **saline w.**, agua salada; **salt w.**,

agua salada; **sea w.**, agua de mar, agua salada; **settled w.**, agua decantada; **slack w.**, agua tranquila; **soft w.**, agua blanda; **soil w.**, agua del suelo; **sour w.**, agua amarga; **spring w.**, agua de manantial; **stagnant w.**, agua estancada; **standing w.**, agua estancada, agua durmiente; **still w.**, agua tranquila, estancada, aguas muertas; **surface w.**, agua superficial; **sweet w.**, agua dulce; **syngenetic w.**, agua singenética; **tap w.**, agua de grifo, agua corriente; **territorial waters**, aguas territoriales; **thermal w.**, agua termal; **unaccounted-for w.**, agua no contabilizada; **waste of w.**, desperdicio (m) de agua, pérdida (f) de agua; **w. (supply) company**, servicio (m) de distribución de agua; **w. abstraction**, extracción (f) de agua; **W. Act (England and Wales)**, ley (f) sobre las aguas; **w. allocation**, reparto (m) del agua; **w. authority (England and Wales)**, confederación (f) hidrológica (Sp); **w. balance**, balance (m) del agua; **w.-bearing formation**, formación (f) acuífera; **w. closet**, water (m), retrete (m); **w. column**, columna (f) de agua; **w. cone**, cono (m) de agua; **w. conservation**, conservación (f) del agua; **w. consumption**, consumo (m) de agua; **w. demand**, demanda (f) de agua; **w. distribution**, distribución (f) de agua; **w. diviner**, zahorí (m), rabdomante (m); **w. duct**, conducto (m), caz (m); **w. encroachment**, avance (m) de agua marginal; **w. film (thin layer)**, película (f) de agua; **w. flow**, corriente (m) de agua; **w. hammer**, golpe (m) de ariete, arietazo; **w. hardness**, dureza (f) del agua; **w. head**, carga (f) hidrostática; **w. ingress**, entrada (f) de agua; **w. leak control**, control (m) de pérdidas de agua; **w. mains**, conductos (m.pl) de aguas potables; **w. mark (Mar)**, nivel (m) de altura de marea; **w. meter**, contador (m) de agua; **w. of crystallization**, agua de cristalización; **w. pipe**, tubería (f) de agua; **w. pump**, bomba (f) de agua; **w. quality criteria**, criterios (m.pl) de calidad de las aguas; **w. quality monitoring**, vigilancia (f) de la calidad de las aguas; **w. rationing**, racionamiento (m) del agua; **w. recirculation**, recirculación (f) del agua; **w. recycling**, reciclaje (m) del agua, reclamo (m) del agua; **w. resource exploration**, aprovechamientos (m.pl) hídricos; **w. resource management**, gestión (f) de recursos hidráulicos; **w. resources**, recursos (m.pl) hídricos; **w. rights**, derechos (m.pl) de aguas; **w. surface**, superficie (f) del agua; **w. table**, nivel (m) de agua freática, capa (f) freática; **w. tank**, tanque (m), aljibe (m) (cistern), depósito (m) de agua; **w. tower**, depósito (m) elevado de agua; **w. trough for animals**, abrevadero (m); **w. utilities**, empresas (f.pl) de agua potable; **w. well**, pozo (m) de agua; **well w.**, agua de pozo.

watercourse, n. cauce (m), curso (m) de agua,

lecho de río (m), corriente de agua (f), rambla (f) (intermittent, southern Spain).

watercress, n. berro (m); **w. bed,** prado (m) de berros.

waterfall, n. cascada (f), catarata (f).

waterfront, a. ribereño/a.

waterline, n. línea (f) de agua; línea de flotación (on a boat).

waterlogged, a. encharcado/a.

waterlogging, n. encharcamiento (m).

watermark, n. nivel (m) de agua.

watermelon, n. sandía (f), melón (m) de agua.

watermill, n. molino (m) de agua.

waterpower, n. energía (f) hidráulica.

waterproof, 1. v. impermeabilizar; 2. a. impermeable.

waterproofing, n. impermeabilización (f).

water-repellent, a. hidrófugo/a; **to make w.-r. (v),** hacer hidrófugo.

watershed, n. divisoria (f) de aguas (GB), divisoria (f) fluvial (GB), punto (m) de divisoria de aguas (GB); cuenca (f) (US); **w. management (US),** gestión (f) de cuenca.

waterspout, n. tromba (f) marina, manga (f) marina.

watertight, a. estanco/a, hermético/a.

watertightness, n. hermeticidad (f), estanqueidad (f).

waterway, n. vía (f) fluvial; **navigable w.,** vía (f) navegable.

waterwell, n. pozo (m); **w. drilling,** perforación (f) para agua.

waterwheel, n. rueda (f) hidráulica; aceña (f) (on river) (Arab); noria (f) (above a well) (Arab); batán (system for irrigating using river flow) (Arab); máquina (f) hidráulica; **w. driven by animal,** noria de sangre, noria a tracción animal.

waterworks, n. planta (f) depuradora (water supply plant).

watt, n. vatio (m), watio (m), watt (m).

wave, n. onda (f) (Phys), ola (f) (of water), oleada (f) (of enthusiasm); ondulación (f) (on surface); **electromagnetic w.,** onda electromagnética; **heat w.,** ola de calor; **large w.,** oleada; **long w.,** onda larga; **longitudinal w.,** onda longitudinal; **P w. (Geol),** onda P; **refractive w.,** onda de refracción; **S w. (Geol),** onda S; **seismic w.,** onda sísmica; **short w.,** onda corta; **sound w.,** onda acústica; **standing w.,** onda estacionaria; **tidal w.,** ola de marea; **transverse w.,** onda transversal; **w. fetch,** alcance (m) de ola; **w. length,** longitud (f) de onda; **w. mechanics,** mecánica (f) ondulatoria; **w. motion,** movimiento (m) ondulatorio.

waveband, n. banda (f) de frecuencia.

wavelength, n. longitud (f) de onda.

wavellite, n. wavellita (f).

wavy, a. ondulado/a.

wax, n. cera (f).

waxing, a. creciente (moon).

wealth, n. riqueza (riches), abundancia (f), profusión (f).

wear away, v. desgastar.

weather, 1. v. meteorizar, exponer a la intemperie (to expose to the weather); curar (wood); desgastarse, erosionarse (Geol); **to w. the storm,** aguantar la tempestad; 2. n. tiempo (m); **w. balloon,** globo (m) sonda, globo piloto; **w. forecast,** previsión (f) del tiempo, previsión sinóptica, pronóstico (m); **w. permitting,** si el tiempo lo permite; **w. vane,** veleta (f).

weathered, a. meteorizado/a.

weathering, n. meteorización (f), alteración (f).

weatherman, n. hombre (m) del tiempo.

web, n. red (f) (net); telaraña (f) (spider's); membrana (f) (of skin); **food w.,** trama (f) trófica, red alimentaria.

web-footed, a. palmípedo/a.

wedge, n. cuña (f); **quartz w.,** cuña de cuarzo; **saline w. (in aquifers),** cuña salina.

weed, n. mala hierba (f).

weed-killer, n. herbicida (m).

weevil, n. gorgojo (m).

weighbridge, n. báscula (f) (for vehicles).

weight, n. peso (m) (heaviness), pesa (f) (unit of measurement); **atomic w.,** peso atómico; **equivalent w.,** peso equivalente; **light w. (a),** de poco peso; **net w.,** peso neto; **specific w.,** peso específico; **weighty argument,** argumento (m) de peso.

weighted, a. en contra (Fig); ponderado/a (Mat); **w. least squares,** cuadrados (m.pl) mínimos ponderados.

weighty, a. pesado/a.

weightless, a. ingrávido/a.

weightlessness, n. ingravidez (f).

weir, n. vertedero (m); azud (m) (for controlling flows); presa (f) (dam); aliviadero (m) (overflow); derramadero (m) (spillway of a dam); **broad-crested w.,** vertedero de pared gruesa; **compound w.,** vertedero compuesto; **drowned w.,** vertedero anegado; **rectangular w.,** vertedero rectangular; **rounded-crested w.,** vertedero curvo; **thin-walled w.,** vertedero de pared delgada; **trapezoidal w.,** vertedero trapecial; **triangular w.,** vertedero triangular; **V-notch w.,** vertedero (m) en V.

weld, 1. v. soldar (metal), unir (to join); 2. n. soldadura (f) (of metal); **butt w.,** soldadura a tope; **spot w.,** soldadura por puntos.

welded, a. soldado/a (metal); aglutinado/a (Geol); **w. tuffs,** tobas (f.pl) soldadas.

welder, n. soldador (m) (person), soldadora (f) (machine).

welding, n. soldadura (f) (of metal); **oxyacetylene w.,** soldadura oxiacetilénica, autógena; **w. torch,** soplete (m) de soldar.

welfare, n. bienestar (m); **social w.,** bienestar social; **w. state,** estado (m) de bienestar social.

well, n. pozo (m); **abstraction w.,** pozo de

abastecimiento; **Abyssinian w.**, pozo abisinio; **artesian w.**, pozo artesiano; **coastal w.**, pozo costero; **collector w.**, captación (f); **drilled w.**, pozo taladrado; **driven w.**, pozo clavado, pozo hincado; **dug w.**, pozo excavado; **flowing w.**, manantial (m), pozo brotante; **gas w.**, pozo de gas; **image w.**, pozo imagen, contrapozo (m); **injection w.**, pozo de inyección; **interception w.**, pozo de intercepción; **large natural w.**, cenote (m) (Mex); **large-diameter w.**, pozo de gran diámetro; **observation w.**, pozo de observación; **oil w.**, pozo de petróleo; **open w.**, pozo abierto; **overflowing w.**, pozo surgente; **sanitary w. protection**, pozo de protección sanitaria; **small-diameter w.**, pozo de pequeño diámetro; **stilling w.**, pozo amortiguador; **test w.**, pozo de prueba; **water w.**, pozo de agua; **wellhead**, puntal (m) del pozo, cabeza (f) de sondeo; **w. logging**, sondeo (m) en el pozo, diagrafía (f); **w. performance**, rendimiento (m) de pozo, desempeño (m) de pozo (LAm); **w. protection**, protección (f) del pozo; **w. with adits/galeries/headings**, pozo con galerías.

wellhead, n. pozo (m) puntal, boca (f) de sondeos, cabeza (f) de pozo.

well-known, a. conocido/a.

wellpoint, n. piezómetro (m).

wellsite, n. emplazamiento (m) de la perforación.

Wenlockian, n. Wenlockense (m).

West, n. occidente (m).

west, a. oeste, del oeste.

western, a. occidental; **the w. world,** el mundo (m) occidental.

Westphalian, n. Westfaliense (m).

wet, a. mojado/a, empapado/a, húmedo/a (damp), lluvioso/a, pluvioso/a; **wetted perimeter (Geog)**, perímetro (m) mojado.

wetland, n. humedal (m), palencia (f), gandulla (f).

wettability, n. humectabilidad (f), mojabilidad (f).

whale, n. ballena (f); **blue w.**, ballena azul; **finback w.**, rorcual (m) común; **killer w.**, orca (f); **sperm w.**, cachalote (m); **w. industry**, industria (f) ballenera.

wharf, n. muelle (m) (jetty, breakwater); embarcadero (m) (for embarcation); desembarcadero (m) (for desembarcation); descargadero (m) (for cargo unloading).

wheat, n. trigo (m); **w. field**, trigal (m); **whole-w. (a)**, integral (a) (of flour).

whetstone, n. muela (f), piedra (f) de afilar.

whirlpool, n. remolino (m), remolino del agua.

whirlwind, n. torbellino (m), remolino (m) del aire; manga (f) de viento; **land w.**, tromba (f) terrestre.

white, n. clara (f) (of an egg).

whitener, n. blanqueador (m), blanqueo (m).

whitening, n. blancura (f).

whiting, n. pescadilla (f), merlán (m), bacadilla (f), plegonero (m) (fish).

WHO, n. OMS (f); **World Health Organization**, Organización (f) Mundial de la Salud.

wholesale, a. al por mayor; **to buy w. (v)**, comprar al por mayor; **to sell w. (v)**, vender al por mayor.

wholesaler, n. mayorista (m, f).

whorl, n. espira (f) (of a shell); verticilo (m) (Bot).

widen, v. ensanchar, extender; **to w. a road**, ensanchar una carretera.

widening, 1. n. ensanche (m), ensanchamiento (m), hinchazón (m); 2. a. crecido/a.

widespread, a. extendido/a.

width, n. anchura (f), ancho (m) (of an object); distancia (f) (the space between); **river w.**, anchura (f) de un río.

wild, a. salvaje (of land, of fierce animals); silvestre.

wilderness, n. yermo (m) (wasteland), desierto (m) (desert), soledad (isolated place), despoblado (m) (depopulated).

wildfowl, n. ave (f) acuática de caza.

wildland, n. yermo (m), páramo (m).

wildlife, n. flora (f) y fauna (f) silvestre; **w. habitat**, hábitat (m) de flora y fauna silvestre.

willow, n. sauce (m); **weeping w.**, sauce llorón.

wilt, v. marchitar (of a plant).

wilting, n. marchitez (m), marchitamiento (m); **permanent w. point**, punto (m) de marchitez permanente.

winch, n. torno (m) (for raising loads), chigre (m) (Mar), cabrestante (m) (windlass).

wind, n. viento (m); **dominant/predominant w.**, viento predominante; **katabatic w.**, viento catabático; **w. farm**, parque (m) eólico; **w. power**, energía (f) del viento; **w. pump**, bomba (f) eólica, molinete (m) eólico.

windbreak, n. protección (f) contra el viento; seto (m) cortaviento (hedge).

windchill, n. enfriamiento (m); **w. factor**, factor (m) de enfriamiento.

winding, n. serpenteo (m) (meandering).

windlass, n. cabrestante (m) (capstan on well), torno (m) (winch).

windmill, n. molino (m) de viento.

window, n. ventana (f); ventanilla (f) (e.g. of vehicle, of a ticket office).

windpipe, n. tráquea (f).

wind-powered, a. impulsado/a por el viento.

windpump, n. bomba (f) eólica, bomba de aeromotor.

windrowing, n. abonado (m) en montones en hilera.

windsurfing, n. windsurf (m).

windward, n. barlovento (m), del lado del viento; **w. side**, banda (f) de barlovento.

wing, n. ala (f); **w.-nut**, mariposa (f).

winter, 1. n. invierno (m); 2. a. de invierno, invernal, hibernal.

winterbourne, n. arroyo (m) intermitente, bourne (m) de invierno, rambla (f) (in southern Spain).

wintry, a. hibernal, invernal, de invierno.

wire, n. hilo (m), alambre (m); **barbed w.**, alambre (m) de púas; **cross wires**, hilos cruzados; **w. fence**, cerca (f) de alambre.

wireline, n. wireline (m); **w. logging**, testificación (f) geofísica con wireline.

withdrawal, n. retirada (f) (from a bank), abandono (m) (abandonment).

wither, v. desecar (de plantas).

witherite, n. witherita (f).

withybed, n. soto (m) para mimbre.

WMO, n. OMM (f); **World Meteorological Organization**, Organización (f) Meteorológica Mundial.

wold, n. hondonada (f) (Geog).

wolf, n. lobo (m).

wolframite, n. wolframita (f).

wollastonite, n. wollastonita (f).

womb, n. útero (m).

wood, n. bosque (m) (copse, forest); madera (f) (timber); **hardwood**, madera dura; **softwood**, madera blanda.

wood-cutter, n. leñador/a (m, f).

wooded, a. arbolado/a, poblado/a; **w. countryside**, paisaje (m) poblado.

wooden, a. de madera.

woodland, n. bosque (m), masa (f) forestal; **seminatural w.**, bosque seminatural.

woodlice, n.pl. cochinillas (f.pl).

woodlouse, n. cochinilla (f).

woodworm, n. carcoma (f).

woody, a. leñoso/a; **w. plant**, planta (f) leñosa.

work, 1. v. trabajar (to be employed), obrar (machine), faenar (to work, e.g. fishing grounds); 2. n. trabajo (m) (Gen); faena (f) (task); esfuerzo (m) (effort); obra (f) (product of labour); **civil works (Eng)**, obras civiles; **detailed w.**, trabajo de detalle; **fieldwork**, trabajo de campo; **hydraulic w.**, obra hidráulica; **irrigation w.**, obra de regadío; **manual w.**, trabajos manuales; **part-time w.**, trabajo de media jornada; **piecework**, trabajo a destajo; **seasonal w.**, trabajo temporal, trabajo estacional; **shift w.**, trabajo por turno; **w. force**, mano (f) de obra.

work out, v. desentrañar (to figure out).

workable, a. que se puede trabajar, explotable (exploitable); **w. mine**, mina (f) explotable; **w. plan**, proyecto (m) viable.

workbench, n. mesa (f) de trabajo.

worker, n. trabajador/a (m, f) (Gen); obrero/a (m, f), operario/a (m, f) (industrial); changador/a (m, f) (Arg, Bol, Ur, odd job man); **seasonal w.**, trabajador/a estacional; **skilled w.**, obrero cualificado, obrero especializado.

workforce, n. mano (f) de obra.

workhorse, n. caballo (m) de tiro.

working, a. obrero/a, laborable, activo/a, que trabaja/funciona; **w. asset**, activo (m) realizable; **w. capital**, capital (m) de trabajo; **w. day**, día (m) laborable; **w. face (Min)**, frente (m) de corte; **w. knowledge**, un conocimiento (m) básico; **w. order**, en estado (m) de funcionamiento, operativo; **w. speed**, velocidad (f) de funcionamiento.

working-class, n. de la clase (f) obrera.

workman, n. trabajador (m), obrero (m) (manual, industrial); artesano (m) (craftsman).

workmanlike, a. concienzudo/a (person); bien hecho/a (job).

workmanship, n. habilidad (f), destreza (f) (craft); artesanía (fine skill).

workplace, n. puesto (m) de trabajo.

works, n.pl. obras (f.pl); **engineering w.**, obras de ingeniería; **public w.**, obras públicas.

workshop, n. taller (m), fábrica (f).

world, 1. n. mundo (m), tierra (f); Universo (m) (universe); **in the scientific w.**, en el mundo científico; **third w.**, tercer mundo; 2. a. mundial; **on a w. scale**, a escala (f) mundial.

worm, n. gusano (m); **earthworm**, lombriz (f); **flatworm**, gusano plano; **intestinal w.**, lombriz (f) intestinal.

worn away, a. desgastado/a.

worn-out, a. gastado/a, usado/a, muy estropeado/a.

worsening, n. empeoramiento (m).

worth, v. valer (to be of value).

wrap, v. envolver; cubrir (to cover).

wrapping, n. envoltura (f) (covering); envase (m), embalaje (m) (packaging).

wreckage, n. restos (m.pl) (of a car, aeroplane); escombros (of a building).

writing, n. redacción (f), escritura (f); **w. of reports**, redacción de informes.

writing-off, n. amortización (f) (of debt).

wrought, a. forjado/a; **w. iron**, hierro (m) forjado.

WWF, n. WWF (m); **World Wildlife Fund**, Fondo (m) Mundial para la Naturaleza.

X

x-ray, n. rayo-X; radiografía (f) (Med, image).
xanthoprotein, n. xantoproteína (f); **xanthopro-**
tein t. (Biol, Chem), test (m) xantoproteico.
xenoblast, n. xenoblasto (m).
xenoblastic, a. xenoblástico/a.
xenolith, n. xenolito (m).
xenolithic, a. xenolítico/a.

xenomophic, a. xenomórfico/a.
xenon, n. xenón (m) (Xe).
xerophyte, n. xerofita (f).
xerosere, n. xeroserie (f).
xylem, n. xilema (m).
xylene, n. xileno (m).

Y

yacht, n. yate (m).
yachting, n. navegación (f) de recreo, navegación a vela.
yam, n. batata (f); boniato (m); camote (m)
(LAm); ñame (m) (yam plant).
yard, n. yarda (f) (measure of length, equiv. to
0.9144 metres); patio (m) (of building),
corral (for animals).
yearbook, n. anuario (m).
yeast, n. levadura (f); espuma (f) (froth).
yellow, 1. n. amarillo (m); **cadmium y.,**
amarillo de cadmio; 2. a. amarillo/a.
yew, n. tejo (m).
yield, n. caudal (m) (Gen); cosecha (f) (crops);
producción (f) (of a factory); rendimiento (m)

(profits); **crop y.,** rendimiento de cultivos,
rendimiento de la cosecha; **drought y.,** caudal
de sequía; **maximum sustainable y.,**
rendimiento máximo sostenible; **minimum**
reliable y., caudal de seguridad mínima;
pumping y., rendimiento del bombeo; **reser-**
voir y., caudal del embalse; **safe y.,** caudal
seguro; **specific y.,** rendimiento específico; **y.-**
drawdown curve, curva (f) de caudal-
descenso.
yolk, n. yema (f) (of an egg); **y. sac,** saco (m)
vitelino.
ytterbium, n. yterbio (m) (Yb).
yttrium, n. itrio (m) (Y).
yucca, n. yuca (f).

Z

Zanclian, n. Zancliense (m).
zebra, n. cebra (f).
zenith, n. cenit (m), zenit (m).
zeolite, n. ceolita (f), zeolita (f).
zero, n. cero (m); **absolute z. (0 deg. K)**, cero absoluto.
zinc, n. zinc (m), cinc (m) (Zn); **z. blende,** blenda (f), esfalerita (f).
zircon, n. circón (m).
zirconium, n. circonio (m) (Zr).
zoisite, n. zoisita (f).
zone, n. zona (f); **abscission z.**, zona de abscisión; **abyssal z.**, zona abisal; **aerobic z.**, zona aerobia; **anaerobic z.**, zona anaerobia; **aphotic z.**, zona afótica; **arid z.**, zona árida; **buffer z.**, zona tampón (Ecol), zona intermediaria; **climate z.**, zona climática; **collapse z.**, zona de hundimientos; **eulittoral z.**, zona eulitoral; **euphotic z.**, zona eufótica; **fault z.**, zona de falla; **fracture z.**, zona de fractura; **intertidal z.**, zona mareal, zona de oscilación de las mareas; **littoral z.**, zona litoral; **nuclear-free z.**, zona desnuclearizada; **photic z.**, zona fótica; **postal z.**, zona postal; **root z.**, zona de raíz; **semi-arid z.**, zona semiárida; **shattered z.**, zona triturada; **shear z.**, zona de cizalla; **soil moisture z.**, zona de humedad del suelo; **subduction z.**, zona de subducción; **tectonic z.**, zona tectonizada; **temperate z.**, zona templada; **tidal z.**, zona mareal; **transition z.**, zona de transición; **urban z.**, zona urbana; **z. of capture**, zona de la cuenca; **z. of saturation**, zona de saturación; **z. of weathering**, zona de meteorización.
zonification, n. zonificación (f).
zoogeography, n. zoogeografía (f).
zoology, n. zoología (f).
zooplankton, n. zooplancton (m).
zoospore, n. zoospora (f).

Part 2

ESPAÑOL–INGLÉS
SPANISH–ENGLISH

A

Aaleniano, m. Aalenian.

abajeño, m. (LAm) lowland; lowland or coastal dweller.

abajino, m. (LAm) lowland; lowland or coastal dweller.

Abajo, m. (Perú, Ch) the northern coast.

abajo, ad. below; underneath (esp. LAm) (posición); down, downwards (dirección); **cuesta (f) a.**, downhill.

abancalamiento, m. terracing.

abancalar, v. to terrace.

abandonar, v. to abandon; **a. una pertenencia**, to abandon a claim.

abandono, m. abandonment, withdrawal.

abanico, m. fan; **a. aluvial**, alluvial cone; **a. de roca**, rock fan; **en forme de a.**, fan-shaped.

abarcar, v. to embrace, to encompass.

abarrotar, v. to overstock.

abasoloa, f. genus of Compositae common in LAm (Bot).

abastecimiento, m. supply, supplying, provisioning; **a. de aguas**, water supply, supply of water; **a. industrial**, industrial supply; **red (f) de a.**, supply network.

abatir, v. to demolish, to knock down (Con); to fell, to cut down (árbol).

abdomen, m. abdomen.

abdominal, a. abdominal.

abedul, m. birch.

abeja, f. bee; **a. maestra**, queen bee; **a. melífera**, honey bee; **a. neutra, a. obrera**, worker bee; **a. reina**, queen bee.

abejón, m. bumblebee.

abejorro, m. bumble bee, humble bee; cockchafer.

abertal, m. unfenced land; ground cracked open as a consequence of drought.

abertura, f. aperture, opening; **a. entre estratos**, gullet.

abeto, m. fir.

abiertola, a. open; **explotación (f) minera a cielo a.**, opencast mining.

abigarradola, a. mottled, multi-coloured, variegated.

abióticola, a. abiotic.

abisal, a. abysmal, abyssal.

abismo, m. abyss, chasm.

ablación, f. ablation.

ablandamiento, m. softening; **a. del agua**, water softening; **a. por intercambio iónico**, softening by ion exchange.

abogacía, f. legal profession.

abogado, m. lawyer (Gen); solicitor (notario); counsel (asesor); barrister, attorney (US) (en un tribunal).

abolladura, f. notch, indentation.

abonadero, m. compost heap.

abonado, m. manure, compost, composting; **a. en montones en hilera**, windrowing; **lodos (m.pl) de a.**, sludge composting.

abonadora, f. composter (máquina).

abonar, v. to fertilize, to dress (Agr); to pay; to take out a subscription for (suscribirse); to vouch for (dar por cierto); to improve (mejorar).

abono, m. fertilizer, compost (Agr); payment, subscription (suscripción); **a. compuesto**, compound fertilizer; **a. de acción lenta/ rápida**, slow-acting/fast-acting fertilizer; **a. fosfatado**, phosphate fertilizer; **a. nitrogenado**, nitrogen fertilizer, nitrate fertilizer; **a. orgánico**, organic fertilizer; **convertir en a.**, to compost.

aboral, a. aboral (más lejos de).

abovedadola, a. arched.

abovedamiento, m. arching.

abovedar, v. to arch.

aboyar, v. to mark with buoys.

abrasión, f. abrasion; **a. eólica**, eolian (wind) abrasion; **a. fluvial**, fluvial abrasion; **a. glacial, a. glaciar**, glacial abrasion; **a. marina**, marine abrasion.

abrasivola, a. abrasive.

ábrego, m. south-west wind.

abrevadero, m. natural watering place, drinking trough, water trough.

abrevar, v. to water (livestock).

abrigo, m. shelter (refugio); coat (ropa); **al a. de**, sheltered from, shielded from.

abrillantador, m. brightening agent (para lavar ropas).

abrir, v. to open; **a. un surco**, to open a trench.

abruptola, a. rugged.

abscisa, f. abscissa.

abscisas, f.pl. abscissae.

abscisina, f. abscisin.

abscisión, f. abscission, leaf-fall.

absentismo, m. absenteeism.

ábside, m. apse.

absolutola, a. absolute; **cero (m) a. (0 deg. K)**, absolute zero; **grados (m.pl) absolutos**, degrees absolute.

absorbencia, f. absorbency, absorptivity.

absorbente, a. absorbent, absorbing (Fig).

absorber, v. to absorb.

absorbible, a. sorptive.

absorción, f. absorption, sorption; **capacidad (f) de a.**, absorptive capacity.

absortivola, a. absorptive.

abstracción, f. abstraction, extraction.

abstraer, v. to abstract, to extract (p.e. agua).

abundamiento, m. abundance.

abundancia, f. abundance.
abundante, a. abundant.
abuso, m. abuse, misuse; **a. ambiental,** environmental abuse.
acacia, f. acacia; **a. blanca, a. falsa,** locust tree.
académicola, a. academic.
Acadiense, m. Acadian.
acampanadola, a. bell-shaped.
acanaladola, a. grooved, furrowed.
acanaladura, f. channelling, groove.
acantilado, m. cliff; **a. marino,** sea cliff.
acantiladola, a. steep, sheer, shelving.
ácaro, m. acarus (pl. acari); **ácaros (Zoo, orden),** Acarina (pl).
acarreador, m. carrier (un transportista).
acarreo, m. haulage, carriage; **gastos (m.pl) de a.,** haulage costs.
acatarrarse, v. to catch a cold (Med).
acceder, v. to access (entrar), to agree to (acordar); to log in (Comp).
acceso, m. access; **vía (f) de a. de exposición (p.e. de contaminantes),** exposure pathway.
accidente, m. accident (siniestro); unevenness (de terreno).
acción, f. action, act; share, stock (Fin); **a. eólica,** wind action; **a. hidráulica,** hydraulic action.
accionariado, m. shareholding.
accionista, m. y f. shareholder.
acebo, m. holly.
aceite, m. oil; **a. combustible,** fuel oil; **a. de linaza,** linseed oil; **a. de parafina,** paraffin oil; **a. diesel,** diesel oil; **a. ligero,** light oil; **a. mineral,** mineral oil; **a. pesado,** diesel fuel; **a. vegetal,** vegetable oil.
aceituna, f. olive.
aceleración, f. acceleration; **a. de la gravedad,** acceleration of gravity.
acelerar, v. to accelerate, to speed up.
acelerómetro, m. accelerometer.
aceña, f. (Arab) water mill, waterwheel on river.
aceptante, m. acceptor (Quím).
acequia, f. unlined irrigation channel.
acera, f. pavement; sidewalk (US).
aceración, f. steelmaking.
acercamiento, m. approach (de un objeto).
acería, f. steelworks.
acero, m. steel; **a. de aleación,** steel alloy; **a. dulce,** mild steel; **a. forjado,** forged steel; **a. inoxidable,** stainless steel; **fundición (f) de a.,** steel casting.
acetaldehido, m. acetaldehyde.
acetamida, f. acetamide.
acetato, m. acetate.
acéticola, a. acetic.
acetileno, m. acetylene.
acetilo, m. acetyl.
acetona, f. acetone.
acíclicola, a. acyclic.
acidez, f. acidity.
acidificación, f. acidification.

acidificar, v. to acidize.
acidimetría, f. acidimetry.
ácido, m. acid; **á. absísico,** abscisic acid; **á. acético,** acetic acid; **á. ascórbico,** ascorbic acid; **á. bórico,** boric acid; **á. bromhídrico,** hydrobromic acid; **á. carbónico,** carbonic acid; **á. carboxílico,** carboxylic acid; **á. cítrico,** citric acid; **á. clorhídrico,** hydrochloric acid; **á. clorsulfónico,** chlorosulphonic acid; **á. desoxirribonucleico (abr. ADN),** deoxyribonucleic acid (abr. DNA); **á. fluorhídrico,** hydrofluoric acid; **á. fórmico,** formic acid; **á. fosfórico,** phosphoric acid; **á. graso,** fatty acid; **á. húmico,** humic acid; **á. maléico,** maleic acid; **á. nítrico,** nitric acid; **á. oxálico,** oxalic acid; **á. ribonucleico (abr. ARN),** ribonucleic acid (abr. RNA); **á. salicílico,** salicylic acid; **á. sulfúrico,** sulphuric acid; **á. úrico,** uric acid.
acidófilola, a. acidophile.
acidogénesis, m. acidogeneisis.
acimut, m. azimuth.
aclaración, f. clarification, explanation.
aclarar, v. to rinse (limpiar); to clarify (explicar); to thin out (selva); to clear (LAm, líquido); to make lighter (color).
aclimatación, f. acclimatization; acclimation (US).
aclimatizar, v. to acclimatize.
acmita, f. acmite.
acondicionador, m. conditioner; **a. de aire,** air conditioner; **a. de suelo,** soil conditioner.
acondicionamiento, m. improvement (mejoramiento); conditioning (de lugar, sustancia); **a. de cauces,** channel improvement; **a. de lodos,** sludge conditioning; **a. del suelo,** soil conditioning.
acondicionar, v. to fit out, to arrange; to improve (p.e. carreteras).
aconsejar, v. to advise, to counsel.
acopio, m. store, stock.
acoplamiento, m. mating.
acortamiento, m. shortening.
acortar, v. to shorten.
acotada, f. bounded land used for tree planting.
acotar, v. to survey, to set out (Ing); to fence in (limitar); to annotate (un libro).
acre, 1. m. acre (equiv. a. 0,4047 ha); **número (m) de acres, superficie (f) en acres,** acreage; 2. a. acrid, pungent.
acrecentamiento, m. accretion.
acreción, f. accretion.
acreditación, f. accreditation.
acre-pie, m. acre-foot (unidad de volumen de agua, 271.200 gálones británicos, 226.100 gálones EEUU o 1233.5 metros cúbicos).
acre-pulgada, m. acre-inch (unidad de volumen de agua, 22.600 gálones británicos, 18.850 gálones EEUU o 102.8 metros cúbicos).
acrílicola, a. acrylic; **resina (f) a.,** acrylic resin.
acrópolis, m. acropolis.

actínido, m. actinide.
actinio, m. actinium (Ac).
actinolita, f. actinolite.
activación, f. activation, speeding up.
activadola, a. activated; **lodos (m.pl) activados**, activated sludge.
actividad, f. activity; **a. comercial**, commercial activity; **a. fabril**, manufacturing activity; **a. humana**, human activity; **a. iónica**, ionic activity; **a. sísmica**, seismic activity; **actividades molestas, insalubres, nocivas y peligrosas**, dangerous and nuisance activities and activities harmful to health.
activista, f. activist.
activola, a. active.
actuación, f. behaviour, action, conduct (conducta); performance, event (suceso).
actuador, m. actuator; **a. neumático**, pneumatic actuator.
actual, a. present, current; **método (m) a.**, present-day method.
actualidad, f. present, present time (tiempo); topicality (tema).
actualización, f. modernization.
actualizar, v. to update, to bring up to date.
actuarial, a. actuarial, actuarian.
actuariola, m. y f. actuary.
acuacultura, f. aquaculture.
acuario, m. aquarium (pl. aquaria, aquariums).
acuáticola, a. aquatic.
acueducto, m. aqueduct.
ácueola, a. aqueous.
acuerdo, m. agreement, understanding; **de a. con**, in compliance with.
acuicludo, m. aquiclude.
acuicultura, f. aquaculture.
acuífero, m. aquifer; **a. aluvial**, alluvial aquifer; **a. anisótropo**, anisotropic aquifer; **a. artesiano**, artesian aquifer; **a. calcáreo**, calcareous aquifer; **a. confinado**, confined aquifer; **a. consolidado**, consolidated aquifer; **a. de doble porosidad**, dual-porosity aquifer; **a. estratificado**, stratified aquifer; **a. explotable**, exploitable aquifer; **a. fisurado**, fissured aquifer; **a. fracturado**, fractured aquifer; **a. freático**, phreatic aquifer; **a. heterogéneo**, heterogeneous aquifer; **a. isótropo**, isotropic aquifer; **a. kárstico**, karstic aquifer; **a. litoral**, coastal aquifer; **a. multicapa**, multi-layered aquifer; **a. múltiple**, multiple aquifer; **a. no confinado**, unconfined aquifer; **a. poroso**, porous aquifer; **a. profundo**, deep aquifer; **a. recargado**, recharged aquifer; **a. semiconfinado**, semi-confined aquifer; **a. sobreexplotado**, over-exploited aquifer; **a. surgente**, overflowing aquifer; **a. volcánico**, volcanic aquifer; **descontaminación (f) de acuíferos**, aquifer decontamination; **propiedades (f.pl) del a.**, aquifer properties; **protección (f) del a.**, aquifer protection; **recarga (f) del a.**, aquifer recharge; **vulnerabilidad (f) del a.**, aquifer vulnerability.

acuíferola, a. aquiferous.
acuífugo, m. aquifuge, leaky aquifer.
acuitardo, m. aquitard.
acumulación, f. accumulation; **a. aluvial**, alluvial accumulation; **a. de arena**, sand drift; **a. glacial**, **a. glaciar**, glacial accumulation.
acumular, v. to accumulate.
acumulativola, a. cumulative; **distribución (f) a. (Mat)**, cumulative distribution.
acuñamiento, m. petering out, pinching out (Geol).
acuñar, v. to coin.
acuosola, a. aqueous; **no acuoso/a**, non-aqueous.
acusadola, a. accused.
acústica, f. acoustics.
acústicola, a. acoustic, acoustical.
achaparradola, a. scrubby, scrub (Bot).
achatamiento, m. flattening.
achicador, m. bailer.
achicar, v. to bale out (agua); to make smaller (reducir); to intimidate (intimidar).
adaptabilidad, f. adaptability, flexibility.
adaptación, f. adjustment, adaptation.
adecuación, f. fitting, adjusting.
adelanto, m. advance, progress, breakthrough.
adelfa, f. oleander.
adelgazamiento, m. petering out, thinning (Geol).
ADENA|WWF, f. **Asociación (f) para la Defensa de la Naturaleza (Esp)**, Association for the Defence of Nature.
adenosina, f. adenosine; **trifosfato (f) de a. (abr. ATP)**, adenosine triphosphate (abr. ATP).
adherencia, f. adherence, adhesion.
adhesión, f. adhesion, adherence.
adiabáticola, a. adiabatic; **velocidad (f) de caída a. (Met)**, adiabatic lapse rate.
adición, f. addition (Mat); acceptance (Jur); bill, check (US) (factura).
adiestramiento, m. training.
aditivo, m. additive; **a. antiincrustante**, anti-incrustation additive.
administración, f. administration; **a. de recursos hidráulicos**, administration of water resources.
admisible, a. admissible.
admixtura, f. admixture.
ADN, m. DNA; **ácido (m) desoxirribonucleico**, deoxyribonucleic acid.
adobe, m. adobe, sun-dried brick.
adobera, f. lake from clay pit for bricks.
adquisición, f. acquisition.
adrenalina, f. adrenalin, adrenaline, epinephrin (US).
adsorber, v. to adsorb.
adsorción, f. adsorption.
aductor, m. adductor.
advección, f. advection.
adversola, a. adverse; **efecto (m) a.**, adverse effect.
adyacente, a. adjacent.

AEET, f. Asociación (f) Española de Ecología Terrestre, Spanish Association of Terrestrial Ecology.

AELC, f. EFTA; Asociación (f) Europea de Libre Comercio, European Free Trade Association.

aéreola, a. aerial; compañía (f) a., airline company; ferrocarril (m) a., elevated railway, overhead railway; fotografía (f) a., aerial photography; tráfico (m) a., air traffic; transportador (m) a., cableway.

aeróbicola, a. aerobic.

aerobiola, a. aerobic; respiración (f) a., aerobic respiration.

aerobús, m. airbus.

aerodeslizador, m. hovercraft.

aerodinámica, f. aerodynamics.

aerodinámicola, a. aerodynamic; túnel (m) a., wind tunnel.

aeródromo, m. aerodrome, airfield, airdrome (US).

aerofotografía, f. aerial photography.

aerofotogrametría, f. aerial photography.

aerolito, m. meteorite.

aeronáuticola, a. aeronautical.

aeronave, m. airship; a. espacial, spacecraft.

aeropostal, m. airmail.

aeropuerto, m. airport.

aerosol, m. aerosol.

áfido, m. aphid.

afiladola, a. tapering.

afinidad, f. affinity.

afloramiento, m. upwelling (Mar); outcrop (Geol).

aflorar, v. to crop out, to expose, to outcrop.

afluencia, f. aflux, inflow, influx; crowd (multitud).

afluente, 1. m. inflow, tributary; 2. a. affluent, copious; tributario (m) a., affluent tributary.

aforador, m. gauging device; a. trapezoidal, trapezoidal flume; a. venturi, venturi meter.

aforamiento, m. gauging, measurement (acción).

aforar, v. to gauge, to measure.

aforestación, f. afforestation.

aforo, m. gauging, measurement; a. con flotadores, float gauging; a. con trazadores radioactivos, gauging with radioactive tracers; a. de caudal, flow gauging; a. por cuchareo, measuring flow with bailer; a. químico, chemical gauging; estación (f) de a., gauging station; recipiente (m) de a., flow measuring tank.

África, f. Africa.

africanola, a. African.

afrolíticola, a. aphrolitic; lava (f) a., aphrolitic lava.

agalla, f. gill (de peces); gall (de plantas); temple, side of head (de aves).

agar, m. agar.

agar-agar, m. agar-agar.

agárico, m. agaric.

ágata, f. agate.

agave, m. agave.

agencia, f. agency; a. de desarrollo, development agency; a. de transportes, haulage company.

agenda, f. agenda.

agente, m. agent; a. contaminante, agent of contamination; a. dispersante, dispersive agent; a. erosivo, agent of erosion; a. mineralizante, agent of mineralization.

agitación, f. agitation.

agitador, m. agitator, shaker.

agitar, v. to shake, to agitate.

aglomeración, f. agglomeration; a. de tráfico, traffic jam; a. urbana, urban agglomeration.

aglomerado, m. agglomerate; a. volcánico, volcanic agglomerate.

agnatos, m.pl. Agnatha (Zoo, superclase).

agorafobia, v. agoraphobia.

agotable, a. exhaustible; recursos (m.pl) agotables, exhaustible resources.

agotadola, a. emptied, depleted, drained.

agotamiento, m. depletion, exhaustion; a. de aguas subterráneas, depletion of underground water resources; a. del agua, dewatering.

agotar, v. to drain, to empty.

agrandamiento, m. enlargement.

agrandar, v. to enlarge.

agrariola, a. agrarian; reforma (f) a., agrarian reform.

agregación, f. aggregation.

agregado, m. aggregate.

agresión, f. aggression.

agresividad, f. aggressiveness, aggressivity; a. del agua, aggressiveness of water.

agresivola, a. aggressive; aguas (f.pl) agresivas, aggressive waters.

agrícola, f. agricultural, farming.

agricultorla, m. y f. farmer.

agricultura, f. agriculture, farming; a. en franjas, strip-farming; a. sostenible, sustainable agriculture; escuela (f) de a., agricultural college.

agrietamiento, m. fissuring.

agrietar, v. to crack, to split.

agrimensor, m. surveyor.

agrimensura, f. land surveying.

agroalimentaria, f. agrifood industry.

agroganadero, m. cattle-raising, stockbreeding.

agroindustria, f. agroindustry.

agronegocios, m.pl. agribusiness.

agronomía, f. agronomy.

agrónomola, 1. m. y f. agronomist; 2. a. agricultural; ingeniero (m) a., agricultural engineer.

agropecuariola, a. agricultural, farming; industrias (f.pl) agropecuarias, farming practices, agricultural industry

agrosilvicultura, f. agroforestry.

agua, f. water; a. ácida, acid water; a. alcalina, alkaline water; a. altamente mineralizada, highly-mineralized water; **a. amarga,** sour water; **a. artesiana,** artesian water; **a. blanda,** soft water; **a. capilar,** capillary water; **a. clara,** clear water; **a. colgada,** perched water; **a. combinada,** combined water; **a. contaminada,** polluted water, contaminated water; **a. corriente,** running water; **a. corrosiva,** corrosive water; **a. desionizada,** deionized water; **a. de arriba,** high water; **a. de baño,** bathing water; **a. de bebida,** drinking water; **a. de cal,** lime water; **a. decantada,** settled water; **a. de campo petrolífero,** oil-field water; **a. de compensación,** compensation water; **a. de cristalización,** water of crystallization; **a. de derretimiento,** melt water; **a. de deshielo,** melt water; **a. de enfriamiento,** cooling water; **a. de fondo,** bottom water; **a. de formación,** connate water; **a. de grifo,** tap water; **a. de infiltración,** infiltration water; **a. de lixiviación,** leachate water; **a. del medio receptor,** receiving water; **a. del suelo,** soil water; **a. de lluvia,** rainwater, stormwater; **a. de manantial,** spring water; **a. de mar,** sea water, seawater; **a. de mina,** mine water; **a. de natación,** bathing water; **a. de percolación,** percolation water; **a. de pozo,** well water; **a. depurada,** treated effluent, final effluent; **a. de regadío,** irrigation water; **a. destilada,** distilled water; **a. dulce,** sweet water, fresh water; **a. dura,** hard water; **a. durmiente,** standing water; **a. embotellada, a. envasada,** bottled water; **a. estancada,** slack water, still water; **a. ferruginosa,** ferruginous water; **a. fluorada,** fluoridated water; **a. fósil,** fossil water; **a. freática,** phreatic water; **a. intersticial,** interstitial water; **a. intersticial,** pore water; **a. kárstica,** karst water; **a. limosa,** muddy water; **a. limpia,** clean water; **a. loja,** muddy water; **a. magmática,** magmatic water; **a. marejada (Mar),** rough water; **a. meteórica,** meteoric water, rainwater; **a. neutra,** neutral water; **a. no contabilizada,** unaccounted-for water; **a. oxigenada,** hydrogen peroxide; **a. pesada,** heavy water; **a. potable,** drinking water, potable water; **a. pura,** pure water; **a. receptora,** receiving water; **a. reciclada,** recycled water, reclaimed water; **a. recirculada,** recirculated water, reclaimed water; **a. salada,** salt water, saline water; **a. salobre,** brackish water; **a. selenitosa,** gypsum-rich water; **a. singenética,** syngenetic water; **a. singénica,** connate water; **a. subterránea artesiana,** artesian groundwater; **a. subterránea colgada,** perched groundwater; **a. subterránea libre,** free groundwater; **a. subterránea,** groundwater; **a. superficial,** surface water; **a. termal,** thermal water; **a. tranquila,** slack water, still water; **a. turbia,** peaty water; **aguas abajo,** downstream; **almacenamiento (m) de**

a. subterránea, groundwater storage; **avance (m) de a. marginal,** water encroachment; **balance (m) del a.,** water balance; **bomba (f) de a.,** water pump; **columna (f) de a.,** water column; **cono (m) de a.,** water cone; **conservación (f) del a.,** water conservation; **consumo (m) de a.,** water consumption; **contador (m) de a.,** water meter; **control (m) de pérdidas de a.,** water leak control; **corriente (f) de a.,** watercourse (curso); water flow (flujo); **curso (m) de a.,** watercourse; **demanda (f) de a.,** water demand; **depósito (m) elevado de a.,** water tower; **desperdicio (m) de a.,** waste of water; **distribución (f) de a.,** water distribution; **divisoria (f) de aguas,** watershed (GB); **dureza (f) del a.,** water hardness; **empresa (f) de a. potable,** water company; **entrada (f) de a.,** water ingress; **extracción (f) de a.,** water supply abstraction; **nivel (m) de a. subterránea,** groundwater level; **película (f) de a. (capa delgada),** water film; **pozo (m) de a.,** water well, waterwell; **racionamiento (m) del a.,** water rationing; **rebote (m) de a. subterránea,** groundwater rebound; **reciclaje (m) del a.,** water recycling; **recirculación (f) del a.,** water recirculation; **red (f) dual de a.,** dual water supply network; **reparto (m) del a.,** water allocation; **salto (m) de a.,** cascade, waterfall; **servicio (m) de distribución de a.,** water (supply) company; **superficie (f) del a.,** water surface; **tubería (f) de a.,** water pipe; **vigilancia (f) de la calidad de las a.,** water quality monitoring.

aguacero, m. cloudburst, downpour; **fuerte a.,** heavy downpour.

aguachar, m. (LAm) small pond.

aguachinal, m. land subject to frequent flooding.

aguadizo, m. land frequently flooded.

aguaje, m. (LAm) downpour.

aguamala, f. jellyfish.

aguamarina, f. aquamarine.

aguanieve, f. sleet, slush; **a. de deshielo glaciar,** glacial meltwater.

aguanoso, m. waterlogged ground.

aguar, v. to water (p.e. diluir el vino).

aguarrás, f. oil of turpentine, thinners.

aguas, f.pl. water, waters (pl); **a. abajo (Mar),** downstream; **a. cloacales,** wastewater, sewage; **a. de aclarado (Trat),** washwater; **a. negras,** sewage, cesspool contents; **a. residuales de curtiembres (LAm),** tannery wastewater; **a. residuales decantadas,** settled sewage; **a. residuales de fábricas de curtidos,** tannery wastewater; **a. residuales de mataderos,** slaughterhouse wastewater, abattoir wastewater; **a. residuales de papeleras,** papermill wastewater; **a. residuales industriales,** industrial wastewater; **a. territoriales,** territorial waters; **aprovechamiento (m) de a. residuales,** wastewater reuse; **conductos**

(m.pl) de a. potables, water mains; **criterios (m.pl) de calidad de las a.,** water quality criteria; **depuración (f) de a. residuales,** wastewater treatment; **derechos (m.pl) de a.,** water rights; **disposición (f) de a. residuales,** wastewater disposal; **evacuación (f) de a. residuales,** wastewater discharge; **hongo (m) de a. residuales (en ríos),** sewage fungus; **reuso (m) de a. residuales (LAm), reutilización (f) de a. residuales,** wastewater reuse; **tratamiento (m) de a. residuales,** wastewater treatment.

aguatocho, m. small bog.

aguazal, m. flooded depression, swamp.

aguja, f. needle (Gen), pointer; spire, steeple (Con); swordfish (Zoo).

agujereadola, a. leaky, full of holes.

agujero, m. hole, rough mine.

aguosola, a. aquatic, watery.

ahogadola, a. drowned (p.e. animales).

ahogar, v. to drown (en agua); to suffocate; to choke.

ahorro, m. saving; thrift (cuidado); **a. de agua,** water saving; **a. energético,** energy saving.

ahuyentar, v. to dispel, to drive away.

aire, m. air; **a. acondicionado,** air conditioned, air conditioning; **a. comprimido,** compressed air; **bolsa (f) de a.,** air lock; **cámara (f) de a.,** air space; **conducto (m) de a.,** air duct; **contaminación (f) del a.,** air contamination, air pollution; **control (m) de la calidad del a.,** air quality control; **corriente (f) de a.,** air current, draught; **filtración (f) de a.,** air filtration; **inclusión (f) de a.,** air entrainment; **monitoreo (m) del a.,** air monitoring; **normas (f.pl) de la calidad del a.,** air quality standards; **polución (f) del a.,** air pollution; **presión (f) del a.,** air pressure; **purificación (f) del a.,** air purification; **purificador (m) de a.,** air cleaner; **recipiente (m) de a.,** air vessel.

aireación, f. aeration, ventilation; **a. con burbujas,** bubble aeration.

aireadola, a. aerated; **no a.,** unaerated.

airear, v. to aerate, to gasify (con gas).

airecito, m. breeze, light wind.

aisladola, a. isolated.

aislamiento, m. insolation, insulation (Elec); **a. acústico,** noiseproofing; **a. térmico,** thermal insulation.

aislante, a. insulating; **material (m) a.,** insulation material.

aislar, v. to isolate (Gen); to separate (separar); to insulate (Elec).

ajuste, m. adjustment, fit; **a. gráfico,** graphical adjustment; **a. por mínimos cuadrados,** least squares adjustment; **prueba (f) de bondad del a., test (m) de buen a.,** goodness-of-fit test.

alabastro, m. alabaster.

álabe, m. (Arab) wooden cog or tooth, bucket of waterwheel (noria).

alabeadola, a. warped.

alabeo, m. warp; **a. hacia abajo,** downwarp; **a. hacia arriba,** upwarp.

ALADI, f. LAATI; **Asociación (f) Latinoamericana de Integración,** Latin American Association for Trade Integration.

ALALC, f. LAFTA; **Asociación (f) Latinomericana de Libre Comercio,** Latin American Free Trade Association.

alambre, m. wire; **a. de espinas, a. de púas, a. espinoso,** barbed wire; **alambres cruzados,** crossed wires; **cerca (f) de a.,** wire fence.

álamo, m. poplar; **á. blanco,** white poplar; **á. negro,** black poplar; **á. temblón,** aspen.

alargadola, a. prolonged, extended (tiempo); lengthened, elongated (de largo).

alargar, v. to elongate.

albañal, m. sewer, drain.

albañil, m. mason, bricklayer.

albañilería, f. masonry (artesanía).

albardilla, f. capping, coping (Con); saddle; mound of earth (en un huerto).

albardón, m. river levée (Esp), raised berm (LAm), bund (dique).

albaricoque, m. apricot.

albaricoquero, m. apricot tree.

albariza, f. saline lake.

ALBE, f. **Asociación (f) de Licenciados de Ciencias Biológicas de España,** Spanish Association of Licensed Biologists.

albedo, m. albedo.

alberca, f. water tank (small reservoir).

albercón, m. large water storage tank.

albergar, v. to house (p.e. una máquina).

Albiense, m. Albian (m).

albina, f. (Arg) dried saltmarsh, salt pan.

albinismo, m. albinism.

albino, m. albino.

albita, f. albite, moonstone.

albufera, f. lagoon (lago pequeño); **a. de marea,** tidal lagoon.

albumen, m. albumen.

albúmina, f. albumin.

albuminoideola, a. albuminoid; **nitrógeno (m) a.,** albuminoid nitrogen.

albura, f. sapwood.

alburno, m. alburnum.

alcacel, m. barley field.

alcachofa, f. artichoke.

álcali, m. alkali.

alcalinidad, f. alkalinity.

alcalinización, f. alkalinization.

alcalinola, a. alkaline; **serie (f) a. cálcica,** calcalkaline series.

alcalinotérreos, m.pl. alkaline earths.

alcaloide, m. alkaloid.

alcance, m. fetch, reach (distancia), range, scope (ámbito); **a. de las olas,** fetch of waves; **a. intercuartil,** interquartile range.

alcanfor, m. camphor.

alcano, m. alkane.

alcantarilla, f. sewer, drain; **a. cuadrada,** box culvert; **a. interceptadora,** intercepting sewer.

alcantarillado, m. sewers (pl), drains (pl), sewerage system; **a. pluvial,** stormwater sewer, stormwater sewerage; **con a.,** sewered, provided with sewerage system; **efluente (m) de a.,** sewer overflow; **sistemas (m.pl) de a. combinado,** combined sewerage systems.

alcoba, f. alcove.

alcohol, m. alcohol.

alcohólicola, a. alcoholic.

alcoholímetro, m. alcoholometer.

alcoholismo, m. alcoholism.

alcornocal, m. cork grove.

alcornoque, m. cork tree, cork oak.

aldehído, m. aldehyde.

aldrín, m. aldrin.

aleación, f. alloy.

aleatoridad, f. randomness.

aleatoriola, a. random (Mat); contingent; **muestreo (m) a.,** random sampling; **ruido (m) a.,** random noise; **variable (f) a.,** random variable.

alegación, f. allegation, assertion.

alejadola, a. distant, far away.

alelo, m. allele.

alemánla, a. German.

Alemania, f. Germany.

alerce, m. larch.

alergia, f. allergy.

alérgicola, a. allergic; **manifestación (f) a.,** allergic reaction.

alerta, f. alert; **sistema (m) de a.,** alarm system.

aleta, f. fin, fluke (de ballena); **aletas pares,** paired fins.

alfa, f. alpha.

alfalfa, f. alfalfa.

alfaque, m. bar, sandbank.

alféizar, m. embrasure (Arq).

alga, f. alga (pl. algae); seaweed (en el mar); **a. marina,** seaweed.

algal, a. algal.

algas, f.pl. algae; **a. pardas,** brown algae; **a. verde-azuladas,** blue-green algae; **a. verdes,** green algae; **floración (f) de a., florecimiento (m) de a.,** algal bloom.

álgebra, f. algebra (Mat); **á. booleana,** Boolean algebra.

algebraicola, a. algebraic.

algébricola, a. algebraic.

algología, f. algology.

algoritmo, m. algorithm.

alguicida, m. algicide.

alhóndiga, f. corn exchange.

alhondiguero, m. corn merchant.

alianza, f. alliance.

alicíclicola, a. alicyclic.

alícuota, f. aliquot.

aliento, m. breath.

alifáticola, a. aliphatic.

aligeramiento, m. reduction, lightening (de carga).

alimentación, f. diet, food; **tubo (m) de a.,** feed pipe, standpipe.

alimentar, v. to feed, to nourish.

alimentariola, a. alimentary, food; **ciclo (m) a.,** food web, food cycle.

alimentarse de, v. to prey on, to prey upon.

alimenticiola, a. food, nutritive; **industria (f) a.,** food industry; **productos (m.pl) alimenticios,** foodstuffs; **valor (m) a.,** nutritive value, food value.

alimento, m. food, nutrients (pl), nourishment; **a. irradiado,** irradiated food; **procesamiento (m) de a.,** food processing.

alineación, f. alignment; **a. de la perforación,** borehole alignment; **fuera (f) de a.,** out of alignment.

alineamiento, m. alignment; **estar fuera de a. (v),** to be out of alignment.

alinear, v. to align.

alisar, m. alder grove.

aliseda, f. alder grove.

aliso, m. alder.

aliviadero, m. overflow channel, relief sewer; **a. de crecida,** storm overflow.

aliviar, v. to alleviate; to mitigate (mitigar).

alivio, m. alleviation; mitigation (mitigación).

aljibe, m. (Arab) cistern (covered domestic water reservoir); **a. abovedado,** vaulted cistern.

almacén, m. warehouse, store; grocery store (LAm); **gran a.,** department store.

almacenadola, a. stored.

almacenaje, m. warehousing, storage dues (pl).

almacenamiento, m. storage, warehousing; **a. de acuífero,** aquifer storage; **a. en el pozo,** well storage; **a. en la ribera (un embalse),** bankside storage; **a. específico,** specific storage; **a. subterráneo,** underground storage; **capacidad (f) de a.,** storage capacity; **embalse (m) por a.,** storage reservoir; **tanque (m) de a.,** storage tank.

almacenar, v. to store.

almadraba, f. tuna fishing (pesca), tunny net (red); tuna grounds (caladero).

almandino, m. almandite.

almarcha, f. wet meadow.

almarge, m. wet meadow.

almarjal, m. much-flooded land.

almazara, f. olive oil press.

almena, f. battlement.

almendra, f. almond (nut).

almendral, m. almond grove.

almendro, m. almond (tree).

almiar, m. haystack.

almidón, m. starch.

Almirantazgo, m. Admiralty.

alo-, pf. allo-.

alóctonola, a. allochthonous; **especie (f) a.,** introduced species, non-native species, allocthonous species.

alofana, f. (LAm) allophane.

alomorfismo, m. alomorphism.

alotropía, f. allotrophy.
alpechín, m. foul smelling seepage from olives pressings.
alpinismo, m. mountaineering.
alqueno, m. alkene.
alquería, f. farm, farmstead, farmhouse.
alquibenzol, m. alkibenzene.
alquil, m. alkyl.
alquimia, f. alchemy.
alquino, m. alkyne.
alquitrán, m. pitch, tar; **a. de hulla**, coal tar.
alquitranado, m. tarpaulin.
alrededores, m.pl. environs.
alteración, f. alteration.
alteradola, a. altered.
alterar, v. to alter.
alterio, m. halteres.
alternador, m. alternator.
alternar, v. to alternate.
alternativa, f. option, choice (opción); alternation (sucesión), alternative; **a. viable**, viable alternative.
alternola, a. alternate; alternating (Elec).
altillo, m. hillock, small hill; attic (LAm).
altímetro, m. altimeter.
altiplanicie, m. high plateau.
altiplano, m. (LAm) high plateau; Andean high plateau.
altitud, f. altitude.
altola, a. high.
altoestrato, m. altostratus.
altozono, m. hill, knoll, hillock (cerro).
altura, f. height; **a. capilar**, capillary height; **a. de caida**, head (of pressure); **a. de impulsión**, head of water; **a. de vuelo**, flight height; **a. hidrostática**, hydrostatic head; **a. manométrica**, manometer height; **a. piezométrica**, piezometric height; **a. topográfica**, topographic height.
alubia, f. bean.
alud, m. landslide, avalanche; **a. de nieve**, snow slip, avalanche; **a. de roca**, rockfall; **a. de tierra**, landslide.
alumbrado, m. lighting (sistema), illumination; **a. público**, street lighting, street lights; **poste (m) de a.**, street light, street lamp.
alumbradola, a. lit, lighted, illuminated.
alumbramiento, m. childbirth (parto); lighting, lighting-up (encendimiento).
alumbre, m. alum.
alúmina, f. alumina.
aluminato, m. aluminate.
aluminio, m. aluminium (Al).
aluminosis, f. aluminosis.
aluvial, a. alluvial; **abanico (m) a.**, alluvial fan.
aluvión, m. alluvium.
aluvionamiento, m. alluviation.
alveoladola, a. honeycombed, pitted.
alveolo, alvéolo, m. alveolus (pl. alveoli).
alzado, m. elevation (de edificio); proud, haughty (LAm); **a. de fachada**, front elevation.

amalgama, f. amalgam.
amalgamación, f. amalgamation.
amanecer, m. daybreak, first light; **al a.**, at daybreak, at dawm.
amapola, f. poppy.
amarguillola, a. salinity.
amarillo, m. yellow; **a. de cadmio**, cadmium yellow.
amarillola, a. yellow.
amarradero, m. berth (para barco).
amarre, m. mooring.
amatista, f. amethyst.
ámbar, m. amber.
ambiental, a. environmental; **ciencia (f) del medio a.**, environmental science; **científico/a (m. y f.) a.**, environmental scientist; **educación (f) a.**, environmental education; **evaluación (f) de impacto a. (abr. EIA)**, environmental impact assessment (abr. EIA); **impacto (m) a.**, environmental impact; **protección (f) a.**, environmental protection; **salud (f) a.**, environmental health.
ambiente, 1. m. environment, atmosphere; **a. acuático**, aquatic environment; **a. humano**, human environment; **a. marino**, marine environment; **a. químico**, chemical environment; **a. reductor**, reducing environment; **a. sedimentario**, sedimentary environment; 2. a. environment, atmosphere; **Agencia (f) de Medio A. (abr. AMA) (Esp)**, Environmental Agency (abr. EA) (England and Wales); **ley (f) del medio a.**, environmental law; **medio (m) a.**, environment; **medio (m) a. controlado**, controlled environment; **temperatura (f) a.**, ambient temperature.
ambigüedad, f. ambiguity.
ámbito, m. scope, ambit, compass, environment.
ameba, f. amoeba (pl. amoebas, amoebae).
ameboide, a. amoeboid; **movimiento (m) a.**, amoeboid movement.
amenaza, f. threat, menace; **a. para la salud**, threat to health.
amenazadola, a. threatened; **especie (f) a.**, threatened species.
América, f. America, Latin America (Nota: en inglés, la palabra America empleada a solas, se refiere normalmente a los Estados Unidos).
América Latina, f. Latin America.
americanola, a. American (del continente), Latin American (de Iberoamérica), American (de Estados Unidos).
amerindiola, 1. m. y f. American Indian, Amerindian; 2. a. American Indian, Amerindian.
amfifílicola, a. amphiphilic.
amida, f. amide.
amileno, m. amylene.
amilo, m. amyl.
amina, f. amine.
aminoácido, m. amino acid.

aminofenol, m. aminophenol.
aminorar, v. to lessen, to decrease.
amnios, m. amnion.
amoniacal, a. ammoniacal; **nitrógeno (m) a.,** ammoniacal nitrogen.
amoníaco, m. ammonia.
amónicola, a. ammonium; **carbonato (m) a.,** ammonium carbonate.
amonificación, f. ammonification.
amonio, m. ammonium.
amonita, f. ammonite.
amontonar, v. to bank.
amorfola, a. amorphous.
amortiguación, f. dampening, softening; buffering (Quím); **capacidad (f) de a.** (Quím), buffering capacity; **estanque (m) de a.,** stilling basin.
amortiguadola, a. muffled, dampened.
amortiguador, m. shock absorber, damper, buffer.
amortiguadorla, a. cushioning, absorbing; **efecto (m) a.,** cushioning effect.
amortización, f. repayment (de préstamo); writing-off (de capital); depreciation (de puesto); amortization (Jur).
amperio, m. ampere, amp (abr. amp).
ampliación, f. enlargement, extension; **a. de capital,** extension of capital.
amplificación, f. amplification.
ampliola, a. extensive, comprehensive.
amplitud, f. width, fullness, amplitude.
anabólicola, a. anabolic.
anabolismo, m. anabolism.
anaeróbicola, a. anaerobic.
anaerobio, m. anaerobe.
anaerobiola, a. anaerobic; **zona (f) a.,** anaerobic zone.
anal, a. anal.
análisis, m. analysis; **a. armónico,** harmonic analysis; **a. bacteriológico,** bacteriological analysis, bacterial analysis; **a. canónico,** canonical analysis; **a. de componentes principales,** principal component analysis (abr. PC); **a. de conglomerados/grupos/agrupamientos,** cluster analysis; **a. de costos-beneficios,** cost-benefit analysis; **a. de datos,** data analysis; **a. de laboratorio,** laboratory analysis; **a. de mercados,** market research, market analysis; **a. de muestras,** sample analysis; **a. de paso crítico,** critical path analysis; **a. de riesgos,** risk analysis; **a. de varianza (abr. ANOVA),** analysis of variance; **a. de viabilidad,** feasibility study; **a. discriminante,** discriminant analysis; **a. espectral,** spectral analysis; **a. estadístico,** statistical analysis; **a. factorial,** factorial analysis; **a. fisicoquímico,** physicochemical analysis; **a. granulométrico,** grain-size analysis; **a. hidrológico,** hydrological analysis; **a. mecánico,** mechanical analysis; **a. microscópico,** microscopic analysis; **a. parasitológico,** parasitological analysis; **a.**

químico, chemical analysis; **a. volumétrico,** volumetric analysis.
analista, m. y f. analyst.
analíticola, a. analytical; **método (m) a.,** analytical technique.
analizador, m. analyzer (máquina).
analizar, v. to analyse.
analógicola, a. analogue, analog (US); **ordenador (m) a.,** analogue computer, analog computer (US).
análogola, m. y f. analogue, analog (US); **a. eléctrico,** electrical analogue.
anastomosadola, a. braided; **río (m) a.,** braided river.
anatasa, f. anatase, octahedrite.
anatomía, f. anatomy; **a. vegetal,** plant anatomy.
anatómicola, a. anatomic, anatomical.
ancla, f. anchor; **a. flotante,** drogue.
anclaje, m. anchorage, anchor (Con); **a. de terreno (Ing),** ground anchor; **bloque (m) de a.,** anchor block; **macizo (m) de a.** (Con), deadman; **perno (m) de a.,** anchor bolt.
ancón, m. (Méx) cove; corner.
anchoa, f. anchovy.
anchura, f. width, breadth; **a. de un río,** river width.
andadura, f. walking (acción), walk, gait.
andalucita, f. andalucite.
andamiaje, m. staging, scaffolding.
andamio, m. scaffolding.
andarivel, m. cable ferry; rope bridge (LAm).
andén, m. platform (ferrocarril); hard shoulder (de autopista); pavement, sidewalk (US); agricultural terrace (Arg, Bol, Perú).
andesita, f. andesite.
andinola, a. Andean.
andro-, pf. andro-.
androceo, m. androecium.
andrógeno, m. androgen.
anea, f. bulrush.
anegacidola, a. land subject to flooding.
anegadola, a. drowned (tierra, animales); **vertedero (m) a.,** drowned weir.
anegamiento, m. flooding.
anegar, v. to drown, to flood; to overwhelm (Fig); **a. un campo,** to flood a field.
anélido, m. annelid; **anélidos (Zoo, filum),** Annelida (pl).
anemia, f. anaemia, anemia (US).
anemómetro, m. anemometer, wind speed indicator.
aneroide, a. aneroid; **barómetro (m) a.,** aneroid barometer.
anfibio, m. amphibian; **anfibios (Zoo, clase),** Amphibia (pl).
anfibiola, a. amphibian.
anfíbol, m. amphibole.
anfibolita, f. amphibolite.
anfiteatro, m. amphitheatre, amphitheater (US).
anfólito, m. ampholyte.

ánfora, f. amphora.
angiosperma, f. angiosperm; angiospermas (Bot, división), Angiospermae (pl).
anglosajón, m. anglosaxon.
anglosajón, anglosajona, a. anglosaxon.
angstrom, m. angstrom (abr. Å, unidad de medida de tamaño atómico).
angstromio, m. angstrom unit.
anguila, f. eel.
anguílula, f. eelworm.
angular, 1. m. angle iron, angle bar; 2. a. angular.
ángulo, m. angle; á. de desplazamiento, angle of slip; á. de inclinación, angle of dip; á. de reposo, angle of repose; junta (f) de á., angle joint.
angulosidad, f. angularity.
anhedral, a. anhedral.
anhídrido, m. anhydride.
anhidrita, f. anhidrite.
anhidrola, a. anhydrous.
anilina, f. anilin, aniline; azul (m) de a., aniline blue.
anillo, m. ring; a. anual (Bot), annual ring.
animal, 1. m. animal; a. de carga, pack animal, beast of burden; a. doméstico, domestic animal; a. salvaje, wild animal; animales inferiores, lower animals; 2. a. animal; reino (m) a., animal kingdom.
anión, m. anion.
Anisiense, m. Anisian.
anisotropía, f. anisotropy.
anisotrópicola, a. anisotropic.
anisotrópola, a. anisotropic.
ano, m. anus.
anochecer, m. sundown, sunset.
anodizar, v. to anodize.
ánodo, m. anode.
anomalía, f. anomaly; a. térmica, thermal anomaly.
anómalola, a. anomalous.
anormalidad, f. abnormality.
anortita, f. anorthite.
anortosa, f. anorthoclase.
anotar, v. to note down, to annotate; to enroll, to register (esp. rg, Bol, Uru).
ANOVA, m. ANOVA; análisis (m) de varianza, analysis of variance.
anóxicola, a. anoxic.
antagonismo, m. antagonism.
antárticola, a. Antarctic.
Antártida, f. the Antarctic.
antecámara, f. antechamber.
antena, f. antenna (pl. antennas, antennae), feeler (Zoo); aerial, antenna (p.e. de radio); a. emisora, transmitting aerial; a. parabólica satellite dish; a. receptora, receiving aerial.
anteproyecto, m. preliminary plan (proyecto); blueprint (esp. dibujos); a. de ley (Jur), draft bill; a. de referencia, base-line survey.
antera, f. anther.
anteridio, f. antheridium (pl. antheridia).

anti-, pf. anti-.
antibiótico, m. antibiotic.
antibióticola, a. antibiotic.
anticiclón, m. anticyclone.
anticiclónicola, a. anticyclonic.
anticlinal, m. anticline; a. aportillado, breached anticline; a. cerrado, closed anticline.
anticlinorio, m. anticlinorium.
anticoagulante, m. deflocculent, anticoagulant.
anticonceptivo, m. contraceptive.
anticontaminante, a. anti-pollution, clean (Fig).
anticorrosivola, a. anticorrosive.
anticorrosivos, m.pl. anticorrosives.
anticuerpo, m. antibody.
antidetonante, a. anti-knocking.
antídoto, m. antidote.
antiespumante, 1. a. anti-foaming; agente (m) a., anti-foaming agent; 2. n. defoamant, anti-foaming agent.
antiforma, f. antiform.
antigénicola, a. antigenic, antigen.
antígeno, m. antigen.
antigüedad, f. antiquity (época antigua), age.
antimoniato, m. antimonate.
antimonio, m. antimony (Sb).
antimonito, m. antimonite.
antioxidante, m. antioxidant.
antipolución, f. anti-pollution; medidas (m.pl) de a., anti-pollution measures.
antiséptico, m. antiseptic.
antisépticola, a. antiseptic.
antisifonaje, f. antisiphonage; tubería (f) de a., antisiphoning pipe.
antozoos, m.pl. anthozoa (pl).
antracita, f. anthracite.
ántrax, m. anthrax.
antro-, pf. anthro-.
antrópicola, a. anthropogenic.
antropocéntricola, a. anthropocentric.
antropógenola, a. anthropogenic.
antropoide, 1. m. anthropoid; antropoides (Zoo, suborder), Anthropoidea (pl); 2. a. anthropoid.
antropoideola, a. anthropoid.
antropología, f. anthropology.
anual, a. annual; planta (f) a., annual, annual plant.
anualidad, f. annuity, annual payment.
anuario, m. yearbook, annual; a. telefónico, telephone directory.
anulador, a. annulated.
anular, v. to annul, to cancel, to repeal.
ánulo, m. annulus (pl. annulata).
anunciar, v. to announce, to advertise (Com); to be a sign of (pronosticar).
anuncio, m. announcement (declaración), advertisement (Com).
anuro, m. anuran (pl. anura); anuros (Zoo, clase), Anura (pl).
añadir, v. to add.
año, m. year; a. de luz, a. luz, light-year.
aorta, f. aorta.

apagadola, a. muffled, muted.

aparato, m. apparatus (pl. apparatuses, apparatus), instrument; **a. bucal**, mouthparts (p.e. de insectos); **a. circulatorio**, circulatory system; **a. de medida**, gauge; **a. digestivo**, digestive system; **a. eléctrico**, electrical appliance.

aparcamiento, m. car park; car lot (US); lay-by (en la carretera).

aparejo, m. bond (de ladrillos), block and tackle (Con), rigging (Mar); **a. a la holandesa (de ladrillos)**, Dutch bond; **a. de poleas**, block and tackle; **a. de tijón (de ladrillos)**, header bond; **a. en espina (de ladrillos)**, herringbone bond; **a. hiriente**, winding gear; **a. tipo jardín (de ladrillos)**, garden bond; **aparejos de pesca**, fishing tackle; **doble a. flamenco (de ladrillos)**, double Flemish bond.

aparente, a. apparent, evident.

apariencia, f. appearance; **a. física**, physical appearance.

apartamento, m. apartment, flat.

apartar, v. to remove, to separate.

apatito, m. apatite.

apelación, f. appeal.

apellido, m. surname, family name, last name; **nombre (m) y apellidos**, full name.

apéndice, m. appendix (pl. appendices, appendixes).

apertura, f. gap, pass; **a. de agua**, water gap; **a. de viento**, wind gap.

apical, a. apical.

ápice, m. apex (pl. apices, apexes), tip; **a. de anticlinal**, apex of anticline.

apicultura, f. apiculture, bee keeping.

apio, m. celery.

apisonadora, f. road roller, steamroller; **a. vibratoria**, vibrating roller.

aplanamiento, m. levelling, leveling (US), flattening; **a. del suelo**, levelling of the ground.

aplastar, v. to crush, to grind.

aplazar, v. to put off, to postpone.

aplicable, a. applicable; enforceable (de leyes).

aplicación, f. application, use; **de múltiples aplicaciones**, multi-purpose.

aplicar, v. to apply, to implement.

apogeo, m. apogee.

aportación, f. contribution; **a. estatal (Com)**, state aid contribution; **a. media anual**, average yearly contribution; **centro (m) de a. de residuos**, centre of waste generation.

aporte, m. (LAm) contribution.

apoyarse, v. to abut, to lean against.

apoyo, m. support (ayuda, protección).

aprendizaje, m. learning (process); apprenticeship, training; **curva (f) de a.**, learning curve.

apresurar, v. to speed up, to quicken.

apretar, v. to squeeze, to compress.

aprobación, f. approval, consent.

aprobar, v. to approve, to authorize.

aprovechamiento, m. exploitation; development

(desarrollo); **a. energético de residuos**, waste-to-energy scheme; **a. hidráulico**, water resource exploitation.

aproximación, f. approximation; **a. de Dupuit-Forscheimer**, Dupuit-Forscheimer approximation; **a. logarítmica de Jacob**, Jacob's logarithmic approximation.

aproximadola, a. approximate.

aproximar, v. to approximate.

Aptiense, m. Aptian.

apto, a. fit, apt; **no a. para el consumo humano**, unfit for human consumption.

apuntadola, a. cuspate (Geog), pointed; **promontorio (m) a.**, cuspate foreland.

apuntalamiento, m. underpinning (acción).

apuntalar, v. to support; to underpin to prop (Con).

Aquitaniense, m. Aquitanian.

árabe, a. Arab, Arabic, Arabian.

arábicola, arábigola, a. Arabic.

arable, a. arable.

arácnido, m. arachnid; **arácnidos (Zoo, clase)**, Arachnida (pl).

arácnidola, a. arachnid.

arada, f. ploughing, plowing (US) (acción); ploughed land, plowed land (US).

arado, m. plough, plow (US) (herramienta); ploughing (acción); **a. por curvas de nivel**, contour ploughing.

aragonita, f. aragonite.

araña, f. spider; **a. de mar**, sea spider; **tela (f) de a.**, spider's web.

arar, v. to plough, to plow (US).

arbitrariola, a. arbitrary.

árbol, m. tree; shaft (Ing); **á. respiratorio**, respiratory system; **árboles talados**, felled trees, trees cut down.

arboladola, a. wooded.

arboleda, f. grove, coppice, spinney.

arbóreola, a. arboreal, tree; **cubierta (f) a.**, tree cover.

arboreto, m. arboretum.

arboricultura, f. arboriculture.

arbotante, m. arch buttress (Con).

arbusto, m. shrub, bush.

arcada, f. arcade, archway.

Arcaico, m. Archean.

arce, m. maple.

archipiélago, m. archipelago.

archivar, v. to archive.

archivo, m. archive.

arcilla, f. clay; **a. para ladrillos**, brick clay; **a. refractaria**, refractory clay, fire clay; **a. tixotrópica**, thixotropic clay; **revestimiento (m) de a.**, clay lining.

arcillosola, a. argillaceous, clayey; **revestimiento (m) a. (p.e. de un vertedero)**, clay lined; **sedimento (m) a.**, argillaceous sediment.

arco, m. arc (Mat), arch (Con); **a. ciego**, blind arch; **a. concéntrico**, concentric arch; **a. de aligeramiento**, relieving arch; **a. de medio**

punto, semicircular arch; **a. de soldadura,** welding arc; **a. eléctrico,** electric arc; **a. invertido,** inverted arch; **a. iris,** rainbow; **a. isla,** island arc; **presa (f) de a.,** arch dam; **riostra (f) en a.,** arch brace; **viga (f) en a.,** arch beam.

arcosa, f. arkose.

arcósicola, a. arkosic.

arder, v. to burn; **a. sin llama,** to smoulder.

área, f. area; **á. abierta de la rejilla,** screen open area; **á. colectora,** drainage area; **á. de bombeo,** pumping area; **á. de captación,** catchment area; **á. de drenaje,** drainage area; **á. de influencia,** area of influence; **á. de recarga,** recharge area; **á. de recepción (de ríos),** drainage basin; **á. urbana,** urban area.

arena, f. sand; **acumulación (f) de a.,** sand drift; **a. angulosa,** sharp sand; **a. de cuarzo,** quartz sand; **a. de sílice,** silica sand; **a. movedizas,** quicksands; **a. volcánica,** volcanic sand; **depósitos (m) de a.,** sand drift; **tormenta (f) de a.,** sandstorm.

arenáceola,, a. arenaceous.

arenero, m. sand pit.

arenícola, a. sand-dwelling.

Arenigiense, m. Arenigian.

arenisca, f. grit, coarse sand, sandstone; **a. de grano grueso,** coarse sandstone; **extracción (f) de a.,** grit removal.

arenita, f. arenite.

arenosola, a. arenaceous, sandy; **barra (f) a.,** sand bar.

areograma, m. pie chart.

areómetro, m. areometer, hydrometer.

arete, m. arête.

Argentina, f. Argentina.

argentinola, 1. m. y f. Argentinian, Argentine; 2. a. Argentinian, Argentine.

argentita, f. argentite.

argilita, f. argillite, claystone.

argón, m. argon (Ar).

aridez, f. aridity, dryness.

áridola, a. arid, barren.

ariete, m. ram (golpe); **a. hidráulico,** hydraulic ram.

arista, f. edge (Mat, p.e. de un cubo), sharp edge; arête (Geog); beard (Bot, p.e. del trigo).

aritmética, f. arithmetic.

armadola, a. reinforced (hormigón); assembled (montado); provided (provisto).

armadura, f. armour, armor (US) (Gen); reinforcement, reinforcing bars (Con).

armazón, f. framework, frame, skeleton; reinforcement (Con).

armónicola, a. harmonic; **serie (f) a.,** harmonic series.

armonizar, v. to harmonize.

ARN, m. RNA; **ácido (m) ribonucleico,** ribonucleic acid.

aromáticola, a. aromatic; **hidrocarburos (m.pl) aromáticos,** aromatic hydrocarbons.

arpón, m. harpoon, gaff.

arqueadola, a. arcuate.

arquegonio, m. archegonium.

arqueología, f. archaeology, archeology.

arqueológicola, a. archaeological.

arqueólogola, m. y f. archaeologist.

Arquímedes, m. Archimedes; **tornillo (m) de A. (para bombear),** Archimedes screw.

arquitecto, m. architect; **a. paisajista,** landscape architect.

arquitectónicola, a. architectonic, architectural.

arquitectura, f. architecture.

arquitectural, a. architectural.

arranque, m. starting, beginning; stripping (Min).

arrastrar, v. to trawl, to drag, to haul, to pull.

arrastre, m. trawling, dragging, pulling (acción); entrainment (hidráulica); silt load (de ríos); tracking (radio); **a. de fondo,** bottom trawl, bottom trawling; **a. de partículas finas (Geog),** transport of fine particles; **a. de suelo,** soil creep; **flota (f) de a.,** trawler fleet; **red (f) de a.,** trawl net.

arrecife, m. reef; **a. artificial,** artificial reef; **a. calcáreo,** limestone reef; **a. coralino,** coral reef; **a. costero en orla,** fringing reef; **a. de coral,** coral reef.

arreglo, m. arrangement, agreement (acuerdo); settlement (de un asunto).

arrestar, v. to confine, to pull in.

arribanzo, m. boggy ground.

arriesgadola, a. hazardous, dangerous.

arrocerola, m. y f. rice grower.

arrojadola, a. ejected.

arrojar, v. to eject, to throw out; to release (desechos).

arrollao, m. intermittent stream through meadow.

arroyo, m. stream, rivulet, watercourse, brook; intermittent stream (US); gutter (en calle); **a. intermitente,** bourne, winterbourne, lavant (GB); **a. seco,** wash.

arroyuelo, m. small creek, brook.

arroz, m. rice.

arrozal, m. rice paddy, paddy field.

arruinar, v. to ruin.

arsenato, m. arsenate.

arsénico, m. arsenic (As).

arseniosis, m. arseniosis.

arseniosola, a. arsenious, arsenous.

arsenito, m. arsenite.

arseniuro, m. arsenide.

arsenolita, f. arsenolite.

arsenopirita, f. arsenopyrite.

arsina, f. arsine.

arte, m. y f. art (Gen); craft, skill (habilidad); **artes tradicionales,** traditional methods.

artefacto, m. artefact.

arteria, f. artery.

artesanía, f. skilled, workmanlike.

artesano, m. craftsman, skilled worker.

artesianola, a. artesian; **acuífero (m) a.,** artesian

aquifer; **altura (f) a.**, artesian head; **cuenca (f) a.**, artesian basin; **flujo (m) a.**, artesian flow; **manantial (m) a.**, artesian spring; **pozo (m) a.**, artesian well; **presión (f) a.**, artesian pressure.
árticola, a. arctic; **Océano (m) á.**, Arctic Ocean.
articulación, f. joint (Anat); articulation (pronunciación); **a. universal**, universal joint.
articuladola, a. articulate, hinged.
artículo, m. article, item.
artrópodo, m. arthropod; **artrópodos (Zoo, filum)**, Arthropoda (pl).
artrópodola, a. arthropod.
arveja, f. pea.
asbesto, m. asbestos; **a.-cemento**, asbestos cement, fibrocement; **techo (m) de a.**, asbestos roofing.
asbestosis, m. asbestosis.
ascáride, m. roundworm.
ascenso, m. rise; **a. capilar**, capillary rise; **a. salino**, saline rise.
ascensor, m. lift; elevator (US).
asensión, f. asension, rise, elevation; **a. continental**, continental rise.
asentamiento, m. settlement, settling; slumping (Geol); **a. humano**, human settlement; **a. rural**, rural settlement; **balsa (f) de a.**, settlement basin.
asentar, v. to settle, to seat.
asépticola, a. aseptic.
aserradero, m. sawmill, timber yard.
aserrador, m. sawyer, sawer.
aserradora, f. power saw.
asesor, m. y f. advisor; **a. jurídico**, legal advisor; **biólogo a.**, biological advisor.
asesoramiento, m. advice, opinion; advising (acción); **a. técnico**, technical advice.
asesorar, v. to counsel, to advise.
asesoría, f. consultancy; **a. de ingeniería, a. ingenieril**, engineering consultancy.
asexual, a. asexual; **reproducción (f) a.**, asexual reproduction.
asfálticola, a. asphalt, asphaltic; **cemento (m) a.**, asphalt cement.
asfalto, m. asphalt; **cubierta (f) de a.**, asphalt roofing.
asfixia, f. asphyxia.
asfixiadola, a. suffocated, asphyxiated.
asfixiante, a. suffocating, asphyxiating; **gas (m) a.**, suffocating gas, suffocating fumes (pl).
asfixiar, v. to asphyxiate.
Ashgiliense, m. Ashgillian.
asiáticola, a. Asiatic.
asignación, f. assignment, allocation; **a. de fondos**, allocation of funds.
asimetría, f. asymmetry, skewness; **coeficiente (m) de a.**, coefficient of skewness.
asimilación, f. assimilation.
asimilar, v. to assimilate.
asimililador, m. digester (Trat); **a. de fangos de alcantarilla**, sewage sludge digester.
asíntota, f. asymptote.

asintóticola, a. asymptotic.
asma, f. asthma.
asmáticola, a. asthmatic.
asnal, a. of asses or donkeys.
asociación, f. association, society; **a. de fauna**, faunal assemblage; **a. profesional**, professional association; **a. vegetal**, plant association.
asociado, m. associate, partner (Com).
asociadola, a. associated, associate.
asociar, v. to associate.
asolamiento, m. devastation.
asolar, v. to ruin, to destroy.
asomo, m. exposure (Geol); **a. granítico**, granitic outlier.
aspecto, m. aspect; **a. económico**, economic aspect; **aspectos culturales**, cultural aspects; **aspectos de la salud**, health aspects; **aspectos sociales**, social aspects; **aspectos socioeconómicos**, socioeconomic aspects.
aspereza, f. roughness (of surface).
asperón, m. grit, sandstone.
aspersión, f. spraying, sprinkling.
aspillera, f. loophole (Mil).
aspiración, f. aspiration, breathing (Anat); aspiration, intake (entrada); **a. de la bomba**, pump aspiration, pump intake.
aspiradorla, m. aspirate; **bomba (f) a.**, suction pump.
aspirante, a. suction (de succión); aspiring (Fig).
aspirar, v. to inhale, to breathe in.
aspirina, f. aspirin.
asquerosola, a. filthy, obscene.
astato, m. astatine (At).
astenosfera, f. asthenosphere.
asteroide, m. asteroid.
asteroideos, m.pl. Asteroidea (Zoo, clase).
astilla, f. splinter.
astillamiento, m. chipping.
astillar, v. to splinter.
astillero, m. boatyard, dockyard.
asunto, m. theme, subject, subject matter.
ataguía, f. cofferdam.
ataque, m. attack; **a. de los silicatos**, silicate attack; **a. químico**, chemical attack.
atascado, m. clogging.
atascar, v. to clog up, to plug.
atascarse, v. to get stuck, to get bogged down.
atasco, m. blockage, obstruction; traffic jam (circulación).
atenuación, f. attenuation; **a. de lixiviado**, leachate attenuation; **a. natural**, natural attenuation.
atenuar, v. to attenuate; **política (f) de a. y dispersar (para vertederos)**, attenuate and disperse policy.
atenuarse, v. to thin out (Geol).
aterrada, m. (Br) waste tip, garbage dump, landfill.
aterrazamiento, m. terracing.
aterrazar, v. to terrace.

ático, m. attic (desván), penthouse (apartamento).
atípicola, a. atypical.
atlánticola, a. Atlantic; **el Océano (m) a.**, the Atlantic Ocean.
atlas, m. atlas.
atmósfera, f. atmosphere.
atmosféricola, a. atmospheric.
atolón, m. atoll; **a. coralino**, coral atoll.
atómicola, a. atomic; **bomba (f) a.**, atom bomb, atomic bomb; **masa (f) a.**, atomic mass; **número (m) a.**, atomic number; **partícula (f) a.**, atomic particle; **unidad (f) de masa a. (abr. uma)**, atomic mass unit (abr. amu).
átomo, m. atom.
átomo-gramo, m. gram-atom.
atorón, m. traffic jam (circulación).
ATP, m. ATP; **trifosfato (m) de adenosina**, adenosine triphosphate.
atrapaniebla, f. fog collector.
atravesar, v. to pass through, to go through.
atrazina, f. atrazine.
atrición, f. attrition.
atrio, m. atrium.
atrofia, f. atrophy.
atún, m. tuna.
audible, a. audible; **zona (f) a.**, audible range.
audición, f. hearing (Med).
audiovisual, m. y a. audiovisual.
auditar, v. to audit.
auditor, m. auditor.
auditoría, f. audit, auditing; **a. ambiental, a. medioambiental**, environmental audit; **pista (f) de a., registro (m) de a.**, audit trail.
augita, f. augite.
aulaga, f. gorse.
aumentar, v. to augment, to increase; **a. en espesor**, to thicken.
aumento, m. increase, rise (incremento); enhancement (mejora); augmentation; **proyecto (m) para el a. del río**, river augmentation scheme.
aura, f. turkey vulture, turkey buzzard.
aureola, f. aureole; **a. de contacto**, contact aureole.
auríferola, a. auriferous, gold-bearing.
aurora, f. aurora; **a. boreal**, aurora borealis, the Northern Lights.
ausencia, f. absence.
ausentismo, m. absenteeism.
austral, m. austral (Arg, unidad monetaria 1985-91).
Australia, f. Australia.
australianola, a. Australian.
autecología, f. autecology.
autillo, m. owl.
auto, m. car, motorcar.
auto-, pf. auto-.
autoabastecimiento, m. self-supply, self-sufficiency.
autoabsorción, f. autoabsorption.
autobús, m. bus.

autoclásticola, a. autoclastic.
autoclasto, m. autoclast.
autoclave, m. autoclave (Quím).
autocorrelación, f. autocorrelation.
autóctonola, a. indigenous (nativo), autochthonous.
autodepuración, f. self-purification (p.e. de ríos); **a. biológica (p.e. en ríos)**, biological self-purification.
autoevaluación, f. self-assessment.
autofecundación, f. self-fertilization.
autofinanciadola, a. self-financing.
autogamia, f. autogamy.
autogeneración, f. autogeneration, self-regulation; **a. y conversión energética**, waste-to-energy.
autolimpiable, a. self-cleaning.
automáticola, a. automatic.
automatización, f. automation.
automatizar, v. to automate.
automización, f. automization.
automóvil, m. automobile, car, motorcar.
autónomola, a. self-governing, autonomous; **comunidad (f) a., (Esp)**, autonomous region.
autopista, f. motorway; freeway; super highway (US); **a. de peaje**, toll-paying motorway.
autopolinización, f. self-pollination.
autoprotección, f. self-protection.
autopsia, f. post-mortem, post-mortem examination.
autoridad, f. authority; **a. portuaria**, port authority; **a. reguladora**, regulatory authority; **tener a. para (v)**, to have authority to.
autoritariola, a. authoritative.
autorización, f. authorization.
autorizadamente, ad. authoritatively, with authority.
autorizadola, a. authorized.
autorizar, v. to authorize, to approve.
autorregulación, f. self-regulation.
autorreguladorla, a. self-regulating.
autostop, m. hitch-hiking; **hacer a.**, to hitch-hike.
autostopista, m. y f. hitch-hiker.
autosuficiencia, f. self-sufficiency.
autotróficola, a. autotrophic.
autótrofo, m. autotroph.
autótrofola, a. autotrophic.
autovía, f. dual-carriageway.
auxiliar, a. auxiliary, ancillary.
avalancha, f. avalanche.
avance, m. advance, breakthrough; **a. gradual eólico (Geog)**, wind encroachment; **el último a.**, the latest development.
avanzo, m. balance sheet (Com).
ave, f. bird; **a. acuática**, water bird; **a. corredora**, flightless birds; **a. de caza**, game fowl, game bird; wildfowl (normalmente acuático); **a. de corral**, poultry, farmyard fowl; **a. de paso**, bird of passage, migratory bird; **a. de presa, a. de rapiña**, bird of prey; **a.**

marina, sea bird; **a. pasajera,** bird of passage, migratory bird; **a. zancuda,** wader; **aves (Zoo, clase),** Aves (pl).

avellana, f. hazelnut.

avellanar, v. to countersink.

avellano, m. hazel (tree).

avena, f. oat, oats (pl).

avenamiento, m. river drainage; **a. dendrítico,** dendritic drainage; **diseño (m) de a.,** drainage pattern; **hoya (f) de a.,** drainage basin.

avenar, v. to drain (Agr).

avenida, f. avenue (calle); flood, spate (de ríos); **a. anual (flujo máximo en un año),** annual flood; **a. de río,** river spate; **a. repentina,** flash flood; **caudal (m) de a.,** flood flow.

avería, f. breakdown; **localización (f) de averías, reparación (f) de averías,** troubleshooting.

averiguar, v. to find out, to ascertain, to check.

avícola, f. poultry.

avifauna, f. avifauna, bird life.

avión, f. airplane.

avisador, m. notifier.

aviso, m. warning, announcement; **a. de inundación,** flood warning; **sin previo a.,** without warning, without notice; **sistema (m) de a.,** warning system.

avispa, f. wasp.

avitaminosis, f. vitamin deficiency.

avituallamiento, m. provisioning, supplying.

avogadro, m. avogadro (Quím).

avulsión, f. avulsion.

axila, f. axil (Bot); armpit (sobaco).

axioma, m. axiom.

axiomáticola, a. axiomatic.

ayllu, m. Amerindian family commune (caserío indio).

ayuda, f. help, assistance; **a. económica,** economic aid.

ayudante, m. assistant.

ayudar, v. to help, to assist.

ayuntamiento, m. town hall (edificio); town council, city council (consejo).

azadón, m. large hoe, mattock.

azafrán, m. saffron.

azagaya, f. spearhead (lanza).

azar, m. chance, random; **al a.,** at random.

azeótropo, m. azeotrope.

azida, f. azide.

azúcar, m. y f. sugar; **a. de caña,** cane sugar; **a. de la fruta,** fruit sugar.

azucarera, f. sugar mill.

azud, m. (Arab) flow retention weir, diversion dam; **a. compuesto,** compound weir; **a. para peces,** fish pass.

azuela, f. adze, adz (hacha).

azufradola, a. sulphonated, sulfonated (US).

azufrar, v. to sulphurate, to sulfurate (US).

azufre, m. sulphur (S).

azufrosola, a. sulphorous, sulfurous (US).

azul, 1. m. blue; **a. índigo,** indigo blue; 2. a. blue.

azulejo, f. glazed tile.

azurita, f. azurite.

B

bacaladero, m. cod-fishing boat, codfisherman.
bacalao, m. cod.
bacaldilla, f. blue whiting.
bacilo, m. bacillus (pl. bacilli).
bacteria, f. bacterium (pl. bacteria).
bacterianola, a. bacterial.
bacterias, f.pl. bacteria; **b. aerobias**, aerobic bacteria; **b. anaerobias**, anaerobic bacteria; **b. de azufre**, sulphur bacteria; **b. del suelo**, soil bacteria; **b. de nitrato**, nitrate bacteria; **b. de nitrito**, nitrite bacteria; **b. desnitrificantes**, denitrifying bacteria; **b. entéricas**, enteric bacteria; **b. fijadoras de nitrógeno**, nitrogen-fixing bacteria; **b. fotosintéticas**, photosynthetic bacteria; **b. nitrificantes**, nitrifying bacteria; **b. purpúreas**, purple bacteria; **b. quimiosintéticas**, chemosynthetic bacteria.
bactericida, 1. m. bactericide; 2. a. bactericidal.
bacteriófago, m. bacteriophage.
bacteriología, f. bacteriology.
bacteriológicola, a. bacteriological.
bacteriólogola, m. y f. bacteriologist.
bacteriostáticola, a. bacteriostatic.
bache, m. hole, pothole (in the road).
baén, m. pasture watered by groundwater seepage.
bahía, f. bay.
bajada, f. alluvial outwash cone or fan.
bajamar, f. low tide, low water.
bajial, m. flood plain, lowland.
bajío, m. shoal, sandbank (en el mar); lowland (LAm).
bajola, a. low, short; **b. calidad**, low grade; **b. relieve**, low relief; **Los Países (m.pl) Bajos**, the Low Countries, the Netherlands; **marea (f) baja**, low tide; **tierras (f) bajas**, lowlands.
Bajociense, m. Bajocian.
baladí, a. trivial, insignificante.
balance, m. balance; balance sheet (Com); **b. de agua en el suelo**, soil water balance; **b. de energía**, energy balance; **b. de humedad**, moisture balance; **b. ecológico**, ecological balance; **b. energético**, energy balance; **b. hídrico**, water balance; **b. hidrodinámico**, hydrodynamic balance; **b. iónico**, ionic balance; **b. químico**, chemical balance; **hacer (v) b.**, to take stock.
balanceo, m. swinging, rocking.
balancín, f. beam (de máquina), seesaw (subibaja); halteres (pl) (Zoo); **máquina (f) de b.**, **motor (m) de b.**, beam engine.
balanza, f. balance, scales (para pesar); **b. analítica**, analytical balance; **b. comercial**, balance of trade; **b. de pagos**, balance of accounts; **b. de resorte**, spring balance.
balasto, m. ballast.

balate, m. terrace field wall in southern Spain.
balaustrada, f. ballustrade.
balboa, m. (Pan) balboa (unidad monetaria).
balcón, m. balcony.
balde, m. bucket, pail; rubbish bin (GB); garbage can (US).
baldío, m. wasteland, uncultivated land.
baldosa, f. floor tile.
balizar, v. to mark with buoys.
balneario, m. spa.
balneariola, a. spa; **pueblo (m) b.**, spa town.
balsa, f. raft; basin, pond, reservoir; ferry (US) (barco de transporte); **b. de decantación**, settling basin; **b. de depuración**, purification basin; **b. de evaporación**, evaporation basin; **b. de infiltración**, infiltration basin; **b. de recarga**, recharge basin.
balsón, m. (LAm) swamp, bog.
baluarte, m. bulwark.
ballena, f. whale; **b. azul**, blue whale.
ballenerola, a. whaling; **industria (f) b.**, whaling industry.
ballicar, m. wet depression in pasture.
bambú, m. bamboo.
bananola, m. y f. banana.
bancal, m. bank of high irrigation channel.
bancarrota, f. bankrupcy (Fin).
banco, m. bank, bed (Geog); bank (Com); shoal (Geog); school (de peces); bench, desk (asiento, mesa); oxbow lake (US); **b. costero**, longshore spit; **b. de arena**, sandbank; **b. de ceniza**, ash bed; **b. de coral**, coral reef; **b. de turba**, peat bed; **b. encorvado (Geog)**, hook; **b. incompetente (Geol)**, incompetent bed; **B. Mundial (Com)**, World Bank; **b. submarino**, submarine bar.
banda, f. band, ribbon; strip (de tierra); lane (de carreteras); **b. de frecuencia**, waveband; **la B. Oriental**, Uruguay (Arg, Chi, Uru).
bandada, f. flock (of birds), shoal (de peces).
bandeadola, a. banded.
bandejón, m. (LAm) floe, ice floe.
banquero, m. banker.
banquisa, f. pack ice.
baña, f. animal watering place.
bañado, m. (LAm) swamp, marshland; extensive semi-permanent wetland (Arg); bath; **b. de remanso**, backwater swamp.
bañil, m. animal watering pool.
baquelita, f. bakelite.
bar, m. bar (unidad de medida de presión).
barandilla, f. guardrail, handrail; sill, apron (de construcción).
barata, f. cockroach (Ch); bargain (Méx).
barba, f. beard, chin (de la cara); barb (Bot); barb (de plumas, de flechas); varve (Geol).

barbacana, f. embrasure, barbican (defensa).

barbadola, a. varved; **arcilla (f) b.,** varved clay.

barbecho, m. fallow land (Agr); in preparation (CSur) (Fig); **dejar en b. (v),** to leave fallow.

barbilla, f. barb; tenon (Ing).

barbitúrico, m. barbiturate.

barcaza, f. barge.

bárcena, f. cultivated riverside.

barchán, m. barchan, barkhan.

barco, m. boat, ship, vessel; shallow depression caused by groundwater piping; **b. aljibe,** tanker; **b. cisterna,** tanker; **b. de carga,** cargo boat, merchant ship; **b. de guerra,** warship; **b. del práctico,** pilot boat; **b. de pasajeros,** passenger liner, liner; **b. de recreo,** pleasure boat; **b. de vela,** sailing boat; **b. mercante,** cargo boat, merchant ship.

bardoma, f. (Arg) flood deposit.

báricola, a. baric; **cloruro (m) b.,** barium chloride.

bario, m. barium (Ba).

barita, f. barite.

barlovento, m. windward; **banda (f) de b.,** windward side.

barniz, m. varnish.

barnizar, v. to varnish.

barógrafo, m. barograph.

barómetro, m. barometer; **b. aneroide,** aneroid barometer; **b. de mercurio,** mercury barometer.

barra, f. bar (Gen); rod, stick (palo); **b. arenosa,** sandbar; **b. colectora (Elec),** bus bar; **b. costera,** sandbar; **b. en cúspide,** cuspate bar; **b. en desembocadura de río,** rivermouth bar; **b. en lazo,** looped bar; **cinta (f) de barras,** bar screen; **playa (f) de b.,** barrier beach.

barranca, f. (Méx) ravine, gully.

barranco, m. barranco, bluff, ravine; chine (garganta), canyon.

barranquera, f. (southern Spain) top end of ravine.

barraquismo, m. (LAm) the problem of shanty towns.

barreal, m. (Arg) alluvial fans at confluence of desert dry valleys (wadis).

barredora, f. sweeper (machine).

barreduras, f.pl. sweepings, rubbish, refuse.

Barremiense, m. Barremian.

barrena, f. drill, drill bit, gimlet.

barrenar, v. to bore, to drill.

barrendero, m. (Bol, C.Rica, ElS, Par, Ur) waste tip operative, garbage man.

barrenderola, m. y f. cleaner, sweeper (persona).

barreno, m. drill (máquina); borehole, bore; blasthole (Min); **b. helicoidal,** auger.

barrera, f. barrier, barricade, roadblock; boom (p.e. a través de puerto); **b. antigás,** gas barrier; **b. de bombeo,** pumping barrier; **b. de contención,** retaining wall; **b. de intrusión marina,** marine intrusion barrier; **b. de inyección,** injection barrier; **b. del sonido,** sound barrier; **b. física,** physical barrier; **b. impermeable,** impermeable barrier.

barrero, m. claypit; **b. lleno de agua,** waterfilled claypit.

barriada, f. (Pan, Perú) slum, shanty town.

barrial, m. (Perú) highly-fertile soils.

barrido, m. sweeping; **b. mecánico,** mechanical sweeping; **b. y limpieza de vías,** street sweeping and cleaning.

barril, m. barrel (igual a 159 litros de petróleo).

barrio, m. district, residential district, area; neighborhood (US); **b. marginal (LAm),** slum, squatter settlement.

barro, m. clay, mud; **b. cocido,** baked earth.

barruntar, m. (central España) ground with water at shallow depth.

Bartoniense, m. Bartonian.

basal, a. basal.

basalto, m. basalt.

basamento, m. basement, base structure; **rocas (f) del b.,** basement rocks.

báscula, f. scales (pl) (balanza); weighbridge (para vehículos).

basculamiento, m. dumping, tipping.

basculante, m. tipper, dumper.

base, f. base, basis; **b. aérea,** air base; **b. de cálculo,** basis of calculation; **b. de comparación,** basis of comparison; **b. de datos,** database; **b. de datos informática,** computer database, computerized database; **b. del acuífero,** base of aquifer; **b. química,** chemical base; **sentar las bases de (v),** to lay the foundation for.

basicidad, f. basicity.

básicola, a. basic.

basidiomiceto, m. basidiomycete.

basifilola, basófilola, a. basiphilic.

basito, m. small spring emerging between rocks.

basura, f. rubbish, refuse; trash, garbage (US); litter (en el campo, en la calle); **b. doméstica,** domestic waste, domestic rubbish; **b. municipal,** municipal waste; **recogida (f) de b.,** rubbish collection, refuse collection, garbage collection (US); **rejilla (f) para la b.,** waste screen, trash screen.

basural, m. (Arg, Ch, Ur) rubbish dump, waste tip.

basureigo, m. (Col) waste tip scavenger.

basurero, m. dustman, garbage man; waste tip operative (Arg, Ch, Méx, Nic, Par, Perú, Ur); rubbish dump (Esp, Bol, Col, ElS, Nic, Par, Ven) (vertedero); litter bin, trash can (US).

basurientola, a. (LAm) full of rubbish.

batán, m. (Arab) waterwheel for raising water using river flow.

bateado, m. panning (Geol); **b. para oro,** gold panning.

batería, f. battery; **b. de acumuladores,** storage

battery; **b. de pozos,** group of wells; **b. seca,** dry battery.

Bathoniense, m. Bathonian.

batial, a. bathyal.

batilimnio, m. bathylimnion.

batimetría, f. bathymetry.

batimétricola, a. bathymetric.

batir, v. to beat, to strike; to beat, to flap (alas de ave); to rinse (LAm) (aclarar la ropa); to strike, to hammer (metal).

batume, m. (Par) masses of floating aquatic vegetation.

bauxita, f. bauxite.

baya, f. berry.

bayou, m. (US) ponded back tributary river.

Beaufort, m. Beaufort; **escala (f) de B.,** Beaufort Scale (para medida del viento).

bebida, f. beverage, drink; **b. refrescante,** soft drink; **b. sin alcohol,** non-alcoholic drink; **bebidas envasadas,** bottled beverages, bottled drinks.

becquerel, m. becquerel (abr. Bq).

BEI, m. IEB; **Banco (m) Europeo de Inversiones,** European Investment Bank.

bejuco, m. reed, rush, cane.

belemnite, m. belemnite.

bellota, f. oak apple.

benceno, m. benzene.

bencenocarbaldehído, m. benzenecarbaldehyde.

bencina, f. petrol; gas, gasoline (US); gasolene (US, Ch).

bencinera, f. petrol station; (US, Ch) gas station.

beneficio, m. benefit, profit, equity; **b. ambiental,** environmental benefit; **b. anual,** yearly profit; **b. bruto,** gross profit; **b. directo,** direct benefit; **b. netos,** net benefit; **b. social,** social benefit; **costo-b. (m),** cost benefit, cost-benefit; **margen (m) de b.,** profit margin.

bentónicola, a. benthic; **flora (f) y fauna (f) bentónicas,** benthic flora and fauna.

bentonita, f. bentonite; **sello (m) de b. en un piezómetro,** bentonite seal in a borehole.

bentos, m. benthos, benthic organisms.

benzaldehído, m. benzaldehyde.

benzoílo, m. benzoyl.

benzol, m. benzene.

berbiquí, m. brace, carpenter's brace; **b. y barrena,** brace and bit.

berenjena, f. aubergine, eggplant.

berenjenal, m. aubergine field, eggplant field.

beriberi, m. beriberi.

berilio, m. beryllium (Be).

berilo, m. beryl.

berma, f. berm, raised river bank.

Berriasense, m. Berriasian.

berro, m. cress, watercress; **prado (m) de berros,** watercress bed, cressbed.

berza, f. cabbage.

berzal, m. cabbage field, cabbage patch.

besugo, m. sea bream.

beta, f. beta.

betarraga, f. (Ch, Perú) beet, beetroot.

betavel, m. (Méx) beetroot.

bi-, pf. bi-.

biáxicola, a. biaxial.

biaxial, a. biaxial.

bicarbonato, m. bicarbonate; **b. de calcio,** calcium bicarbonate; **b. de sodio,** sodium bicarbonate.

bichicome, m. (Ur) waste tip scavenger.

bichos, m.pl. vermin, pests (plagas); bugs (Zoo, hemípteros).

bichucon, m. (LAm) solid waste segregator.

bicicleta, f. bicycle, cycle; bike (Fam).

biconvexola, a. biconvex.

bicromato, m. dichromate.

BID, m. IDB; **Banco (m) Interamericano de Desarrollo,** Inter-American Development Bank.

bidimensional, a. two-dimensional.

bidimensional, a. two-dimensional.

bidón, m. can, drum (envase).

bienal, a. biennial; **planta (f) b.,** biennial, biennial plant.

bienestar, m. welfare, well-being; **b. social,** social welfare; **estado (m) de b. social,** welfare state.

bifurcación, f. bifurcation, diversion; **relación (f) de b.,** bifurcation ratio.

bifurcadola, a. bifurcated.

bilarciasis, f. bilharzia.

bilateral, a. bilateral; **simetría (f) b.,** bilateral symmetry.

bilineal, a. bilinear.

billón, m. billion (millón de millones). (Note: en los EEUU y GB billón es equivale a mil millones).

bimodal, a. bimodal.

binariola, a. binary; **zona (f) b.,** binary scale.

binoculares, f.pl. binoculars.

binomial, a. binomial; **distribución (f) b.,** binomial distribution; **ecuación (f) b.,** binomial equation.

binómica, a. binomial; **distribución (f) b.,** binomial distribution.

bio-, pf. bio-.

bioabsorción, f. biosorption.

bioacumulación, f. bioaccumulation.

bioaireación, f. bio-aeration.

biobarrera, f. bio-barrier.

biocenología, f. biocenology.

biocenosis, f. biocoenosis, biocenosis.

biocéntricola, a. biocentric.

biocida, m. biocide.

biocilindro, m. biodisc (Trat).

bioclásticola, a. bioclastic.

bioclasto, m. bioclast.

biodegradabilidad, f. biodegradability.

biodegradable, a. biodegradable.

biodegradación, f. biodegradation; **b. anaerobia,** anaerobic degradation.

biodigestor, m. biodigester.

biodisco, m. biodisc (Trat).

biodisponibilidad, f. bioavailability.
biodiversidad, f. biodiversity.
bioenergética, f. bioenergetics.
bioensayo, m. bioassay, biological assay.
biofacies, f.pl. biofacies.
biofilm, m. biofilm.
biofiltro, m. biofilter.
biofísica, f. biophysics.
biofloculación, f. bioflocculation.
biogás, m. biogas.
biogénesis, f. biogenesis.
biogenéticola, a. biogenetic.
biogénicola, a. biogenic.
biogeografía, f. biogeography.
bioindicador, m. bioindicator.
biología, f. biology; **b. marina,** marine biology.
biológicola, a. biological; **capa (f) b.** (Trat), biological film; **control (m) b.** (Agr), biological control; **filtro (m) b.** (Trat), biological filter.
biólogola, m. y f. biologist.
bioluminescente, a. bioluminescent.
bioluminiscencia, f. bioluminescence.
bioma, m. biome.
biomagnificación, f. biomagnification.
biomasa, f. biomass.
biometría, f. biometry.
bioquímica, f. biochemistry.
bioquímicola, 1. m. y f. biochemist; 2. a. biochemical; **atenuación (f) b.,** biochemical attenuation; **demanda (f) b. de oxígeno (abr. DBO),** biochemical oxygen demand (abr. BOD).
bioreactor, m. bioreactor.
bioremediación, f. bioremediation; **b. en el sitio,** in situ bioremediation, on site bioremediation; **b. intrínseca,** intrinsic bioremediation.
biorritmo, m. biorhythm.
biosfera, f. biosphere; **Reservas (f.pl) de la B.,** Biosphere Reserves.
biosistema, m. biosystem, ecosystem.
biostratigrafía, f. biostratigraphy.
biostromo, m. biostrome.
biota, f. biota.
biotecnología, f. biotechnology.
bióticola, a. biotic; **factor (m) b.,** biotic factor; **índice (m) b.,** biotic index.
biotipo, m. biotype.
biotita, f. biotite.
biótopo, m. biotope.
biotransformación, f. biotransformation.
biotratamiento, m. biological treatment.
bioturbación, f. bioturbation.
bioventilación, f. bioventing.
birrefringencia, f. birefringence.
bisector, m. bisector, bisectrix.
bisectriz, f. bisector, bisectrix, bisecting line;.
bisel, m. bevel, bevel edge.
biseladola, a. bevelled.
bisexual, a. bisexual.
bismuto, m. bismuth (Bi).
bisulfato, m. bisulphate.

bitumen, m. bitumen.
bituminosola, a. bituminous.
bivalente, a. bivalent.
bivalvo, m. bivalve.
bivalvola, a. bivalve.
blancura, f. whitening.
blandola, a. soft; **agua (f) b.,** soft water.
blanqueador, m. brightening agent, brightener, whitening agent, whitener; **b. óptico,** optical brightener; **b. químico,** chemical brightening agent.
blanquear, v. to bleach.
blanqueo, m. whitener (sustancia); bleaching (acción); **b. de papel,** paper bleaching.
blastoide, m. blastoid.
blenda, f. zinc blende, sphalerite.
bloque, m. block; **b. de apartamentos,** block of flats (GB), apartment block (US); **b. de falla,** block fault; **b. de pisos (apartments),** blocks of flats; **b. errático de arenisca silicea,** sarsen, sarsen stone; **b. fallado,** block-faulted; **b. rift,** rift block; **el b. asiático,** the Asiatic block; **en b.,** en bloc.
bloque, m. block; **relieve (m) de bloques,** block relief.
boa, f. boa, boa constrictor.
boca, f. mouth (de un animal, de un río); pincers (pl) (de crustáceos); lip (de un vasija); jaws (Zoo, Ing); opening (de un horno); **ancho de b.,** wide-mouthed (p.e. de un vaso); **b. de alcantarilla,** sewer inlet; **b. de incendio,** fire hydrant; **b. de metro,** metro/tube/underground entrance, subway entrance (US); **b. de riego,** hydrant; **b. de río,** river mouth, mouth of river.
bocamina, f. pithead, mine entrance.
bocana, f. inlet (de río), estuary; **b. de marea,** tidal inlet.
bocio, m. goiter (Med).
bodonal, m. boggy ground.
bofadal, m. (Arg) pastureland with lakes.
bohío, m. (CAm) shanty town.
boletín, m. bulletin, certificate; **b. informativo,** news bulletin.
boliche, m. small furnace; dragnet (red); skittles (juego de bolos); snack bar CSur).
bolichera, f. (Perú) fishing boat.
bolita, f. pellet; **b. fecal,** faecal pellet.
bolívar, m. (Ven) bolivar (unidad monetaria).
Bolivia, f. Bolivia.
boliviano, m. (Bol) boliviano (unidad monetaria).
bolivianola, 1. m. y f. Bolivian; 2. a. Bolivian.
bolsa, f. bag, purse; stock exchange, stock market (Com); **b. de aire,** air pocket; **b. de basura,** rubbish bag, rubbish sack; **b. de cereales,** corn exchange; **b. subterránea,** underground pocket of water (aquifer).
bolsada, f. ore pocket (Geol).
bolsón, m. (US) evaporite deposit in fault-block depression.
bomba, f. pump (p.e. para agua), bomb

(explosiva); **b. aspirante,** suction pump; **b. atómica,** atom bomb, atomic bomb; **b. centrífuga,** centrifugal pump; **b. contra incendios,** fire pump; **b. de aire comprimido,** air-lift pump; **b. de barrido,** scavenger pump; **b. de desplazamiento positivo,** positive-displacement pump; **b. de diafragma,** diaphragm pump; **b. de doble efecto,** double-action pump, reciprocating pump; **b. de flujo axial,** axial flow pump; **b. de hidrógeno,** hydrogen bomb; **b. de inyección,** injection pump; **b. de mano,** hand pump; **b. de refuerzo,** booster pump; **b. de relojería,** time bomb; **b. de succión,** suction pump; **b. de superficie,** surface pump; **b. de triple efecto,** triple action pump; **b. dúplex,** duplex pump; **b. eólica,** windpump; **b. H,** H-bomb; **b. oscilante,** reciprocating pump; **b. peristáltica,** peristaltic pump; **b. por fuerza,** ram pump; **b. rotatoria,** rotary pump; **b. sumergible,** submersible pump; **b. volcánica,** volcanic bomb; **casa (f) de bombas,** pumphouse.

bombeo, m. pumping (p.e. de agua); bulge, convexity (convexidad); warp (alabeo); **b. a caudal constante,** constant-rate pumping; **b. intermitente,** intermittent pumping; **b. por aire comprimido,** pumping with compressed air; **b. por interferencia,** interference pumping; **ensayo (m) de b.,** pumping test; **estación (f) de b.,** pumping station; **etapas (f.pl) de b.,** **niveles (m.pl) de b.,** pumping steps.

bombero, m. fireman; **coche (m) de bomberos,** fire engine; **cuerpo (m) de bomberos,** fire brigade (GB), fire department (US).

bombilla, f. electric light bulb.

bongo, m. dug-out canoe, barge.

boniato, m. sweet potato, yam.

bonito, m. bonito, Atlantic bonito, striped tunny.

booleanola, a. boolean (Mat); **álgebra (f) b.,** boolean algebra.

boquera, f. (southern Spain) rock weir to divert riverflow into irrigation channel.

boquerón, m. anchovy.

boquete, m. hole (agujero); narrow opening (paso estrecho); gap, breach (brecha).

borano, m. borane.

borato, m. borate.

bórax, m. borax.

borde, m. boundary; **al b. de,** on the edge of; **b. del mar,** seaside, seashore; **b. delgado,** feather-edge; **b. de placa,** plate boundary; **b. enfriado (Geol),** chilled margin; **b. rectangular,** rectangular boundary.

bordo, a. board (Mar); **ordenador (m) de a b. (Mar),** onboard computer.

bóricola, a. boracic, boric.

bornita, f. bornite.

boro, m. boron (B).

borrador, m. draft (de escribir).

borrasca, f. squall, rough weather.

borreguil, m. mountain peat bog.

bosque, m. wood, woodland; forest (selva); **b. de clima templado,** temperate forest; **b. denso,** forest, dense woodland; **b. esclerófilo,** sclerophyllous forest; **b. seminatural,** seminatural woodland; **b. virgen,** virgin forest.

bosquecillo, m. copse, coppice.

botadero, m. rubbish dump; ford (Bol, Col, C.Rica, Ec, ElS, Nic, Pan, Perú, Ven); **b. a cielo abierto,** open dump.

botánica, f. botany.

botánicola, a. botanical.

botanista, m. y f. botanist.

botar, v. to launch (a boat).

bote, m. jump, leap (salto); can, tin can; boat (Mar).

boteal, m. area with abundant springs.

botella, f. bottle; **b. Winckler,** Winckler bottle.

botellero, m. (Col) waste tip scavenger.

botulismo, m. botulism.

bóveda, f. canopy (de un bosque); vault, cavity (Con).

bóveda, f. vault; **b. de arista,** cross-vaulting; **b. de cañón,** barrel arch; **b. de medio punto,** barrel vault.

bovino, m. bovines (vacas, etc.).

boya, f. buoy.

boyante, a. buoyant; **lo b. (m),** buoyancy.

boyar, v. to float.

bráctea, f. bract.

branquia, f. branchia (pl. branchiae); gill (de peces).

branquial, a. branchial.

braquiópodo, m. brachiopod.

Brasil, m. Brazil.

brasileñola, m. y f. Brazilian.

brasileñola, a. Brazilian.

brasilerola, m. y f. Brazilian.

brasilerola, a. Brazilian.

braza, f. fathom (la braza inglesa es 1,83m, la braza española es 1,671m).

brazo, m. arm, limb, branch (of a river); **b. de mar,** arm of sea; **b. de río,** branch of river.

brea, f. pitch, tar.

brecha, f. breccia; **b. de falla,** fault breccia; **b. volcánica,** volcanic breccia.

breña, f. scrub (Bot).

Bretaña, f. Brittany (Francia); **Gran B.,** Great Britain.

breva, f. fig.

brezal, m. heath, moor.

brezo, m. heather, heath.

brillantez, f. brilliance, radiance.

brillo, m. sheen, lustre, shine; **b. del sol,** sunshine; **b. metálico,** metallic lustre; **b. vitreo,** vitreous lustre.

briofita, f. bryophyte; **briofitas (Bot, división),** Bryophyta (pl).

briología, f. bryology.

briozoo, m. bryozoan; **briozoos (Zoo, filum),** Bryozoa (pl).

briozoos, m.pl. bryozoa.

brisa, f. breeze; **b. del mar,** sea breeze; **b. fresca,** fresh breeze; **b. fuerte,** strong breeze; **b. leve,** light breeze.

británicola, 1. m. y f. British citizen; 2. a. British.

brocal, m. curb (p.e. de un pozo).

brochal, m. header beam, trimmed joist.

bromación, f. bromination.

bromato, m. bromate.

bromo, m. bromine (Br).

bromoclorometano, m. bromochloromethane.

bromuro, m. bromide.

bronce, m. bronze.

bronquio, m. bronchus (pl. bronchii).

bronquiolo, m. bronchiole.

bronquitis, f. bronchitis.

brotar, v. to spring, to bud (Bot); to gush, to flow (water).

brote, m. sprout, shoot, bud (de plantas); gushing (agua); outbreak (p.e. de peste).

broza, f. leaf litter, dead wood (mantillo); undergrowth, thicket (matorral); rubbish (basura).

brújula, f. compass (para navegación); **b. prismática,** prismatic compass.

brumosola, a. misty, foggy.

bruscola, a. sharp, sudden; **una curva (f) b.,** a sharp curve.

brutola, a. crude, rough (materiales); gross (peso, Com); uncut (piedra); **hierro (m) b.,** pig iron, crude iron; **peso (m) en b.,** gross weight; **petróleo (m) b.,** crude oil; **producto (m) nacional b. (abr. PNB),** gross national product (abr. GNP).

buceador, m. diver, skin-diver.

bucear, v. to dive, to swim under water.

buceo, m. diving, skindiving; **equipo (m) de b.,** dive team, diving team.

buche, m. crop (de aves); maw (de animales).

bueyada, f. (LAm) drove of oxen or bullocks.

búfalo, m. buffalo.

bufete, m. practice (clientela), lawyer's office (de abogado), writing desk (mesa).

buffer, m. buffer (Quím); **solución (f) b.,** buffer solution.

búho, m. owl; **b. chico,** Long-eared Owl; **b. real,** Eagle Owl.

bujeo, m. boggy ground.

bujía, f. candle (vela); candle power (Fís); spark plug (automóvil).

bulbo, m. bulb.

bulldozer, m. bulldozer, excavator.

bungalow, m. bungalow, chalet.

buque, m. ship, boat; **b. aljibe,** tanker, oil tanker; **b. de cabotaje,** coastal vessel, coaster; **b. de guerra,** warship; **b. de vapor,** steamship; **b. faro,** lightship; **b. mercante,** merchant vessel; **b. transatlántico,** ocean-going vessel.

burbuja, f. bubble.

burbujeo, m. sparging; **b. con aire,** air sparging.

Burdigaliense, m. Burdigalian.

bureta, f. burette, buret (US).

buril, m. burin (Arq, herramienta).

burocracia, f. bureaucracy.

bursátil, a. stock market, stock-exchange.

buscar, v. to search for, to look for; to prospect for (Geol).

buseta, f. (LAm) small bus.

búsqueda, f. search, inquiry.

butadieno, m. butadiene.

butaldehído, m. butaldehyde.

butanal, m. butanal.

butano, m. butane.

butanol, m. butanol.

buteno, m. butene.

butileno, m. butylene.

butilo, m. butyl.

buzamiento, m. dip (Geol); **b. aparente,** apparent dip; **b. hacia abajo,** down-dip; **b. hacia arriba,** up-dip; **b. regional,** regional dip; **b. verdadero,** true dip.

buzandola, a. plunging, dipping (Geol).

buzar, v. to dip, to plunge (Geol).

buzo, m. (C.Rica) waste tip scavenger.

buzón, m. box, mailbox, chamber; **b. de visita,** inspection manhole.

byte, m. byte.

C

CA, f. AC; **comunidad (f) autónoma (Esp)**, autonomous region; **corriente (f) alterna**, alternating current.

cabalgamiento, m. thrust (Geol); **c. de alto ángulo**, high-angle thrust.

caballa, f. mackerel.

caballito, m. small horse; bridge of the nose (US); nappy (US) (diaper); **c. de diablo**, dragonfly.

caballo, m. horse; **c. de vapor (de vehículo), c. inglés**, horsepower.

caballón, m. ridge (Agr, entre surcos).

cabaña, f. cabin, hut; livestock (ganado); **c. nacional**, the nation's livestock.

cabeceo, m. lurch, jolt, shaking (de la cabeza); pitch (Geol).

cabecera, f. head (Gen); seat of honour; headline; headwater; **c. de río**, headwater of river; **obras (f.pl) de c. (Con)**, headworks.

cabeza, f. head; source (LAm) (de un río); **c. de biela (Ing)**, big-end; **c. de línea**, terminus; **c. de partido**, county town; **c. sonora**, recording head, tape head; **dolor (m) de c.**, headache.

cabida, f. space, capacity (Gen); area, extent (terreno).

cabina, f. cabin (p.e. de avión); **c. de medición**, instrument cabin, recorder hut.

cable, m. cable, lead (Elec) (unidad de longitud, 185,19 m); **c. armado**, armoured cable; **c. de alta tensión**, high tension cable; **c. de remolque**, towline; **c. eléctrico**, electrical cable; **c. principal**, main cable, trunking; **c. toma a tierra, c. toma de tierra (Elec)**, earth.

cableado, m. cabling, rigging.

cabo, m. cape, headland (Geog); end, extremity; **C. de Buena Esperanza**, Cape of Good Hope; **C. de Hornos**, Cape Horn.

cabotaje, m. coastal trade.

cabra, f. goat; **c. montés**, mountain goat.

cabrestante, m. capstan, windlass (en un pozo), winch (torno).

cabria, f. shear legs for hoist (Ing).

cacahuate, m. groundnut, peanut.

cacahuete, m. groundnut, peanut.

cacao, m. cacao (árbol y grano); cocoa (en el comercio); chocolate (LAm).

cacaotal, m. cocoa plantation, cacao plantation.

cacería, f. hunting.

cacto, m. cactus.

cactus, m. cactus.

cachalote, m. sperm whale.

cachurero, m. (Ch) waste tip scavenger.

cadañego, m. plant which bears abundant fruit every year.

cadáver, m. corpse, cadaver.

cadena, f. chain; **c. alimentaria**, food chain; **c. de fabricación**, production line; **c. de Markov (Mat)**, Markov chain; **c. de montaje**, production line; **c. de montañas**, mountain chain, mountain range; **c. trófica**, food chain; **llave (f) de c.**, chain tongs; **reacción (f) en c.**, chain reaction.

cadmio, m. cadmium (Cd).

cadozo, m. pothole, deep hole (en río o laguna).

caducidad, f. expiry; **c. del permiso**, expiry of permit, licence expiry; **fecha (f) de c.**, expiry date.

caducifoliola, a. deciduous, broadleaved.

caducola, a. deciduous (de plantas, de árboles); **de hoja (f) c.**, deciduous, broadleaved.

café, m. coffee.

cafeína, f. caffeine.

cafetal, m. coffee plantation.

cahozo, m. (Valencia, Esp) semi-permanent pool, usually forming a chain along a river bed.

caída, f. drop, fall; **c. de hielo**, ice fall; **c. de roca**, rock fall.

caja, f. box, case, crate; **c. de arena**, sand box; **c. de conexión**, junction box; **c. de registro (sobre contador)**, manhole; **c. de velocidades**, gearbox; **c. registradora**, cash till.

cajón, m. crate, chest; caisson (Con).

cal, f. lime; **c. apagada, c. muerta** slaked lime; **c. viva**, quick lime; **horno (m) de c.**, lime kiln.

cala, f. cove, creek; prospecting pit (Min); core (de sondeos).

calabacín, m. marrow.

calabaza, f. pumpkin, gourd.

caladero, m. fishing grounds (pl), fishery.

calafateadura, f. caulking.

calafateo, m. caulking.

calamar, m. squid (pl. squid).

calambre, m. electric shock; cramp (Med).

calamina, f. calamine.

calcarenita, f. calcarenite.

calcáreola, a. calcareous.

calcedonia, f. chalcedony.

cálcicola, a. calcic; **carbonato (m) c.**, calcium carbonate.

calcícola, f. calcicole, lime-lover.

calcíferola, a. calciferous.

calcífuga, f. calcifuge, lime-hater.

calcilutita, f. calcilutite.

calcinar, v. to calcine, to burn, to roast.

calcio, m. calcium (Ca).

calcirrudita, f. calcirudite.

calcita, f. calcite.

calcíticola, a. calcitic.

calco, m. tracing (acción de calcar); copy

(copia) (Fig); **c. de corriente,** flow cast; **c. internal,** internal cast; **papel de c.,** tracing paper.

calcocita, f. chalcosite.

calcodolomita, f. calcdolomite.

calcopirita, f. chalcopyrite.

calculadorla, m. y f. calculator (persona).

calculadora, f. calculator, calculating machine.

calcular, v. to calculate.

cálculo, m. calculation; estimation (conjetura); calculus (Mat); **c. estadístico,** statistical calculation; **c. integral,** calculus; **hoja (f) de c. (Comp),** spreadsheet.

caldas, f.pl. hot springs.

caldera, f. caldera (Geol), boiler, cauldron; **c. de vapor,** steam boiler.

calefacción, f. heating; **c. central,** central heating; **c. solar,** solar heating.

calendario, m. calendar.

calentador, m. heater.

calentamiento, m. heating, warming; **c. geotérmico,** geothermal heating; **c. global,** global warming.

calentar, v. to heat up.

calentura, f. fever, temperature; **tener c. (Med),** to have a temperature.

calera, f. limekiln, lime pit.

caleta, f. small bay, cove, creek.

calgón, m. calgon.

calibración, f. calibration; **ensayo (m) de c.,** calibration test.

calibrador, m. gauge, caliper; **c. micrométrico,** vernier calliper.

calibrar, v. to calibrate.

calibre, m. calibre, caliber (US), bore (tamaño).

caliche, m. caliche, saltpetre, saltpeter (US).

calidad, f. quality; **c. bacteriológica,** bacterial quality, bacteriological quality; **c. de vida,** quality of life; **c. físico-química,** physical-chemical quality; **c. industrial,** industrial quality; **control (m) de c.,** quality control; **norma (f) de c.,** quality standard.

cálidola, a. warm (climate).

caliente, a. warm, hot.

calificación, f. qualification, evaluation, assessment.

calina, f. mist, haze.

cáliz, m. calyx.

caliza, f. limestone; **c. bioclástica,** shelly limestone; **c. coralina,** coral limestone.

callampa, f. (Ch) shanty town.

calma, f. calm, calmness; **c. chicha,** dead calm; **calmas tropicales,** horse latitudes; **zona (f) de calmas ecuatoriales,** doldrums.

calor, m. heat; **c. de combustión,** heat of combustion; **c. de reacción,** heat of reaction; **c. específico,** specific heat; **c. latente de evaporación,** latent heat of evaporation; **c. radiante,** radiant heat; **c. solar,** solar heat.

caloría, f. calorie (abr. cal).

caloríficola, a. calorific; **potencia (f) c., valor (m) c.,** calorific value.

calorífugola, a. heat-resistent.

calorimetría, f. calorimetry.

calorzo, m. (Valencia, Esp) series of intermittent pools along watercourse.

calurosola, a. warm, hot; **un verano (m) c.,** a hot summer.

calvero, m. clearing, glade (en bosque).

calzada, f. road, road surface, carriageway.

callapa, f. (Ch) slum, shanty town.

calle, f. street, road; lane (de autovía); **c. ciega,** dead end street, cul-de-sac; **c. de sentido único,** one-way street; **c. mayor,** high street, main street (US).

callejón, m. alley, narrow street; **c. sin salida,** dead end street, blind alley, cul-de-sac.

Calloviense, m. Callovian.

camada, f. litter (de animales), brood (de aves); layering (Geog); **c. de flujo (Geol),** flow layering.

camalotal, m. (Arg) floating island on river.

camanchacas, f.pl. fog.

cámara, f. chamber, room; camara (cine y televisión); **c. alta/baja,** upper/lower chamber; **c. de combustión,** combustion chamber; **C. de los Comunes,** House of Commons; **C. de los Representantes,** House of Representatives; **c. legislativa,** legislative chamber; **c. séptica (LAm),** septic tank, septic chamber; **cámaras y pilares (Min),** pillar and stall (mining).

cambio, m. change, exchange; **agente (m) de c. y bolsa,** stockbroker; **c. catiónico,** cationic exchange; **c. climático,** climate change; **c. de base,** base exchange; **c. de facies,** change of facies; **c. de política,** change of policy; **c. eustático,** eustatic change; **c. isotópico,** isotopic change; **tipo (m) de c.,** exchange rate.

Cámbrico, m. Cambrian.

cambur, m. banana.

camellón, m. central reservation (GB); median strip (of road) (US).

camino, m. road, route, way, path; **c. crítico,** critical path; **c. de acceso,** approach road; **c. de herradura,** bridleway, bridle path; **c. de hierro (LAm),** railway; **c. de sirga (LAm),** canal towpath; **c. de tierra,** dirt track, dirt road; **c. forestal,** forest track; **c. sin firme,** dirt track, dirt road; **c. trillado (Fig),** the beaten track.

camión, m. lorry, truck; coach, bus (US); **c. cisterna,** road tanker; **c. de carga pesada,** heavy goods vehicle; **c. volquete,** dumptruck.

camioneta, f. van; coach, bus (US).

camote, m. yam, sweet potato.

campaña, f. campaign; countryside (campo).

Campaniense, m. Campanian.

campero, m. jeep, four-wheel drive.

campesino, m. farmer, small farmer.

cámping, m. campsite, camping site; campground (US).

campiña, f. countryside, cultivated land.

campito, m. property.

campo, m. field; country, countryside (país); camping; **biología (f) de c.,** field biology; **c. a través,** cross-country; **c. abierto,** open country, countryside; **c. aurífero,** goldfield; **c. de actividad,** field of activity; **c. de aterrizaje,** landing strip; **c. de aviación,** airfield; **c. de batalla,** battlefield; **c. de bombeo,** pumping field; **c. de deportes,** sports ground, sports field; **c. de filtración,** cultivated fields or pasture irrigated with wastewater; **c. de golf,** golf course, golf links; **c. de hielo,** ice field; **c. de pozos,** well-field; **c. de refugiados,** refugee camp; **c. de trigo,** field of wheat, field of corn (US); **c. electromagnético,** electromagnetic field; **c. petrolífero,** oil field; **c. recreativo,** leisure park; **c. santo,** cemetary, churchyard; **c. visual,** field of vision; **capacidad (f) del c.,** field capacity; **libreta (f) de c.,** fieldbook; **trabajo (m) de c.,** fieldwork.

campus, m. campus.

canal, m. canal, channel, strait (Geog, de agua); duct, tract (Anat); guttering (Con); watercourse; **c. aleviadero,** flood relief channel; **c. de alcantarillado,** drainage canal; **c. de drenaje,** drainage canal; **c. de erupción,** volcanic vent; **c. de marea,** tidal channel; **C. de Panamá,** Panama Canal; **c. de riego,** irrigation channel; **c. suplementario,** relief channel; **canales (Con),** guttering; **capacidad (f) del c.,** channel capacity; **el C. de la Mancha,** the English Channel; **el túnel (m) del C. de la Mancha,** the Channel Tunnel.

canaleta, f. flume.

canalización, f. canalization, channelling, channeling (US); **redes (f.pl) de c.,** pipe networks.

canalizar, v. to channel, to canalize.

cáncer, m. cancer.

cancerígeno/a, a. cancerigenic, cancerogenic.

cancha, f. dump (Min).

canchal, m. bouldery ground, scree.

caneca, f. pot; can, tin can (lata); rubbish bin, trash bin (cubo de basura) (LAm).

cangilón, m. bucket (de rueda hidraúlica).

cantegriles, m.pl. (Ur) shanty town, slum.

cantera, f. quarry; **c. de arena,** sand pit; **c. de caliza,** limestone quarry; **explotación (f) de canteras,** quarrying.

cantería, f. stonework, masonry; quarrying.

cantero, m. quarryman.

cantidad, f. quantity, sum.

cantiléver, a. cantilever; **grúa (f) de c.,** cantilever crane; **puente (m) de c.,** cantilever bridge.

canto, m. edge, corner, rim; pebble (guijarro); **al c.,** in support; **c. rodado,** pebble, boulder.

caña, f. reed, cane; **c. de azúcar,** sugar cane.

cañabrava, f. bamboo.

cañada, f. ravine, gulley, gulch, flume.

caño, m. pipe, conduit (de una fuente); watercourse through mudflats (Ven).

cañaveral, m. reed-bed (Bot), sugar-cane plantation (de azúcar).

cañería, f. piping (tubería), pipeline (tubo); **c. principal (para agua),** trunk main; **fugas en cañerías,** water main leaks.

cañón, m. canyon; cannon, gun barrel; pipe; flue (de chimenea); **c. de vertido,** discharge pipe; **c. submarino,** submarine canyon.

caoba, f. mahogany.

caolín, m. kaolin.

caolinita, f. kaolinite, china clay.

caolinización, f. kaolinization.

caos, m. chaos; **teoría (f) del c.,** chaos theory.

capa, f. layer, band, cover (Geol); **c. arable,** topsoil; **c. de árboles (Ecol),** tree layer; **c. de arbustos (Ecol),** shrub layer; **c. de arcilla (Geol),** clay pan; **c. de carbón,** coal seam; **c. de hielo,** sheet ice; **c. de transición (Geol),** passage bed; **c. dura,** hardpan; **c. ferruginosa (Geol),** iron pan; **c. guía,** marker bed; **c. herbácea (Ecol),** herb layer; **c. límite,** boundary layer; **sección (f) de c.,** bed profile.

capacidad, f. capacity; **c. de campo,** field capacity; **c. de carga (Ecol),** carrying capacity; **c. de drenaje,** drainage capacity; **c. de infiltración,** infiltration capacity; **c. del pozo,** well capacity; **c. de recarga,** recharge capacity; **c. específica,** specific capacity; **c. financiera,** financial status.

capacitancia, f. capacitance (Elec).

capariche, m. rubbish tip operative, garbage man.

capataz, m. foreman.

capear, v. to brave, to ride out, to dodge; **c. el temporal,** to weather the storm.

capilar, a. capillary; **acción (f) c.,** capillary action; **ascenso (m) c.,** capillary rise; **franja (f) c.,** capillary fringe; **presión (f) c.,** capillary pressure; **tubo (m) c.,** capillary tube; **zona (f) c.,** capillary zone.

capilaridad, f. capillarity.

capital, 1. m. capital (finanzas); **c. (m) inicial,** initial capital; **fondos (m.pl) de c. riesgo,** venture capital; 2. f. capital (ciudad); **ciudad (f) c.,** capital city.

capitalización, f. capitalization (Fin).

capitel, m. capital (de edificio).

capítulo, m. chapter.

caprinola, a. of goats.

cápsula, f. capsule.

capsulado, m. capping (Con).

capsuladora, f. capping machine.

capsular, v. to cap, to cover over (Geol).

captación, f. catchment, harnessing (de aguas), abstraction; **c. de neblina,** mist catchment; **c. ilegal de agua,** illegal water abstraction; **zona (f) de c.,** catchment.

captar, v. to harness, to capture (potencia del agua).

captura, f. capture; **c. de neutrones,** neutron capture; **c. de ríos,** river capture; **c. fluvial,** river capture.

capturar, v. to capture.

capullo, m. cocoon.

cárabo, m. Tawny Owl.

caracol, m. snail (gasterópodo terrestre); winkle (gasterópoda marino); seashell (concha de mar).

caracolla, m. y f. seashell.

característica, f. characteristic, feature; **características de las aguas residuales,** wastewater characteristics; **características de los residuos peligrosos,** hazardous waste characteristics; **características del aquífero,** aquifer characteristics; **características del suelo,** soil characteristics; **características hidráulicas,** hydraulic characteristics.

característicola, a. characteristic, typical.

caracterización, f. characterization; **c. de muestras (Quím),** sample characterization, sample profiling; **c. de sitio,** site characterization.

caracterizadola, a. characterized.

caracterizar, v. to characterize.

Caradocense, m. Caradocian.

carámbano, m. icicle.

carancho, m. owl (Perú), vulture (GB), turkey buzzard (US).

caraota, f. bean.

carapazón, m. carapace.

carbaríl, m. carbaryl.

carbohidrato, m. carbohydrate.

carbón, m. coal; **c. activo,** activated carbón; **c. bituminoso,** biuminous coal; **c. de leña,** charcoal; **c. lignítico,** lignite; **c. natural,** bone black; **c. orgánico,** organic carbon; **c. pardo,** lignite, brown coal; **cuenca (f) de c.,** coal basin; **frente (m) de c.,** coalface; **industria (f) de c.,** coal industry; **mina (f) de c.,** coal mine; **yacimiento (m) de c.,** coalfield.

carbonáticola, a. carbonate; **dureza (f) c.,** carbonate hardness.

carbonato, m. carbonate; **c. de calcio,** calcium carbonate; **c. de sodio,** sodium carbonate.

carboneo, m. charcoal burning.

carbónicola, a. carbonic.

carbonización, f. carbonization.

carbono, m. carbon (C) (elemento); **c. catorce,** carbon 14; **datación (f) por c. radioactivo,** carbon dating; **relación (f) c.-nitrógeno (abr. relación C/N),** carbon-nitrogen ratio, C/N ratio; **tetracloruro (m) de c.,** carbonte-trachloride.

carbonosola, a. carbonaceous; **oxidación (f) c.,** carbonaceous oxidation.

carborundo, m. carborundum; **piedra (f) de c.,** carborundum stone.

carboxílicola, a. carboxylic.

carburante, m. fuel.

carburo, m. carbide; **c. hidrogenoclorofluorado,** chlorofluorocarbon (abr. CFC).

cárcava, f. gully, ravine.

carcinógeno, m. carcinogen.

carcinógenola, a. carcinogenic.

carcinoma, m. carcinoma.

carcoma, f. woodworm.

cardíacola, a. cardiac, heart.

cardinal, a. cardinal.

cardo, m. thistle (planta espinosa); cardoon (planta comestible); **c. borriquero,** cotton thistle.

carencia, f. lack, shortage, scarcity; deficiency (Med).

careo, m. aquifer recharge in mountain areas through fissures or by spreading.

carga, f. load (peso, Elec); head (presión); duty (Elec); **c. de un vehículo,** vehicle load; **c. eléctrica, c. en el circuito,** electrical load; **c. hidráulica,** hydraulic load; **c. hidrostática,** water head.

cargadora, f. loader (máquina); **pala (f) c. de ataque frontal,** front-end loader.

cargamento, m. shipment.

cargar, v. to load, to fill.

caribe, 1. m. y f. Carib; 2. a. Caribbean; **Mar c.,** Caribbean Sea.

carneola, f. carnelian.

carneril, a. livestock.

Carniense, m. Carnian.

carnívoro, m. carnivore.

carnívorola, a. carnivorous.

carpa, f. carp.

carpelo, m. carpel.

carpintería, f. joinery or carpentry shop.

carpintero, m. carpenter, joiner.

carrascal, m. holm-oak forest; stony ground (LAm) (pedregal).

carretera, f. main road, highway; **c. comarcal,** B road (GB), secondary road; **c. de acceso,** approach road; **c. de circunvalación,** ring road, bypass; **c. de doble calzada,** dual carriageway; **c. general,** class A road, main road, arterial highway (US); **c. nacional,** A road, main road, arterial highway (US); **c. secundaria,** class B road (GB), secondary road; **c. sin firme,** unmade road, unmetalled road, unsurfaced road; **red (f) de carreteras,** road network.

carretero, a. for vehicles; **camino c.,** cart track.

carretilla, f. truck; **c. elevadora,** forklift truck.

carrier, m. carrier (Med, vector).

carrizal, m. reed-bed.

carrizo, m. reed.

carro, m. car, motorcar.

carroña, f. carrion.

carroñero, m. scavenger.

carta, f. letter (correspondencia); map, chart (Geog); menu (restaurantes); **c. certificada,** registered letter; **c. comercial,** business letter; **c. de navegación,** navigation chart; **c. de**

pago, receipt; **c. de venta,** bill of sale; **c. marítima,** chart, navigation chart; **c. náutica,** navigation chart, nautical chart; **c. urgente,** special delivery letter, express delivery letter.

cartel, m. poster, placard.

cartesianola, a. Cartesian; **coordenadas (f.pl) cartesianas,** Cartesian coordinates.

cartografía, f. cartography, mapping, map making; **c. digital,** digital mapping.

cartográficola, a. cartographic.

cartógrafola, m. y f. cartographer.

cartón, m. cardboard; **c. corrugado,** corrugated cardboard.

cartonero, m. (Arg, Ch) waste tip scavenger.

cartón-fibra, f. fibreboard.

cartón-yeso, m. plasterboard.

casa, f. house, flat, apartment; household; home (hogar); **c. comercial,** business firm; **C. de la Villa,** Town Hall.

cascada, f. cascade, waterfall.

cascajo, m. debrís, rubble.

cáscara, f. shell (de huevo, de fruta seca); skin, peel (de fruta); bark (de troncos); husk (de cereales).

cascote, m. rubble.

caserío, m. homestead, country house.

caserola, a. household.

caseta, f. stall, booth, cubicle.

casiterita, f. cassiterite.

casquete, m. skullcap, helmet; **c. polar,** ice cap.

castaña, f. chestnut.

castañar, m. chestnut grove.

castañeda, f. chestnut grove.

castaño, m. chestnut tree; **c. de Indias,** horse chestnut.

castellanola, a. Spanish, Castilian.

castro, m. hill fort, Iron-Age settlement (Arq).

cata, f. tasting, sampling; **c. de ensayo,** taste test.

catabólicola, a. catabolic.

catabolismo, m. catabolism.

catador, m. taster, sampler (persona); (Br) waste tip scavenger.

Catalán, m. Catalan (lengua y persona).

catalánla, a. Catalan, Catalonian.

catálisis, f. catalysis.

catalíticola, a. catalytic.

catalizador, m. catalyst; catalytic converter (vehículos).

catalizar, v. to catalyze.

catalogación, f. cataloguing.

catálogo, m. catalogue, catalog (US); **c. de biblioteca,** library catalogue.

Cataluña, f. Catalonia.

catamina, f. (Ch, Perú) corrugated iron.

catarata, f. cataract (del ojo); waterfall (de agua), cataract (flujo muy fuerte).

catarómetro, m. katharometer.

catastro, m. census, official land register.

catástrofe, f. catastrophe; **c. natural,** natural disaster.

catastróficola, a. catastrophic; **inundación (f) c.,** catastrophic flood.

cateando, m. prospecting (Min) (LAm).

catear, v. to prospect (Min) (LAm).

categoría, f. category, grade.

categril, m. (Ur) slum, shanty town.

cateo, m. claim, prospect (Min) (LAm).

catión, m. cation.

catiónicola, a. cationic.

catódicola, a. cathodic; **protección (f) c.,** cathodic protection.

cátodo, m. cathode.

cauce, m. watercourse, river bed, channel; **c. (f) de un río,** river channel; **c. efímero,** ephemeral stream; **c. enterrado,** buried channel; **caudal (m) de c. lleno,** bankfull discharge.

cauchal, m. rubber plantation.

caucho, m. rubber.

caudal, 1. m. flow, volume, discharge, yield (de agua); **características (f.pl) del c.,** flow characteristics; **c. crítico,** critical flow; **c. de aguas de lluvia (Trat),** storm weather flow; **c. de avenida,** flood flow; **c. de bombeo,** pumped output (de un pozo); **c. de cauce lleno,** bankfull discharge; **c. de crecida,** flood flow; **c. de estación seca,** dry weather flow; **c. del embalse,** reservoir yield; **c. de seguridad mínima,** minimum reliable yield; **c. del río,** river flow; **c. de sequía,** drought yield; **c. explotable,** exploitable flow; **c. fluvial,** river flow; **c. instantáneo,** instantaneous flow; **c. perenne,** perennial flow; **c. seguro,** safe yield; **c. surgente,** overflow; **curva (f) de c.-descenso,** yield-drawdown curve; **índice (m) de c. base,** baseflow index; 2. a. caudal.

caudalímetro, m. current meter, flow meter (para ríos).

causa, f. cause; **c. y efecto,** cause and effect.

causal, 1. f. reason, grounds; 2. a. causal.

cautivola, a. captive.

cava, f. rill (Geog).

cavador, m. digger (Arq).

cavadura, f. excavation, digging.

cavar, v. to excavate, to scoop out.

caverna, f. cavern.

cavernícola, a. cave-dwelling; **hombre (m) c.,** caveman, troglodyte.

cavernosola, a. cavernous; **caliza (f) c.,** cavernous limestone.

cavidad, f. cavity; **c. de disolución,** solution cavity.

cavitación, f. cavitation.

caz, m. millrace, leat (de molino de agua), water duct.

caza, f. hunting, shooting (con armas de fuego); trapping (con trampas); game (animales que se cazan); open season, hunting season (temporada); **ave (f) de c.,** game bird; **c. furtiva,** poaching; **coto (m) de c.,** game reserve; **licencia (f) de c.,** permiso

(m) de c., hunting licence, hunting license (US); **vedado (m) de c.**, game reserve.

cazadorla, m. y f. hunter; **c. recolector**, hunter-gatherer.

CC, f. DC; **corriente (f) continua (Elec)**, direct current.

CCAA, f.pl. **Comunidades (f.pl) Autónomas (Esp)**, the seventeen autonomous regions of Spain.

CD, m. CD; **CD-Rom**, CD-Rom (abbr. of Compact Disc Read-Only-Memory).

CE, f. EC; **Comunidad (f) Europea**, European Community.

cebada, f. barley.

cebadola, a. fattened (ganado).

cebo, m. feed (para ganado); bait (para pescar); fuse (Elec); detonator, primer (explosivos); **c. vivo (pesquería)**, live bait.

cebolla, f. onion.

cebollar, m. onion patch, onion field.

cebolleta, f. chives (pl), spring onions (pl).

cebra, f. zebra.

CECA, f. ECSC; **Comunidad (f) Europea del Carbón y del Acero**, European Coal and Steel Community.

cedazo, m. sieve, sifter.

cedro, m. cedar.

CEE, f. EEC; **Comunidad (f) Económica Europea**, European Economic Community.

cefalón, m. cephalon.

cefalópodo, m. cephalopod; **cefalópodos (Zoo, clase)**, Cephalopoda (pl).

ceguera, f. blindness; **c. de río**, river-blindness.

ceja, f. eyebrow (Anat); rim, flange (Con); brow, crown (Geog).

celda, f. cell (Quím).

celentéreo, m. coelenterate; **celentéreos (Zoo, filum)**, Coelenterata (pl).

celestina, f. celestine.

celofán, m. celophane.

celsius, a. Celsius, centigrade.

celtíberola, a. Celto-Iberian, Celtiberian.

célula, f. cell; **c. vegetal**, plant cell.

celular, a. cellular, cell; **cultivo (m) c.**, tissue culture; **división (f) c.**, cell division; **membrana (f) c.**, cell membrane; **pared (f) c.**, cell wall.

celuloide, f. celluloid.

celulosa, f. cellulose.

celulósicola, a. cellulose.

cellisca, f. sleet.

cementación, f. cementation; **c. penetrante**, grouting.

cementar, v. to cement.

cementera, f. cement works.

cementerio, m. cemetery, graveyard.

cemento, m. cement; **cañón (m) de c.**, cement gun; **c. Portland**, Portland cement; **mortero (m) de c.**, cement mortar; **pasta (f) de c.**, cement paste.

CEN, m. **Consejo (m) de Economía Nacional (Esp)**, National Economic Coucil.

cenagal, m. bog, swamp, marsh, quagmire (pantano); nasty business (Fig).

cenagosola, a. swampy, muddy.

cenefa, f. edging, border (Con).

cenit, m. zenith.

ceniza, f. ash; **c. de combustible pulverizado**, pulverized fuel ash (abr. PFA); **c. volcánica**, volcanic ash, colcanic dust; **cenizas en suspensión**, fly ash; **lluvia (f) de cenizas**, ash shower.

Cenomaniense, m. Cenomanian.

cenote, m. (Méx) large natural well.

Cenozoico, m. Cenozoic.

censo, m. census (empadronamiento); ground rent (para una casa); pension (renta); **c. electoral**, electoral roll, electoral register.

centavo, m. hundredth, hundredth part; centavo (unidad monetaria).

centeno, m. rye.

centésimo, m. hundredth, hundredth part; centesimo (unidad monetaria).

centi-, pf. centi- (10.E2).

centigradola, a. centigrade, Celsius.

centímetro, m. centimetre.

céntimo, m. hundredth, hundredth part; centimo (C.Rica, Haiti, Par, Ven) (unidad monetaria).

central, f. head office (sede); station, plant (Con); power station; **c. azucarera**, sugar mill; **c. de correos**, general post office; **c. eléctrica**, **c. energética**, power station; **c. hidráulica**, hydroelectric power (abr. HEP) station; **c. nuclear**, nuclear power station; **c. térmica**, power station.

centralizar, v. to centralize.

centrífuga, f. centrifuge.

centrifugación, f. centrifuging.

centrifugadora, f. centrifuge.

centrifugar, v. to centrifuge.

centrífugola, a. centrifugal; **bomba (f) c.**, centrifugal pump.

centro, m. centre, middle; **c. de gravedad**, centre of gravity; **c. de interpretación**, information centre.

Centroatlántica, f. Central Atlantic.

centroatlánticola, a. Central Atlantic.

centella, f. spark (chispa); flash of lightening (rayo).

centelleante, m. scintillator.

centelleo, m. scintillation, sparkling, flashing; **contador (m) de c.**, scintillation counter.

ceolita, f. zeolite.

CEPAL, f. ECLA; **Comisión (f) Económica para América Latina**, Economic Commission for Latin America.

CEPE, f. EEC; **Comisión (f) Económica Europea**, European Economic Commission.

CEPIS, m. CEPIS; **Centro (m) Panamericano de Ingeniería Sanitaria y Ciencias de Ambiente (LAm)**, Pan American Center for Sanitary Engineering and Environmental Sciences.

cera, f. wax.
cerámica, f. china, ceramics.
cerámicola, a. ceramic; **pieza (f) c.**, ceramic tile.
cercado, m. enclosure.
cercadola, a. enclosed; **c. de tierra**, landlocked.
cercanía, f. proximity; **c. de la costa**, foreshore; **cercanías**, outskirts of a city.
cercanola, a. nearby, neighbouring, neighboring (US), close; **c. a la orilla**, inshore, nearshore.
cerco, m. ring, circle; **c. de jareta (pesca) (Galicia, Esp)**, purse seine net.
cerdito, m. piglet.
cereal, m. cereal, grain; **cereales (pl)**, cereal, grain.
cerebro, m. brain, cerebrum.
cereza, f. cherry (fruta); coffee bean (café).
cerezo, m. cherry (tree).
cerio, m. cerium (Ce).
cerniduras, f.pl. screenings, siftings, riddlings (Min).
cero, m. zero, nought; **c. absoluto (0 deg. K)**, absolute zero.
cerrillo, m. small hill.
cerrojo, m. bolt (de puerta).
certificación, f. certification.
certificadola, a. certified, registered; **carta (f) c.**, registered letter, registered post.
cerusita, f. cerussite.
cese, m. cessation, discontinuation, cut-off.
cesio, m. caesium (Cs).
césped, m. lawn; **c. artificial**, astroturf.
cestón, m. gabion.
cetáceo, m. cetácean; **cetáceos (Zoo, orden)**, Cetacea (pl).
cetona, f. ketone.
cetrino, m. citrine.
CFC, m. CFC; **clorofluorocarbono (m)**, chlorofluorocarbon.
cianamida, f. cyanamide.
cianato, m. cyanate.
cianita, f. kyanite, cyanite.
cianobacteria, f.pl. cyanobacteria.
cianofita, f. cyanophyte; **cianofitas (Bot, división)**, Cyanophyta (pl).
cianógeno, m. cyanogen.
cianuro, m. cyanide.
cibernética, f. cybernetics.
cicatriz, f. scar.
cíclicola, a. cyclic, cyclical.
ciclo, m. cycle; **c. alimentario**, food web, food cycle; **c. celular**, cell cycle; **c. de agua**, water cycle, hydrological cycle; **c. de carbono**, carbon cycle; **c. de nitrógeno**, nitrogen cycle; **c. de oxígeno**, oxygen cycle; **c. económico**, economic cycle; **c. geoquímico**, geochemical cycle; **c. hidrológico**, hydrological cycle; **c. lunar**, lunar cycle; **c. menstrual**, menstrual cycle; **c. sedimentario**, sedimentary cycle; **c. solar**, solar cycle; **ciclos por segundo**, cycles per second.

cicloalcano, m. cycloalkane.
cicloalquino, m. cycloalkene.
ciclohexano, m. cyclohexane.
ciclohexanol, m. cyclohexanol.
ciclón, m. cyclone.
ciclónicola, a. cyclonic.
cicloparafinas, f.pl. cycloparaffins.
cicloterma, f. cyclotherm.
cicloturismo, m. cycling tourism.
cidaroide, m. cidaroid.
cielo, m. sky; ceiling (techo); heaven (religión); **a c. abierto (Min)**, open cast; **c. encapotado**, overcast sky; **claro de c.**, bright interval.
ciempiés, m. centipede.
ciénaga, f. (Méx, US) swamp, extensive boggy ground, mudflats.
ciencia, f. science; **c. ambiental**, environmental science; **ciencias del medio**, environmental sciences.
cieno, m. mud, mire, slime, sludge, ooze; **c. activado**, activated sludge; **c. sin tratar**, raw sludge.
científicola, 1. m. y f. scientist; **c. ambiental**, environmental scientist; 2. a. scientific.
cierre, m. closure, shutdown; **c. de fábrica**, factory closure; **c. definitivo**, decommissioning; **c. estructural**, structural closure; **c. industrial**, industrial shutdown.
cifra, f. figure, number, numeral; code, cipher; bench mark (Ing); **c. global**, lump sum; **cifras significativas (Mat)**, significant figures; **en c.**, in code.
cigala, f. Dublin Bay prawn.
cigüeñal, m. (Arab) crankshaft, lever for raising water.
CIJ, f. ICJ; **Corte (f) Internacional de Justicia**, International Court of Justice.
ciliado, m. ciliate; **ciliados (Zoo, clase)**, Ciliata (pl).
ciliadola, a. ciliate.
ciliar, a. ciliary.
cilíndricola, a. cylindrical.
cilindro, m. cylinder.
cilio, m. cilium (pl. cilia).
CIMA, f. **Comisión (f) Interministerial del Medio Ambiente (Esp)**, Interministerial Committee for the Environment.
cima, f. summit, top, peak.
cimacio, m. ogee (Con); **arco (m) c.**, ogee arch.
cimbra, f. soffit (centro del arco).
cimentación, f. foundation, foundations, laying of foundations (acción).
cimientos, m.pl. foundation (de un edificio); **c. de una balsa**, raft foundation; **c. en los pilotes**, piled foundation; **echar los c. de un edificio (v)**, to lay the foundations of a building.
cinabrio, m. cinnabar.
cinc, m. zinc (Zn).
cincadola, a. of zinc; **c. electrolítico**, zinc plating.
cine, m. cinema; movies (US).

cinegéticola, a. hunting.

cinemática, f. kinematics, cinematics; **c. de fluidos,** fluid kinematics.

cinéticola, a. kinetic; **altura (f) c.,** velocity head, kinetic head.

cinéticola, a. kinetic; **energía (f) c.,** kinetic energy.

cinta, f. band, ribbon, belt, tape; **c. adhesiva,** adhesive tape; **c. aislante,** insulating tape; **c. de goma, c. elástica,** rubber band, elastic band; **c. magnetofónica,** magnetic tape, recording tape; **c. métrica,** tape measure, measuring tape; **c. perforada,** punched tape; **c. transportadora,** conveyor belt.

cinturón, m. belt.

circadianola, a. diurnal; **ritmo (m) c.,** diurnal rhythm; **variación (f) c.,** diurnal variation.

circo, m. circus, amphitheatre, amphitheater (US); **c. glaciar,** glacial cirque, corrie, cwm (País de Gales).

circón, m. zircon.

circonio, m. zirconium (Zr).

circuito, m. circuit; **c. integrado,** integrated circuit; **placa (f) de c. impreso,** printed circuit board; **televisión (f) por c. cerrado,** closed-circuit television (abr. CCTV).

circulación, f. circulation; traffic (tráfico); **arterias (f.pl) de c. vial,** principal traffic routes; **c. atmosférica,** atmospheric circulation; **c. del tráfico,** road traffic, vehicular traffic; **c. inversa,** reverse circulation.

circular, 1. v. to circulate; 2. a. circular, round.

círculo, m. circle (Mat); group, faction; **C. Antártico,** Antarctic Circle; **C. Polar,** Arctic Circle; **en círculos comerciales,** in business circles.

circumpolar, a. circumpolar.

circunferencia, f. circumference.

circunscritola, a. circumscribed.

circunstancia, f. circumstance.

circunvalación, f. bypassing; **vía (f) de c.,** bypass, ring road (GB).

circunvolución, f. convolution.

cirros, m.pl. cirrus; **cirroestratos,** cirrostratus.

cirtocónicola, a. cirtoconical.

ciruela, f. plum; **c. claudia,** greengage; **c. damascena,** damson; **c. pasa,** prune.

ciruelo, m. plum tree.

ciruja, m. y f. (Arg, Ur) waste tip scavenger.

cisterna, f. cistern, tank; cistern, flush (del lavabo).

citación, f. citation (Jur).

citadinola, 1. m. y f. city dweller, citizen; 2. a. (LAm) urban (urbano).

citólisis, f. cytolysis.

citología, f. cytology.

citoplasma, m. cytoplasm.

citoquímica, f. cytochemistry.

cítricola, a. citrus, citric; **ácido (m) c.,** citric acid.

ciudad, f. town, city; **c. satélite,** satellite town;

c. universitaria, university campus; **las afueras (f) de la c.,** suburbs.

ciudadanola, m. y f. city dweller, citizen; **ciudadanos,** townspeople (pl).

ciudadela, f. citadel, fortress.

civil, a. civil; **derecho (m) c.,** civil law.

cizalla(s), f. o f.pl. shears, metal clippers.

cizalladura, f. shearing.

cizallamiento, m. shearing; **zona (f) de c.,** shear zone.

cizallar, v. to cut, to shear.

clara, f. white (de un huevo).

claraboya, f. skylight.

claridad, f. clarity.

clarificación, f. clarification, explanation.

clarificador, m. clarifier; **c. de flujo ascendente,** upward-flow clarifier; **c. primario,** primary clarifier.

clase, f. class, grade, sort.

clasificación, f. classification, grading; **c. de suelos,** soil classifiaction; **c. química,** chemical classification.

clasificador, m. grader, sorter (máquina).

clasificadorla, m. y f. classifier (persona).

clasificar, v. to grade, to classify.

clásticola, a. clastic; **sedimento (m) c.,** clastic sediment.

clasto, m. clast.

clausura, f. recess, closure.

clavar, v. to nail.

clave, m. key; keystone (Con).

clavo, m. nail (de metal).

clima, m. climate; **c. artificial,** air conditioning; **c. marítimo,** maritime climate; **c. templado,** temperate climate.

climácicola, a. climax; **vegetación (f) c.,** climax vegetation.

climáticola, a. climatic, climate; **cambio (m) c.,** climate change.

climatización, f. air-conditioning.

climatizadola, a. air-conditioned.

climatología, f. climatology.

clímax, m. climax; **bosque (m) c. (Ecol),** climax forest; **comunidad (f) c. (Ecol),** climax community.

clina, f. cline.

cline, m. cline.

clinoformola, a. clinoform.

clinográfico, m. clinograph.

clinómetro, m. clinometer.

clinopiroxenita, f. clinopyroxenite.

cliserie, f. clisere.

clivaje, m. cleavage, foliation.

cloaca, f. drain, sewer; **c. de aguas negras,** foul sewer; **c. maestra,** main sewer.

clon, m. clone.

cloración, f. chlorination; **c. de agua,** water chlorination; **c. rotura,** break-point chlorination.

cloradola, a. chlorinated; **disolvente (m) c.,** chlorinated solvent; **hidrocarburo (m) c.,** chlorinated hydrocarbon.

clorador, m. chlorinator (m).
cloramina, f. chloramine.
clorato, m. chlorate; **c. sódico**, sodium chlorate.
clorhídricola, a. hydrochloric.
clóricola, a. chloric.
clorinación, f. chlorination.
clorinador, m. chlorinator.
clorinización, f. chlorination; **límite (m) de c.**, break-point chlorination.
clorita, f. chlorite.
cloritoide, m. chloritoid.
cloro, m. chlorine (Cl); **c. libre**, free chlorine; **c. residual**, residual chlorine; **tratar con c. (v)**, to chlorinate.
clorobenzol, m. chlorobenzene.
clorobifenol, m. chlorobiphenol.
clorodibromometano, m. chlorodibromomethane.
cloroficeas, f.pl. Chlorophyceae (Bot, class).
clorofila, f. chlorophyl, chlorophyll.
clorofito, m. chlorophyte.
clorofluorocarbono, m. chlorofluorocarbon (abr. CFC).
cloroformo, m. chloroform.
cloroplasto, m. chloroplast.
clorosis, f. chlorosis.
cloruro, m. chloride; **c. cálcico**, calcium chloride.
clostridium, m. clostridium.
clypeasteroide, m. clypeasteroid.
CMCC, f. CARICOM; **Comunidad (f) y Mercado Común del Caribe**, Caribbean Community and Common Market.
coadaptación, f. coadaptation.
coagulación, f. coagulation.
coagulante, m. coagulant.
coagular, v. to coagulate.
coágulo, m. coagulum, clot.
coalescencia, f. coalescence.
coalescente, a. coalescent.
cobálticola, a. cobaltic.
cobalto, m. cobalt (Co).
cobertura, f. coverage.
cobijadura, f. cover, shelter; nappe (Geol) (Lm).
cobijo, m. shelter, cover.
cobre, m. copper (Cu); **c. amarillo**, brass.
cocaína, f. cocaine.
cocción, f. boiling (acción de hervir), cooking (acción de cocer), baking (de pan, etc.).
coche, m. car, motorcar, automobile; coach, carriage, car (ferrocarril).
cochinilla, f. woodlouse (pl. woodlice); cochineal (tinte).
cociente, m. quotient; **c. respiratorio**, respiratory quotient (abr. RQ).
cocolito, m. coccolith.
codal, m. strut (Con).
código, m. code, rules; **c. civil**, civil code; **c. de barras**, bar code; **c. de buena práctica**, code of practice; **c. de conducta**, code of conduct, code of behaviour; **c. de leyes**, statute book;

c. de máquina (Comp), machine code; **c. penal**, penal code; **c. postal**, post code.
codisposición, f. codisposal.
coeficiente, m. coefficient; **c. de Coriolis**, Coriolis coefficient; **c. de expansión**, coefficient of expansion; **c. de fricción**, coefficient of friction; **c. de permeabilidad**, coefficient of permeability; **c. de regresión**, regression coefficient; **c. de selección**, coefficient of sorting; **c. térmico**, temperature coefficient.
coeliminación, f. co-disposal.
coenzima, f. coenzyme.
coevolución, f. coevolution.
cogeneración, f. cogeneration.
cognadola, a. cognate.
coherente, a. coherent.
cohesión, f. cohesion.
cohesivola, a. cohesive.
cohete, m. rocket.
cohorte, f. cohort.
coihué, m. Southern false beech (Bot).
coincidencia, f. coincidence; conformity (Mat); **c. de curvas**, curve-fitting.
col, f. cabbage (planta); col (Geog); **c. de Bruselas**, Brussels sprout; **c. glacial, c. glaciar**, glacial col.
cola, f. tail; **c. de caballo (Bot)**, horsetail; **c. de milano, c. de pato**, dovetail joint; **c. de pescado**, fish glue; **c. de tráfico**, tailback; **viento (m) de c.**, tailwind.
colaboración, f. collaboration.
colaborar, v. to collaborate, to cooperate (cooperar).
colacionar, v. to collate, to compare.
colada, f. washing (lavado); outflow (Geol).
colador, m. strainer, colander.
colágeno, m. collagen.
colas, f.pl. tailings (Min).
coleccionista, m. collector.
colectar, v. to collect.
colectiva, f. (LAm) bus.
colectivo, m. collective taxi, small bus.
colector, m. collector (persona); main sewer (alcantarilla); sump, trap (pozo); **c. de drenaje**, drainpipe; **c. de un edificio**, house drain, house sewer; **red (f) de c. público**, main sewerage.
colegio, m. college; **C. de Geólogo (Esp)**, Association of Professional Geologists.
coleóptero, m. coleopteran; **coleópteros (Zoo, orden)**, Coleoptera (pl).
cólera, f. cholera.
colgadola, a. perched; **freático (m) c.**, perched groundwater.
coliflor, f. cauliflower.
coliforme, m. coliform; **c. fecal**, faecal coliform, fecal coliform (US); **coliformes totales**, total coliforms; **organismo (m) c.**, coliform organism.
colimación, f. collimation; **error (m) de c.**, collimation error.
colina, f. hill.

colmatación, f. infilling, clogging; **c. de charcas,** silting up of ponds (con sedimentos).

colmena, f. hive, beehive.

colmillo, m. tusk (p.e. de un elefante); fang, canine tooth, eye-tooth (diente);.

colocación, f. employment, employment office (empleo); positioning (sitio); investment (dinero).

colocar, v. to place, to put.

coloidal, a. colloidal.

coloide, m. colloid.

coloideola, a. colloidal.

Colombia, f. Colombia.

colombianola, 1. m. y f. Colombian; 2. a. Colombian.

colon, m. colon (Anat).

colón, m. colon (LAm) (unidad monetaria).

colonia, f. colony, housing development; residential district (LAm) (barrio); **cuenta (f) de c.** (Biol), colony count.

colonización, f. colonization, settlement.

color, m. colour, color (US); **trazador (m) de c.,** dye tracer.

coloración, f. coloration.

coloradola, a. coloured, colored (US).

colorante, m. dye, colouring.

colorimetría, f. colorimetry.

colorimétricola, a. colorimetric.

columna, f. column; **c. de agua,** head of water; **c. de extracción (Quím),** extraction column; **c. geológica,** geological column.

columnata, f. colonnade.

coluvial, a. colluvial.

coluvión, f. colluvium.

collado, m. fell (Geog).

coma, f. comma.

comarca, f. region, area; county (US).

comarcal, a. regional.

comba, f. bend (Gen), warp, sag.

combadura, f. camber (in road), curve, bend.

combar, v. to sag.

combinación, f. combination.

combustible, 1. m. fuel; **c. fósil,** fossil fuel; **c. sólido,** solid fuel; 2. a. combustible.

combustión, f. combustion; **instalación (f) de c.,** incineration plant.

combustóleo, m. fuel-oil, heavy oil.

comején, m. termite (insecta).

comensalismo, m. commensalism.

comercial, a. business, commercial; **centro (m) c.,** shopping centre, shopping center (US), mall (US); **parque (m) c.,** business park; **trato (m) c.,** business deal.

comercialización, f. commercialization, marketing (mercadeo).

comercializador, m. merchant.

comercializar, v. to commercialize.

comerciante, m. y f. dealer, merchant, tradesman/tradeswoman; shopkeeper (tendero).

comercio, m. commerce; **Cámara (f) de C.,** Chamber of Trade; **código de c.,** commercial law; **c. al por mayor,** wholesale trade; **c. al por menor,** retail trade; **c. de exportación,** export trade; **c. exterior,** foreign trade, overseas trade; **c. interior,** domestic trade; **libre c.,** free trade; **Ministerio (m) de C.,** Department of Trade.

comezcla, f. (Agr) admixture.

comisaría, f. commissariat.

comisión, f. commission; **C. Europea,** European Commission.

compacidad, f. compactness.

compactación, f. compaction; **proporción (f) de c. (en vertederos),** compaction ratio.

compactador, m. compactor.

compactola, a. compact.

compañía, f. company; **c. de seguros,** insurance company.

comparación, f. comparison; **c. de costos,** comparison of costs; **en c. con,** in comparison with.

compartimiento, m. compartment; **c. estanco,** watertight compartment, cofferdam.

compás, m. compass; pair of compasses (Mat); **c. magnético,** magnetic compass.

compatibilidad, f. compatibility.

compendio, m. compendium (pl. compendia, compendiums).

compensación, f. compensation, reparation, reward.

competencia, f. competence, ability.

competente, a. competent; **autoridad (f) c.,** competent authority.

competición, f. competition.

competidorla, m. y f. competitor.

competitividad, f. competitiveness.

compiladorla, m. y f. compiler.

compilar, v. to compile.

complejación, f. complexation; **reacciones (f.pl) de c.,** complexation reactions.

complejidad, f. complexity.

complejo, m. complex; **c. basal,** basal complex; **c. industrial,** industrial complex.

complejola, a. complex; **número (m) complejo,** complex number.

completar, v. to complete, to finish.

complexional, a. complexing.

complexométricola, a. complexometric.

componente, 1. m. component; 2. a. component.

componer, v. to compose, to form; to mend (arreglar); to compound (Quím).

comportamiento, m. behaviour, behavior (US).

composición, f. composition; **c. del agua,** composition of water; **c. de lugar,** stocktaking; **c. química,** chemical composition.

compostado, m. compost.

compostador, m. composter (máquina).

compostaje, m. composting.

compostificación, f. (LAm) composting; **planta (f) de c. (LAm),** composting plant.

compra, f. purchase.

compradorla, m. y f. buyer.

comprar, v. to buy, to purchase; **c. la parte de,** to buy out.

compraventa, f. dealing, buying and selling.

comprender, v. to understand, to comprehend.

comprensivola, a. understanding.

compresibilidad, f. compressibility; **c. del acuífero,** aquifer compressibilty.

compresión, f. compression; **barra (f) de c.** (Ing), compression bar; **c. diferencial,** differential compression; **ensayo (m) de c.,** compression test; **filtro (m) de c.,** compression filter; **rotura (f) por c.,** compression failure.

compresividad, f. compressiveness.

compresivola, a. compressive; **fuerza (f) de c.,** compressive strength.

compresor, m. compressor (máquina); **c. de aire,** air compressor.

comprimidola, a. compressed; **aire (m) c.,** compressed air.

comprimir, v. to compress; to condense (líquidos).

comprobable, a. checkable, verifiable.

comprobación, f. check, verification; proof (datos).

comprobar, v. to check.

comprometer, v. to compromise, to jeopardize; to implicate, to involve.

comprometidola, a. involved, endangered; **especie (f) c.,** endangered species.

compuerta, f. sluice, floodgate, penstock; **c. de esclusa,** sluice gate, lock gate.

compuestas, f.pl. Compositae (pl) (Bot, familia).

compuesto, m. compound (Quím); **c. químico,** chemical compound; **compuestos orgánicos volátiles,** volatile organic compounds (abr. VOCs).

compuestola, a. composite, compound; **material (m) c.,** composite material; **revestimiento (m) c. (para vertedero),** composite liner.

computación, f. computación (f); calculación (esp. LAm) (f).

computadorla, m. y f. computer; **c. central,** mainframe computer.

computadora, f. computer terminal.

computar, v. to compute, to calculate.

computerización, f. computerization.

computerizar, v. to computerize.

cómputo, m. computation, calculation.

común, a. common, commonplace, shared.

comunicación, f. communication.

comunicado, m. bulletin, communiqué (declaración).

comunicar, v. to communicate, to convey, to inform.

comunidad, f. community; **c. autónoma (Esp),** autonomous region; **c. climácica, c. climax, (Ecol),** climax community; **c. de bienes,** community property; **c. de proprietarios,** owners' association (de casas); **C. Económica Europea (abr. CEE),** European Economic Community (abr. EEC); **C. Europea (abr.**

CE), European Community (abr. EC); **c. pionera (Ecol),** pioneer community; **c. final (Ecol),** climax community.

cóncavola, a. concave.

concejal, m. town councillor.

concentración, f. concentration; **c. de fondo,** background concentration; **c. iónica,** ionic concentration; **c. máxima permisible,** maximum permissible concentration; **c. por evaporación,** evaporative concentration.

concentrar, v. to concentrate.

concéntricola, a. concentric.

concepto, m. concept.

conceptual, a. conceptual; **modelos (m.pl) conceptuales,** conceptual models.

concesión, f. concession.

concesionario, m. licence or concession holder, licensee (p.e. de una mina).

concesionariola, 1. m. y f. licence holder, licensee; 2. a. concessionary.

conciencia, f. consciousness; awareness (conocimiento).

concienzudola, a. workmanlike.

conclusión, f. conclusion; **llegar a la c. de que (v),** to come to the conclusion that.

concoidal, a. concoidal.

concordancia, f. conformity, agreement.

concreción, f. concretion.

concreto, m. concrete (LAm); **c. armado,** reinforced concrete.

concha, f. shell (caracol); peel, bark (Ven).

condensación, f. condensation.

condensado, m. condensate.

condensador, m. condenser, capacitor (Elec).

condensar, v. to condense, to compact.

condición, f. condition; **condiciones climáticas,** climatic conditions; **condiciones del caso más desfavorable,** worst-case conditions; **condiciones de trabajo,** working conditions; **condiciones iniciales,** initial conditions; **en las condiciones más favorables,** under optimum conditions.

condominio, m. flat, apartment, condominium.

condón, m. condom.

conducción, f. transport, transportation (transporte); piping (de líquidos); wiring (por cable); management (Com); leading (acción); **c. abiertas,** open conduits; **c. de desagües,** sewage conduits.

conductancia, f. conductance; **c. eléctrica,** electrical conductance.

conductividad, f. conductivity; **c. eléctrica,** electrical conductivity; **c. hidráulica,** hydraulic conductivity; **c. térmica,** thermal conductivity.

conductivímetro, m. conductivimeter.

conducto, m. duct (Ing, Anat); conduit, pipe; **c. abierto,** open conduit.

conectadola, a. connected; switched on (encendido); on-line (Comp).

conejera, f. rabbit burrow.

conexión, f. connection, communication; **c.**

cruzada, crossed connection; **c. domiciliaria,** household connection; **c. hidráulica entre ríos y acuíferos,** river-aquifer hydraulic connection; **c. ilícita,** illegal connection.

conexo/a, a. connected, related, allied.

confederación, f. confederation; **C. Hidrográfica (Esp),** Hydrological Confederation; **c. hidrológica (Esp),** river board, river catchment board.

conferencia, f. conference, meeting; lecture, talk; **c. de prensa,** press conference.

conferenciante, m. y f. lecturer, speaker.

confiabilidad, f. reliability, dependability.

confianza, f. confidence; **en c.,** in confidence; **límites (m) de c.,** confidence limits.

configuración, f. configuration.

confinado/a, a. confined; **acuífero (m) c.,** confined aquifer; **no confinado/a,** unconfined.

confinamiento, m. confinement.

confirmación, f. confirmation.

confiscación, f. confiscation, sequestration (de propiedad).

confluencia, f. confluence; **c. de los ríos,** river confluence.

conformidad, f. conformity, agreement; compliance (con reglas).

congelación, f. freezing; **punto (m) de c.,** freezing point.

congelado/a, a. frozen.

congelador, m. freezer.

congelar, v. to freeze.

conglomerado, m. conglomerate; **c. basal,** basal conglomerate; **c. de cantos bien seleccionados,** well-sorted conglomerate.

congreso, m. congress, meeting, convention (asamblea).

congrio, m. conger eel.

congruencia, f. congruence.

congruente, a. congruous.

Coniacense, m. Coniacian.

conífera, f. conifer; **coníferas (Bot, orden),** Coniferophyta, Coniferae, Coniferales (pl).

conjugado/a, a. conjugate.

conjunto, m. collection, group, whole; **visión (f) de c.,** overview.

conjunto/a, a. combined, joined.

conminutador, a. comminutor.

conmutador, m. commutator.

cono, m. cone; **c. de cenizas volcánicas,** ash cone; **c. de escorias volcánicas,** cinder cone; **C. Sur,** Southern Cone (Arg+Ch+Ur).

conocer, v. to know.

conocido/a, a. known, well-known.

conocimiento, m. knowledge.

conodonto, m. conodont.

consanguíneo, a. similar; cosanguineous (by blood).

consecución, f. attainment, realization.

consecuencia, f. consequence; **c. grave,** grave consequence.

consejería, f. regional government ministry.

consejo, m. advice; council, board (Reg); **c. de**

administración, board of directors; **C. de Europa,** Council of Europe.

consenso, m. consensus.

conserva, f. preserving (proceso), preserve (producto); **conservas alimenticias,** tinned (US, canned) goods; **conservas de carnes,** meat preserves; **conservas de pescados,** fish preserves.

conservación, f. conservation; **c. de la naturaleza,** nature conservation; **estado (m) de c.,** conservation status.

conservacionista, m. y f. conservationist; **grupo (m) c.,** conservation group.

conservar, v. to preserve, to conserve.

conservas, f.pl. canned products.

consiguiente, a. consequential.

consolidación, f. consolidation.

consolidar, v. to consolidate, to strengthen.

constante, m. constant; **c. arbitraria,** arbitrary constant; **c. de desintegración,** decay constant; **c. dieléctrica,** dielectric constant.

constelación, f. constellation.

constitución, f. constitution (Jur).

constitutivo/a, a. constituent.

construcción, f. construction; **c. metálica,** steel-frame construction.

constructivo/a, a. constructional, constructive.

constructor/a, 1. m. y f. builder (de edificios); manufacturer (de coches); 2. a. building, construction; **empresa (f) c.,** building company, construction company.

construir, v. to construct.

consuetudinario/a, a. customary (Jur); **derecho (m) c.,** customary law, common law (GB).

consulta, f. consultation; **c. pública,** public consultation; **obra (f), libro (m) de c.,** reference book.

consultor/a, m. y f. consultant; **biólogo (m) c.,** biological consultant, consultant biologist.

consultoría, f. consulting.

consumidor/a, m. y f. consumer; **c. primario (Ecol),** primary consumer; **c. secundario (Ecol),** secondary consumer.

consumo, m. consumption; **bienes (m.pl) de c.,** consumer goods; **c. energético,** energy consumption; **c. industrial de agua,** industrial water consumption; **c. no contabilizado de agua,** unaccounted-for water consumption; **c. total,** consumptive use; **no apto para el c. humano,** unfit for human consumption; **sociedad (f) de c.,** consumer society.

contabilidad, f. accounting.

contabilizado, m. entry (en un registro).

contable, m. y f. accountant, bookkeeper.

contacto, m. contact.

contactor, m. contactor; **c. rotatorio (Trat),** rotary contactor.

contador, m. meter, counter; accountant (LAm); **c. de agua,** water meter.

contagiar, v. to infect, to spread disease.

contaminación, f. contamination; pollution (de agua, aire); **c. atmosférica,** atmospheric contamination; **c. bacteriológica,** bacterial contamination; **c. biológica,** biological contamination; **c. de agua,** water contamination; **c. de costas,** coastal pollution; **c. del aire,** air pollution; **c. de las aguas subterráneas,** groundwater contamination; **c. de playas,** beach contamination; **c. de ríos,** river pollution; **c. difusa,** diffuse pollution (de cauces); **c. directa (p.e. de acuíferos),** direct pollution, point source pollution; **c. fecal,** faecal contamination; **c. indirecta,** indirect contamination; **c. industrial,** industrial contamination; **c. por hidrocarbonos,** contamination by hydrocarbons; **c. por pesticidas,** contamination by pesticides; **c. radioactiva,** radioactive contamination; **carga (f) de c.,** pollution load.

contaminadola, a. contaminated; **altamente c.,** highly contaminated; **limpieza (f) de suelos contaminados,** contaminated land clean-up; **recuperación (f) de suelo c.,** contaminated land recovery; **suelo (m) c.,** contaminated land.

contaminador, m. pollutant (sustancia que contamina); polluter (persona/empresa).

contaminante, 1. m. pollutant, contaminant; **c. acuático,** aquatic pollutant; **c. biológico,** biological contaminant; **c. conservativo,** conservative contaminant; **c. no biodegradable,** non-biodegradable contaminant; **c. químico,** chemical contaminant; **c. radioactivo,** radioactive contaminant; **eliminación (f) de contaminantes,** contaminant removal; **transporte (m) de contaminantes,** contaminant transport; 2. a. polluting.

contaminar, v. to pollute, to contaminate.

contemporáneola, a. contemporaneous.

contención, f. containment, restraint (control); **c. de precios,** price control; **depósito (m) de c.,** impounding reservoir; **lugar (m) de c.,** containment site; **muro (m) de c.,** retaining wall.

contenedor, m. container, receptacle; container ship (buque); **c. de escombros,** skip; **c. de vidrio,** bottle bank.

contenerización, f. containerization.

contenido, m. content (cantidad contenido); **c. en carbón,** carbon content.

contextualizar, v. to put into context.

continental, a. continental; **deriva (f) c.,** continental drift.

continente, m. continent, mainland (Geog); container (recipiente); **el c. europeo,** the Continent of Europe.

contingencia, f. contingency.

contingente, a. contingent.

continuar, v. to continue.

continuidad, f. continuity; **ecuación (f) de la c.,** equation of continuity.

continuola, a. continuous, continual.

contorno, m. contour.

contorsión, f. contortion.

contorsionadola, a. contorted.

contrabalancear, v. to counterbalance.

contracción, f. contraction, shrinkage; **c. diferencial,** differential contraction.

contracepción, f. contraception.

contracorriente, f. cross-current (agua); eddy (torbellino); surging (Ing).

contradiagonal, m. counterbracing.

contrafuerte, m. spur (de una montaña); buttress (Con); abutment (Ing).

contraincendio, m. fire fighting measure, fire precaution.

contramedida, f. countermeasure.

contramolde, m. cast.

contrapartida, f. balancing entry (Fin).

contrapesar, v. to counterbalance.

contrapeso, m. counterweight, counterbalance.

contrapozo, m. counter well, image well.

contrapresión, f. counterpressure.

contrarradiación, f. counter-radiation.

contrarrestar, v. to counteract, to block, to resist.

contraste, m. contrast; **en c. con,** in contrast with.

contratación, f. hiring, contract; **c. de un consultor,** hiring of a consultant.

contratante, a. contracting; **las partes (f.pl) contratantes,** the contracting parties.

contratar, v. to contract (hacer un contrato).

contratiempo, m. setback, difficulty.

contratista, m. y f. contractor; **c. de obras,** building contractor.

contrato, m. contract; **c. para construcción,** construction contract.

contravenir, v. to contravene.

contraventana, f. shutter (en ventana).

contribución, f. contribution; **c. urbana (Esp),** urban tax contribution, rates (GB).

control, m. control, check; **bajo c.,** under control; **compuerta (f) de c. (Ing),** control gate; **c. ambiental,** environmental monitoring; **c. automático,** automatic control; **c. biológico (Agr, Ecol),** biological control; **c. de calidad,** quality control; **c. de fugas,** leakage control; **c. de la natalidad,** birth control; **c. de nivel freático,** control of water level; **c. de pH,** pH control; **c. de proceso,** process control; **c. químico,** chemical control; **fuera de c.,** out of control; **reglas (f.pl) de c.,** control rules (pl); **sala (f) de c.,** control room; **torre (m) de c.,** control tower.

controlable, a. controlable (US), controllable.

conuco, m. (Ven) smallholding, plot of land.

conurbación, f. conurbation.

convección, f. convection; **corriente (f) de c.,** convection current.

convectivola, a. convective.

convención, f. convention (custom, usage).

convencional, a. conventional; **energía (f) c.,** conventional energy.

convenio, m. agreement (acuerdo).

conventillo, m. (LAm) squatter settlement.

convergencia, f. convergence; **c. óptica,** optical convergence.

convergir, v. to converge.

conversión, f. conversion; **c. energética,** energy conversion.

convexidad, f. convexity, curvature.

convexo/a, a. convex.

convoluto/a, a. convoluted; **plegamiento (m) c.,** convoluted folding.

cooperación, f. cooperation, collaboration; **c. económica,** economic cooperation; **c. técnica,** technical cooperation.

cooperar, v. to cooperate, to collaborate.

cooperativo/a, a. cooperative.

coordenada, f. coordinate; **c. cilíndrica,** cylindrical coordinates; **c. esférica,** spherical coordinate; **coordenadas geográficas,** geographical coordinates.

coordinación, f. coordination.

coordinado/a, a. coordinated.

coordinar, v. to coordinate.

copa, f. cup, glass (vaso); crown (de un árbol).

coparticipación, f. joint interest.

copartícipe, m. y f. joint partner.

copépodo, m. copepod; **copépodos (Zoo, subclase),** Copepoda (pl).

copo, m. flake (de nieve), tuft (de lana), lump (de harina); raincloud (US) (nube).

copolimerización, f. copolimerization.

coprecipitación, f. coprecipitation.

copropiedad, f. joint ownership.

cópula, f. copulation (Biol); conjunction.

coque, m. coke (carbón).

coquización, f. coking.

coquizadora, f. coking plant (planta coquizadora).

coral, m. coral.

coraliforme, a. corallitic.

coralino/a, a. coral, coralline; **arrecife (m) c.,** coral reef; **caliza (f) c.,** coral limestone.

corazón, m. heart.

corcho, m. cork.

cordado, m. chordate; **cordados (Zoo, filum),** Chordata (pl).

cordierita, f. cordierite.

cordillera, f. cordillera, mountain chain, mountain range.

cordillerano/a, a. (CSur) Andean.

córdoba, m. (Nic) cordoba (unidad monetaria).

corindón, m. corundum.

Coriolis, m. Coriolis; **efecto (m) de C.,** Coriolis effect.

cormo, m. corm.

cornamenta, f. antlers (pl).

corneana, f. hornfel.

cornisa, f. cornice; **la C. Cantábrica,** the Cantabrian coast.

cornubianita, f. hornfel.

corola, f. corolla.

corolario, m. corollary.

corona, f. crown; **c. luminosa (Met),** luminous corona; **c. para sondeos,** core bit.

corporación, f. corporation.

corpúsculo, m. corpuscle; **c. sanguíneo,** blood corpuscle.

corral, m. yard, courtyard; pen (para ganado), run (para animales).

corrección, f. correction; **c. de niveles,** level correction; **c. de pH,** pH correction; **c. paisajística y medioambiental,** environmental and landscape improvement; **factor (m) de c.,** correction factor.

correctivo/a, a. corrective, remedial; **tomar las medidas correctivas (v),** to take corrective action.

corrector/a, a. corrective, remedial; **medidas (f.pl) correctoras,** corrective measures.

corredor, m. agent, broker; runner (deportes); corridor (pasillo); **c. de ribera (Ecol),** river corridor; **c. de bolsa,** stockbroker; **c. de fincas,** estate agent, real estate broker (US); **c. de fincas rurales,** land agent; **c. natural (Ecol),** wildlife corridor.

corregir, v. to correct, to amend.

correlación, f. correlation; **coeficiente (m) de c.,** correlation coefficient, coefficient of correlation; **c. estratigráfica,** stratigraphical correlation; **c. múltiple,** multiple correlation; **grado (m) de c.,** degree of correlation.

correlacionar, v. to correlate.

correlograma, m. correlogram, correlation graph.

correntada, f. (Arg, Ur, Ch) rapids, strong flow.

correntera, f. (LAm) rapids.

correo, m. post, mail (cartas); courier (mensajero); **casa (f) de correos, correos,** post office.

correr, v. to run; to rush, to hurry; to speed (conductor o coche); to flow (agua); to pass (tiempo).

correspondencia, f. correspondence; letters, post (cartas); communications (ferrocarril etc.).

corriente, 1. f. current (Elec, agua), stream (agua); current month (mes); tendency (tendencia); **c. alterna,** alternating current; **c. costera longitudinal,** longshore current; **c. cruzada,** cross-flow; **c. de barro,** mudflow, lahar; **c. de Humboldt,** Humboldt current; **c. de marea,** tidal current; **c. de resaca,** rip current, undercurrent, undertow; **c. de suspensión,** suspension current; **c. de turbidez,** turbidity current; **c. directa,** direct current; **c. litoral,** coastal current; **c. marginal,** longshore current; **c. profunda,** undercurrent; **c. trifásica,** three-phase current; **c. turbulenta,** turbulent flow; 2. a. running, flowing; **agua (f) c.,** running water, watercourse.

corrimiento, m. landslide, slipping; thrust (Geol); **c. hacia arriba,** upthrust.

corroboración, f. corroboration.

corroborar, v. to corroborate.

corroer, v. to corrode.

corrosión, f. corrosion, rust; **c. bimetálica,** bimetallic corrosion; **c. electrolítica,** electrolitic corrosion; **c. electroquímica,** electrochemical corrosion; **c. por desgaste,** fatigue corrosion.

corrosividad, f. corrosivity.

corrosivola, a. corrosive.

corrugación, f. corrugation.

cortadora, f. cutter, cutting machine.

cortafuego, m. fire gap (en bosque).

corte, 1. m. cut; cutting, felling (de árboles); **c. de corriente,** power cut; **c. transversal,** cross-section; 2. f. royal court, law court; **C. Suprema,** Supreme Court.

corte, f. royal court, law court; **Cortes (Esp),** the Spanish Parliament.

corteza, f. bark (de árboles); crust; **la c. terrestre,** the Earth's crust.

cortical, a. crustal.

cortijo, m. farmhouse.

cortisona, f. cortisone.

cortocircuito, m. short circuit (Elec).

corza, f. doe (hembra).

corzo, m. roe deer (genérico); roe buck (macho).

cos, m. cos (Mat, abr. para coseno).

cosec, m. cosec (m) (Mat, abr. para cosecante).

cosecante, m. cosecant (Mat, abr. cosec).

cosecha, f. crop, yield, harvest; **c. actual,** cover crop; **c. continua,** continuous cropping; **c. permanente,** standing crop.

cosechadora, f. cropper (máquina).

cosechar, v. to harvest, to reap (cereals), to pick (fruit).

coseno, m. cosine (Mat, abr. cos).

cosméticola, a. cosmetic.

cosmopolita, a. cosmopolitan.

cosolvente, m. cosolvent.

costa, f. coast, coastline (del mar); riverbank, lakeside (orilla); cost, price, costs (gasto); **c. baja,** low coast; **c. dentada,** indented coast; **c. marina,** seaboard; **hacia la c.,** shoreward.

costado, m. side, flank; **c. del valle,** valley side.

costalero, m. (Col) waste tip scavenger.

Costa Rica, f. Costa Rica.

costarricense, 1. m. y f. Costa Rican; 2. a. Costa Rican.

costarriqueñola, m. y f. Costa Rican.

costarriqueñola, a. Costa Rican.

coste, m. cost, price; **base (m) del c. más beneficio,** cost-plus-profit basis; **base (m) del c. más comisión,** cost-plus-commission basis; **c. marginal,** marginal cost; **c. beneficio,** cost-benefit; **c. de energía,** energy cost; **c. de instalación,** installation cost; **contrato (m) por el c. más honorarios fijos,** cost-plus-fixed-fee contract; **costes variables,** variable costs; **efectividad (f) del c.,** cost-effectiveness; **índice (m) del c. de la vida,** cost-of-living index; **reducción (f) en el c.,** cost-cutting.

costear, v. to pay for, to finance; to pasture (CSur).

costeñola, 1. m. y f. (LAm) person who lives near the coast; 2. a. coastal.

costera, f. side; slope (Geog); fishing season, nearshore fishing fleet.

costerola, a. coastal.

costo, m. cost; **cálculo (m) de costos,** costing; **contrato (m) al c. más beneficio,** cost-plus contract; **contrato (m) por el c. más un porcentaje,** cost-plus-percentage contract; **c.-eficacia,** cost-effectiveness; **costos de fabricación,** manufacturing costs; **costos-beneficios,** cost-benefit; **depreciación (f) de c. y mantenimiento,** cost-plus-maintenance depreciation; **recuperación (f) de c.,** cost recovery; **reducción (f) de costos,** cost-reduction.

costra, f. crust; **c. continua,** hardpan; **c. de hielo,** ice crust.

COT, m. TOC; **carbono (m) orgánico total,** total organic carbon.

cota, f. height, elevation, height above sea level.

cotangente, m. cotangent (Mat, abr. cotan).

cotiledón, m. cotyledon.

cotización, f. quotation, price; **precio (m) de c.,** quotation price.

coto, m. enclosure (Agr); reserve, hunting ground (caza); goitre, goiter (US) (LAm, Med); **c. cerrado,** closed shop (de sindicatos); **c. de caza,** game reserve, hunting ground.

Couviniense, m. Couvinian.

COV, m. VOC; **compuesto (m) orgánico volátil,** volatile organic compound.

covadera, f. (LAm) guano deposit.

covalencia, f. covalence.

covalente, a. covalent.

covarianza, f. covariance.

cpue, f. cpue; **captura (f) por unidad de esfuerzo (pesquería, caza),** catch per unit effort.

cráneo, m. cranium.

craqueo, m. cracking (Quím).

cráter, m. crater; **c. lateral,** lateral crater; **c. meteorítico,** meteorite crater.

cratón, m. craton.

crecer, v. to grow, to increase (Gen); to rise (prices).

crecida, f. spate, flood; **alerta (f) de crecidas,** flood warning; **canal (m) de desagüe de crecidas,** flood channel; **c. anual,** annual flood; **margen (m) de crecidas,** flood banks; **obras (f.pl) de protección contra las crecidas,** flood protection works; **previsión (f) de crecidas,** flood forecast; **protección (f) contra crecidas,** flood relief; **regulación (f) de crecidas,** flood routing; **riesgo (m) de c.,** flood risk.

crecidola, a. swollen (de río); full-grown (person); large (grande).

creciente, 1. m. crescent (p.e. de la luna); **C. Fértil (Geog),** Fertile Crescent; 2. a. growing; waxing (de la luna).

crecimiento, m. growth, increase; flooding,

rising (de un río); waxing (de la luna); **anillo (m) de c. (Bot)**, growth ring; **c. bacterial,** bacterial growth; **c. de la población,** population growth; **c. demográfico,** population growth; **c. económico sostenido,** sustained economic growth; **c. excesivo,** overgrowth; **c. irruptivo,** irruptive growth; **c. secundario,** secondary growth; **curva (f) de c.,** growth curve; **rapidez (f) de c.,** rate of growth.

crédito, m. credit (Fin), goodwill, standing; **cuenta (f) de c.,** credit account.

crematorio, m. waste tip, rubbish dump, landfill (EIS).

crenuladola, a. crenulate.

crepuscular, a. crepuscular.

cresa, f. maggot, larva (pl. larvae).

cresta, f. crest, ridge, comb (Zoo, de aves); tuft (Zoo, copete); **c. dentada,** serrated ridge; **c. de presión,** pressure ridge; **c. morrénica,** morainic ridge.

creta, f. chalk; **c. nodulosa,** nodular chalk; **pudelado (m) con c. (p.e. para revestimiento de cauces),** chalk puddling.

Cretácico, m. Cretaceous.

cría, f. breeding, rearing (acción); litter (camada), brood (de aves); **c. de ganado,** cattle rearing, stock raising; **c. intensiva,** intensive breeding.

criadero, m. nursery (de plantas), breeding place (de animales), hatchery (de peces); orebody (Geol); **c. de ostras,** oyster bed.

criba, f. screen, latticework; **c. de cinta,** band screen; **c. de tambor,** drum screen.

cribado, m. sieving.

cribaduras, f.pl. screenings, siftings (Min).

cribar, v. to sieve.

crinoide, m. crinoid.

criolita, f. cryolite, criolite.

crioscopia, f. cryoscopy.

criosfera, f. criosphere.

criptocristalinola, a. cryptocrystalline.

crisálida, f. chrysalis.

crisis, f. crisis (pl. crises); **control (m) de c.,** crisis management.

crisol, m. crucible, melting pot.

crisolita, m. chrysolite.

cristal, m. glass (vidrio), crystal (fino), window (ventana), lens (de gafas); **c. de seguridad,** safety glass; **c. reforzado,** reinforced glass; **c. líquido,** liquid crystal.

cristalinola, a. crystalline; **enrejado (m) c.,** crystal latice.

cristalización, f. crystallization; **c. fraccionada,** fractional crystallization.

cristalografía, f. crystallography.

cristobalita, f. cristobalite.

criterio, m. criterion (pl. criteria).

crítica, f. criticism, critique, review.

críticola, a. critical; **método (m) del camino c.,** critical path method.

cromado, m. chromium plating.

cromáticola, a. chromatic.

cromato, m. chromate.

cromatografía, f. chromatography; **c. de gases,** gas chromatography; **c. de gases-líquidos,** gas-liquid chromotography.

cromita, f. chromite.

cromitita, f. chromitite.

cromo, m. chromium (Cr).

cromosoma, m. chromosome.

crónicola, a. chronic.

cronología, f. chronology; **c. por el método de isótopos de carbono,** radio-carbon dating.

cronológicola, a. chronological.

cronostratigrafía, f. chrono-stratigraphy.

cruce, m. crossing (acción de cruzar); cross-roads (de caminos); crossbreed (animal híbrido); crossing, crossbreeding (hibridación); short-circuit (Elec); **c. de peatones (US),** pedestrian crossing, crosswalk; **c. giratorio,** roundabout (GB).

crucero, m. cleavage.

crudo, m. crude oil.

crudola, a. crude (aceite), raw (no cocido), unripe (no maduro), harsh (cruel); **desagües (m.pl) crudos,** crude sewage.

Crustácea, f.pl. Crustacea (pl) (Zoo, clase).

crustáceo, m. crustacean; **crustáceos (Zoo, clase),** Crustacea (pl).

cruz, f. cross; **C. Roja,** Red Cross.

cruzadola, 1. m. y f. cross, crossbreed, hybrid; 2. a. cross; **fertilización (f) c.,** cross-fertilization; **polinización (f) c.,** cross-pollination.

cruzamiento, m. crossing, crossbreeding (de híbridos); **c. consanguineo,** inbreeding; **c. prueba,** back-cross.

cruzar, v. to cross; to crossbreed (de hibridismo).

ctg, m. cotan (Mat, abr. para cotangente).

cuadrado, m. square; cuadrat (m) (Bot); **mínimos (m.pl) cuadrados (Mat),** least squares.

cuadrante, m. quadrant.

cuadráticola, a. quadratic; **ecuación (f) c.,** quadratic equation.

cuadrícula, f. grid, cross-ruling.

cuadriculado, m. grid, cross ruling.

cuadriculadola, a. grid; **papel (m) c.,** graph paper, squared paper.

cuadrilátero, m. four-sided figure.

cuadripolo, m. quadripole.

cuadrivalente, a. quadrivalent, tetravalent.

cuadro, m. square (Mat, polígono); table, diagram (Mat, gráfico); bed, plot (Agr); team (equipo).

cuajada, f. curdling, clotting; setting (Bot, frutas).

cuajo, m. rennet.

cualitativola, a. qualitative.

cuantía, f. quantity (cantidad), extent (extensión).

cuántica, a. quantum; **teoría (f) c.,** quantum theory.

cuantificable, a. quantifiable; **no c.,** non-quantifiable.

cuantificación, f. quantification.

cuantil, m. quantile, fractile.

cuantitativola, a. quantitative.

cuanto, m. quantum (pl. quanta).

cuarcíferola, a. quartziferous.

cuarcita, f. quartzite.

cuartil, m. quartile; **c. más bajo,** lower quartile; **c. superior/inferior,** upper/lower quartile.

cuarzo, m. quartz; **c. ahumado,** smoky quartz; **c. rosado,** rose quartz; **que contiene c. (a),** quartz-bearing.

cuarzosola, a. quartzose.

Cuaternario, m. Quaternary.

Cuba, f. Cuba.

cubanola, 1. m. y f. Cuban; 2. a. Cuban.

cubeta, f. keg, small cask (tonel); bucket, pail (cubo); **c. de precipitación (Quím),** beaker.

cúbicola, a. cubic; **raíz (f) c. (Mat),** cube root.

cubierta, f. cover, overburden; **c. arbórea,** tree cover; **c. de suelo,** soil cover; **c. rocosa,** rock cover.

cubiertola, a. covered; overcast (el cielo).

cubo, m. cube (polígono); bucket (balde), tank (tanque); **c. de basura,** rubbish bin (GB), garbage can (US), trash can (US); **elevar al c. (v) (Mat),** to cube.

cucaracha, f. cockroach.

cuchara, f. spoon; bailer, auger (Ing), excavator scoop (en una excavadora); **c. hidráulica,** hydraulic grab.

cucharada, f. spoonful.

cuenca, f. catchment, basin (GB); watershed (US) (Geog, de ríos); eye socket (Anat); **c. experimental,** experimental catchment; **c. artesiana,** artesian basin; **c. colectora,** basin catchment; **c. de alabeada,** down-warped basin; **c. de captación,** drainage area; **c. de drenaje,** drainage basin, river basin; **c. hidrográfica,** drainage basin, hydrographic basin; **c. oceánica,** ocean basin; **c. sedimentaria,** sedimentary basin; **gestión (f) de c.,** catchment management; **la c. del Ebro,** the Ebro basin.

cuenta, f. account, counting (acción), bill (factura); **c. bancaria,** bank account; **c. corriente,** current account; **cuentas de gestión,** management accounts.

cuerno, m. horn, antler.

cuerpo, m. body; **c. igneo,** igneous body; **c. mineralizado,** ore body.

cuesta, f. slope, hill; **c. abajo,** downhill; **c. arriba,** uphill; **en c.,** sloping, on a slope.

cuestión, f. question, issue, matter.

cuestionario, m. questionaire.

cueva, f. cave (Geog); vault, cellar (Con).

cuidado, m. care; **c. parental (Zoo),** parental care.

culebra, f. snake.

culebrina, f. forked lightening.

culminación, f. culmination.

culombio, m. coulomb (abr. C, unidad de carga eléctrica).

culombiómetro, m. coulombmeter.

cultivable, a. cultivable, arable; **tierra (f) cultivable,** arable land.

cultivador, m. y f. farmer, cultivator, grower (de plantas).

cultivar, v. to cultivate, to grown, to till (labrar la tierra).

cultivo, m. cultivation, farming (Agr); culture (crecimiento); **caldo (m) de c. (Biol),** culture medium; **c. celular,** tissue culture; **c. de microorganismos,** culture of microorganisms; **c. hidropónico,** hydroponics; **c. in vitro,** tissue culture.

cultura, f. culture; **la c. popular,** popular culture.

cumbre, f. summit, top, hilltop; **conferencia (f) c.,** summit conference.

cumplir, v. to comply.

cumulativola, a. cumulative.

cúmulo, m. cumulus (nube); accumulation (montón).

cúmulola, a. cumulous.

cumulonimbos, m. cumulonimbus.

cuneta, f. ditch; road bank, hard shoulder (borde de la carretera); gutter (en calle).

cuña, f. wedge; **c. de cuarzo,** quartz wedge; **c. salina (en aquíferos),** saline wedge.

cuota, f. quota, share.

cupo, m. quota, share.

cúpricola, a. cupric.

cuprita, f. cuprite.

cuproníquel, m. cupronickel.

cuprosola, a. cuprous.

cúpula, f. cupola.

curie, m. curie (abr. Ci, unidad de medida de radiactividad).

currículum, m. curriculum; **c. vitae,** curriculum vitae.

curso, m. course, direction (rumbo); watercourse (cauce); **c. de agua,** watercourse; **el c. de un río,** the course of the river; **moneda (f) de c. legal,** legal tender, legal money.

curtido, m. tanning; **aguas (f.pl) residuales de planta de curtidos,** tannery wastewater; **industria (f) de curtidos,** tanning industry.

curtiduría, f. tannery.

curva, f. curve; **coincidencia (f) de curvas,** curve-fitting; **c. acumulativa,** cumulative curve; **c. característica de una bomba,** pump characteristic curve; **c. de caudales en función del nivel,** stage-discharge curve; **c. de crecimiento de población,** population growth curve; **c. de descenso-tiempo,** time-drawdown curve; **c. de doble masa,** double-mass curve; **c. de escape,** breakaway curve; **c. de frecuencia,** frequency curve; **c. de intensidad de lluvia-duración,** rainfall intensity duration curve; **c. de nivel,** contour line;

c. de pozo, well curve; **c. de recesión,** recession curve; **c. de regresión,** regression curve; **c. de remanso,** backwater curve; **c. de secado,** drying curve; **c. de volumen embalsado,** reservoir storage curve; **c. envolvente,** envelope curve; **c. gausiana,** gaussian curve; **c. hipsométrica,** hypsometric curve; **c. logarítmica,** logarithmic curve (abr. log curve); **c. tipo,** type curve.

curvatura, f. curvature.

cúspide, f. cusp (Anat, Geog); summit, pinnacle, zenith, peak; apex (Mat).

custodia, f. custody, care, safekeeping.

custodio, m. custodian.

cutáneola, a. cutaneous.

cutícula, f. cuticle.

cutis, m. skin.

cutrero, m. (LAm) solid waste segregator.

CH

chabola, f. hut, shack, shanty.
chacal, m. jackal.
chacarero, m. (LAm) smallholder (dueño), smallholding (terreno).
chaflán, m. bevel, chamfer.
chagra, m. y f. (Ec) peasant, farm.
chagüe, m. (CAm) bog, swamp.
chagüite, m. (CAm, Méx) swamp.
chalibita, f. chalybite.
chambero, m. (Ec) waste tip scavenger.
champa, f. (Arg) clod.
chancaca, f. (US) loaf sugar (azúcar).
changadorla, m. y f. odd job man.
chapa, f. plate, sheet (de metal); plywood (contrachapado); lock (LAm) (cerradura); bottle top (de una botella); c. de acero, steel sheet; c. ondulada, corrugated iron.
chapapote, m. bitumen, asphalt.
chaparral, m. chaparral, thicket of dwarf oaks.
chaparro, m. dwarf oak, holm oak (encina), kermes oak (coscoja).
chaparrón, m. shower (de lluvia); c. violento, cloudburst, downpour.
chapatal, m. waterlogged ground.
chapopote, m. pitch, tar.
chapote, m. pitch, tar; asphalt (CAm, Carib, Méx).
chapulín, m. grasshopper.
charapa, f. (Perú) jungle dweller.
charca, f. small pond, puddle.
charco, m. pond, pool; c. de agua estancada, stagnant pond.
charcón, m. large pond.
charnela, f. hinge, hinge line; c. del anticlinal, anticlinal crest.
charnokita, f. charnokite.
charnoquita, f. charnokite.
chatarra, f. scrap (de metal, etc.); c. de metal, scrap metal; c. de vidrio, cullet; parque (m) de c., scrapyard.
Chattiense, m. Chattian.
check-list, m. check-list.
chepeiro, m. (Br) waste tip scavenger.

chequeo, m. check-up (revisión), check; comparison (comparación).
chernozem, m. (Arg) chernozem.
chícharo, m. pea.
chiflón, m. (LAm) incline; c. de descarga, chute.
chihuido, m. (LAm) volcanic neck (tapón volcánico).
Chile, m. Chile.
chilenola, m. y f. Chilean.
chilenola, a. Chilean.
chimenea, f. chimney, vent (Geol); cañón (m) de la c., chimney stack; c. central, central vent; salida (f) de la c., chimney stack.
China, f. China.
chinampa, f. (Méx) pre-Colombian method of water meadow cultivation.
chinche, f. bug (hemíptero).
chip, m. chip (Comp).
chirimoya, m. chirimoya, cherimoya, custard apple (fruta).
chirivía, f. parsnip.
chirriar, v. to stridulate (de insectos).
chirrido, m. stridulation (de insectos).
choclo, m. corn (US, Arg, Bol, Ch, Ec), maize (GB).
chopera, f. poplar grove.
chopo, m. poplar.
choque, m. blow, impact.
chorizo, m. rubbish tip operative, garbage man.
chorro, m. blast, jet; spray (de mar); c. de arena, sandblast; c. de granalla, shotblast; limpieza (f) con c. de agua a presión, jetting.
chortal, m. springhead of intermittent river (barranco).
chubasco, m. shower (of rain).
chumbera, f. prickly pear plant.
chumbola, m. y f. prickly pear (fruta).
chupón, m. y f. sucker, shoot (Bot).
chupónla, a. sucking.
churequero, m. (Nic) waste tip scavenger.

D

DAC, m. CAD; **diseño (m) asistido por ordenador,** computer-aided design.
dacita, f. dacite.
dacrón, m. dacron.
damasco, m. apricot (fruta); apricot tree (LAm) (árbol).
Daniense, m. Danian.
daño, m. damage; **evaluación (f) de daños,** damage evaluation.
DAO, m. CAD; **diseño (m) asistido por ordenador,** computer-aided design.
darro, m. small stream used for village drainage.
datación, f. dating; **d. con radiocarbono,** radiocarbon dating; **d. con tritio,** tritium dating; **d. geológica,** geological dating.
datar, v. to date.
dátil, m. date (fruit).
datilero, m. date tree.
dato, m. datum (pl. data).
datos, m.pl. data (pl); **análisis (m) de d.,** data analysis; **conjunto (m) de d.,** data set; **d. a tratar,** raw data; **d. categorizados,** ranked data; **d. climáticos,** climatic data; **d. de la prueba,** test data; **d. disponibles,** available data; **d. en bruto,** raw data; **d. estadísticos,** statistical data; **d. experimentales,** experimental data; **d. no analizados, d. no evaluados, d. sin procesar,** raw data; **d. ordenados por sus rangos,** ranked data; **procesamiento (m) de d.,** data processing; **registrador (m) automático de d.,** data logger; **toma (f) de d.,** data collection.
daya, f. depression with abundant vegetation in arid areas.
DBO, f. BOD; **demanda (f) bioquímica de oxígeno,** biochemical oxygen demand.
DDT, m. DDT; **diclorodifenil-tricloroetano (m),** dichlorodiphenyltrichloroethane.
debate, m. debate; discussion, argument (discusión).
debilitamiento, m. debilitation, weakening.
decaimiento, m. decay, die-off; **d. bacteriano,** bacterial die-off.
decantación, f. decanting, decantation; **lavado (m) por d.,** elutriation.
decantador, m. decanter (Trat).
decantar, v. to decant.
decapado, m. cleaning, scouring; pickling, descaling (adobado); **cuba (f) de d.,** pickling tank.
decapar, v. to pickle (metal); **instalación (f) para d.,** pickling plant.
decápodo, m. decapod; **decápodos (Zoo, orden),** Decapoda (pl).
decarbonatación, f. decarbonation.

decenio, m. decade.
decibelio, m. decibel.
decíduola, a. deciduous.
decimal, 1. m. decimal; 2. a. decimal.
decisión, f. decision; **proceso (m) de toma de decisiones,** decision making process.
declaración, f. statement; **D. de Impacto Ambiental (Esp, abr. DIA),** Environmental Impact Statement (GB, abr. EIS); **d. obligatoria (a),** notifiable.
declinación, f. declination; **d. magnética,** magnetic declination.
decoloración, f. discolouration, discoloration (US), fading.
decolorante, m. bleach, bleaching agent.
decolorar, v. to discolour, to discolor (US), to fade.
decoración, f. decoration, ornamentation.
decrecimiento, m. decrease, lessening.
decreto, m. decree, order; enactment (para poner en ejecución las leyes); **d. ley,** decree-law; **real d. (Esp, abr. RD),** royal decree; **real d. legislativo (Esp, abr. RDL),** legislative decree.
dedalera, f. foxglove.
dedo, m. finger (del mano); toe (del pie).
deducción, f. deduction, inference.
defectivola, a. defective, imperfect.
defecto, m. defect.
defectuosola, a. defective, deficient.
defensa, f. defence, defense (US); **d. contraincendio,** fire defence, fire defense (US).
defensivola, a. defensive.
defensor, m. defender; **D. del Pueblo (Jur),** ombudsman.
deficiencia, f. deficiency, defect; deficiency disease (Med, enfermedad).
déficit, m. deficit; **d. de humedad en el suelo,** soil moisture deficit; **d. de saturación,** saturation deficit.
definición, f. definition.
definidola, a. definite, defined; **bien d.,** well-defined.
definir, v. to define.
deflación, f. deflation.
deflector, m. deflector, baffle, baffle plate.
defluorización, f. defluoridation.
defluoruración, f. defluoridation.
deforestación, f. deforestation.
deforestar, v. to deforest.
deformación, f. deformation; **d. elástica,** elastic deformation; **d. geométrica,** geometric distortion.
deformar, v. to deform, to distort.
deforme, a. shapeless.

degeneración, f. degeneration (proceso); degeneracy (estado).

degradable, a. degradable.

degradación, f. degradation.

degradar, v. to degrade.

degustación, f. tasting, sampling.

degustar, v. to taste, to sample.

dehesa, f. pasture with oak trees; **d. carneril,** dehesa grazed by ruminants; **d. potral,** dehesa used for growing on foals.

dehiscente, a. dehiscent.

delegación, f. delegation; council; local government, municipal district (LAm).

delegar, v. to delegate.

deleznable, a. weak, fragile, feeble.

delfín, m. dolphin.

delgadola, a. thin, slim.

delicuescencia, f. deliquescence.

delicuescente, a. deliquescent.

delimitación, f. delimiting, delimitation.

delimitar, v. to delimit.

delineación, f. contouring.

delineante, m. draughtsman (GB), draftsman (US).

delinear, v. to delineate, to trace.

delito, m. offence, offense (US), crime, misdeed; **d. ambiental,** environmental abuse.

delta, m. delta; **d. cuspidada,** cuspate delta; **d. encorvado,** arcuate delta; **d. triangular,** cuspate delta.

deltáicola, a. deltaic.

deltoide, m. deltoid.

demanda, f. demand; **d. bioquímica de oxígeno (abr. DBO),** biochemical oxygen demand (abr. BOD); **d. de oxígeno,** oxygen demand; **d. elástica,** fluctuating demand; **d. industrial,** industrial demand; **d. oscilante,** fluctuating demand; **d. química de oxígeno (abr. DQO),** chemical oxygen demand.

demografía, f. demography.

demográficola, a. demographic, population; **crecimiento (m) d.,** population increase; **explosión (f) d.,** population explosion.

demoler, v. to demolish (un edificio).

demolición, f. demolition.

demostración, f. demonstration, show; **planta (f) de d.,** demonstration plant.

demostrar, v. to demonstrate, to show.

dendríticola, a. dendritic.

dendrocronología, f. dendrochronology.

dendrocronólogola, m. y f. dendrochronologist.

dendroide, m. dendroid.

denitrificar, v. to denitrify.

densidad, f. density; **d. absoluta,** absolute density; **d. aparente,** apparent density; **d. de agua salina,** density of saline water; **d. de masa,** bulk density; **d. de población,** population density; **d. de probabilidad,** probability density; **d. espectral,** spectral density; **d. relativa,** relative density.

densímetro, m. densimeter.

densola, a. dense, thick; **humo (m) d.,** thick smoke.

dentadola, a. jagged.

dental, a. dental.

denudación, f. denudation.

denuncia, f. denunciation (Jur), accusation.

departamento, m. department, section (sección administrativa); region, province (provincia); compartment (p.e. de una caja); **D. de Comercio,** Department of Trade.

depauperación, f. impoverishment, starvation.

deporte, m. sport; **deportes acuáticos,** water sports.

deposición, f. deposition; **d. sedimentaria,** sedimentary deposition.

deposicional, a. depositional.

depositar, v. to deposit, to settle.

depósito, m. deposit; **d. aluvial,** alluvial deposit; **d. de aguas someras,** shallow-water deposits; **d. de estuario,** estuary deposit; **d. de relleno de cauce,** channel-filled deposit; **d. de seguridad (de desechos),** safe repository; **d. eólico,** wind-blown deposit; **depósitos de arrastre,** outwash.

depreciación, f. depreciation.

depreciar, v. to depreciate.

depredación, f. predation.

depredador, m. predator.

depredadorla, a. predatory.

depresión, f. depression (Gen, Met), hollow (hueco); dip (en carretera); drop, fall; recession (en economía); **d. de bombeo,** pumping depression; **d. estructural (Geol),** saddle; **d. kárstica (Geog),** sink.

depuración, f. purification, cleansing; debugging (Comp); **d. biológica,** biological treatment; **d. de agua,** water treatment; **d. de aguas residuales,** sewage treatment; **d. previa,** preliminary water purification; **d. primaria,** primary water purification; **d. secundaria,** secondary water purification; **d. terciaria,** tertiary water purification.

depuradorla, a. purifying, treating; **estación (f) d.,** treatment plant, purifying plant.

depuradora, f. purifying plant; **d. de agua,** water treatment plant; **d. de aguas residuales,** sewage works, sewage farm, wastewater treatment works.

depurar, v. to cleanse, to purify (Gen); to scrub (gas); to debug (Comp).

derecho, m. law (ley); right, claim (pretensión); tax, duty (impuestos); fees (cargas); **d. civil,** civil law; **d. consuetudinario,** customary law, common law (GB); **d. de paso,** right of way; **d. de prescripción (establecido por uso),** prescriptive right; **d. marítimo,** maritime law; **derechos civiles,** civil rights; **derechos del autor,** copyright; **por d.,** by right.

deriva, f. drift; **arte (m) de d. (pesca),** drift net; **d. continental,** continental drift; **d. de los continentes,** contiental drift; **d. de los polos (Geol),** polar wandering; **d. genética,** genetic drift; **d. litoral,** longshore drift; **red (f) de deriva,** drift net.

derivada, f. derivative (Mat); by-product; **d. lácteo,** milk products; **d. parcial,** partial derivative; **ecuación (f) d. parcial,** partial derivative equation.

derivado, m. derivative, by-product (Quím); **d. del petróleo,** petroleum derivative.

derivar, v. to derive.

derivativo, m. derivative (Mat).

dermatitis, f. dermatitis.

dermatología, f. dermatology.

dérmicola, a. dermal, skin.

dermis, f. dermis.

derogación, f. derogation (of the law).

derogar, v. to repeal, to annul, to abolish.

derramadero, m. spillway of dam.

derramamiento, m. outpouring, spilling.

derramar, v. to spill (rebosar); to pour (verter); to scatter (esparcir).

derrame, m. overflow, spillage, leakage; **d. de llanura aluvial,** flood-plain outflow; **d. de petróleo,** oil spill.

derretimiento, m. thawing (de nieve), melting (Gen).

derretir, v. to thaw (de nieve), to melt (Gen).

derrocamiento, m. demolition; overturning (p.e. de un gobierno).

derrocar, v. to knock down, to demolish.

derrubiar, v. to scour, to erode.

derrubio, m. scree, drift; **d. glacial,, d. glaciar,** glacial drift; **pendiente (m) de derrubios,** scree slope.

derrumbadola, a. caved.

derrumbamiento, m. collapse (desplome), demolition, plunge (caída), tipping; **d. de tierra,** landslide; **d. ilegal,** fly tipping.

derrumbar, v. to knock down, to demolish.

derrumbarse, v. to collapse, to cave in.

derrumbe, m. slide, collapse.

desabastecimiento, m. shortage.

desacuerdo, m. disagreement.

desaguable, a. drainable.

desaguadero, m. drain; outfall (de efluente); **d. de pie (de una presa),** toe drain; **d. marino,** marine outfall.

desaguar, v. to dewater, to drain (p.e. a mine).

desagüe, m. drainage (drenaje); drain, outlet (tubo, canal); **d. de piso,** field drain; **d. interior,** house drain; **desagües domésticos,** domestic sewage, municipal wastewater.

desaireación, f. deaeration.

desalar, v. to desalinar, to desalt.

desalinar, v. to desalinate.

desalinización, f. desalination; **capacidad (f) instalada de d.,** desalination installed capacity; **d. nuclear,** nuclear desalination; **d. por hidratación,** desalination by hydration; **d. por intercambio iónica,** desalination by ion exchange; **d. por ósmosis inversa,** desalination by reverse osmosis; **sistemas (m.pl) de d.,** desalination systems.

desalinizadorla, a. desalinating; **capacidad (f) d.,** desalinating capacity; **planta (f) d.,** desalinating plant.

desalinizar, v. to desalinate.

desamortización, f. disentailment, freeing of land (Jur).

desaparecer, v. to disappear.

desaparición, f. disappearance.

desaprobación, f. disapproval.

desarenador, m. grit remover, desander (Trat, máquina).

desarraigar, v. to uproot, to dig up; **d. un árbol,** to uproot a tree.

desarraigo, m. uprooting; **d. social,** social uprooting.

desarreglo, m. disorder, mess.

desarrollador, m. developer.

desarrollo, m. development; growth, expansion (Com); **agencia (f) de d.,** development agency; **banco (m) de d.,** development bank; **corporación (f) de d.,** development corporation; **d. del acuífero,** aquifer development; **d. del pozo por sobrebombeo,** well development by overpumping; **d. efectivo,** effective development; **d. rural,** rural development; **d. sostenible,** sustainable development; **índice (m) de d. (Com),** growth rate; **país (m) en vías de d.,** developing country; **plan (m) de d.,** development plan; **polo (m) de d., zona (f) de d.,** development area.

desasimilación, f. dissimilation.

desasimilar, v. to dissimilate.

desastre, m. disaster; **d. natural,** natural disaster; **d. provocado por el hombre,** man-made disaster.

desautorizadola, a. unauthorized.

desautorizar, v. to remove authority, to repudiate.

desbancar, v. to supplant, to replace (suplantar).

desbarbador, m. cyclone separator.

desbaste, m. smoothing, polishing, planing.

desbloquear, v. to release, to unblock, to free.

desbordamiento, m. overflowing.

desbordante, a. overflowing.

desbordar, v. to overflow.

desbrozar, v. to clear (basura o matorral).

descableado, m. uncabling.

descamación, f. peeling (Geol).

descansar, v. to relax; to rest on (Geol).

descanso, m. rest (rato); break (en trabajo), relaxation (relajamiento); toilet (LAm, retrete).

descantillar, v. to chip.

descarga, f. discharge, outlet, spill, unloading; **d. de agua,** flushing; **d. de agua subterránea,** groundwater discharge; **d. de manantiales,** spring discharge; **d. de residuos peligrosos al mar,** disposal of hazardous waste to sea; **d. de vapor,** steam flushing; **d. inducida,** induced discharge; **d. natural,** natural discharge.

descargador, m. docker.

descargar, v. to unload, to discharge.

descascaramiento, m. flaking.

descascarar, v. to flake, to peel.

descascarillado, m. scaling, peeling-off, flaking.

descenso, m. drop, fall (caida); descent (bajada); drawdown (bajada del nivel de agua etc.); down-gradient (cuesta abajo); **d. de nivel de agua subterránea,** groundwater drawdown; **d.-distancia,** distance drawdown; **d. específico,** specific drawdown; **d. observado,** observed drawdown; **d. piezométrico,** piezometric drawdown; **d. residual,** residual drawdown; **d. teórico,** theoretical drawdown; **d.-tiempo,** time-drawdown.

descentralización, f. decentralization.

descentralizar, v. to decentralize.

descifrar, v. to solve, to decipher, to figure out.

descloración, f. dechlorination.

desclorinización, f. dechlorination.

desclorización, f. dechlorination.

descolmatación, f. desilting.

descolorar, v. to bleach (to remove colour).

descomponer, v. to decay, to decompose, to rot.

descomposición, f. decay, decomposition; **d. aerobia,** aerobic decomposition.

descompresión, f. decompression.

desconexión, f. disconnection.

descongelar, v. to thaw.

descongestión, f. decongestion.

descongestionar, v. to relieve congestion.

descontaminación, f. decontamination; cleanup (Trat, p.e. de suelo).

descontaminar, v. to decontaminate.

descontrol, m. lack of control.

descoordinación, f. lack of coordination.

descortezar, v. to debark (un árbol).

descubrimiento, m. discovery.

descubrir, v. to discover.

descuento, m. discount; **tipo (m) de d. (Fin),** discount rate.

descuidadola, a. careless, negligent.

descuido, m. negligence, oversight.

desecación, f. drying, desiccation, withering (de plantas).

desecadola, a. desiccated, dried.

desecar, v. to drain, to dry up, to desiccate; to wither (de plantas).

desechar, v. to throw out, to reject, to scrap.

desecho, m. waste, discharge; **aprovechamiento (m) de desechos,** waste recovery; **d. comercial,** commercial waste, trade waste; **d. de declaración obligatoria,** notifiable waste; **d. de demolición,** demolition waste, building waste; **d. doméstico,** domestic waste; **d. especial,** special waste; **d. industrial,** industrial discharge; **d. inerte,** inert waste; **d. tóxico,** toxic waste; **desechos radioactivos,** radioactive waste; **producto (m) de d.,** waste

product; **productor (m) de d.,** waste producer; **recuperación (f) de desechos,** waste recovery; **transporte (m) de desechos,** waste transport; **transportista (m) de desechos,** waste carrier.

desembarcadero, m. pier, jetty, landing stage.

desembocadura, f. outfall, river mouth; **d. de aguas residuales,** sewage outfall; **d. del río,** river mouth; **d. en el mar,** long sea outfall.

desembocar, v. to flow out of.

desembolso, m. outlay.

desempeño, m. performance test.

desempleadola, a. unemployed.

desempleados, m.pl. unemployed people (pl).

desempleo, m. unemployment; **d. estacional,** seasonal unemployment.

desengrasado, m. degreasing.

desengrasador, m. degreaser (Trat, máquina).

desengrasar, v. to degrease.

desenlodamiento, m. desludging.

desenterrar, v. to unearth.

desentrañar, v. to work out; to figure out (US).

desenvolvimiento, m. development (of events).

desérticola, a. desert, barren (erial); deserted (despoblado).

desertificación, f. desertification (por las actividades humanos).

desertización, f. desertification (procesos naturales).

desestabilizar, v. to destabilize.

desfasadola, a. out of phase.

desfasaje, m. phase shift, phase difference.

desfasar, v. to phase out.

desfase, m. phase shift.

desfavorecidola, a. underprivileged.

desfibradora, f. shredder; **d. de desecho,** waste shredder.

desfiladero, m. narrow pass, defile.

desgarrar, v. to tear.

desgasificación, f. degassing.

desgasificadola, a. degassed.

desgasificar, v. to degas.

desgastadola, a. worn away.

desgastar, v. to wear away, to wear down.

desgaste, m. scouring, wearing away, erosion; **d. fluvial,** river scouring.

desglose, m. breakdown, cutting, editing; **d. de los costos,** breakdown of costs.

desgrasado, m. (LAm) degreasing.

deshabitadola, a. uninhabited, unoccupied.

deshabitar, v. to depopulate.

deshalogenación, f. dehalogenation.

deshelar, v. to thaw.

deshidratación, f. dewatering, dehydration; **d. de los fangos residuales,** sewage sludge dewatering.

deshidratadola, a. dehydrated.

deshidratar, v. to dehydrate.

deshielo, m. thaw, thawing.

deshumidificador, m. dehumidifier.

deshumidificar, v. to dehumidify.

desierto, m. desert; **d. de arena,** sand desert, erg; **d. de baja latitud,** low-latitude desert.

desigualdad, f. inequality (Mat); roughness, unevenness (rugosidad); inconsistency (inconsistencia).

desilicificación, f. desilicification.

desilificación, f. desilification.

desincentivización, f. dissuasion.

desincrustación, f. deincrustation; **d. química,** chemical deincrustacion.

desinfección, f. disinfection; **d. por luz UV (Trat),** UV disinfection.

desinfectante, m. disinfectant.

desinfectar, v. to disinfect.

desinformación, f. disinformation, false information.

desintegración, f. disintegration, breaking up.

desintegrador, m. disintegrator.

desintegrarse, v. to disintegrate.

desintoxicación, f. detoxification.

desintoxicar, v. to detoxify.

desionización, f. deionization.

desionizar, v. to deionize.

desistir, v. to desist, to give up.

deslinde, m. delineating, defining.

deslizamiento, m. slip, slide; creep (Geol); **d. de masas,** mass displacement; **d. de suelo,** soilslip; **d. de tierra,** landslip; **d. rotacional,** rotational slide; **superficie (f) de d.,** slip surface.

deslustradola, a. tarnished.

deslustrar, v. to tarnish.

deslustre, m. dullness, tarnish.

desmantelamiento, m. dismantling.

desmenuzador, m. (LAm) shredder (máquina).

desmineralización, f. demineralization.

desmineralizar, v. to demineralize.

desmontar, v. to strip (Min), to level.

desmonte, m. cutting; clearing (de árboles); soil stripping; **d. y relleno (Ing), d. y terraplén (Ing),** cut and fill.

desmonterado, m. stripping (Min).

desmoronamiento, m. slumping, settling; **d. submarino,** marine slumping.

desnatadola, a. skimmed (de leche).

desnaturalizar, v. to denature.

desnitrificación, f. denitrification.

desnitrificante, a. denitrifying.

desnivel, m. unevenness; bench (Geol).

desnutrición, f. malnutrition.

desodorización, f. deodourization. deodorization (US).

desorción, f. desorption.

desoxi-, pf. deoxy-.

desoxidación, f. deoxidation, deoxidization.

desoxidante, 1. m. deoxidant; 2. a. deoxidizing.

desoxidar, v. to deoxidize.

desoxigenación, f. deoxygenation.

desoxigenar, v. to deoxygenate.

despeñadero, m. precipice, cliff.

desperdicio, m. waste, waste products; **desperdicios (m.pl),** rubbish, refuse, garbage (US).

despido, m. dismissal (de un empleado).

despilfarrar, v. to waste, to squander.

despilfarro, m. wasting, squandering, wastefulness.

desplazamiento, m. displacement, slip; **círculo (m) de d.,** slip circle; **d. costero,** longshore drift; **d. de falla,** fault slip.

desplazar, v. to displace, to move.

desplegar, v. to display, to unfold.

despliegue, m. deployment, show, display.

desplome, m. collapse, fall (caida); batter (Con, fachada inclinada).

despoblación, f. depopulation.

despoblado, m. wilderness, open country, uninhabited place.

despobladola, a. uninhabited, deserted, desolate; **d. de árboles,** treeless.

despoblar, v. to depopulate; **d. de árboles,** to deforest.

despojo, m. plundering (robo); plunder, booty (botín); debris, rubble (escombro).

despojos, m.pl. remains (pl); dispossession (desposeimiento); booty, spoils (botín); **d. mortales,** mortal remains.

desportillamiento, m. chipping.

desportillar, v. to break open, to break down.

despreciable, a. negligible (insignificante); worthless (de poco valor); despicable (de una persona).

desprendimiento, m. detachment, separation (separación); emission, shedding, escaping (p.e. de gases); **d. de tierras,** landslide.

desproporcionadola, a. misfit; **río (m) d.,** misfit river.

desprotegidola, a. unprotected.

desregulación, f. deregulation.

desregular, v. to deregulate.

destacar, v. to highlight, to detail.

destilación, f. distillation; **d. al vacío,** vacuum distillation; **d. de vapor,** steam distillation; **d. fraccional,** fractional distillation; **d. instantánea,** flash distillation; **d. multiefecto,** multistage distillation; **d. por termocompresión,** distillation by thermocompression.

destilación, f. distillation; **d. solar,** solar distillation.

destilado, m. distillate.

destiladola, a. distilled.

destilar, v. to distil.

destino, m. destiny, fate; **d. de contaminantes,** fate of contaminants.

destrozo, m. breakage, ruin.

destructivola, a. destructive.

desulfurización, f. desulphurization, desulfurization; **d. del gas de combustión,** flue-gas desulphurization.

desván, m. attic.

desviación, f. diversion, deviation; **canal (m) de d.,** diversion channel; **d. estándar,** standard deviation; **d. uniforme (Mat),** standard deviation; **presa (f) de d.,** diversion dam.

desvío, m. deviation, deflection (de rumbo);

diversion (en una carretera); **d. del río,** river diversion.

detallar, v. to detail.

detalle, m. detail.

detección, f. detection; **límite (f) de d.,** detection limit.

detector, m. detector, scanner; **d. direccional,** direction scanner.

detención, f. stoppage, standstill, detention; **período (m) de d. (Trat),** detention time.

detergente, m. detergent; **d. aniónico,** anionic detergent; **d. sintético,** synthetic detergent.

deterioro, m. deterioration, impairment, damage; **d. en la calidad de las aguas,** deterioration in water quality.

determinación, f. determination; **d. de campo,** field determination; **d. de laboratorio,** laboratory determination.

determinadola, a. definite, specific.

determinante, m. determinand.

determinista, a. deterministic; **modelo (m) d.,** deterministic model.

determinísticola, a. deterministic.

detonador, m. detonator, blasting cap.

detonar, v. to detonate, to blow up, to explode.

detrimento, m. detriment, damage, harm; **en d. de,** to the detriment of.

detríticola, a. detrital.

detritívoro, m. detritivore.

detrito, m. detritus, debris.

detritos, m. cuttings (Min); **d. de perforación,** drill cuttings.

deuterio, m. deuterium.

devastación, f. devastation.

devastar, v. to devastate.

devitrificación, f. devitrification.

Devónico, m. Devonian.

dextrina, f. dextrin.

DGMA, f. **Dirección (f) General de Medio Ambiente (Esp),** Environment Board of the Ministry of Public Works.

DIA, f. EIS; **Declaración (f) de Impacto Ambiental (Esp),** Environmental Impact Statement (GB).

diabasa, f. diabase.

diaclasa, f. joint; **d. de tracción,** tension joint; **d. principal,** major joint; **d. transversal,** cross joint; **familia (f) de diaclasas,** joint set; **superficie (f) de diaclasas,** joint surface.

diaclasamiento, m. jointing.

diacrónicola, a. diachronous.

diafragma, m. diaphragm; **d. de contención,** diaphragm wall.

diagénesis, f. diagenesis.

diagenéticola, a. diagenetic.

diagnosticar, v. to diagnose.

diagnóstico, m. diagnosis.

diagnósticola, a. diagnostic.

diagonal, 1. f. diagonal; 2. a. diagonal.

diagrafía, f. well logging, downhole logging (en sondeos).

diagrama, m. diagram; **d. bidimensional,** two-

dimensional diagram; **d. circular,** circular diagram; **d. de dispersión,** scatter diagram; **d. de esfuerzos cortantes (Fís),** shear diagram; **d. de flujo,** flow diagram; **d. de frecuencia,** contour diagram; **d. logarítmico,** logarithmic diagram; **d. triangular,** triangular diagram; **d. tridimensional,** tridimensional diagram.

dial, m. dial (p.e. de un radio).

diálisis, f. dialysis.

dializar, v. to dialyze.

diamante, m. diamond.

diámetro, m. diameter; **d. de la bomba,** pump diameter; **d. del pozo,** well diameter; **d. eficaz,** effective diameter.

diapausa, f. diapause.

diapiro, m. diapir (Geol).

diapositiva, f. diapositive, colour slide.

diario, m. diary, record, log; **d. de barco,** ship's log; **d. de vuelo,** flight log.

diarrea, f. diarrhoea, diarrhea.

DIAS, f. IWSD; **Década (f) Internacional del Agua y Saneamiento,** International Water and Sanitation Decade.

diatomáceola, a. diatomaceous.

diatomea, f. diatom; **diatomeas (f.pl) (Bot, clase),** Diatomaea (pl).

diatomita, f. diatomite, diatomaceous earth.

dibujadola, a. outlined.

dibujo, m. drawing; **d. a escala,** scale drawing.

diccionario, m. dictionary.

diciclicola, a. dicyclical.

dicloretano, m. dichloroethane.

diclorobenzol, m. dichlorobenzene.

dicotiledónea, f. dicotyledon; **dicotiledóneas (Bot, clase),** Dicotyledonae (pl).

dicotiledóneola, a. dicotyledonous.

dicotomía, f. dichotomy.

dictamen, m. opinion (opinión), report (informe), legal opinion (Jur).

dichograpto, m. dichograptid.

didácticola, a. didactic, guide; **guía (f) didáctica,** instruction manual, users' guide.

diductor, m. diductor.

didymograptido, m. didymograptid.

dieldrín, m. dieldrin.

diente, m. tooth (Zoo, Bot); cog (Ing, de rueda dentada); serration (serrado); **d. de león (flor),** dandelion.

dieta, f. diet.

difenilo, m. diphenyl.

diferencia, f. difference; **d. de nivel (Ing),** head difference; **d. finita,** finite difference.

diferenciación, f. differentiation.

diferencial, a. differential; **ecuación (f) d.,** differential equation; **hundimiento (m) d.,** differential settlement; **presión (f) d.,** differential pressure.

diferenciar, v. to differentiate.

difracción, f. diffraction.

difteria, f. diphteria.

difundir, v. to diffuse, to spread.

difusión, f. diffusion; **coeficiente (m) de d.,**

coefficient of diffusion; **d. de fluido,** fluid diffusion; **d. molecular,** molecular diffusion.

difusividad, f. diffusivity; **coeficiente (m) de d.,** coefficient of diffusivity; **d. hidráulica,** hydraulic diffusivity; **d. molecular,** molecular diffusivity.

difusor, m. diffuser, diffusor.

difusorla, a. diffusive, diffusing, disseminating; **domo (m) d.,** dome diffuser.

digerir, v. to digest.

digestión, f. digestion; **d. aerobia,** aerobic digestion; **d. anaerobia,** anaerobic digestion; **d. bacteriana,** bacterial digestion; **d. de fango,** sludge digestion; **d. mesofílica,** mesophilic digestion; **d. por ácido,** acid digestion; **fango (m) de d.,** digested sludge; **tanque (m) de d.,** digestion tank.

digestivola, a. digestive, digested; **aparato (m) d.,** digestive system; **tubo (m) d.,** alimentary canal.

digestor, m. digester; **d. de fangos de alcantarilla,** sewage sludge digester.

digitación, f. digitation.

digitalización, f. digitalization.

digitalizar, v. to digitize, to digitalize.

dígito, m. digit.

dilatación, f. dilation, expansion, enlargement; **d. termal,** thermal expansion.

dilatar, v. to dilate.

dilema, m. dilemma.

dilución, f. dilution.

diluir, v. to dilute; **política (f) de d. y dispersar (para vertederos),** dilute and disperse policy.

diluvio, m. deluge (de lluvia).

dimensión, f. dimension, size; **d. del grano,** grain-size.

dimensionamiento, m. sizing (de una instalación).

dimetilamina, f. dimethylamine.

dimetilbenzol, m. dimethylbenzene.

diminutivola, a. diminutive.

dinámica, f. dynamics; **d. de fluidos,** fluid dynamics; **d. poblacional,** population dynamics.

dinámicola, a. dynamic.

dinosaurio, m. dinosaur.

dintel, m. lintel (de una puerta).

diodo, m. diode.

dioicola, a. dioecious.

diópsido, m. diopside.

diorita, f. diorite.

dioxano, m. dioxane.

dióxido, m. dioxide; **d. de azufre,** sulphur dioxide; **d. de carbono,** carbon dioxide; **d. de titanio,** titanium dioxide.

dioxina, f. dioxin.

diploide, a. diploid.

diplópodo, m. diplopod; **diplópodos (Zoo, clase),** Diplopoda (pl).

dipolo, m. dipole.

díptero, m. dipteran; **dípteros (Zoo, orden),** Diptera (pl).

dípterola, a. dipteran.

diputado, m. representative, delegate; member of the Spanish Parliament (Esp), member of Parliament (GB), Congressman (US).

dique, m. dyke, dike (US), breakwater (terraplén etc.); dock (dársena); **d. anular,** ring dyke; **d. de contención,** dam; **d. en dársena,** dry dock; **d. flotante,** floating dock; **d. holandés,** Dutch dyke; **d. radial,** radial dyke; **d. seco,** dry dock; **enjambre (m) de diques, haz (m) de diques,** dyke swarm; **poner diques (v),** to embank (ríos).

dirección, f. address (señas); direction (gestión, tendencia); steering (Ing, de vehículo); course, route (Mar); management (gestión); **d. de obras (Ing),** works management; **d. horizontal (Geol),** strike; **d. preferente,** preferred direction.

directiva, f. directive (instrucciones); board of directors (consejo de administración); **d. comunitaria (Jur), d. europea,** European Community Directive, EC Directive.

directivo, m. director, board member.

directola, a. direct; **en d.,** live (Comm); on-line (Comp).

director, m. director, manager, head; **d. comercial,** business manager; **d. gerente,** managing director.

directorio, m. directive (asamblea directiva); directory (libro de instrucciones).

directriz, f. directive, guideline, instruction.

disco, m. disk, disc; **d. biológico (Trat),** biological disc; **d. compacto (abr. CD),** compact disc (abr. CD); **filtro (m) de d. (Trat),** disk filter.

discoidal, a. discoidal.

discontinuidad, f. discontinuity.

discordancia, f. unconformity; **d. angular,** angular unconformity.

discrepancia, f. discrepancy, disagreement.

discretización, f. discritization; **d. espacial,** spatial discritization.

discretizar, v. to discritize.

discretola, a. discrete, low-profile.

discriminación, f. discrimination.

discriminante, a. discriminant.

disecación, f. dissection.

disecar, v. to dissect.

disección, f. dissection.

diseminación, f. dissemination.

diseminar, v. to disseminate, to spread.

disentería, f. dysentery.

diseñadorla, m. y f. designer.

diseño, m. pattern, design; **d. de avenamiento,** drainage pattern; **fallo (m) en el d.,** design fault.

disgregarse, v. to disintegrate.

disimetría, f. disymmetry.

disipar, v. to dissipate (p.e. duda).

disjunción, f. jointing; **d. columnar,** columnar jointing.

diskette, m. diskette, floppy disk (Comp).

dislocación, f. dislocation, fault, break-up.
dislocadola, a. dislocated.
disminución, f. decrease, reduction, drop, depletion.
disminuir, v. to diminsh, to decrease; to degrade (Geol).
disociación, f. dissociation.
disociar, v. to dissociate.
disoluble, a. soluble, dissoluble, dissolvable.
disolución, f. dissolution.
disolvente, m. solvent, dissolvent; **d. clorado.**, chlorinated solvent; **recuperación (f) de d.**, solvent recovery.
disolver, v. to dissolve.
dispar, a. disparate, unlike.
dispersante, m. dispersant.
dispersar, v. to disperse; **política (f) de atenuar y d.**, **(para vertederos)**, attenuate and disperse policy.
dispersión, f. dispersion, dispersal; **d. hidrodinámica**, hydrodynamic dispersion; **d. isotrópica**, isotropic dispersion; **d. lateral**, lateral dispersion; **d. longitudinal**, longitudinal dispersion; **d. molecular**, molecular dispersion.
dispersividad, f. dispersivity; **coeficiente (m) de d.**, coefficient of dispersivity.
dispersivola, a. dispersive; **modelo (m) d.**, dispersive model.
dispersola, a. dispersed, scattered.
disponibilidad, f. availability.
disponible, a. available (Gen); on hand, disposable (dinero); **oxígeno (m) d.**, available oxygen.
disposición, f. layout, arrangement; **d. del avenamiento**, drainage pattern.
dispositivo, m. device, mechanism (Ing); arrangement, array (disposición); **d. de salida (Comp)**, output device; **d. Schlumberger**, Schlumberger array; **d. Wenner**, Wenner array.
disprosio, m. dysprosium (Dy).
disquete, m. floppy disk, floppy disc (Comp).
disquetera, f. disc drive.
distancia, f. distance; **d. entre curvas de nivel**, contour interval.
distanciamiento, m. distancing, spacing.
distorsionar, v. to distort.
distribución, f. distribution; **d. binomial**, binomial distribution; **d. de frecuencias**, frequency distribution; **d. de Gaus**, Gaussian distribution; **d. de Poisson**, Poisson distribution; **d. de t de Student**, Student's t distribution; **d. esporádica**, patchy distribution; **d. estadística**, statistical distribution; **d. gamma**, gamma distribution; **d. logarítmiconormal**, log-normal distribution; **panel (m) de d.**, distribution panel.
distribuidola, a. distributed; **d. ampliamente**, widespread.
distribuidor, m. distributor, dealer; **d. automático**, vending machine.

distrito, m. district; circuit (Jur); **d. municipal**, municipal district.
distróficola, a. dystrophic.
disturbio, m. disturbance.
disuasión, f. dissuasion.
disuasoriola, a. disuasive.
disueltola, a. dissolved; **flotación (f) por aire d.**, dissolved-air flotation; **oxígeno (m) d.**, dissolved oxygen; **sales (f.pl) disueltas**, dissolved salts; **sólidos (m.pl) disueltos**, dissolved solids.
diurético, m. diuretic.
diuréticola, a. diuretic.
divergencia, f. divergence; **d. de opiniones**, divergence of opinions.
divergente, a. divergent.
diversidad, f. diversity; **d. biológica**, biological diversity; **d. de especies**, species diversity; **d. espacial**, spatial diversity; **índice (m) de d.**, diversity index.
diversificación, f. diversification.
diversificar, v. to diversify.
dividendo, m. dividend (Fin).
dividir, v. to divide, to separate.
división, f. division.
divisoria, f. divide; **d. de aguas (Geog)**, watershed (GB), water divide, rainshed (US).
divisoriola, a. dividing, separating; **línea (f) d.**, dividing line; **línea d. de aguas (Geog)**, watershed (GB), water divide, rainshed (US).
divulgación, f. disclosure, divulging, revelation.
doblamiento, m. doubling.
doble, a. double; **bombeo (m) con acción d.**, double-action pump.
documentación, f. documentation.
dodecaedro, m. dodecahedron.
dólar, m. dollar.
dolerita, f. dolerite.
dolina, f. doline, (depresión kárstica).
dolmen, m. dolmen (Arq).
dolomita, f. dolomite.
dolomíticola, a. dolomitic; **caliza (f) d.**, dolomitic limestone.
dolomitización, f. dolomitization.
dolor, m. ache, pain; **d. de cabeza**, headache.
domésticola, a. domestic; **basura (f) d.**, domestic waste, domestic rubbish.
dominante, a. dominant; prevailing (de viento).
dominicanola, 1. m. y f. Dominican; 2. a. Dominican.
dominio, m. property, authority, control; domain (territorio); **d. público**, public property.
domo, m. dome; **d. de exfoliación**, exfoliation dome; **d. de inyección**, injection dome; **d. de sal**, salt dome; **d. periclinal**, periclinal dome; **d. volcánico**, volcanic dome; **formación (f) de domos (p.e. en vertederos)**, doming.
donación, f. donation, endowment; **d. de sangre**, donation of blood.

donante, m. y f. donor.

donar, v. to donate.

donativo, m. donation.

dorada, f. sea bream.

dormancia, f. dormancy.

dormina, f. dormin.

dorsal, a. dorsal.

dosificación, f. dosage, dosing; titration (Quím); **sifón (m) de d.**, dosing siphon; **vasija (f) de d.**, dosing chamber.

dosificador, m. dosimeter.

dosificar, v. to dose.

dosímetro, m. dosimeter.

dosis, f. dose (Med); proportion (Quím); **d. de radiación,** radiation dose; **d. permisible,** permissible dose.

dotar, v. to provide, to endow, to equip; **d. de alcantarillado,** to sewer, to install sewerage.

DQO, f. COD; **demanda (f) química de oxígeno,** chemical oxygen demand.

dracunculosis, m. dracunculosis.

draga, f. dredge, dredger, dredging machine.

dragado, m. dredging; **zona (f) de vertido de productos de d. (Mar),** spoil area, spoil dump.

dragalina, f. dragline.

dragar, v. to drag (un río), to dredge (un puerto).

dren, m. drain.

drenable, a. drainable.

drenado/a, a. drained; **d. con tubos de barro cocido,** tile drained.

drenaje, m. drainage; **colector (m) de d. (para aguas negras),** main drainage; **densidad (f) de la red de d.,** drainage network density; **d. agrícola,** agricultural drainage; **d. anastomosado,** braided drainage; **d. de carreteras,** road drainage; **d. del terreno,** land drainage; **d. diferido,** delayed drainage; **d. gravitacional,** gravitational drainage; **d. interno,** internal drainage; **d. litoral,** coastal drainage; **d. para tormentas,** storm sewer, storm drain; **d. por un río,** drainage by a river; **d. poroso,** tile drain, field drain; **d. superficial,** surface drainage; **tubo (m) de d.,** drainage pipe.

drenar, v. to drain away.

droga, f. drug.

drumlin, m. drumlin (loma glacial).

drupa, f. drupe.

dualidad, f. duality.

dúctil, a. ductile.

dueñola, m. y f. owner.

dumper, m. dumptruck (US).

duna, f. dune; **d. costera,** coastal dune; **d. de arena,** sand dune; **d. longitudinal,** longitudinal dune.

dunar, a. dune; **formaciones (f.pl) dunares,** dune formations.

durabilidad, f. durabilty.

duración, f. duration; **corta d.,** short-lived duration; **curva de d.-caudal,** flow duration curve; **d. de la vida,** life-span.

duramen, m. heartwood.

duraznero, m. peach tree.

durazno, m. peach.

dureza, f. hardness, toughness; **d. bicarbonatada,** bicarbonate hardness; **d. del agua,** hardness of water; **d. del carbonato,** carbonate hardness; **d. permanente,** permanent hardness; **d. temporal,** temporary hardness; **d. total,** total hardness.

durola, a. hard, tough.

E

ebanista, m. carpenter, cabinetmaker.
ebanistería, f. joinery, cabinet making.
ebonita, f. ebonite.
ebullición, f. boiling (de líquidos); **entrar en e.** **(v)**, to come to the boil; **punto (m) de e.**, boiling point.
ebuloscopio, m. ebulloscope.
ecalafón, m. list, table, roll (p.e. de empleados).
ecdisis, f. ecdysis, moult, molt (US).
echar, v. to eject, to expel; **e. brotes (Bot)**, to sprout.
eclogita, f. eclogite.
eco, m. echo.
eco-, pf. eco-.
ecocéntricola, a. ecocentric.
ecodesarrollo, m. ecodevelopment.
ecoequilibrio, m. ecobalance.
ecoetiqueta, f. ecolabelling.
eco-etiquetado, m. eco-labelling, ecolabelling.
ecohidrológicola, a. ecohydrologic.
ecología, f. ecology; **e. acuática**, aquatic ecology.
ecológicola, a. ecological; **sucesión (f) e.**, ecological succession.
ecologismo, m. environmentalism.
ecologista, m. y f. environmentalist.
ecólogola, m. y f. ecologist.
economía, f. economy; **e. de mercado**, market economy.
económicamente, ad. economically; **e. viable**, economically viable.
económicola, a. economic; **año (m) e., ejercicio (m) e.**, financial year, fiscal year; **prosperidad (f) e.**, economic prosperity.
economista, m. y f. economist.
economizar, v. to economize.
ecosensible, a. ecosensitive.
ecosfera, f. ecosphere.
ecosistema, m. ecosystem; **e. amazónico**, Amazon ecosystem; **e. andino**, Andean ecosystem.
ecosonda, m. echo sounder.
ecosondador, m. echo sounder.
ecotipo, m. ecotype.
ecotono, m. ecotone.
ecotoxicología, f. ecotoxicology.
ecotoxicológicola, a. ecotoxicological.
ecoturismo, m. ecotourism.
ectoparásito, m. ectoparasite.
ECU, f. ECU; **Unidad (f) Monetaria Europea**, European Currency Unit.
ecuación, f. equation; **e. de conservación de masa**, equation of conservation of mass; **e. de continuidad**, continuity equation; **e. de Laplace**, Laplace equation; **e. de primer grado**, simple equation; **e. de segundo grado**, second order equation; **e. de transporte de masa**, mass transport equation; **e. diferencial**, differential equation; **e. no lineal**, non-linear equation; **e. parabólica**, parabolic equation.
Ecuador, m. Ecuador; **El E.**, Ecuador.
ecuador, m. equator.
ecuatorianola, a. Ecuadoran, Ecuadorian.
eczema, m. eczema.
edad, f. age; **E. de Bronce**, Bronze Age; **E. de Cobre**, Copper Age; **E. de Hierro**, Iron Age; **E. de Hielo**, Ice Age; **E. de Piedra**, Stone Age; **e. geológica**, geological age; **e. real**, actual age.
edáficola, a. edaphic; **factor (m) e.**, edaphic factor.
edafología, f. pedology, soil science, edaphology.
EDAR, f. STW; **estación (f) depuradora de aguas residuales**, sewage treatment works (abr. STW), sewage farm.
edema, m. oedema.
edificación, f. construction, building; **patología (f) de la e.**, building pathology.
edificio, m. building; **e. protegido**, protected building, listed building (GB).
edómetro, m. oedometer, consolidometer.
EEUU, m.pl. US, USA; **Estados (m.pl) Unidos, Estados Unidos de América**, United States, United States of America.
efectividad, f. effectiveness; **e. de tratamiento**, effectiveness of treatment.
efectivola, a. effective; **lluvia (f) e.**, effective rainfall; **porosidad (f) e.**, effective porosity.
efecto, m. effect; **e. barométrico**, barometric effect; **e. de las temperaturas diurnas**, diurnal temperature effect; **e. invernadero**, greenhouse effect; **e. osmótico**, osmotic effect; **e. pelicular (flujo del agua)**, skin effect.
efervescencia, f. effervescence.
eficacia, f. efficacy.
eficiencia, f. efficiency; **e. barométrica**, barometric efficiency; **e. de pozo**, well efficiency; **e. energética**, energy efficiency; **e. relativa**, relative efficiency.
efímerola, a. ephemeral; **cauce (m) e.**, ephemeral stream.
eflorescencia, f. efflorescence.
eflorescente, a. efflorescent.
efluente, m. effluent; **calidad (f) de e. (Trat)**, effluent quality; **descarga (f) de e. (Trat)**, effluent discharge; **e. crudo**, raw effluent; **e. doméstico**, domestic effluent; **e. tratado**, treated effluent.
efusión, f. outpouring, effusion; **e. basalto, e. de**

basalto, basalt flow; **e. por grietas,** fissure flow.

egiptología, f. egyptology (Arq).

eglefino, m. haddock.

EIA, m. y f. EIA; **Estudio (m) de Impacto Ambiental, evaluación (f) de impacto ambiental,** Environmental Impact Assessment.

Eifeliense, m. Eifelian.

eje, m. axis; axle (Ing); **e. de inclinación,** axis of tilt; **e. de la hélice,** propeller shaft; **e. delantero,** front axle; **e. del anticlinal,** anticlinal axis; **e. de la Tierra,** Earth's axis; **e. de levas,** cam shaft; **e. del pliegue,** fold axis; **e. de simetría,** axis of symmetry; **e. motor,** drive shaft; **e. trasero,** rear axle; **e. vertical,** vertical axis; **e. x,** x-axis; **e. y,** y-axis.

ejecución, f. carrying out, execucution, undertaking; **e. de un esquema,** undertaking of a scheme.

ejecutar, v. to execute, to undertake (un orden); **e. un plan de acciones,** to execute an action plan.

ejecutivo/a, 1. m. y f. executive; 2. a. executive.

ejecutorio/a, a. enforceable (de un contrato).

ejemplo, m. example.

ejido, m. common land.

ejote, m. green bean, string bean.

elaborar, v. to elaborate.

elasticidad, m. elasticity, resilience (de un objeto); **coeficiente (m) de e.,** coefficient of elasticity.

elástico/a, a. elastic, flexible.

elastómero, m. elastomer.

eldrín, m. eldrin.

electricidad, f. electricity.

electricista, 1. m. y f. electrician; 2. a. electrical; **ingeniero/a (m, f) e.,** electrical engineer.

eléctrico/a, a. electric (corriente, aparato); electrical (Ing); **corriente (f) e.,** electric current; **estufa (f) e.,** electric heater; **ingeniero/a (m, f) e.,** electrical engineer.

electrificación, f. electrification.

electrificar, v. to electrify (p.e. sistema de ferrocarriles).

electroanálisis, m. electroanalysis.

electrobomba, f. electric pump; **e. sumergida,** submersible electric pump.

electrocardiograma, m. electrocardiogram (Med).

electrocromatografía, f. electrochromatography.

electrodiálisis, m. electrodialysis.

electrodinámico/a, a. electrodynamic.

electrodo, m. electrode; **e. conductivimétrico,** conductivity meter electrode; **e. de hidrógeno,** hydrogen electrode; **e. negativo,** negative electrode; **e. positivo,** positive electrode.

electrodoméstico, m. electrical appliance, electrical household appliance.

electroencefalograma, m. electroencephalogram (Med).

electroforesis, f. electrophoresis.

electrógeno/a, a. generator, generating; **grupo (m) e.,** generator set.

electrogravimetría, f. electrogravimetry.

electroimán, m. electromagnet.

electrólisis, f. electrolysis.

electrolítico/a, a. electrolytic; **proceso (m) e.,** electrolytic process.

electrolito, m. electrolyte.

electrolizar, v. to electrolyze.

electromagnético/a, a. electromagnetic; **campo (m) e.,** electromagnetic field.

electrón, m. electron.

electronegativo/a, a. electronegative.

electrónica, f. electronics.

electrónico/a, a. electronic; **correo (m) e.,** electronic mail.

electrón-voltio, m. electron-volt (abr. eV).

electroósmosis, f. electroosmosis.

electroosmótico/a, a. electroosmotic.

electropesca, f. electrofishing.

electroplastia, f. electro-plating.

electropositivo/a, a. electropositive.

electroquímica, f. electrochemistry.

electroquímico/a, a. electrochemical.

electrostático/a, a. electrostatic; **precipitador (m) e.,** electrostatic precipitator.

elefante, m. elephant.

elemento, m. element; **e. tóxico,** toxic element; **e. traza,** trace element.

elevación, f. elevation, height, rise; uplift (Geol); **e. continental,** continental rise; **e. salina,** saline rise; **e. suave (Mar),** swell.

elevador, m. lift; elevator (US); step-up transformer (Elec); **e. de voltaje (Elec),** booster; **e. hidráulico,** hydraulic ramp.

elevar, v. to lift, to raise (subir); to step up (producción); to boost (Elec); **e. a la enésima potencia (Mat),** to raise to the power n; **e. al cubo (Mat),** to cube; **e. a una potencia (Mat),** to raise to a power.

elevarse, v. to rise, to aggrade (Geol).

eliminación, f. elimination, removal; **e. de fosfato,** phosphate removal; **e. de residuos,** waste disposal.

elipse, f. ellipse.

elipsoide, m. ellipsoid; **e. de deformación,** strain ellipsoid; **e. de esfuerzo,** stress ellipsoid.

elíptico/a, a. elliptical, elliptic.

elisión, f. elision.

élitro, m. elytron o elytrum (pl. elytra).

El Niño, m. El Niño (Mar).

elongación, f. elongation.

elote, m. maize (GB), corn (US).

El Salvador, m. El Salvador.

elución, f. elution.

eluir, v. to elute (Quím).

elutriación, f. elutriation.

eluviación, f. eluviation.

eluvial, a. eluvial.

eluviar, v. to elute, to leach (Quím).

eluvión, f. eluvium.

eluyente, m. eluent (Quím).

elvan, m. (LAm) elvan (pórfido feldespático).

elvanita, f. elvan.

emanación, f. emanation.

emanar, v. to emanate.

embalaje, m. packaging; **envases y embalajes**, packaging.

embalar, v. to bale (p.e. basuras).

embalsado, m. (LAm) large floating island.

embalsamiento, m. impounding.

embalsar, v. to dam up (un río), to store (agua).

embalsar, v. to impound, to dam.

embalse, m. reservoir, dam, damming-up; **e. cubierto**, covered reservoir, service reservoir; **e. de almacenamiento**, storage reservoir; **e. de almacenamiento bombeado**, pumped-storage reservoir; **e. de compensación**, balancing reservoir, compensation reservoir; **e. de contención**, impounding reservoir; **e. de regulación**, regulating reservoir; **e. de servicio**, service reservoir; **e. para control de crecidas**, flood-control reservoir; **gestión (f) de e.**, reservoir management; **operación (f) de e.**, reservoir operation; **revestimiento (m) de e.**, reservoir lining.

embarazada, a. pregnant (de una mujer).

embarazo, m. pregnancy (de mujeres).

embarcadero, m. pier, jetty, wharf, landing stage.

embargo, m. embargo (Gen), seizure (Jur).

embeber, v. to imbibe, to absorb.

embestida, f. uprush.

embocadura, f. mouth; **e. de un río**, mouth of river.

embotelladola, a. bottled.

embotellar, v. to bottle.

embrague, m. clutch.

embrión, m. embryo.

embrionariola, a. embryonic.

embudo, m. funnel (para el trasiego de líquidos); crater (de bomba, etc.); **e. cardo (Quím)**, thistle funnel; **e. de filtro**, filter funnel; **e. de separación**, separation funnel.

emergencia, f. emergence (acción de emerger); emergency (crisis); **e. de fitoplancton**, phytoplankton bloom; **en caso (m) de e.**, in case of emergency.

emergente, a. emergent, resultant.

emético, m. emetic.

eméticola, a. emetic.

emigrar, v. to migrate.

eminencia, f. height, rise (Geog).

emisario, m. discharge, emission; emissary (person); **e. submarino**, submarine discharge.

emisión, f. emission (Gen); effluent (agua sucia); output, issue (Com, de acciones); broadcast, programme, program (US) (radio, TV); **criterio (m) de e.**, emission standard; **e. fugitiva**, fugitive emission;

emisiones de chimeneas, stack emissions; **emisiones gaseosas**, gaseous emissions; **límite (m) de e. fijo**, fixed emission limit, limit value; **norma (f) de e.**, emission standard.

emisor, m. emitter; **e. alpha/beta/gamma**, alpha/beta/gamma emitter.

emisora, f. transmitter, radio station; **estación (f) e.**, broadcasting station.

emitir, v. to emit; **e. un jucio**, to express an opinion.

empalizada, f. stockade.

empalme, m. joint, connection; **e. de enchufe y cordón**, spigot and socket joint.

empañarse, v. to tarnish, to steam up.

empapadola, a. wet, soaked.

empapamiento, m. soakage.

empapar, v. to soak, to absorb.

empaque, m. packing; **e. de gravas**, gravel packing.

empaquetadola, a. packed.

empaquetador, m. packer (e.g. for well testing).

empaquetar, v. to pack.

empedradola, m. cobbled.

empeoramiento, m. deterioration, worsening.

empíricola, a. empirical.

empirismo, m. empiricism.

emplazamiento, m. installation, emplacement, siting; **e. de la perforación**, wellsite; **e. profundo**, deep-seated.

empleadola, m. y f. employee, clerk, office worker; **e. de limpieza (Méx)**, waste tip operative, garbage man.

empleo, m. employment, use, work; **condiciones (f.pl) de e.**, conditions of employment; **e. comunitario**, community work; **e. de trazadores**, use of tracers; **modo (m) de e.**, working method; **pleno e.**, full employment; **sin e.**, unemployment.

empobrecer, v. to impoverish.

empobrecimiento, m. impoverishment.

emprender, v. to undertake (una tarea).

empresa, f. compamy, firm (compañía), business, enterprise, venture (negocio); **e. privada**, private enterprise.

empresariola, m. y f. entrepreneur (Com).

empuje, m. thrust, pressure, push.

emulsión, f. emulsion.

emulsionador, m. emulsifier.

emulsionante, a. emulsifying; **agente (m) e.**, emulsifying agent.

emulsionar, v. to emulsify.

enarenar, v. to sand.

encajamiento, m. insertion (inserción); culvert (de cauces); gouging (Geol).

encalladero, m. shoal, sandbank.

encañar, v. to pipe (agua), to drain (tierra).

encapotadola, a. overcast (Met).

encapsulación, f. encapsulation.

encarar, v. to confront, to face.

encauzamiento, m. flood routing, channelling,

channeling (US); **e. de la inundación,** flood routing.

encauzar, v. to embank, to channel (un río).

encerrada/a, a. enclosed.

encerrar, v. to enclose, to shut in.

encina, f. evergreen oak, holm oak, holly oak.

enclavar, v. to interlock.

enclave, m. enclave.

encofrado, m. formwork, timberwork.

encofrar, v. to plank, to timber.

encontrar, v. to find, to meet; **e. agua,** to find water; **e. una solución,** to find a solution.

encorvada/a, a. curved.

encorvamiento, m. arching.

encrinita, f. encrinite.

encristalado, m. glazing.

encuesta, f. inquiry (Gen), investigation; sounding (public opinion, Geol); **e. por muestreo,** sampling survey.

encharcada/a, a. waterlogged, swamped.

encharcamiento, m. flooding, waterlogging, inundation.

encharcar, v. to inundate, to flood.

endémico, m. endemic.

endémica/a, a. endemic.

endibia, f. endive.

endo-, pf. endo-.

endoesqueleto, m. endoskeleton.

endógena/a, a. endogenetic, endogenous; **procesos (m.pl) e.,** endogenetic processes.

endoparásito, m. endoparasite.

endospermo, m. endosperm.

endotérmica/a, a. endothermic; **reacción (f) e.,** endothermic reaction.

endurecedor, m. stiffener, hardener.

endurecida/a, a. hardened, indurated.

enebro, m. juniper.

energética/a, a. energetic; **eficiencia (f) e.,** energy efficiency.

energéticos, m.pl. fuels.

energía, f. energy; **abastecimiento (m) de e.,** energy supply; **e. alternativa, e. blanda,** alternative energy; **e. calorífica,** heat energy; **e. cinética,** kinetic energy; **e. convencional,** conventional energy; **e. de olas,** wave energy; **e. dura,** conventional energy; **e. eléctrica,** electrical energy; **e. eólica,** wind energy; **e. fósil,** fossil energy; **e. geotérmica,** geothermal energy; **e. hidráulica,** water-power, hydropower; **e. hidroeléctrica,** hydroelectric energy; **e. hidrotérmica,** hydro-thermal energy; **e. mareal,** tidal energy; **e. mecánica,** mechanical energy; **e. nuclear,** nuclear energy; **e. química,** chemical energy; **e. radiante,** radiant energy; **e. renovable,** renewable energy; **e. solar,** solar energy; **e. térmica,** thermal energy; **suministro (m) de e.,** energy supply; **utilización (f) de e.,** energy use.

enésima/a, a. n, nth (Mat); **elevar a la e. potencia (v),** to raise to the power n.

énfasis, m. emphasis; **e. especial,** special emphasis.

enfatizar, v. to emphasize.

enfermedad, f. disease, illness; **e. del sueño,** sleeping sickness; **e. de origen hídrico,** waterborne disease; **e. gastrointestinal,** gastrointestinal disease; **e. parasitaria,** parasitic disease; **e. respiratoria,** respiratory disease; **e. transmisible,** communicable disease.

enferma/a, a. diseased.

enfisema, f. emphysema.

enfoque, m. focusing, focussing; **e. ecológico,** ecological focusing.

enfrentar, v. to confront, to face up to.

enfriamiento, m. cooling, chilling; **balsas (f.pl) de e.,** cooling ponds; **factor (m) de e.,** windchill factor; **lagunas (f.pl) de e.,** cooling ponds.

engastar, v. to embed, to imbed, to set, to mount.

englobar, v. to include, to comprise, to lump together.

engolfamiento, m. (LAm) embayment.

engranada/a, a. meshed.

engranaje, m. gear, cogs (pl), gear teeth (pl); **diente (m) de e.,** sprocket.

engranar, v. to mesh.

engrasar, v. to grease, to oil (lubricar); to manure (Agr).

engrosar, v. to enlarge, to swell.

enjambre, m. swarm.

enjuiciamiento, m. prosecution, lawsuit, trial.

enlace, m. connection, linkage; bond, bonding (Quím); **e. atómico,** atomic bond; **e. covalente,** covalent bond; **e. doble,** double bond; **e. iónico,** ionic bond; **e. químico,** chemical bond; **e. simple,** single bond; **e. triple,** triple bond; **único e.,** single bond.

enladrillado, m. brick paving.

enladrillador, m. bricklayer.

enlame, m. silting.

enlatados, m.pl. canned products.

enlosado, m. flagstone paving.

enlucido, m. plaster, coat of plaster.

enlucida/a, a. plastered.

enmaderada/a, a. timbered (p.e. un techo).

enmaderar, v. to timber (p.e. un techo).

en masa, a. en masse.

enmendar, v. to amend (p.e. la ley).

enmienda, f. amendment, correction.

en pendiente, a. sloping, inclined.

enrarecimiento, m. rarefaction.

enrasar, v. to make flush or level.

enrase, m. flush, level.

enredadera, f. creeper (planta), bindweed; **e. de campanillas, e. de campo,** bindweed.

enrejado, m. railings (pl) (verja); bars (pl) (de jaula); **e. de alambre,** wire fencing.

enriquecer, v. to enrich.

enriquecimiento, m. enrichment; **e. en hierro,** iron enrichment.

enrollada/a, a. rolled, convoluted.

enrollar, v. to convolute, to roll.

ensanchamiento, m. swelling, widening; enlargement (p.e. de una ciudad).

ensanchar, v. to widen, to extend; **e. una carretera**, to widen a road.

ensanche, m. widening.

ensayar, v. to test, to try, to assay (Quím).

ensaye, m. assay (de un metal).

ensayo, m. test, testing, trial; **aparato (m) de e.**, test kit; **datos (m.pl) de los ensayos**, test data; **e. a largo plazo**, long-term test; **e. ambiental**, environmental test; **e. a nivel constante**, constant-level test; **e. con trazadores**, tracer test; **e. de acuífero**, aquifer test; **e. de bombeo**, pumping test; **e. de compresión**, compression test; **e. de criba (Geol)**, screen test; **e. de cuchareo**, bailer test; **e. de dureza**, hardness test; **e. de eficiencia**, efficiency test; **e. de formación**, formation test; **e. de funcionamiento seguro**, reliabilty test; **e. de interferencia**, interference test; **e. de inyección**, injection test; **e. de la t de Student**, Student's t test; **e. de medida del caudal**, flow test; **e. de molinete**, vane test; **e. de permeabilidad**, permeability test; **e. de recuperación (Geol)**, recovery test; **e. de rejilla**, screen test; **e. de sellador (Geol)**, packer test; **e. destructivo**, test to destruction; **e. escalonado (Geol)**, step test; **e. in situ**, field test; **e. mecánico de roca**, mechanical test of rock; **e. termonuclear**, thermonuclear test; **e. triaxial**, triaxial test; **instalación (f) de e.**, pilot plant.

ensenada, f. inlet, cove.

ensilaje, m. silage; **pinza (f) de e.**, silage clamp.

ensisadura, f. (Aragón, Esp) floodwater channel.

entablado, m. boarding (Con).

entablar, v. to plank, to board over.

entalpía, f. enthalpy.

entarquinamiento, m. silting.

entero, m. integer.

enterobácter, m. enterobacter.

enterobacteria, f. enterobacterium (pl. enterobacteria).

enterococos, m.pl. enterococci.

enterovirus, m. enterovirus.

enterramiento, m. burial, tomb.

entibación, f. timbering, shuttering, shoring (de madera).

entibado, m. timbering (en una mina).

entibadola, a. timbered (en una mina).

entibar, v. to timber, to install shuttering, to prop (a mine).

entomófilola, a. entomophilous.

entomología, f. entomology.

entomólogola, m. y f. entomologist.

entorno, m. environment, surroundings; **e. socioeconómico**, socio-economic environment.

entrada, f. entry, entrance; beginning (principio); influx (of people); admission (of people); input (Comp); **entrada (f)/salida (f) (Comp)**, input/output (abr. I/O).

entreabiertola, a. half-open.

entrecruzadola, a. (LAm) cross-bedded.

entrecruzamiento, m. (LAm) cross-bed.

entredicho, m. prohibition, ban; injunction (Jur).

entrelazarse, v. to interlock.

entremezclar, v. to intermix.

entrepaño, m. low pasture between sown fields.

entresuelo, m. mezzanine (piso).

entropía, f. entropy.

entubado, m. casing, lining, tubing; **e. provisional**, temporary casing, temporary lining; **e. telescópico**, telescopic lining.

entubar, v. to pipe, to tube.

enturbiar, v. to make cloudy.

enumeración, f. enumeration.

enumerar, v. to enumerate.

envasadora, f. packer (máquina).

envejecimiento, m. ageing.

envenenamiento, m. poisoning; **e. de la sangre**, blood-poisoning.

envenenar, v. to poison; to pollute (el aire).

envoltura, f. wrapping, covering.

envolvente, a. surrounding, enveloping; **línea (f) e.**, enveloping line.

envolver, v. to wrap, to wrap up, to envelop.

enzima, f. enzyme.

eólicola, a. eolian.

eolinita, f. eolinite.

eolo, m. wind.

eón, m. eon.

eosina, f. eosin.

epi-, pf. epi-.

epicentro, m. epicentre, epicenter (US).

epidemia, f. epidemic.

epidémicola, a. epidemic.

epidemiología, f. epidemiology.

epidermis, f. epidermis.

epidiorita, f. epidiorite.

epidota, f. epidote.

epífito, m. epiphyte.

epigénesis, m. epigenesis.

epiginola, a. epigenetic, epigenic.

epilimnio, m. epilimnion.

epitermal, a. epithermal.

época, f. epoch, era; **é. de reproducción**, mating season.

épsilon, f. epsilon.

equi-, pf. equi-.

equidad, f. equity, fairness.

equidistancia, f. equidistance, equal distance.

equidistante, a. equidistant.

equigranular, a. even-grained.

equiláterola, a. equilateral.

equilibrar, v. to balance, to counterbalance.

equilibrio, m. equilibrium (pl. equilibria, equilibriums); **condiciones (f.pl) de e.**, equilibrium conditions; **e. agua-roca**, rock-water equilibrium; **e. dinámico**, dynamic equilibrium; **e. estable**, stable equilibrium; **e.**

hidráulico, hydraulic equilibrium; e. inestable, unstable equilibrium; e. químico, chemical equilibrium.

equinoccio, m. equinox; e. de otoño, autumn equinox; e. de primavera, vernal equinox, spring equinox.

equinodermo, m. echinoderm; equinodermos (Zoo, filum), Echinodermata (n.pl).

equinoide, m. echninoid; equinoideos (Zoo, clase), Echinoidea.

equipamiento, m. equipping, fitting out.

equiparar, v. to compare.

equipo, m. team (p.e. deportes), shift (de trabajadores); equipment (herramientas); bienes (m.pl) de e., capital goods; e. de bombeo, pumping equipment; e. de sondeo, down-hole instrumentation; e. eléctrico, electrical equipment; e. perforador, drilling rig.

equipotencial, a. equipotential; línea (f) e., equipotential line.

equivalencia, f. equivalence.

equivalente, a. equivalent; peso (m) e., equivalent weight.

era, f. era, age; threshing ground (Agr); crushing ground (Min); patch, bed (para flores); e. atómica, atomic age.

erbio, m. erbium (Er).

erg, m. erg (Fís); sandy desert.

ergio, m. erg (Fís).

ergonomía, f. ergonomics.

ergot, m. ergot.

erguir, v. to raise up, to lift up (Con).

erial, m. wasteland, uncultivated land.

erigir, v. to erect, to build, to raise.

erosión, f. erosion; e. en mantos, sheet errosion; e. eólica, wind erosion; e. glacial, e. glaciar, glacial erosion; e. pluvial, rainwash; e. por saltación, saltation.

erosionadola, a. eroded.

erosionar, v. to erode.

erosivola, a. erosive; agente (m) e., agent of erosion.

erradicación, f. eradication.

erradicar, v. to eradicate; to uproot (plantas).

erróneola, a. erroneous, mistaken.

error, m. error, mistake (equivocación); e. analítico, analytical error; e. de truncación, truncation error; e. humano, human error; e. residual, residual error; e. sistemático, systematic error; manipulación (f) de errores, tratamiento (m) de errores (Comp), error handling.

erupción, f. eruption; e. volcánica, volcanic eruption.

eruptivola, a. eruptive.

escabrosola, a. rugged, tough (tierra); uneven (superficie).

escafópodo, m. escaphopod.

escala, f. scale (graduación); range (rama); size (amplitud de un proyecto); extent (de un evento); a gran e. (a), large-scale; a pequeña e. (a), small-scale; aumentar e. (v), to scale up; dibujar a e. (v), to scale, to draw to scale; e. centígrada, centigrade scale; e. de Beaufort (para la medición del viento), Beaufort scale; e. de Fahrenheit, Fahrenheit scale; e. de magnitud, scale of magnitude; e. de pH, pH scale; e. de Richter, Richter Scale; e. de tiempo, timescale; e. graduada, scale of measurement; e. para peces, fish ladder; dibujo (m) a e. scale drawing; hecho/a a e. (a), drawn to scale; modelo (m) a e., scale model; reducir a e. (v), to scale down (un plano).

escalar, a. scalar; dibujo (m) e., scale drawing.

escalera, f. staircase, stairway; e. de caracol, spiral staircase; e. de incendios, fire escape; e. de mano, ladder; e. de tijera, stepladder, steps; e. mecánica, escalator.

escalilla, f. gauge board (para la medición del nivel de un río).

escalón, m. step, stair (peldaño); rung (de escalera de mano); e. de bombeo, pumping step, pumping stage.

escalonar, v. to phase, to plan in stages (poner en fase).

escama, f. scale (de peces, de serpientes, de piel), flake; e. de mica, mica flakes.

escamosola, a. scaly.

escandio, m. scandium (Sc).

escapada, f. breakaway.

escape, m. escape (huida); leak, leakage, escape (de gases); exhaust (Ing); curva (f) de e. breakaway curve (Mat); gases (m.pl) de e., exhaust fumes, exhaust gases; tecla (f) e. (Comp), escape key; tubo (m) de e., exhaust pipe, tail-pipe (US).

escarabajo, m. beetle.

escaramujo, m. dog rose.

escarbadora, f. scraper; máquina (f) e., mechanical scraper, earth scraper.

escarcha, f. ground frost (en la tierra).

escariación, f. scouring; e. fluvial, stream scouring.

escariado, m. reaming (ensanchando).

escariador, m. reamer.

escariar, v. to ream (ensanchar).

escarificador, m. scarifier.

escarificar, v. to scarify.

escarpadola, a. steep.

escarpadura, f. escarpment, steep slope.

escarpe, m. escarpment, scarp; e. de creta, chalk escarpment; e. de falla, fault scarp; pendiente (f) del e., scarp slope.

escasez, f. scarcity, scarceness, shortage; famine (LAm) (hambruna).

escenario, m. scenario.

Escitiense, m. Scythian.

esclerófilola, a. sclerophyllous; bosque (m) e., sclerophyllous forest.

esclusa, f. sluice, floodgate; lock (en un canal); e. de aire, airlock; e. para el aire (en sub-

marinos), airlock; **sistema (m) de esclusas**, lockage.

escollera, f. breakwater, jetty.

escollo, m. reef (arrecife), rock (roca); pitfall, stumbling block (obstáculo).

escombrera, f. dump, tip, rubbish dump; **e. de carbón**, coal tip, slag heap; **e. de escoria**, slag heap.

escombros, m.pl. rubble, rubbish, debris, spoil (p.e. roca); ruins (de un edificio).

esconder, v. to conceal, to hide.

escoria, f. slag; **e. de alto horno**, blastfurnace slag.

escoriáceola, a. scoriaceous, rough; **lava (f) e.**, scoriaceous lava.

escorial, m. slag heap.

escorrentía, f. runoff, flow; **coeficiente (m) de e.**, runoff coefficient; **e. estacional**, seasonal runoff; **e. superficial**, overland flow.

escrobiculadola, a. pitted.

escudo, m. shield; **el E. Laurentiano**, the Laurentian Shield.

escuela, f. school.

esencia, f. essence.

esencial, a. essential (imprescindible); chief, main (principal).

esfagno, m. sphagnum.

esfalerita, f. sphalerite, zinc blende.

esfena, m. sphene, titanite.

esfera, f. sphere; **e. de actividad**, sphere of activity; **e. de influencia**, sphere of influence.

esfericidad, f. sphericity.

esféricola, a. spherical.

esferita, f. spheryte.

esferoide, m. spheroid; **e. oblato**, oblate spheroid.

esfuerzo, m. stress, effort (Fís); **e. cizallante**, shear stress; **e. de pesca**, fishing effort; **e. de tensión**, tensile stress.

esker, m. esker.

eskip, m. skip (vasija para basura).

esmaltado, m. enamelling; **e. protector**, protective enamelling.

esmalte, m. enamel.

esmectita, f. smectite.

esmeralda, f. emerald.

esmeril, m. emery; **tela (f) de e.**, emery cloth.

esmeriladola, a. ground (lustrado); **papel (m) e.**, emery paper.

esmerilar, v. to grind (lustrar).

esmog, m. smog (Met).

espacial, a. spatial.

espaciamiento, m. spacing, distancing.

espacio, m. space; **e. natural**, natural area, wildlife park, nature reserve; **e. natural protegido**, protected wildlife area; nature park, nature reserve.

espaciosola, a. spacious, roomy.

espadaña, f. bulrush.

España, f. Spain.

español, m. Spanish language.

españolla, 1. m. y f. Spaniard (person); 2. a. Spanish.

esparcidola, a. interspersed.

esparcimiento, m. spreading (dispersión); relaxation (descanso).

esparcir, v. to space, to spread out.

espárrago, m. asparagus.

esparto, m. esparto (Bot); **fibra (f) de e.**, esparto grass fibre.

espato, m. spar (Geol); **e. de Islandia**, Iceland Spar; **e. satinado**, Satin Spar.

especia, f. spice.

especiación, f. speciation; **e. acuosa**, aqueous speciation.

especialidad, f. specialization.

especialista, m. y f. specialist.

especialización, f. specialization.

especializar, v. to specialize.

especie, f. species (pl. species); **e. accidental**, accidental species; **e. amenazada**, threatened species; **e. característica**, characteristic species; **e. en peligro de extinción**, endangered species; **e. extinguida**, extinct species; **e. nominal**, type species; **e. protegida**, protected species; **e. tipo**, type species; **e. vulnerable**, vulnerable species; **riqueza (f) de especies**, species richness.

especificación, f. specification; **e. de construcción**, building specification.

específicamente, ad. specifically, expressly.

especificar, v. to specify.

específicola, a. specific, definite; **capacidad (f) e.**, specific capacity; **nombre (m) e.**, specific name; **rendimiento (m) e.**, specific yield; **retención (f) e.**, specific retention.

espectral, a. spectral; **análisis (m) e.**, spectral analysis.

espectro, m. spectrum; **e. ultravioleta**, ultraviolet spectrum; **e. visible**, visible spectrum.

espectrofotocolorímetro, m. spectrophotocolorimeter.

espectrofotometría, f. spectrophotometry.

espectrofotómetro, m. spectrophotometer; **e. de flama**, flame spectrophotometer.

espectrografía, f. spectrography.

espectrográficola, a. spectrographic.

espectrógrafo, m. spectograph.

espectrometría, f. spectrometry; **e. de masas**, mass spectrometry.

espectrómetro, m. spectrometer; **e. de masas**, mass spectrometer.

espectroscopio, m. spectroscope.

especulación, f. speculation; **e. bursátil**, speculation on the stock exchange; **e. inmobiliaria**, property speculation.

especuladorla, m. y f. speculator; **e. urbanizador**, developer, property developer.

espejismo, m. mirage.

espejo, m. mirror; **e. de falla (Geol)**, slickenside.

espeleología, f. speleology, potholing.

espera, f. wait, waiting.

esperanza, f. expectation; **e. de vida,** life expectancy.

esperma, f. sperm.

espermatofita, f. spermatophyte; **espermatofitas (Bot, división),** Spermatophyta (pl).

espesador, m. thickener; **e. de fangos, e. de lodos,** sludge thickener.

espesamiento, m. thickening; **tanque (m) de e.,** thickening tank.

espesar, v. to thicken.

espesartita, f. spessartite.

espesola, a. dense, thick.

espesor, m. thickness; **e. del acuífero,** aquifer thickness; **e. no saturado,** unsaturated thickness; **e. saturado,** saturated thickness.

espesura, f. denseness, thickness (de un flúido).

espícula, f. spicule; **e. de una esponja,** sponge spicule.

espiga, f. ear, spike (Bot); spit (Geog); tenon (de madera); pin, peg (macho); masthead (Mar); **e. marginal,** longshore spit.

espigón, m. breakwater (muelle); ear (Bot); spike (espinazo).

espilita, f. spilite.

espina, f. spine (espinazo); thorn (de plantas); bone (de peces); **e. dorsal,** spine, backbone.

espinaca, f. spinach.

espinal, a. spinal; **médula (f) e. (Anat),** spinal cord.

espinapez, f. herringbone; **sistema (m) de e. (para drenaje),** herringbone system.

espinazo, m. spine (Anat); hogsback (Geog, cresta larga).

espinela, f. spinel.

espino, m. hawthorn; **e. artificial,** barbed wire.

espinosola, a. prickly, thorny.

espira, f. spire; spiral, whorl (de una concha).

espiral, 1. f. spiral; hairspring (de un reloj); **la e. inflacionaria,** the inflationary spiral; **moverse (v) en e.,** to spiral; 2. a. spiral.

espirifido, m. spirifid.

espliego, m. lavender.

espolón, m. spur (de aves, de montañas, de edificios); levee, levée (de ríos); groyne, groin; seawall (malecón); perch (percha de aves).

esponja, f. sponge.

espontaneidad, f. spontaneity.

espora, f. spore.

esporádicola, a. sporadic.

espuma, f. foam; scum (espuma con impurezas), froth, spume; **fraccionamiento (m) por e.,** foam fractionation.

espumado, m. skimming; **tanque (m) de e.,** skimming tank.

espumar, v. to froth, to foam; to skim off (quitar espuma).

espumosola, a. foaming.

espueje, m. cutting, slip (de una planta).

esqueleto, m. skeleton.

esquema, m. scheme, diagram, outline.

esquemáticola, a. schematic.

esquematizadola, a. schematized.

esquiladorla, m. y f. sheepshearer (person).

esquilar, v. to shear (ovejas), to cut

esquileo, m. sheepshearing.

esquilmar, v. to overuse, to overcrop (terreno),to overgraze (pastoro), to overexploit (recursos); **e. el suelo,** to exhaust the soil.

esquilmo, m. impoverishment, exhaustion (p.e. de suelo).

esquina, f. corner; coign, coin (Con).

esquirla, f. chip (de piedra).

esquisto, m. schist; **e. arcilloso,** shale; **e. bituminoso,** oil shale.

esquistosidad, f. schistosity.

esquistosola, a. schistose.

estabilidad, f. stability; **e. de la pendiente,** slope stability; **e. numérica (Mat),** numerical stability.

estabilización, f. stabilization; **e. de niveles,** stabilization of levels; **e. de presiones,** stabilization of pressures; **e. de residuos,** waste stabilization; **estanque (m) de e.,** stabilization tank; **laguna (f) de e.,** stabilization pond.

estabilizador, m. stabilizer; **e. de formación,** formation stabilizer.

estabilizar, v. to stabilize.

estable, a. stable.

establecer, v. to establish, to set up, to found.

establecimiento, m. establishment, institution, institute; settlement (de gente).

estaca, f. stake, post.

estación, f. station (Gen); season (temporada); **e. climatológica,** climate station; **e. de aforo,** gauging station; **e. de ferrocarril,** railway station; **e. de investigación,** research station; **e. depuradora,** treatment plant; **e. de tratamiento de aguas potables,** water treatment plant, water treatment works (abr. WTW); **e. de vigilancia,** monitoring station, monitoring point; **e. meteorológica,** weather station; **e. pluviométrica,** rainfall station; **las cuatro estaciones,** the four seasons.

estacional, a. seasonal.

estacionamiento, m. car park.

estacionariola, a. stationary, stabilized; **estado (m) e. (Mat),** steady state.

estadista, m. statesman, statistician.

estadística, f. statistics (ciencia), statistic (data); **e. paramétrica/no paramétrica,** parametric/non-parametric statistics; **según las estadísticas,** statistically, as far as the statistics show.

estadísticola, m. y f. statistician.

estadísticola, a. statistical.

estado, m. state, status; stage (Geol); **e. actual de la tecnología,** state of the art; **e. de conservación,** conservation status; **e. de emergencia,** state of emergency; **e. equilibrio, e. estacionario,** steady state; **e. físico,** physical state; **e. inicial,** initial state; **e. juvenil,** youth stage; **e. nacional,** national status; **e. senil,** old age stage; **e. sólido,** solid state; **e. transitorio,** transitory state.

estadounidense, a. American, North American.

estalactita, f. stalactite.

estalagmita, f. stalagmite.

estambre, m. stamen.

estanato, m. stannate.

estancadola, a. stagnant.

estancamiento, m. stagnation, standstill, checking of flow.

estancia, f. (LAm) ranch.

estancola, a. watertight; state monopoly (monopolio); tobacconist's; cigar shop (US) (tienda).

estándar, m. standard; **e. de calidad de las aguas,** water quality standards, water quality goals (US); **Estándares británicos,** British Standards.

estandardización, f. standardization.

estandardizar, v. to standardize.

estandarización, f. standardization.

estandizar, v. to standardize.

estanina, f. stannite.

estanque, m. tank, small reservoir (artificial); pool, pond (natural); **e. de estabilización,** balancing tank; **e. de evaporación,** evaporation tank; **e. de sedimentación,** settling tank.

estanqueidad, f. watertightness.

estañadera, f. soldering, tin-plating.

estañado, m. tin-plating, soldering.

estañar, v. to solder, to tin-plate.

estaño, m. tin (Sn).

estatal, a. state; **sector (m) e.,** state sector.

estáticola, a. static; **presión (f) e.,** static head.

estatutariola, a. statutory.

estatuto, m. statute, statutory law; **e. de ciudad,** bylaw; **estatutos (m.pl),** charter; **estatutos sociales,** articles of association.

estaurolita, f. staurolite.

estavella, f. underground stream in karstic terrain.

estearato, m. stearate; **e. de sodio,** sodium stearate.

esteatita, f. steatite, soapstone.

Estefaniense, m. Estefanian.

estepa, f. steppe, steppes (pl).

estequiometría, f. stoichiometry.

éster, m. ester.

estercolar, v. to manure (el suelo).

estereograma, m. stereogram.

estereoquímica, f. stereochemistry.

estereoscópicola, a. stereoscopic; **par (m) e.,** stereoscopic pair.

estereoscopio, m. stereoscope.

estéril, 1. m. mine spoil, gangue (Min); 2. a. sterile, barren.

esterilización, f. sterilization.

esterilizar, v. to sterilize.

esterizante, m. sterilant.

esterlina, f. sterling; **la libra (f) e.,** the pound sterling.

estero, m. estuary; marsh, swamp (LAm); stream (Ch) (favela).

esteroide, m. steroid.

estéticola, a. aesthetic, esthetic (US).

estiaje, m. low tide, low water.

estiércol, m. manure, muck, dung.

estigma, m. stigma.

estilo, m. style.

estimación, f. estimate, approximation; **e. de la población,** population estimate; **e. puntual (Mat),** point estimation.

estimador, m. y f. valuer, appraiser (US) (Fin).

estimulación, f. stimulation.

estimulante, m. stimulant.

estimular, v. to stimulate.

estímulo, m. stimulus; stimili (pl).

estireno, m. styrene.

estirpe, f. stock, lineage.

estivación, f. aestivation, estivation (US).

estocásticola, a. stochastic.

estolón, m. stolon.

estoma, m. stoma (pl. stomata).

estómago, m. stomach.

estragón, m. tarragon.

estrategia, f. strategy.

estratégicola, a. strategic; **desarrollo (m) e.,** strategic development; **planificación (f) e.,** strategic planning.

estratificación, f. stratification, bedding (Geol); **e. cruzada,** cross-bedding; **e. fina,** thin-bedded (a); **e. gradada,** graded bedding; **e. gruesa,** thick-bedded (a); **e. rizada,** hassock bedding; **e. termal,** **e. térmica,** thermal stratification; **plano (m) de e.,** bedding plane.

estratificadola, a. stratified, bedded; **no e.,** unstratified.

estratigrafía, f. stratigraphy.

estratigráficola, a. stratigraphical.

estrato, m. stratum (pl. strata); bed (Geol); strato (m) (Met); **e. competente (Geol),** competent bed; **e. delgado,** thin bed; **e. fosilífero,** fossiliferous bed; **e. herbáceo (Ecol),** herb layer.

estratocúmulo, m. stratocumulus.

estratosfera, f. stratosphere.

estratosféricola, a. stratospheric.

estrecho, m. narrows, straits (pl) (Geog); **E. de Gibraltar,** Straits of Gibraltar; **e. de mar,** firth (p.e. Firth of Forth).

estreptococo, m. streptococcus (pl. streptococci).

estreptomicina, f. streptomycin.

estría, f. striation, groove.

estriadola, a. striated.

estribación, f. spur (Geog); **estribaciones (pl),** foothills.

estribo, m. bridge abutment (puente); buttress (contrafuerte); foundation (Fig, base).

estricnina, f. strychnine.

estro, m. oestrus.

estrógeno, m. oestrogen.

estromatolito, m. stromatolite.

estroncianita, f. strontianite.
estroncio, m. strontium (Sr).
estropajado, m. scouring, swabbing.
estropear, v. to damage, to ruin, to spoil.
estructura, f. structure, frame; **e. anticlinal,** anticlinal structure; **e. columnar,** columnar structure; **e. de dirección,** management structure; **e. molecular,** molecular structure; **e. organizativa,** organizational structure.
estructural, a. structural.
estruendosola, a. thunderous.
estuarinola, a. estuarine.
estuario, m. estuary.
estuco, m. stucco.
estudiar, v. to study.
estudio, m. study, investigation; **e. de factibilidad,** feasibility study; **e. de planificación,** planning study; **e. de recursos hidráulicos,** water resources study; **e. económico,** economic study; **e. geotécnico,** geotechnical study; **e. preliminar, e. previo,** preliminary study.
estupefaciente, a. narcotic.
etano, m. ethane.
etanol, m. ethanol.
etapa, f. stage; **dos etapas,** two-stage; **etapas múltiples,** multi-stage; **por etapas,** in stages.
eteno, m. ethene.
éter, m. ether; **e. etílico,** ethyl ether.
etilo, m. ethyl.
etilobenzol, m. ethylbenzene.
etino, m. ethyne.
etiqueta, f. label; **e. ecólogica,** ecolabel, green labelling.
etiquetado, m. labelling.
étnicola, a. ethnic.
etnográficola, a. ethnographic.
etología, f. ethology.
etoxi-, pf. ethoxy-.
eucalipto, m. eucalyptus (pl. eucalyptuses, eucalypti).
eucariota, f. eucaryote, eukaryote; **eucariotas (Zoo, Bot, grupo),** Eucaryota, Eukaryota (pl).
eufóticola, a. euphotic; **zona (f) e. (Biol),** euphotic zone.
euroasiáticola, a. Eurasian.
Europa, f. Europe.
europeola, 1. m. y f. European (de nacionalidad); 2. a. European; **Comisión (f) E.,** European Commission; **Comunidad (f) Económica E., (abr. CEE),** European Economic Community (abr. EEC); **Comunidad (f) E. (abr. CE),** European Community (abr. EC); **Unión (f) E. (abr. UE),** European Union (abr. EU).
europio, m. europium (Eu).
eustasia, f. eustacy.
eustáticola, a. eustatic.
eutécticola, a. eutectic.
eutróficola, a. eutrophic.
eutrofización, f. eutrophication.

eutrofizar, v. to eutrophicate.
eV, m. eV (electrón-voltio).
evacuación, f. evacuation, removal; **e. de las aguas residuales,** sewage disposal.
evaluación, f. evaluation, assessment; **e. del impacto ambiental (abr. EIA),** environmental impact assessment (abr. EIA); **e. de riesgo,** risk assessment.
evaluar, v. to evaluate, to assess.
evaporación, f. evaporation; **cubeta (f) de e.,** evaporation pan; **e. potencial,** potential evaporation; **e. real,** actual evaporation; **e. solar,** solar evaporation; **tanque (m) de e.,** evaporation tank.
evaporador, m. evaporator.
evaporar, v. to evaporate.
evaporígrafo, m. evapograph.
evaporímetro, m. evaporimeter; **e. de balanza,** balancing evaporimeter.
evaporita, f. evaporite.
evapotranspiración, f. evapotranspiration; **e. potencial,** potential evapotranspiration; **e. real,** actual evapotranspiration.
evapotranspirómetro, m. evapotranspirometer.
evidencia, f. evidence (Jur).
evolución, f. evolution; **teoría (f) de e.,** theory of evolution.
evolucionar, v. to evolve.
evolutivola, a. evolutionary.
exactitud, f. exactness, accuracy.
examen, m. examination, study, survey; **e. de documentación,** literature review.
excavación, f. excavation, dig, digging; **sobre e. (Min),** overbreak.
excavadorla, m. y f. excavator, digger (p.e. par Arq).
excavadora, f. excavator (máquina); **e. de retrodescarga,** back-actor excavator.
excavar, v. to dig, to excavate.
exceder, v. to exceed.
excentricidad, f. eccentricity.
excéntricola, a. eccentric.
excitación, f. excitation.
excreción, f. excretion.
excremento, m. faeces (pl), feces (pl) (US), excrement; dung, animal manure (Agr).
excursión, f. excursion, trip, outing.
excursionismo, m. outings (pl), excursions (pl), hiking.
excursionista, m. y f. hiker, tripper.
exención, f. exemption; **e. jurídico,** legal exemption.
exentola, a. exempt, free (de obligación); **e. de aduanas,** duty-free; **e. de toda obligación,** free of obligation.
exfiltración, f. exfiltration.
exfoliación, f. exfoliation.
exfoliar, v. to exfoliate.
exhalar, v. to exhale, to breathe out.
exhondamiento, m. gouging, channel scouring.
exhumar, v. to exhume.
exigencia, f. requirement, demand.

exo-, pf. exo-.

éxodo, m. exodus; **é. rural**, rural depopulation.

exoesqueleto, m. exoskeleton.

exógenola, a. exogenous.

exotérmicola, a. exothermic; **reacción (f) e.**, exothermic reaction.

expandir, v. to expand, to extend.

expansión, f. expansion, enlargement; **e. del fondo oceánico**, sea floor spreading; **e. termal**, thermal expansion; **junta (f) de e.**, expansion joint.

expediente, m. dossier, file, record (fichero); case, proceedings (Jur); inquiry (investigación); **e. profesional**, professional record; **e. sancionador (urbanismo)**, cessation order.

expeler, v. to expel, to eject.

experiencia, f. experience.

experimentación, f. experimentation.

experimental, a. experimental.

experimentar, v. to experiment, to test.

experimento, m. experiment; **hacer experimentos (v)**, to experiment.

expertola, m. y f. expert.

explanación, f. explanation.

explicar, v. to explain.

explicativola, a. explanatory.

explícitola, a. explicit; **función (f) e.**, explicit function.

exploración, f. exploration; prospecting (Min).

explorar, v. to explore.

exploratoriola, a. exploratory; **sondeo (m) e.**, exploratory drilling.

explosímetro, m. explosimeter.

explosión, f. explosion; **e. demográfica**, population explosion; **e. nuclear**, nuclear explosion; **e. subterránea**, underground explosion.

explosivo, m. explosive.

explosivola, a. explosive; **goma (f) e.**, plastic explosive; **residuos (m.pl) e.**, explosive wastes.

explotabilidad, f. exploitability.

explotable, a. exploitable.

explotación, f. exploitation, development; **e. a cielo abierto (Min)**, opencast working; **e. de canteras**, quarrying; **e. de recursos**, exploitation of resources; **e. económica**, economic development; **e. forestal**, forestry; **e. óptima**, optimum exploitación; **e. por fajas aisladas**, cut-and-fill mining; **e. por galerías**, drift mining; **e. por socavón**, adit mining; **e. sostenible**, sustainable development; **gastos de e.**, operating costs.

explotadola, a. exploited; **no e.**, unexploited.

explotar, v. to explode, to blow up.

exponencial, a. exponential; **función (f) e.**, exponential function.

exportación, f. export trade, exportation; **artículos (m.pl) de e., bienes (m.pl) de e.**, export goods; **comercio (m) de e.**, export trade.

exportar, v. to export.

expresadola, a. expressed.

expresamente, ad. expressly, specifically.

expresar, v. to express.

expresión, f. expression; **e. matemática**, mathematical expression.

expropiación, f. expropriation.

expropiar, v. to expropriate.

expulsar, v. to expel.

extender, v. to extend, to spread.

extendidola, a. widespread; **ampliamente e.**, widely distributed.

extensímetro, m. extensometer.

extensión, f. extension (ampliación); extent (superficie); duration (tiempo); range, scope (de conocimientos, de proyectos), spreading; **e. urbana**, urban sprawl.

extensivola, a. extendible; extensive (Agr); **cultivo (m) e.**, extensive cultivation.

extensola, a. extensive, large; widespread (de conocimientos).

exterior, a. exterior; overseas, foreign (Com); overseas, abroad (en los extranjeros).

extinción, f. extinction.

extinguidola, a. extinct (p.e. una raza), extinguished (p.e. un fuego); wiped out, obliterated (exterminado); **especie (f) e. (Bot, Zoo)**, extinct species.

extinguir, v. to extinguish.

extintola, a. extinct (una raza, un volcán), extinguished (un fuego).

extintorla, m. y f. extinguisher; **extintor (m) de fuego**, fire extinguisher.

extracción, f. extraction; **e. de disolvente**, solvent extraction; **e. de vapor del suelo**, soil vapour extraction; **e. por vacío**, vacuum extraction.

extractar, v. to summarize, to abridge.

extractivola, a. extractive; **actividades (f.pl) extractivas**, extractive industry.

extracto, m. extract.

extraer, v. to extract, to abstract (abstraer).

extrapolación, f. extrapolation.

extrapolar, v. to extrapolate.

extrarradio, m. suburbs (pl).

extremidad, f. extremity.

extrusivola, a. extrusive; **roca (f) e.**, extrusive rock.

exudación, f. exudation.

exudar, v. to exude.

eyectar, v. to eject.

eyector, m. ejector; **e. neumático**, pneumatic ejector.

F

fábrica, f. fabric; factory, mill (p.e. de textiles, papel); masonry; **f. de cerveza**, brewery; **f. de deformación**, **f. deformacional (Geol)**, deformation fabric; **f. harina**, flour mill; **marca (f) de f.**, trademark.

fabricación, f. manufacture, fabrication; **costos (m.pl) de f.**, manufacturing costs; **f. de acero**, steelmaking.

fabricante, m. manufacturer.

fabricar, v. to manufacture.

fabril, a. manufacturing.

faceta, f. facet.

fachada, f. facade, frontage, front (de edificio); **alzado (m) de f.**, front elevation.

facies, f. facies (pl. facies); **f. conchíferas**, shelly facies; **f. lacustres**, lacustrine facies; **f. marinas**, marine facies.

facilidad, f. facility; **f. para el almacenamiento de combustible**, fuel-storage facility.

factibilidad, f. feasibility, practicality; **análisis (m) de f.**, feasibility study.

factible, a. feasible (de un plan).

factor, m. factor; **f. biótico**, biotic factor; **f. climático**, climatic factor; **f. de crecimiento**, growth factor; **f. edáfico**, edaphic factor; **f. limitador**, **f. limitante**, limiting factor; **f. político**, political factor; **f. social**, social factor; **f. socioeconómico**, socioeconomic factor.

factorial, a. factorial.

factorización, f. factorization.

factura, f. invoice, bill; sales check (US).

facturación, f. invoicing, billing.

facturar, v. to invoice.

facultativo/a, a. facultative; **parasitario (m) f.**, facultative parasite; optional, facultative (no obligatorio).

faenar, v. to work in, to fish.

faenero, m. (Ch) farm worker.

fago, m. phage, bacteriophage.

Fahrenheit, m. Fahrenheit.

faid, m. (Arab) hillside spring in mountains.

faja, f. strip, band, fringe (Gen); zone, belt (Geog); **f. capilar**, capillary fringe; **f. de meandros**, meander belt, meander zone.

fajeado/a, a. streaky.

falda, f. skirt (ropa); flap, fold (pliegue), foothill (Geog).

falsificación, f. falsification.

falsificar, v. to falsify.

falso/a, a. false; **f. estratificación (f)**; false bedding.

falla, f. fault, defect; **bloque (m) de f.**, fault block; **brecha (f) de f.**, fault breccia; **complejo (m) de f.**, fault complex; **escarpe (m) de f.**, fault scarp; **f. de alto ángulo**, high-angle fault; **f. de arrastre**, drag fault; **f. de cizalla**, tear fault; **f. de empuje**, thrust fault; **f. de gradería**, step fault; **f. de rasgadura**, tear fault; **f. en escalón**, step fault; **f. normal**, normal fault; **flanco (m) de f.**, limb of fault; **límite (m) de f.**, fault boundary; **línea (f) de f.**, fault line; **salto (m) vertical de una f.**, throw of fault; **zona (f) de f.**, fault zone.

fallado/a, a. faulted.

familia, f. game, sport (diversión); set (conjunto); **f. de diaclasas (Geol)**, joint set.

Fammeniense, m. Fammenian.

fanega, f. grain measure (Esp. 1,58 bushels, Méx. 2,57 bushels, CSur. 3,89 bushels); area measure (Esp. 1,59 acres, Carib. 1,73 acres).

fanerita, f. phanerite.

fangal, m. bog, quagmire.

fangar, m. boggy land subject to flooding.

fango, m. mud, muck, ooze; sewage sludge (Trat); **deshidratación (f) de fangos**, sludge dewatering; **destrucción (f) de f.**, sludge disposal; **digestión (f) de f.**, sludge digestion; **f. activado**, activated sludge; **f. calcáreo**, calcareous ooze; **f. digestivo**, digested sludge; **f. dragado**, dredging, dredged mud; **f. mixto**, mixed sludge; **fangos cloacales**, sewage sludge; **incinerador (m) de los fangos de alcantarilla**, sludge incinerator; **licor (m) de f.**, sludge liquor; **secado (m) de fangos**, sludge drying; **tratamiento (m) de f.**, sludge treatment.

fangolita, f. mudstone.

faradio, m. farad (abr. F, unidad de medida de capacidad eléctrica).

farallón, m. headland, bluff, rocky peak (Geog); outcrop (Geol).

farmacéutico/a, a. pharmaceutical; **industria (f) f.**, pharmaceutical industry.

farmacia, f. pharmacy; chemist's, chemist's shop; drugstore (US) (tienda).

farmacología, f. pharmacology.

faro, m. lighthouse (torre), beacon (señal luminosa); headlight (en un coche); small atoll (Arg); **buque (m) f.**, lightship.

fase, f. phase, stage (etapa); **desfasaje (m)**, **desfase (m)**, phase difference; **estar en f. (v)**, to be in phase; **f. avanzada**, advanced phase; **f. inmiscible**, immiscible phase; **f. libre**, free-phase.

FAT, f. FOE; **Federación (f) de Amigos de la Tierra (Esp)**, Federation of Friends of the Earth.

fatal, a. fatal, inevitable.

fatiga, f. fatigue; **f. del metal**, metal fatigue.

fauna, f. fauna; **f. acuática**, aquatic fauna; **f. conchífera**, shelly fauna.

faunal, a. faunal.
favela, f. (LAm) squatter settlement.
favorable, a. favourable, propitious.
fayd, m. (Arab) hillside spring in mountains.
febril, a. fevered.
fecal, a. faecal, fecal (US); **streptococos (m.pl) fecales**, faecal/fecal streptococci.
fecundación, f. fertilization.
fecundar, v. to fertilize (de reproducción).
fecundidad, f. fecundity.
fecundola, a. fertile, fecund (de organismos).
fecha, f. date (tiempo); **f. de caducidad**, expiry date; **f. límite**, **f. tope**, deadline.
FED, m. EDF; **Fondo (m) Europeo de Desarrollo**, European Development Fund.
FEDER, m. ERDF; **Fondo (m) Europeo de Desarrollo Regional**, European Regional Development Fund.
federación, f. federation; **F. Europea de Geólogos (abr. FEG)**, European Federation of Geologists (abr. EFG).
federal, a. federal; **entidad (f) f.**, federal body.
feedback, m. feedback; **f. negativo**, negative feedback; **f. positivo**, positive feedback.
feldespáticola, a. feldspathic.
feldespato, m. feldspar, felspar.
femeninola, a. female.
feniciola, m. y f. Phoenician.
fénicola, a. carbolic.
fenilamina, f. phenylamine.
fenileno, m. phenylene.
fenilo, m. phenyl.
fenocristal, m. phenocryst.
fenol, m. phenol.
fenolftaleína, f. phenolphthalein.
fenómeno, m. phenomenon (pl. phenomena).
fenotipo, m. phenotype.
feofíceas, f.pl. brown algae, Phaeophyta (n.pl) (Bot, class).
FEPMA, m. **Fundación (f) para la Ecología y la Protección del Medio Ambiente (Esp)**, Foundation for Ecology and Environmental Protection.
feral, a. feral.
feraz, a. fertile.
fermentación, f. fermentation; **f. anaerobia**, anaerobic fermentation.
fermentar, v. to ferment.
fermento, m. ferment.
feromona, f. pheromone.
ferrato, f. ferrate.
férreola, a. iron-like, ferreous, ferrous; **línea (f) f.**, **vía (f) f.**, railway, railroad (US).
férricola, a. ferric.
ferrita, f. ferrite.
ferrobacteria, f. iron bacterium (pl. iron bacteria).
ferrocarril, m. railway; **f. de cercanías**, city line.
ferrocemento, m. ferrocement, reinforced concrete.
ferrocianuro, m. ferrocyanide.
ferromagnésianola, a. ferromagnesian.

ferromagnéticola, a. ferromagnetic.
ferromanganeso, m. ferromanganese.
ferroníquel, m. ferronickel.
ferrosola, a. ferrous; **no ferrosola**, non-ferrous.
ferruginosola, a. ferruginous.
fértil, a. fertile (de suelo).
fertilidad, f. fertility.
fertilización, f. fertilization; **f. cruzada**, cross-fertilization.
fertilizante, 1. m. fertilizer; **f. orgánico**, organic fertilizer; **f. químico**, chemical fertilizer; **f. sintético**, synthetic fertilizer; 2. a. fertilizing.
fertilizar, v. to fertilize (de suelo).
fetal, a. foetal, fetal (US).
fétidola, a. foetid, fetid (US), foul-smelling.
feto, m. foetus, fetus (US).
fiabilidad, f. reliability; **estudio (m) de f.**, reliability study.
fianza, f. deposit (Fin, anticipo); surety, security (Fin).
fiberglass, f. fibreglass, fiberglass (US).
fibra, f. fibre, fiber (US); grain (de madera); **f. alimenticia**, dietary fibre; **f. artificial**, man-made fibre; **f. de vidrio**, fibreglass, fiberglass (US), glass fibre, glass fiber (US); **f. dietética**, dietary fibre; **f. óptica**, optical fibre; **f. sintética**, synthetic fibre; **f. textil**, textile fibre.
fibrocemento, m. fibrocement, asbestos cement.
fibrolita, f. fibrolite.
fibrosola, a. fibrous.
ficomiceto, m. phycomycete; **ficomicetos (Zoo, Bot, clase)**, Phycomycetes (pl).
ficha, f. index card (de archivo); token (juegos etc.); form, dossier (expediente).
fichero, m. file, record (documento); file (Com); filing cabinet (mueble); card index (archivo); **f. de reserva (Comp)**, back-up file.
FIDA, m. IFAD; **Fondo (m) Internacional de Desarrollo Agrícola**, International Fund for Agricultural Development.
fidedignola, a. reliable, trustworthy.
fiebre, f. fever; **f. aftosa**, foot and mouth disease; **f. amarilla**, yellow fever; **f. tifoidea**, typhoid fever.
fierro, m. (LAm) iron.
fijación, f. fixing, fixation; **f. del nitrógeno**, nitrogen fixation; **f. química**, chemical fixation.
fijador, m. fixer; **f. de nitrógeno**, nitrogen fixer.
fijar, v. to fix (por bacteria).
fijola, a. fixed.
filamento, m. filament; **f. branquial**, gill filament.
filamentosola, a. filamentous; **organismo (m) f.**, filamentous organism.
filita, f. phyllite.
filón, m. seam, vein (Min); **f. capa**, sill; **f. cuarcífero**, quartz vein; **f. de carbón**, coal seam; **f. estrato (Geol)**, sill; **f. mineralizado**, lode.
filoncillo, m. veinlet (Min).
filonita, f. phyllonite.
filosofía, f. philosophy.

filoxera, f. phylloxera.
filtrabilidad, f. filtrability.
filtración, f. filtration; leak (Fig); **alimentador (m) por f.** (Zoo), filter feeder; **f. en prensa,** filtration press; **f. mecánica,** mechanical filtration; **f. por gravedad,** gravity filtration; **f. previa,** prefiltration; **f. rápida por arena,** rapid sand filtration; **proceso (m) de f. extendida,** extended-filtration process.
filtrado, m. filtrate, leachate.
filtrador/a, a. filter, filtering; **colchón (m) f.** (Trat), filter blanket; **prensa (f) f. de membrana,** membrane filter press.
filtradora, f. filter.
filtrante, a. filtering; **medio (m) f.,** filter medium.
filtrar, v. to filter; to leak (información).
filtrarse, v. to seep, to filter through.
filtro, m. filter; **f. a presión,** pressure filter; **f. arenoso de flujo ascendente,** upward-flow sand filter; **f. biológico aireado,** biological-aerated filter; **f. biológico rotativo,** rotating biological filter; **f. biológico,** biological filter; **f. de aire,** air scrubber; **f. de compresión,** compression filter; **f. de goteo,** percolating filter; **f. de grava,** gravel filter; **f. de mallas anchas/pequeñas,** coarse-/fine-mesh filter; **f. de manga,** bag filter; **f. de membrana,** membrane filter; **f. de nilón,** nylon filter; **f. de pie (p.e. de una presa),** toe filter; **f. de puentecillos,** bridge-slotted well screen; **f. en rejilla,** slotted well screen; **f. para grasa,** grease filter; **f. percolador,** trickling filter, percolating filter; **f. por vacío,** vacuum filter; **f. prensa,** press filter; **f. rápido de arena por gravedad,** rapid gravity sand filter; **filtros verdes** (Trat), grass plot treatment (GB, tratamiento de pastizal), reed-bed treatment (tratamiento de cañaveral); **prensa (f) de f.,** filter press; **torta (f) de f.** (Trat), filter cake.
finalización, f. finalization.
finalizar, v. to finalize, to complete, to finish (acabar).
financiación, f. financing.
financiamiento, m. (LAm) financing.
financiero, m. financier; **intermediario (m) f.,** broker.
financiero/a, a. financial; **análisis (m) f.,** financial analysis; **capacidad (f) f.,** financial status; **ejercicio (m) f.,** financial year; **estado (m) f.,** financial statement; **respaldo (m) f.,** financial backing; **retorno (m) f.,** financial return.
finanzas, f.pl finance, funds (pl); **director/a (m, f) de f.,** financial director.
finca, f. (LAm) country house, farm.
finito/a, a. finite; **aproximación (f) de diferencias finitas implícitas,** implicit finite-difference approximation; **modelo (m) por diferencias finitas,** finite-difference model; **modelo (m) por elementos finitos,** finite-element model.
finos, m.pl. fines (Min).

finura, f. fineness.
fiordo, m. fjord.
firme, a. firm, secure, stable, steady.
firmeza, a. firmness.
fiscal, m. public prosecutor (Jur, GB), district attorney (Jur, US); treasury official (empleado de Hacienda).
fiscal, a. fiscal, treasury.
fisibilidad, f. fissibilty.
física, f. physics.
físico/a, 1. m. y f. physicist; 2. a. physical; **modelo (m) f.,** physical model.
físico-químico/a, a. physico-chemical.
fisil, a. fissile.
fisiográfico/a, a. physiographic.
fisiología, f. physiology.
fisión, f. fission; **f. nuclear,** nuclear fission.
fisura, f. fissure.
fisuración, f. jointing.
fisurado/a, a. fissured; **acuífero (m) f.,** fissured aquifer.
fito-, pf. phyto-.
fitófago/a, a. phytophagous.
fitogenética, f. phytogenetics.
fitogeografía, f. phytogeography.
fitómetro, m. phytometer.
fitoplancton, m. phytoplankton; **emergencia (f) de f.,** phytoplankton bloom.
fitosanitario/a, a. plant health, phytosanitary; **productos (m.pl) fitosanitarios,** phytosanitary products.
fitosociología, f. phytosociology, plant sociology.
fitotoxicidad, f. phytotoxicity.
fitotóxico/a, a. phytotoxic.
flanco, m. side, flank, limb; **f. de anticlinal,** flank of anticline; **f. de un pliegue,** flank of a fold.
flash, m. flash, flashlight (luz); newsflash (noticias).
flecha, f. arrow; **punta (f) de f. de sílex** (Arq), flint arrowhead.
flete, m. freight.
flexibilidad, f. flexibility, adaptability.
flexión, f. flexion, inflexion, bending; **momento (m) de f.,** bending moment.
floc, m. floc.
floculación, f. flocculation; **agente (m) de f.,** flocculating agent; **f. mecánica,** mechanical flocculation.
floculador, m. flocculator; **f. mecánico,** mechanical flocculator.
floculante, 1. m. flocculant; 2. a. flocculating; **agente (m) f.,** flocculating agent.
flocular, v. to flocculate.
flóculo, m. floc.
floema, m. phloem.
flor, f. flower, bloom, blosssom; **en f.,** in blossom.
flora, f. flora (pl. flora); **f. acuática,** aquatic flora; **f. endémica,** endemic flora.
floración, f. bloom, flowering; **f. de plancton,** algal bloom.

flota, f. fleet; **f. congeladora,** factory fishing fleet; **f. de bajura, f. de fresco,** inshore fishing fleet, short-haul fishing fleet; **f. pesquera de altura,** deep sea fishing fleet, ocean-going fishing fleet.

flotabilidad, f. buoyancy, flotation.

flotación, f. flotation; **f. por aire disuelto,** dissolved-air flotation.

flotador, m. float (Gen); **interruptor (m) operado por un f.,** float-operated switch; **válvula (f) operada por un f.,** float-operated valve.

flotante, a. floating; **dique (m) f.,** floating dock; **pontón (m) f.,** floating bridge.

fluctuación, f. fluctuation, variation; **f. diurna,** diurnal fluctuation; **f. estadística,** statistical fluctuation.

fluctuar, v. to fluctuate, to oscillate.

fluente, a. flowing.

fluidez, f. liquidity, fluidness; **índice (m) de f. (Ing),** liquidity index.

fluidificación, f. fluidization.

fluidificadola, a. fluidized.

fluido, m. fluid.

fluidola, a. fluid, fluidized; **incinerador (m) de lecho f.,** fluidized-bed incineration.

flujo, m. flow, stream (caudal); flux (transporte); swing (de volantes); **estructura (f) de f.,** flow structure; **f. basal,** base flow; **f. bidimensional,** two-dimensional flow; **f. calorífico,** heat flow; **f. de agua subterránea,** groundwater flow; **f. de base,** base flow; **f. en canales abiertos,** open-channel flow; **f. específico,** specific flow; **f. inferior,** underflow; **f. kárstico,** karstic flow; **f. laminar,** laminar flow; **f. multifase,** multiphase flow; **f. regional,** regional flow; **f. subcrítico,** subcritical flow; **f. transversal,** cross-flow; **f. turbulento,** turbulent flow; **hidrograma (m) de f.,** flow hydrograph; **línea (f) de f.,** flow line; **plano (m) de f. cero (Geog),** zero flux plane.

flúor, m. fluorine (F).

fluorar, v. to fluorate.

fluoresceína, f. fluorescene.

fluorescencia, f. fluorescein.

fluorescente, a. fluorescent; **ser f. (v),** to fluoresce; **trazador (m) f.,** fluorescent tracer.

fluorita, f. fluorite, fluorspar.

fluorización, f. fluoridation, fluoridization.

fluorómetro, m. fluorometer.

fluoruro, m. fluoride.

fluvial, a. fluvial; **meandros (m.pl) fluviales,** river meandering; **vía (f) f.,** waterway.

fluviodeltaicola, a. fluvio-deltaic.

fluvioglaciar, a. fluvioglacial.

fluviolacustre, a. fluvio-lacustrine.

fluviomarinola, a. fluvio-marine.

fluyente, a. flowing.

flysch, m. flysch.

foca, f. seal (mamífero); sealskin (piel); **f. monje,** monk seal.

foco, m. focus (pl. foci, focuses); center (US).

fofadal, m. (Arg) bog, quagmire.

foliación, f. foliation, cleavage (Geol).

foliadola, a. foliaceous (Geol).

folíolo (m), m. foliole.

follaje, m. foliage.

fomentar, v. to encourage, to foster.

fondeo, m. anchoring, anchorage (Mar).

fondo, m. bottom (base); funds, finance (Com); **concentración (f) de f.,** background concentration; **de f.,** long-distance; **de medio f.,** middle-distance; **f. de valle,** valley floor; **f. del mar,** bottom of the sea, ocean floor; **F. Monetario Internacional,** International Monetary Fund; **f. oceánico,** ocean floor; **fondos de capital riesgo,** venture capital; **fondos de inversión,** unit trust; **fondos propios,** equity; **mar (m) de f.,** groundswell.

fonocaptor, m. sound pickup (de sonido).

fontanería, f. plumbing.

foraminíferos, m.pl. Foraminifera (pl) (Zoo, grupo).

foráneola, a. foreign, alien (extranjero).

forestación, f. forestation, afforestation.

forestal, a. forest; **incendio (m) f.,** forest fire; **producción (f) f.,** timber production.

forjadola, a. wrought, forged; **acero (m) f.,** forged steel.

forjar, v. to forge.

forma, f. form, shape; **de f. triangular,** triangular shape; **en f. de U,** U-shaped; **en f. de V,** V-shaped; **f. de abanico,** fan-shaped; **f. de bloques,** blocky; **f. de pago,** method of payment; **f. del terreno, f. de relieve, f. fisiográfica,** landform; **formas arriñonadas,** boudinage.

formación, f. formation; **f. acuífera,** water-bearing formation; **f. blanda,** soft formation; **f. calcárea,** calcareous formation; **f. compacta (Geol),** tight formation; **f. de cavidades,** cavitation; **f. de estrías (Geol),** rodding; **f. de montañas,** mountain building; **f. estratificada,** stratified formation; **f. fisurada,** fissured formation; **f. kárstica,** karstic formation; **f. no consolidada,** unconsolidated formation.

formaldehído, m. formaldehyde.

formalina, f. formalin.

formato, m. format; **f. de disco (Comp),** disk format.

fórmicola, a. formic.

fórmula, f. formula; **f. empírica,** empirical formula; **f. matemática,** mathematical formula; **f. molecular,** molecular formula; **f. química,** chemical formula.

formulación, f. formulation.

formular, v. to formulate.

formulario, m. form, application form.

forraje, m. fodder, forrage, foliage.

forro, m. lagging, lining.

fortalecimiento, m. strengthening.

fortaleza, f. strength (fuerza); fort, stronghold (p.e. castillo).

fortificación, f. fortification.

fortín, m. small fort, bunker, blockhouse.
fortuitola, a. by chance, haphazard; fortuitous.
forzadola, a. forced, hard.
fosa, f. basin, trough (Geog); grave (tumba); cavity (Anat); fossa (Arq); **f. común,** common grave; **f. de infiltración,** infiltration basin; **f. de recarga,** recharge basin; **f. marina,** deep sea trough; **f. oceánica,** oceanic trench; **f. séptica,** septic tank.
fosfatar, v. to phosphatize.
fosfáticola, a. phosphatic.
fosfato, m. phosphate; **eliminación (f) de f.,** phosphate removal; **f. abono,** phosphate fertilizer.
fosfina, f. phosphine.
fosfito, m. phosphite.
fosfolípido, m. phospholipid.
fosforescencia, f. phosphorescence.
fosforescer, v. to phosphoresce.
fosforita, f. phosphorite.
fósforo, m. phosphorous (P).
fosgeno, m. phosgene.
fósil, m. fossil; **energía (f) f.,** fossil energy.
fosiliferola, a. fossiliferous.
fosilización, f. fossilization.
foso, m. ditch (zanja); hole, pit (hoya); **f. de reparaciones (para coches),** inspection pit; **f. negro,** cesspool.
fóticola, a. photic; **zona (f) f. (Geog),** photic zone.
foto-, pf. photo-.
fotocolorímetro, m. photocolorimeter.
fotoeléctricola, a. photoelectric; **célula (f) f.,** photoelectric cell, photo cell.
fotogeología, f. photogeology.
fotografía, f. photography (técnica); photograph, photo (foto); **f. aérea,** aerial photography.
fotográficola, a. photographic.
fotogrametría, f. photogrammetry.
fotoheterótrofo, m. photoheterotroph, photoheterotrophe.
fotoíndice, m. photo-index.
fotointerpretación, f. photointerpretation.
fotólisis, m. photolysis.
fotoluminescencia, f. photoluminescence.
fotometría, f. photometry.
fotomontaje, m. photomontage.
fotomosaico, m. photomosaic.
fotomultiplicador, m. photomultiplier.
fotón, m. photon.
fotoperiodicidad, f. photoperiodism.
fotoperiodismo, m. photoperiodism.
fotoquímicola, a. photochemical.
fotosíntesis, f. photosynthesis.
fotosintéticola, a. photosynthetic.
fotosintetizar, v. to photosynthesize.
fototaxia, f. phototaxis.
fototropismo, m. phototropism.
fotovoltaicola, a. photovoltaic, solar power; **célula (f) f.,** photoelectric cell; **tecnología (f) f.,** solar-power technology.

FPNE, m. **Fondo (m) para el Patrimonio Natural Europeo,** European Fund for Natural Heritage.
FR, f. RF; **radiofrecuencia (f),** radio frequency.
fracaso, m. failure.
fracción, f. fraction (f) (Mat); part, portion (parte); faction (grupo); **f. representativa,** representative fraction.
fraccionadola, a. fractional; **cristalización (f) f.,** fractional crystalization.
fraccionamiento, m. fractionation, fractionating; (Méx) housing development, housing estate; **columna (f) de f.,** fractionating column; **f. por espuma,** foam fractionation; **torre (m) de f.,** fractionating tower.
fractal, a. fractal.
fractura, f. fracture.
fracturación, f. fracturing, faulting; **f. hidráulica,** hydraulic fracturing, hydro-fracturing.
fracturadola, a. fractured.
frágil, a. fragile.
fragmentadola, a. crumbly, fragmented.
fragmentariola, a. fragmentary.
fragmento, m. fragment; **f. encorvado,** shard; **fragmentos de rocas (Min),** brash (GB), mush (US).
fragmocono, m. phragmocone.
fraguado, m. setting (of cement).
frambuesa, f. raspberry.
frambueso, m. raspberry bush, raspberry cane.
francéslfrancesa, a. French, Frenchman/Frenchwoman.
Francia, f. France.
francio, m. francium (Fr).
francobordo, m. freeboard (de buques).
franja, f. fringe, border (Gen); strip (Geog); **f. capilar,** capillary fringe; **f. central,** central reservation (GB), median strip (US); **f. de interferencia,** interference fringe; **f. subtropical,** subtropical fringe; **f. urbana,** urban fringe; **la f. de Gaza,** the Gaza Strip.
franquicia, f. franchise, exemption from.
frasco, m. flask, small bottle; **f. de Mariotte,** Mariotte bottle.
Frasniaense, m. Frasnian.
freático, m. unconfined groundwater; **freático (m) colgado,** perched groundwater.
freáticola, a. phreatic.
freatofita, f. phreatophyte.
frecuencia, f. frequency; **distribución (f) de f.,** frequency distribution; **f. de los ríos,** stream frequency; **f. de red,** mains frequency; **f. granulométrica (Geol),** size frequency; **radiofrecuencia (f) (abr. FR),** radio frequency (abr. RF).
frejol, m. (esp. Perú) bean.
frente, m. front; face (Min); **f. aluvial,** alluvial front; **f. atmosférico,** atmospheric front; **f. cálido,** warm front; **f. de agua recargada,** recharge front; **f. de carbón,** coal face; **f. de corte (Min),** working face; **f. de deslizamiento,** slip face; **f. de desplazamiento,**

displacement front; **f. de ribera,** shore face; **f. frío,** cold front; **f. ocluido,** occluded front; hacer **f. (v),** to face up to, to confront.

freón, m. freon.

fresa, f. strawberry.

fresal, m. strawberry patch, strawberry field.

frescal, m. groundwater-fed meadow.

frescola, a. cool, fresh; **viento (m) f.,** cool wind.

fresneda, f. soils with high water table.

fresno, m. ash (Bot).

fresón, m. cultivated strawberry.

freza, f. spawn (huevos); spawning (desove); spawning season (temporada).

friabilidad, f. friability.

friable, a. friable.

frialdad, f. coldness, chilliness (tiempo); indifference (indiferencia).

fricción, f. friction; **pérdida (f) por f.,** friction loss.

frigorífico, m. refrigerator.

frigoríficola, a. refrigerated; **conservación (f) en cámara f.,** cold storage.

frijol, m. bean.

frío, m. cold; **coger f. (v),** to catch cold; **hacer f. (v) (Met),** to be cold; **ola (f) de f.,** cold spell, cold snap.

fríola, a. cold; **de sangre (f) f.,** cold-blooded; **frente (m) f.,** cold front.

fronda, f. frond (hoja); foliage (follaje).

frondosidad, f. luxuriance (de vegetación); foliage, leafiness, leaves (pl) (follaje).

frondosola, a. leafy (de una planta); luxuriant, thick (de vegetación).

frontal, a. frontal.

frontera, f. frontier, border, borderland.

fronterizola, a. frontier, border; **zona (f) f.,** border zone.

frotis, m. smear (Med).

fructosa, f. fructose.

fruta, f. fruit; **f. propia,** fresh fruit, fruit of the season.

frutal, a. fruit-bearing; **árbol (m) f.,** fruit tree.

fruticultorla, m. y f. fruit grower, fruit farmer.

fruticultura, f. fruit farming.

fruto, m. fruit; **dar f. (v),** to bear fruit.

ftaleína, f. phtalein.

fucosa, f. fucose.

fuego, m. fire; light (p.e. para cigarrillos); beacon (Mar); burner (estufa).

fuel, m. fuel-oil; **f.-oil (m),** fuel-oil.

fuelle, m. bellows (pl).

fuel-oil, m. fuel-oil.

fuente, f. spring, source, fountain; **de f. fidedigna,** from a reliable source; **f. de agua,** source of water; **f. de alimentos,** source of food; **f. de beber,** drinking fountain; **f. de energía,** source of energy; **f. de suministro,** source of supply; **f. difusa (de contaminación),** diffuse source (of pollution); **f. fija (de contaminación),** point source (of pollution);

f. pública de agua, public standpost (US), public standpipe (GB); **f. puntual,** point source; **f. radioactiva,** radioactive source.

fuerte, a. forceful, strong.

fuerza, f. strength, force, power; **f. centrífuga,** centrifugal force; **f. cizallante, f. de corte,** shear strength; **f. de un ácido,** acid strength; **f. hidráulica,** hydro power; **por f.,** by force.

fuga, f. flight (vuelo); escape (huida); spill (reboso); leak (gotera); **control (m) de fugas,** leakage control; **f. de distribución,** distribution leak; **f. de embalse,** reservoir leak.

fugitivola, a. fugitive; **emisión (f) f.,** fugitive emission.

fulminante, 1. m. fuse (para detonator); 2. a. fulminant.

fumarola, f. fumarole.

fumigación, f. fumigation; **f. con plaguicidas, f. de los cultivos,** crop spraying.

fumigante, m. fumigant.

fumigar, v. to fumigate.

función, f. function, performance; **f. de error,** error function; **f. de pozo,** well function; **f. de probabilidad,** probability function; **f. de transferencia,** transfer function; **f. matemática,** mathematical function; **funciones de Bessel,** Bessel functions.

funcional, a. functional.

funcionamiento, m. functioning, working; **sociedad (f) en f.,** going concern.

funcionariola, m. y f. official, civil servant; **f. público,** public official.

fundación, f. foundation.

fundamental, a. fundamental.

fundente, m. flux (de metales), melting.

fundición, f. foundry; smelting, melting; **f. de acero,** steel works; **f. de hierro,** iron foundry.

fundida, f. casting (Ing).

fundido, m. melted; **en estado f.,** molten.

fundidola, a. melted; molten (roca, metal); smelted (Min, para extraer metal); **roca (f) f.,** molten rock.

fundidor, m. smelter.

fundir, v. to fuse, to melt (líquido); to smelt (hierro); to cast (moldear); to join, to merge (unir); to blow (soplar); **f. hierro,** to smelt iron; **fundirse un fusible, fundirse un plomo (v) (Elec),** to blow a fuse.

fundo, m. ranch.

fungicida, m. fungicide.

funidraga, f. dragline.

furano, m. furan.

fusible, m. fuse (Elec); **caja (f) de fusibles,** fuse box.

fusiforme, a. fusiform.

fusión, f. fusion, melting; **en f.,** melted, molten; **f. nuclear,** nuclear fusion.

fusionar, v. to join, to fuse together.

fuste, m. log, timber; shaft (de lanza, de chimenea).

G

gabarra, f. barge, flat-bottomed boat.
gabinete, m. study, library (sala); cabinet (política); **g. de consulta**, consulting room.
gablete, m. gable.
gabro, m. gabbro.
gacela, f. gazelle.
gadolinio, m. gadolinium (Gd).
galardón, m. reward.
galaxia, f. galaxy.
galena, f. galena.
galenita, f. galenite.
galería, f. gallery (Min), heading, adit; qanat, canat, khanat (Arab); **bosque (m) de g. (LAm)**, riverside woodland; **g. de desagüe**, drainage level; **g. de infiltración**, infiltration gallery; **g. de pozo**, well heading, well adit.
galga, f. millstone (de molino); hub brake (freno); gauge (calibrador).
gálibo, m. gauge (galga), template, pattern.
galio, m. galium (Ga).
galón, m. gallon (medida de volumen, 4,55 litros en GB, 3,79 litros en US).
galvánicola, a. galvanitic; **pila (f) g.**, galvanitic cell.
galvanizadola, a. galvanized; **lámina (f) g.**, galvanized sheet.
galvanizar, m. to galvanize.
galvanómetro, m. galvanometer.
galvanoplásticola, a. electroplating; **industria (f) g.**, electroplating industry.
galvanotecnia, f. electroplating (technology).
gallinazo, m. turkey vulture; turkey buzzard (US); **g. sin pluma**, waste tip operative, garbage man.
gallito, m. (Perú) prominent hill at meander.
gallo, m. cock, rooster (ave); dory, john dory (pez).
gama, m. range, scale, gamut.
gamba, f. prawn; shrimp (US).
gameto, m. gamete.
gamma, f. gamma; **rayos (m.pl) g.**, gamma rays.
ganadería, f. stockbreeding, cattle raising.
ganado, m. cattle, livestock, stock; **g. lanar**, sheep; **g. menor**, sheep, goats, pigs; **g. ovejuno**, sheep; **g. porcino**, pigs, swine; **g. vacuno**, cattle.
ganat, m. (Arab) ganat, qanat, khanat; gallery, heading, adit.
ganchero, m. (Par) waste tip scavenger.
gándara, f. reed-bed in silted up lake.
gandulla, f. wetland.
ganga, f. gangue, rock waste, mine spoil (Min); bargain (price).
gangrenarse, v. to go gangrenous, to putrefy.
gangrenosola, a. gangrenous.

ganister, m. gannister.
garaje, m. garage.
garantía, f. guarantee, surety; **certificado (m) de g.**, guarantee certificate; **g. de suministro**, guarantee of supply.
garantizar, v. to guarantee, to assure.
garbanzal, m. chick-pea field.
garbanzo, m. chick-pea.
garfio, m. grapple, hook (gancho), grapnel (rezón).
gargallón, m. spring issuing from rock.
garganta, f. ravine, narrow pass (Geog); throat, gullet (Anat).
gari, m. (Br) rubbish tip operative, garbage man.
garra, f. claw (Zoo, uña); talon (de aves de rapiña).
garrapata, f. mite, tick.
garriga, f. garigue.
garúa, f. (LAm) drizzle.
gas, m. gas; **contador (m) de g.**, gas meter; **cromatografía (f) de gases-espectrometría de masas**, gas chromatography-mass spectrometry; **desulfurización (f) del g. de combustión**, flue-gas desulphurization; **detector (m) de gases**, gas detector; **extracción (f) de gases**, gas extraction; **g. de cloaca**, sewer gas; **g. de combustión**, flue-gas; **g. de los pantanos**, marsh gas; **g. de origen bioquímico**, biogas; **g. de vertido controlado**, landfill gas; **g. disuelto**, dissolved gas; **g. inerte**, inert gas; **g. natural**, natural gas; **gases de efecto invernadero**, greenhouse gases; **gases nobles**, noble gases; **gases raros**, rare gases; **instalación (f) lavadora del g. de combustión**, flue-gas washing plant; **lavado (m) de gases**, gas scrubbing; **lavador (m) del g. de combustión**, flue-gas scrubber; **licor (m) de gases**, gas liquor; **saturación (f) con g.**, gas saturation; **ventilación (f) de gases**, gas ventilation.
gaseoducto, m. gas pipeline.
gaseosola, a. gaseous, aerated; **agua (f) g.**, aerated water; **emisiones (f.pl) gaseosas**, gaseous emissions.
gasfitería, f. (LAm) plumbing.
gasificación, f. gasification.
gasificar, v. to gasify, to aerate.
gasohol, m. gasohol.
gasolina, f. petrol; gas, gasoline (US); gasolene (US, Ch); **surtidor (m) de g.**, petrol pump, gas pump (US).
gasolinera, f. petrol station, gas station (US) (estación de gasolina); motorboat (lancha).
gasómetro, m. gasholder, gasometer.
gastadola, a. worn, worn-out.

gasterópodo, m. gastropod; **gasterópodos (Zoo, clase)**, Gastropoda (pl).

gasto, m. spending, expenditure (acción); outlay; consumption (consumo); waste (desgaste).

gastos, m.pl. costs, charges, expenses, expenditure; **g. anuales de administración**, annual administration costs; **g. de acarreo**, transport charges; **g. de depreciación**, depreciation costs; **g. de energía**, energy costs; **g. de flete**, freight charges; **g. de mantenimiento**, maintenance costs; **g. de tramitación**, handling charge; **g. financieros**, financial costs; **g. generales**, overheads; **g. indirectos**, indirect costs; **g. menores (de caja)**, petty cash; **ingresos (m.pl) y g.**, income and expenditure.

gástricola, a. gastric.

gastrópodo, m. gastropod; **gastrópodos (Zoo, clase)**, Gastropoda (pl).

gato, m. cat (Gen), tomcat (macho) (Zoo); jack (para levantar cargas); **g. hidráulico**, hydraulic jack.

GATT, m. GATT; **acuerdo (m) general sobre aranceles aduaneros**, General Agreement on Tariffs and Trade.

gaussianola, a. gaussian; **distribución (f) g.**, gaussian distribution.

gaviota, f. gull.

gea, f. non-living natural resources.

geco, m. gecko.

Gedinnense, m. Gedinnian.

GEF, m. GEF; **Fondo (m) para el Medio Ambiente Mundial**, Global Environment Facility.

Geiger, m. Geiger; **contador (m) G.**, Geiger counter.

gel, m. gel.

gelatina, f. gelatine.

gelatinosola, a. gelatinous.

gelignita, f. gelignite.

gemelo, m. twin; **gemelos (de campo)**, binoculars, field glasses.

gemelola, a. twinned; **cristal (m) g.**, twinned crystal.

gemelos, m.pl. twins; binoculars (prismáticos); **g. de campo**, field glasses, binoculars.

gen, m. gene.

gene, m. gene; **pool (m) génico**, gene pool.

generación, f. generation; **g. de datos**, data generation.

generador, m. generator (Elec); **g. eléctrico**, electrical generator.

general, a. general; common (corriente).

generalidad, f. generality.

Generalitat, f. Catalan autonomous government.

generalizadola, a. generalized; **modelo (m) lineal g.**, generalized linear model.

generatriz, f. generatrix.

genéricola, a. generic; **nombre (m) g. (Biol)**, generic name.

género, m. race (raza); type, sort; genus (pl. genera) (grupo taxonómico).

genética, f. genetics.

genéticola, a. genetic; **intercambio (m) g.**, genetic exchange; **variación (f) g.**, genetic variation.

génicola, a. gene; **pool (m) g.**, gene pool.

genoma, m. genome.

genotipo, m. genotype.

geo-, pf. geo-.

geoarqueología, f. geoarchaeology.

geoda, f. geode.

geodesia, f. geodesy.

geodésicola, a. geodesic, geodetic.

geodinámicola, a. geodynamic.

geoeléctricola, a. geoelectrical.

geoestadística, f. geostatistics.

geofísica, f. geophysics.

geofísicola, 1. m. y f. geophysicist; 2. a. geophysical.

geófono, m. geophone.

geogénicola, a. geogenic.

geografía, f. geography.

geográficola, a. geographical.

geógrafola, m. y f. geographer.

geohidrología, f. geohydrology.

geohidrológicola, a. geohydrological.

geohidrólogola, m. y f. geohydrologist.

geohidroquímica, f. geohydrochemistry.

geohidroquímicola, a. geohydrochemical.

geoide, m. geoid.

geología, f. geology.

geológicola, a. geological; **Instituto (m) G. y Minero de España (Esp)**, British Geological Survey (GB).

geólogola, m. y f. geologist.

geomembrana, f. geomembrane.

geometría, f. geometry.

geométricola, a. geometric; **progresión (f) g.**, geometric progression.

geomorfología, f. geomorphology.

geomorfológicola, a. geomorphological.

geomorfólogola, m. y f. geomorphologist.

geopolítica, f. geopolitics.

geoquímica, f. geochemistry.

geoquímicola, m. y f. geochemist.

geoquímicola, a. geochemical.

Georgiense, m. Georgian.

georiesgo, m. geohazard.

georreferenciado, m. georeference.

geosfera, f. geosphere.

geosinclinal, 1. m. geosyncline; 2. a. geosynclinal.

geosintéticola, a. geosynthetic; **revestimiento (m) g. (Trat, en un vertedero)**, geosynthetic lining.

geotecnia, f. geotechnics.

geotérmicola, a. geothermal; **reacción (f) g.**, geothermal reaction.

geotextil, 1. m. geotextile; 2. a. geotextile; **revestimiento (m) g. (Trat, en un vertedero)**, geotexile lining.

geotropismo, m. geotropism.
gerente, m. manager, director.
germanio, m. germanium (Ge).
germen, m. germ (Bot); germ (microorganismo).
germinación, f. germination.
germinal, a. germ.
germinar, v. to germinate.
gestión, f. step, measure (trámite); management, conduct (administración); **g. ambiental**, environmental management; **g. integrada**, integrated management; **g. sostenible**, sustainable management.
gestoría, f. stewardship; **g. ambiental**, environmental stewardship.
geyser, m. geyser.
giardia, f. giardia.
giba, f. hump.
giga-, pf. giga- (10.E9).
gigabyte, m. gigabyte.
gigantescola, a. gigantic, mountainous.
gimnosperma, f. gymnosperm; **gimnospermas (Bot, división)**, Gymnospermae (pl).
gipsita, f. gypsite.
girasol, m. sunflower.
girder, m. girder.
giro, m. turn, rotation.
girocono, m. gyrocone.
gis, m. chalk.
gitanola, m. y f. gypsy.
Givetiense, m. Givetian.
glaciación, f. glaciation; **que ha sufrido g.**, glaciated.
glacial, m. glacial.
glaciar, 1. m. glacier; **g. de piedemonte, g. de pedemontano (LAm)**, piedmont glacier; **g. de valle**, valley glacier; 2. a. glacial, icy; **casquete (m) g.**, icecap.
glaciar, a. glacial, icy; **período (m) g.**, ice age.
glaciofluvial, a. glaciofluvial.
glándula, f. gland.
glandular, a. glandular.
glaucófana, f. glaucophane.
glauconita, f. glauconite.
gley, m. (LAm) gley (suelo arcilloso).
gliceraldehído, m. glyceraldehyde.
glicérido, m. glyceride.
glicerina, f. glycerine.
glicerol, m. glycerol.
glicol, m. glycol.
global, a. global; overall, total (total); comprehensive (p.e. un estudio); total, aggregate (cantidad); lump (suma); **calentamiento (m) g.**, global warming.
globo, m. globe, sphere, balloon; **g. aerostático**, balloon; **g. cautivo**, observation balloon; **g. del ojo**, eyeball; **g. dirigible**, airship; **g. ocular**, eyeball; **g. piloto, g. sonda**, sonde, weather balloon, trial balloon; **g. terráqueo**, the Globe (la Tierra).
glóbulo, m. globule; corpuscle (Zoo); **g. blanco sanguíneo**, white blood cell, white blood

corpuscle; **g. rojo sanguíneo**, red blood cell, red blood corpuscle.
globulosola, a. globular, blistered (Geol).
glorieta, f. square (plaza); roundabout (cruce giratorio); bower (cenador).
glosario, m. glossary; **g. de términos**, glossary of terms.
glucógeno, m. glycogen.
glucólisis, f. glycolysis.
glucosa, f. glucose.
glucosamina, f. glucosamine.
gluma, f. glume.
glutamina, f. glutamine.
gluten, m. gluten.
gneis, m. gneiss.
gobernadorla, 1. m. y f. governor, ruler; **g. civil**, civil governor; **g. provincial**, provincial governor; 2. a. governing; **junta (f) de gobierno**, governing board.
gobierno, m. government (de país, estado), management (gestión); **g. civil**, provincial government.
goethita, f. goethite.
golf, m. golf; **campo (m) de g.**, golf course, golf links.
golfo, m. gulf; **el g. de México/Méjico**, the Gulf of Mexico.
golpe, m. blow, knock, bump.
golpetear, v. to beat, to hammer; to blast; **g. con arena**, to sandblast.
golpeteo, m. tapping, beating, drumming; **g. de arena**, sandblasting.
goma, f. gum, glue; rubber (caucho); **g. arábica**, gum arabic; **g. explosiva**, plastic explosive.
goma-espuma, f. foam rubber.
gomera, f. gum tree.
gomero, m. rubber tree; rubber plantation worker (LAm).
gónada, f. gonad.
góndola, f. goods wagon, freight car (US) (ferrocarril); bus (Bol, Ch, Perú).
goniatita, f. goniatite.
gonitado, m. grout curtain.
gorgojo, m. weevil, grub (insecto); runt, dwarf (Fig).
gorgón, m. (LAm) concrete.
gota, f. drop (de líquido); **g. de agua**, drop of water; **g. de lluvia**, raindrop.
gotita, f. droplet.
gozne, m. hinge.
GPS, m. global positioning system (sistema de positionamiento terráqueo).
grabación, f. recording (p.e. de sonido); **g. en cinta magnetofónica**, tape recording.
grabadola, a. recorded (sonido etc.), on tape.
grabadorla, m. y f. tape recorder.
grabar, v. to record (sonido etc.), to engrave (p.e. un dibujo).
graben, m. graben.
grada, f. slip, slipway (Mar); step, stair (peldaño); row, tier (en un estadio).

gradiente, m. gradient, slope; **g. de concentración,** concentration gradient; **g. geotérmico,** geothermal gradient; **g. hidráulico,** hydraulic gradient; **g. piezométrico,** piezometric gradient; **g. umbral,** threshold gradient.

grado, m. degree (en una escala), grade, rank, quality; **g. absoluto,** degree absolute; **g. alto/bajo,** high/low grade; **g. Celsius,** degree Celsius; **g. centígrado,** degree centigrade; **g. de incertidumbre,** degree of uncertainty; **g. de latitud,** degree of latitude; **g. de libertad,** degree of freedom; **g. de saturación,** degree of saturation; **g. de selección (Geol),** degree of sorting; **g. Fahrenheit,** degree Fahrenheit; **g. metamórfico,** metamorphic grade.

graduación, f. grading, graduation (Mat).

gráfico, 1. m. graph, plot; **g. de caudal-descenso,** yield-drawdown graph; **g. de descenso-tiempo,** time-drawdown graph; **g. de niveles-tiempos,** water level-time graph; **g. logarítmico,** logarithmic plot; 2. a. graphic, graphical; **proceso (m) de información g.,** graphic data processing, computer graphics.

grafito, m. graphite.

Gram, a. Gram; **G. negativo/a,** Gram-negative; **G. positivo/a,** Gram-positive; **tinción (f) G.,** Gram's stain.

grama, f. pasture, grass.

Graminae, f.pl. Graminae (pl) (Bot, familia).

gramíneas, f.pl. Graminae (pl), Graminaceae (pl) (Bot, familia).

gramíneola, a. graminaceous, gramineous.

gramo, m. gramme, gram.

granada, f. pomegranate (fruit).

granate, m. garnet.

grande, a. large, big.

granero, m. barn (for grain), grain store.

granitización, f. granitization.

granito, m. granite.

granizada, f. hailstorm.

granizar, v. to hail.

granizo, m. hail.

granja, f. farm, farmhouse (cortijo), dairy (lechería); **g. avícola,** poultry farm; **g. colectiva,** collective farm.

granjerola, m. y f. farmer.

grano, m. grain, bean, seed; **g. fino,** fine-grained (a); **g. grueso,** coarse-grained (a); **tamaño (m) de g.,** grain-size.

granodiorita, f. granodiorite.

granófiro, m. granophyre.

granudola, a. grainy.

granular, a. granular.

granulita, f. granulite.

gránulo, m. granule.

granulometría, f. granulometry.

granzas, f.pl. screenings, siftings, riddlings.

grapa, f. clamp.

graptolita, f. graptolite.

grasa, f. grease; **grasas y aceites,** fats, oils and greases (Trat, abr. FOG), oil and grease (Ing); **interceptor (m) de grasas,** grease trap.

grauvaca, f. greywacke, graywacke.

grava, f. gravel; **cubierta con g. (carretera),** surfaced, metalled; **empaque (m) de gravas,** gravel pack; **g. de crecida,** flood gravel; **g. de playa,** beach gravel.

gravedad, f. gravity (Fís); seriousness (seriedad); **centro (m) de g.,** centre of gravity; **drenaje (m) por g.,** gravity drainage; **filtro (m) por g.,** gravity filter; **sistema (m) de g. (p.e. el suministro de agua),** gravity system.

gravimetría, f. gravimetry.

gravitación, f. gravitation.

gravitacional, a. gravitational.

gravitar, v. to gravitate.

gredal, m. claypit.

greisen, m. greisen.

gremial, a. trade-union.

grieta, f. crack, fissure, crevice; crevasse (Geog, en un glaciar); **g. de desecación, g. de retracción,** sun crack.

grifo, m. tap (p.e. de agua); faucet (US); petrol station, gas station (US) (gasolinera); **g. del agua,** water tap.

grillo, m. cricket (insecto); shoot (de una planta).

gripa, f. (LAm) flu, influenza.

gripe, f. flu, influenza.

griposola, a. of influenza; **estar g. (v),** to have the flu.

grisalla, f. scrap metal.

grisú, m. firedamp (Min).

grosella, f. currant; **g. espinosa,** gooseberry; **g. negra,** blackcurrant.

grosellero, m. currant bush.

grosería, f. coarseness.

groserola, a. coarse, crude.

grosularita, f. grossularite.

grúa, f. crane; **g. de torre,** tower crane.

gruesola, a. thick, rough, heavy; **mar (m) g.,** heavy sea.

grulla, f. crane.

grullera, f. shallow wetland supporting wading birds.

grupo, m. group; **g. de árboles,** clump of trees; **g. de presión,** pressure group; **g. sanguíneo,** blood group.

gruta, f. cavern, cave.

guacal, m. gourd.

guad, m. (Arab) wadi (desert valley or ravine).

guadal, m. (LAm) swamp.

guagua, f. (LAm) bus.

guaico, m. (LAm) basin, hollow (hondonada); garbage dump (US) (vertedero).

guano, m. guano.

guaraní, m. (Par) guarani (lengua, unidad monetaria).

guardabosque, m. game warden (GB).

guardacosta, m. coastguard.

guardamonte(s), m. game warden.

guardapesca, f. fishery protection; **buque (m) g.**, fishery protection vessel.

guatal, m. (CAm) hillock.

guatemaltecola, 1. m. y f. Guatemalan; 2. a. Guatemalan.

guate, m. maize; corn (US); maize plantation (CAm).

Guatemala, f. Guatemala.

Guayana, f. Guyana, Guiana.

gubernamental, a. government, governmental; state (US).

guía, 1. m. y f. leader, guide (persona); 2. f. guide, guidance; handbook, manual (p.e. libro para mantener máquinas); directory (teléfono); prompt (Comp); guidebook (libro turístico); **g. de teléfono, g. telefónica,** telephone directory.

guijarral, m. stony place; shingle beach, pebble beach (playa).

guijarro, m. shingle, pebble, cobblestone; **g. aplanado,** flat pebble; **playa (f) de guijarros,** shingle beach.

guijarrosola, a. gravelly.

guijo, m. cobble (canto rodado).

guijón, m. (Arg) cobble.

güinche, m. (LAm) crane, derrick.

guineo, m. small banana.

guisante, m. pea.

gunita, f. gunite, grout.

gunitado, m. grout curtain, grouting.

gunitadola, a. grouted.

gunitar, v. to grout.

gusano, m. worm; **g. de seda,** silkworm; **g. plano,** flatworm.

gusto, m. taste, flavour (sabor); pleasure, liking (gusto).

Gymnophyta, f.pl. Gymnophyta (Zoo, división).

H

haba, f. bean.
habichuela, f. bean.
habilidad, f. workmanship.
habitante, m. inhabitant (Gen); resident (vecino).
habitat, hábitat, m. habitat.
hacendado, m. (LAm) landowner, rancher.
hacia, p. toward; **h. abajo,** downward, downwards; **h. arriba,** upward, upwards; **h. la costa,** shoreward, shorewards.
hacienda, f. large farm, country estate; ranch (LAm); cattle, livestock (LAm, esp. Arg); **Ministerio (m) de H.,** Exchequer (GB), Treasury (US), Ministry of Finance (los demas paises).
hacha, f. axe, hatchet; **h. de piedra,** stone axe.
hachís, m. hashish.
hafnio, m. hafnium (Hf).
Haití, m. Haiti.
haitianola, 1. m. y f. Haitian; 2. a. Haitian.
halita, f. halite.
hallazgo, m. discovery, finding.
halo, m. halo (Met).
halofana, f. allophane.
halófilola, a. halophilic.
halófito, m. halophyte.
halogenación, f. halogenation.
halogenadola, a. halogenated; **disolvente (m) h.,** halogenated solvent; **hidrocarburo (m) h.,** halogenated hydrocarbon.
halogénesis, m. halogenesis.
halógeno, m. halogen.
halogenuro, m. halogen, halogenated compound; **h. de carbono,** halocarbon; CFC (chlorofluorocarbon).
haluro, m. halide; **h. volátil,** volatile halide.
hambre, m. hunger; famine (escasez); starvation (inanición).
haploide, a. haploid.
haptotropismo, m. haptotropism, geotropism.
hardware, m. hardware (Comp).
harina, f. flour, meal, powder; **h. de hueso,** bonemeal; **h. de pescado,** fishmeal.
hastial, m. gable end.
hato, m. cattle ranch.
Hauteriviense, m. Hauterivian.
haya, f. beech.
haz, f. bundle, bunch (fajo); face, surface (superficie); **h. de curvas,** family of curves; **h. de luz,** beam of light; **h. de rayos,** bundle of rays.
haza, f. plot of arable land.
HCB, m. HCB; **hexaclorobenzol,** hexachlorobenzene.
HCBD, m. HCBD; **hexaclorobutadieno,** hexachlorobutadiene.

heces, f.pl. faeces (pl), feces (pl) (US), excrement.
hectárea, f. hectare.
hecto-, pf. hecto-.
hectogramo, m. hectogramme, hectogram (US).
hectolitro, m. hectolitre, hectoliter (US).
hectómetro, m. hectometre, hectometer (US) (unidad de volumen : 10.E4 metros cúbicos).
HCH, m. HCH; **hexaclorociclohexano,** hexachlorocyclohexane.
hecho, m. action (acción); event (acto); fact (dato); **hechos innegables,** hard facts.
hechola, a. made, built; **h. de encargo,** purpose-built, purpose-made.
helada, f. frost, frosty weather; **acción (f) de la h.,** frost action.
helarse, v. to ice up, to ice over.
helecho, m. fern.
hélice, f. helix (espiral); propeller, airscrew (de buque o aeroplano).
helio, m. helium (He).
heliofíticola, a. heliophytic.
heliógrafo, m. heliograph; **h. de Campbell-Stokes (registro de las horas de sol),** Campbell-Stokes heliograph.
helión, m. helium nucleus.
heliotropo, m. heliotrope, bloodstone.
hematita, f. haematite, hematite (US).
hematites, m. haematite, hematite.
hematología, f. haematology, hematology (US).
hembra, f. female (Zoo); mujer (woman); nut (Ing).
hemíptero, m. hemipteran (pl. hemiptera, hemipterans); hemipteron (US).
hemípterola, a. hemipteran, hemipterous.
hemisferio, m. hemisphere; **h. norte,** northern hemisphere; **h. sur,** southern hemisphere.
hemocianina, f. haemocyanin, hemocianina (US).
hemofilia, f. haemophilia, hemophilia (US).
hemofílicola, m. y f. haemophiliac, hemophiliac (US).
hemoglobina, f. haemoglobin, hemoglobin (US).
hemorragia, f. haemorrhage, hemorrhage (US).
henar, m. hayfield, meadow.
hendidura, f. cleft, split.
hendimiento, m. splitting, cleavage.
henequén, m. henequen, agave.
heno, m. hay; **fiebre (f) del h.,** hay fever.
hepática, f. liverwort; **hepáticas (Bot, clase),** Hepaticae (pl).
hepáticola, a. hepatic, liver.
hepatitis, f. hepatitis.

heptágono, m. heptagon.
heptaldehído, m. heptaldehyde.
heptano, m. heptane.
heptanol, m. heptanol.
hepteno, m. heptene, heptylene.
heptileno, m. heptylene, heptene.
heptino, m. heptyne.
heptocloro, m. heptachlor.
herbácea, f. grass.
herbáceola, a. herbaceous.
herbicida, 1. m. herbicide; **2. a.** herbicide.
herbívorola, 1. m. y f. herbivore; **2. a.** herbivorous.
hereditariola, a. hereditary.
heridola, a. injured.
hermafrodita, 1. m. y f. hermaphrodite; **2. a.** hermaphrodite.
hermeticidad, f. watertightness, airtightness.
herméticola, a. watertight, airtight.
heroína, f. heroin (droga).
herramental, m. tools (pl.), set of tools, tool kit.
herramienta, f. tool; **caja (f) de herramientas,** tool box, tool kit; **conjunto (m) de herramientas,** tools, set of tools; **h. de corte,** cutting tool; **h. de perforación,** drilling tool.
herrera, f. sea bream.
herrumbre, f. rust (orín).
hervir, v. to boil.
hetero-, pf. hetero-.
heterocíclicola, a. heterocyclic.
heterocigóticola, a. heterozygous.
heterocigoto, m. heterozygote.
heterogeneidad, f. heterogeneity.
heterogéneola, a. heterogeneous.
heterólisis, m. heterolysis.
heterosexual, a. heterosexual.
heterotróficola, a. heterotrophic.
heterótrofo, m. heterotroph, heterotrophe.
Hettangiense, m. Hettangian.
heurísticola, a. heuristic.
hexa-, pf. hexa-.
hexaclorobenzol, m. hexachlorobenzene (abr. HCB).
hexaclorobutadieno, m. hexachlorobutadiene (abr. HCBD).
hexaclorociclohexano, m. hexachlorocylcohexane (abr. HCH); **alfa/beta/gamma/épsilon h.,** alpha/beta/gamma/epsilon hexachlorocyclohexane.
hexacloroetano, m. hexachloroethane.
hexacloruro, m. hexachloride; **h. de benceno,** hexachlorobenzene (abr. HCB).
hexagonal, a. hexagonal.
hexágono, m. hexagon.
hexaldehído, m. hexaldehyde.
hexametilenotetramina, f. hexamethylenetetramine, hexamine.
hexamina, f. hexamine.
hexanal, m. hexanal.
hexano, m. hexane.
hexanol, m. hexanol.

hexápodo, m. hexapod.
hexavalente, a. hexavalent.
hexeno, m. hexene.
hexileno, m. hexilene.
hialita, f. hyalite, hydrophane.
hialófana, f. hyalophane.
hiato, m. hiatus.
hibernación, f. hibernation.
hibernal, a. wintry, winter (p.e. de tiempo); hibernal (durante el invierno).
hibernar, v. to hibernate.
hibridización, f. hybridization.
híbridola, a. hybrid; **vigor (m) h.,** hybrid vigour, hybrid vigor (US).
hidra, f. hydra.
hidracina, f. hydrazine.
hidrante, m. fire hydrant.
hidratación, f. hydration.
hidratadola, a. hydrated.
hidrato, m. hydrate.
hidráulica, f. hydraulics; **h. de canal a cielo abierto,** open-channel hydraulics; **h. de pozos,** well hydraulics; **ariete (m) h.,** hydraulic ram; **carga (f) h.,** hydraulic loading; **construcción (f) h.,** hydraulic structure; **energía (f) h.,** hydraulic energy; **gato (m) h.,** hydraulic jack; **gradiente (m) h.,** hydraulic gradient; **junta (f) h.,** hydraulic joint; **obras (f.pl) hidráulicas,** hydraulic works; **presión (f) h.,** hydraulic pressure; **radio (m) h.,** hydraulic radius; **salto (m) h.,** hydraulic jump; **válvula (f) h.,** hydraulic valve.
hídricola, a. of water, hydric; **el medio (m) h.,** the water environment; **recursos (m.pl) hídricos,** water resources.
hidro-, pf. hydro-.
hidroavión, m. seaplane, flying boat.
hidrobiología, f. hydrobiology.
hidrocarburo, m. hydrocarbon; **h. clorado,** chlorinated hydrocarbon; **h. halogenados,** halogenated hydrocarbons; **h. volátil,** volatile hydrocarbon.
hidroclásticola, a. hydroclastic.
hidrocraqueo, m. hydrocracking.
hidrodiálisis, m. hydrodialysis.
hidrodinámica, f. hydrodynamics.
hidrodinámicola, a. hydrodynamic; **modelo (m) h.,** hydrodynamic model.
hidroeléctricola, a. hydroelectric; **proyecto (m) h.,** hydroelectric scheme.
hidroenergía, f. hydropower.
hidrofílicola, a. hydrophilic.
hidrófilola, a. hydrophylic.
hidrofita, m. y f. hydrophyte.
hidrofóbicola, a. hydrophobic.
hidrófobola, a. hydrophobic.
hidrofugación, f. damp-proofing.
hidrófugola, a. water-repellent, waterproof; **hacer h. (v),** to make water-repellent.
hidrogenación, f. hydrogenation.
hidrogenar, v. to hydrogenate.

hidrógeno, m. hydrogen (H).
hidrogenocarbonato, m. hydrogencarbonate; **h. cálcico**, calcium hydrogencarbonate.
hidrogeofísica, f. hydrogeophysics.
hidrogeología, f. hydrogeology.
hidrogeológicola, a. hydrogeological.
hidrogeólogola, m. y f. hydrogeologist.
hidrogeomorfología, f. hydrogeomorphology.
hidrogeoquímica, f. hydrogeochemistry.
hidrografía, f. hydrograph, hydrography.
hidrográficola, a. hydrographic; **confederación (f) h. (Esp)**, water board.
hidrograma, m. hydrograph; **análisis (m) de hidrogramas**, hydrograph analysis; **h. complejo**, complex hydrograph; **h. compuesto**, compound hydrograph; **h. de caudales**, flow hydrograph; **h. unitario**, unit hydrograph; **separación (f) de hidrogramas**, hydrograph separation.
hidrólisis, m. hydrolysis.
hidrolizar, v. to hydrolyze.
hidrología, f. hydrology; **h. de superficie**, surface hydrology; **h. kárstica**, karstic hydrology; **h. subterránea**, groundwater hydrology; **h. superficial**, surface hydrology; **h. urbana**, urban hydrology.
hidrológicola, a. hydrological; **confederación (f) h. (Esp)**, river authority, river catchment board.
hidrólogola, m. y f. hydrologist.
hidromecánica, f. hydromechanics (pl.).
hidrometeorología, f. hydrometeorology.
hidrometría, f. hydrometry.
hidrométricola, a. hydrometric; **esquema (m) h.**, hydrometric scheme.
hidrómetro, m. hydrometer, areometer.
hidroneumática, f. hydropneumatics.
hidropesía, f. dropsy.
hidroponia, f. hydroponics.
hidropónicola, a. hydroponic; **cultivo (m) h.**, hydroponics.
hidroscopia, f. dowsing, water divining; **varita (f) de avellano para h.**, divining rod.
hidroscopista, m. y f. dowser, water diviner.
hidroserie, f. hydrosere.
hidrosfera, f. hydrosphere.
hidrosola, a. hydrous.
hidrostática, f. hydrostatics.
hidrostáticola, a. hidrostatic; **nivel (m) h.**, hydrostatic level.
hidrotermal, a. hydrothermal; **manantial (m) h.**, hydrothermal spring.
hidrotérmicola, a. hydrothermal; **energía (f) h.**, hydrothermal energy.
hidróxido, m. hydroxide; **h. cálcico**, calcium hydroxide; **h. sódico**, sodium hydroxide.
hidroxilión, m. hydroxide ion (OH-).
hidróxilo, m. hydroxyl.
hidrozoos, m.pl. hydrozoa.
hidruro, m. hydride.
hiedra, f. ivy.
hiel, f. bile, gall.

hielo, m. ice; **banco (m) de h., h. de pack (LAm)**, pack ice, ice field; **bloqueado por el h.**, icebound (carretera); **brecha (f) de h.**, ice breccia; **campo (m) de h.**, ice field; **h. de pack (LAm)**, pack ice; **h. desgajado**, broken ice; **h. fragmentado**, brash (GB), mush (US)
hielo, m. ice; **h. grasiento**, slush, slush ice; **h. podrido**, rotten ice; **h. polar**, polar ice; **h. seco**, dry ice; **h. troceado (Met)**, brash (GB), mush (US); **manto (m) de h.**, ice sheet; **paquete (m) de h. (Med)**, icepack.
hiena, f. hyena, hyaena.
hierba, f. grass; herb, medicinal plant; **h. mate**, maté, Paraguay tea; **malahierba (f), mala h.**, weed.
hierbal, m. grassland.
hierro, m. iron (Fe); **h. disuelto**, dissolved iron; **h. dulce**, soft iron; **h. en lingotes**, pig iron; **h. forjado**, wrought iron; **h. fundido**, cast iron; **reducción (f) de h.**, iron reduction; **solubilización (f) de h.**, iron solubilization.
hifa, f. hypha (pl. hyphae).
hígado, m. liver.
higiene, f. hygiene; **h. ambiental**, environmental hygiene.
higiénicola, a. hygienic.
higienizar, v. to clean up.
higo, m. fig; **h. chumbo**, prickly pear.
higro-, pf. hygro-.
higrometría, f. hygrometry.
higrómetro, m. hygrometer.
higroscopicidad, f. hygroscopicity.
higroscópicola, a. hygroscopic.
higroscopio, m. hygroscope.
higuera, f. fig tree.
hija, f. daughter.
hijuela, f. legal document detailing the corresponding part to each heir.
hilo, m. thread, wire; **hilos cruzados**, cross wires.
himenóptero, m. hymenopteran (pl. hymenopterans, hymenoptera), hymenopteron (US); **himenópteros (Zoo, orden)**, Hymenoptera (pl).
hincapié, m. firm footing; **hacer h. en (v)**, to insist on, to emphasize.
hincar, v. to drive in, to insert; **h. un piezómetro**, to drive in a piezometer.
hinchadola, a. swollen (inflamado); vain (vanidoso).
hinchamiento, m. swelling (Gen).
hinchar, v. to inflate, to swell, to blow up (con aire); **h. con una bomba**, to inflate with a pump.
hinchazón, f. swelling, lump, bump; arrogance, vanity (vanidad).
hinterland, m. hinterland.
hiper-, pf. hyper-.
hipérbola, f. hyperbola.
hiperbólicola, a. hyperbolic.
hipermetropía, f. longsightedness.
hipersensibilidad, f. hypersensitivity.

hipertónicola, a. hypertonic.
hipo-, pf. hypo-.
hipocausto, m. hypocaust (Arq).
hipoclorito, m. hypochlorite.
hipofosfito, m. hypophosphite.
hipolimnio, m. hypolimnion.
hipoteca, f. mortgage (Fin).
hipotermal, a. hypothermal.
hipotermia, f. hypothermia.
hipótesis, f. hypothesis (pl. hypotheses); **h. nula**, null hypothesis.
hipotéticola, a. hypothetical, hypothetic.
hipotónicola, a. hypotonic.
hipsométricola, a. hypsometric.
hirviendo, a. boiling; **agua (f) h.**, boiling water.
hirviente, a. boiling.
hispanola, m. y f. Hispanic person.
hispanola, a. Hispanic.
Hispanoamérica, f. Hispano-America (los países de Norteamérica, Centroamérica y Suramérica donde se habla el español como la lengua materna), Latin America.
hispanoamericanola, 1. m. y f. Latin-American; 2. a. Spanish-American, Hispanoamerican, Latin-American.
histamina, f. histamine.
histéresis, m. hysteresis.
histograma, m. histogram.
histología, f. histology.
histológicola, a. histological.
historiadorla, m. y f. historian.
hocico, m. snout, muzzle.
hocino, m. glen (valle).
hogar, m. hearth, fireplace (chimenea); home (Fig); furnace (de caldera).
hoja, f. leaf (pl. leaves); sheet, leaf (p.e. de papel), blade (p.e. de cuchillo); **de h. caduca**, deciduous; **de h. perenne**, evergreen; **h. superpuesta**, overlay.
hojarasca, f. rubbish, litter, dead leaves, fallen leaves (Bot, hojas).
hojuela, f. flake, chip; leaflet (Bot); **h. de roca**, stone flake.
Holanda, f. Holland, The Netherlands.
holandéslholandesa, a. Dutch.
holasteroide, m. holasteroid.
holísticola, a. holistic.
holmio, m. holmium (Ho).
holo-, pf. holo-.
Holoceno, m. Holocene.
holocristalinola, a. holocrystalline.
holofíticola, a. holophytic.
holozoicola, a. holozoic.
hollín, m. soot.
hombre, m. man, mankind; **hombres (pl)**, manpower.
homeopatía, f. homeopathy.
homeopáticola, a. homeopathic.
homeostasis, f. homeostasis.
homeotermola, a. homeothermic.
homo-, pf. homo-.
homocigóticola, a. homozygous.

homogeneidad, f. homogeneity.
homogeneización, f. homogenization.
homogéneola, a. homogeneous.
homografía, f. homography.
homólisis, m. homolysis.
homologación, f. confirmation, acknowledgement.
homología, f. homology.
homólogola, a. homologous.
homosexual, a. homosexual.
homotáxicola, a. homotaxial.
homotaxis, m. homotaxis.
hondola, a. deep, profound.
hondonada, f. hollow, dip (depresión), gulley; lowland, wold (Geog).
Honduras, f. Honduras.
hondureñola, 1. m. y f. Honduran; 2. a. Honduran.
hongo, m. fungus (pl. fungi), toadstool, mushroom.
hontanar, m. spring, group of springs.
hora, f. hour; time; **cincuenta kilómetros por h.**, fifty kilometres per hour, fifty kilometres an hour; **h. de Greenwich**, Greenwich Mean Time; **h. punta**, rush hour.
horadar, v. to bore through, to drill, to pierce, to perforate.
horario, m. timetable; **huso (m) h.**, time zone.
horizontal, a. horizontal.
horizonte, m. horizon; **h. de suelo**, soil horizon; **h. estratigráfico**, stratigraphical horizon; **línea del h.**, skyline.
hormiga, f. ant.
hormigón, m. concrete; **bloques (m.pl) de h.**, concrete blocks; **h. aireado**, aerated concrete; **h. armado**, reinforced concrete; **h. bombeado**, shotcreting; **h. en masa**, mass concrete; **h. postensado**, post-tensioned concrete; **h. preamasado**, ready-mixed concrete; **h. prefabricado**, precast concrete; **h. pretensado**, prestressed concrete; **muro (m) de h.**, concrete wall.
hormigonera, f. concrete mixer.
hormona, f. hormone; **h. de crecimiento**, growth hormone; **h. sexual**, sex hormone; **h. vegetal**, plant hormone.
hormonal, a. hormonal.
hornblenda, f. horneblende.
hornblendita, f. hornblendite.
horno, m. oven, kiln, furnace; **h. alto**, blast furnace; **h. crematorio**, crematorium; **h. de reverbero**, reverberatory furnace.
horst, m. horst (bloque tectónico elevado).
hortaliza, f. vegetable; market-garden, truck-garden (US).
hortícola, a. horticultural.
horticultorla, m. y f. horticulturalist.
horticultura, f. horticulture.
hortifrutícola, a. fruit and vegetable; **industria (f) h.**, market gardening, truck gardening (US).

hospital, m. hospital; **incinerador (m) de h.**, hospital incinerator.

hospitalariola, a. hospital; **residuos (m.pl) hospitalarios**, hospital wastes.

hostelería, f. hotel trade.

hotel, m. hotel.

hoya, f. hole, pit; grave (tumba); vale, valley (Geog); **h. glaciar (Geog)**, kettle hole; **h. hidrográfica**, watershed (US).

hoyada, f. hollow.

hoyanco, m. (Méx) pothole.

hoyo, m. hollow, pit; **h. de poste (Arq)**, posthole.

hoyuelo, m. dimple, pit.

hoyuelola, a. dimpled, pitted.

huaiqueria, f. (LAm) badlands.

huaso, m. (LAm) peasant.

huayco, m. (Perú, Ch) landslide.

hueco, m. hollow (cavidad), hole (agujero); empty space (sitio libre); vesicle (Geol).

huecola, a. hollow.

huella, f. track (de un animal, persona, vehículo); trace, imprint; **h. de corriente**, flow cast; **h. de desgaste**, scour mark; **h. de lluvia**, rain imprint.

huerta, f. orchard and vegetable garden, irrigated land; cultivated plain (Valencia y Murcia, Esp).

huerto, m. small orchard, vegetable garden, garden patch.

hueso, m. bone; stone, pit (of fruit); **harina (f) de huesos**, bonemeal.

huesosola, a. bony.

huéspedla, m. y f. host (Zoo, Bot, p.e. de parásito).

huevo, m. egg (p.e. de ave); **huevos (p.e. de anfibios)**, spawn.

huincha, f. measuring tape.

hular, m. (Méx) rubber plantation.

hule, m. (LAm) rubber, oilcloth, oilskin; **h.-espuma**, foam rubber.

hule, m. rubber (caucho).

hulla, f. soft coal; **h. bituminosa**, bituminous coal; **h. sub-bituminosa**, semi-bituminous coal; **mina (f) de h.**, coal mine.

humanola, a. human (p.e. el cuerpo humano), humane (p.e. un enfoque humano a los animales); **ambiente (m) h.**, human environment; **ser (m) h.**, human being.

humear, v. to fume (Quím).

humectabilidad, f. wettability.

humectación, f. humidification, moistening.

humectador, m. humidifier.

humedad, f. humidity; **grado (m) de h.**, degree of humidity; **h. absoluta**, absolute humidity; **h. del aire**, air moisture; **h. del suelo**, soil moisture; **h. específica**, specific humidity; **h. relativa**, relative humidity.

humedal, m. wetland.

humedecedor, m. humidifier.

humedecer, v. to moisturize.

humedecimiento, m. moistening, humidifying.

húmedola, a. humid.

húmicola, a. humic; **ácido (m) h.**, humic acid.

humo, m. smoke, steam, vapour (vapor, US), fumes (pl); **campana (f) de h.**, fume cupboard; **echar h. (v)**, to fume; **h. de las fábricas**, factory fumes; **humos nocivos**, noxious fumes.

humus, m. humus.

hundidola, a. sunken.

hundimiento, m. sinking, subsidence, settlement; **h. de oxígeno**, oxygen sag.

hundirse, v. to sink (en aqua, arena etc.); to collapse (edificio etc.); to subside (terreno).

huracán, m. hurricane.

hurgador, m. (Ur) waste tip scavenger.

Huronense, m. Huronian.

huso, m. spindle; fuselage (de los aviones); **h. horario**, time zone.

I

I & D, m. R & D; **investigación (f) y desarrollo (m)**, research and development.

Iberoamérica, f. Latin America, Iberoamerica (excepto Brasil).

iberoamericano/a, m. y f. Latin-American.

iceberg, m. iceberg.

ICONA, m. **Instituto (m) para la Conservación de la Naturaleza (Esp)**, Institute for Nature Conservation.

ictericia, f. jaundice, icterus.

ictérico/a, a. jaundiced.

identificación, f. identification.

ídolo, m. idol.

idoneidad, f. suitability, fitness.

idóneo/a, a. suitable; capable.

ígneo/a, a. igneous; **cuerpo (m) í.**, igneous body; **intrusión (f) í.**, igneous intrusion; **masa (f) í.**, igneous body.

ignición, f. ignition, combustion, burning; **temperatura (f) de i.**, ignition temperature.

ignifugación, f. fire resistance, fireproofing.

ignífugo/a, a. fireproof, fire-resistant.

ignimbrita, f. ignimbrite.

ignitable, a. ignitable, flammable.

iguana, f. iguana.

ilegal, a. illegal.

íleon, m. ileum.

ilíaco/a, a. iliac.

ilmenita, f. ilmenite.

iluminación, f. illumination, lighting; **i. indirecta**, artificial lighting.

iluminación, f. illumination, lighting (sistema); **i. con focos**, floodlighting.

iluminar, v. to illuminate.

ilustración, f. illustration; example (ejemplo).

ilustrar, v. to illustrate.

illita, f. illite.

imagen, f. image; **i. estereoscópica**, stereoscopic image.

imago, m. imago.

imán, m. magnet.

imbibición, f. imbibation, soaking up, absorption.

imbricación, f. overlapping, superposition, imbrication.

imbricar, v. to overlap.

impacto, m. impact, blow; **evaluación (f) de i. ambiental (abr. EIA)**, environmental impact assessment (abr. EIA); **i. ambiental**, environmental impact; **i. de meteorito**, meteorite impact; **i. visual**, visual impact.

impar, a. odd, uneven; **día (m) i.**, odd-numbered day; **número (m) i.**, odd number.

IMPE, m. **Instituto (m) de la Mediana y Pequeña Empresa (Esp)**, institute for small and medium businesses.

impedancia, f. impedance.

impedir, v. to impede, to prevent, to hinder, to obstruct.

impeler, v. to propel.

impenetrable, a. impenetrable.

imperfecto/a, a. imperfect, defective.

impermeabilidad, f. impermeability.

impermeabilización, f. waterproofing, impermeabilization.

impermeabilizar, v. to waterproof.

impermeable, a. impermeable, impervious; waterproof; **límite (m) i.**, impermeable boundary.

imperturbado/a, a. undisturbed.

ímpetu, m. impetus, momentum.

implantación, f. implantation; introduction (ideas).

implementabilidad, f. implementability.

implementar, v. to implement, to put into effect.

implemento, m. implement, tool (herramienta).

implicación, f. implication, inference.

implícito/a, a. implicit; **función (f) i.**, implicit function.

impoluto/a, a. unpolluted, uncontaminated.

importación, f. import; **artículos (m.pl) de i.**, **bienes (m.pl) de i.**, import goods, imported goods.

importar, v. to import (a/de un país); to amount to, to be worth, to cost (Com); to matter, to be important (importancia).

imprecisión, f. imprecision, lack of precision, vagueness.

impregnación, f. impregnation.

impregnar, v. to impregnate.

imprescindible, a. indispensable, essential.

impresión, f. impression, imprint; **i. digital**, fingerprint.

impreso/a, a. printed.

impresor/a, m. y f. printer; **impresora láser**, laser printer.

imprevisto/a, a. unforeseen; **riesgos (m.pl) imprevistos**, unforeseen hazards.

imprimir, v. to imprint, to impress (Gen); to output, to print out (Comp).

improductivo/a, a. unproductive.

impulsión, f. impulsion.

impulso, m. impulse; **i. nervioso**, nerve impulse.

impulsor, m. impeller; **i. de una bomba**, pump impeller.

inactividad, f. inactivity; **estado (m) de i. (máquinas)**, standby.

inactivo/a, a. dormant; lazy (perezoso).

inadmisible, a. inadmissible, impermissible.

inalterado/a, a. unweathered.

inarticulado/a, a. inarticulate.

inauguración, f. inauguration.
incandescencia, f. incandescence, glow.
incandescente, a. incandescent.
incendiarse, v. to catch fire, to burst into flames.
incendio, m. fire; **abra (f) de i.**, fire gap; **defensas (f.pl) contra los incendios**, fire precautions, fire fighting measures; **i. forestal**, forest fire; **riesgo (m) de i.**, fire hazard.
incentivo, m. incentive.
incertidumbre, f. uncertainly, doubt; **i. científica**, scientific uncertainty.
incidencia, f. incidence (f); incident (suceso); **ángulo (m) de i.**, angle of incidence.
inciertola, a. uncertain.
incineración, f. incineration; **planta (f) de i.**, incineration plant.
incinerador, m. incinerator; **i. de basura**, waste incinerator; **i. de fangos de alcantarilla**, sewage sludge incinerator.
incineradora, f. incinerator, incineration plant (para residuos urbanos).
incinerar, v. to incinerate.
inclinación, f. inclination; hade (Geol, en fallas).
inclinadola, a. inclined, sloping, tilted.
inclinar, v. to incline, to dip.
incluidola, a. included.
incluir, v. to include, to enclose; to merge (fusionar).
inclusión, f. inclusion.
incógnita, f. unknown quantity (Mat).
incoherente, a. incoherent.
incombustible, a. incombustible, fire-proof.
incompetente, a. incompetent.
incomprensible, a. incomprehensible.
incompresibilidad, f. incompressibility.
inconformidad, f. nonconformity, unconformity.
inconsistente, a. inconsistent.
incontaminadola, a. uncontaminated, unpolluted.
incremento, m. increment.
incrustación, f. incrustation, encrustation; **i. carbonática**, carbonate incrustation; **i. de hierro**, iron incrustation; **i. de sílice**, silica incrustation.
incubación, f. incubation.
incubadora, f. incubator; **ensayo (m) de i.**, incubator test.
incumplimiento, m. non-compliance.
indegradabilidad, f. persistence, non-degradability.
indemnización, f. indemnification.
indemnizar, v. to indemnify.
independiente, a. independent; self-sufficient (autosuficiente); self-contained (unidad); stand-alone (Comp).
indicación, f. indication.
indicador, m. indicator, guide; **i. de contaminación**, pollution indicator; **i. de nivel**

de agua, water level indicator; **i. de polución**, pollution indicator; **i. de presión del aceite**, oil pressure gauge; **i. de temperatura**, temperature indicator; **organismos (m.pl) indicadores**, indicator organisms; **especie (f) i.**, indicator species.
índice, m. index; rate; forefinger (Anat); **í. biótico**, biotic index; **í. de calidad**, quality index; **í. de cambio de bases**, base-exchange index; **í. de diversidad**, diversity index; **í. de fertilidad**, fertility rate; **í. del coste de la vida**, cost-of-living index; **í. de mortalidad**, mortality rate; **í. de pH**, pH scale; **í. de refracción**, refractive index; **í. de riesgo**, hazard index; **í. de saturación**, saturation index.
índice, m. index; rate; forefinger (Anat); **Í. General de Calidad (Esp, de aguas superficial, abr. IGC)**, General Quality Index; **í. granulométrico (Geol)**, sorting index; **í. hidroquímico**, hydrochemical index.
indígena, a. indigenous, native.
indio, m. indium (In).
indirectola, a. indirect.
indisoluble, a. indissoluble.
indo-arábiga, a. Indoarabian.
índole, f. nature (manera).
inducción, f. induction.
industria, f. industry; **i. agroalimentaria**, agrifood industry; **i. automovilística**, car industry, motor industry; **i. clave**, key industry; **i. curtidora**, tanning industry; **i. de curtiembres**, tanning industry; **i. de procesamiento de alimentos**, food processing industry; **i. de pulpa y papel**, pulp and paper industry; **i. enlatadora**, canning industry; **i. láctea**, dairy industry; **i. olivarera**, olive industry; **i. pesada**, heavy industry; **i. petrolera**, oil industry, petroleum industry; **i. siderúrgica**, steel industry.
industrial, 1. m. manufacturer, industrialist; 2. a. industrial; **acción (f) i.**, industrial action, strike; **agua (f) i.**, **desechos (m.pl) industriales**, industrial discharge(s).
industrialización, f. industrialization.
ineficiencia, f. inefficiency.
inelasticidad, f. inelasticity.
inelásticola, a. inelastic.
inercia, f. inertia.
inerte, a. inert, passive; **i. químicamente**, chemically inert.
inertización, f. stablilization, making inert.
inestabilidad, f. instability.
inestabilizar, v. to destabilize.
inestable, a. unstable.
infección, f. infection.
infeccionar, v. to infect.
infecciosola, a. infectious.
infectar, v. to infect.
infectola, a. infected, contaminated.
inferencia, f. inference, implication.

inferior, a. inferior, lower, lesser; **el lado i.**, the underneath; **i. extremo**, lowermost.

infiltración, f. infiltration; **balsa (f) de i.**, infiltration basin; **capacidad (f) de i.**, infiltration capacity; **eficacia (f) de i.**, infiltration efficiency; **galería (f) de i.**, infiltration gallery; **i. anual**, annual infiltration; **i. eficaz**, infiltration efficiency; **i. inducida**, induced infiltration; **velocidad (f) de i.**, infiltration rate.

infiltrarse, v. to infiltrate, to seep.

infiltrómetro, m. infiltrometer.

infinidad, f. infinity.

infinitesimal, a. infinitesimal.

infinitola, a. infinite.

inflamable, a. flammable, inflammable.

inflamación, f. inflammation.

inflar, v. to inflate.

inflexión, f. inflexion; **punto (m) de i.**, inflexion point.

influencia, f. influence; **cono (m) de i.**, cone of influence; **ejercer una i. sobre (v)**, to have an influence upon; **i. del bombeo**, pumping influence.

influente, a. influent; **río (m) i.**, influent river.

influenza, f. influenza, flu.

información, f. information; **autopista (f) de la i.**, information highway; **tecnología (f) de la i.**, information technology.

informar, v. to inform, to report; **informarse**, to find out.

informática, f. data processing, computer science.

informe, 1. m. report (documento); piece of information (dato); file, dossier (expediente); **i. anual**, annual report; **i. diario del sondista**, driller's log; 2. a. shapeless.

infracción, f. infringement, breach, offence; **i. de la ley**, breach of the law; **i. de las leyes federales (US)**, federal offense.

infradesarrollo, m. infrastructure.

infraestructura, f. infrastructure; **i. viaria**, traffic infrastructure.

infrarrojola, a. infrared.

infravaloración, f. underestimation.

infrayacente, a. underlying (Geol).

infrayacer, v. to underlie (Geol).

infundir, v. to infuse.

infusión, f. infusion.

ingeniería, f. engineering; **i. agrícola**, agricultural engineering; **i. ambiental**, environmental engineering; **i. civil**, civil engineering; **i. eléctrica**, electrical engineering; **i. litoral**, coastal engineering; **i. medioambiental**, environmental engineering.

ingeniero, m. engineer; **i. agrónomo**, agricultural engineer; **i. civil**, civil engineer; **i. consultor**, consulting engineer; **i. de caminos**, highway engineer; **i. de caminos, canales y puertos (Esp)**, civil engineer; **i. de carreteras**, highway engineer; **i. de minas**, mining engineer; **i. de proyecto**, project engineer; **i. eléctrico**, electrical engineer; **i. hidráulico**,

water engineer; **i. mecánico**, mechanical engineer; **i. medioambiental**, environmental engineer; **i. municipal**, municipal engineer; **i. sanitario**, sanitary engineer.

ingerir, v. to ingest.

ingestión, f. ingestion.

Inglaterra, f. England.

ingléslinglesa, 1. m. y f. English (persona); 2. a. English.

ingreso, m. ingress, admission; **i. de agua**, ingress of water.

ingresos, m.pl. revenue, income; **i. y gastos**, income and expenditure.

inhalar, v. to inhale.

inherente, a. inherent; **error (m) i.**, inherent error.

inhibición, f. inhibition.

inhibidor, m. inhibitor; **i. antidetonante**, anti-knocking agent (en gasolina); **i. de corrosión**, corrosion inhibidor; **i. de crecimiento (Bot, Zoo)**, growth inhibitor.

inhomogeneidad, f. inhomogeneity.

iniciativa, f. initiative, leadership; **i. privada**, private enterprise; **obrar por propia i. (v)**, to use one's initiative.

injertar, v. to graft (Bot).

injerto, m. graft (Bot).

inmediaciones, f.pl. neighbourhood, neighborhood (US), environs (pl); **en las i. de**, in the neighbourhood of.

inmersión, f. immersion, submersion.

inmovilización, f. immobilization.

inmovilizado, m. fixed asset (Fin).

inmundicia, f. squalor, filth; **inmundicias**, rubbish.

inmune, a. immune; **sistema (m) i.**, immune system.

inmunidad, f. immunity.

inmunitoriola, a. immune; **sistema (m) i.**, immune system.

inmunización, f. immunization.

inmunizar, v. to immunize.

inmunoensayo, m. immunoassay.

inmunología, f. immunology.

inmunológicola, a. immunological.

innovación, f. innovation; **i. técnica**, technical innovation; **i. tecnológica**, technological innovation.

innovadorla, a. innovative.

inoculación, f. inoculation.

inocular, v. to inoculate.

inodoro, m. flush toilet.

inorgánicola, a. inorganic.

inoxidable, a. stainless; **acero (m) i.**, stainless steel.

input, m. input (Comp); **i. para ordenadores**, computer input.

insalubre, a. unhealthy, insalubrious; **actividades (f.pl) molestas, insalubres, nocivas y peligrosas**, dangerous and nuisance activities and activities harmful to health.

insalubridad, f. unhealthiness.

inscripción, f. inscription.
insecticida, 1. m. insecticide; **accuon (f) i.** insecticidal action; **i. de contacto**, contact insecticide; **i. respiratorio**, respiratory insecticide; 2. a. insecticidal.
insectívoro, m. insectivore; **insectívoros (Zoo, orden)**, Insectivora (pl).
insectívorola, a. insectivorous.
insecto, m. insect; **insectos (Zoo, clase)**, Insectos (pl).
insegurola, a. unsafe, insecure (peligroso); uncertain (que lleva incertidumbre).
inselberg, m. inselberg.
inseminación, f. insemination; **i. artificial**, artificial insemination.
inseminar, v. to inseminate.
inserción, f. insertion.
insignificante, a. insignificant, negligible.
in situ, m. in situ, on site.
insolación, f. insolation, sunshine.
insoluble, a. insoluble.
insolvencia, f. insolvency (bancarrota).
insonorización, f. soundproofing, noiseproofing; **cabina (f) de i.**, soundproof chamber.
insonorola, a. soundproof.
inspección, f. inspection, examination; **cámara (f) de i.**, inspection chamber; **i. sanitaria**, sanitary inspection.
inspiración, f. inspiration.
instalación, f. installation; **i. de calefacción**, heating installation; **i. de fontanería**, plumbing installation; **i. de la perforación**, well installation; **i. eléctrica**, electrical installation; **i. para la recuperación (de desechos)**, (waste) recovery facility.
instantáneola, a. instantaneous.
instar, m. instar.
instauración, f. establishment, installation, setting up; **i. de un régimen de gestión**, establishment of a management regime.
instigación, f. instigation.
instigadorla, m. y f. instigator.
instintivola, a. instinctive.
instinto, m. instinct.
institución, f. institution, establishment.
instituto, m. institute, institution; **I. de Hidrología de España**, Spanish Hydrological Institution; **I. Geológico de España (IGE)**, Spanish Geological Institute; **I. Nacional de Meteorología (Esp)**, Meteorological Office (GB).
instrucción, f. instruction.
instrumentación, f. instrumentation; **i. de sondeo**, down-hole instrumentation.
instrumento, m. instrument; **tablero (m) de instrumentos**, instrument panel.
insuficiencia, f. insufficiency, inadequacy; lack, shortage (falta).
insulina, f. insulin.
insumo, m. input (Com).
intactola, a. unbroken, intact.

integración, f. integration.
integradola, a. integrated; **circuito (m) i.**, integrated circuit.
integrador, m. integrator.
integral, f. integral; **cálculo (m) i.**, integral calculus.
integrante, a. integral.
integumento, m. integument.
inteligencia, f. intelligence; **i. artificial**, artificial intelligence.
intemperie, f. weather, elements; **dejar a la i. (v)**, to leave out in the elements.
intención, f. intention.
intencionalidad, f. intention.
intendente, m. (CSur) governor (de estado, municipio).
intensidad, f. intensity (Gen), strength; **i. de la luz**, light intensity; **i. máxima pluvial**, maximum rainfall intensity.
intensivola, a. intensive.
inter-, pf. inter-.
interacción, f. interaction.
interaccionar, v. to interact.
intercalación, f. intercalation, interbedding.
intercambiador, m. exchanger; **i. de calor**, heat exchanger.
intercambio, m. exchange, interchange; **i. de bases**, base exchange; **i. iónico**, ionic exchange.
intercelular, a. intercellular.
intercepción, f. interception.
interceptógrafo, m. interceptograph.
interceptor, m. interceptor.
intercomunicarse, v. to intercommunicate.
interconectar, v. to interconnect.
interconexión, f. interconnection.
intercontinental, a. intercontinental.
intercotidal, a. (LAm) intertidal; **zona (f) i. (LAm)**, intertidal zone.
intercrecimiento, m. intergrowth (Geol).
intercuartil, a. interquartile; **alcance (m) i.**, interquartile range.
interdependencia, f. interdependence.
interdigitación, f. interfingering, interdigitation.
interdisciplinariola, a. interdisciplinary; **estudio (m) i.**, interdisciplinary study.
interés, m. interest, appeal, concern; **i. actualizado (Com)**, actualized interest; **i. compuesto (Com)**, compound interest; **i. público**, public interest; **i. simple (Com)**, simple interest; **tener i. por (v)**, to take an interest in; **un i. de un diez por ciento (Com)**, ten per cent interest.
interesante, a. interesting.
interespecíficola, a. interspecific.
interfase, f. interface; ecotone (Ecol); **i. agua dulce-agua marina**, freshwater-seawater interface; **i. aire-agua**, air-water interface; **i. de equilibrio**, equilibrium interface; **i. dinámica**, dynamic interface; **profundidad (f) de la i.**, depth of interface.

interferencia, f. interference; **i. de pozos**, well interference.
interflujo, m. interflow.
interfluvio, m. interfluve.
interglacial, a. interglacial.
interglaciar, a. interglacial.
interior, 1. m. interior; domestic, home, national (Com); inside (parte interior); interior, hinterland (Geog); 2. a. interior, internal; **comercio (m) i.**, domestic trade, home trade.
intermareal, a. intertidal; **zona (f) i.**, intertidal zone.
intermediario, m. middleman; **i. financiero (Com)**, broker.
intermitente, a. intermittent; **descarga (f) i.**, intermittent discharge; **manantial (m) i.**, intermittent spring.
INTERNET, m. INTERNET.
internola, a. internal, interior.
interpolación, f. interpolation.
interpretación, f. interpretation; **i. fotográfica**, photographic interpretation; **i. geofísica**, geophysical interpretation.
interpretar, v. to interpret.
interrelación, f. inter-relation.
interrumpir, v. to interupt, to disrupt; to cut off (una provisión).
interrupción, f. interruption.
interruptor, m. switch, cut-out; **i. de disparo**, trip switch; **i. de nivel mínimo**, low-level cut-out; **i. eléctrico automático**, automatic time switch; **i. protector**, circuit breaker.
intersección, f. road intersection (carreteras).
intersticial, a. interstitial; **fluido (m) i.**, interstitial fluid.
intersticio, m. interstice.
interstratificación, f. interstratification.
interstratificadola, a. interbedded, interstratified.
intervalo, m. interval; **i. de ajuste**, adjustment interval; **i. de tiempo**, time interval.
intervención, f. intervention; audit (de cuentas).
intestino, m. intestine; **i. ciego**, caecum; **i. delgado**, small intestine; **i. grueso**, large intestine.
intestinola, a. intestinal, intestine.
inti, m. (Perú) inti (unidad monetaria).
intracelular, a. intracellular.
intraespecíficola, a. intraspecific.
intraplegamiento, m. convoluted folding.
intrínsecola, a. intrinsic; **permeabilidad (f) i.**, intrinsic permeability.
introducción, f. introduction.
intruir, v. to intrude.
intrusión, f. intrusion; **i. ígnea**, igneous intrusion; **i. marina**, marine intrusion; **i. salina**, saline intrusion; **i. visual**, visual intrusion.
intumescencia, f. swelling, intumescence, surging; **tanque (m) de i. (suministro de agua)**, surge tank.
inundable, a. floodable.

inundación, f. flood, flooding, inundation; **control (m) de inundaciones**, flood control; **pronóstico (m) de inundaciones**, flood forecast.
inundadola, a. flooded.
inundar, v. to flood.
inútil, a. useless.
invariante, a. invariant.
inventario, m. inventory, list, stocktaking; **i. de emisiones**, emissions inventory; **i. de material**, quantity survey.
invernada, f. (LAm) winter pasture.
invernadero, m. greenhouse; **efecto (m) i.**, greenhouse effect; **gases (m.pl) de efecto i.**, greenhouse gases.
invernal, a. winter, wintry.
inversión, f. inversion, reversal; investment (Fin); **i. de salinidad**, salinity inversion; **i. de temperatura**, temperature inversion.
inversionista, m. investor.
inversola, a. inverse, reverse; **función (f) i.**, inverse function.
inversorla, m. y f. investor.
invertebrado, m. invertebrate; **invertebrados (Zoo, grupo)**, Invertebrata (pl).
invertido, m. invert.
invertidola, a. inverted.
invertir, v. to invest (Com).
investigación, f. investigation; **i. científica**, scientific research; **i. de campo**, field investigation; **i. y desarrollo (abr. I&D)**, research and development (abr. R&D).
investigador, m. y f. investigator.
invierno, m. winter; **de i. (a)**, winter, wintry.
involucrar, v. to introduce into.
involutola, a. involuted.
inyección, f. injection; **i. a presión**, pressure injection; **i. de aire comprimido**, injection of compressed air; **i. de cemento**, cement injection; **i. del trazador**, injection of tracer; **i. de mortero de cemento a presión**, pressure cement grouting; **i. de pozo en profundidad**, deep well injection; **i. de vapor**, steam injection; **i. en sondeo**, borehole injection; **i. química**, chemical injection; **i. subterránea de aguas residuales**, wastewater subsurface injection; **i. subterránea de residuos peligrosos**, hazardous waste deep well injection.
inyectadola, a. injected.
inyectar, v. to inject.
ioduro, m. iodide.
ión, m. ion; **i. calcio**, calcium ion; **i. complejo**, complex ion; **i. menor**, minor ion; **i. metálico**, metallic ion.
ión-gramo, m. gram-ion.
iónicola, a. ionic; **balanza (f) i.**, ionic balance; **cambio (m) i.**, ion exchange.
ionización, f. ionization; **energía (f) de i.**, ionization energy.
ionizante, a. ionizing; **radiación (f) i.**, ionizing radiation.
ionizar, v. to ionize.

ionosfera, f. ionosphere.
IPCC, m. IPCC; **Panel (m) Internacional sobre el Cambio Climático**, International Panel on Climatic Change.
iridio, m. iridium (Ir).
Irlanda, f. Ireland; **I. del Norte**, Northern Ireland; **República (f) de I.**, Republic of Ireland.
irlandés/irlandesa, a. Irish, Irish person; **el Mar i.**, the Irish Sea.
irradiación, f. irradiation.
irradiar, v. to radiate, to irradiate.
irreductible, a. irreducible.
irregular, a. irregular.
irregularidad, f. irregularity.
irrigación, f. irrigation.
irrigar, v. to irrigate.
irritación, f. irritation.
irruptivo/a, a. irruptive; **crecimiento (m) i.**, irruptive growth.
IRYDA, m. **Instituto (m) para la Reforma y el Desarrollo Agrario (Esp)**, Institute for Land Reform and Development.
Islandia, f. Iceland.
isleta, f. islet, small island.
islote, m. small island, islet.
ISO, f. ISO; **Organización (f) Internacional de Normalización**, International Standards Organization.
iso-, pf. iso-.
isobara, f. isobar.
isobárico/a, a. isobaric.
isobutano, m. isobutane.
isobutileno, m. isobutylene.
isocíclico/a, a. isocyclic.
isoclinal, a. isoclinal.
isócrono/a, a. isochronous.
isodrín, m. isodrin.
isolínea, f. isoline.

isomería, f. isomerism.
isomerización, f. isomerization.
isómero, m. isomer.
isomórfico/a, a. isomorphic.
isomorfismo, m. isomorphism.
isopaca, f. isopach, isopachyte.
isopleta, f. isopleth.
isopreno, m. isoprene.
isopropilo, m. isopropyl.
isóptero, m. isopteran; **isópteros (Zoo, orden)**, Isoptera (pl).
isósceles, a. isosceles.
isostasia, f. isostacy.
isostático/a, a. isostatic.
isoterma, f. isotherm.
isotérmico/a, a. isothermic.
isotónico/a, a. isotonic.
isotopía, f. isotopy.
isotópico/a, a. isotopic; **cambio (m) i.**, isotopic change.
isótopo, m. isotope; **i. radioactivo**, radioactive isotope.
isotropía, f. isotropy.
isotrópico/a, a. isotropic.
isoyeta, f. isohyet; **mapa (m) de isoyetas**, isohyetal map.
isoyetal, a. isohyetal.
istmo, m. isthmus (pl. ismthmuses) (Geog).
Italia, f. Italy.
italiano/a, a. Italian.
iteración, f. iteration.
iterar, v. to iterate, to repeat.
iterativo/a, a. iterative.
itinerario, m. itinerary; **i. de la naturaleza**, nature trail; **i. urbana**, town trail.
itrio, m. yttrium (Y).
IVA, m. VAT; **impuesto (m) sobre el valor añadido/agregado (Esp)**, value added tax (GB).

J

jábega, f. seine, dragnet, sweep net; **pesca (f) con j.**, seine fishing.

jabón, m. soap; **j. de sastre**, soapsone, steatite, French chalk; **j. en polvo**, soap powder, washing powder.

jacal, m. (Méx) shanty town.

jade, m. jade.

jadeíta, f. jadeite.

jalón, m. rod, ranging rod.

Jamaica, f. Jamaica.

jamba, f. post, jamb (de una puerta).

jaqueca, f. migraine (dolor de cabeza).

jardín, m. garden; **j. zoológico**, zoo, zoological park.

jardinaje, m. gardening.

jardinería, f. gardening.

jardinerola, m. y f. gardener; **j. paisajista**, landscape gardener.

jareta, f. netting (red); hem (dobladillo); rope, cable (Mar); **cerco (m) de j. (pesca)**, purse seine net.

jarosita, f. jarosite.

jaspe, m. jasper, jaspilite.

jaspilita, f. jaspilite, jasper.

jaula, f. cage (Gen, Min); lockup (carcel).

jeep, m. jeep.

jefa, f. boss, head, manageress.

jefatura, f. leadership, management.

jefe, m. boss, head, manager; **j. de gobierno**, head of government; **j. de servicio**, department manager; **j. de taller**, foreman.

JEN, f. AEA (GB), AEC (US); **Junta (f) de Energía Nuclear (Esp)**, Atomic Energy Authority (GB), Atomic Energy Commission (US).

jerarquía, f. hierarchy.

jerárquicola, a. hierarchial; **orden (m) j.**, hierarchial order.

jerarquizadola, a. hierarchical.

jeringa, f. syringe.

jeringuilla, f. syringe.

jeroglífico, m. hieroglyph (Arq).

ji, m. chi; **distribución (f) ji-cuadrada**, Chi-squared distribution.

jirafa, f. giraffe.

joint-venture, f. joint venture.

jornalerola, f. farm worker, farm labourer (laborer, US).

jote, m. turkey vulture, turkey buzzard.

judía, f. bean; **j. verde**, runner bean, string bean.

judicatura, f. judiciary.

judicial, a. judicial.

judiciario, m. judiciary.

juego, m. game, sport (diversión); set (conjunto); **j. de tamices**, set of sieves.

jugo, m. juice; **jugos digestivos**, digestive juices; **jugos gástricos**, gastric juices.

juicio, m. judgement, judgment; trial (Jur, pleito); common sense (sentido común); **j. de peritos**, expert opinion.

julio, m. joule.

juncal, m. rush-bed.

junco, m. rush, reed; **estera (f) de juncos**, rush matting.

jungla, f. jungle.

junquera, f. rush, bulrush.

junta, f. meeting, assembly (asamblea); board, council (consejo); **celebrar j. (v)**, to hold a meeting; **j. general de accionistas**, annual general meeting (of shareholders); **j. general extraordinaria**, extraordinary general meeting.

Júpiter, m. Jupiter.

jurel, m. horse mackerel, jack mackerel, scad, jural, saurel (pez).

jurídicola, a. judicial, legal (establecido por la ley), lawful (dentro la ley); **asesor (m) j.**, legal advisor; **procedimiento (m) j.**, legal procedure.

jurisdicción, f. jurisdiction.

jurisprudencia, f. jurisprudence.

justificación, f. justification.

juzgar, v. to judge, to deem.

K

kanat, m. (Arab) khanate, qanat, ganat; gallery, heading, adit.

kanato, m. (Arab) khanate, qanat, ganat; gallery, heading, adit.

karst, m. karst; **desarrollo (m) de k.**, karst development.

kársticola, a. karstic; **acuífero (m) k.**, karstic aquifer; **sistema (m) k.**, karstic system.

karstificación, f. karstification.

Kelvin, m. Kelvin (Quím, grados absolutos, abr. K).

keroseno, m. kerosene, paraffin.

khanat, m. (Arab) khanate, qanat, ganat; gallery, heading, adit.

kilo-, pf. kilo-.

kilobyte, m. kilobyte.

kilocaloría, f. kilocalorie (abr. kCal).

kilociclo, m. kilocycle.

kilogramo, m. kilogramme, kilogram (US) (abr. kg).

kilometraje, m. distance travelled in kilometres.

kilómetro, m. kilometre, kilometer (US).

kilovatio, m. kilowatt.

kilovatio-hora, f. kilowatt-hour (abr. KWh, unidad de consumición eléctrica).

kilovoltio, m. kilovolt.

Kimmeridgense, m. Kimmeridgian.

Kjeldahl, m. Kjeldahl; **nitrógeno (m) K.**, Kjeldahl nitrogen.

kriptón, m. krypton (Kr).

L

labio, m. lip, rim (Gen); labium (pl. labia) (Zoo); **l. bajo (Geol)**, downthrow side; **l. levantado (Geol)**, upthrow side.

labor, f. job, work (tarea); working (funcionamiento de una máquina); **caballo (m) de l.**, workhorse; **l. a rajo abierto**, open-cast working; **l. de campo**, farm work; **l. minera**, mine working; **tierra (f) de l.**, arable land.

laboratorio, m. laboratory; **análisis (m) de l.**, laboratory analysis.

labra, f. cutting, carving (de piedra, madera).

labrado, m. cutting, carving (de piedra, madera); **l. tosco**, rustication.

labradorla, m. y f. farm worker, farmhand, peasant farmer.

labrantío, m. arable land.

labrantíola, a. arable.

labranza, f. arable farmland, arable land (tierra); farming, cultivation (cultivación); **instrumentos (m) de l.**, farm implements, farm tools.

labrar, v. to work, to plough (to plow, US), to till (arar); to carve (madera, piedra).

labriegola, m. y f. farmhand.

labro, m. labrum (pl. labra).

laburo, m. (LAm) work, job.

laca, f. lacquer.

lacolito, m. laccolith.

lactación, f. lactation.

lactancia, f. lactation.

lactar, v. to breast-feed, to bottle-feed (niños); to lactate, to suckle (animales).

lácteola, a. milk, milky; **dieta (f) l.**, milk diet.

lactosa, f. lactose, milk-sugar.

lacustre, a. lacustrine.

ladeadola, a. tilted.

ladear, v. to tilt.

ladera, f. hillside, slope; **l. abajo**, downhill; **l. arriba**, uphill; **l. de valle**, valley wall.

lado, m. side.

ladrillera, f. brickworks.

ladrillo, m. brick; **almacén (m) de ladrillos**, **fábrica (f) de ladrillos**, brickyard; **l. prensado**, engineer's brick; **l. refractario**, refractory brick, firebrick; **l. silicocalcáreo**, silica brick; **l. vidriado**, glazed brick; **pared (f) de l.**, brick facing; **tapiar con ladrillos (v)**, to brick up, to brick in.

lag, m. lag, delay.

lagartija, f. small lizard.

lagarto, m. lizard.

lago, m. lake, loch; mere (GB, lago superficial); **fluctuaciones (f) de l.**, lake fluctuations; **l. de agua salada**, saltwater lake.

lagomorfo, m. lagomorph; **lagomorfos (Zoo, orden)**, Lagomorpha (pl).

lágrima, f. tear.

laguna, f. lagoon, small lake; turlough (Ir, lago kárstico); **l. aireada (Trat)**, aerated lagoon; **l. de maduración (Trat)**, maturation pond; **l. de oxidación (Trat)**, oxidation lagoon; **l. en forma de semiluna**, ox-bow lake; **l. glaciar**, cirque lake; **l. para verter las aguas residuales**, sewage lagoon.

lagunajo, m. small lake left after flooding.

lagunosola, a. swampy, marshy.

lahar, m. lahar (flujo de lodo).

laja, f. (LAm) flagstone; rock.

lajosola, a. slabby; flaggy (Geol).

lama, f. mud, slime, ooze (lodo); mould, mold (US) (LAm); crushed ore (Min); lama (Zoo).

lamedero, m. evaporite soils.

lamelibranquio, m. lamellibranch; **lamelibranquios (Zoo, clase)**, Lamellibranchia pl.

Lamellibranchia, f.pl. Lamellibranchia (pl) (Zoo, clase).

lámina, f. lamella (capa); gill (de hongos); section (Anat); **l. delgada**, thin section.

laminación, f. lamination; **l. cruzada**, cross-bed.

laminadola, a. laminated.

laminar, a. laminar; **flujo (m) l.**, laminar flow; **velocidad (f) l.**, laminar velocity.

lamprea, f. lamprey.

lancha, f. launch; **l. motora**, motor launch.

lanchón, m. barge (buque).

Landiniense, m. Landinian.

Landsat, m. Landsat (satélite artificial que describe 14 órbitas diarias alrededor de la tierra).

langosta, f. lobster.

langostina, f. prawn; shrimp (US).

Languiense, m. Langhian.

lanolina, f. lanolin, lanoline.

lantánido, m. lanthanide.

lantano, m. lanthanum (La).

lanza, f. spearhead.

lanzamiento, m. launching, flotation (e.g. de una compañía).

lanzar, v. to launch (un proyectil); to throw (echar).

laña, f. clamp (grapa); green coconut (coco).

lápida, f. stone, memorial tablet; **l. mortuoria**, **l. sepulcral**, gravestone.

lapilli, m. lapilli.

lapislázuli, m. lapis lazuli.

largavistas, f.pl. (LAm) binoculars (prismáticos).

laringe, f. larynx (pl. larynxes, larynges).

laríngeola, a. laryngeal.

laringitis, f. laryngitis.

larva, f. larva (pl. larvae), grub.

larval, a. larval.

LASCA 246 *LEVANTAMIENTO*

lasca, f. chip of stone.
láser, m. laser.
lastrado, m. ballasting, weighing down.
lastrar, v. to ballast, to weigh down.
lastre, m. ballast.
lata, f. can, tin, tin can.
latente, a. latent; dormant (Biol); **calor (m) l.**, latent heat.
lateral, a. lateral, side of; **línea (f) l. (de peces)**, lateral line.
laterita, f. laterite.
lateríticola, a. lateritic; **suelo (m) l.**, latosol.
látex, m. latex (pl. latexes, latices).
latido, m. beat (of pulse).
latifundio, m. latifundium (pl. latifundia), large estate (tierra).
Latinoamérica, f. Latin America.
latinoamericanola, a. Latin-American.
latita, f. latite.
latitud, f. latitude; **grados (m) de l.**, degrees of latitude.
latitudinal, a. latitudinal.
latón, m. brass.
lava, f. lava; **campo (m) de l.**, lava field; **colada (f) de l.**, lava flow; **cono (m) de l.**, lava cone; **l. almohadilla**, pillow lava; **l. pahoehoe**, pahoehoe lava; **meseta (f) de l.**, lava plateau.
lavado, m. washing (p.e. ropa), scrubbing (gases); **lavado (m) de gases**, gas scrubbing.
lavadora, f. washing machine.
lavaje, m. washing; **l. pluvial (LAm)**, rainwash.
lavamano, m. wash basin.
lavanda, f. lavender.
lavandería, f. laundry; **l. química**, dry cleaners.
lavazas, f.pl. washings (agua sucia).
lazurita, f. lazurite.
lecitina, f. lecithin.
lectorla, m. y f. reader (que lee); assistant professor; **l. de cinta perforada**, punched tape reader.
lechada, f. grout, paste, whitewash; **l. para inyección**, grout.
leche, f. milk; **l. completa**, full-cream milk; **l. desnatada**, skimmed milk; **l. entera**, full-cream milk; **l. en polvo**, powdered milk, milk powder.
lechería, f. dairy, dairy farm.
lecho, m. bed (Gen); bed, layer (Geol); bed, bedding (Agr); river bed (de un río); **l. (m) bacteriano, l.**, filter bed; **l. de río**, riverbed; **l. de roca**, bedrock; **l. filtrante (Trat)**, filter bed; **l. fluidificado**, fluidized bed; **l. percolador (Trat)**, percolating filter bed.
lechón, m. piglet, suckling pig.
lechuza, f. owl; **l. campestre**, Short-eared Owl; **l. común**, Barn Owl.
legal, a. legal, lawful; **contrato (m) l.**, legal contract.
légamo, m. mud, bog, slime.
legislación, f. legislation; **l. ambiental**, environmental legislation; **l. autonómica (Esp)**, legislation enacted by an autonomous

community; **l. comunitaria europea**, EC legislation, European Community legislation; **l. estatal**, state legislation; **l. mediombiental**, environmental legislation; **l. nacional**, national legislation; **l. privada**, private legislation; **l. propuesta**, proposed legislation; **l. sobre contaminación**, pollution legislation.
legisladorla, m. y f. legislator (persona).
legislar, v. to legislate.
legislativo, m. assembly which approves legislation (Jur).
legislativola, a. legislative.
legumbre, f. legume (planta con vainas); vegetable (verdura); **legumbres secas**, dried legumes, dried pulses; **legumbres verdes**, green vegetables.
leguminosa, f. pulse (legumbre), legume (Agr); **leguminosas (Bot, familia)**, Leguminosae (pl).
leguminosola, a. leguminous.
lejanía, f. distance.
lejía, f. bleach.
lempira, m. (Hond) lempira (unidad monetaria).
lengua, f. tongue; **la l. española**, the Spanish language; **la l. inglesa**, the English language; **l. materna**, mother tongue; **l. nativa**, native tongue.
lenguado, m. sole.
lenguaje, m. language (Gen), idiom, speech; **l. científico**, scientific language; **l. de máquina (Comp)**, machine language.
leníticola, a. lentic.
lente, m. lens; **l. de aumento**, magnifying glass.
lenteja, f. lentil.
lenticular, a. lenticular.
leña, f. firewood.
leñadorla, m. y f. wood-cutter.
leñosola, a. woody, ligneous; **planta (f) l.**, woody plant.
león, m. lion.
lepidolita, f. lepidolite.
lepidóptero, m. lepidopteran (pl. lepidoptera, lepidopterans), lepidopteron (US); **lepidópteros (Zoo, orden)**, Lepidoptera (pl).
lepidópterola, a. lepidopteran.
lepra, f. leprosy.
leprosola, m. y f. leper.
leptospirosis, m. leptospirosis.
letal, a. lethal, deadly.
letargo, m. dormancy (Zoo); lethargy (Gen).
letrina, f. latrine; **l. de hoyo seco**, pit latrine; **l. seca**, dry latrine.
leucita, f. leucite.
leucitófiro, m. leucitophyre.
leucocito, m. leukocyte, leukocyte, white blood cell.
levadura, f. yeast; **l. química**, baking powder.
levantamiento, m. raising, lifting, hoisting (elevación); drawing up plans (preparación de planes); survey (Geog, agrimensura, etc.);

l. aéreo, aerial survey; **l. del terreno,** ground heave; **l. de planos,** mapping; **l. topográfico,** land mapping, survey.

levante, m. east, east wind.

ley, f. law; act, bill (en las Cortes); rule (regla); statute (de una asamblea); purity (de un metal); quality (calidad), weight (peso), dimension (medida); **al margen de la l.,** outside the law; **anteproyecto (m) de l.,** draft legislation; **aprobar una l. (v),** to pass a bill; **contra la l.,** against the law; **decreto-l.,** decree law; **fuera de la ley,** outside the law; **l. de acción de masas (Fís),** law of mass action; **l. de aguas,** water legislation; **L. de aguas (Esp),** Water Act (GB); **L. de Costas,** Coastal (Planning) Law (Esp); **l. de decrecientes,** law of diminishing returns; **l. de extracción de agua,** water abstraction law; **l. de la naturaleza,** law of nature; **l. de la oferta y la demanda,** law of supply and demand; **l. del Congreso (EE UU),** Act of Congress; **l. del medio ambiente,** environmental law; **l. del parlamento,** act of Parliament (GB); **L. del Suelo (Esp),** Land Protection Law; **l. de superposición (Geol),** law of superposition; **l. de un asamblea,** statute law; **l. estatal,** state law; **l. física de la conservación de la materia y la energía,** physical law of conservation of mass and energy; **l. orgánica,** constitutional law; **l. seca,** prohibition law; **l. vigente,** law in force; **leyes de la física,** laws of physics, physical laws; **proyecto (m) de l.,** bill, measure.

leyenda, f. legend (en mapas).

liana, f. liana, liane.

Liásico, m. Lias.

libélula, f. damselfly.

liberalización, f. liberalization.

liberar, v. to free, to liberate.

libertad, f. freedom; **l. de información,** f. of information (absence of censure).

libre, a. free (Gen); vacant (sitio); **l. cambio (m), l. comercio (m),** free trade; **radical (m) l. (Quím),** free radical.

libreta, f. notebook, memorandum; **l. de campo,** fieldbook.

licencia, f. licence, license (US), permission, approval; **l. para pescar,** fishing licence.

licitación, f. bid (a un subasta).

lícito/a, a. lawful, legal, licit.

licor, m. liquor (Quím); **l. de gases,** gas liquor; **l. mixto,** mixed liquor; **licores (m.pl) residuales (Trat),** waste liquors.

licuación, f. liquefaction.

licuar, v. to liquify.

licuarse, v. to deliquesce.

licuefacción, f. liquefaction.

licuefacer, v. to liquefy, to liquify.

LIDEMA, f. **Liga (f) de Defensa de Medio Ambiente (LAm),** Environment Defence League.

lienzo, m. fabric, material; stretch (de pared) (Arq).

liga, f. flux, adhesive.

ligamento, m. ligament.

lignígrafo, m. water level recorder (para flujos de un río).

lignina, f. lignin.

lignito, m. lignite.

lima, f. lime (fruta); file (herramienta).

limbo, m. leafblade (Bot, de hojas); limb (Mat, Bot).

limero, m. lime tree.

limitación, f. limitation, restriction.

limitada, a. limited; **sociedad (f) (de responsibilidad) l. (abr. s.l.),** limited company, limited liability company (abr. ltd).

límite, m. limit, end, boundary, border; **l. de detección,** limit of detection; **l. de drenaje,** drainage boundary; **l. de emisión,** fixed emission limit, uniform emission standard; limit value; **l. del acuífero,** aquifer boundary; **l. de la nieve perenne,** snow line; **l. del área de captación,** catchment boundary; **l. de la vegetación arbórea, l. del bosque,** tree-line, timber-line; **l. del hielo,** ice limit; **l. de los árboles,** tree-line, timber-line; **l. de nivel constante,** constant-head boundary; **l. de recarga,** recharge boundary; **l. de tolerancia,** tolerance limit; **l. impermeable,** impermeable boundary; **l. líquido,** liquid limit; **l. plástico,** plastic limit; **l. superior,** upper limit.

limítrofe, a. bordering, neighbouring, neighboring (US); **país (m) l.,** bordering country.

limnígrafo, m. limnigraph.

limnigrama, m. limnigram.

limnímetro, m. limnimeter.

limnología, f. limnology.

limo, m. silt; lime tree (Bot).

limolita, f. siltstone.

limón, m. lemon.

limonar, m. lemon grove.

limonero, m. lemon tree.

limonita, f. limonite.

limoso/a, a. silty, slimy.

limpia, f. cleansing, cleaning.

limpiada, f. (LAm) clean-up.

limpiador/a, a. cleansing, purifying.

limpieza, f. cleaning, cleansing, clean-up (de contaminación), cleanliness; **l. de banco (Trat),** bed cleaning; **l. en seco,** dry cleaning.

limpio/a, a. clean.

lince, m. lynx.

lindano, m. lindane.

línea, f. line, cable; **en l. (Comp),** on-line; **l. aérea,** airline; **l. central,** centre line, center (US) line; **l. de agua,** waterline; **l. de base,** base line; **l. de carga máxima (Mar),** Plimsoll mark, Plimsoll line; **l. de corriente,** streamline; **l. de costa,** coastline; **l. de drenaje,** drainage line; **l. de falla,** fault line; **l. de flotación,** waterline; **l. de flujo,** flow line; **l. de plomada,** plumb line, vertical line; **l. de**

puntos, dotted line; **l. de reticulado,** gridline; **l. de rompientes,** line of breakers; **l. de vuelo,** flight line; **l. divisoria,** dividing line; **l. equipotencial,** equipotential line; **l. férrea,** railway line; **l. isócrona,** isochrone; **l. recta,** straight line.

lineación, f. lineation.

lineal, a. linear; **dibujo (m) l.,** line drawing; **ecuación (f) no l.,** non-linear equation.

linearización, f. linearization.

linfa, f. lymph.

linfáticola, a. lymphatic; **sistema (m) l.,** lymphatic system.

linfocito, m. lymphocyte.

lingote, m. ingot.

lino, m. flax (planta), linen (tela).

liofílicola, a. lyophilic; **coloide (m) l.,** lyophilic colloid.

liofilización, f. freeze-drying.

liofilizadola, a. freeze-dried.

liofóbicola, a. lyophobic; **coloide (m) l.,** lyophobic colloid.

lípido, m. lipid.

lipólisis, f. lipolysis.

liquen, m. lichen.

líquido, m. liquid; **el l. elemento,** water.

lisímetro, m. lysimeter; **l. humectado,** irrigated lysimeter.

lista, f. list.

listado, m. listing (Comp), list.

listadola, a. striped, banded.

listón, m. batten, lathe, strip (pieza de madera); **colocación (f) de listones,** battening.

literatura, f. literature; **búsqueda (f) sistemática de la l.,** literature search.

líticola, a. lithic.

litificación, f. lithification.

litigio, m. litigation.

litio, m. lithium (Li).

litoclasa, f. lithoclase.

litofacies, f. lithofacies.

litófilola, a. lithophilous.

litogénesis, m. lithogenesis.

litología, f. lithology.

litológicola, a. lithological.

litoral, 1. m. littoral, seaboard, coast; 2. a. littoral, coastal; **deriva (f) l.,** longshore drift.

litoserie, f. lithosere.

litosfera, f. lithosphere.

litosféricola, a. lithospheric; **placas (f) l.,** lithospheric plates.

litostáticola, a. lithostatic.

litostratigráficola, a. lithostratigraphic.

litro, m. litre, liter (US).

lixao, m. (Br) waste tip, garbage dump, landfill.

lixiviación, f. leaching, lixiviation.

lixiviado, m. leachate; **recolección (f) de l.,** leachate collection; **tratamiento (m) de l.,** leachate treatment.

lixiviar, v. to leach, to elute.

loam, m. (LAm) loam.

lobby, m. lobby.

lobo, m. wolf (pl. wolves); **l. marino,** seal.

lóbulo, m. lobe; **l. de solifluxión,** solifluction lobe.

local, 1. m. place (Gen), premises (oficinas, etc.); **en los locales,** on the premises; 2. a. local.

localidad, f. locality; **l. tipo,** type locality.

localización, f. siting of, locating, location; **l. de embalses,** siting of reservoirs.

locomotora, f. locomotive.

lodar, m. pond to collect moisture (dew pond).

lodazal, m. bog.

lodo, m. mud; sewage sludge (Trat); **descarga (f) de lodos al mar,** sludge disposal to sea; **digestión (f) de lodos,** sludge digestion; **incineración (f) de lodos,** sludge incineration; **l. activado,** activated sludge; **l. bentonítico,** bentonite mud; **l. biodegradable,** biodegradable mud; **l. de depuración,** sewage sludge; **l. de perforación,** drilling mud; **l. orgánico,** organic mud; **lodos activados por aireación extendida,** extended aeration activated sludge.

lodolita, f. mudstone.

loess, m. loess.

log, m. log (Mat, abr. de logaritmo).

logarítmicola, a. logarithmic; **fase (m) l.,** log phase, logarithmic phase.

logaritmo, m. logarithm.

loma, f. hillock, low ridge; **l. en forma de lomo de ballena,** whaleback ridge.

lomada, f. hilly ridge.

lo más bajo, lowermost.

lombriz, f. earthworm; **l. intestinal,** intestinal worm.

lona, f. canvas, sackcloth, sailcloth (Mar); **l. alquitranada,** tarpaulin.

longar, m. embanked field to collect groundwater.

longitud, f. length; longitude (Geog); **grados (m.pl) de l.,** degrees of longitude; **l. de onda,** wave length.

longitudinal, a. longitudinal.

lopolito, m. lopolith.

losa, f. slab.

lubina, f. bass (pez).

lubricante, m. lubricant.

lubricar, v. to lubricate; to oil, to grease (con aceite).

Ludloviense, m. Ludlovian.

lugar, m. place, spot (sitio); room (espacio); village (pueblo); passage (en un libro); **l. común,** commonplace; **l. de desarrollo,** development site; **l. geométrico (Mat),** locus.

luminiscencia, f. luminescence.

luminosidad, f. luminosity.

luna, f. moon; **iluminadola (a) por la l.,** moonlit; **l. llena,** full moon; **l. nueva,** new moon; **luz (f) de l.,** moonlight; **media l.,** half moon.

lunar, a. lunar.

lupa, f. magnifying glass, lens.
lustrar, v. to shine.
lustre, m. lustre, luster (US).
lustro, a. lustre; luster (US); 5-year period.
lustrosola, a. lustrous.
Luteciense, m. Lutetian.
lutecio, m. lutetium (Lu).
lutita, f. lutite, shale.
lutíticola, a. shaly.
luxullianita, f. luxullianite.

luz, f. light; lamp, electricity; span (Con); **a la l. del día,** in the light of day; **dar l. verde (v),** to give the green light to; **en plena l.,** in broad daylight; **l. de luna,** moonlight; **l. del sol,** sunlight; **l. eléctrica,** electric light; **l. intermitente,** flashing light; **l. natural,** natural light; **l. polarizada,** polarized light; **l. solar,** sunlight; **l. y sombra,** light and shade; **media l.,** half-light; **primera l.,** daybreak, dawn; **salir a la l. (v),** to come to light.

LL

llamada, f. call; **ll. de fondo,** upwelling; **ll. de socorro,** distress signal, SOS.

llampo, m. (LAm) pulverized ore.

llanada, f. plain, flat ground.

llanca, f. (LAm) copper ore.

Llandeilo, m. Llandeilian.

Llandoveriense, m. Llandoverian.

llano, m. flat, plain, savannah; **Los Llanos (Ven),** Venezuelan Plains.

llanura, f. flat, plain, savannah; **ll. aluvial,** floodplain; **ll. costanera, ll. costera,** coastal plain; **ll. de fangos,** mud flat; **ll. de inundación,** floodplain, floodable land; **ll. de pedimentos,** pediplain; **ll. litoral,** coastal plain; **ll. mareal,** tidal flat.

Llanvirniense, m. Llanvirnian.

llave, f. key; spanner, wrench (US) (para tuercas); switch (interruptor); **ll. de paso,** stopcock; **planta (f) ll. en mano (Fab),** turnkey installation.

lloradero, m. (Arg) intermittent wetland.

llover, v. to rain.

llovida, f. (LAm) rain, shower.

llovizna, f. drizzle; **ll. helada,** freezing drizzle.

lloviznar, v. to drizzle.

lluvia, f. rain, rainfall (cantidad); **agua (f) de ll.,** rainwater; **distribución (f) de la ll.,** rainfall distribution; **estimulación (f) de lluvias,** rain seeding; **gota (f) de ll.,** raindrop; **ll. abundantes,** heavy rains; **ll. ácida,** acid rain; **ll. anual,** annual rainfall; **ll. copiosas,** heavy rains; **ll. diaria,** daily rainfall; **ll. efectiva,** effective rainfall; **ll. ligera,** drizzle; **ll. media,** average rainfall; **lluvias caídas,** rainfall; **medición (f) de la ll., medida (f) de la ll.,** rainfall measurement; **recarga (f) por ll.,** rainfall recharge; **siembra (f) para ll. artificial,** artificial rain seeding; **sombra (f) de ll.,** rain shadow.

lluviosola, a. rainy, wet, showery; **temporada ll.,** rainy season.

M

mln, f. **moneda (f) nacional (Arg, Méx)**, local currency.

Maastrichtense, m. Maastrichtian.

macerador, m. macerator (máquina).

macho, m. male.

macizo, m. mass, lump; massif (Geog); bed, plot (de plantas); **m. de hormigón**, mass of concrete; **m. volcánico**, volcanic mass.

macizola, a. solid, massive, stout; **asunto (m) m.**, solid case; **roca (f) m.**, solid rock.

macro-, pf. macro-.

macroartefacto, m. macroartifact.

macroeconomía, f. macroeconomics.

macrofito, m. macrophyte.

macroinvertebrado, m. macroinvertebrate; **m. bentónico**, benthic macroinvertebrate.

macronutriente, m. macronutrient.

macroscópicola, a. macroscopic.

madera, f. wood, timber, lumber; **de m. (a)**, wooden; **industria (f) de m.**, lumber industry; **m. blanda**, softwood; **m. contrachapada**, plywood; **m. dura**, hardwood; **m. escuadrada**, balk; **m. prensada**, hardboard; **negociante (m) de m.**, timber merchant.

maderaje, m. shuttering, timbering (Con), timberwork.

maderamen, m. shuttering, timbering (para construcción), timberwork.

maderería, f. lumber yard, timber yard.

madero, m.pl. log, piece of timber; **maderos**, timber, lumber

madre, f. mother; **m. de un río**, river bed, river source; **m. vieja (Geog)**, ox-bow lake (abandoned meander).

madrejón, m. (Arg) old river channel.

madrepórica, a. madreporic; **placa (f) m.**, madreporic plate.

madreselva, f. honeysuckle.

madriguera, f. burrow, hole (de conejos), earth, lair, den (zorros, tejones).

maduración, f. ripening, maturing, maturation; **laguna (f) de m. (Trat)**, maturation pond.

madurez, f. ripeness, maturity; wisdom (conocimiento); **m. de un río**, river maturity; **m. sexual**, sexual maturity.

madurola, a. mature.

máficola, a. mafic.

maganeso, m. manganese (Mn).

magistrado, m. magistrate.

magma, f. magma.

magmáticola, a. magmatic.

magnésicola, a. magnesic.

magnesio, m. magnesium (Mg).

magnesita, f. magnesite.

magnéticola, a. magnetic; **anomalía (f) m.**, magnetic anomaly; **campo (m) m.**, magnetic field; **cinta (f) m.**, magnetic tape; **disco (m) m.**, magnetic disc; **norte (m) m.**, magnetic north.

magnetismo, m. magnetism.

magnetita, f. magnetite, lodestone.

magnetómetro, m. magnetometer.

magnitud, f. magnitude; **(cálculo de) magnitudes geométricas**, mensuration (Mat).

maicillo, m. (Ch) road gravel.

maíz, m. maize.

maizal, m. maize field, cornfield (US).

malahierba, f. weed.

malaquita, f. malachite.

malaria, f. malaria.

malatión, m. malathion.

maleabilidad, f. malleability.

maleable, a. malleable.

malecón, m. pier, jetty.

maleza, f. scrub (arbustos); undergrowth (broza); weed, weeds (malas hierbas); brambles (zarzas); **lleno (m) de m. (Ecol)**, undergrown, choked.

malignola, a. malignant.

malnutrición, f. malnutrition.

maloliente, a. smelly.

malsanola, a. sickly, unhealthy.

malta, f. malt (cebada), malt (Quím); **fábrica (f) de m.**, malthouse.

maltaje, m. malting.

maltasa, f. maltase.

malteadola, a. malted; **leche (f) m.**, malted milk.

maltear, v. to malt.

maltería, f. malthouse.

maltosa, f. maltose.

malla, f. mesh, network; swimsuit (US); **m. cuadrada**, square mesh; **m. cuadriculada**, grid mesh; **m. de capacidades y resistencias**, capacitor-resistor mesh; **m. poligonal**, polygonal mesh; **m. rectangular**, rectangular mesh.

mallín, m. (Arg, Ch) large-scale depression caused by groundwater discharge.

mama, f. mamma.

mamífero, m. mammal; **mamíferos (Zoo, clase)**, Mammalia (pl).

mamíferola, a. mammalian, mammiferous.

mampara, f. bulkhead.

mampostería, f. masonry, stonework; **m. en seco**, dry rubble; **m. seca**, dry masonry, dry stonework

mana, f. (LAm) spring, fountain.

manada, f. herd; **llevar en m. (v)**, to herd (animals).

manantial, m. spring, source, fountain; **m. de agua termal**, hot spring; **m. de ebullición,** boiling spring; **m. gaseoso,** gaseous spring; **m. gravitacional,** gravity spring; **m. intermitente,** intermittent spring; **m. kárstico,** karst spring; **m. mineral,** mineral spring; **m. perenne,** perennial spring; **m. termal,** thermal spring; **surgencia (f) de un m.,** spring discharge.

mancha, f. stain, mark, flaw; **La M.,** region in central Spain; **m. de petróleo,** oil slick; **m. solar,** sun spot.

mandarina, f. mandarin, tangerine.

mandarino, m. mandarine tree, tangerine tree.

mandato, m. mandate; **poner bajo el m. de (v),** to mandate.

mandíbula, f. mandible; jaw, jawbone.

manejo, m. handling, running; operation (de máquina).

manera, f. manner; **m. de presentarse,** mode of occurrence.

manga, f. sleeve (de vestido); hose, hosepipe (para agua); windsock (en un campo de aviación); stretch (recto de un río); crowd (LAm) (multitud); **m. de riego,** garden hosepipe; **m. de ventilación,** ventilation shaft; **m. de viento,** whirlwind; **m. marina (Met),** waterspout.

mangada, f. narrow embanked field fed by groundwater.

manglar, m. mangrove swamp.

mangle, m. mangrove.

manguera, f. hose, garden hosepipe.

manguito, m. sleeve (Ing), joint; **m. roscado (Ing),** threaded sleeve.

maní, m. peanut, groundnut.

manifestación, f. manifestion; **m. hidrotérmica,** hydrothermal manifestation.

manifestar, v. to declare, to express.

manifiesto, m. manifest (lista).

manifiestola, a. manifest, evident (evidente).

manigua, f. swampy scrubland (pantano), thicket.

manipulación, f. manipulation.

manipular, v. to manipulate.

manómetro, m. manometer.

mansiega, f. fen, prone to drying and burning.

mantenimiento, m. maintenance; **gastos (m.pl) de m.,** maintenance costs; **m. de bombas,** pump maintenance; **m. de máquinas,** machine maintenance; **m. de planta,** plant maintenance; **m. preventivo,** preventive maintenance.

mantillo, m. farmyard manure (estiércol); vegetable mould, vegetable mold (US), litter (capa de suelo).

manto, m. mantle, overburden (Geol); cloak; **m. de corriente,** nappe; **m. de hielo,** ice sheet; **m. rocoso,** rock mantle.

manual, m. handbook, manual (libro).

manubrio, m. crank, handle.

manufacturar, v. to manufacture.

manufacturerola, a. manufacturing.

manutención, f. maintenance (de un edificio).

manzana, f. apple (fruta); block (building).

manzanar, m. apple orchard.

manzano, m. apple tree.

MAPA, m. MAFF (GB); **Ministerio (m) de Agricultura, Pesca y Alimentación (Esp),** Ministry of Agriculture, Fisheries and Food (GB).

mapa, m. map, chart; **m. aéreo,** aerial map; **m. base,** base map; **m. de Almirantazgo,** Admiralty chart; **m. de curvas de nivel,** contour map; **m. de isopacas,** isopachyte map; **m. de subsuelo,** subsurface map; **m. del Ministerio de Marina,** Admiralty chart; **m. estructural,** structural map; **m. hidrográfico,** hydrographic chart; **m. piezométrico,** piezometric map; **m. superpuesto,** map overlay; **m. topográfico,** topographic map.

mapeo, m. mapping, digital mapping.

maquiladora, f. (LAm) assembly plant.

máquina, f. machine, engine, locomotive; **cuarto (m) de máquinas,** engine room; **m. automática,** vending machine; **m. de cálcular,** calculating machine, calculator; **m. de escribir,** typewriter; **m. de vapor,** steam engine; **m. herramienta,** machine tool, lathe; **m. niveladora, m. para igualar terreno,** grader; **sala de máquinas,** engine room.

maquinaria, f. machinery.

maquis, m. maquis.

mar, m. y f. sea; **m. agitado,** rough sea; **m. de fondo,** groundswell; **m. gruesa,** heavy sea; **m. rizada,** choppy sea.

marca, f. mark; **m. de fábrica, m. patentada, m. registrada,** trademark; **m. terrestre,** landmark.

marcasita, f. marcasite.

marchal, m. (southern Spain) spring water.

marchitamiento, m. wilting.

marchitar, v. to wilt (de una planta).

marchitez, m. wilting; **punto (m) de m. permanente,** permanent wilting point.

marea, f. tide; **alcance (m) de la m., amplitud (f) de la m.,** tidal range; **constituyente (m) armónico de la m. (Mat),** tidal constituent, tidal harmonic; **de la m. (a),** tidal; **energía de la m., línea (f) de la m. alta,** tidemark; **m. alta,** high tide; **m. baja,** low tide, low water; **m. creciente,** rising tide; **m. descendente,** outgoing tide; **m. entrante,** incoming tide; **m. equinoccial,** equinoctial tide; **m. menguante,** ebb tide; **m. muerta,** neap tide; **m. negra,** oil slick; **m. viva,** spring tide; **tablas (f.pl) de mareas,** tide tables.

mareal, a. tidal; **zona (f) m.,** tidal zone.

marejada, f. swell (en el mar).

maremoto, m. sea quake.

marga, f. marl.

margalita, f. (LAm) marlstone.

margen, m. y f. margin, edge; **m. continental,** continental margin; **m. de error,** margin of error; **m. de río,** river bank, riverside; **m. de**

seguridad, safety margin; **m. marino,** seashore.

marginal, a. marginal.

margosola, a. marly.

mariguana, f. marijuana, cannabis, hachis, pot.

marihuana, f. marijuana, cannabis, hachis, pot.

marijuana, f. marijuana, cannabis, hachis, pot.

marina, f. navy, marines (US) (Mil); coast (litoral); navigation (navegación); **m. mercante,** merchant navy.

marinero, m. sailor.

marinerola, a. seaworthy (de barco); seafaring (gente).

marino, m. seaman, sailor; **m. mercante,** merchant seaman.

marinola, a. marine; **milla (f) m.,** nautical mile.

mariposa, f. butterfly (insecto); butterfly-nut, wing-nut (Ing); **m. nocturna,** moth.

mariscadorla, m. y f. shellfisherman.

mariscos, m.pl. shellfish (pl. shellfish).

marisma, f. salt marsh; **m. de marea,** tidal marsh.

marítimola, a. maritime.

marjal, m. fen, marsh, bog.

márketing, m. marketing.

marlita, f. (LAm) marl, marlite.

marmita, f. pot (cooking); **m. de gigante (Geol),** pothole.

mármol, m. marble.

marmoladola, a. marbling; **efecto (m) m.,** marbling effect.

marquesina, f. canopy (cobertiza).

marsupial, 1. m. marsupial; 2. a. marsupial.

Marte, m. Mars.

martillo, m. hammer, striker; **m. de aire comprimido,** air hammer; pneumatic drill, jackhammer; **m. neumático,** pneumatic drill, air hammer; jack hammer; **m. pilón,** pile driver.

martinete, m. pile driver.

más, ad. plus sign (Mat, +); times as (Mat); **2 más 2 son 4,** 2 plus 2 is 4; **4 veces m. grande,** 4 times as big.

masa, f. mass (Fís), bulk, volume, quantity; **m. de mineral,** ore body; **m. terrestre,** land mass; **unidad (f) de m. atómica (abr. u.m.a.),** atomic mass unit (abr. a.m.u.).

masculinola, a. masculine, male.

masificación, f. over-crowding (p.e. of beaches).

masilla, f. putty (para ventanas).

mástil, m. mast, spar, stem (Ing).

mata, f. bush, shrub; clump, grove (de árboles); forest, jungle (LAm); **matas,** thicket, scrub.

matadero, m. slaughterhouse; **desechos (m.pl) de m.,** slaughterhouse wastes.

mate, a. frosted, dull, matt (delustrado/a).

matemáticas, f.pl. mathematics (abr. maths).

matemáticola, 1. m. y f. mathematician; 2. a. mathematical; **expresión (f) m.,** mathematical expression; **modelo (m) m.,** mathematical model.

materia, f. matter, subject matter; **m. carbonosa,** carbonaceous matter; **m. disuelta,** dissolved material; **m. en suspensión,** matter in suspension, suspended matter; **m. insoluble,** insoluble matter; **m. orgánica,** organic matter; **m. prima,** raw material; **m. viviente,** living matter.

material, 1. m. material; **m. de construcción,** building material; **m. de oficina,** office equipment; **m. de partida,** parent material; **m. de relleno,** filler; **m. en fusión,** molten material; **m. granular,** particulate material; **m. particulado,** particulate material; 2. a. material; **bienes (m.pl) materiales,** material assets.

materialización, f. materialization.

mates, f.pl. maths, math (US) (abr. de mathematics).

matorral, m. scrub, bush, thicket.

matraz, m. flask (Quím); **m. cónico,** conical flask; **m. filtro,** filter flask; **m. graduado,** graduated flask.

matriz, f. matrix (Mat, Min); womb, uterus (Anat); die, mould (troquel); master copy, original (de documentos); nut (tuerco); **m. de roca,** rock matrix.

mausoleo, m. mausoleum.

maxilar, m. maxilla (pl maxillae, maxillas); **m. superior,** maxilla.

máximo, m. maximum (pl. maxima, maximums).

máximola, a. maximum.

mayor, a. bigger, larger; **comprar al por m. (v),** to buy wholesale; **vender al por m. (v),** to sell wholesale.

mayoría, f. majority; **en la m. de los casos,** in most cases, in the majority of cases.

mayorista, m. y f. wholesaler, merchant.

mazorca, f. (LAm) corn cob, maize ear.

MCE, m. ECM; **Mercado (m) Común Europeo,** European Common Market.

meandro, m. meander; **con meandros,** meandering; **m. abandonado,** abandoned meander, ox-bow lake; **m. del río,** river meander; **m. excavado,** incised meander; **que forma meandros,** meandering; **rizo (m) de m.,** meander loop; **zona (f) de meandros,** meander belt.

mecánica, f. mechanics; **m. de rocas,** rock mechanics; **m. de suelos,** soil mechanics, geotechnics; **m. ondulatoria,** wave mechanics.

mecánico, m. mechanic, mechanist.

mecánicola, a. mechanical.

mecanización, f. mechanization.

mecanizar, v. to mechanize.

mecanografiadola, a. typed, typewritten; **texto (m) m.,** typescript.

mechero, m. lighter, cigarette lighter, burner; **m. Bunsen,** Bunsen burner.

media, f. mean, average (Mat); **m. anual,** annual average; **m. aritmética,** arithmetic mean; **m. geométrica,** geometric mean; **m. móvil,** moving average.

mediación, f. mediation.

mediana, f. median (Mat, término central).

medianola, a. medium, average, mediocre.

medianoche, f. midnight.

medicamentosola, a. medicinal.

medicina, f. medicine, health; **m. de trabajo,** occupational health.

medicinal, a. medicinal; **aguas (f.pl) medicinales,** medicinal waters.

medición, f. measuring, gauging; **estación (f) de m. de caudales,** gauging station.

medida, f. measure, means, measurement; **m. correctiva,** corrective measure; **m. de la humedad del suelo,** soil moisture measurement; **m. de la insolación,** insolation measurement; **m. de la lluvia,** rainfall measurement; **m. de la presión,** pressure measurement; **m. de la temperatura,** air temperature measurement; **m. de niveles de agua,** water level measurement; **m. de superficie,** square measure; **m. de tratamiento,** means of treatment; **m. de volumen,** cubic measure, volumetric measure; **m. preventiva,** preventative measures.

medidor, m. y f. recorder, meter (instrumento); measurer (persona); **m. de cantidades de obra,** quantity surveyor; **m. de carga de agua,** domestic flow meter; **m. de cinta perforada,** punched tape recorder; **m. de cinta,** tape measure; **m. de corriente,** current meter, flow meter; **m. de flotador,** float recorder; **m. de Venturi,** Venturi meter; **m. neumático,** pneumatic recorder.

medio, m. half (mitad); middle, centre, center (US) (centro); means (pl) (método); atmosphere (atmósfera); environment, medium (Biol); **m. ambiente,** environment; **m. anaerobio,** anaerobic medium; **m. anisótropo,** anisotropic medium; **m. de cultivo,** culture medium; **m. de vida,** half-life; **m. filtrante,** filter medium; **m. heterogéneo,** heterogeneous medium; **m. isotrópico,** isotropic medium; **m. poroso,** porous medium; **m. rural,** rural environment; **m. semipermeable,** semi-permeable medium; **medios (Comm),** media (pl).

mediola, a. mean, average (Mat), medium; **temperatura (f) m.,** average temperature; **término (m) m.,** median, middle term.

medioambiental, a. environmental; **ciencia (f) m.,** environmental science.

medioambientalismo, m. environmentalism.

medioambientalista, m. y f. environmentalist.

medioambiente, m. environment; **ley (f) de m.,** environmental law.

mediodía, m. noon, midday; **a m.,** at midday.

medios, m.pl. media (comunicaciones).

medir, v. to measure.

Mediterráneo, m. Mediterranean sea.

mediterráneola, a. Mediterranean.

médula, f. pith, marrow (Bot); medulla (Anat); **m. espinal,** nerve cord; **m. ósea,** bone marrow.

medusa, f. medusa; jellyfish.

mega-, pf. mega- (10.E6).

megabyte, m. megabyte.

megalíticola, a. megalithic (Arq).

megalitismo, m. the building of megaliths (Arq).

megalito, m. megalith (Arq).

megalópolis, m. megacity, very large city (p.e. Ciudad de México); megalopolis (US).

megatón, m. megaton.

megavatio, m. megawatt.

megavoltio, m. megavolt.

mejicanola, m. y f. Mexican.

mejicanola,, a. mexican.

Méjico, m. Mexico; Mexico City (capital city); **ciudad (f) de M.,** Mexico City.

mejora, f. improvement, amelioration.

mejoramiento, m. improvement.

mejorar, v. to improve, to ameliorate.

melanina, f. melanin.

melanismo, m. melanism; **m. industrial,** industrial melanism.

melaza, f. molasses; treacle.

melocotón, m. peach, peach tree.

melón, m. melon.

melonar, m. melon patch, melon field.

melonerola, m. y f. melon grower, melon dealer.

melladola, a. jagged, chipped.

membrana, f. membrane; **m. celular,** cell membrane; **m. filtrante,** filter membrane; **m. hidrófuga,** damp-proof membrane; **m. mucosa,** mucous membrane; **m. semipermeable,** semi-permeable membrane.

memoria, f. memory; memo, memorandum (documento); memoirs (libro); **m. de acceso aleatorio (Comp),** random-access memory (abr. RAM); **m. de acceso directo (Comp),** direct-access memory; **m. del ordenador,** computer memory; **m. de sólo lectura,** read-only memory (abr. ROM).

mena, f. ore.

mengua, f. diminution, lessening, decrease.

menguadola, a. diminished, decreased.

menguante, 1. m. ebb tide, low water; 2. a. decreasing, diminishing; waning (la luna); ebb (marea).

menguar, v. to decrease, to diminish; to ebb (marea); to wane (luna).

menisco, m. meniscus (pl. menisci, meniscuses).

menor, a. smaller; **al por m.,** retail.

menos, p. minus, less.

menoscabo, m. lessening, reduction; damage (daño); **con/en m. de,** to the detriment of; **sin m.,** unimpaired.

mensaje, m. message.

mensuración, f. mensuration.

mentalizar, v. to persuade (convencer); to brainwash (lavar el cerebro).

mentol, m. menthol.

mercadeo, m. trading (comercio); marketing (marketing).

mercadería, f. commodity; **mercaderías**, goods, merchandise.

mercado, m. market; **análisis (m) de mercado(s)**, market analysis; **demanda (f) del m.**, market demand; **equilibrio (m) del m.**, market equilibrium; **estudios (m.pl) de m.**, market research; **fuerzas (f.pl) del m.**, market forces; **libre m.**, free market; **líder (m) del m.**, market leader; **m. exterior**, overseas market; **m. interior, m. nacional**, domestic market, home market; **m. único**, single market.

mercadotecnia, m. market research.

mercancía, f. commodity; **mercancías**, goods, merchandise.

mercantil, a. mercantile, commercial; **sociedad (f) m.**, trading company.

mercaptano, m. mercaptan.

mercúricola, a. mercuric.

mercurio, m. mercury (Hg); Mercury (la planeta).

mercuriosola, a. mercurous.

meridiano, m. meridian.

meridianola, a. midday, noon (hora); crystal-clear (Fig).

meridional, a. meridional, southern; **América (f) m.**, South America.

meristemo, m. meristem; **m. apical**, apical meristem.

merluza, f. hake.

mero, m. grouper (pez).

mesofílicola, a. mesophylic.

mesófilola, a. mesophilic; **digestión (f) m. (Trat)**, mesophilic digestion.

mesofitola, m. y f. mesophyte.

mesosfera, f. mesosphere.

Mesozoico, m. Mesozoic.

Messinense, m. Messinian.

meta-, pf. meta-.

metabólicola, a. metabolic; **actividad (f) m.**, metabolic activity.

metabolismo, m. metabolism; **m. basal**, basal metabolic rate, base metabolism.

metabolito, m. metabolite.

metacuarcita, f. metaquartzite.

metafosfato, m. metaphosphate.

metahemoglobinamia, f. methaemoglobinaemia.

metal, m. metal; **m. no ferroso**, non-ferrous metal; **m. pesado**, heavy metal.

metaldehído, m. methaldehyde.

metalero, m. (Pan) waste tip scavenger.

metálicola, a. metallic.

metalistería, f. metalwork.

metaloide, m. metalloid.

metamórficola, a. metamorphic.

metamorfismo, m. metamorphism.

metamorfosear, v. to metamorphose, to metamorphosize.

metamórfosis, f. metamorphosis (pl. metamorphoses).

metano, m. methane, marsh gas.

metanogénesis, f. methanogenesis.

metanol, m. methanol.

metanotróficola, a. methanotrophic; **bacterias (f.pl) m.**, methanotrophic bacteria.

metasedimento, m. metasediment.

metasomatismo, m. metasomatism.

metazoo, m. metazoan (pl. metazoa, metazoans); **metazoos (Zoo, subreino)**, Metazoa (pl).

meteorito, m. meteorite.

meteorización, f. weathering.

meteorizadola, a. weathered; **no m.**, unweathered.

meteorizar, v. to weather.

meteoro, m. meteor.

meteorología, f. meteorology.

meteorológicola, a. meteorological.

meteorólogola, m. y f. meteorologist.

metilamina, f. methylamine.

metilbenzol, m. methylbenzene.

metileno, m. methylene; **azul (m) de m.**, methylene blue.

metilmercurio, m. methylmercury.

metilo, m. methyl; **naranja (f) de m.**, methyl orange.

metilparathión, m. methylparathion.

metódicola, a. methodical.

método, m. method; **m. analógico**, analogue method; **m. de cuchara**, bailer method; **m. de diferencias finitas**, finite-difference method; **m. de dilución**, dilution method; **m. de inversión de matrices**, method of matrix inversion; **m. de las imágenes**, method of images; **m. de optimización**, optimization method; **m. de recuperación**, recovery method; **m. de relajación**, relaxation method; **m. de ruta crítica (LAm)**, critical path method; **m. de separación de variables**, method of separation of variables; **m. de superposición**, method of superposition; **m. de vía crítica**, critical path method; **m. explícito**, explicit method; **m. gráfico**, graphical method; **m. gravimétrico**, gravimetric method; **m. implícito**, implicit method; **m. iterativo**, iterative method; **m. logarítmico**, logarithmic method.

metodología, f. methodology.

metoxicloro, m. methoxychlor.

métricola, a. metric; **cinta (f) m.**, tape measure, measuring tape; **sistema (m) m.**, metric system.

metro, m. metre, meter (US); meter (instrumento de medidas); (US) underground, tube, subway (trenes); **m. cúbico**, cubic metre.

metrópoli, f. metropolis.

metropolitanola, a. metropolitan.

mexicanola, 1. m. y f. Mexican; 2. a. Mexican.

México, m. Mexico; Mexico City (capital city); **ciudad (f) de M.**, Mexico City.

mezcla, f. mixture, blend.

mezclador, m. mixer (máquina).

mezclar, v. to mix, to blend.

mhos, m. mhos (anterior unidad de conductancia eléctrica).

mica, f. mica.

micacita, f. (LAm) mica schist.

micaesquisto, m. mica schist.

micelio, m. mycelium (pl. mycelia).

micófito, m. mycophyte.

micología, f. mycology.

micorriza, f. mycorrhiza.

micra, f. micron, micrometre, micrometer (US).

micrinita, f. micrinite.

micro-, pf. micro- (Mat, 10.E-6).

microanálisis, m. microanalysis.

microartefacto, m. microartifact (Arq).

microbianola, a. microbial, microbic; **contaminación (f) m.**, microbial contamination.

microbio, m. microbe; germ.

microbiología, f. microbiology.

microblasto, m. microblast.

microbus, f. small bus.

microclima, m. microclimate.

microclimatología, f. microclimatology.

microclina, f. microcline.

microcribado, m. microstraining.

microcristalinola, a. microcrystalline.

microchip, m. microchip.

microdepurador, m. small-scale purification plant.

microfaradio, m. microfarad (unidad de capacitancia eléctrica).

microfiltración, f. microfiltration.

microfisuración, f. microfissuring.

microfósil, m. microfossil.

microgota, f. droplet, aerosol.

micrografía, f. micrograph; **m. electrónica**, electron micrograph.

microhm, m. microhm (unidad de resistencia eléctrica).

microhmio, m. microhm (unidad de resistencia eléctrica).

microlog, m. microlog.

micrometeorología, f. micometeorology.

micrométricola, a. micrometric.

micrómetro, m. micrometre, micrometer (US).

micrón, m. micron, micrometre, micrometer (US).

micronutriente, m. micronutrient.

microonda, f. microwave.

microorganismo, m. microorganism.

microplaqueta, f. microchip (Comp).

microprocesador, m. microprocessor.

microregistro, m. microlog.

microscopia, f. microscopy.

microscópicola, a. microscopic; **examen (m) m.**, microscopic examination.

microscopio, m. microscope; **m. de luz polarizado**, polarizing microscope; **m. electrónico**, electron microscope; **m. simple**, simple microscope.

microsiemen, m. microsiemen (unidad de conductancia eléctrica).

microsónicola, a. microsonic.

microtamiz, m. microscreen, microstrainer.

microtamizado, m. microscreening.

microterremoto, m. microquake, small tremor.

mid-atlánticola, a. mid-Atlantic.

miedo, m. fear; **tener m. de (v)**, to be frightened of.

miel, f. honey.

mielga, f. lucerne.

miembro, m. member; limb (Anat); **Estado (m) m.**, member state; **los países (m.pl) miembros**, member countries; **m. de Parlamento**, Member of Parliament (GB).

migmatita, f. migmatite.

migmatización, f. migmatization.

migración, f. migration; **m. de lixiviado**, leachate migration; **m. hacia arriba de gases**, upward gas migration; **m. lateral de gases**, lateral gas migration.

migraña, f. migraine.

migratoriola, a. migratory, migrant; **trabajador (m) m.**, migrant worker.

mijo, m. millet.

mildeu, m. mildew.

mildiu, mildiú, m. mildew.

mili-, pf. milli- (10.E-3).

miliamperio, m. miliampere (unidad de corriente eléctrica).

milibar, m. millibar (unidad de presión atmosférica).

miligramo, m. milligram, milligramme.

mililitro, m. millilitre, milliliter (US).

milímetro, m. millimetre, millimeter (US); **m. de mercurio**, millimetre of mercury, millimeter of mercury (US).

milonita, f. mylonite.

milonitización, f. mylonitization.

milpa, f. field of corn (US); maize field (GB, LAm).

milpero, m. (CAm) maize grower.

milpiés, m. millipede.

milla, f. mile (1609 metros); **m. marina (1852 metres)**, nautical mile.

mimetismo, m. mimicry.

mina, f. mine; **campo (m) de minas**, minefield; **m. abandonada**, abandoned mine; **m. de carbón**, coal mine.

minador, m. (LAm) solid waste segregator.

mineral, 1. m. ore, mineral; **cuerpo (m) de m. (LAm)**, ore body; **m. de hierro**, iron ore; **m. esencial (Med)**, essential mineral; **masa (f) de m.**, ore body; **minerales pesados**, heavy minerals; 2. a. mineral; **aceite (m) m.**, mineral oil; **agua (f) m.**, mineral water; **sales (m.pl) minerales**, mineral salts.

mineralización, f. mineralization.

mineralizar, v. to mineralize.

mineralogía, f. mineralogy.

mineralometría, f. mineralometry.

minero, m. miner (obrero), mine owner (propietario), mine operator (explotador).

minifundio, m. smallholding, small farm.

Minimata, f. Minimata; **enfermedad (f) de M.,** Minimata disease.

minimización, f. minimization.

minimizar, v. to minimize.

mínimo, m. miminum (pl. minima, minimums).

mínimola, a. minimum.

ministerio, m. ministry; **M. de Cultura (Esp),** Ministry of Culture; **M. de Industria y Comercio,** Ministry of Industry and Commerce; **M. de la Vivienda,** Ministry of Housing; **M. del Medio Ambiente,** Ministry for the Environment; **M. de Obras Públicas,** Ministry of Public Works; **M. del Interior,** Home Office (GB), Department of State (US).

ministro, m. minister; **Primer M.,** Prime Minister.

minorista, m. y f. retailer, retail dealer.

minuciosidad, f. thoroughness (meticulosidad); detailed nature (detalle); minuteness (de tamaño pequeño).

mio-, pf. myo-.

Mioceno, m. Miocene.

miope, a. myopic, short-sighted, near-sighted.

miopía, f. myopia, short-sightedness, near-sightedness.

mirador, m. vantage point, panoramic view (vista); bay window (ventana).

miscibilidad, f. miscibility.

mitad, f. half; **a m. de precio,** at half price.

mitigación, f. mitigation; alleviation (alivio).

mitigar, v. to mitigate; to alleviate (aliviar).

mixomiceto, m. myxomycete, slime fungus.

mixtola, a. mixed.

mixtura, f. mixture.

mobilario, m. furniture; **m. urbano (p.e. alumbrado público),** street furniture.

moco, m. mucus.

mochuelo, m. little owl, owlet.

moda, f. mode (Mat).

modal, a. modal.

modalidad, f. modality.

modelado, m. modelling, modeling (US).

modeladorla, a. modelling, modeling (US).

modelar, v. to model.

modelista, m. y f. modeller, modeler (US).

modelización, f. modelling; modeling (US) (Comp).

modelo, m. model, pattern (patrón); **m. analítico,** analytical model; **m. analógico,** analogue model; **m. autoregresivo,** autoregressive model; **m. bidimensional,** two-dimensional model; **m. conceptual,** conceptual model; **m. de advección,** advection model; **m. de aproximación de diferencias finitas implícitas,** implicit finite-difference approximation model; **m. de cambio climático,** global climate model (abr. GCM); **m. de difusión,** diffusion model; **m. de parámetro concentrado,** lumped-parameter model; **m.**

dependiente del tiempo, time-dependent model; **m. determinista,** deterministic model; **m. de transporte,** transport model; **m. digital,** digital model; **m. estocástico,** stochastic model; **m. físico,** physical model; **m. híbrido,** hybrid model; **m. hidrodinámico,** hydrodynamic model; **m. matemático,** mathematical model; **m. por diferencias finitas,** finite-difference model; **m. probabilístico,** probabilistic model; **m. transitorio,** transient model; **m. unidimensional,** one-dimensional model; **validación (f) de un m.,** model validation.

módem, m. modem.

modernola, a. modern.

modificación, f. modification.

modo, m. mode, manner, way.

modular, a. modular; **flujo (m) m.,** modular flow.

módulo, m. modulus (Mat); module; **m. de elasticidad,** modulus of elasticity.

MOF, m. **Ministerio (m) de Fomento (Esp),** Ministry of Development.

moheda, f. land subject to flooding.

Moho, m. (abr. discontinuidad Mohorovicic), Moho (Geol); (abr. Mohorovicic discontinuity); **capa (f) de M.,** Moho layer.

moho, m. mould, mold (US), mildew; rust (orín); verdigras (cubre); **criar m. (v),** to go mouldy; **m. mucoso,** slime fungus.

mohosola, a. mouldy, moldy (US) (p.e. frutas); rusty (hierro); **ponerse m. (v),** to go mouldy (p.e. frutas); to rust (hierro).

mojabilidad, f. wettability.

mojadola, a. wet, damp; **perímetro (m) m. (Geog),** wetted perimeter.

mojar, v. to dampen, to wet.

mojeda, f. land subject to flooding.

mojón, m. landmark; boundary stone (piedra).

mol, m. mole (Quím, abr. mol).

molal, a. molal.

molalidad, f. molality.

molar, a. molar; **solución (f) m.,** molar solution.

molaridad, f. molarity.

moldura, f. moulding, molding (US).

molécula, f. molecule.

molécula-gramo, m. gram-molecule.

molecular, a. molecular.

moler, v. to grind, to crush.

molestia, f. nuisance, annoyance, bother.

molibdato, m. molybdate.

molibdeno, m. molybdenum (Mo).

molienda, f. grinding, milling, crushing.

molinero, m. miller.

molinete, m. windmill, vane, ventilator (de un ventana); current meter, flow meter (para la medición de la velocidad del flujo); **m. eólico,** wind pump; **micro-m.,** small or micro current meter.

molino, m. mill; **m. de agua,** watermill; **m. de**

martillo (para la pulverización de desechos) (Ing), hammermill; **m. de viento,** windmill.

molusco, m. mollusc, mollusk (US); **moluscos (Zoo, filum),** Mollusca (pl).

molusquicida, m. molluscicide.

momia, f. mummy (Arq).

monacita, f. monazite.

mondar, m. water-filled claypit.

monel, m. monel (metal).

moniliforme, m. (LAm) boudinage.

monitor, m. monitor (Comp).

monitoreo, m. monitoring; **m. del ambiente,** environmental monitoring.

monitorización, f. monitoring.

mono, m. ape, monkey.

mono-, pf. mono-.

monocarril, m. monorail.

monocíclicola, a. monocyclic, monocyclical.

monoclinal, 1. m. monocline; **bloque (m) m.,** monoclinal block; 2. a. monoclinal.

monocotiledónea, f. monocotyledon; **monocoti-ledóneas (Bot, clase),** Monocotyledonae (pl).

monocultivo, m. monoculture, single-crop farming.

monodimensional, a. one-dimensional.

monoespecie, f. monospecies.

monofilamento, m. monofilament; **m. de nilón (pesquera),** nylon monofilament.

monografía, f. monograph, special edition.

monograptido, m. monograptid.

monohidrato, m. monohydrate.

monoicola, a. monoecious.

monolito, m. monolith.

monómero, m. monomer.

monosacárido, m. monosaccharide.

monotrofía, f. monotrophy.

monovalente, a. monovalent.

monóxido, m. monoxide; **m. de carbono,** carbon monoxide.

montacarga, m. hoist, lift, good lift, service lift; freight elevator (US); **m. de horquilla,** forklift.

montaje, m. assembly, assembling, putting together; **planta (f) de m. (Con),** assembly plant.

montaña, f. mountain; **cadena (f) de montañas,** mountain chain.

montante, m. upright, stanchion (soporte), mullion (de ventana).

montaña, f. mountain; **ladera (f) de m.,** mountain side.

montañismo, m. climbing, mountaineering.

montañosola, a. hilly, mountainous.

monte, m. mountain, forest; **Ley (f) de Montes (Esp),** Forest Law; **m. bajo,** undergrowth; **m. submarino,** seamount.

montecillo, m. hillock.

Monteniense, m. Montenian.

montículo, m. motte (Arq); **m. y banqueta,** motte and bailey.

montmorillonita, f. montmorillonite.

montón, m. pile, heap (p.e. de arena).

monumento, m. monument; memorial (de conmemoración).

monzón, m. monsoon.

monzonita, f. monzonite.

MOPU, m. **Ministerio (m) de Obras Públicas y Urbanismo (Esp),** Ministry of Public Works and Urban Planning.

mora, f. mulberry, blackberry.

moratoria, f. moratorium.

morbosidad, f. morbidity.

morena, f. moray eel.

morfina, f. morphine.

morfología, f. morphology; **m. foliar (Bot),** leaf morphology.

morfológicola, a. morphological.

morón, m. hummock.

morrena, f. moraine (Geol); moray (pez); **m. basal,** basal moraine; **m. central,** central moraine; **m. terminal,** terminal moraine.

morro, m. snout, nose (Zoo); pier, jetty (malecón).

mortalidad, f. mortality; **m. infantil,** infant mortality.

mortero, m. mortar (cemento).

mosca, f. fly (Biol).

moscovita, f. moscovite.

mosquito, m. mosquito (pl. mosquitos, mos-quitoes).

mosto, m. must, grape must.

mota, f. mottle, speckle.

moteadola, a. mottled, speckled.

motel, m. motel.

motil, a. motile.

motocultivo, m. mechanized farming.

motoniveladora, f. bulldozer.

motor, m. motor, engine; **impulsado por un m.,** motor-driven; **m. de arranque,** starter motor; **m. de balancín,** beam engine; **m. de com-bustión interna,** internal combustion engine; **m. de gasolina,** petrol engine; **m. de reacción,** jet engine; **m. de vapor,** steam engine; **m. diesel,** diesel engine; **m. eléctrico,** electric motor; **m. fuera borda,** outboard motor.

motora, f. motorboat.

motriz, a. motor.

movidola, a. moved; **m. hacia abajo,** down-throw.

movilidad, f. mobility.

movilización, f. mobilization.

movilizar, v. to mobilize.

movimiento, m. movement, motion; **m. de bloques,** block movement; **m. de caja (Com),** transactions; **m. de contaminantes,** move-ment of contaminants; **m. de desplazamiento en dirección,** strike-slip movement; **m. del agua subterránea,** movement of underground water; **m. de las olas,** wave movement; **m. de la tierra,** earth movement; **m. de mercancías (Com),** turnover, volume of business; **m. rotacional,** rotational movement; **m. sísmico,** earth tremor.

mucílago, m. mucilage.

mucosa, f. mucosa (pl. mucosas, mucosae); mucous membrane.

mucosola, a. mucous; **membrana (f) m.,** mucous membrane.

mucus, m. mucus.

muchedumbre, f. crowd (de personas), spectators (espectadores), flock (de gente).

muda, f. moult, molt (US); moulting season, molting season (US) (temporada); slough (piel de una culebra).

mudar, v. to moult, to molt (US) (piel); to mutate (de génes).

muela, f. molar, molar tooth; whetstone, grinding stone.

muelle, m. wharf, dock, pier (malecón); embankment (de un río); freight platform (de ferrocarril); spring (de un mecanismo).

muerte, f. death; **m. negra,** the Black Death, plague, bubonic plague.

muertola, 1. m. y f. dead person; corpse (cuerpo); **muertos,** victims, casualties of an accident; 2. a. dead; **cal (m) m.,** slaked lime; **estar m. (v),** to be dead; **punto (m) m.,** dead centre.

muesca, f. notch, nick, mortice.

muestra, f. sample, specimen; **m. aleatoria,,** random sample; **m. de agua,** water sample; **m. de invertebradas,** invertebrate sample; **m. de mano,** hand specimen; **m. de río,** river sample; **m. inalterada,** unaltered sample; **m. representativa,** representative sample; **m. tomada al azar,** random sample; **m. tomada por cubeta,** grab sample; **m. tomada por draga,** grab sample (by dredger).

muestreadorla, m. y f. sampler (máquina); **m. automático,** automatic sampler.

muestrear, v. to sample.

muestreo, m. sampling; sampler, driller; **punto (m) de m.,** sampling point.

mugre, m. filth, dirt, grime.

mujer, f. woman, female.

muladar, m. dump, rubbish dump; midden (Arq); dungheap (estercolero).

mulching, m. mulching.

mullidar, m. area with alkaline peats.

multa, f. fine, penalty.

multi-, pf. multi-.

multicelular, a. multicellular.

multidisciplinariola, a. multidisciplinary.

multiespectral, a. multispectral.

multifase, f. multiphase.

multimedia, a. multimedia.

multinacional, a. multinational.

multiparamétricola, a. multiparametric.

múltiple, a. multiple; **regresión (f) m.,** multiple regression.

multiplicación, f. multiplication.

múltiplo, m. multiple; **mínimo común m.,** lowest common multiple.

múltiplola, a. multiple.

multivariante, a. multivariate.

mundial, a. world, global; universal; **a escala (f) m.,** on a world scale.

mundo, m. world; **en el m. científico,** in the scientific world; **tercer m.,** third world.

municipal, a. municipal; town, local (concejo); public (p.e. biblioteca).

municipalidad, f. municipality, township.

municipio, m. township, municipality.

muralla, f. city wall, town wall, rampart; **m. almenada,** battlement.

murciélago, m. bat (mamífero).

muro, m. wall; **anclaje (m) para m.,** wall anchor; **m. cortina,** curtain wall; **m. de contención,** retaining wall, load-bearing wall; **m. de defensa,** groyne; **m. de pie,** toe wall.

musaraña, f. shrew.

músculo, m. muscle; **m. liso,** smooth muscle.

musgo, m. moss.

musgosola, a. mossy.

musulmán, musulmana, m. y f. Moslem.

mutación, f. mutation; **m. génica,** gene mutation, genetic mutation.

mutagenicidad, f. mutagenicity.

mutagénicola, a. mutagenic; **agente (m) m.,** mutagenic agent, mutagen.

mutágeno, m. mutagen.

mutante, a. mutant.

mutualismo, m. mutualism.

N

nabo, m. turnip.

nácar, m. mother-of-pearl.

nacimiento, m. birth, hatching (Biol); spring, source of river (Geog); origin, beginning (origen); **n. de un río,** source of river.

nacionalización, f. nationalization.

nafta, f. petrol, gas (US).

naftalina, f. naphthalene.

Namuriense, m. Namurian.

nano, pf. nano- (10.E-9).

nanoplancton, m. nanoplankton.

naranja, f. orange; **n. de metilo (Quím),** methyl orange.

naranjo, m. orange tree.

narcóticola, a. narcotic.

nariz, f. nose; **n. del anticlinal,** anticlinal nose.

natación, f. swimming, bathing; **agua (f) de n. (EurU),** bathing water; **pileta (f) de n. (LAm),** swimming pool.

natalidad, f. birth; **índice (m) de n., tasa (f) de n.,** birth rate.

natatorio, m. (LAm) swimming pool.

nativola, 1. m. y f. native (habitante); 2. a. native, indigenous; **oro (m) n.,** native gold; **suelo (m) n.,** homeland, native soil.

natural, a. natural; **parque (m) n.,** nature park, nature reserve; **recursos (m.pl) naturales,** natural resources; **selección (f) n.,** natural selection.

naturaleza, f. nature; **la madre (f) n.,** Mother Nature; **leyes (f.pl) de la n.,** laws of nature.

náuticola, a. nautical, marine; **carta (f) n.,** nautical chart.

nautilocono, m. nautilocone.

nautiloide, m. nautiloid.

nava, f. plain between mountains.

nave, f. ship, vessel, craft (buque); nave (de una inglesa); shop; old dunes with xerophytic vegetation; **n. espacial,** spacecraft, spaceship.

navegable, a. navigable.

navegación, f. navigation; **n. a vela,** yachting, sailing; **n. costera, n. de cabotaje,** coastal navigation; **n. de recreo,** yachting, sailing; **n. fluvial,** river navigation.

naviera, f. shipping.

navierola, a. shipping.

navío, m. vessel (barco).

neárticola, a. nearctic.

neblina, f. mist, fog; **captación (f) de n.,** mist catchment; **captador (m) de n., colector de n. (LAm),** mist collector.

nebulosola, a. foggy, misty.

necesidad, f. necessity, need.

necesitar, v. to need.

necrófago, m. scavenger.

necromasa, f. dead organic matter.

necrópolis, m. necropolis.

necrosis, f. necrosis (pl. necroses).

néctar, m. nectar.

necton, m. nekton.

nefelina, f. nepheline.

nefelinita, f. nephelite.

nefrita, f. nephrite.

negativola, a. negative; **carga (f) n.,** negative charge; **electrodo (m) n.,** negative electrode; **ión (m) n.,** negative ion; **polo (m) n.,** negative pole; **taxia (f) n.,** negative taxis; **tropismo (m) n.,** negative tropism.

negligencia, f. negligence, neglect, carelessness; **gran n.,** gross negligence.

negligente, a. negligent, careless.

negociante, m. y f. merchant, dealer; **n. al por mayor,** wholesale merchant.

negocio, m. business (comercio); deal (transacción); **hombre (m) de negocios,** businessman (pl. businessmen); **montar un n. (v),** to set up a business; **mujer (f) de negocios,** businesswoman (pl. businesswomen); **n. redondo,** profitable business, profitable deal; **oportunidad (f) de n.,** business opportunity; **poner un n. (v),** to set up a business.

negruzcola, a. blackish.

neis, m. gneiss.

nematoblásticola, a. nematoblastic.

nemátodo, m. nematode; **nemátodos (Zoo, filum),** Nematoda (pl).

neodimio, m. neodynium (Nd).

Neógeno, m. Neogene.

neolítico, m. Neolithic.

neón, m. neon (Ne).

neopreno, m. neoprene.

neptunio, m. neptunium (Np).

neptunismo, m. neptunism.

neptunita, f. neptunite.

neríticola, a. neritic.

nervadura, f. venation.

nervio, m. nerve (Anat); rib, vein (p.e. de hojas); **n. óptico,** optic nerve.

nerviosola, a. nervous; **célula (f) n.,** nerve-cell; **centro (m) n.,** nerve centre; **impulso (m) n.,** nerve-impulse; **sistema (m) n. central (abr. SNC),** central nervous system (abr. CNS); **terminación (f) n.,** nerve-ending.

netola, a. net; **cantidad (f) n.,** net amount; **peso (m) n.,** net weight.

neumático, m. tyre, pneumatic tyre.

neumáticola, a. pneumatic.

neumatolisis, m. pneumatolysis.

neumoconiosis, m. pneumoconiosis.

neural, a. neural.

neurología, f. neurology.

neurona, f. neurone; **n. motora,** motor neurone.

neuróptero, m. neuropteran (pl. neuroptera, neuropterans); **neurópteros (Zoo, orden),** Neuroptera (pl).

neurotoxicidad, f. neurotoxicity.

neurotóxicola, a. neurotoxic; **gas (m) n.,** nerve gas.

neutral, a. neutral.

neutralización, f. neutralization.

neutralizar, v. to neutralize.

neutrola, a. neuter; neutral (Quím, Elec).

neutrón, m. neutron.

neutrón-gamma, m. neutron-gamma; **sonda (f) de n.-g.,** neutron-gamma sonde.

neutrón-neutrón, m. neutron-neutron; **sonda (f) de n.-n.,** neutron-neutron sonde.

nevada, f. snowfall.

nevado, m. (LAm) snow-capped mountain.

nevadola, a. snow-covered, snow-capped, snowy.

nevar, v. to snow.

nevasca, f. snowstorm.

nevazón, f. (LAm) snowstorm.

nevero, m. snowfield, perennial snowcap.

nevisca, f. light snowfall.

neviscar, v. to snow lightly.

newton, m. newton (unidad de fuerza).

nexo, m. link, bond.

Nicaragua, f. Nicaragua.

nicaragüense, 1. m. y f. Nicaraguan; 2. a. Nicaraguan.

nicotina, f. nicotine.

nicho, m. niche; **n. ecológico,** ecological niche.

nidada, f. brood (de aves), clutch (de gallinas).

nido, m. nest; ore pocket (LAm).

niebla, f. fog, mist (Met); mildew (Bot); **n. helada,** freezing fog; **hay n. (v),** it is foggy.

nieve, f. snow; **copo (m) de n.,** snowflake; **límite (m) de las nieves perpetuas,** snow line; **medidor (m) de nieves,** snow gauge; **n. acumulada, n. a la deriva,** snowdrift.

nilón, m. nylon.

nimboestrato, m. nimbostratus.

ninfa, f. nymph.

niobio, m. niobium (Nb).

níquel, m. nickel (Ni).

níquel-hierro, m. nickel-iron.

nitidez, f. clarity, clearness (de líquidos).

nitrato, m. nitrate; **n. de Chile,** Chilean nitrate; **n. potásico,** saltpeter.

nítricola, a. nitric.

nitrificación, f. nitrification.

nitrificante, a. nitrifying; **bacterias (f.pl) nitrificantes,** nitrifying bacteria.

nitrificar, v. to nitrify.

nitrilo, m. nitryl.

nitrito, m. nitrite.

nitroanilina, f. nitroaniline.

nitrobenzol, m. nitrobenzene.

nitrocelulosa, f. nitro-cellulose, nitrocellulose.

nitrogenación, f. nitrogenation.

nitrogenadola, a. nitrogenated; **abono (m) n.,** nitrogen fertilizer.

nitrógeno, m. nitrogen (N); **ciclo (m) de n.,** nitrogen cycle; **eliminación (f) de n.,** nitrogen removal; **fijación (f) de n.,** nitrogen fixation; **n. albuminoideo,** albuminoid nitrogen.

nitrogenosola, a. nitrogenous.

nitroglicerina, f. nitroglycerine.

nitrosilo, m. nitrosyl.

nitroso, m. nitrous.

nitruro, m. nitride.

nivel, m. level (superficie); standard (categoría); **n. cero,** datum, ordnance datum (GB); **n. de agua freática,** water table; **n. de agua subterránea,** groundwater level; **n. de base,** base level; **n. de burbuja de aire,** spirit level; **n. de datum,** datum, ordnance datum (GB); **n. de inundación,** flood level; **n. del agua,** water level, water mark; **n. de la marea,** tide level; **n. del mar,** sea level; **n. del suelo,** ground level; **n. fosilífero,** fossil horizon; **n. freático,** water table; **n. hidrostático,** hydrostatic level; **n. medio del mar,** mean sea level; **n. piezométrico,** piezometric level.

nivelación, f. flattening, levelling, leveling (US).

nivelar, v. to level.

nocividad, f. noxiousness.

nocivola, a. noxious; **actividades (f.pl) molestas, insalubres, nocivas y peligrosas,** dangerous and nuisance activities and activities harmful to health.

no confinado, a. unconfined; **acuífero (m) no c.,** unconfined aquifer.

nodo, m. node.

nodular, a. nodular.

nódulo, m. nodule; **n. radical (Bot),** root nodule.

nogal, m. walnut tree.

nombre, m. name; noun (gramática); **n. corriente,** trivial name; **n. de pila,** first name, Christian name; **n. específico (Biol),** specific name; **n. genérico (Biol),** generic name; **n. propio,** proper name, proper noun; **n. sistemático,** systematic name; **n. y apellidos,** full name.

nomenclatura, f. nomenclature.

nominal, a. nominal.

nomograma, m. nomogram.

noria, f. (Arab) waterwheel over a well turned by horizontal spar; **n. a tracción animal, n. de sangre,** waterwheel driven by animal pulling a spar.

Noriense, m. Norian.

norita, f. norite.

norma, f. rule, norm (regla); **n. británica,** British Standard (GB, Reg); **n. de calidad ambiental,** environmental quality standard; **n. de calidad del agua,** water quality standard.

normal, a. normal; perpendicular (Mat); **escuela (f) n.,** teacher training college, normal school (US).

normalidad, f. normality (Quím).

normalización, f. normalization; **Instituto (m)**

Británico de Normalización, British Standards Institute (abr. BSI).

normativa, f. rules (pl), regulations (pl); **n. vigente,** existing legislation, legislation currently in force; **Normativas Urbanísticas (Esp),** town-planning regulations.

norte, 1. m. north; **América (f) del N.,** North America; **en dirección (f) n.,** in a northerly direction; **n. de la malla,** grid north; **n. magnético,** magnetic north; **n. verdadero, rumbo n. (LAm),** true north; 2. a. north, northerly.

norteamericanola, a. North American.

norteñola, a. northern.

notación, f. notation.

notariola, m. y f. notary, solicitor; commissioner for oaths; notary public (US); **n. público,** public notary.

notificador, m. notifier, advisor.

notificar, v. to notify.

novísimola, a. latest (very new).

NOX, m. NOX (abr. para los óxidos de nitrógeno).

NPK, m. NPK (abr. para nitrógeno, fósforo y potasio); **abono (m) NPK,** NPK fertilizer.

nubarrón, m. rain cloud.

nube, f. cloud; **n. ardiente,** nueé ardent.

nubladola, a. cloudy, overcast.

nublar, v. to cloud over.

nubosidad, f. cloudiness, nubosity.

nubosola, a. cloudy.

nucleación, f. nucleation.

nuclear, a. nuclear; **energía (f) n.,** nuclear energy; **fisión (f) n.,** nuclear fission; **fusión (f)** n., nuclear fusion; **residuos (m.pl) nucleares,** nuclear waste.

núcleo, m. nucleus; core (Elec); **n. atómico,** atomic nucleus; **n. de población,** population centre, centre of population; **n. laminado,** laminated core; **n. residencial,** residential area, housing estate.

nucleón, m. nucleon.

nucleótido, m. nucleotide.

nuclídico, m. nuclide.

nudo, m. knot (velocidad, lazo), node; **n. de carreteras,** road intersection; **n. de la malla (Mat),** mesh node; **n. llano,** slipknot.

nudosola, a. knotted, knarled.

Nueva Zelanda, f. New Zealand.

Nueva Zelandia, f. (LAm) New Zealand.

nuez, f. nut (fruta); walnut; **n. moscada,** nutmeg.

numeración, f. numeration.

numerador, m. numerator.

numéricola, a. numerical; **modelo (m) n.,** numerical model.

número, m. number; **n. atómico,** atomic number; **n. cardinal,** cardinal number; **n. complejo,** complex number; **n. de referencia,** reference number; **n. ordinal,** ordinal number; **n. primo,** prime number; **n. redondo,** round number.

nunatak, m. nunatak.

nutrición, f. nutrition.

nutriente, m. nutrient; **nutrientes disponibles,** available nutrients.

nutrimiento, m. nutrient.

nutrir, v. to nourish.

nutritivola, a. nourishing, nutritious.

Ñ

ñacurutú, m. night bird; owl (LAm) (lechuza).

ñadi, f. (Ch) waterlogged woodland.

ñame, m. yam plant.

ñocle, m. quicksand.

ñucuruto, m. (LAm) owl.

O

OAA, f. FAO; **Organización (f) de las Naciones Unidas para la Alimentación y la Agricultura**, Food and Agriculture Organization.

oasis, m. oasis.

obesidad, f. obesity.

objetividad, f. objectivity.

objetivo, m. objective; **o. primordial, o. principal**, primary objective.

objetivola, a. objective; **especie (m) no o.** (Agr), non-target species; **población (f) o.** (Agr), target population.

oblatola, a. oblate; **esferoide (m) o.**, oblate spheroid.

oblicuola, a. oblique, slanting.

obligación, f. obligation; **exento (m) de toda o.**, free of obligation.

obligadola, a. obligate; **parásito (m) o.**, obligate parasite.

obligar, v. to oblige, to force, to compel.

obligatoriola, a. obligatory, compulsory.

obra, f. labour, work; **al pie (m) de la o.**, delivered on site; **mano (f) de o.**, work force, labour force; **mano (f) de o. especializada**, skilled labour; **o. de construcción**, construction site; **o. de regadío**, irrigation work; **o. hidráulica**, hydraulic work; **obras civiles**, civil works; **obras de ingeniería**, engineering works; **obras públicas**, public works.

obrerola, 1. m. y f. worker, labourer; **o. cualificado**, skilled worker; **o. de aseo urbano (LAm)**, waste tip worker, garbage man; **o. de la baja policía (Perú)**, waste tip scavenger; **o. especializado**, skilled worker; 2. a. working.

obscenola, a. obscene.

observadola, a. observed.

observar, v. to observe, to watch.

obsidiana, f. obsidian.

obsolescencia, f. obsolescence.

obsolescente, a. obsolescent, out of date.

obstaculizar, v. to obstruct.

obstáculo, m. obstacle, hindrance.

obtención, f. obtaining, collection; **o. de datos**, data collection.

obtener, v. to obtain, to get; to acquire (adquirir).

obturador, m. seal, packer, stopper (en sondeos).

obtusola, a. obtuse.

occidental, a. western; **el mundo (m) o.**, the western world.

occidente, m. the West, Occident.

OCDE, f. OECD; **Organización (f) para la Cooperación y el Desarrollo Económico**, Organization for Economic Cooperation and Development.

oceánicola, a. oceanic; **placa (f) o.** (Geol), oceanic plate.

océano, m. ocean; **O. Atlántico/Índico/Pacífico**, Atlantic/Indian/Pacific Ocean.

oceanografía, f. oceanography.

oceanógrafola, m. y f. oceanographer.

oceanología, f. oceanology.

ocio, m. leisure, idleness; **ocios**, spare time (tiempo libre); pastime (hobby); **ratos (m.pl) de o.**, leisure time.

ocle, f. kelp.

ocluidola, a. occluded; **frente (m) o.** (Met), occluded front.

oclusión, f. occlusion.

ocre, m. ochre, ocher (US).

octaedrita, f. octahedrite, anatase.

octaedro, m. octahedron.

octanaje, m. octane rating.

octanal, m. octanal.

octano, m. octane.

octanol, m. octanol.

octeno, m. octene.

octino, m. octyne.

octocorales, m.pl. octocorals.

ocupación, f. occupation; occupancy (of a dwelling).

ODECA, f. **Organización (f) de los Estados Centroamericanos**, Organization of Central American States.

OEA, f. OAS; **Organización (f) de Estados Americanos**, Organization of American States.

OECE, f. OEEC; **Organización (f) Europea para la Cooperación Económica**, Organization for European Economic Cooperation.

oedómetro, m. oedometer.

oferta, f. offer, quotation; **hacer una o. (v)**, to make an offer.

oficialismo, m. (LAm) government authorities.

oficina, f. office; **edificio (m) de oficinas**, office block; **horas (f.pl) de o.**, office hours, business hours; **o. al pie de la obra**, site office; **o. central**, head office, central office; **o. principal**, head office.

oficio, m. official letter, communiqué (comunicado); profession, trade (profesión); job, post (sitio).

ofiolita, f. ophiolite.

ofíticola, a. ophitic.

ohm, m. ohm (unidad de corriente eléctrica); **o.-metro**, ohm-metre.

ohmio, m. ohm (unidad de corriente eléctrica); **o.-metro**, ohm-metre.

oído, m. ear; **o. externo**, outer ear; **o. interno**, inner ear; **o. medio**, middle ear.

ojo, m. eye; **o. (de agua subterranea)**, upwelling

(of groundwater); **o. compuesto,** compound eye; **o. de tigre (Geol),** tiger's eye; **o. simple,** simple eye.

ola, f. wave; **alcance (m) de o.,** wave fetch; **o. de calor,** heat wave; **o. de marea,** tidal wave; **o. larga,** rollers; **o. sísmica,** tsunami.

oleada, f. big wave, ocean swell (Mar); surge of people (muchedumbre).

oleaje, m. sea swell, surge (Mar); **o. oceánico,** ocean swell.

olefina, f. olefine.

oleoducto, m. oil pipeline.

oler, v. to smell.

olfacción, f. olfaction.

olfatear, v. to smell.

olfato, m. sense of smell.

Oligoceno, m. Oligocene.

oligoclasa, f. oligoclase.

oligoelemento, m. accessory element (Geol); trace element (Quím, Biol).

oligoqueto, m. oligochaete; **oligoquetos (Zoo, clase),** Oligochaetae (pl).

oligotróficola, a. oligotrophic.

oliva, f. olive.

olivar, m. olive-grove.

olivina, f. olivine.

olivinita, f. olivinite.

olivo, m. olive tree.

olla, f. pot; **o. porosa,** porous pot.

ollar, m. nostril.

olmedola, m. y f. elm grove.

olmo, m. elm.

olor, m. odour, smell; **eliminación (f) de o.,** elimination of odour.

ombligo, m. navel; centre (Fig).

OMI, f. IMO; **Organización (f) Marítima Internacional,** International Maritime Organization.

OMM, f. WMO; **Organización (f) Meteorológica Mundial,** World Meteorological Organization.

ómnibus, m. bus.

omnívoro, m. omnivore.

omnívorola, a. omnivorous.

OMS, f. WHO; **Organización (f) Mundial de la Salud,** World Health Organization.

onda, f. wave; **longitud (f) de o.,** wavelength; **o. acústica,** sound wave; **o. corta,** short wave; **o. de refracción,** refractive wave; **o. electromagnética,** electromagnetic wave; **o. estacionaria,** standing wave; **o. explosiva,** shock wave, blast; **o. larga,** long wave; **o. longitudinal,** longitudinal wave; **o. P, o. S (Geol),** P wave, S wave; **o. sísmica,** seismic wave; **o. transversal,** transverse wave; **ondas de alta frecuencia,** high-frequency waves; **ondas de baja frecuencia,** low-frequency waves.

ondulación, f. undulation; wrinkling (arrugas), waviness (rizos); **ondulaciones de arena,** sand ripples.

onduladola, a. wavy, undulating (tierra);

uneven (en carreteras), corrugated; **cartón (m) o.,** corrugated cardboard; **chapa (f) o.,** **placa (f) o.,** corrugated iron.

ondulatoriola, a. undulatory; **mecánica (f) o.,** wave mechanics; **movimiento (m) o.,** wave motion.

ondulita, f. ripple mark.

ONG, f. NGO (US), quango (GB); **organización (f) no gubernamental,** non-governmental organization (US).

ónice, m. onyx.

onitar, m. evaporite soil layer rich in nitrates.

onix, m. onyx.

on-line, a. on-line (Comp).

Ontariense, m. Ontarian.

ontrón, m. pond covered with abundant vegetation.

ONU, f. UN; **Naciones (f.pl) Unidas, Organización (f) de las Naciones Unidas,** United Nations Organization.

onza, f. ounce (unidad de peso, 28,35 gramos).

oolíticola, a. oolitic.

oolito, m. oolite, oolith.

opacidad, f. opacity.

opacímetro, m. opacimeter.

opacola, a. opaque.

ópalo, m. opal.

opción, f. option, choice; **o. económica,** economic choice.

operando, m. operand.

operariola, m. y f. operative, worker; **o. de limpieza (Nic),** waste tip o., garbage man.

opérculo, m. operculum (pl. opercula).

opinión, f. opinion; **o. pública,** public opinion; **ser de la o. que (v),** to be of the opinion that; **sondeo (m) de la o. pública,** public opinion poll.

opio, m. opium.

oportunidad, f. opportunity.

oportunista, a. opportunist.

oposición, f. opposition; **o. fuerte,** strong opposition.

optativola, a. optional.

ópticola, a. optical; **disco (m) ó. (Comp),** optical disc.

optimización, f. optimization.

optimizar, v. to optimize.

óptimola, a. optimum, optimal.

oquerosola, a. cavernous, hollowed; **caliza (f) o.,** cavernous limestone.

oral, a. oral.

órbita, f. orbit; eye socket (Anat).

orbital, a. orbital.

orca, f. killer whale.

orden, m. y f. order, command (mando); writ, mandate (Jur); **o. alfabético,** alphabetical order; **o. del día,** agenda; **o. de magnitud,** order of magnitude; **o. de reacción,** order of reaction; **o. fluvial,** stream order; **o. judicial,** court order; **por o. cronológico,** chronological order.

ordenación, f. order, arrangement; **o. regional,**

regional planning; **o. rural,** rural planning; **Planes (m.pl) Generales de O. (Esp),** General Land-Use Plans.

ordenada, f. ordinate (Mat).

ordenador, m. computer; **o. personal (abr. PC),** personal computer (abr. PC); **o. potente,** powerful computer.

ordenamiento, m. planning, putting tidy; **o. ambiental,** environmental planning; **o. de los recursos hídricos,** water resources planning.

ordenanza, f. ordinance (Jur).

ordinal, a. ordinal.

Ordovícico, m. Ordovician.

oreja, f. ear.

orgánicola, a. organic; **crecimiento (m) o.,** organic growth; **materia (f) o.,** organic material.

organigrama, m. chart, diagram; **o. funcional,** flow chart.

organismo, m. organism (Bot, Zoo), organization, body (institución); **o. de control,** regulatory organization; **o. vivo,** living organism.

organización, f. organization; **o. no gubernamental (abr. ONG)** non-government organization (US, abr. NGO).

organizar, v. to organize.

órgano, m. organ; **ó. de sentido,** sense organ; **ó. reproductivo,** reproductive organ.

organocloradola, a. organochlorine; **pesticidas (m.pl) organoclorados,** organochlorine pesticides.

organolépticola, a. organoleptic.

orientación, f. orientation, direction.

oriental, a. eastern, oriental.

orientar, v. to orient, to orientate; to direct (dirigir).

oriente, 1. m. the East; **Extremo O., Lejano O.,** Far East; **O. Medio,** Middle East; 2. a. east.

orificio, m. orifice, outlet; **vertedero (m) de o. (para medida de caudal),** orifice plate.

origen, m. origin.

orilla, f. edge, border (Gen); bank, riverside (de un río); shore (litoral).

orín, m. rust; **orines,** urine.

orina, f. urine.

orinar, v. to urinate.

orla, f. border, fringe.

ornamentación, f. ornamentation, decoration.

ornitología, f. ornithology.

ornitológicola, a. ornithological.

ornitólogola, m. y f. ornithologist.

oro, m. gold (Au); **campo (m) de o.,** gold field; **o. de aluvión,** alluvial gold; **o. de los tontos,** fool's gold.

orogénesis, m. orogenesis, orogeny.

orogenia, f. orogeny.

orogénicola, a. orogenic; **cinturón (m) o.,** orogenic belt.

orografía, f. orography.

orográficola, a. orographic; **lluvia (f) o.,** orographic rain.

orquídea, f. orchid.

orto-, pf. ortho-.

ortoclasa, f. orthoclase.

ortocónicola, a. orthoconical.

ortofosfato, m. orthophosphate.

ortogonal, a. orthogonal.

ortopiroxenita, f. orthopyroxenite.

ortóptero, m. orthopteran; **ortópteros (Zoo, orden),** Orthoptera (pl).

ortosa, f. orthoclase.

oruga, f. caterpillar (Zoo); caterpillar tread (máquina); **tractor (m) o.,** caterpillar tractor.

orza, f. glazed earthenware jar.

osario, m. ossuary, ossuaria (para huesos).

oscilación, f. oscillation, fluctuation; **o. climática,** climatic oscillation; **o. de período largo,** long-period oscillation; **o. sinusoidal,** sinusoidal oscillation.

oscilante, a. reciprocating, oscillating.

oscilar, v. to oscillate, to fluctuate.

óseola, a. bone, bony; **médula (f) ó.,** bone marrow.

osículo, m. ossicle.

osificación, f. ossification.

osificar, v. to ossify.

osmio, m. osmium (Os).

osmómetro, m. osmometer.

osmoregulación, f. osmoregulation.

osmosis, ósmosis, f. osmosis; **o. inversa,** reverse osmosis.

osmóticola, a. osmotic; **potencial (m) o.,** osmotic potencial; **presión (f) o.,** osmotic pressure.

osmotróficola, a. osmotrophic.

oso, m. bear; **o. pardo,** Brown Bear.

ostracodo, m. ostracod.

OTAN, f. NATO; **Organización (f) del Tratado de Atlántico Norte,** North Atlantic Treaty Organization.

otoño, m. autumn.

output, m. output (Comp).

ovario, m. ovary.

oveja, f. sheep (pl. sheep); **cría (f) de ovejas,** sheep farming.

ovejería, f. sheep farming; sheep (Ch).

ovinola, a. of sheep.

ovíparola, a. oviparous.

ovipositor, m. ovipositor.

óvulo, m. ovule.

oxalato, m. oxalate.

oxálicola, a. oxalic.

Oxfordiense, m. Oxfordian.

oxhidrilola, a. hydroxyl.

oxi-, pf. oxy-.

oxiacetilénicola, a. oxyacetylene; **soldadura (f) o.,** oxyacetylene welding.

óxicola, a. oxic.

oxidadola, a. oxidized; rusty (mohoso).

oxidante, 1. m. oxidant; 2. a. oxidizing; **medio (m) o.,** oxidizing environment.

oxidar, v. to oxidize.

óxido, m. oxide; **o. de aluminio,** alumina; **o. de calcio,** lime.

oxidorreducción, f. oxidation-reduction (abbr. redox); **potencial (m) de o. (abr. potencial redox)**, oxidation-reduction potential (abr. redox potential).

oxigenación, f. oxygenation; **capacidad (f) de o.**, oxygenation capacity.

oxigenadola, a. oxygenated.

oxigenar, v. to oxygenate.

oxígeno, m. oxygen (O); **hundimiento (f) de o.**, oxygen sag.

oxihidróxido, m. oxyhydroxide.

oxi-reducción, f. redox, oxidation-reduction; **condiciones (f.pl) de oxi-r.**, redox conditions.

ozonar, v. to ozonize.

ozónido, m. ozonide.

ozonización, f. ozonization.

ozonizador, m. ozonizer.

ozonizar, v. to ozonize.

ozono, m. ozone; **agujero (m) de o.**, ozone hole; **capa (f) de o.**, ozone layer.

ozonosfera, f. ozonosphere.

P

PAC, f. CAP; **Política (f) Agraria Común (EurU)**, Common Agricultural Policy.
pacer, v. to graze, to pasture.
pacíficola, a. Pacific; **el Océano (m) P.**, the Pacific Ocean.
pacil, m. saltings in Donaña delta (Esp) subject to flooding.
padre, m. father (m); parent (m. o f.).
padúl, m. boggy ground.
pagar, v. to pay; to pay for (costear).
pago, m. payment (Com); region, district (CSur).
pagoda, f. pagoda.
PAHO, f. PAHO; **Organización (f) Panamericana de la Salud**, Pan American Health Organization.
país, m. country (nación); land, region (tierra); **p. en vías de desarrollo**, developing country; **P. Vasco**, Basque Country.
paisaje, m. countryside, landscape, scenery; **rasgo (m) residual de p.**, relic landscape feature.
paisajismo, m. landscaping.
paisajísticola, a. landscape; **valor (m) p.**, landscape value.
paisanola, m. y f. peasant.
Países Bajos, m.pl. Holland, The Netherlands.
paja, f. straw, chaff, thatch; **cubrir con p. (v)**, to thatch; **techo (m) de p.**, thatched roof.
pájaro, m. small bird; **p. carpintero**, woodpecker.
pajote, m. mulch (Agr).
pala, f. shovel, shovelful (contenido de la pala), paddle (de noria), spade (de jardinero); **p. cargadora, p. mecánica**, mechanical digger, excavator.
paladio, m. palladium (Pd).
palanca, f. lever, crowbar.
palencia, f. wetland.
paleo-, pf. palaeo-, paleo- (esp. US).
paleobotanía, f. palaeobotany, paleobotany (esp. US).
paleobotánica, f. palaeobotany, paleobotany (esp. US).
paleobotánicola, m. y f. palaeobotanist.
Paleoceno, m. Paleocene.
paleoclima, f. paleoclimate.
paleoclimatología, f. paleoclimatology.
Paleógeno, m. Paleogene.
paleogeografía, f. paleogeography.
paleografía, f. paleography.
paleógrafola, m. y f. paleographer.
paleolimnología, f. paleolimnology.
Paleolítico, m. Palaeolithic; **P. inferior**, Lower Palaeolithic; **P. medio**, Middle Palaeolithic; **P. superior**, Upper Palaeolithic.

paleontología, f. palaeontology, paleontology.
paleontológicola, m. y f. paleontologist.
Paleozoico, m. Palaeozoic, Paleozoic.
paleta, f. shovel (pala); vane (de una hélice); blade (de una turbina); bucket (de noria); shoulder blade (Anat).
paletada, f. shovelful.
paliar, v. to paliate, to alleviate (aliviar).
palillo, m. small stick; rod and line (para la pesca).
palinogénesis, m. palinogenesis.
palinología, f. pollen analysis, palynology.
palmera, f. palm, date palm.
palmeral, m. palm grove.
palmípedola, a. web-footed, palmiped.
palo, m. pole, stick.
palúdicola, a. malarial.
paludismo, m. malaria.
pampa, f. prairie; pampas (LAm).
Panamá, m. Panama.
panameñola, 1. m. y f. Panamanian; 2. a. Panamanian.
panamericanola, a. pan-american.
páncreas, m. pancreas.
pandeamiento, m. buckling, warping.
pandemia, f. pandemic.
pandémicola, a. pandemic.
panel, m. panel; **p. solar**, solar panel.
panorama, m. panorama, view, scene.
panorámicola, a. panoramic.
pantalla, f. shade, screen; **p. impermeable (Geog)**, impermeable cover.
pantano, m. marsh, bog, wetland (humedal); small artificial reservoir (tanque); **gas (m) de los pantanos**, marsh gas.
pantanosola, a. boggy, marshy.
paño, m. cloth, material.
papa, f. (LAm) potato.
papel, m. paper; role (función); **p. cartón**, cardboard; **p. cuadriculado**, graph paper; **p. de filtro**, filter paper; **p. de lija**, abrasive paper; **p. de periódico**, newsprint; **p. doble logarítmico**, double-logarithmic paper; **p. madera (LAm)**, cardboard; **p. probalístico-logarítmico**, logarithmic-probability paper; **p. reciclado**, recycled paper; **p. semi-logarítmico**, semi-logarithmic paper; **p. tornasol**, litmus paper; **p. usado, papeles viejos**, waste paper.
papeleo, f. red tape (Fig); **p. administrativo**, paper work.
papera, f. goitre, goiter (US) (bocio); **paperas**, mumps (enfermedad).
papila, f. papilla (pl. papillae); **p. gustativa**, taste-bud.
papiro, m. papyrus.

paquete, m. package; computer package (Comp, lote de programa).
par, m. pair, couple (Gen); **p. de bases,** base pair; **p. estereoscópico,** stereo pair; **p. térmico,** thermocouple.
para-, pf. para-.
parábola, f. parabola (Mat).
parabólicola, a. parabolic.
paradola, a. unemployed.
paradoja, f. paradox.
parafina, f. paraffin (Quím).
paragénesis, m. paragenesis.
Paraguay, m. Paraguay.
paraguayola, 1. m. y f. Paraguayan; 2. a. Paraguayan.
paralaje, f. parallax.
paraldehído, m. paraldehyde.
paralelo, m. parallel (p.e. latitud).
paralelola, a. parallel; **proceso (m) en p.,** parellel processing.
paralelogramo, m. parallelogram.
paramétricola, a. parametric; **técnicas (f.pl) no paramétricas (Mat),** non-parametric techniques.
parámetro, m. parameter.
páramo, m. moor, heath, wildland.
pararrayos, m.pl. lightening conductor, lightening rod (US).
parasitariola, a. parasitic, parasitical.
parasitismo, m. parasitism.
parásito, m. parasite; **p. facultativo,** facultative parasite; **p. obligado,** obligate parasite.
parásitola,, a. parasitic, parasitical.
parasitología, f. parasitology.
parata, f. (southern Spain) stone wall across mountain stream to catch soil wash.
paratión, m. parathion.
parcela, f. plot (terreno), smallholding; particle (átomo).
parcial, a. partial.
parcialmente, ad. partly, partially.
pared, f. wall; **p. basal (Geog),** footwall; **p. celular,** cell wall; **p. colgante (Geog),** hanging wall; **p. de valle,** valley wall.
pareja, f. pair (par); couple (hombre y mujer); **parejas reproductoras,** breeding pairs.
parental, a. parental; **cuidado (m) p. (Zoo),** parental care.
paridad, f. parity.
parking, m. car park, car lot (US); lay-by (en la carretera);.
parlamento, m. parliament (asamblea); negotiation (negociación); **P. Europeo,** European Parliament.
paro, m. stoppage, shutdown (de trabajo); strike (huelga); unemployment (desempleo); **índice (m) de p.,** level of unemployment; **p. biológico,** "biological stoppage" (term for fishing moratorium called to permit recovery of fish stocks); **p. del sistema,** system shutdown; **p. encubierto,** underemployment; **p. estacional,** seasonal unemployment.

párpado, m. eyelid.
parque, m. park, public gardens; **p. de atracciones,** amusement park, fun fair; **p. nacional,** national park; **p. zoológico,** zoological park, zoo.
parqueadero, m. (Col, Pan) car park.
parra, f. vine, grapevine, climbing vine.
parral, m. vine; vine arbor (US).
parrilla, f. grating, grill.
parte, 1. m. report (informe); **p. meteorológico,** weather report; 2. f. part (Gen); portion (sección); share (participación); **p. destacable,** highlights; **p. sensible,** sensitive spot.
partenogénesis, f. parthenogenesis.
partición, f. partition; **coeficiente (m) de p.,** partition coefficient.
particionamiento, m. partitioning.
participación, f. participation; **p. en beneficios,** profit-sharing.
partícula, f. particle; **p. alfa/beta/gamma,** alpha/beta/gamma particle; **p. suspendida,** suspended particle; **p. viable en el aire,** airborne particle.
particuladola, a. particulate; **material (m) p.,** particulate material.
particular, a. particular, special.
particularidad, f. particularity.
partidola, a. divided, split.
partidor, m. distributor, distribution network.
parto, m. childbirth, delivery, parturition.
pasadizo, m. catwalk.
pasaje, m. passage, passing (acción); voyage, crossing (Mar); fare (tarifa).
pasajerola, a. fleeting, transient, intermittent.
pasarela, f. footbridge, gangway; **p. de servicio,** catwalk; **p. de vía urbana,** highway footbridge.
paseo, m. stroll, walk, promenade; **p. marítimo,** seaside promenade.
paso, m. step, pace; walk (paseo); passage (derecho), clearing (de un obstáculo), strait, straits (pl) (Geog, estrecho); **P. de Calais,** Straits of Dover (GB); **p. elevado,** flyover; **p. inferior,** underpass; **p. protegido,** right-of-way; **p. subterráneo,** subway; **p. superior,** flyover, over-crossing; **prohibido el p.,** no entry.
pasta, f. paste, pulp; **p. colada,** slurry; **p. de papel/madera,** paper/wood pulp.
pastal, m. (LAm) pasture, grazing land.
pasterización, f. pasteurization.
pasterizadola, a. pasteurized.
pasterizar, v. to pasteurize.
pasteurización, f. pasteurization.
pasteurizadola, a. pasteurized.
pastinaca, f. parsnip.
pastizal, m. pastureland, rangeland; **p. adehesado,** pasture with trees.
pasto, m. pasture; pasture land, grazing land (tierra); grazing (acción); fodder, feed

(alimento); grass, lawn (LAm); **p. fertilizado,** fertilised pasture.

pastor, m. shepherd, sheepherder (US); **perro (m) p.,** sheep dog.

pastora, f. shepherdess.

pastoreo, m. grazing, shepherding, pastoral farming.

pata, f. leg (Gen); hoof (de caballo, oveja, vaca etc.); paw (gato, oso etc.); foot (pie de personas); duck (hembra de pato).

patata, f. potato.

patatal, m. potato field.

patatar, m. potato field.

patela, f. patella (pl. patellae, patellas).

patente, f. patent (de invención).

patio, m. patio, yard; **"en el p. trasero de mi casa, ¡no!",** "not in my back yard" (vease NIMBY); **p. rectanglar,** quadrangle.

pato, m. duck; drake (pato macho).

patógeno, m. pathogen.

patógenola, a. pathogenic; **agente (m) p.,** pathogenic agent.

patología, f. pathology.

patrimonio, m. heritage, patrimony; **p. forestal del Estado,** state forests, crown forests (GB); **p. nacional,** national heritage; **p. real,** crown land.

patrocinadorla, m. y f. sponsor.

patrocinio, m. sponsorship.

paulatinola, a. gradual; **enfriado (a) p. (Geol),** slow-cooling.

pauta, f. rule, guide (regla); guideline (guía); model (modelo); **marcar la p. (v),** to lay down guidelines; **servir de p. (v),** to act as a model for.

pavimento, m. pavement; **p. de caliza,** limestone pavement; **p. del desierto,** desert pavement.

PC, m. PC; **ordenador (m) personal,** personal computer.

PCB, m. PCB; **policlorobifenilo,** polychlorbiphenyl.

PCD, f. LCD; **pantalla (f) de cristal líquido,** liquid crystal display.

peatónla, m. y f. pedestrian; **paso (m) de peatones,** pedestrian crossing (GB), crosswalk (US).

peatonal, a. pedestrian; **zona (f) p.,** pedestrian area.

peatonalización, f. pedestrianization.

pecinal, m. stagnant pool.

peciolo, pecíolo, m. petiole.

pecios, m.pl. flotsam, wreckage (de un barco).

pectina, f. pectin.

pectoral, a. pectoral.

pecuariola, a. livestock.

pechblenda, f. pitchblende, uraninite.

pecho, m. chest, breast; **dar el p. (v),** to breastfeed.

pedernal, m. flint, silex; **utensilios (m.pl) de p.,** flint implements.

pedestal, m. pedestal.

pediatría, f. paediatrics, pediatrics (US).

pedido, m. order (Com); **hacer un p. (v),** to place an order, to order.

pedimento, m. pediment.

pedología, f. pedology, soil science.

pedómetro, m. pedometer.

pedostratigráficola, a. pedostratigraphic.

pedregosola, a. stony, rocky.

pedreplén, m. hardcore, rap, rough stone; riprap (US).

pedrera, f. quarry.

pedrero, m. scree (Geog).

pedrisco, m. large hailstone.

peduncular, a. peduncular.

pegajosola, a. sticky.

pegmatita, f. pegmatite.

peinador, m. (Bol) waste tip scavenger.

pelágicola, a. pelagic.

película, f. film (de fotos); film, movie (US) (cine); pellicle (Biol, capa); **p. de agua,** water film; **p. infrarroja,** infrared film.

peligro, m. danger; **especies (f.pl) en p.,** endangered speices.

peligrosidad, f. danger, riskiness.

peligrosola, a. dangerous, hazardous (arriesgado/a); **actividades (f.pl) molestas, insalubres, nocivas y peligrosas (Esp),** dangerous and nuisance activities and activities harmful to health.

pelita, f. pelite.

pelíticola, a. pelitic.

pelo, m. hair; **p. radical (Bot),** root hair.

pélvicola, a. pelvic.

pelvis, f. pelvis.

pellet, m. pellet (aglomerado).

pelletización, f. pelletization; **planta (f) de p.,** pelletization plant.

PEN, m. **Plan (m) Energético Nacional (Esp),** National Energy Plan.

penacho, m. plume, tuft, crest (Biol).

penalización, f. sanction (castigo).

pendiente, f. slope, incline; pitch (de techo); **en p.,** sloping, inclined; **p. ascendente,** upslope; **p. del plano de fallas,** hade; **p. del tejado,** pitch of a roof; **p. descendente,** downslope; **p. estructural suave,** dip slope; **p. regional,** regional dip.

péndulo, m. pendulum.

pene, m. penis.

penecontemporáneola, a. penecontemporaneous.

peneplanización, f. peneplanation.

penetrabilidad, f. penetrability.

penetrable, a. penetrable.

penetración, f. penetration; **p. del pozo,** well penetration; **p. marina,** marine penetration; **p. salina,** saline penetration.

penetrómetro, m. penetrometer.

penicilina, f. penicillin.

penillanura, f. peneplain.

península, f. peninsula.

peniplanicie, f. peneplain, peneplane; **p. disectada**, dissected peneplain.

penta-, pf. penta-.

pentaclorobifenol, m. pentachlorobiphenol.

pentaclorofenol, m. pentachlorophenol (abr. PCP).

pentano, m. pentane.

pentanol, m. pentanol.

pentavalente, a. pentavalent.

penteno, m. pentene.

pentino, m. pentyne.

pentóxido, m. pentoxide.

peñascal, m. rocky place, rocky ground.

peñasco, m. crag.

peñón, m. wall of rock, crag.

peón, m. farm labourer (laborer); farm laborer, farm worker (US); rubbish dump operative, garbage man (que trabaja en un vertedero) (Col, C.Rica, Ec, Par, Pan).

peonaje, m. workgang, group of labourers, group of laborers (US).

pepenadorla, m. y f. rubbish sifter; garbage sifter (US).

pepino, m. cucumber.

pepita, f. pip (of fruit), pit (US); nugget (Min).

pepsina, f. pepsin.

péptido, m. peptide.

peptona, f. peptone.

pequeñola, a. small.

per-, pf. per-.

pera, f. pear.

perca, f. perch (pez).

per cápita, a. per capita.

percentil, m. percentile.

perceptible, a. perceptible.

perclorato, m. perchlorate.

percolación, f. percolation.

percolado, m. percolate.

percolar, v. to percolate.

percolímetro, m. percolation gauge.

percusión, f. percussion; **perforación (f) por p.**, percussion drilling.

percha, f. pole, support; perch (para aves);.

pérdida, f. loss, waste, leakage; **p. de agua**, waste of water, water loss; **p. de carga**, head loss; **p. en cabeza**, head losses; **p. en la red de distribución**, distribution losses; **p. en las bombas**, pump losses; **p. por fricción**, friction loss; **pérdidas en tuberías**, water main losses; **pérdidas y ganancias**, profit and loss.

perdidola, a. lost.

perdiz, f. partridge.

perenne, a. perennial (de plantas); evergreen (de hojas, de plantas); **manantial (m) p.**, perennial spring; **planta (f) p.**, perennial, perennial plant.

perfeccionar, v. to perfect.

perfil, m. profile (Gen); outline (silueta); section, cross-section (corte); **de p.**, in profile; **p. de humedad del suelo**, soil moisture profile; **p. de ribera**, shore profile; **p. de salinidad**, salinity profile; **p. geofísico**, geophysical profile; **p. geológico**, geological section; **p. longitudinal**, longitudinal profile; **p. tipo**, type profile; **p. transversal**, profile section, transverse section.

perfiladola, a. profiled.

perfilaje, m. (LAm) logging, profiling.

perfilar, v. to outline (Gen); to streamline (aerodinamizar); to log, to record (registrar un perfil).

perforación, f. borehole (para petróleo, agua); boring, drilling (acción); **p. a percusión**, percussion drilling; **p. con aire comprimido**, air drilling; **p. con circulación directa/inversa**, direct/reverse circulation drilling; **p. con corona de diamantes**, diamond drilling; **p. de pequeño diámetro**, small-diameter drilling; **p. para agua**, waterwell drilling; **p. rotativa**, rotary drilling; **registro (m) del p.**, drilling log.

perforadola, a. perforated, pierced; bored, drilled (Min).

perforador, m. driller.

perforar, v. to perforate, to pierce; to drill, to bore (Min).

perfumería, f. perfumery.

pergelisol, m. (LAm) permafrost.

pérgola, f. pergola.

periantio, m. perianth.

pericia, f. expertise, expertness.

periclinal, m. pericline.

peridotita, f. peridotite.

periféricola, a. peripheral; **drenaje (m) p.**, peripheral drainage.

perigeo, m. perigee.

periglacial, a. periglacial.

periglaciar, a. periglacial.

perímetro, m. perimeter; **p. de protección de agua subterránea**, groundwater protection area; **p. mojado**, wetted perimeter.

periodicidad, f. periodicity.

periódicola, a. periodic.

período, m. period; **p. de gestación**, gestation period; **p. de latencia**, latent period; **p. de lluvias**, pluvial period; **p. de recuperación**, recovery period; **p. de retorno**, return period; **p. de semidesintegración**, half-life period; **p. glacial, p. glaciar**, glacial period; **p. menstrual**, menstrual period, period; **p. pluvial**, pluvial period.

peristálticola, a. peristaltic; **bomba (f) p.**, peristaltic pump.

peristaltismo, m. peristalsis.

peritación, f. expertness; surveying.

peritaje, m. specialist's report (informe); expert's fee (honorario); professional training (estudios), expert opinion.

peritola, a. expert; **juicio (m) de peritos**, expert opinion.

perjudicial, a. prejudicial, detrimental; harmful (que hace daño).

perjuicio, m. harm, damage, injury.

perla, f. pearl.

perlita, f. perlite.

permafrost, m. permafrost.

permanecer, v. to remain, to stay.

permanganato, m. permanganate; **índice (m) de p.**, permanganate index; **p. potásico**, potassium permangonate.

permeabilidad, f. permeability; **p. de Darcy**, Darcy permeabilty; **p. horizontal**, horizontal permeability; **p. intrínseca**, intrinsic permeabilty; **p. no saturada**, unsaturated permeabilty; **p. primaria**, primary permeability; **p. secundaria**, secondary permeability.

permeable, a. permeable, pervious; **formación (f) p.**, permeable formation.

permeómetro, m. permeameter; **p. de carga fija**, fixed-head permeameter; **p. de carga variable**, falling-head permeameter; **p. diferencial**, differential permeameter.

Pérmico, m. Permian.

permisible, a. permissible, allowable.

permiso, m. permission; permit, licence, license (US) (licencia).

permutación, f. permutation.

perno, m. bolt; **p. en U**, U bolt.

peroxidación, f. peroxidation.

peróxido, m. peroxide; **p. de hidrógeno**, hydrogen peroxide.

perpendicular, a. perpendicular.

persistencia, f. persistence.

persistente, a. persistent.

personal, m. personnel, staff, employees (pl).

personero, m. (LAm) government official.

perspectiva, f. perspective.

perspex, m. perspex.

pertenencia, f. claim, ownership (Min).

perthosita, f. perthosite.

pertita, f. perthite.

perturbación, f. disturbance.

perturbadola, a. disturbed.

Perú, m. Peru.

peruanola, m. y f. Peruvian.

pesar, v. to weigh; **p. más que**, to outweigh.

pesca, f. fishing (actividad); catch, haul (redada); **artes (m.pl) de p.**, fishing gear, fishing tackle; **barco (m) de p.**, fishing boat; **p. con caña**, angling; **p. de altura**, deep sea fishing; **p. de bajura, p. de litoral**, inshore fishing, coastal fishing; **p. deportiva**, game fishing (GB, de salmón, trucha), coarse fishing (GB de los demás); **p. submarina**, underwater fishing; **red (f) de p.**, fishing net; **técnicas (f.pl) de p.**, fishing techniques; **zona (f) de p.**, fishing grounds.

pescado, m. fish (pl. fish).

pescador, m. fisherman (pl. fishermen).

pescar, v. to fish, to fish for; **caña (f) de p.**, fishing rod; **licencia (f) para p.**, fishing licence.

pesebre, m. trough (para animales).

peso, m. weight, heaviness, stone (unidad de peso, 6,350 kilos); peso (Arg, Ch, Col, Carib, Ur) (unidad monetaria); **argumento (m) de p.**, weighty argument; **de poco p. (a)**, light weight; **p. atómico**, atomic weight; **p. equivalente**, equivalent weight; **p. específico**, specific weight; **p. neto**, net weight.

pesquera, f. fishing, fishing ground.

pesquero, m. fishing boat; **p. de altura**, long-haul fishing boat, deep sea fishing boat; **p. de arrastre**, fishing trawler.

pesquerola, a. fishing; **puerto (m) p.**, fishing port.

pestaña, f. flange, rim (Ing); ledge (Geog); eyelash (Anat).

peste, f. plague; **p. bubónica**, bubonic plague; **p. pneumónica**, pneumonic plague.

pesticida, m. pesticide; **p. organoclorado**, organochlorine pesticide; **residuos (m.pl) de p.**, pesticide residues.

pestilencia, f. pestilence, plague (plaga); stench, stink (mal olor).

pétalo, m. petal.

petición, f. petition.

pétreola, a. stony, rocky.

petrificación, f. petrification.

petrogénesis, m. petrogenesis.

petrografía, f. petrography.

petrográficola, a. petrographic.

petróleo, m. oil, petroleum; **industria (f) del p.**, oil industry; **p. crudo**, crude oil; **p. lampante (LAm)**, paraffin, kerosene; **pozo (m) de p.**, oil well; **reservas (f.pl) de p.**, oil reserves, petroleum reserves.

petrolero, m. oil tanker.

petrolíferola, a. oil-bearing, petroliferous (Min); of oil; **campo (m) p.**, oilfield; **contaminación (f) p.**, oil pollution; **terminal (f) p.**, oil terminal.

petroquímica, f. petrochemical industry.

petroquímicola, a. petrochemical.

pez, m. fish (pl. fish); **criadero (m) de peces**, fish farm; **escalera (f) para peces**, fish ladder; **p. de agua dulce**, freshwater fish; **p. de agua salada**, saltwater fish; **p. marino**, marine fish; **población (f) de peces**, fish population; **rampa (f) para peces**, fish ladder; **vivero (m) de peces**, fish farm.

pezón, m. teat, nipple (Zoo, Ing); knob (protuberancia).

PGB, m. GDP; **Producto (m) Geográfico Bruto (Ch)**, Gross Domestic Product.

pH, m. pH (concentración de ión hidrógeno); **pH de campo**, field pH.

phi, a. phi; **escala (f) p.**, phi-scale.

phtanita, f. (LAm) chert.

pi, f. pi (Mat).

Piacenciense, m. Piacenzian.

PIB, m. GDP; **Producto (m) Interior Bruto (Esp)**, Gross Domestic Product.

picacho, m. peak, summit.

picadola, a. pricked, perforated (material); pitted (superficie); choppy (el mar).

picante, a. pungent, spicy (comida), sour (vino).

picapedrero, m. stonecutter, quarryman (cantero).

picar, v. to prick, to hew, to shred; **p. piedra**, to knap (p.e. sílex).

piclaje, m. pickling.

pico, m. bill, beak (de aves); beak (de insectos); pick, pickaxe (piqueta); peak (de montañas); **p. verde**, green woodpecker.

pico-, pf. pico- (10.E-12).

picrato, m. picrate.

picrita, f. picrite.

pie, m. foot (Gen) (pl. feet); trunk, stem (Bot); standing (posición); foot (Mat, abr. ft, unidad de longitud igual a 0,3048m); **a p.**, on foot, by foot; **al p. de**, at the foot of; **p. cuadrado**, square foot; **p. de la montaña**, foot of mountain; **seis pies de largo**, six feet long.

piedemonte, m. piedmont.

piedra, f. stone; **p. córnea**, hornstone; **p. de afilar**, oilstone; **p. luna**, moonstone, albite.

piel, f. skin, fur, pelt; leather (producto de la curtiduría).

pienso, m. fodder, feed.

pierna, f. leg (of a person).

piezometría, f. piezometry.

piezométricola, a. piezometric; **nivel (m) p.**, piezometric surface.

piezómetro, m. piezometer, wellpoint; **p. costero, p. litoral**, coastal piezometer.

pigmento, m. pigment.

pila, f. heap, pile (de arena); battery, cell (Elec).

pilar, m. pillar, column (columna); pier (de un puente); pillar, prop (apoyo); **anclaje (m) de p.**, column anchorage.

píldora, f. pill; **la p. (Med, anticonceptivo)**, the pill.

pilón, m. basin (de fuente); trough (bebedero); mortar (mortero); pillar, post (columna); **martillo (m) p.**, drop hammer.

pilotaje, m. piles (pl) (Con); pilotage (Mar).

pilote, m. pile (Con), stake (palo).

pilotola,, a. pilot; **proyecto (m) p.**, pilot scheme.

pimienta, f. pepper.

pimiental, m. pepper patch, pepper field.

pimiento, m. pepper.

pinacoide, m. pinacoid.

pináculo, m. pinnacle.

pinadola, a. pinnate.

pinnadola, a. pinnate.

pino, m. pine.

pinsapo, m. Spanish fir.

pinta, f. pint (medida de líquido igual a 0,568 litros en GB y a 0,473 litros en EEUU).

pintura, f. paint, painting, picture; **p. anti-corrosiva**, anticorrosive paint; **p. con pistola**, spray paint; **p. rupestre**, cave painting.

pinza, f. pincers (pl), claws (pl) (Zoo); pincer, jaw (Ing).

piña, f. pineapple (ananás); pinecone (de pino); cone (de otros árboles); **p. de América**, pineapple.

piojo, m. louse (pl. lice).

pionera, f. pioneer.

pionerola, a. pioneer; **comunidad (f) p. (Ecol)**, pioneer community.

pipeta, f. pipette.

pipote, m. rubbish bin, trash can (US).

piqueta, f. pickaxe, pick.

piramidal, a. pyramidal.

pirámide, f. pyramid.

piranómetro, m. pyranometer.

pirata, m. y f. pirate; hacker (Comp); cowboy operator (Com).

pirca, f. drystone wall.

piretro, m. pyrethrum.

pirgeómetro, m. pyrgeometer.

pirheliógrafo, m. pyrheliograph.

pirheliómetro, m. pyrheliometer.

piridina, f. pyridine.

piridoxina, f. pyridoxine (vitamina B6).

pirita, f. pyrite, pyrites; **p. blanca de hierro**, white pyrites; **p. de cobre**, copper pyrites; **p. de estaño**, stannite, tin pyrites; **p. de hierro**, iron pyrite, iron pyrites.

piritización, f. pyritization.

piro-, pf. pyro-.

piroclásticola, a. pyroclastic.

piroclasto, m. pyroclast.

pirofosfato, m. pyrophosphate.

pirólisis, f. pyrolysis.

pirolusita, f. pyrolusite.

pirómetro, m. pyrometer.

piroxenita, f. pyroxenite.

piroxeno, m. pyroxene.

pisada, f. footprint, footstep.

piscícola, a. piscine, fish.

piscicultura, f. fish farming, pisciculture.

piscifactoría, f. fish farm, fish hatchery.

piscina, f. swimming pool (para bañarse); fishpond (para peces).

piscívorola, m. y f. piscivore.

piso, m. floor, storey; stage (Geol, cronología); **p. entresuelo**, mezzanine floor.

pisolita, f. pisolite.

pisolíticola, a. pisolitic.

pista, f. track, trail (Gen); carriageway (carreteras); **p. de aterrizaje**, runway, landing strip; **p. de tierra**, dirt track.

pistón, m. piston.

pita, f. agave, sisal.

pitón, m. plug (Geol), spout.

pitot, m. pitot; **tubo de p., tubo (m) p.**, pitot tube.

pivote, m. pivot.

pizarra, f. slate, blackboard.

pizarreñola, a. slaty, slatey.

pizarrosidad, f. cleavage, foliation, slaty cleavage.

pizarrosola, a. slaty, slatey, slated; **foliación (f) p.**, slaty cleavage.

pizca, f. pinch, trace amount (cantidad); harvest (cosecha).

placa, f. plate, plaque (Biol); **p. continental**,

continental plate; **p. de apoyo (Con)**, backplate; **p. de asiento**, base plate, bearing plate; **p. de presión**, pressure plate; **p. litosférica**, Earth's plate; **p. oceánica**, oceanic plate; **p. petri**, Petri dish; **recuenta (f) de p.**, plate count; **tectónica (f) de placas**, plate tectonics.

placebo, m. placebo.

placenta, f. placenta (pl. placentas, placentae).

placer, m. gold field, placer (Min); **p. aluvial**, alluvial placer; **p. metalífero**, ore placer.

plaga, f. pest, blight (Bot); plague (enfermedad); **control (m) de plagas**, pest control.

plagioclasa, f. plagioclase.

plagioclímax, m. plagioclimax.

plagionita, f. plagionite.

plaguicida, m. pesticide; **residuos (m.pl) de plaguicidas**, pesticide residues.

plan, m. plan, scheme (proyecto); idea, intention (intención); level (Geog, nivel); **p. comercial**, business plan; **p. de desarrollo**, development plan; **p. de protección**, protection plan; **P. General de Ordenación (Esp)**, general land use plan, general development plan); **p. maestro**, master plan; **p. trienial/ quinquenal**, three-/five-year plan; **p. urbanístico**, town plan.

planario, m. planarian; **planarios (Zoo, clase)**, Planaria (pl).

plancton, m. plankton.

plancha, f. metal plate; iron (para planchar ropas).

planeación, f. planning; **p. urbana**, town-planning.

planeamiento, m. planning.

planicie, f. plain, flat area, level ground; **p. de inundación**, flood plain.

planificación, f. planning; **p. de arriba abajo**, topdown planning; **p. familiar**, family planning; **p. rural**, rural planning.

planificador/a, m. y f. planner.

planificar, v. to plan (producción).

planímetro, m. planimeter.

plano, m. plane; **p. axial**, axial plane; **p. de diaclasa**, joint plane; **p. de estratificación**, bedding plane; **p. de falla**, fault plane; **p. de pedimentos**, pediplane; **p. de referencia**, datum level; **primer p.**, foreground.

planta, f. plant (vegetación); plant (fábrica); floor, storey (piso); plantation (terreno); **p. anual**, annual, annual plant; **p. bienal**, biennial, biennial plant; **p. de demostración**, demonstration plant; **p. de depuración de aguas residuales**, sewage works, sewage treatment plant; **p. de desalinización, p. desalinizadora, p. desaladora (LAm)**, desalination plant; **p. de oxidación (Trat)**, oxidation plant; **p. de pruebas**, pilot plant; **p. depuradora (Trat)**, waterworks, water supply plant; **p. de tratamiento**, treatment plant, processing plant; **p. guía**, characteristic species; **p. hidroeléctrica**, hydroelectric plant; **p. incineradora**, incineration plant; **p.**

industrial, industrial plant; **p. perenne**, perennial, perennial plant; **p. pilota**, pilot plant; **p. potabilizadora**, water treatment plant; **plantas vasculares (Bot)**, vascular plants.

plantación, f. plantation; **p. de bosques**, afforestation.

plantilla, f. template, pattern; insole (Med); payroll (Com); **p. para montaje (para construcción)**, jig.

plantón, m. seedling.

plántula, f. seedling.

plaquita, f. platelet.

plasma, m. plasma.

plasmólisis, f. plasmolysis.

plasticidad, f. plasticity.

plástico, m. plastic.

plásticola, a. plastic.

plasticultura, f. agriculture under plastic.

plata, f. silver (Ag).

plataforma, f. platform; **p. continental**, continental shelf; **p. de perforación**, drilling platform, drilling rig.

platanal, m. banana plantation.

platanar, m. banana plantation.

platanera, f. banana plantation.

platanero, m. banana dealer, banana tree.

plátano, m. banana (fruit).

plateadola, a. silver-plated, silvered.

platelminto, m. platyhelminth, flatworm; **platelmintos (Zoo, filum)**, Platyhelminthes (pl).

platino, m. platinum (Pt).

platinoide, m. platinoid.

plato, m. plate, dish.

playa, f. beach; salt lake bed (Méx); **p. de arena**, sand beach; **p. de tormenta**, storm beach; **p. elevada**, raised beach.

playo, f. foreshore.

playola, a. (Arg, Méx) shallow.

pleamar, f. high tide, high water.

plecóptero, m. plecopteran; **plecópteros (Zoo, clase)**, Plecoptera (pl).

plegadola, a. folded; **p. hacia abajo**, downfolded; **p. hacia arriba**, up-folded.

plegamiento, m. folding; **p. por flexodeslizamiento**, flexure-slip folding.

plegar, v. to fold.

Pleistoceno, m. Pleistocene.

pleito, m. lawsuit, case.

plexiglás, m. plexiglass.

pliegue, m. fold, crease; **eje (m) del p.**, fold axis; **p. buzante (LAm)**, plunging fold; **p. cruzado**, cross fold; **p. de arrastre**, drag fold; **p. isoclinal**, isoclinal fold; **p. por cabalgamiento**, overthrust fold; **p. recumbente**, recumbent fold; **p. secundario**, minor fold; **p. tumbado**, plunging fold.

Pliensbaquiense, m. Pliensbachian.

plinto, m. plinth.

Plioceno, m. Pliocene.

Pliocuaternario, m. Plio-Quaternary.

plomería, f. (LAm) plumbing.

plomo, m. lead (Pb); lead weight (peso), lead shot (de fusil), sinker (de red); fuse (Elec); **envenenamiento (m) por p.**, lead poisoning; **sin p. (gasolina)**, unleaded (petrol).

pluma, f. feather, plume (Biol); plume (p.e. de humo); derrick (Ing, Mar); tap (LAm, grifo); pen (para escribir); **p. de contaminación**, contamination plume; **p. de desagües**, sewage plume; **p. de fondo**, down feather; **p. de humo**, plume of smoke; **p. de lixiviado**, leachate plume; **p. primaria/secundaria**, primary/secondary feather; **p. térmica**, thermal plume.

plumbocalcita, f. plumbocalcite.

pluricelular, a. multicellular.

pluridisciplinariola, a. multi-disciplinary.

plusvalía, f. gain, increase in value.

Plutón, m. Pluto.

plutón, m. pluton.

plutónicola, a. plutonic.

plutonio, m. plutonium (Pu).

pluvial, a. pluvial.

pluviógrafo, m. pluviograph, rainfall recorder; **p. de cangilón**, tilting-bucket rainfall recorder; **p. de sifonación automática**, automatic siphoning rainfall recorder.

pluviometría, f. pluviometry, rainfall measurement; **p. eficaz**, effective rainfall; **p. umbral**, rainfall threshold.

pluviómetro, m. raingauge.

pluviosidad, f. pluviosity, precipitation.

pluviosola, a. rainy, pluvious, wet.

PNB, m. GNP; **Producto (m) Nacional Bruto**, Gross National Product.

pneumonía, f. pneumonia.

PNN, m. **Producto (m) Nacional Neto**, net national product.

PNUD, m. **Programa (m) de Naciones Unidas para el Desarrollo**, United Nations Programme for Development.

PNUMA, m. **Programa (m) de Naciones Unidas para el Medio Ambiente**, United Nations Programme for the Environment.

pobeda, f. depression occupied by poplar trees (álamos).

población, f. population (personas); town, village (ciudad, pueblo); **aumento (m) de la p.**, increase in population; **núcleo (m) de p.**, population centre, centre of population; **p. activa**, working population; **p. rural**, rural population.

poblado, m. built-up area; town (ciudad); centre of population (núcleo urbano).

pobladola, a. populated, inhabited; wooded (arbolado); **densamente p.**, densely populated; **paisaje (m) p.**, wooded countryside.

pocerón, m. (CAm, Méx) large pool.

poceta, f. sump; **p. de drenaje**, drainage sump.

pocilga, f. piggery, pigsty (porquerizo).

pocosin, m. (US) acid shrub-covered peat.

poda, f. pruning (acción); pruning season (época).

poder, m. power (autoridad), possession (posesión), capacity (capacidad), strength (fuerza); **p. adquisitivo**, purchasing power; **p. ejecutivo**, executive power; **plenos poderes**, full powers.

podridola, a. rotten, putrefied.

podsol, m. podsol; **suelo (m) p.**, podsol.

poiquilotermo, m. poikilotherm.

poiquilotermola, a. poikilothermic.

poise, m. poise (unidad de viscosidad).

polar, a. polar; **casquete (m) p.**, polar cap; **círculo (m) p.**, polar circle.

polarimetría, f. polarimetry.

polarización, f. polarization.

polarizadorla, a. polarizing; **microscopio (m) p.**, polarizing microscope.

polder, m. polder.

polea, f. pulley; **p. de cable de perforación**, cable rig pulley.

polen, m. pollen; **grano (m) de p.**, pollen grain.

poli-, pf. poly-.

poliamida, f. polyamide.

policíclicola, a. polycyclic; **hidrocarburo (m) p. aromático (abr. HPA)**, polycyclic aromatic hydrocarbon (abr. PAH).

policloradola, a. polychlorinated.

policultivo, m. mixed farming.

poliedro, m. polyhedron.

polielectrolito, m. polyelectrolyte.

poliéster, m. polyester.

poliestireno, m. polystyrene.

polieteno, m. polythene, polyethylene.

polietileno, m. polyethylene, polythene.

polifacéticola, a. many-sided; versatile (Fig).

polifosfato, m. polyphosphate.

polígamola, a. polygamous.

poligonal, a. polygonal.

polígono, m. polygon (Mat); area (zona); building site (solar); **p. de frecuencia**, frequency polygon; **p. de Thiessen**, Thiessen polygon; **p. industrial**, industrial estate.

polihalita, f. polyhalite.

polilla, f. clothes moth (en ropa).

polimerización, f. polimerization.

polímero, m. polymer.

polínicola, a. pollen; **recuento (m) p.**, pollen count.

polinización, f. pollination; **p. cruzada**, cross-pollination.

polinizar, v. to pollinate.

polinomial, a. polynomial.

polinómicola, a. polynomial.

polio, f. polio, poliomyelitis.

poliomielitis, f. poliomyelitis.

polipasto, m. pulley, hoist block.

polipéptido, m. polypeptide.

poliploide, a. polyploid.

pólipo, m. polyp.

polipropeno, m. polypropene, polipropylene.

polipropileno, m. polypropylene, polypropene.

poliqueto, m. polychaete; **poliquetos (Zoo, clase)**, Polychaeta.

polisacárido, m. polysaccharide.
polisapróbicola, a. polysaprobic.
política, f. policy (programa); politics; **p. agraria,** agricultural policy; **p. ambiental,** environmental policy; **p. ecologista,** green politics; **p. respetuosa con el medio ambiente,** environmental policy; **p. verde,** green politics.
poliuretano, m. polyurethane.
polivinilo, m. polyvinyl.
polo, m. pole (Elec); zone, area; **p. de desarrollo,** development area; **P. Norte/Sur,** North/South Pole; **polos iguales/contrarios,** like/opposite poles.
polonio, m. polonium (Po).
polución, f. pollution.
polutola, a. polluted.
polvareda, f. dust cloud.
polvo, m. dust, powder; **nieve p.,** powdery snow; **p. de carbón,** coal dust.
pollerola, m. y f. poulterer, chicken farmer; gambler (LAm).
pollo, m. chicken, chick, young bird.
pomelo, m. grapefruit.
pómez, f. pumice; **piedra (f) p.,** pumice stone.
ponderadola, a. steady, balanced; **cuadrados (m.pl) mínimos ponderados (Mat),** weighted least squares.
poney, m. pony.
pongo, m. gorge; canyon (LAm); orangutan (mono).
poniente, 1. m. westerly wind (viento); 2. a. west, western.
pontón, m. pontoon; **puente (m) de pontones,** pontoon bridge.
popular, a. popular (conocido); of the people, public; **protesta (f) p.,** public outcry.
por, prep. by; times, multiplied by (Mat); **2 p. 3 son 6,** 2 times 3 is 6.
porcelana, f. porcelain.
porcelanita, f. porcellanite.
porcentaje, m. percentage; **p. de defunciones,** death rate; **p. de peso seco,** percentage dry weight.
porcentual, a. percentage, per cent.
por ciento, m. percent, per cent; **un aumento (m) del veinte p. c.,** a twenty per cent increase.
porcinola, a. pig, swine; **ganado (m) p.,** pigs, swine.
porcinos, m.pl. pigs, swine.
porche, m. porch (de una casa), arcade (soportal).
porfídicola, a. porphyritic.
pórfido, m. porphyry, porphyrite; **p. diorítico,** diorite porphyry; **p. feldespático,** elvan; **p. granítico,** granite porphyry.
porfidoblasto, m. porphyroblast.
porfidoclasto, m. porphyroclast.
porfiroide, m. porphyroid.
porífero, m. porifer; **poríferos (Zoo, filum),** Porifera (pl).

pormenor, m. detail.
pormenorizadola, a. itemized, detailed.
pormenorizar, v. to detail.
poro, m. pore; **presión (f) de p. intersticial,** pore water pressure; **tamaño (m) de poros,** pore size.
porosidad, f. porosity; **p. efectiva, p. eficaz,** effective porosity; **p. primaria/secundaria,** primary/secondary porosity.
porosidad, f. porosity; **p. teórica,** theoretical porosity.
porosimetría, f. porosimetry.
porosola, a. porous.
porotal, m. (LAm) bean field.
poroto, m. (LAm) bean.
porquería, f. filth, rubbish (suciedad); nastiness, obscenity (Fig).
porqueriza, f. pigsty.
portacontenedor, m. container ship.
portadorla, m. y f. carrier (Zoo, Med).
portal, m. doorway, entranceway.
portátil, a. portable.
portavoz, m. y f. spokesman, spokeswoman.
portezuelo, m. (LAm) pass (Geog).
pórtico, m. portal (Con), gantry (de grúa).
portillo, m. narrow pass (Geog), gap, opening.
Portlandiense, m. Portlandian.
portorriqueñola, a. Puerto Rican.
portuario, m. port, harbour, harbor (US); dock (muelle).
Portugal, m. Portugal.
portugués, portuguesa, m. y f. Portuguese.
portués/potuguesa, a. Portugese.
posglacial, a. post-glacial.
posgrado, m. second degree (educación).
posibilidad, f. possibility.
posición, f. position.
posido, m. meadow within more elevated arable land.
positivola, a. positive; **carga (f) p.,** positive charge; **electrodo (m) p.,** positive electrode; **ión (m) p.,** positive ion; **polo (m) p.,** positive pole; **taxia (f) p.,** positive taxis; **tropismo (m) p.,** positive tropism.
positrón, m. positron.
Postdamiense, m. Postdamian.
poste, m. post, pillar, pylon; **p. indicador,** signpost; **p. telegráfico,** telegraph pole.
posterior, a. posterior, subsequent.
postglaciar, a. post-glacial.
potabilidad, f. potability; **análisis (m) de p.,** potability analysis; **p. bacteriológica,** bacterial potability; **p. química,** chemical potability.
potable, a. potable, drinkable; **agua (f) p.,** potable water.
potamología, f. potamology.
potasa, f. potash.
potásicola, a. potassic; **cloruro (m) p.,** potassium chloride.
potasio, m. potassium (K).
potencia, f. power (Fis, Mat); **elevar a la p. de**

dos (v) (Mat), to square, to raise to the power two; **elevar a la p. tres (v),** to cube, to raise to the power of three.

potencial, m. potential; **p. de reducción-oxidación,** reduction-oxidation potential; **p. humano,** manpower; **p. redox, p. de reducción-oxidación,** redox potential, reduction-oxidation potential; **energía (f) p.,** potential energy; **evaporación (f) p.,** potential evaporation; **vegetación (f) p.,** climax vegetation.

potenciometría, f. potentiometry.

potenciométrica, a. potentiometric; **superficie (f) p.,** potentiometric surface.

potenciómetro, m. potentiometer.

potente, a. powerful.

potrero, m. (LAm) fenced pasture; (LAm, CAm) cattle ranch, stock farm.

potro, m. foal.

poza, f. large puddle (charca), natural well (cenote).

pozo, m. well; **desempeño (m) de p. (LAm),** well performance; **cabeza (f) de p.,** wellhead; **p. abierto,** open well; **p. abisinio,** Abyssinian well; **p. aireado,** air shaft; **p. amortiguador,** stilling well; **p. artesiano,** artesian well; **p. brotante,** flowing well; **p. cárstico,** swallow hole; **p. clavado,** driven well; **p. colector,** collector well; **p. con galerias,** well with headings; **p. con respiración,** air shaft; **p. costero,** coastal well; **p. de abastecimiento,** abstraction well; **p. de agua,** water well, waterwell; **p. de desagüe,** drainage shaft; **p. de gas,** gas well; **p. de gran diámetro,** large-diameter well; **p. de intercepción,** interception well; **p. de inyección,** injection well; **p. de observación,** observation well; **p. de pequeño diámetro,** small-diameter well; **p. de petróleo,** oil well; **p. de protección sanitaria,** sanitary well protection; **p. de prueba,** test well; **p. de ventilación,** air shaft; **p. excavado,** dug well; **p. hincado,** driven well; **p. imagen,** image well; **p. negro,** cesspit, cesspool; **p. perforado,** bored well; **p. surgente,** flowing well, overflowing well; **p. taladrado,** drilled well; **protección (f) del p.,** well protection; **puntal (m) del p.,** wellhead; **rendimiento (m) de p.,** well performance.

PPC, f. **Política (f) Pesquera Común (EurU),** Common Fishing Policy.

ppm, m.pl. ppm; **partes (m.pl) por millón,** parts per million.

práctica, f. practice; experience (conocimiento); method (método); **buena p.,** good practice; **código (m) de buena p.,** code of good practice; **código (m) de p.,** code of practice; **p. de verter basuras,** landfilling.

practicar, v. to practise; to practice (US).

prácticola, a. practical.

pradera, f. prairie, meadow; **praderas encharcadizas,** glades (p.e. Everglades, US).

prado, m. meadow, field, pasture, grassland; lawn (LAm).

pragmáticola, a. pragmatic.

pragmatismo, m. pragmatism.

praseodimio, m. praseodynium (Pr).

pre-, pf. pre-.

preacabadola, a. prefinished.

preaireación, f. preaeration.

precalentamiento, m. preheating.

Precámbrico, m. Pre-Cambrian.

precargo, m. preloading.

precarios, m.pl. (Méx, C.Rica) slum, shanty town.

precaución, f. precaution; **con p. (a),** precautionary; **precauciones contraincendios,** fire precautions.

precaver, v. to take precautions, to guard against.

precavidola, a. cautious, wary.

preceptivola, a. obligatory, mandatory, prescriptive.

preciclaje, m. precycling.

precio, m. price, value, worth; **p. de coste,** cost price; **p. neto,** net price; **p. por unidad, p. unitario,** unit price; **subida (f) de p.,** price rise.

precipicio, m. precipice, cliff.

precipitación, f. precipitation, rainfall; **captación (f) de precipitaciones,** rainwater catchment, rainfall catchment; **p. ciclónica,** cyclonic precipitation; **p. convectiva,** convective precipitation; **p. eficaz,** effective precipitation; **p. frontal,** frontal precipitation; **p. orográfica,** orographic precipitation; **p. química,** chemical precipitation; **relación (f) p.-escorrentía,** rainfall-runoff relationship.

precipitado, m. precipitate; **p. químico,** chemical precipitate.

precipitador, m. precipitator; **p. electrostático,** electrostatic precipitator.

precipitar, v. to precipitate (Quím), to deposit.

precisión, f. precision, accuracy, exactness.

predicción, f. prediction.

predominante, a. prevailing; **viento (m) p.,** prevailing wind.

preexistente, a. pre-exisiting.

prefabricadola, a. prefabricated.

prefabricar, v. to prefabricate.

preferencia, f. preference.

preferente, a. preferred; preferable (preferible); prior (derecho); **opción (f) p.,** preferred option.

prefijo, m. prefix.

prefiltación, f. prefiltration, prescreening.

prefiltro, m. prefilter.

prehistóricola, a. prehistoric.

preliminar, a. preliminary.

prensil, a. prehensile.

preñadola, a. pregnant (Zoo).

preñez, f. pregnancy (Zoo).

preparación, f. preparation; training (entrenamiento); **estado (m) de p.,** preparedness.

presa, f. dam (embalse); mill-race, flume (canal para agua); prey (Zoo); capture, seizure (captura); **ave (f) de p.**, bird of prey; **p. con contrafuertes**, buttress dam; **p. de arco**, arch dam; **p. de contención**, barrage; **p. de gravedad**, gravity dam; **p. de mampostería**, masonry dam; **p. de relleno**, rockfill dam; **p. de retención**, impounding dam; **p. de tierra**, earth dam; **p. encofrada**, coffer dam.

presbicia, f. longsightedness.

prescripción, f. specification, prescription.

presencia, f. presence, attendance; **p. de ánimo**, presence of mind.

presentación, f. presentation.

preservación, f. preservation, protection.

preservativola, 1. m. condom; 2. a. preservative.

presidente, m. president; **p. del gobierno**, president of the government.

presión, f. pressure; **a toda p.**, at full pressure; **cresta (f) de p.**, pressure ridge; **filtro (m) de p.**, filter pressure; **grupo (m) de p.**, pressure group; **p. absoluta**, absolute pressure; **p. alta**, high pressure; **p. arterial**, blood pressure; **p. atmosférica**, atmospheric pressure; **p. baja**, low pressure; **p. capilar**, capillary pressure; **p. crítica**, critical pressure; **p. de agua**, pressure head of water; **p. de columna estática**, pressure head; **p. de confinamiento**, confining pressure; **p. del agua**, water pressure; **p. del aire**, air pressure, atmospheric pressure; **p. hidrostática**, hydrostatic pressure; **p. interior**, back pressure; **p. intersticial**, interstitial pressure; **p. osmótica**, osmotic pressure; **p. parcial del vapor**, partial vapour pressure; **p. por inyección**, injection pressure; **p. reducida del agua**, reduced pressure head; **p. sanguínea**, blood pressure; **presiones ambientales**, environmental lobbying; **transductor (m) de p.**, pressure transducer.

préstamo, m. loan.

prestar, v. to lend; **acción (f) de p.**, lending.

presuponer, v. to presuppose.

presupuesto, m. budget, presupposition; **equilibrar el p. (v)**, to balance the budget; **hacer un p. (v)**, to make an estimate.

presurizar, v. to pressurize.

pretensadola, a. prestressed; **hormigón (m) p.**, prestressed concrete.

pretensión, f. claim, pretension; **tener pretensiones de (v)**, to lay claim to.

pretil, m. parapet (de una construcción).

pretratamiento, m. pretreatment; **p. de aguas residuales**, wastewater pretreatment; **p. de residuos**, pretreatment of waste.

prevención, f. prevention; **p. de inundaciones**, flood prevention.

preventivola, a. preventive, preventative, precautionary; **medidas (f) p.**, preventive measures.

previamente, a. previously, beforehand; **p. necesario**, prerequisite.

previsible, a. foreseeable, forecast; **legislación (f) p.**, proposed legislation.

previsión, f. forecast; **p. de avenidas**, flood forecast; **p. de demanda de agua**, water demand forecast; **p. del tiempo**, weather forecast.

Priaboniense, m. Priabonian.

primarias, f.pl primary feathers, primaries (pl).

primariola, a. primary; **clarificador (m) p.**, primary clarifier; **color (m) p.**, primary colour; **enseñanza (f) p.**, primary education; **escuela (f) p.**, primary school; **plumas (f) primarias**, primary feathers, primaries; **tanque (m) para sedimentación p.**, primary sedimentation tank; **tratamiento (m) p.**, primary treatment.

primate, 1. m. primate; **primates (Zoo, orden)**, Primates (pl); 2. a. primate.

primavera, f. spring.

primola, a. prime (Mat); **materia (f) p.**, raw material; **número (m) p.**, prime number.

principio, m. principle; **al p.**, at the start, at the beginning; **p. de Arquímedes**, Archimedes principle; **p. de "quien contamina paga"**, "polluter pays" principle; **p. de superposición**, principle of superposition.

prioridad, f. priority; **alta p.**, high priority.

priorizar, v. to prioritize.

prisma, f. prism; **p. de cuarzo**, quartz prism; **p. de nicol**, nicol prism.

privadola, a. private; **aguas (f.pl) privadas**, private waters.

privatización, f. privatization.

privilegio, m. privilege.

probabilidad, f. probability, likelihood; **distribución (f) de probabilidad(es)**, probability distribution.

probar, v. to prove, to demonstrate.

problema, m. problem.

problemáticola, a. problematical.

probóscide, f. proboscis (pl. proboscises, proboscides).

procariota, f. procaryote, prokaryote; **procariotas (Zoo, Bot, superreino)**, Procaryota, Prokaryota (pl).

procedencia, f. provenance.

procedimiento, m. method, procedure (Gen); process (sistema); means (método).

procesador, m. processor; **p. de palabras**, word processor.

procesamiento, m. processing; prosecution (Jur); **auto (m) de p.**, indictment; **p. de alimentos**, food processing; **p. de datos**, data processing.

procesar, v. to process (data); to prosecute (Jur).

proceso, m. process; trial (Jur); **p. anaerobio**, anaerobic process; **p. Bessemer (para fundición de acero)**, Bessemer process; **p. bioquímico**, biochemical process; **p. de aireación extendida**, extended-aeration process; **p. de depuración**, treatment process; **p. de diagénesis**, diagenetic process; **p. de karsti-**

ficación, karstification process; **p. de la vida,** life process; **p. de modelado (Mat),** modelling process; **p. de sobre-relajación,** over-relaxation process; **p. endógeno,** endogenic process; **p. en paralelo/en serie,** parallel/serial processing; **p. iterativo,** iterative process; **p. mecánico,** mechanical process; **p. termal, p. térmico,** thermal process.

producción, f. production; **p. primaria,** primary production.

producir, v. to produce; to make (hacer); to cause (motivar); to yield (rendir); **p. en serie,** to mass-produce.

productividad, f. productivity (de gente), output (de máquina).

producto, m. product; yield, profit (Com); **p. agroquímico,** agrochemical; **p. disuelto,** dissolved product; **p. final,** end product; **p. industrial,** industrial by-product; **p. interno bruto (Esp, abr. PIB),** gross domestic product; **p. libre (Quím),** free product; **p. nacional bruto (abr. PNB),** gross national product (abr. GNP); **p. secundario,** industrial by-product; **productos manufacturados,** manufactured products; **productos primarios,** primary products.

productor/a, 1. m. y f. producer; worker (trabajador); 2. a. producing.

proenzima, f. proenzyme.

profesión, f. profession.

profesional, a. professional.

profiláctico, m. prophylactic, protective (US, condón).

profilácticola, a. prophylactic.

profundidad, f. depth; **diez metros de p.,** ten metres in depth; **poca p.,** shallow, little depth.

profundizadola, a. deepened, incised (Geog).

profundizar, v. to deepen, to downcut.

profundola, a. deep.

progenie, f. progeny, offspring.

progesterona, f. progesterone.

programa, m. programme, program (US); program (Comp); schedule (plan de trabajo); **p. autocopiado,** computer virus; **p. de aplicación(es) (Comp),** applications program; **p. de desarrollo,** development programme.

programable, a. programmed, programmable.

programación, f. programming, programing (US); **p. dinámica;** dynamic programming; **p. en ordenador, p. informática,** computer programming; **p. lineal,** linear programming.

programadorla, m. y f. programmer (p.e. de ordenadores).

progresión, f. progression; **p. geométrica,** geometric progression.

prohibir, v. to prohibit.

proliferación, f. proliferation.

proliferar, v. to proliferate.

prólogo, m. prologue, foreword (del informe, libro, etc.).

prolongación, f. prolongation, extension.

prolongar, v. to prolong, to extend.

promedio, m. average, mean (Mat); **p. móvil,** moving average.

prometer, v. to promise.

prometio, m. promethium (Pr).

promoción, f. promotion, advancement.

promontorio, m. promontory, headland, foreland.

promotorla, m. y f. promoter.

promulgación, f. promulgation.

promulgadorla, a. promulgating.

promulgar, v. to enact, to promulgate (ley).

pronoto, m. pronotum (pl. pronota).

prontitud, f. promptness.

propagación, f. propagation, spreading; **p. de contaminantes,** contaminant spread; **p. vegetativa,** vegetative propagation.

propagadorla, m. y f. propagator.

propaganda, f. propaganda.

propagar, v. to propagate (Biol); to spread, to disseminate.

propágulo, m. propagule.

propanal, m. propanal.

propano, m. propane.

propanol, m. propanol.

propata, f. proleg.

propiciola, a. propitious, favourable.

propiedad, f. property (objeto); possession, ownership (pertenencia); accuracy (p.e. de una copia); **derechos (m.pl) de p. intelectual,** intellectual property rights; **p. física,** physical property; **p. industrial,** patent rights; **p. intelectual,** copyright; **p. particular,** private property; **propiedades del suelo,** soil properties.

propietariola, m. y f. owner, landowner, landlord (de terreno, casas).

propileno, m. propylene, propene.

propilo, m. propyl.

propino, m. propyne.

proponer, v. to propose; **p. un proyecto (v),** to propose a plan.

proporción, f. proportion, ratio; **p. del carbono,** carbon ratio; **p. de vacío,** void ratio.

proporcionadola, a. proportionate, in proportion.

proporcional, a. proportional; **inversamente p.,** inversely proportional.

proporcionar, v. to supply, to provide.

proposición, f. proposition, proposal.

propuesta, f. proposal; **hacer una p. (v),** to make a proposal.

prospección, f. prospection.

prospectando, m. prospecting (Min).

prospectar, v. to prospect, to search for (Min).

prospector, m. prospector (m) (Min); cateador (m) (LAm) (Min).

pros y contras, m.pl. pros and cons.

protactinio, m. protactinium (Pr).

protección, f. protection; **área p. de aguas,** water protection area; **medidas (f) de p.,**

protection measures; **p. ambiental,** environmental protection; **p. católica,** cathodic protection; **p. contra la corrosión,** protection against corrosion; **p. de acuíferos,** aquifer protection; **p. de aguas subterráneas,** groundwater protection.

protector/a, a. protective; **revestimiento (m) p.,** protective coating.

proteína, f. protein; **p. soluble,** soluble protein.

proteolisis, f. proteolysis.

Proterozoico, m. Proterozoic.

protesta, f. protest, complaint, objection.

protio, m. protium.

proto-, pf. proto-.

protocolo, m. protocol.

protón, m. proton.

prototipo, m. prototype.

protozoarios, m.pl. protozoa.

protozoo, m. protozoan (pl. protozoans, protozoa); **protozoos (Zoo, subreino o filum),** Protozoa (pl).

protuberancia, f. protuberance, protusion.

provecho, m. advantage, benefit; profit (Com).

proveedor/a, m. y f. supplier.

proveer, v. to provide, to supply.

provincial, a. provincial.

proximidad, f. proximity.

proyección, f. projection; **p. de mapas,** map projection; **p. estereográfica,** stereographic projection.

proyectista, m. y f. planner, designer, draftsman (US).

proyecto, m. project, plan, scheme; draft (versión); **estar en p. (v),** to be in the planning stage; **p. de ley (Jur),** bill; **p. hidroeléctrico,** hydroelectric scheme.

prudente, a. prudent.

prueba, f. test, trial (ensayo); proof, evidence (Jur); **banco (m) de pruebas,** bench test; **datos de las pruebas,** test data; **equipo (m) de pruebas,** test panel; **p. ambiental,** environmental test; **p. de bondad de ajuste (Mat),** goodness-of-fit test; **p. de compresión,** compression test; **p. de idoneidad,** fitness test; **p. de jarras (Quím),** jar test; **p. del laboratorio,** laboratory test; **p. de pala (Ing),** vane test; **p. de rendimiento,** performance test; **p. destructiva,** test to destruction; **p. ji-cuadrada (Mat),** Chi-squared test; **p. nuclear,** nuclear test; **p. rutinaria,** routine check; **p. U de Mann Whitney (Mat),** Mann-Whitney U test.

psammita, f. (LAm) psammite.

psammíticola, a. (LAm) psammitic.

pseudo-, pf. pseudo-.

pseudobrecha, f. pseudobreccia.

pseudomonas, f. pseudomonas.

pseudomórficola, a. pseudomorphic.

pseudomorfo, m. pseudomorph.

psicología, f. psychology.

psicrofílicola, a. psychrophylic.

psicrómetro, m. psychrometer.

psiquiatría, f. psychiatry.

PTB, m. GDP; **Producto (m) Territorial Bruto (Perú),** Gross Domestic Product.

pteridofita, f. pteridophyte; **pteridofitas (Bot, división),** Pteridophyta (pl).

púa, f. tine (de tenedor), prong, spike; quill (Biol); **alambre (m) de púas,** barbed wire.

pubertad, f. puberty.

publicación, f. publication.

públic/a, a. public; **obras (f.pl) públicas,** public works.

pudelado, m. puddling; **p. con creta,** chalk puddling.

pudinga, f. pudding stone.

pudridero, m. midden (Arq), rubbish heap.

pudrir, v. to rot, to putrefy.

puente, m. bridge; **estribo (m) de p.,** abutment; **p. colgante,** suspension bridge; **p. de Wheatstone (Elec),** Wheatstone bridge; **p. ferroviario,** railway bridge; **p. giratorio,** swing bridge; **p. levadizo,** drawbridge; **p. transportador,** transporter bridge; **tender un p. sobre (un río) (v),** to bridge.

puerco, m. swine, pig; **p. espín,** porcupine.

puerro, m. leek.

puerto, m. port, harbour, harbor (US); haven (Fig).

Puerto Rico, m. Puerto Rico.

puesta, f. putting, setting; **p. en marcha,** starting up (de una máquina); beginning (de un proyecto).

puesto, m. place, position (lugar); seat (sitio); job, post (empleo); **p. de trabajo,** post, job.

pulgada, f. inch (medida de longitud igual a 2,54 cm).

pulgón, m. plant louse.

pulimentar, v. to polish, to shine.

pulimento, m. polish, polishing, shine.

pulir, v. to polish.

pulmón, m. lung.

pulmonar, a. pulmonary.

pulmonía, f. pneumonia.

pulsación, f. pulsation, beat.

pulsar, v. to press, to push (un teclado, un interruptor); to beat, to pulse (batir).

pulverización, f. spraying, atomization; **p. con pesticidas,** pesticide spraying.

pulverizador, m. spray (recipiente para espolvorear líquidos); pulverizer (máquina).

pulverizar, v. to pulverize.

pumita, f. pumice.

puna, f. high Andean plateau.

punta, f. arrowhead; **p. de flecha de sílex (Arq),** flint arrowhead; point, tip, peak; **p. de avenida,** flood peak.

puntal, m. strut, prop, brace (Con); support (sostén); foundation, base (Con, elemento fundamental).

punto, m. spot, dot (Gen); point (tanto); place (Geog); point (Mat); **p. de comprobación,** control point, reference point; **p. de congelación,** freezing point; **p. de control,** control point, reference point; **p. de divisoria de aguas,** watershed (GB), rainshed (US); **p. de equivalencia (Quím),** end-point; **p. de estancamiento (Mat),** stagnation point; **p. de fusión,** melting point; **p. de inflamación,** flash point; **p. de inflexión,** point of inflexion; **p. de manantial,** springhead; **p. de marchitez permanente,** permanent wilting point; **p. de observación,** observation point; **p. de origen (Mat),** focus; **p. de referencia,** reference point (Gen), landmark (Geog), benchmark (Ing); **p. de rocío,** dew point; **p. de rotura,** breakpoint; **p. focal,** focal point; **p. nodal,** nodal point.

puntual, a. precise, accurate; punctual (de tiempo); **fuente (m) p. de contaminación,** point source of pollution.

punzón, m. punch; **p. para madera,** awl.

pupa, f. pupa (pl. pupae) (insecto); pimple (grano); scab (postilla).

pupila, f. pupil (Anat).

puquío, m. (LAm) spring, fountain.

Purbeckiense, m. Purbeckian.

pureza, f. purity, pureness; **p. atmosférica,** atmospheric purity; **p. del agua,** water purity.

purgación, f. purging, cleaning.

purgar, v. to purge (limpiar).

purificación, f. purification; **p. de agua,** water treatment.

purificadorla, 1. m. y f. purifier, purifying plant; 2. a. purifying, cleansing; **planta (f) p.,** purifying plant.

purificar, v. to purify, to cleanse.

púrpura, f. purple.

purpúreola, a. purple.

pus, m. pus.

putrefacción, f. putrefaction, rot.

putrefactola, a. putrefied, rotten.

putrescente, a. putrescent.

putrescible, a. putrescible; **desechos (m.pl) p.,** putrescible waste.

pútridola, a. putrid, rotten.

PVC, m. PVC; **polivinilcloruro (m),** polyvinylchloride.

PWR, m. PWR; **reactor (m) de agua a presión,** pressurized water reactor.

pyrex, m. pyrex.

Q

qanat, m. (Arab) qanat, ganat, khanate; gallery, heading, adit.

quanta, m.pl. quanta; **teoría (f) de los quanta,** quantum theory.

quanto, m. quantum (pl. quanta); **mecánica (f) cuántica,** quantum mechanics.

quark, m. quark.

quebrada, f. gulley, ravine; brook, mountain stream; gulch (US).

quebradizola, a. fragile.

quebrado, m. fraction (Mat); bankrupt (Com); **línea (f) q.,** broken line.

quebradola, a. broken.

quebrar, v. to break.

quechua, a. quechuan (idioma).

quelante, a. chelating; **agente (m) q.,** chelating agent.

quemador, m. burner; **q. de gas,** gas burner.

quemar, v. to burn; to incinerate (basura).

queratina, f. keratin.

querosén, m. (LAm) kerosene.

queta, f. chaeta (pl. chaetae).

quetzal, m. quetzal (ave); quetzal (Guat, unidad monetaria).

quietud, f. calm, quiescence.

quijada, f. jaw, jawbone.

química, f. chemistry.

químicola, 1. m. y f. chemist; 2. a. chemical.

quimioautótrofo, m. chemoautotroph, chemoautotrophe.

quimioheterótrofo, m. chemoheterotroph, chemoheterotrophe.

quimiosíntesis, f. chemosynthesis.

quimiotaxia, f. chemotaxis.

quimiotropismo, m. chemotropism.

quimo, m. chyme.

quincho, m. (CSur, Ch) mud hut.

quinina, f. quinine.

quintal, m. quintal (unidad de peso, aprox. = hundredweight); **q. (castilla) (46 kg), q. métrico (100 kg),**

quironómido, m. chironomid; **quironómidos (Zoo, clase),** Chironomidae (pl).

quiróptero, m. bat, chiropteran.

quitanieves, m.pl. snowplough, snowplow (US).

quitina, f. chitin.

R

rabanito, m. radish.
rábano, m. radish.
rabdomancia, f. dowsing, water divining.
rabdomante, m. dowser, water diviner.
rabia, f. rabies.
rábicola, a. rabid.
rabil, m. yellow fin tuna.
rabión, m. rapids (pl), riffle (ríos).
rabo, m. tail (de animal); stalk, stem (de hojas, de frutas).
racimo, m. raceme.
racionalización, f. rationalization.
racionalizar, v. to rationalize, to streamline.
rada, f. bight (Geog).
radar, m. radar; **pantalla (f) de r.**, radar screen.
radiación, f. radiation; **r. alfa/beta**, alpha/beta radiation; **r. de onda corta**, short-wave radiation; **r. gamma**, gamma radiation; **r. ionizante**, ionizing radiation; **r. solar**, solar radiation; **r. ultravioleta**, ultraviolet radiation.
radiactividad, f. radioactivity.
radial, a. radial.
radiante, a. radiant.
radical, m. radical, root; **r. ácido**, acid radical; **r. libre**, free radical; **medidas (f) radicales**, radical measures; **pelo (m) r.**, root hair; **presión (f) r.**, root pressure.
radícula, f. radicle.
radio, m. radium (Ra); radius (Mat, pl. radii, radiuses); radio (comunicación); **r. de curvatura**, radius of curvature; **r. de influencia**, radius of influence; **r. hidráulico**, hydraulic radius.
radioactividad, f. radioactivity.
radioactivola, a. radioactive; **emisiones (f.pl) radioactivas**, radioactive emissions; **residuos (m.pl) radioactivos de baja actividad**, low level radioactive waste.
radiocarbono, m. radiocarbon.
radiocomunicación, f. radiocommunication.
radiodiagnósticola, a. radiodiagnostic.
radiodifusora, f. (LAm) radio station.
radiofaro, m. radio beacon.
radiofrecuencia, m. radio frequency.
radioisótopo, m. radioisotope.
radiola, f. radiola.
radiolarios, m.pl. radiolaria.
radiolarita, f. radiolarite.
radiómetro, m. radiometer; **r. de radiación neta**, radiometer of net radiation.
radionucleido, m. radionuclide.
radionúclido, m. radionucleus.
radioquímica, f. radiochemistry.
radiorreceptor, m. radio receiver, wireless.
radioseguimiento, m. radio tracking.

radiosonda, f. radiosonde.
radioteléfono, m. radio telephone.
radón, m. radon (Ra).
rádula, f. radula (pl. radulae).
raedera, f. scraper; trowel (llana).
raedura, f. scraping (acción).
ráfaga, f. gust of wind, squall.
raíz, f. root; **constante (m) de r.**, root constant; **echar raíces (v) (Agr)**, to take root; **r. cuadrada (Mat)**, square root; **raíces adventicias**, adventitious roots; **raíces caulogenas**, buttress roots; **raíces fibrosas**, fibrous roots; **raíces zancos**, buttress roots; **zona (f) de r.**, root zone.
rajadura, f. split, crack.
RAM, f. RAM; **memoria (f) de acceso aleatorio (Comp)**, random access memory.
rama, m. branch (de un árbol, de una compañía), field (sector).
ramada, f. (LAm) shelter with branches.
rambla, f. gulley, intermittent stream, bourne, winterbourne, lavant (GB, flujo invernal).
RAMINP, m. **Reglamento (m) de Actividades Molestas, Insalubres, Nocivas y Peligrosas (Esp)**, Regulation concerning dangerous activities, nuisance and activities harmful to health.
ramita, f. bunch, spray (de flores).
rampa, f. ramp, slope; **r. para peces**, fish ladder.
Ramsar, m. Ramsar; **Humedales (m.pl) R.**, Ramsar sites (pl), Wetlands of International Importance.
ranchada, f. (LAm) shed, shanty hut.
ranchería, f. workers quarters.
rancherío, m. (Ur) slum, shanty town.
rancherola, m. y f. (LAm) rancher, farmer; peasant (Méx).
ranchitos, m.pl. (Ven) shanty town.
rancho, m. (Méx) ranch, large farm; shanty town (Col).
randomización, f. randomization, randomizing.
rango, m. rank (clase); luxury (LAm); **r. de variación (Mat)**, range of variation.
ranura, f. furrow.
rapaz, a. predatory (Zoo); rapacious; **aves (f.pl) rapaces**, birds of prey, raptors.
rape, m. angler-fish.
rapeta, f. (Galicia, Esp) dragnet.
rapidez, f. speed, rapidity; **r. de respuesta**, speed of response.
rápidos, m.pl. rapids.
rapiña, f. robbery, theft; **animal (m) de r.**, predator; **ave (f) de r.**, bird of prey.
raquis, m. rachis.
raquitismo, m. rickets, rachitis.
rarefacción, f. rarefaction.

rareza, f. rarity, scarcity; oddity (peculiaridad).
rarola, a. rare.
ras, m. level with; **a r. de tierra**, on level ground; **r. con r.**, level.
rasa, f. wetland.
rascacielos, m.pl. skyscraper.
rascador, m. scraper.
rasgo, m. feature, characteristic; **r. estructural**, structural feature; **r. geográfico**, geographical feature.
raso, m. treeless plain.
raspador, m. scraper.
raspadura, f. scratch.
rastrajero, m. stubble field.
rastrero, m. runner, stolon (Bot).
rastrerola, a. creeping (Bot); **tallo (m) r. (Bot)**, runner.
rastro, m. course (curso), trace (traza).
rastrojo, m. stubble (Agr); **quema (f) de r.**, stubble burning.
ratificación, f. ratification, confirmation.
ratón, m. mouse (Zoo, Comp).
ratonero, m. buzzard (género Buteo).
raya, f. line, streak; **r. de puntos**, dotted line.
rayo, m. ray, beam; streak of lightning; **r. de sol,,** sunbeam; **r. láser**, laser beam; **r. solar**, sunbeam; **rayos cósmicos**, cosmic rays; **rayos infrarrojos**, infrared rays; **rayos ultravioletas**, ultraviolet rays; **rayos X**, X-rays.
rayón, m. rayon.
rayo-X, m. x-ray.
raza, f. race.
razón, f. reason, cause; ratio, proportion (Mat).
razonamiento, m. reasoning.
RDP, m.pl. **residuos (m.pl) tóxicos y peligrosos**, toxic and dangerous wastes.
reabastecimiento, m. replenishment.
reabsorción, f. resorption.
reacción, f. reaction; **r. de oxidación-reducción**, oxidation-reduction reaction; **r. exotérmica**, exothermic reaction; **r. nuclear**, nuclear reaction; **r. química**, chemical reaction; **r. redox**, redox reaction; **r. termonuclear**, thermonuclear reaction.
reaccionar, v. to react.
reactante, m. reactant (Quím).
reactivación, f. reactivation, rejuvenation.
reactivo, m. reagent.
reactivola, a. reactive.
reactor, m. reactor; **r. nuclear**, nuclear reactor.
reaireación, f. reaeration.
real, m. (Bra) real (unidad monetaria).
realineación, f. realignment.
realineamiento, m. realignment.
realización, f. fulfilment, realization; happening (actuación); broadcast (Comm).
realizar, v. to carry out, to accomplish.
reanimación, f. resuscitation; **aparato (m) de r.**, resuscitation apparatus.
reanimar, v. to resuscitate.
rebajamiento, m. lowering; planation (Geog).

rebanada, f. slice.
rebaño, m. flock (of sheep, goats); herd (of other animals).
rebasamiento, m. overflow (p.e. de embalse), surcharge (Geog).
rebosadero, m. spillway (de una presa).
rebose, m. overflow, spill.
rebote, m. rebound; **r. de agua subterránea**, groundwater rebound.
rebuscador, m. (Ven) waste tip scavenger.
recalcar, v. to emphasize, to stress.
recalentamiento, m. warming; **r. atmosférico**, atmospheric warming.
recarga, f. recharge; **capacidad (f) de r.**, recharge capacity; **r. artificial**, artificial recharge; **r. con aguas residuales**, wastewater recharge; **r. de aguas subterráneas**, groundwater recharge; **r. del acuífero/río**, aquifer/river recharge; **r. inducida**, induced recharge; **r. litoral**, coastal recharge; **r. natural**, natural recharge; **r. neta**, net recharge; **r. por lluvia**, rainfall recharge.
recargar, v. to raise, to increase; to overload (sobrecargar); to recharge (una batería etc.).
recaudación, f. takings, collection (de impuestos, de basura); **r. de basura**, rubbish collection.
recencio, m. fresh breeze.
receptor, m. receptor.
receptorla, a. receiving; **agua (f) del medio r.**, **cauce r. (Trat)**, receiving water; **estación (f) r.**, receiving station.
recesión, f. recession; **curvas (f) de recesión**, recession curves.
recesivola, a. recessive.
reciclable, a. recyclable.
reciclado, m. recycling.
reciclador, m. recycler, recycling machine.
reciclaje, m. recycling; **planta (f) de r.**, recycling plant.
reciclar, v. to recycle.
reciente, a. recent.
recinto, m. enclosure; **r. cercado**, compound.
recipiente, m. recipient (persona); receptacle (vaso); **r. de aforo**, flow measuring tank.
recíproco/a, a. reciprocal.
recirculación, f. recirculation.
recircular, v. to recirculate.
reclamación, f. reclamation.
recobrar, v. to recover.
recocido, m. annealing.
recodo, m. river bend.
recogedor, m. collector; **r. de partículas (Trat)**, grit collector, grit trap, grit arrester (US).
recoger, v. to collect, to gather, to harvest.
recogida, f. collection, harvest; **r. de basura**, rubbish collection.
recolección, f. collection (Gen); harvesting (of vegetables); harvest time; compilation, summary (resumen); **r. de muestras**, collection of samples, sampling.
recolectar, v. to collect.

recolector, m. (Ur) waste tip scavenger.
recolonizar, v. to recolonize.
recomendación, f. recommendation.
recomendar, v. to recommend.
reconocer, v. to recognize (p.e. una persona), to identify (distinguir); to survey (terreno); to check (averiguar).
reconocimiento, m. recognition; reconnaissance (Mil); **r. aéreo**, aerial survey; **r. médico**, medical check-up.
reconvertir, v. to reconvert, to retrain.
recopilación, f. summary (resumen); compilation (colección).
recopilar, v. to summarize (resumir); to compile (reunir).
recorrer, v. to travel through, to tour, to cover.
recorrido, m. journey, route; distance covered (distancia); overhaul (arreglo).
recorte, m. cuttings; cutting out (acción); **recortes de periódicos**, newspaper cuttings.
recristalización, f. recrystallization.
recta, f. straight line; **r. de descenso-tiempo**, time-drawdown curve; **r. de regresión**, regression curve.
rectangular, a. rectangular; **vertedero (m) r. (medidor de flujo)**, rectangular weir.
rectángulo, m. rectangle.
rectificación, f. rectification.
rectificadola, a. rectifying, correcting.
rectificar, v. to rectify, to put right.
rectilíneola, a. rectilinear.
recto, m. straight line (Mat); straight stretch (ríos, caminos etc.); rectum (Anat).
rectola, a. straight (derecho), upright (vertical); **ángulo (m) r.**, right angle.
recubrimiento, m. coating, covering; **r. electrolítico**, metal plating; **r. metálico**, metal plating.
recuperación, f. recovery, remediation, rehabilitation; **r. biológica in situ**, in situ biological remediation; **r. de agua subterránea**, groundwater remediation; **r. de confinamiento**, landfill remediation; **r. de desechos**, waste-to-energy; **r. de disolventes**, solvent recovery; **r. de energía**, waste-to-energy; **r. del acuífero**, aquifer rehabilitation; **r. de la tierra al mar**, land reclamation; **r. de suelo contaminado**, contaminated land remediation; **r. en el sitio**, on-site remediation; **r. informativa**, information retrieval; **r. in situ**, on-site remediation; **tecnologías (f.pl) de r.**, remediation technologies; **tiempo (m) de r.**, recovery time.
recuperador, m. recuperator.
recuperar, v. to recover (un objeto).
recurrencia, f. recurrence; **intervalo (m) de r.**, recurrence interval.
recurrente, a. recurrent.
recursivola, a. recursive.
recurso, m. resource; means (medio).
recursos, m.pl. resources; **gestión (f) de r. hidráulicos**, water resource management; **r. de agua**, water resources; **r. de agua**

subterránea, groundwater resources; **r. disponibles**, available resources; **r. financieros**, financial resources; **r. genéticos**, genetic resources; **r. hídricos**, water resources; **r. humanos**, human resources; **r. minerales**, mineral resources; **r. naturales**, natural resources; **r. no renovables**, non-renewable resources; **r. pesqueros**, fish stocks; **r. renovables**, renewable resources.
red, f. net, network, mesh; **r. alimentaria**, food web; **r. cristalina**, crystal lattice; **r. de arrastre**, drag net; **r. de autovías**, motorway network, highway network; **r. de canales**, channel network; canal network (de canales artificales); **r. de carreteras**, road network; **r. de distribución**, distribution network; **r. de drenaje**, drainage network; **r. de enmallado (pesca)**, gill net; **r. de estaciones de aforos**, flow gauging network; **r. de ferrocarril**, rail network; **r. de flujo**, **r. de percolación (aguas freáticas)**, flow net; **r. de vigilancia**, monitoring network; **r. sísmica**, seismic network.
redacción, f. writing (acción); **equipo (m) de r.**, editors; **primera r.**, first copy, first draft; **r. de informes**, writing of reports.
redactar, v. to draft, to draw up.
redeposición, f. redeposition.
redepositadola, a. redeposited.
redepositar, v. to redeposit.
redesarrollo, m. redevelopment.
redil, m. sheep pen.
redondez, f. roundness.
redox, m. redox; **condiciones (f.pl) de r.**, redox conditions; **potencial (m) r.**, redox potential (abr. Eh); **reacción (f) r.**, redox reaction.
reducción, f. reduction; **r. bacteriana**, bacterial reduction; **r. de hierro**, iron reduction; **r. de sulfatos**, sulphate reduction.
reductible, a. reducible.
reductor, m. reducer (Quím).
reductorla, a. reducing; **medio (m) r.**, reducing environment.
reemplazar, v. to replace, to substitute.
reemplazo, m. replacement.
referencia, f. reference, cross-reference, marker; **nivel (m) de r.**, reference datum; **plano (m) de r.**, reference plane.
refinación, f. refining.
refinado, m. refining; **r. de petróleo**, oil refining.
refinamiento, m. refining; **r. de petróleo**, oil refining; **r. por fundición**, smelting.
refinería, f. refinery; **r. de petróleo**, oil refinery.
reflejo, m. reflex; **r. condicionado/incondicionado (Biol)**, conditioned/unconditioned reflex; **r. pasivo (Biol)**, passive reflex; **acción (f) r.**, reflex action.
reflexión, f. reflection.
reflujo, m. ebb tide (Mar); backflow, reflux; **el flujo (m) y r.**, ebb and flow; **válvula (f) de r.**, reflux valve.
reforestación, f. reforestation.

reforma, f. reform; **r. agraria**, land reform; **r. legislativa**, legislative reform.

reforzar, v. to reinforce, to strengthen.

refracción, f. refraction; **índice (m) de r.**, refractive index; **r. de ondas**, wave refraction.

refractariola, a. refractory, heat-resistent.

refrigeración, f. refrigeration; **r. por aire**, air cooling; **torre (m) de r.**, cooling tower.

refrigerador, m. fridge, refrigerator.

refrigerante, 1. m. refrigerant; 2. a. refrigerant.

refucilo, m. flash of lightening.

refuerzo, f. reinforcement, support.

refugio, m. refuge, shelter; harbourage, harborage (US) (Mar).

regadera, f. watering device; small irrigation channel; sprinkler (pulverizador).

regadío, m. irrigation, irrigated land; **cultivo (m) de r.**, crops grown on irrigated land; **tierra (f) de r.**, irrigated land.

regajío, m. stagnant pond.

regar, v. to irrigate, to water.

regata, f. irrigation channel.

regeneración, f. regeneration; **r. de playa**, beach replenishment; **r. industrial**, industrial regeneración.

regeneradorla, a. regenerative.

regenerar, v. to regenerate.

regenerativola, a. regenerative; **capacidad (f) r.**, regenerative capacity.

régimen, m. regime, rules, system; diet (food); **r. alimenticio**, dietary regime; **r. de flujo**, flow regime; **r. estacionario**, steady-state system; **r. hidráulico**, hydraulic system; **r. no permanente**, non steady-state system; **r. turbulento**, turbulent system.

región, f. region, area; **r. perturbada**, disturbed region.

regional, a. regional; **desarrollo (m) r.**, regional development.

registrador, m. recorder, register; **r. acústico**, acoustic recorder; **r. automático**, automatic recorder.

registrar, v. to search (buscar); to examine (examinar); to register (anotar); to occur (suceder).

registro, m. register, registration, recording, log; **r. analógico**, analogue recorder; **r. de conductividad**, conductivity log; **r. de datos numéricos**, digital data recording; **r. de gamma-gamma**, gamma-gamma log; **r. de gamma natural**, natural gamma log; **r. de la propiedad**, land registry; **r. de perforación**, boring log, borehole log; **r. de potencial espontáneo**, spontaneous potential log; **r. de pruebas**, test log; **r. de resistividad**, resistivity log; **r. de salinidad**, salinity log; **r. de temperatura diferencial**, differential temperature log; **r. digital**, digital recorder; **r. electoral**, electoral register; **r. geofísico**, geophysical log; **r. geológico**, geological record; **r. secuencial de datos**, data logging; **r. sísmico**, seismic log.

regla, f. rule, regulation; ruler (utensilo).

reglamentación, f. regulation, regulations (pl); rules (pl) (reglas).

reglamentariola, a. prescribed, obligatory, required; **flujo (m) r.**, prescribed flow.

reglamento, m. regulation, regulations (pl), rules (pl); **R. de la Unión Europea**, EU Regulation.

regolito, m. regolith.

regresión, f. regression; **r. lineal**, linear regression; **r. marina**, marine regression; **r. múltiple**, multiple regression; **r. polinómica**, polynomial regression.

reguera, f. small irrigation channel.

regulación, f. regulation, control; **r. de cuenca**, basin regulation; **r. fluvial**, river regulation.

regulador, 1. m. regulator; 2. a. regulatory.

regularidad, f. regularity.

rehabilitación, f. rehabilitation.

rehabilitar, v. to reinstate.

reina, f. queen; **abeja (f) r.**, queen-bee.

reino, m. kingdom; **r. animal**, animal knigdom; **R. Unido**, United Kingdom (abr. UK); **r. vegetal**, plant kingdom.

reintroducción, f. reintroduction.

reinversión, f. reinvestment.

reinvertir, v. to reinvest.

reiteración, f. reiteration, follow-up.

reja, f. grille, grid.

rejalgar, m. realgar.

rejilla, f. screen, latticework; **r. de pozo de persiana**, slotted well screen; **r. de pozo de puentecillos**, bridge-slotted screen.

rejuntado, m. pointing (de ladrillos).

rejuvenecimiento, m. rejuvenation.

relación, f. relation, relationship; ratio (Mat, razón), proportion (Mat, proporción); account, report (relato); list (lista); **r. carbono-nitrógeno (abr. r. C/N)**, carbon-nitrogen ratio, (abr. C/N ratio); **r. costobeneficio**, cost-benefit ratio; **r. de agua dulce-agua salada**, freshwater-saltwater ratio; **r. de almacenamiento/transmisividad**, storativity/transmissivity ratio (abr. S/T ratio); **r. de helio-argón**, helium-argon ratio; **r. de storatividad/transmisividad**, storativity/transmissivity ratio (abr. S/T ratio); **r. de sumersión**, submergence ratio; **r. hidráulica**, hydraulic ratio; **r. isotópica**, isotopic ratio.

relámpago, m. lightning, flash of lightning; **r. difuso**, sheet lightning.

relativola, a. relative; **densidad (f) r.**, relative density.

relaves, m.pl. tailings (Min).

relé, m. relay (Elec).

relicto, m. relict (Geol).

relieve, m. relief (Geog); importance (importancia); **bajo r. (Geog)**, low relief; **mapa (m) de r.**, relief map.

reliquia, f. relic; after-effect (Med).

rellano, m. landing (de escalera), shelf (estante).

rellenar, v. to pack, to fill (un hoyo).
relleno, m. filling, packing; fill (Min); landfill (LAm) (vertedero); **r. arcilloso,** clay fill; **r. de seguridad,** secure landfill.
remachar, v. to rivet.
remache, m. rivet, riveting (acción).
remanal, m. land with spring flows.
remanso, m. backwater; **bañado (m) de r. (LAm),** backwater swamp; **r. hidráulico,** backwater.
remediación, f. remediation; **medidas (f.pl) de r.,** remedial measures.
remediador/a, a. remedial.
remediar, v. to remedy, to remediate.
remedio, m. remedy.
remojo, m. soaking, drenching, steeping.
remolacha, f. beet; **r. azucarera,** sugar-beet.
remolino, m. swirl, eddy; whirlpool (en un río); whirlwind (viento).
remolque, m. trailer (detrás de un vehículo); caravan (de turismo).
remover, v. to stir, to disturb (mezclar).
removilización, f. remobilization.
renacuajo, m. tadpole (Zool).
rendija, f. chink.
rendimiento, m. efficiency, performance, yield, output; **r. de bombeo,** pumping output; **r. de cultivos,** crop yield; **r. decreciente,** diminishing return; **r. de la cosecha,** crop yield; **r. específico,** specific yield; **r. máximo sostenible,** maximum sustainable yield.
rendzina, f. rendzina.
renegociación, f. renegotiation.
renegocionar, v. to renegotiate.
renina, f. rennin.
renio, m. rhenium (Re).
renovable, a. renewable; **energía (f) r.,** renewable energy; **recursos (m.pl) renovables,** renewable resources.
renovación, f. renovation, renewal; **r. urbana,** urban renewal.
renovar, v. to renew.
rentabilidad, f. profitability.
rentabilizar, v. to make profitable.
rentable, a. profitable; **hacer r. (v),** to make profitable; **poco r.,** unprofitable.
renunciar, v. to renounce, to waive (un derecho).
reología, f. rheology.
reorganización, f. reorganization.
reoxigenación, f. reoxygenation.
reparador/a, a. remedial, reparative; **medidas (f.pl) reparadoras,** remedial measures.
repartición, f. division, distribution.
reparto, m. sharing-out, distribution; **r. de costos,** sharing-out of costs.
repercusión, f. repercussion, reverberation.
reperforación, f. redrilling.
replantear, v. to lay out (un edificio nuevo).
replanteo, m. layout (de un edificio).
replegamiento, m. refolding (Geol).

réplica, f. replica (Zoo, Bot); answer (respuesta).
replicación, f. replication.
replicar, v. to replicate (Bot, Zoo); to answer back (contestar).
repoblación, f. repopulation; **r. forestal,** reforestation.
reportaje, m. report, article (Comm).
represa, f. dam (en un río), millpond (de molino).
represamiento, m. damming, impounding (acción).
represar, v. to dam, to dam up.
representación, f. representation; **r. doble logarítmica,** double log representation; **r. gráfica,** graphical representation; **r. regional,** regional office (de una organización); **r. semilogarítmica,** semi-logarithmic representation.
representante, m. y f. salesman (m), saleswoman (f), salesperson (m, f).
representativo/a, a. representative; **muestra (f) r.,** representative sample.
reprocesamiento, m. reprocessing; **planta (f) de r.,** reprocessing plant.
reproducción, f. reproduction; recurrence (Med); **derechos (m.pl) de r.,** copyright; **r. asexual/sexual,** asexual/sexual reproduction; **r. vegetativa,** vegetative reproduction.
reproducir, v. to reproduce.
reproductividad, f. reproductivity.
reptación, f. creep, creeping; **r. del suelo,** soil creep.
reptar, v. to creep (p.e. suelos).
reptil, 1. m. reptile; **reptiles (Zoo, clase),** Reptilia (pl); 2. a. reptile, reptilian.
República Dominicana, f. Dominican Republic.
requerimiento, m. injunction, summons (Jur).
requerir, f. to urge, to beg, to request.
requisito, m. requisite, requirement; **r. previo,** prerequisite.
resaca, f. undertow, undercurrent (de marea); backlash (reacción); hangover (Med).
resalte, m. ledge, projection; **r. de pendiente,** nickpoint.
resbalamiento, m. slip, slide; creep (Geol); **r. de suelo,** soil slip.
rescatador, m. (PRico) waste tip scavenger.
resero, m. cowboy, herdsman.
reserva, f. reserve, reservation, stock; **absoluta r.,** strictest confidence; **de r.,** standby (a); **fichero (m) de r. (Comp),** back-up file; **r. de la naturaleza,** nature reserve; **Reservas de la Biosfera,** Biosphere Reserves (pl); **reservas de petróleo,** oil reserves, petroleum reserves; **reservas minerales,** mineral reserves; **sustancias (f.pl) de r. (Bot),** cell sap.
reservas, f.pl. stockpile.
reservorio, m. (LAm) reservoir.
resfriado/a, a. cooled, cold; **estar r. (v) (Med),** to have a cold; **r. común (Med),** common cold.

residencial, a. residential.
residual, a. residual.
residuo, m. residue; remainder (Mat); **r. químico,** chemical residue; **r. radioactivo,** radioactive residue; **r. seco,** dry residue; **r. tóxico,** toxic residue.
residuos, m.pl. waste, residues; **almacenamiento (m) de r. peligrosos,** hazardous waste storage; **disposición (f) de r. peligrosos,** hazardous waste disposal; **disposición (f) de r. sólidos,** solid waste collection; **gestión (f) de r.,** waste management; **inyección (f) subterránea de r. peligrosos,** hazardous deep well injection; **procesamiento (m) de r. peligrosos,** hazardous waste processing; **recogida (f) y tratamiento (m) de r.,** waste collection and disposal; **recolección (f) de r. sólidos,** solid waste collection; **recuperación (f) de r. sólidos,** solid waste recovery; **r. atmosféricos,** fallout; **r. de alta peligrosidad,** special category wastes; **r. de minas de carbón,** colliery waste, colliery spoil; **r. de parque de chatarra,** scrapyard waste; **r. domésticos,** household waste; **r. especiales,** special wastes; **r. hospitalarios,** hospital wastes; **r. municipales,** domestic waste, city garbage (US); **r. municipales y asimilables,** domestic and recyclable waste; **r. nucleares,** nuclear waste; **r. peligrosos,** dangerous waste, hazardous waste; **r. plásticos,** waste plastic; **r. radioactivos (de baja/alta actividad),** (low/high level) radioactive waste; **r. sólidos urbanos (abr. RSU),** urban refuse; **r. tóxicos y peligrosos (abr. RDP),** hazardous and dangerous wastes; **r. venenosos,** poisonous wastes; **reutilización (f) de r. sólidos,** solid waste reuse; **segregador (m) de r. sólidos,** solid waste segregator; **separación (f) de r. sólidos,** solid waste segregation; **transporte (m) de r. peligrosos,** hazardous waste transport; **tratamiento (m) de r. peligrosos,** hazardous waste treatment.
resiliencia, f. resilience.
resina, f. resin; **r. epoxi,** epoxy resin.
resinar, v. to tap (sangrar un árbol).
resistencia, f. resistence; strength (fuerza, dureza); drag (aerodinámico); stamina (aguante); **r. a la compresión, r. al aplastamiento,** compressive strength; **coeficiente (m) de r.,** drag coefficient; **r. a la ruptura,** rupture strength.
resistividad, f. resistivity; **r. aparente,** apparent resistivity; **r. de terreno,** ground resistivity; **r. eléctrica,** electrical resistivity.
resistivímetro, m. resistivimeter.
resolución, f. resolution.
resolver, v. to resolve, to solve (solucionar).
resonancia, f. resonance.
resorción, f. resorption.
resorte, m. spring (muelle).
respiración, f. respiration; **r. aerobia,** aerobic respiration; **r. anaerobia,** anaerobic respiration; **r. artificial,** artificial respiration; **r. tisular,** tissue respiration.
respiradero, m. ventilator, air vent; **r. en una mina,** ventilation shaft.
respirador, m. respirator.
respirar, v. to breathe.
respirómetro, m. respirometer.
resplandor, m. brightness, brilliance.
responsabilidad, f. responsibility; liability (Jur); **r. civil,** public liability; **r. conjunta,** joint liabilty.
responsabilización, f. placing responsibility (Jur).
respuesta, f. response; **tiempo (m) de r.,** response time.
resquebrajadola, a. breached, cracked open.
resquebrajar, v. to break open.
restar, v. to take away, to subtract (Mat); to remain (quedar).
restauración, f. restoration; **r. de lugares contaminados,** site restoration; **r. de terrenos,** land restoration.
restaurar, v. to restore, to rehabilitate.
restingua, f. (Perú) wide river valleys subject to flooding.
resto, m. rest, remainder (Mat); balance (de una cuenta).
restos, m.pl. wreckage (de una máquina etc.), remains; **r. humanos,** human remains.
restricción, f. restriction, limitation; **r. del consumo de agua,** water restriction.
restructuración, f. restructure, restructuring.
resumen, m. résumé, summary, abstract.
resumir, v. to summarize; to sum up (recapitular).
retamal, m. field of broom.
retardadola, a. delayed; **drenaje (m) r.,** delayed storage, delayed yield.
retardo, m. lag, delay; **coeficiente (m) de r.,** retardation coefficient.
retención, f. retention; **período (m) de r. (Trat),** retention time; **r. de avenidas,** flood retention.
reticular, a. reticular; **red (f) r.,** reticular network.
Retiense, m. Rhaetian, Rhaetic.
retina, f. retina (pl. retinas, retinae).
retoque, m. retouching, touching up.
retornable, a. returnable.
retorno, m. return, exchange; **período (m) de r.,** return period; **r. de aguas residuales,** waste-water return; **r. financiero,** financial return.
retorta, f. retort (Quím).
retrabajadola, a. reworked.
retracción, f. retraction.
retranqueo, m. offset (Con).
retrasar, v. to delay, to put off; to slow down (retardar).
retraso, m. lag, delay, slowness; **r. de tiempo,** time-lag.
retrete, m. latrine, water closet; **r. seco,** earth closet, dry latrine.

retroalimentación, f. feedback.
retrocavador, m. back-acting digger.
retroceso, m. retreat, recession; backwash (Ing); **r. de agua subterránea,** groundwater recession; **r. glaciar,** glacial recession.
retrocruzamiento, m. back-cross (genética).
reubicación, f. resiting, repositioning.
reuma, reúma, f. rheumatism.
reumatismo, m. rheumatism.
reutilización, f. reuse, reutilization; **r. de aguas residuales,** wastewater reuse.
revegetación, f. revegetation.
revelado, m. developing (fotografía).
revestimiento, m. lining, liner, coating, panelling, paneling (US), sheeting; **r. arcilloso (Trat, en un vertedero),** clay liner; **r. compuesto (Trat, en un vertedero),** composite liner; **r. del pozo,** borehole relining; **r. de membrana (Trat, en un vertedero),** membrane liner; **r. de vertedero,** landfill liner; **r. tosco,** roughcast.
revisión, f. revision, check, review; overhaul (de una máquina); **r. de cuentas,** audit (of accounts); **r. de precios,** price review.
revitalizar, v. to revitalize.
revocar, v. to revoke, to cancel.
revolución, f. revolution; **r. industrial,** industrial revolution; **r. verde,** green revolution; **revoluciones por minuto,** revolutions per minute (abr. rpm).
revoque, m. rendering, resurfacing, plastering.
rezumar, v. to seep, to ooze, to exude.
rezume, m. seepage, leak.
Rh, a. Rh; **Rhesus (m),** Rhesus.
rhabdosoma, f. rhabdosome.
Rhesus, a. Rhesus; **factor (m) R.,** Rhesus factor.
rhizópodo, m. rhizopod.
rhynconélido, m. rhynchonellid.
ría, f. ria, estuary, river mouth.
riachuelo, m. brook, stream.
riada, f. flood.
riatillo, m. runnel.
ribazo, m. foreshore.
ribera, f. riverbank, riverside, bank (de río); beach, shore (del río o del mar); **almacenamiento (m) en la r.,** bankside storage; **bosque (m) de r.,** riverside woodland.
ribereño, m. y f. person living alongside a river.
ribereñola, a. riparian, waterfront; **bosque (m) r.,** riparian woodland, riverside woodland; **derechos (m.pl) ribereños,** riparian rights.
riboflavina, f. riboflavin (vitamina B2).
riebeckita, f. riebeckite.
riego, m. irrigation; **r. agrícola,** agricultural irrigation, crop irrigation; **r. con aguas residuales,** wastewater irrigation; **r. de aspersión,** sprinkler irrigation; **r. gota a gota,** dripfeed irrigation, trickle irrigation; **r. por aspersión;** spray irrigation, overhead irrigation, sprinkler irrigation; **r. por compartimientos,** basin irrigation; **r. por goteo,** dripfeed irrigation,

trickle irrigation; **r. por surcos,** channel irrigation.
riesgo, m. risk, danger; **análisis (m) de riesgos,** risk analysis; **de alto r.,** high-risk; **de bajo r.,** low-risk; **evaluación (f) de riesgos,** risk assessment; **gestión (f) de riesgos,** risk management; **r. para la salud,** health risk; **valoración (f) de r.,** risk assessment.
rift, m. rift; **bloque (m) de r.,** rift block.
rigidez, f. rigidity, stiffness.
rigidizador, m. stiffener.
rígidola, a. rigid; inflexible inelastic (inflexible).
rigor, m. severity (severidad); rigour, rigor (US) (exactitud); precision (meticulosidad).
rigurosola, a. rigorous, strict; accurate (preciso).
ringlera, f. row, swath.
rinoceronte, m. rhinoceros.
riñón, m. kidney.
río, m. river; **boca (f) de r.,** river mouth; **captación (f) en ríos,** river catchment; **cauce (m) de un r.,** river channel; **contaminación (f) de los ríos,** river pollution; **cuenca (f) del r.,** river basin; **desembocadura (f) del r.,** river mouth; **lecho (m) del r., madre (m) del r.,** river bed; **margen (m) de r.,** river bank; **parte (f) recta de un r.,** reach of a river; **r. anastomosado,** braided river, braided stream; **r. desproporcionado,** misfit river, misfit stream; **r. en equilibrio,** graded river; **r. estacional, r. intermitente,** intermittent river, intermittent stream, bourne (GB); **r. truncado,** truncated river, truncated stream; **tramo (m) recto de un r.,** reach of a river.
riolita, f. rhyolite.
riostra, f. brace, strut (Con).
ripariola, a. riparian.
ripícola, f. river-dwelling.
ripio, m. rubble, broken stone, rap, filling; riprap (US).
riqueza, f. wealth, riches.
risco, m. crag, cliff; **r. y cola (Geog),** crag and tail.
ritmo, m. rhythm; rate, pace (velocidad); **r. circadiano, r. (m) diurno,** diurnal rhythm; **r. de trabajo,** rate of work.
rizadola, a. ripply, curly.
rizoesfera, f. rhizosphere.
rizoide, m. rhizoid.
rizoma, m. rhizome.
roble, m. oak; **agalla (f) de r.,** oak apple.
robledal, m. oak grove.
robótica, f. robotics.
robustecimiento, m. strengthening, fortifying.
roca, f. rock, boulder; **caída (f) de r.,** rock fall; **cubierta (f) de r.,** rock cover, regolith; **r. almacén,** reservoir rock; **r. alterada,** altered rock; **r. de aureola de contacto,** rimstone; **r. ferruginosa,** ironstone; **r. firme,** bedrock; **r. ignea,** igneous rock; **r. jabón,** soapstone; **r. madre,** country rock; **r. metamórfica,** metamorphic rock; **r. productiva,** reservoir rock;

r. sedimentaria, sedimentary rock; **rocas verdes,** greenstone; **sal (f) de r.,** rock salt.
roce, m. friction (fricción); attrition, rubbing (fregamiento).
rociada, f. spray (líquido), spraying (acción).
rociar, v. to spray, to sprinkle.
rocío, m. dew; **punto (m) de r.,** dew point.
rocosola, a. rocky; **desperdicio (m) r.,** rock waste.
rodadola, a. on wheels, vehicular; **tráfico (m) r.,** road traffic, vehicular traffic.
rodamiento, m. roller bearing.
rodamina, f. rhodamine.
rodear, v. to surround, to enclose.
rodenticida, m. rodenticide.
rodilla, f. knee.
rodio, m. rhodium (Rh).
rodocrosita, f. rhodochrosite.
rodofícea, f. red alga (pl algas), rhodophyte; **rodofíceas (Bot, clase),** Rhodophyta (pl).
roedor, m. rodent.
roentgen, m. roentgen; **r. equivalente para el hombre,** roentgen equivalent man (abr. REM).
rogar, v. to beg, to request.
rojola, a. red; **Libro (m) R. (Ecol),** Red Book.
rollo, m. roll (de papel); coil (de cuerda).
ROM, f. ROM; **memoria (f) de sólo lectura (Comp),** read-only memory.
romanola, a. Roman.
romboide, m. rhomboid (Mat).
romeral, f. field of rosemary.
romola, a. blunt, dull.
rompehielos, m.pl. icebreaker.
rompeolas, m.pl. breakwater, seawall, sea defences.
rompiente, m. reef, shoal; **rompientes,** surf, breakers.
ronda, f. rampart, parapet walk.
ronza, f. flood-transported vegetation, rack, flotsam.
ropalócero, m. butterfly.
roquedal, m. rocky place.
rorcual, m. finback whale (ballena).
rosa, f. rose (flor, en mapa); **r. de los vientos (Met),** wind rose.
rosal, m. rose bush; **r. silvestre,** wild rose, dog rose.
roscadola, a. threaded, spiral-shaped.
rostro, m. rostrum, face.
rotación, f. rotation, turn; **perforación (f) por r.,** rotary drilling; **r. de cultivos, r. de la cosecha,** crop rotation.
rotacional, a. rotational; **deslizamiento (m) r.,** rotational slide.
rotavirus, m. rotavirus.
rotola, a. broken, shattered.

rotopercusión, f. rotary percussion.
rotor, m. rotor.
rótula, f. kneecap, patella (de rodilla); ball and socket joint (Gen).
rótulo, m. label, tag (etiqueta); title, heading (título); poster (cartel).
rotura, f. breaking, smashing; fracture (quiebra); failure (fracaso); **r. por cizalla,** shear failure; **r. por tracción,** tension failure.
roturación, f. rotavation.
roturar, v. to plough, to plow (US) (arar), to rotovate.
roya, f. mildew, rust, blight (en plantas).
roza, f. clearance (de terreno).
rozadura, f. abrasion.
rozamiento, m. friction, rubbing.
rpm, f.pl. rpm; **revoluciones (f.pl) por minuto,** revolutions per minute.
RS, m.pl. **residuos (m.pl) sólidos,** solid wastes.
RSU, m.pl. **residuos sólidos urbanos (Esp),** solid urban waste.
rubéola, f. rubella, German measles.
rubi, m. ruby.
rubidio, m. rubidium (Rb).
rudáceola, a. rudaceous.
ruderal, a. ruderal.
rudita, f. rudite.
rueda, f. wheel; circle (círculo); **r. de molino,** millstone; **r. dentada,** cog, cogwheel; **r. hidráulica,** waterwheel; **r. libre,** freewheel.
rugosidad, f. rugosity, roughness; **coeficiente (m) de r.,** coefficient of rugosity.
ruido, m. noise; sound (sonido); **contaminación (f) por r.,** noise pollution; **control (m) de r.,** noise control; **r. callejero,** road noise; **r. de fondo,** background noise; **r. irritante,** noise nuisance; **reducción (f) de r.,** noise reduction; **ruidos y vibraciones,** noise and vibration.
ruina, f. ruin(s), remains.
rumbo, m. direction, course, bearing.
rumen, m. rumen.
rumiante, m. ruminant.
Rupeliense, m. Rupelian.
rupestre, a. rupestrian; **pintura (f) r.,** cave painting.
rupícola, a. rock-dwelling.
ruptura, f. rupture, fracture.
rural, a. rural; **abastecimiento (m) r. de agua,** rural water supply; **desarrollo (m) r.,** rural development; **éxodo (m) r.,** rural depopulation; **planificación (f) r.,** rural planning; **población (f) r.,** rural population.
rutenio, m. ruthenium (Ru).
rutilo, m. rutile.
rutinariola, a. routine; **prueba (f) r.,** routine check.

S

s.a., f. ltd., plc, Corp. (US), Inc. (US); **sociedad (f) anónima de capital variable**, limited company (GB), Incorporated (US); **sociedad (f) anónima**, public limited company (GB), joint stock company, corporation (abr. Corp, US).

sabana, f. savanna, savannah.

sabandija, f. bug (bicho); **sabandijas**, vermin.

sabinio, m. sabin (unidad de absorción acústica).

sabor, m. flavour, flavor (US); taste (sentido).

sacáridola, a. saccharide.

sacarina, f. saccharine.

sacarita, f. (Perú) channel-cutting meander.

sacaromiceto, m. saccharomycete.

sacarosa, f. saccharose.

sacatestigos, m.pl. core barrel.

saco, m. sack, bag; sac (Bot, Zoo); **s. aéreo**, air sac; **s. polínico**, pollen sac; **s. vitelino**, yolk sac.

sacudida, f. shaking (p.e. de un terremoto); strike, blow (golpe).

sacudir, v. to shake, to agitate.

saetín, m. millrace, flume, leat (de molino de agua).

sal, f. salt; **s. común**, common salt; **s. de Epsom**, Epsom Salts; **s. gema**, rock salt; **sales disueltas**, dissolved salts; **sales minerales**, mineral salts.

salada, f. (Méx, US) small salt lake.

saladar, m. salt marsh.

saladola, a. salty.

saladura, f. salting.

salamandra, f. salamander.

salamanquesa, f. salamander.

salar, v. to salt.

salario, m. salary, wage.

salazón, m. salting; **s. en salmuera**, brining.

salgüero, m. mineralized soil with hardpan.

salicilato, m. salicylate.

salida, f. exit, outlet (Geog), departure, leak (Ing).

salina, f. salt works, salt mine.

salinidad, f. salinity.

salinización, f. salinization; **s. de suelo**, soil salinization.

salinola, a. saline; **cuña (f) s. (Geol)**, saline wedge.

salita, f. salt pan.

salitre, m. Chilean nitrate (nitrato de Chile), saltpetre.

salitrosa, a. rich in mineral salts.

saliva, f. saliva.

salmón, m. salmon.

salmonete, m. mullet.

salmuera, f. brine, pickle.

salobre, a. brackish.

salpicadura, f. splash (mancha), splashing (acción).

saltamontes, m. grasshopper.

salto, m. jump, leap, vault; **s. vertical de una falla**, throw of fault.

saltos, m.pl. falls, waterfall.

salud, f. health; **atención (f) primaria de la s.**, primary health care; **riesgo (m) para la s.**, health risk; **s. ocupacional**, occupational health.

salvado, m. bran.

salvadoreñola, 1. m. y f. Salvadoran, Salvadorian; 2. a. Salvadoran, Salvadorian.

salvaguardar, v. to safeguard.

salvaguardia, f. safeguard.

salvaje, a. wild (de tierra, de animales feroces).

salza, f. (US) small saline spring discharge.

samario, m. samarium (Sm).

samita, f. psammite.

samíticola, a. psammitic.

sanción, f. sanction.

sandía, f. watermelon.

saneamiento, m. drainage (de terreno); sanitation; stabilization (de moneda); clean-up (limpieza); **Planes (m.pl) de S. (Esp)**, clean-up plans; **política (f) de s.**, sanitation policy; **s. de emergencia**, emergency sanitation; **s. de la vivienda**, housing sanitation; **s. del terreno**, land drainage; **s. rural**, rural sanitiation.

sangre, f. blood; **de s. caliente**, warm-blooded; **de s. fría**, cold-blooded; **envenenamiento (m) de la s.**, blood poisoning; **pura s.**, thoroughbred (caballo).

sangüeño, m. areas of saline water discharge.

sanguijuela, f. leech.

sanguíneola, a. blood; **glóbulo (m) rojo s.**, red blood cell; **grupo (m) s.**, blood group; **transfusión (f) s.**, blood transfusion.

sanidad, f. health; **s. pública**, public health.

sanitario, m. water closet.

sanitariola, a. sanitary, health; **atención (f) s.**, health care; **ingeniería (f) s.**, sanitary engineering; **instalación (f) s.**, sanitary installation; **sistema (m) de alcantarillado s.**, sanitary sewerage system.

Santoniense, m. Santonian.

santuario, m. sanctuary, shrine.

saponificación, f. saponification.

saprobióticola, a. saprobiotic; **zona (f) s.**, saprobiotic zone.

saprófito, m. saprophyte.

saprófitola, a. saprophytic.

saprolito, m. saprolith, saprolite.

saprozoicola, a. saprozoic.

sarampión, m. measles; **s. alemán,** German measles.

sarcoma, m. sarcoma.

sardina, f. sardine, pilchard.

sardinel, m. English bond (ladrillos).

sarna, f. scabies; mange (de animales).

sarta, f. string, series; **s. de varillas de perforación,** drilling string.

satélite, m. satellite; **ciudad (f) s.,** satellite town; **s. de telecomunicaciones,** telecommunications satellite; **s. espacial,** space satellite; **teledetección (f) por s.,** satellite remote sensing.

saturación, f. saturation; **grado (m) de s.,** degree of saturation; **índice (m) de s.,** saturation index; **s. crítica,** critical saturation.

saturadola, a. saturated; **no s.,** unsaturated; **zona (f) s.,** saturated zone.

Saturno, m. Saturn.

sauce, m. willow; **s. llorón,** weeping willow.

saúco, m. elder.

savia, f. sap.

saxícola, a. rock-dwelling.

schlieren, m. schlieren.

SDT, m. TDS; **sólidos (m.pl) disueltos totales,** total dissolved solids.

sebáceola, a. sebaceous; **secreción (f) s.,** sebum.

sebkha, f. (Arab) saline coastal lake.

sebo, m. tallow (para velas, jabón); sebum (Anat); grease, fat (grasa).

sec, m. sec (Mat, abr. de secante).

secador, m. dryer; **banda (f) s.,** band dryer.

secano, m. dry land, dry region, unirrigated land.

secante, 1. m. secant (Mat, abr. sec); 2. a. drying.

secar, v. to dry; to drain (p.e. ropas).

secarse, v. to dieback, to wither.

sección, f. section; **s. típica,** type section; **s. transversal,** cross-section.

secola, a. dry.

secreción, f. secretion.

secretar, v. to secrete.

sector, m. sector, area (zone); **el s. energético,** the energy sector; **s. estatal,** public sector; **s. industrial,** industrial sector; **s. privado,** private sector; **s. público,** public sector.

secuencia, f. sequence, order.

secundario, m. secondary.

secundariola, a. secondary; **escuela (f) s.,** secondary school; **producto (m) s.,** by-product; **sedimentador (m) s.,** secondary settling tank; **tratamiento (m) s.,** secondary treatment.

seda, f. silk (tela); bristle (cerda); **gusano de s.,** silkworm.

sedante, m. sedative.

sedativo, m. sedative.

sedativola, a. sedative.

sede, f. headquarters (de una organización); seat (de gobierno);; **s. social,** head office (de un sociedad).

sedentariola, a. sedentary.

sedimentación, f. sedimentation.

sedimentador, m. settling tank.

sedimentar, v. to deposit, to settle out.

sedimentariola, a. sedimentary.

sedimentarse, v. to settle.

sedimento, m. deposit, sediment; **carga (f) de s.,** sediment load; **s. costanero,** shore deposit; **s. de fondo,** bottom deposit; **s. no consolidado,** unconsolidated sediment; **s. someras,** shallow-water deposit; **sedimentos aluviales,** alluvial deposit, flood deposit; **transporte (m) de s.,** sediment transport.

sedimentología, f. sedimentology.

segadorla, m. y f. harvester (persona).

segadoratrilladora, f. combine harvester.

segar, v. to reap, to cut, to mow.

segmento, m. segment.

segregación, f. segregation.

segregador, m. (Bol, Par, Perú) waste tip scavenger.

seguia, f. irrigation channel.

seguimiento, m. monitoring, tracking (acción); continuation (continuación).

seguridad, f. security; safety (ausencia de peligro); **cristal (m) de s.,** safety glass; **dispositivo (m) de s.,** safety device; **equipo (m) de s.,** safety equipment; **lámpara (f) de s.,** safety lamp; **margen (m) de s.,** safety margin; **medidas (f.pl) de s.,** safety measures, safety precautions; **s. social,** social security, National Health Service (GB); **válvula (f) de s.,** safety valve; **vidrio (m) de s.,** safety glass.

seguro, m. insurance; safety device, security device; **compañía (f) de s.,** insurance company; **póliza (f) de s.,** insurance policy; **prima (f) de s.,** insurance premium; **s. a todo riesgo,** fully comprehensive insurance; **s. social,** social insurance, National Insurance (GB).

segurola, a. safe, secure.

seísmo, m. earthquake.

selección, f. selection; **s. artificial (Biol),** artificial selection; **s. natural (Biol),** natural selection.

seleccionadola, a. sorted, selected; **s. granulométricamente,** size-sorted.

seleccionar, v. to sort, to select.

selectividad, f. selectivity.

selenio, m. selenium (Se).

selenita, f. selenite.

selva, f. forest, rainforest; **la s. amazónica,** the Amazon rainforest; **s. tropical,** rainforest, tropical rainforest; **s. virgen,** virgin forest.

selváticola, a. forest, woodland.

selvícola, a. woodland-dwelling, forest-dwelling.

selladola, a. sealed, stamped.

sellador, m. seal, packer; **ensayo (m) de s.,** packer test (in boreholes).

sellante, m. sealant.

sellar, v. to seal, to stamp, to brand.

semáforos, m.pl. traffic lights.
sembrado, m. sown field.
sembradora, f. seed drill.
sembradura, f. sowing.
sembrar, v. to sow, to seed.
sembrío, m. (LAm) sown field.
semejanza, f. likeness, resemblance.
semen, m. semen, sperm.
semental, m. stud (animal).
semestral, a. biannual, half-yearly.
semiáridola, a. semiarid.
semiconductor, m. semiconductor.
semiconfinadola, a. semi-confined.
semiconfinamiento, m. semi-confinement.
semiconsolidadola, a. semi-consolidated.
semicuerda, f. semichord.
semiempíricola, a. semi-empirical.
semiinfinitola, a. semi-infinite.
semillero, m. nursery (plantas).
seminario, m. seminar.
seminatural, a. seminatural; **ecosistema (m) s.,** seminatural ecosystem.
sen, m. sin (Mat, abr. por sen).
senado, m. senate.
senda, f. trail, path, footpath; **s. ecológica,** nature trail.
senderismo, m. hiking (andadura).
sendero, m. track, path, footpath.
senescencia, f. senescence.
senilidad, f. senility.
seno, m. sinus (Anat); breast (Zoo); sine (Mat, abr. sin).
sensibilidad, f. sensitivity (de aparatos, etc.); awareness (perceptibilidad); **análisis (m) de s. (Mat),** sensitivity analysis; **s. ambiental,** environmental awareness.
sensible, a. sensitive (responsivo); perceptible, tangible, noticeable (diferencia); **diferencia (f) s.,** perceptible difference; **zona (f) menos s. (EU),** less sensitive waters.
sensor, m. sensor; **s. remoto,** remote sensor.
sentadola, a. seated.
sentido, m. sense, meaning; direction (rumbo); **calle (f) de s. único,** one-way street; **en s. contrario,** in the opposite direction; **órgano (m) del s.,** sense organ; **s. común,** common sense, good sense.
sentina, f. bilge; **agua (f) de s.,** bilge water.
señal, f. sign (Gen); symptom, signal; **s. de tráfico,** traffic sign, traffic signal; **s. electromagnética,** electromagnetic signal; **s. mecánica,** mechanical signal.
señalar, v. to mark, to point out; to indicate (indicar).
señas, f.pl. address.
SEO, f. **Sociedad (f) Española de Ornitología (Esp),** Spanish Ornithological Society.
sépalo, m. sepal.
separación, f. separation, detachment.
separadola, a. separated, divided.
separador, m. separator; **s. ciclónico,** cyclone separator; **s. con platos inclinados (Trat),** inclined plate separator; **s. de grasa,** grease trap; **s. de vórtice,** vortex separator; **s. lamella,** lamella separator.
separar, v. to separate, to space, to spread out.
sepiolita, f. sepiolite.
septentrional, a. northern.
septicemia, f. septicaemia, septicemia (US).
septicidad, f. septicity.
sépticola, a. septic; **fosa (f) s., tanque (m) s.,** septic tank.
septo, m. septum.
sepultura, f. grave, tomb (tumba); burial (enterramiento).
sequedad, f. dryness.
sequía, f. drought; **secuencia (f) de sequías, serie (m) de sequías,** drought sequence.
ser, m. being; **s. humano,** human being; **s. vivo,** living being.
serac, m. serac.
sereno, m. night dew.
sericita, f. sericite.
sericultura, f. silk culture.
serie, f. series (pl. series); sere (f) (Ecol); **s. alcalina cálcica,** calc-alkali series; **s. armónica,** harmonic series; **s. continua,** continuous series; **s. cronológica,** chronological series; **s. de suelos,** soil series; **s. determinista,** deterministic series; **s. estocástica,** stochastic series; **s. sintética de tiempo,** synthetic time series.
serie, f. series (pl. series); sere (f) (Ecol); **series temporales,** time series.
serpentear, v. to meander (un río).
serpenteo, m. winding, meandering; **s. de río,** river meandering.
serpentina, f. serpentine.
serpentinita, f. serpentinite.
serpiente, f. snake, serpent.
serradola, a. serrated, serrate (Biol).
serranía, f. mountainous area, mountain range.
serranola, a. highland, hilly.
Serravaliense, m. Serravallian.
serrería, f. sawmill, lumber yard, timber yard.
serrín, m. sawdust.
servicio, m. service; **puesta (f) en s. (de una instalación),** commissioning; **s. de reparaciones,** repair service; **s. permanente,** 24-hour service; **s. público,** public service; **s. social,** social service; **servicios (pl) (en restaurantes, etc.),** toilets.
servidumbre, f. obligation (obligación); servitude (servicio); right (derecho); easement (acuerdo para acceso sobre terreno); **s. de paso,** right of way, easement.
servomecanismo, m. servomechanism.
servosistema, m. servosystem.
sésamo, m. sesame.
sesgadola, a. skewed; **distribución (f) s.,** skewed distribution; **estadísticas (f) sesgadas,** skewed statistics.
sésil, a. sessile.
sesquióxido, m. sesquioxide.

seta, f. mushroom.
seto, m. hedge; **s. cortavientos**, windbreak.
seudo-, pf. pseudo-.
se vende, v. "for sale".
severidad, f. severity.
sexo, m. sex.
sexual, a. sexual; **hormona (f) s.**, sex hormone.
shock, m. shock (Med); **estado de s. (Med)**, state of shock.
SIAL, f. SIAL (Geol).
sicomoro, sicómoro, m. sycamore.
sicula, f. sicula.
SIDA, m. AIDS; **síndrome (m) de inmunodeficiencia adquirida**, aquired immune deficiency syndrome.
siderita, f. siderite.
siderometalurgía, f. ferrous metallurgy.
siderurgia, f. iron and steel industry.
siderúrgicola, a. iron and steel; **fábrica (f) s.**, **siderúrgica (f)**, iron and steel works; **industria (f) s.**, iron and steel industry.
siega, f. harvesting, reaping, mowing.
Siegeniense, m. Siegenian.
siembra, f. sowing (de plantas), sowing time (época); **s. de nubes**, cloud seeding.
sienita, f. syenite.
sierra, f. saw (herramienta); mountain range (montañas); **s. abrazadera**, pit saw; **s. de cinta**, band saw; **s. mecánica**, power saw.
sífilis, f. syphilis.
sifón, m. siphon, syphon (de líquido); trap, U-bend (trapa); **s. invertido**, inverted siphon/syphon.
sifonación, f. siphoning.
sifonamiento, m. siphoning.
SIG, m. GIS; **Sistema (m) de Información Geográfica**, Geographical Information System.
sigmoideola, a. sigmoid.
significativola, a. significant, important; **cifras (f) s.**, significant figures.
silano, m. sylane.
silenciador, m. silencer, muffler (US) (en un coche).
sílex, m. silex, flint, chert.
silexita, f. silexite, flintstone.
silicato, m. silicate.
sílice, f. silica; **arena (f) de s.**, silica sand.
silíceola, a. siliceous.
silicificación, f. silicification.
silicio, m. silicon (Si); **carburo (m) de s.**, silicon carbide.
silicosis, f. silicosis.
silo, m. silo, grain silo (para cereales).
siloxano, m. siloxane.
silueta, f. silhouette, profile, outline.
Silúrico, m. Silurian.
silvanita, f. sylvanite.
silvestre, a. wild (flora y fauna).
silvicultura, f. forestry, silviculture, sylviculture.
silla, f. chair, seat; saddle (Geog).
sillar, m. ashlar (piedra).

sillimanita, f. sillimanite.
sima, f. chasm (m, f) (Geog); sink (Geog, depresión); swallow hole (Geol, de cárstica); depths (Fig).
SIMA, f. SIMA (Geol).
simazina, f. simazine.
simbiosis, f. symbiosis.
simbióticola, a. symbiotic.
símbolo, m. symbol; **s. en mapa**, map symbol; **s. igual/menos/más**, equal/minus/plus symbol; **s. topográfico**, topographic symbol.
simetría, f. symmetry; **s. bilateral**, bilateral symmetry; **s. radial**, radial symmetry.
simétricola, a. symmetrical.
simplificación, f. simplification; **sobresimplificación**, over-simplification.
simposio, m. symposium.
simulación, f. simulation; **modelo (m) de s.**, simulation model; **s. numérica**, numerical simulation.
simular, v. to simulate.
simúlido, m. simulid; **simúlidos (Zoo, clase)**, Simulidae (pl).
simultaneidad, f. simultaneity.
simultáneola, a. simultaneous.
sinapsis, f. synapse.
sinclinal, 1. m. syncline; 2. a. synclinal.
sinclinorio, m. synclinorium.
sindicato, m. trade union.
síndrome, m. syndrome; **s. de inmunodeficiencia adquirida (abr. SIDA)**, acquired immune deficiency syndrome (abr. AIDS); **s. del edificio enfermo**, sick building syndrome.
Sinemuriense, m. Sinemurian.
sinergia, f. synergy.
sinérgicola, a. synergistic.
sinergismo, m. synergism, synergy.
sinfín, m. countless number.
singenéticola, a. syngenetic.
siniestralidad, f. accident rate; loss, casualty.
siniestro, m. catastrophe, accident (catástrofe); fire (incendio); **s. forestal**, forest fire.
sinistrorsola, a. counterclockwise, anticlockwise.
sinópticola, a. synoptic; **cuadro (m) s.**, chart, diagram.
sintaxis, f. syntax.
sinter, m. sinter.
síntesis, m. synthesis.
sintéticola, a. synthetic, man-made; **detergente (m) s.**, synthetic detergent; **generación (f) de datos sintéticos**, synthetic data generation; **serie (f) s. de tiempo**, synthetic time series.
sintetizar, v. to synthesize, to synthetize.
sinuadola, a. sinuate.
sinuosidad, f. sinuosity.
sinuosola, a. sinuous.
sinusoidal, a. sinusoidal.
sirimiri, m. drizzle.
siroco, m. sirocco.
sisal, m. sisal.
sismicidad, f. seismicity.

sísmicola, a. seismic; **fenómeno (m) s.**, seismic phenomena; **red (f) s.**, seismic network.

sismo, m. earthquake, earth tremor.

sismograma, m. seismograph.

sismología, f. seismology.

sismómetro, m. seismometer.

sismorresistente, a. seismic resistant.

sistema, m. system; **s. administrativo**, administrative system; **s. agua-vapor**, water-vapour system; **s. circulatorio**, circulatory system; **s. conjugado**, conjugated system; **s. decimal**, decimal system; **s. de diaclasas (Geol)**, joint system; **s. digestivo**, digestive system; **s. fluvial**, river system; **s. inmunitario**, immune system; **s. métrico**, metric system; **s. montañoso**, mountain system; **s. multiacuífero**, multi-aquifer system; **s. nervioso**, nervous system; **s. solar**, solar system; **s. vascular**, vascular system.

sistemática, f. systematics, taxonomy.

sistemáticola, a. systematic.

sistémicola, a. systemic.

sitio, m. place, site; **caracterización (f) de s.**, site characterization; **en el s.**, in situ, on site; **remediación (f) en el s.**, on-site remediation; **s. de construcción**, building site; **s. de vertedero**, landfill site; **s. específico**, site-specific.

situación, f. situation, location, site.

skarn, m. skarn.

s.l., f. ltd; **sociedad (f) limitada, sociedad de responsabilidad limitada**, limited company, limited liability company.

slickenside, m. slickenside.

smoothing, m. smoothing (Mat).

SNC, m. CNS; **sistema (m) nervioso central**, central nervous system.

soberanía, f. sovereignty.

sobrebombeo, m. over-pumping.

sobrecarga, f. overload, overburden; **s. eléctrica**, electrical overload.

sobrecargar, v. to overload.

sobrecolgante, m. overhang.

sobreconsolidación, f. overconsolidation (de arcillas).

sobreconsumo, m. overconsumption.

sobrecorrimiento, m. (LAm) overthrust.

sobredimensionamiento, m. oversizing; **s. de obras**, oversizing of works.

sobreexplotación, f. overexploitation; **s. de acuíferos**, overexploitation of aquifers.

sobreexplotadola, a. overexploited, overdrawn; **acuífero (m) s.**, overexploited aquifer.

sobreimponer, v. to superimpose.

sobrenadante, a. supernatent; **licor (m) s.**, supernatant liquor.

sobrepasar, v. to exceed, to surpass; **s. los límites de calidad**, to exceed the quality standards.

sobrepastoreo, m. overgrazing.

sobrepesca, f. overfishing.

sobrerelajación, f. over-relaxation.

sobrerriego, m. overwatering.

sobresalir, v. to overhang.

sobresimplificación, f. over-simplification.

socavación, f. undermining.

socavadola, a. undercut.

socavón, m. excavation; hole (en la carretera); subsidence (Con); gallery (para minería); adit, heading (para agua).

social, a. social; **asistente (m, f) s.**, social worker.

sociedad, f. company (Com), society; **s. anónima (abr. s.a.)**, public limited company (GB, abr. plc), joint stock company, corporation (US, abr. Corp), incorporated (US, abr. Inc); **s. anónima de capital variable**, limited company (GB, abr. Ltd), incorporated (US, abr. Inc); **s. comanditaria**, limited partnership; **s. cooperativa**, cooperative, cooperative society; **s. de acciones**, stock company; **s. de consumo**, consumer society; **s. de responsabilidad limitada (abr. s.l.)**, limited liability company (abr. Ltd); **s. en comandita**, limited partnership; **s. limitada (abr. s.l.)**, limited liability company (abr. Ltd); **s. mercantil**, trading company.

socioeconómicola, a. socio-economic; **planificación (f) s.**, socio-economic planning.

sociólogola, m. y f. sociologist.

socovación, f. stoping (Min); **s. magmática**, magmatic stoping.

sodalita, f. sodalite.

sódicola, a. sodic; **cloruro (m) s.**, sodium chloride, common salt.

sodio, m. sodium (Na).

sofito, m. soffit (centro de arco).

sofocación, f. suffocation, asphyxiation.

sofocadorla, a. suffocating.

sofocante, a. suffocating.

sofocar, v. to suffocate, to smother.

sofoco, m. suffocation.

sofosión, f. suffusion, piping.

software, m. software; **s. de aplicación**, application software; **soporte (m) s.**, software support.

sol, m. sun; sunlight, sunshine (luz de sol); sol (Quím, coloide); **nuevo s.**, nuevo sol (Perú, unidad monetaria).

solapa, f. flap (aletazo); pretext (Fig).

solapado, m. overlap.

solar, 1. m. lot, plot (terreno); building site (para construcción); 2. a. solar; **energía (f) s.**, solar power.

solarímetro, m. solarimeter.

soldadola, a. welded; **tobas (f.pl) soldadas**, welded tuffs.

soldador, m. welder (persona).

soldadura, f. weld, welding (de metal), welder (máquina); **s. a tope**, butt weld; **s. oxiacetilénica**, oxyacetylene welding; **s. por puntos**, spot weld.

soldar, v. to solder, to weld (metal); **soplete (m) de s.**, welding torch.

solevantamiento, m. (LAm) upthrust, high-angle reverse fault.

solicitud, f. care, diligence (cualidad); request, application form (petición).

solidificación, f. solidification.

sólido, m. solid; **contenido (m) de sólidos secos,** dry solids content; **sólidos suspendidos,** suspended solids.

solidola, a. solid.

solifluxión, f. solifluction, soil creep; **lóbulo (m) de s.,** solifluction lobe.

solsticio, m. solstice; **s. de invierno,** winter solstice; **s. de verano,** summer solstice.

soltar, v. to release, to let go.

solubilidad, f. solubility; **s. de gases,** gas solubility.

solubilización, f. solubilization.

solubilizar, v. to solubilize.

soluble, a. soluble; **s. en agua,** water-soluble.

solución, f. solution (Quím, de problema); answer (explicación); **s. normal,** normal solution; **s. saturada,** saturated solution; **s. sobresaturada,** oversaturated solution; **s. tampón (Quím),** buffer solution.

solucionar, v. to solve, to resolve.

soluto, m. solute.

solvencia, f. solvency (Fin).

sombra, f. shade; shadow (propio de un objeto).

somerola, a. shallow, superficial.

sonda, f. sounding (acción); sonde (instrumento); bore, drill, probe (perforación); **s. acústica,** echo sounder.

sondear, v. to drill, to bore (perforar); to make soundings (explorar).

sondeo, m. drilling (acción de perforación); borehole (pozo); sounding (encuesta, investigación); **cabeza (f) de s.,** wellhead; **registrado (m) del s.,** down-hole log, borehole log, well log; **s. acústico,** echo sounding; **s. a percusión,** percussion borehole; **s. a rotación,** rotary borehole; **s. de ensayo,** test borehole; **s. de exploración,** exploratory borehole; **s. de observación,** observation borehole; **s. de pequeño diámetro,** small-diameter borehole; **s. de reconocimiento,** investigatory borehole; **s. de rotopercusión,** rotary percussion drilling; **s. profundo,** deep borehole; **s. surgente,** overflowing borehole; **testificación (f) del s.,** down-hole logging.

sondista, m. driller, drill operator; **parte (m) del s.,** driller's log.

sonido, m. sound.

sonorola, a. loud; **nivel (m) s.,** noise level.

soplar, v. to blow (Gen); **s. racheado,** to gust (con lluvia o viento).

soplete, m. blowtorch, blowlamp.

sopletear, v. to blow, to puff; **s. con arena,** to sandblast.

sorgo, m. sorghum.

soro, m. sorus (pl. sori).

SOS, f. SOS, distress call; **captar un SOS (v),** to pick up an SOS.

sosa, f. soda (Quím); saltwort (Bot); **s. cáustica,** caustic soda.

sosero, m. black alkaline soil.

sósten, m. support (ayuda); support, prop (Con); food, sustenance (alimento).

sostenibilidad, f. sustainability.

sostener, v. to sustain.

sostenible, a. sustainable; **desarrollo (m) s.,** sustainable development.

sostenidola, a. sustained.

sostenimiento, m. sustainability.

sótano, m. basement, cellar (bodega).

sotavento, m. lee, leeside.

soterramiento, m. burial.

soto, m. thicket, grove, copse; riparian vegetation (al lado de un río); **s. para mimbre,** withybed.

sotobosque, m. understorey, undergrowth.

spatangoide, m. spatangoid.

spilita, f. spilite.

status quo, m. status quo.

stock, m. stock, reserves (la pesca).

stoping, m. (LAm) stoping (Min).

storatividad, f. storativity.

stripping, m. stripping (Con); **s. de sulfuro,** sulphur stripping, sulfur stripping (US).

suavizante, m. softener.

subacuáticola, a. subaqueous, submarine.

subaéreola, a. subaerial.

subasta, f. auction.

subatómicola, a. subatomic; **partículas (f.pl) subatómicas,** subatomic particles.

subcomité, m. subcommittee.

subconsumación, f. underconsumption.

subcontratadola, a. subcontracted.

subcortical, a. subcrustal.

subcutáneola, a. subcutaneous.

subdesarrolladola, a. underdeveloped.

subdesarrollo, m. underdevelopment.

subdesérticola, a. subdesert.

subdimensionamiento, m. undersizing; **s. de obras,** undersizing of works.

subdren, m. underdrain.

subducción, f. subduction.

subestación, f. substation.

subestructura, f. substructure.

subglacial, a. subglacial.

subglaciar, a. subglacial.

subgrado, m. subgrade (Trat).

subhúmedola, a. subhumid.

subir, v. to go up, to rise, to ascend.

súbitola, a. sudden.

subjetividad, f. subjectivity.

subjetivola, a. subjective.

sublimación, f. sublimation.

submareal, a. subtidal.

submarinista, m. y f. skin-diver, submariner.

submarino, m. submarine.

submarinola, a. submarine, underwater.

submuestra, f. subsample.

subordinadola, a. subordinate, auxiliary, ancillary.

subóxido, m. suboxide.

subpiso, m. substage (cronología).

subproducto, m. by-product.

subrayar, v. to underline; **s. la importancia de algo**, to underline the importance of something.

subreino, m. subkingdom (Biol).

subsidiariola, a. subsidiary, secondary.

subsidio, m. subsidy, grant.

subsiguente, a. subsequent.

subsistema, m. sub-system.

subsistencia, f. subsistence.

substancia, f. substance; **s. coloidal**, colloidal substance; **s. dañina (LAm)**, noxious substance; **s. en la lista ámbar**, amber-list substance; **s. en la lista gris/negra/roja (UE)**, grey/black/red-list substance; **s. orgánica**, organic substance; **s. peligrosa**, hazardous substance; **s. radioactiva**, radioactive substance; **s. tóxica**, toxic substance.

substitución, f. substitution.

substrato, m. substratum (Gen), bedrock (Geol).

subsuelo, m. subsoil.

subterráneo, m. subway, metro.

subterráneola, a. underground; **drenaje (m) s.**, underground drainage.

subtropical, a. subtropical.

subtrópicola, a. subtropical.

suburbios, m.pl. slum, shanty town.

subvención, f. subsidy.

subvencionar, v. to subsidize.

subyacente, a. underlying.

subyacer, v. to underlie.

succión, f. suction (de líquido).

sucesión, f. succession; **s. ecológica**, ecological succession.

suceso, m. occurrence, event, happening.

suciedad, f. dirt, dirtiness.

suciola, a. dirty.

sucre, m. (Ec) sucre (unidad monetaria).

suculento, m. succulent (Bot).

suculentola, a. succulent.

sucursal, f. branch (oficina de una compañía); subsidiary company (compañía).

Sudamérica, f. South America.

sudamericanola, a. South American.

sudamina, f. prickly heat.

sudar, v. to sweat.

sudor, m. sweat, perspiration.

suelo, m. soil, ground, surface; **acidez (f) del s.**, soil acidity; **conservación (f) del s.**, land conservation; **contaminación (f) del s.**, soil contamination; **déficit (m) de humedad del s.**, soil moisture deficit; **erosión (f) del s.**, soil erosion; **estructura (f) del s.**, soil structure; **hundimiento (m) del s.**, ground subsidence; **mecánica (f) de suelos**, soil mechanics; **ordenación (f) del s.**, land use planning; **perfil (m) del s.**, soil profile; **reptación (f) del**

s., soil creep; **s. de pradera**, prairie soil; **s. de tundra**, tundra soil; **s. franco**, loamy soil; **s. laterítico**, laterite soil; **s. no urbanizable (Esp, urbanismo)**, rural land, land which can not be built on; **s. pardo**, brownearth; **s. podsol**, podsol; **s. residual**, residual soil; **s. rojo desértico**, red desert soil; **s. saturado**, saturated soil; **s. turboso**, boggy soil; **s. urbanizable (Esp)**, buildable land, designated urban land; **s. vegetal**, topsoil; **suelos agrícolas**, farmland; **uso (m) del s.**, land use.

sueltola, a. loose.

suero, m. serum (pl. sera, serums); **s. sanguíneo**, blood serum.

sufragar, v. to help, to pay, to defray; **s. los gastos**, to defray the costs.

sujetar, v. to fasten (to attach).

sujeto, m. subject (tema).

sulfamida, f. sulphamide, sulfamide (US).

sulfanato, m. sulphonate, sulfanate (US).

sulfano, m. sulphane.

sulfato, m. sulphate, sulfate (US); **s. reducción**, sulphate reduction.

sulfito, m. sulphite, sulfite (US).

sulfocloruro, m. sulphochloride, sulfochloride (US).

sulfonación, f. sulphonation, sulfonation (US).

sulfonamida, f. sulphonamide, sulfonamide (US).

sulfurar, v. to sulphurate; to sulfurate (US).

sulfúricola, a. sulphuric; sulfuric (US).

sulfuro, m. sulphide, sulfide (US); **s. de hidrógeno**, hydrogen sulphide.

suma, f. sum, total; addition (Mat); summary (resumen).

sumatorio, m. summation.

sumergible, a. submersible.

sumergidola, a. submerged.

sumergir, v. to submerge.

sumersión, f. submergence, immersion; **relación (f) de s.**, submergence ratio.

sumidero, m. drain, sewer (cloaca); cesspool (letrina); sump (Con); sinkhole (Geol); **s. ciego (Trat)**, soakaway; **s. de topo**, mole drain; **s. francés**, french drain; **s. inverso**, reverse drain; **s. por desplome**, collapse sinkhole.

suministrar, v. to supply, to provide.

suministro, m. supply (acción de abastecimiento), provision.

superdesarrolladola, a. overdeveloped.

superdesarrollo, m. overdevelopment.

superestructura, f. superstructure.

superficial, a. superficial; **aguas (f.pl) superficiales**, surface waters.

superficie, f. surface (Gen), outside (exterior); **área (f) de la s.**, surface area; **en una extensa s.**, over a wide area; **ruta (f) de s.**, surface route; **s. a nivel**, level surface; **s. de erosión**, surface of erosion; **s. de rodadura (Eng)**, tread (de un neumático); **s. freática**, phreatic surface; **s. inferior**, underside; **s. libre**, free surface; **s. nivelada,** level surface; **s. piezométrica,** piezometric surface; **s. rocosa,** rock surface; **s. terrestre,** Earth's surface; **s. topográfica,** topographic surface.

superfosfato, m. superphosphate.

superior, a. superior; upper, higher.

supermercado, m. supermarket.

superóxido, m. superoxide.

superpoblación, f. overpopulation.

superpobladola, a. overpopulated.

superponer, v. to overlie, to superimpose.

superposición, f. superposition.

superproducción, f. overproduction.

supervisión, f. supervision.

supervivencia, f. survival; **lucha (f) por la s.,** the fight for survival; **s. de los más aptos, s. de los mejor dotados,** survival of the fittest.

superviviente, 1. m. y f. survivor; 2. a. surviving.

suplantar, v. to replace, to supplant (desbancar).

supracortical, a. supracrustal.

suprayacente, a. overlying.

suprayacer, v. to overlie.

supresión, f. supression, elimination.

sur, m. South; **América (f) del S.,** South America.

suramericanola, a. South American.

surcadola, a. furrowed, grooved.

surco, m. furrow.

surfactante, m. surfactant.

surgencia, f. outflow, surgence, spring, issue; **s. submarina,** submarine spring; **s. termal,** thermal spring.

surgente, a. overflowing, upwelling.

surgir, v. to spurt, to issue forth.

susceptibilidad, f. susceptibility.

susceptible, a. susceptible.

suspensión, f. suspension; **s. coloidal,** colloidal suspension; **sólidos (m.pl) en s.,** suspended solids.

sustancia, f. substance; essence (esencia); matter (materia); **s. húmica,** humic substance.

sustentabilidad, f. sustainability.

sustentar, v. to support; to sustain, to nourish (of food).

sustento, m. sustenance, food, nourishment.

sustraer, v. to subtract.

sustrato, m. substrate; substratum (pl. substrata); **s. rocoso,** bedrock.

sutura, f. suture.

Swaziano, m. Swazian.

T

tabaco, m. tobacco.

tabique, m. partition (muro delgado).

tabla, f. plank, board (de madera); slab (de piedra); sheet (de metal); flood plain (Geog); table (Mat); **t. de calibración,** calibration table; **t. de consulta,** look-up table; **t. de logaritmos,** logarithm tables; **t. para piso de madera,** batten plate; **T. Periódica de los Elementos,** Periodic Table of Elements.

tablero, m. board, panel; blackboard (encerado); switchboard (Elec); **t. de anuncios,** notice board, bulletin board; **t. de dibujo,** drawing board; **t. de gráficos,** graph pad; **t. de instrumentos,** instrument panel.

tablestaca, f. sheet pile.

tablestacado, m. sheet piling.

tablón, m. plank; notice board, bulletin board (US).

tabulación, f. tabulation.

tabular, v. to tabulate.

tacana, f. (LAm) cultivated hillside terrace.

táctil, a. tactile.

tactismo, m. taxis; **fototactismo,** phototaxis.

tacto, m. touch (sentido).

tacho, m. bucket.

tahuampa, f. (Perú) lake in a wood.

taiga, f. taiga.

tajamar, m. breakwater.

tala, f. tree felling (de árboles); pruning (poda); destruction (destrucción).

taladola, a. felled, cut down; **árboles (m.pl) talados,** felled trees, timber.

taladradorla, m. y f. driller, borer.

taladradora, f. drill (máquina), boring machine; **t. neumática,** pneumatic drill.

taladrar, v. to bore, to drill, to auger; to pierce (perforar).

taladro, m. drill, drill bit.

talar, v. to fell, to cut down.

talco, m. talc, soapstone; **polvo (m) de t.,** talcum powder.

talio, m. thallium (Tl).

talo, m. thallus.

talud, m. talus; **estabilización (f) de taludes,** talus stabilization; **t. continental,** continental slope.

taller, m. workshop, garage, shop; **t. mecánico (para vehículos),** garage.

tallo, m. stem, stalk; **t. bulboso,** corm, bulb.

tamaño, m. size; **de t. natural (a),** life-size, life-sized; **t. de grano,** grain size; **t. natural,** life size.

tambo, m. (Arg, Par, Ur) small dairy farm.

tambor, m. drum (cilindro); **t. de aceite,** oil drum.

tamiz, m. sieve; **juego (m) de tamices,** set of sieves; **t. vibratorio,** vibrating screen.

tampón, m. ink pad; tampon (Med), plug; buffer (Quím); **solución (f) t. (Quím),** buffer solution; **zona (f) t.,** buffer zone.

tamponamiento, m. clogging.

tamponar, v. to buffer (Quím).

tangente, m. tangent (Mat, abr. tan).

tanino, m. tannin.

tanque, m. tank, reservoir; **t. con desviadores,** baffle tank; **t. de aforo,** gauge tank; **t. de aireación,** aeration tank; **t. de almacenamiento,** storage tank; **t. de compensación,** equalizing tank, balancing tank; **t. de descenso constante,** constant-head tank; **t. de ruptura de carga,** surge tank; **t. de tratamiento,** digester tank.

tántalo, m. tantalum (Ta).

tanteo, m. approximate calculation, rough estimate, guesstimate, guestimate.

tapar, v. to cover, to put the lid on.

tapia, f. mud wall (de adobe).

tapisca, f. (CAm, Méx) maize harvest, corn harvest.

tapiscar, v. (CAm, Méx) to harvest.

tapón, m. inkpad, pad; buffer (Quím); tampon (Med); plug, stopper (tapón); **capacidad (f) t. (Quím),** b. capacity; **disolución (f) t. (Quím),** buffer solution; **t. de bentonita en un piezométro,** bentonite seal in a piezometer.

taponamiento, m. plugging, stopping up.

taqué, m. stopper, plug.

taquímetro, m. tachymetry.

tarajal, m. ground with deep-rooted water-seeking plants.

tardanza, f. delay (retraso); slowness (lentitud).

tarifa, f. tariff, rate, price list.

tarjeta, f. card; **t. de crédito,** credit card; **t. perforada,** punched card.

tarro, m. jar, can; **t. de mermelada,** jam jar.

tarsianola, a. tarsal; **hueso (m) t.,** tarsal, tarsal bone.

tarso, m. tarsus (de insectos); tarsal (Med, hueso tarsiano); **t. parpebral,** tarsus (Med, tobillo).

tártaro, m. tartar (Quím).

tartrato, m. tartrate.

tasa, f. appraisal (valoración); tax (impuesta); measure (medida); rate (índice); **t. de interés,** rate of interest; **t. de natalidad,** birth rate.

tasación, f. appraisal, valuation (valoración), calculation (cálculo).

tasador, m. y f. valuer, valuator, appraiser (US) (Fin).

taxia, f. taxis; **t. negativa,** negative taxis; **t. positiva,** positive taxis.

taxón, m. taxon.

taxonomia, f. taxonomy; **t. numérica,** numerical taxonomy.

taxonómicola, a. taxonomic; **unidad (f) t.,** taxonomic unit.

té, m. tea.

teca, f. teak (árbol, madera); theca (Bot, Zoo).

teclado, m. keyboard (Comp).

tecnecio, m. technetium (Tc).

técnica, f. technique, method; technology (tecnología); engineering (ingenería); **t. aeroespacial,** aerospace technology; **t. hidráulica,** hydraulic engineering.

tecnicidad, f. technicality.

tecnicismo, m. technicality.

técnicola, 1. m. y f. technician; engineer (reparador); 2. a. technical; **evaluación (f) t.,** technical assessment; **terminología (f) t.,** technical terminology.

tecnología, f. technology; **t. alternativa,** alternative technology; **t. apropiada,** appropriate technology; **t. blanda,** soft technology; **t. de bajo costo,** low-cost technology; **t. de bajo impacto,** low impact technology; **t. intermedia,** intermediate technology; **t. suave,** soft technology; **tecnologías limpias,** clean technologies.

tectónica, f. tectonics; **t. de placas,** plate tectonics.

tectónicola, a. tectonic.

techo, m. ceiling, roof (tejado); **t. corredizo,** sliding roof, sun roof; **vigueta (f) de t.,** ceiling joist.

techumbre, m. roofing, roof.

teflón, m. teflon.

tegumento, m. tegument, integument (piel); seed-coat (Bot).

teja, f. roof tile; **t. de caballete,** ridge tile; **t. flamenca,** pantile.

tejado, m. roof, tile roof; **inclinación (f) de t.,** roof pitch; **tirante (m) de t.,** roof tie.

tejamaní, m. (LAm) shingle (en tejado).

tejamanil, m. (LAm) shingle (en tejado).

tejido, m. tissue; **t. adiposo,** adipose tissue; **t. conectivo,** connective tissue; **t. vascular,** vascular tissue.

tejo, m. yew (árbol); quoit, ring (aro).

tejocote, m. hawthorn.

tela, f. cloth, fabric; **t. metálica,** wire netting.

telecomunicación, f. telecommunication.

teledetección, f. remote sensing; **t. espacial, t. por satélite,** remote sensing by satelite.

teledirección, f. telemetry, remote control.

teledirigidola, a. remote-controlled.

telefónicola, a. telephone, telephonic; **central (m) t.,** telephone exchange.

teléfono, m. telephone.

telegráficola, a. telegraph, telegraphic.

telégrafo, m. telegraph, telegraph pole.

telemetría, f. telemetry.

telemétricola, a. telemetric.

teleósteo, m. teleost; **teleósteos (Zoo, taxón),** Teleostei (pl).

teleósteola, a. teleost.

televisión, f. television; **t. (f) por circuito cerrado,** closed-circuit television (abr. CCTV).

televisor, m. television set; **t. en sondeos,** downhole television.

telón, m. curtain, screen; **t. de lechada,** grout curtain.

teluro, m. tellurium (Te).

tema, m. subject, theme, topic.

temblor, m. (LAm) tremor; earthquake; **t. de tierra,** earth tremor.

témpano, m. floe; **t. de hielo,** ice floe.

temperatura, f. temperature; **t. absoluta,** absolute temperature; **t. ambiente,** ambient temperature, room temperature; **t. de congelación,** freezing point; **t. de inflamabilidad,** flash point; **t. de saturación,** saturation point, dew point; **t. media anual,** annual mean temperature.

tempestad, f. tempest, storm.

templadola, a. temperate; **zona (f) t.,** temperate zone.

templo, m. temple.

temporada, f. season, period, time; **alta/baja t.,** high/low season.

temporal, 1. m. severe storm; **capear el t. (v),** to weather the storm; 2. a. temporal (de tiempo); temporary, seasonal (provisional, estacional).

tenaz, a. adhesive, sticky.

tenca, f. tench.

tendencia, f. trend, tendency; **la t. al alza, la t. al aumento,** rising trend; **medida (f) de la t. central (Mat),** measure of central tendency.

tender, v. to spread, to spread out.

tendido, m. construction, building (edificio); laying (de un cable); washing (ropa puesta a secarse); coat of plaster (capa de yeso); **t. de redes (LAm),** pipe laying; **tendidos eléctricos,** power cables.

tendidola, a. spread out.

tendón, m. tendon; **t. de Aquiles,** Achilles tendon.

tenería, f. tannery.

tenia, f. tapeworm.

tensiómetro, m. tensionometer.

tensión, f. tension, stress (Med); **cable (m) de alta t., línea (f) de alta t.,** high-tension cable; **t. capilar,** capillary pressure; **t. de régimen,** voltage rating; **t. intergranular,** intergranular pressure; **t. nerviosa (Med),** stress; **t. superficial,** surface tension.

tensoactivola, a. tensoactive.

tensor, m. tensor.

tentaculito, m. tentaculite.

tentáculo, m. tentacle.

tentativa, f. attempt, endeavour.

teodolito, m. theodolite.

teorema, m. theorem.

teoría, f. theory; **t. atómica,** atomic theory; **t. cuántica, t. de los quanta,** quantum theory.

teóricamente, ad. theoretically, in theory.

teóricola, a. theoretical.

tera-, pf. tera- (10.E12).

terbio, m. terbium (Tb).

Terciario, m. Tertiary.

terileno, m. terylene.

terma, f. hot spring.

termal, a. thermal.

térmicola, a. thermal; **estratificación (f) t.,** thermal stratification; **imagen (f) t.,** thermal image; **valor (m) t. medio,** average temperature.

terminación, f. termination, ending, finish.

terminal, 1. m. terminal (Elec); 2. f. terminal (lugar); **t. aérea,** air terminal; **t. de un ordenador,** computer terminal; 3. a. terminal.

término, m. end, term; **en términos de,** in terms of; **glosario (m) de términos,** glossary of terms; **t. técnico,** technical term.

terminología, f. terminology; **t. técnica,** technical terminology.

termistor, m. thermistor.

termita, f. termite (insecta); thermite (Quím).

termo-, pf. thermo-.

termoclina, f. thermocline.

termodinámica, f. thermodynamics; **primera ley (f) de la t.,** first law of thermodynamics.

termodinámicola, a. thermodynamic.

termoesfera, f. thermosphere.

termófilo, m. thermophile.

termófilola, a. thermophilic; **digestión (f) anerobia t.,** thermophilic anerobic digestion.

termografía, f. thermography; **t. infrarroja,** thermal imaging; infrared thermography.

termógrafo, m. thermograph.

termómetro, m. thermometer; **t. de las máximas y mínimas,** maximum and minimum thermometer; **t. de mercurio,** mercury thermometer.

termonuclear, a. thermonuclear; **prueba (f) t.,** thermonuclear test.

termopar, m. thermocouple.

termoplástico, m. thermoplastic.

termoplásticola, a. thermoplastic.

termopluviometría, f. temperature and rainfall measurement.

termoquímicola, a. thermochemical.

termostato, m. thermostat.

terpeno, m. terpene.

terracota, f. terracotta.

terral, m. (LAm) dust cloud.

terrapene, m. terrapin.

terraplén, m. embankment, earthwork, causeway.

terraplenación, f. terracing.

terraplenado, m. backfill.

terraplenar, v. to level, to bank, to bank up (para carreteras).

terraza, f. terrace; **t. aluvial,** alluvial terrace; **t.** elevada, raised beach; **t. fluvial,** river terrace; **t. ribereña,** shore terrace.

terremoto, m. earthquake, earth tremor.

terreno, m. ground, terrain; **corrección (f) de terrenos,** land restoration, land reclamation; **restauración (f) del t.,** land restoration; **t. abandonado,** derelict land; **t. arenoso,** sandy terrain; **t. bajo,** low land; **t. contaminado,** contaminated land; **t. fisurado,** fissured terrain; **t. horizontal,** level land; **t. incoherente,** broken terrain; **t. kárstico,** karstic terrain; **t. pantanoso,** fen land; **t. pedregoso,** stony ground; **t. plano,** flat land; **terrenos ganados al mar,** land reclaimed from the sea.

terrestre, a. terrestrial.

terrestrificación, f. land reclamation (from the sea).

territorialidad, f. territoriality.

territorio, m. territory; region (LAm); district (LAm) (barrio).

terrón, m. clod of earth.

terrosola, a. earthy.

tesauro, m. thesaurus.

tesis, f. thesis.

tesoro, m. treasure.

test, m. test; **t. de Ames (Biol, Quím),** Ames test; **t. del anillo marrón (Biol, Quím),** brown ring test; **t. de Lassaigne (Biol, Quím),** Lasaigne test.

testa, f. testa.

teste, m. testis (pl. testes).

testículo, m. testicle.

testigo, m. witness (Jur); core (Geol); **caja (f) de testigos,** core box; **t. de perforación,** drilling core; **t. ocular,** eyewitness; **t. orientado,** oriented core.

testosterona, f. testosterone.

tétano, m. tetanus, lockjaw.

tétanos, m. tetanus.

tetilla, f. teat, nipple.

tetra-, pf. tetra-.

tetrabratúlido, m. tetrabratulid.

tetraclorobifenol, m. tetrachlorobiphenol.

tetracloroetileno, m. tetrachloroethylene.

tetracloruro, m. tetrachloride; **t. de carbono,** carbon tetrachloride.

tetraedro, m. tetrahedron.

tetragraptido, m. tetragraptid.

tetrahidrato, m. tetrahydrate.

tetramicina, f. tetramycin.

tetranitrometano, m. tetranitromethane.

tetrápodo, m. tetrapod.

tetrápodola, a. tetrapod.

tetravalente, a. tetravalent, cuadrivalent.

textil, a. textile; **industria (f) t.,** textile industry.

texto, m. text; **t. mecanografiado,** typescript.

textura, f. texture; **t. de avenamiento,** drainage texture; **t. homogénea,** even texture.

tg, m. tan (Mat, abr. de tangente).

Thanetiense, m. Thanetian.

tiamina, f. thiamine (vitamina B1).

tibia, f. tibia (pl. tibias, tibiae).

tiburón, m. shark.

tiempo, m. time (Gen); period, epoch (época); weather (Met); **hace buen t.**, the weather is fine; **limitación (f) de t.**, time limit; **motor (m) de dos tiempos**, two-stroke engine; **previsión (f) del t.**, weather forecast; **retardo (m) de t.**, time lag; **si el t. lo permite**, weather permitting; **t. corrido (Comp)**, run time; **t. de arranque (máquinas)**, start time; **t. de concentración (Geog)**, time of concentration; **t. de contacto (Trat)**, contact time; **t. de estancia**, residence time; **t. de llegada**, arrival time; **t. de reacción**, reaction time; **t. de renovación**, recovery time; **t. de residencia**, residence time; **t. de respuesta**, response time, turnaround time; **t. de retención**, retention time; **t. de tránsito**, transit time; **t. geológico**, geological time; **T. Medio de Greenwich (abr. TMG)**, Greenwich Mean Time (abr. GMT); **t. real**, real time.

tienda, f. shop, store; tent (camping).

tierra, f. earth, ground, soil; the Earth (La Tierra); **bajo t.**, underground; **equipo (m) para el movimiento de tierras**, earthmoving equipment; **mejoramiento (m) de tierras**, land improvement; **t. adentro**, inland; **t. agrícola**, cropland, arable land; **t. alta/baja**, highland/lowland; **t. cultivable**, cultivable land, arable land; **t. de aluvión**, alluvial land; **t. de cultivo**, arable land; **t. de diatomeas**, diatomaceous earth; **t. de labranza**, arable land; **t. de regadío**, irrigated land; **t. de secano**, unirrigated land; **t. desigual**, rough terrain (uneven); **t. diatomácea**, diatomaceous earth; **t. estéril**, sterile land; **t. fértil**, fertile land; **t. firme**, dry land; **t. floja**, loose earth; **t. negra**, black earth; **t. parda**, brown earth; **t. suelta**, loose earth; **tierras estériles, tierras malas (LAm)**, badlands; **toma (f) de t.**, earth (Elec).

tiesto, m. potsherd, piece of pottery, shard.

tifoideola, a. typhoid; **tifoidea (f), fiebre (f) t.**, typhoid fever.

tifón, m. typhoon.

tifus, m. typhus.

tigmotropismo, m. thigmotropism.

tijera, f.pl. scissors (pl);; **tijeras**, scissors, tongs.

tijón, m. header (Con).

till, m. till (Geol); **t. glacial, t. glaciar**, glacial till.

tillita, f. tillite.

tímpano, m. eardrum.

tinción, f. stain, staining; **t. Gram**, Gram's stain.

tintal, m. (Méx) wooded area in karst country.

tiña, f. ringworm.

tio-, pf. thio-.

tiocianato, m. thiocyanate.

tiosulfato, m. thiosulphate.

tipischa, f. (Perú) oxbow, abandoned river meander.

tipo, m. type, kind (clase); pattern (patrón); build, figure (Anat); type, typeface (tipografía); rate (Com); **t. de cambio**, exchange rate; **t. de cultivo**, type of cultivation; **t. salvaje (Ecol)**, wild type.

tipológicola, a. typological.

tiradero, m. (US, Méx) rubbish tip/dump, garbage tip/dump.

tirante, m. tie beam, brace, strut (Con); **t. de cercha**, tie beam; **t. falso**, collar beam.

tiro, m. throw (lanzamiento), load (carga), range (sitio); **t. de falla**, throw of fault.

titanio, m. titanium (Ti); **dióxido (m) de t.**, titanium dioxide.

titanita, f. titanite, sphene.

titración, f. titration.

titulación, f. titration (Quím); degrees and diplomas (enseñanza); **curva (f) de t.**, titration curve.

título, m. title, degree, qualification; **t. universitario**, university degree.

tixotropía, f. thixotropy.

tixotrópicola, a. thixotropic.

tiza, f. chalk.

tizne, m. soot, grime, dirt.

TMG, m. GMT; **Tiempo (m) Medio de Greenwich**, Greenwich Mean Time.

Toarciense, m. Toarcian.

toba, f. tuff, tufa; **tobas soldadas**, welded tuffs.

tobáceola, a. tuffaceous.

tobar, m. place with abundant tufa.

tobillo, m. ankle.

tobogán, m. slide, chute.

todoterreno, m. four-wheel-drive vehicle, all-terrain vehicle (US); jeep (US).

toldo, m. awning, canvas cover; sunshade (en la playa).

toleita, f. tholeiite.

tolerancia, f. tolerance, toleration; **límite (m) de t.**, tolerance limit.

tolueno, m. toluene.

tolva, f. hopper, chute.

tolvanera, f. dust cloud, dust storm.

tolla, f. peat bog.

toma, f. taking, capture; dose; inlet (de aire); ditch (acequia); wire, cable, cord (US) (Elec); **t. de muestras**, sampling, sample collection.

tomate, m. tomato (pl. tomatoes), tomato plant.

tomatera, f. tomato plant.

tomaterola, m. y f. tomato dealer, tomato grower.

tomillar, m. field of thyme.

tomillo, m. thyme.

tonalita, f. tonalite.

tonelada, f. ton; **t. corta**, ton (US) (907.18 kg); **t. larga**, ton (GB) (1.106 kg); **t. métrica**, metric ton, tonne (1.000 kg).

tonsilitis, f. tonsillitis.

topacio, m. topaz.

topadora, f. (LAm) bulldozer.

tope, m. end; **al t.**, end to end; **fecha (f) t.**, deadline date, closing date.

topillo, m. vole.

topo, m. mole.

topografía, f. topography; **t. de bloques fallados**, block-faulted topography; **t. submarina**, submarine topography.

topógrafola, m. y f. surveyor, topographer.

topográficola, a. topographical.

topología, f. topology.

topónimo, m. place name.

torácicola, a. thoracic.

tórax, m. thorax.

torbellino, m. whirlwind (Met), twister (US) (Met), maelstrom; vortex (Fís).

torio, m. thorium (Th).

tormenta, f. storm (Met); upheaval (trastorno); **aviso (m) de t.**, storm warning; **t. de arena**, sandstorm; **t. de polvo**, dust storm; **t. eléctrica**, electric storm.

tormentosola, a. thundery.

tornado, m. tornado.

tornasol, m. sunflower (Bot); litmus (Quím); **papel (m) de t.**, litmus paper; **prueba (f) de t.**, litmus test.

tornillo, m. screw, small lathe; **t. de Arquímedes**, Archimedes screw; **t. sin fin**, worm gear, endless screw.

torno, m. winch (for raising loads), windlass (sobre un pozo), lathe (máquina herramienta).

toronja, f. grapefruit.

torr, m. torr (unidad de medida de presión).

torre, f. tower, mast, pylon; headframe (Min); **t. de control**, control tower; **t. de extracción**, oil derrick (de petróleo); **t. del homenaje (de un castillo)**, keep; **t. de pisos**, tower block; **t. de refrigeración**, cooling tower.

torrencial, a. torrential.

torrente, m. torrent; **a torrentes**, in torrents; **t. de lodo**, mudflow.

torrentosola, a. torrential.

torsión, f. torsion, twisting, torque.

Tortoniense, m. Tortonian.

tortuga, f. tortoise; **t. acuática**, terrapin; **t. marina**, turtle.

tortuosidad, f. tortuosity.

tos, f. cough; **t. ferina**, whooping cough.

tosca, f. hardpan; **t. arcillosa**, claypan.

toscal, m. area with hardpan soils.

toscola, f. rough, coarse.

totora, f. rush, reed.

Tournasiense, m. Tournaisian.

toxicidad, f. toxicity.

tóxico, m. poison, toxin, toxicant.

tóxicola, a. toxic, poisonous; **desecho (m) t.**, toxic waste; **substancia (f) t.**, toxic substance.

toxicología, f. toxicology.

toxina, f. toxin.

traba, f. bond, tie; obstacle (Fig).

trabajadorla, m. y f. worker, labourer, laborer (US); **t. estacional**, seasonal worker.

trabajar, v. to work, to work at, to strive.

trabajo, m. work, labour, labor (US) (tarea); effort (esfuerzo); **mesa (f) de t.**, workbench; **t. de campo**, field work; **t. de detalle**, detailed work; **t. de equipo**, teamwork; **t. de explosivos**, blasting; **t. de media jornada**, part-time employment; **t. estacional**, seasonal work; **t. manual**, manual work, manual labour; **t. por turno**, shift work; **t. temporal**, temporary work, seasonal work.

tracción, F. traction, tension.

tractor, m. tractor (Agr); **t. oruga**, caterpillar tractor.

tráfico, m. traffic; trade, business (Com); **accidente (m) de t.**, road accident; **atasco (m) de t.**, traffic jam; **centro (m) de control de t.**, traffic control centre; **circulación (f) del t.**, traffic flow; **señal (f) de t.**, road sign, traffic sign; **t. marítimo**, maritime traffic; **t. rodado**, road traffic, vehicular traffic.

tragaluz, m. skylight.

tragante, m. flue (de horno); mouth, opening; **instalación (f) extractora de polvo del t.**, flue dust removal plant.

trainera, f. trawler (barco).

trama, f. web, screen, frame; **t. trófica**, food web.

tramadel, m. quagmire.

tramitación, f. transaction, negotiation; **t. de un asunto**, business transaction.

tramitar, v. to take steps.

trámite, m. step, procedure (procedimiento); **hacer trámites para hacer alguna cosa (v)**, take steps to do something; **t. legal**, legal procedure.

tramo, m. stretch, section (de carretera); span (bridge); flight (de escalera); plot (terreno); **t. de carretera**, stretch of road.

trampa, f. trap (caza); trapdoor (en el suelo), hatch (abertura); cheating, fraud.

trampal, m. area with soft ground.

tranquilidad, f. calmness.

tranquilola, a. calm, quiet; **mar (m) t.**, calm sea.

transductor, m. transducer.

transecto, m. transect; **t. lineal**, linear transect.

transferencia, f. transfer, transference; **estación (f) de t. (Trat)**, transfer station; **función (f) de t. (Mat)**, transfer function; **t. de agua entre cuencas**, inter-basin transfer; **t. de tecnología**, technology tranfer.

transformación, f. transformation; **t. de Fourier**, Fourier transformation; **t. de Laplace**, Laplace transformation.

transformadola, a. transformed.

transformador, m. transformer.

transfronterizola, a. transboundary.

transfusión, f. transfusion; **t. sanguínea**, blood transfusion.

transgresión, f. transgression; **t. marina**, marine transgression.

transistor, m. transistor.

transitorio/a, a. transient, transitory; **modelo (m) t.**, transient model.
translocación, f. translocation.
transmisibilidad, f. transmissibility.
transmisión, f. transmission.
transmisividad, f. transmissivity.
transparencia, f. transparency.
transparente, a. transparent.
transpiración, f. transpiration (Bot); perspiration (sudor).
transportable, a. transportable.
transportador, m. conveyer, transporter; protractor (Mat); **t. aéreo**, cableway.
transportador/a, a. transporting; **banda (f) t.**, conveyor belt; **puente (m) t.**, transporter bridge.
transportar, v. to transport, to carry (llevar); **t. mercancías**, to transport goods.
transporte, m. transport, transportation (acción); freight (carga); **buque (m) de t.**, transport ship; **gastos (m.pl) de t.**, transport costs; **medio (m) de t.**, means of transport; **Ministerio (m) de t.**, Ministry of Transport; **red (f) de t.**, transport network; **t. colectivo**, public transport; **t. de mercancías**, transport of goods; **t. por ferrocarril**, rail transport; **t. público**, public transport; **t. terrestre**, overland transport.
transportista, m. haulier, carrier, transporter.
transvasar, v. to transfer (p.e. agua de un río a otro); to decant (líquidos).
transversal, a. transverse.
transversola, a. transverse.
trapecial, a. trapezoidal.
trapecio, m. trapezium, trapezoid (US).
trapezoidal, a. trapezoidal; **vertedero (m) t.**, trapezoidal weir.
trapezoide, a. trapezoid.
tráquea, f. trachea (pl. tracheas, tracheae); windpipe (Anat).
traquiandesita, f. trachyandesite.
traquibasalto, m. trachybasalt.
traquita, f. trachyte.
traslapar, v. to overlap.
traslape, m. overlap (Geol); **t. regresivo (Geol)**, offlap.
traspaso, m. transfer, sale, conveyance (de una propiedad).
trasplante, m. transplant.
trasvase, m. transfer; large-scale water transfer (Ing, grandes conducciones); decanting (decantación).
tratado, m. treatise (tesis); treaty (acuerdo); **t. de matemáticas**, mathematical treatise.
tratadola, a. treated.
tratamiento, m. treatment, processing; **aguas (f.pl) de t.**, process liquors (GB); **facultad (f) de t.**, treatability; **t. aerobio**, aerobic treatment; **t. anaerobio**, anaerobic treatment; **t. biológico**, biological treatment; **t. de aguas residuales**, wastewater treatment, sewage treatment; **t. de cañaveral**, reed-bed treat-

ment; **t. de la información**, data processing; **t. de materias primas**, processing of raw materials, treatment of raw materials; **t. de pastizal**, grass-plot treatment; **t. del agua**, water treatment; **t. desoxidante (piclaje)**, pickling; **t. en la fuente**, treatment at source; **t. en origen**, on site treatment, treatment at source; **t. por etapas**, stage treatment; **t. preliminar**, pretreatment, preliminary treatment; **t. primario/secundario/terciario**, primary/secondary/tertiary treatment.
tratar, v. to treat, to process (material); to deal with, to discuss (una tema).
trato, m. treatment, manners (pl), behaviour; **t. inhumano**, inhuman treatment; **t. terrestre**, ground treatment.
travertino, m. travertine.
través, m. crossbeam, slant, reverse; **t. del campo**, cross-country.
travesaño, m. crosspiece, transom, sleeper (de ferrocarril).
traviesa, f. sleeper, crosstie (US) (de ferrocarril); crossbeam (Con).
trayectoria, f. trajectory, path; **t. profesional**, career.
traza, f. tracer; trace, plan; **elementos (m.pl) trazas**, trace elements; **t. inorgánica**, inorganic trazer; **t. metálica**, metallic tracer; **t. orgánica**, organic trazer; **trazas**, trace amount.
trazado, 1. m. layout (disposición); plan, design (de un edificio etc.); drawing (dibujo); outline (contorno); course, route (de un camino); 2. a. laid out (p.e. una carretera), designed.
trazador, m. tracer (diseñador, perseguidor); plotter (de gráficos); **t. artificial**, artificial tracer; **t. colorante**, dye tracer; **t. de dilución**, dilution tracer; **t. isotópico**, isotopic tracer; **t. químico**, chemical tracer; **t. radioactivo**, radioactive tracer.
trazador/a, a. tracer.
trazar, v. to draw up; to outline (bosquejar); to trace; **t. un programa**, to draw up a programme; **t. la diagonal**, to diagonalize.
trazo, m. line, trace; **t. rectilíneo**, straight line.
trébol, m. clover.
tremadal, m. quagmire, morass, mountain peat bog.
Tremadociense, m. Tremadocian.
Trematoda, f.pl. Trematoda (pl) (Zoo, clase).
tremátodo, m. trematode, fluke; **tremátodos (Zoo, clase)**, Trematoda (pl).
trementina, f. turpentine.
tremolita, f. tremolite.
trépano, m. bit (para perforación).
tri-, pf. tri-.
triangulación, f. triangulation.
triangular, 1. v. to triangulate; 2. a. triangular; **vertedero (m) t.**, t. weir, V-notch weir.

triángulo, m. triangle; **t. equilátero,** equilateral triangle; **t. isósceles,** isosceles triangle.

Triásico, m. Triassic.

triaxial, a. triaxial; **ensayo (m) t.,** triaxial test.

triazina, f. triazine.

tribromometano, m. tribromomethane.

tributación, f. levy, taxation, payment.

tributario, m. tributary.

tríceps, m. triceps.

tricloroacetaldehído, m. trichloroacetaldehyde.

tricloroetileno, m. trichloroethene, trichloroethylene.

triclorometano, m. trichloromethane.

tricono, m. drilling bit, tricone bit.

triconola, a. tricone; **trépano (m) t.,** tricone bit.

tricóptero, m. trichopteran, caddis fly; **tricópteros (Zoo, clase),** Trichoptera (pl).

tridimensional, a. three-dimensional, tridimensional.

trifásica, a. three-phase; **corriente (f) t.,** three-phase current.

trifenilmetano, m. triphenylmethane.

trigal, m. cornfield (GB), wheatfield (US, GB).

trigo, m. wheat (US, GB), corn (GB) (cereal).

trigonometría, f. trigonometry.

trihalometano, m. trihalomethane.

trilobites, m.pl. trilobite.

trilla, f. threshing.

trilladora, f. threshing machine; **t. segadora,** combine harvester.

trillar, v. to thresh.

trimestral, a. quarterly, three-monthly.

trimetilamina, f. trimethylamine.

trinchera, f. trench.

trinitrofenol, m. trinitrophenol.

trinitroglicerina, f. trinitroglycerine.

trinitrotolueno, m. trinitrotoluene.

trióxido, m. trioxide.

trípode, m. tripod.

trismo, m. lockjaw.

tritio, m. tritium; **correlación (f) por t.,** tritium correlation; **datación (f) por t.,** tritium dating; **distribución (f) por t.,** tritium distribution.

tritón, m. newt.

triturador, m. grinder.

triturar, v. to crush, to grind.

trivalente, a. trivalent.

trivial, a. trivial.

triza, f. shard, small piece.

troctolita, f. troctolite.

tróficola, a. trophic; **nivel (m) t.,** trophic level.

troglodita, m. y f. troglodyte, cave-dweller.

tromba, f. whirlwind; **t. de agua,** violent downpour; **t. marina,** waterspout; **t. terrestre,** land whirlwind.

tromel, m. drum screen (Trat).

trompa, f. proboscis (pl. proboscises, proboscides); snout (hocico), horn (cuello).

trona, f. (esp. US) highly alkaline evaporite deposit.

tronada, f. heavy shower; **t. con chubascos,** thunder shower.

tronar, v. to thunder.

tronco, m. trunk (p.e. de árbol).

tropero, m. cattle drover.

trópicola, a. tropical, tropic.

Trópicos, m.pl. the Tropics.

tropismo, m. tropism; **t. negativo/positivo,** negative/positive tropism.

troposfera, f. troposphere.

troquiforme, a. trochiform.

trozo, m. fragment.

trucha, f. trout.

trueno, m. thunder.

truncación, f. truncation.

truncadola, a. truncated.

truncamiento, m. truncation, rounding off (Mat).

tse-tse, tsé-tsé, a. tsetse; **mosca (f) t.,** tsetse fly.

tsunami, m. tsunami.

tubérculo, m. tuber (Bot); tubercle (Zoo, Bot).

tuberculosis, f. tuberculosis.

tubería, f. piping, pipes (pl) (conjunto de tubos); tube, pipe (un tubo); **conducto (m) de t.,** pipe duct; **pérdidas (f.pl) en tuberías,** water mains losses; **t. con brida,** flange pipe; **t. de acero,** steel pipe; **t. de acero estirado,** ductile iron pipe; **t. de bambú,** bamboo pipes; **t. de cobre,** copper pipe; **t. de conducción de agua,** water supply pipe; **t. de fibrocemento,** asbestos cement pipe; **t. de gas,** gas main; **t. de hormigón armado,** reinforced concrete pipe; **t. de impulsión,** rising main; **t. de PVC,** PVC pipe; **t. de revestimiento,** lining tube, casing; **t. de servicio,** service pipe; **t. emisora,** discharge pipe; **t. perforada,** perforated pipe; **t. principal,** water main (para agua); **t. provisional,** temporary pipeline; **tendido (m) de t.,** pipe-laying.

tubo, m. tube, pipe; **brida (f) de t.,** pipe flange; **envuelta (f) de tubos,** pipe wrapping; **revestimiento (m) de tubos,** pipe relining; **t. burbujeador,** sparger; **t. capilar,** capillary tube; **t. de agua,** water pipe; **t. de desagüe,** drain pipe, waste pipe; **t. de drenaje,** drainage pipe; **t. de ensayo,** test tube; **t. de escape,** exhaust pipe, tailpipe (US); **t. de gas,** gas pipe; **t. de Pitot,** Pitot tube; **t. de rayos catódicos,** cathode ray tube; **t. de rayos-X,** X-ray tube; **t. galvanizado,** galvanized pipe; **t. para aguas de alcantarilla,** sewage pipe; **t. piezométrico,** piezometric tube; **t. Visking (Biol),** Visking tubing.

túbulo, m. tubule.

tugurio, m. (Col, ElS) squatter settlement.

tulio, m. thulium (Tm).

tumba, f. tomb, grave.

tumor, m. tumour, tumor (US); **t. benigno,** benign tumour, benign tumor (US); **t. maligno,** malignant tumour, malignant tumor (US).

túmulo, m. tumulus, barrow (Arq); **t. largo,** long barrow.

tumultosola, a. tumultous.

tundra, f. tundra.

túnel, m. tunnel; **t. aerodinámico,** wind tunnel.

tunelización, f. tunnelling.

tungsteno, m. tungsten (W); **carburos (m.pl) de t.,** tungsten carbide.

tungstita, f. tungstite.

turba, f. peat, turf; **capa (f) de t.,** peat bed; **pantano (m) de t.,** peat bog.

turbalización, f. (LAm) peat formation.

turbidez, m. turbidity, cloudiness (en líquidos); **corriente (f) de t. (Geog),** turbidity current.

turbidímetro, m. turbidimeter.

turbidita, f. turbidite.

túrbidola, a. muddy.

turbiedad, f. turbidity, cloudiness (en líquidos).

turbina, f. turbine; **t. de condensación,** condensation turbine; **t. de contrapresión,** counterpressure turbine.

turbiniforme, a. turbiniform.

turbiola, a. cloudy, turbid (agua), blurred (vista).

turbobomba, f. turbine pump.

turbonada, f. squall (Met).

turbulencia, f. turbulence; **t. del viento,** wind turbulence.

turbulentola, a. turbulent; **flujo (m) t.,** turbulent flow.

turgente, a. turgid.

turgor, m. turgor.

turismo, m. tourism, tourist trade, sightseeing; private car, motorcar.

turísticola, a. tourist.

turmalina, f. tourmaline.

turno, m. shift, turn; **t. diurno,** day shift; **t. nocturno,** night shift.

Turoniense, m. Turonian.

turquesa, f. turquoise.

U

uadi, m. wadi.
ubicación, f. position, situation, location.
ubicar, v. to locate, to place, to site.
ubicuola, a. ubiquitous.
ued, m. (Arab) intermittent watercourse in desert regions.
UEM, f. EMU; **Unión (f) Económica y Monetaria,** European Monetary Union.
UICN, f. IUCN; **Unión (f) Internacional para la Conservación de la Naturaleza y Recursos Naturales,** International Union for the Conservation of Nature and Natural Resources.
úlcera, f. ulcer.
ulceración, f. ulceration.
ulja, f. (Arab) low terrace of a desert valley.
ultrabásicola, a. ultrabasic.
ultracentrífuga, f. ultracentrifuge.
ultrafiltración, f. ultrafiltration.
ultramáficola, a. ultramaphic.
ultrasónicola, a. ultrasonic.
ultrasonido, m. ultrasound.
ultravioleta, a. ultraviolet.
UMA, f. AMU; **unidad (f) de masa atómica,** atomic mass unit.
umbela, f. umbel.
umbelífera, f. umbellifer; **umbelíferas (Bot, familia),** Umbelliferae (pl).
umbo, m. umbo.
umbral, m. threshold; **u. de tolerancia,** tolerance threshold.
umbríola, a. shady.
umbrosola, a. shady.
unanimidad, f. unanimity.
UNEP, f. UNEP; **programa (m) ambiental de la ONU,** UN Environment Programme.
únicola, a. unique (incomparable); only, sole (solo).
unidad, f. unit, unity; harmony; **u. central de proceso (abr. UCP) (Comp),** central processing unit (abr. CPU); **u. de masa atómica (abr. u.m.a.),** atomic mass unit (abr. a.m.u.).
unidimensional, a. one-dimensional.
uniforme, a. uniform; level, smooth (superficie); steady (velocidad).
uniformidad, f. uniformity; **coeficiente (m) de u.,** coefficient of uniformity.
uniformitarismo, m. uniformitarianism; **teoría (f) del u. (Geol),** theory of uniformitarianism.
unión, f. union (una organización), meeting of, joining (acción).
unir, f. to join, to link; **u. carreteras,** to link up roads.

unisexual, a. unisexual.
unitariola, a. unitary.
univalvo, m. univalve.
universidad, f. university.
universo, m. universe; **el U.,** the Universe.
uña, f. nail, claw (Anat).
UPA, f. PAU; **Unión (f) Panamericana,** Panamerican Union.
UPC, m. CPU; **Unidad (m) de Procesamiento Central (Comp),** Central Processing Unit.
uralita, f. uralite.
uraninita, f. uraninite, pitchblende.
uranio, m. uranium (U).
Urano, m. Uranus.
urbanismo, m. town planning.
urbanista, m. y f. town planner.
urbanísticola, a. urban, city centre (plan, entorno); town-planning, city planning (US) urbanismo); **conjunto (m) u.,** building development (Gen); housing development, housing estate (casas, pisos).
urbanización, f. urbanization (extensión de ciudades); town planning, city planning (US) urbanismo); building development (construcción); new town (ciudad nueva); housing estate (viviendas).
urbanizadola, a. built-up.
urbanola, a. urban; **extensión (f) u.,** urban sprawl; **población (f) u.,** urban population; **renovación (f) u.,** urban renewal.
urea, f. urea.
uréter, m. ureter.
uretra, f. urethra.
urinación, f. urination.
urinario, m. urinal.
urinariola, a. urinary.
urubú, m. turkey vulture, turkey buzzard.
Uruguay, m. Uruguay.
uruguayola, 1. m. y f. Uruguayan; 2. a. Uruguayan.
usina, f. (CSur) factory, plant; power station.
uso, m. use, usage; **u. posterior,** afteruse.
usurpación, f. impingement, encroachment.
utensilio, m. utensil, tool (herramienta).
útero, m. uterus (pl. uteri), womb.
útil, a. useful, handy.
utilidad, f. usefulness.
utilización, f. utilization; **de u. sencilla (Comp),** user-friendly.
utilizar, v. to use, to utilize; to reclaim (desechos); **fácil de u. (Comp),** user-friendly.
utillaje, m. tolls (pl); equipment.
uva, f. grape; **racimo (m) de uvas,** bunch of grapes; **u. espina,** gooseberry; **u. pasa,** raisin.

V

vaciadero, m. dumping-ground.
vaciamiento, m. emptying.
vaciar, v. to empty; to drain (de líquido).
vacío, m. void, vacuum; **espacio (m) de v.**, void space; **extracción (f) por v.**, vacuum extraction; **filtro (m) v.**, vacuum filter; **proporción (f) de v.**, void ratio.
vacíola, a. empty.
vacuna, f. vaccine.
vacunación, f. vaccination.
vacunar, v. to vaccinate.
vacunola, a. bovine; **carne (f) de v.**, corned beef; **ganado (m) v.**, cattle.
vacuola, f. vacuole.
vadear, v. to ford, to wade across (de pie).
vado, m. ford (a través de un río).
vagón, m. wagon; **v. cerrado**, covered wagon; **v. de mercancías**, freight car (US); **v. para ganado**, cattle truck.
vagoneta, f. wagonette; **v. de mina**, mine wagon, mine skip.
vaguada, f. thalweg (Geog), trough (Met).
vaina, f. sheath; pod, bool (Bot); **v. de mielina**, myelin sheath, medullary sheath.
vajilla, f. crockery, china.
Valanginense, m. Valangian.
valencia, f. valency, valence.
valer, v. to be worth; **v. más que**, to outweigh, to be worth more.
validación, f. validation; **v. del modelo**, model validation.
validar, v. to validate.
validez, f. validity.
válidola, a. valid; robust, strong (fuerte); lawful (Jur).
valor, m. value, worth (Gen); price (precio); **impuesto (m) al v. añadido/agregado (abr. IVA)**, value added tax (abr. VAT); **v. comercial**, commercial value; **v. de reemplazamiento**, replacement value; **v. ecológico**, ecological value; **v. límite imperativo (Reg, de directivas europeas)**, mandatory value (GB); **v. medio**, average value; **v. propio (Mat)**, eigenvalue; **v. residual**, residual value; **v. térmico medio (Mat)**, average temperature; **v. umbral**, threshold value; **valores extremos máximos y mínimos**, extreme values; **valores normales**, background values, background levels.
valoración, f. valuation; appraisal, assessment; titration (Quím); **curva (f) de v. (Ing)**, rating curve; **v. a punto final (Quím)**, end point titration; **v. de la calidad**, quality assessment.
valorador, m. valuer; land agent (de fincas).
válvula, f. valve; **v. de admisión**, inlet valve, intake; **v. de clapeta**, clack valve, non-return valve; **v. de compuerta**, gate valve; **v. de control**, control valve; **v. de control de flujo**, flow-reducing valve; **v. de descarga**, discharge valve; **v. de pie**, foot valve; **v. de purga**, blow-off valve; **v. de purga de aire**, air blow-off valve; **v. de reflujo**, reflux valve; **v. de regulación**, regulator valve; **v. de retención**, retention valve, reflux valve; **v. de seguridad**, safety valve, pressure-relief valve; **v. de tres vías**, three-way valve; **v. mariposa**, butterfly valve; **v. piezorreductora**, pressure-reducing valve.
valvulería, f. valving.
valle, m. valley; **fondo (m) de v.**, valley bottom; **v. ahogado**, drowned valley; **v. aluvial**, alluvial valley; **v. anegado**, drowned valley; **v. anticlinal**, anticlinal valley; **v. antiguo**, old age valley; **v. antigúo**, old age valley; **v. ciego**, blind valley; **v. colgante**, hanging valley; **v. de bloques de rift**, rift-block valley; **v. de drenaje**, drainage valley; **v. de línea de falla**, fault-line valley; **v. enterrado**, buried valley; **v. estrecho entre cerros**, combe, coomb; **v. fluvial**, river valley; **v. juvenil**, youthful valley; **v. seco**, dry valley; **v. senil**, old age valley; **v. sepultado**, buried valley.
vanadio, m. vanadium (V).
vano, m. opening, bay (hueco).
VAO, m. **vehículos (m.pl) de alta ocupación (abr. VAO)**, passenger vehicles, public transport.
vapor, m. vapour, vapor (US), steam (vaho); **caldera (f) de v.**, steam boiler; **de v. (a)**, steam; **extracción (f) de vapores en suelos**, soil vapour extraction; **máquina (f) de v.**, steam engine; **presión (f) de v.**, vapour pressure; **tensión (f) de v.**, vapour tension; **v. de agua**, water vapour.
vaporador, m. (LAm) vaporizer.
vaporizador, m. vaporizer.
vaquería, f. cowshed (establo); dairy (lechería).
vaquero, m. cowman; cowboy (US, LAm); **vaqueros (pl) (ropa)**, jeans.
varadero, m. dry dock.
variabilidad, f. variability.
variable, f. variable; **análisis (m) de múltiples variables**, multivariate analysis; **v. aleatoria**, random variable; **v. continua**, continuous variable; **v. discreta**, discrete variable.
variación, f. variation, change; **sin v. (a)**, unchanging, constant; **v. climática**, climatic variation; **v. de caudal de bombeo**, variation in pumped output; **v. de espesor saturado**, variation in saturated thickness; **v. estacional**, seasonal variation; **v. genética**,

genetic variation; **v. magnética**, magnetic variation.

variadola, a. assorted.

variante, 1. f. variant; 2. a. variant.

varianza, f. variance (Mat); **análisis (m) de v.**, analysis of variance (abbr. ANOVA); **v. de la muestra**, sample variance; **v. espectral**, spectral variance; **v. explícita**, explicit variance; **v. total**, total variance.

varicela, f. chicken-pox, varicella.

variedad, f. variety.

varillaje, m. ribbing, ribs (pl).

variola, a. various (unos cuantos); diverse, different (diferente).

variolíticola, a. variolitic.

varita, f. small stick, rod.

varón, m. man, boy, male.

varva, f. varve; **arcilla (f) de v. (LAm)**, varved clay.

vascular, a. vascular; **planta (f) v.**, vascular plant.

vasectomía, f. vasectomy.

vaselina, f. vaseline (R).

vasija, f. vessel, pot, jar; skip (envase por basura); **v. de barro**, earthenware pot.

vaso, m. vessel (recipiente); glass (copa); **v. sanguíneo**, blood vessel.

vasoconstricción, f. vasoconstriction.

vasodilación, f. vasodilation.

vasodilatación, f. vasodilation.

vástago, m. shoot (Biol); offspring (Biol); piston, rod (Con); **v. de perforación**, drill stem, drilling rod.

vatio, m. watt (unidad de consumo eléctrico).

vazadouro, m. (Br) waste tip, garbage dump, landfill.

vecinola, a. neighbouring, neighboring (US); **país (m) v.**, neighbouring country, neighboring country (US).

vector, m. vector; **v. propio (Mat)**, eigenvector.

vectorial, a. vector; **análisis (m) v.**, vector analysis.

vega, f. flat fertile land; hillside fed by groundwater (Ch); backswamp (US).

vegetación, f. vegetation; **v. antrópica**, antropic vegetation; **v. natural/semi-natural**, natural/semi-natural vegetation.

vegetal, a. plant, vegetable; **reino (m) v. (Bot)**, plant kingdom.

vegetativola, a. vegetative; **reproducción (f) v.**, vegetative reproduction.

veguilla, f. productive soils in mountains.

vehículo, m. vehicle; **v. ligero**, light vehicle; **v. pesado**, heavy goods vehicle (GB, abr. HGV); **vehículos de alta ocupación (abr. VAO)**, passenger vehicles.

vejiga, f. bladder; **v. de aire**, air bladder, gas bladder (de peces); **v. de la bilis**, gall bladder; **v. natatoria**, swim bladder; **v. urinaria**, bladder.

vela, f. sail (de un barco); sailing (deporte); vigil (sobre el enfermo); candle (bujía); **barco (m) de v.**, sailing boat.

velero, m. sailing boat.

veleta, f. weather vane.

velocidad, f. velocity, speed; **v. angular**, angular velocity; **v. crítica**, critical velocity; **v. de flujo**, flow velocity; **v. de la luz**, speed of light; **v. del viento**, wind velocity, wind speed; **v. de ondas sísmicas**, velocity of seismic waves; **v. de pulso**, pulse rate; **v. de reacción**, rate of reaction; **v. de sonido**, speed of sound; **v. en el canal**, channel velocity; **v. terminal**, terminal velocity.

vena, f. vein.

venación, f. venation; **v. paralela**, parallel venation.

vendaval, m. gale; **v. de fuerza 10**, gale force 10.

vendedorla, m. y f. seller, salesman/saleswoman; shop assistant (en una tienda).

vender, v. to sell; **"se vende"**, **"for sale"**.

vendimia, f. grape harvest.

veneno, m. poison, venom.

venenosola, a. poisonous (Gen); venemous (p.e. un serpiente); tóxic (p.e. drogas).

venera, f. scallop (marisco).

venero, m. intermittent spring.

venezolanola, 1. m. y f. Venezuelan; 2. a. Venezuelan.

Venezuela, m. Venezuela.

venosola, a. venous.

venta, f. sale; **precio (m) de v.**, sale price.

ventana, f. window; nostril (Zoo, de nariz).

ventanilla, f. window.

ventarrón, m. gale.

ventilación, f. ventilation.

ventilador, m. aerator, ventilator; **v. cónico**, cone aerator.

ventisca, f. blizzard, snowstorm.

ventiscar, v. to blow a blizzard.

ventisquear, v. to blow a blizzard.

ventisquero, m. blizzard, snowstorm.

ventolada, f. (LAm) gale, strong wind.

ventolera, f. gust of wind.

ventolina, f. (LAm) sharp gust of wind.

ventosa, f. sucker (Zoo).

ventral, a. ventral.

ventrículo, m. ventricle.

Venturi, m. Venturi (Ing); **canal (m) V.**, Venturi flume; **contador (m) V.**, Venturi meter.

Venus, f. Venus.

verde, a. green (color); unripe (fruta); unseasoned (madera); premature (plan); foliage (Bot); green (política); **libra (f) v.**, green pound (esterlinas); **Los Verdes, Partido V. (política)**, The Greens, Green Party; **política (f) v.**, green politics; **temas (m.pl) verdes**, green issues; **zona (f) v.**, green belt.

verdear, v. to turn green (plantas).

verdura, f. vegetable.

vereda, m. path, lane (senda); pavement, sidewalk (US) (andén).

verificable, a. verifiable, checkable.
verificación, f. verification.
verificar, v. to verify.
verja, f. grating, grill, railings (pl).
vermiculita, f. vermiculite.
vernalización, f. vernalization.
verosimilitud, f. likelihood, probability.
vértebra, f. vertebra (pl. vertebrae, vertebras).
vertebrado, m. vertebrate; **vertebrados (Zoo, subfilum),** Vertebrata (pl).
vertebral, a. vertebral; **columna (f) v.,** vertebral column.
vertedero, m. rubbish dump, tip (de basuras); overflow, gauging weir (para agua); **gas (m) de v. controlado,** landfill gas; **v. compuesto,** compound weir; **v. controlado,** licensed tip (GB), waste management facility (US); **v. curvo,** rounded-crested weir; **v. de basuras (casas),** rubbish chute; **v. de pared gruesa,** broad-crested weir; **v. de superficie,** open tip/dump; **v. en V,** V-notch weir; **v. ilegal,** illegal wastetip; **v. incontrolado,** illegal tip/dump; **v. público,** municipal landfill, public waste tip.
verter, v. to spill, to empty, to dump.
vertido, m. spill, spillage, waste, discharge; **cañón (m) de v.,** discharge pipe; **v. accidental de petróleo,** accidental oil spill; **v. controlado,** landfill; **v. de alpechín,** spillage of foul smelling liquid which runs from olive waste; **v. directo,** direct discharge; **v. en terraplén,** landfill; **v. indirecto,** indirect discharge; **vertidos controlados,** controlled waste.
vertiente, f. slope; spring, fountain (LAm); **en la v. norte,** on the north slope; **v. del lado del viento,** windward slope.
vesícula, f. vesicle, bladder; **v. biliar,** gall bladder.
vesiculación, f. blistering.
vestigial, a. vestigial.
vestigio, m. trace, relic.
vesubiana, f. vesuvianite.
veta, f. vein, seam (Min); land in marshes above flood level (Geog).
veterinario/a, 1. m. y f. vet, veterinary doctor, veterinary surgeon, veterinarian (US); **veterinaria (f),** veterinary science, veterinary medicine; 2. a. veterinary.
vía, f. route, track, way; **limpieza (f) de v.,** street cleaning; **v. del ferrocarril,** railway, railroad (US).
viabilidad, f. viability, feasibility; **estudio (m) de v.,** feasibility study, viability study.
viable, a. viable.
viaducto, m. viaduct.
viajar, v. to journey, to travel.
viaje, m. journey, trip.
Vía Láctea, f. Milky Way.
viaria, f. of roads, of traffic; **red (f) viaria,** transport network.
vibración, f. vibration.
vibrador, m. vibrator, shaker.

vibrar, v. to vibrate.
vibroflotación, f. vibroflotation.
viciadola, a. contaminated, polluted (aire); stuffy (atmósfera).
vid, f. vine, grapevine; vine arbour (US).
vida, f. life; **análisis (m) ciclo v.,** life cycle assessment; **duración (f) de la v.,** life span; **nivel (m) de v.,** standard of living; **v. media,** half-life; **v. útil,** useful life; **v. vegetal (Bot),** plant life; **valoración (f) del ciclo de v.,** life cycle assessment.
vidriado, m. glaze; glazed earthenware (cerámica).
vidriadola, a. glazed.
vidriar, v. to glaze (cerámica).
vidrio, m. glass (material), glazing (p.e. para ventanas); **contenedor (m) de v.,** container glass; **desperdicios (m.pl) de v.,** cullet (para reciclaje), scrap glass; **fibra (f) de v.,** glass fibre, glass fiber (US), fibreglass, fiberglass (US); **v. esmerilado,** ground glass.
viento, m. wind; **energía (f) del v.,** wind power; **protección (f) contra el v.,** windbreak; **v. catabático,** katabatic wind; **v. predominante,** dominant/predominant wind.
vientre, m. belly (estómago); bowels (intestino); womb (matriz); guts (de animal muerto).
viga, f. beam, rafter, girder; **v. de hormigón armado,** reinforced concrete beam; **v. en H,** H beam; **v. maestra, v. principal,** main beam; **v. transversal,** crossbeam; **v. voladizo,** cantilever.
vigente, a. valid, applicable, in force; **normativa (f) v.,** existing legislation, legislation currently in force.
vigilancia, f. monitoring, surveillance; **estación (f) de v.,** monitoring station, monitoring point; **red (f) de v.,** monitoring network; **v. de la calidad de las aguas,** water quality monitoring.
vigor, m. vigour, vigor (US), force; **entrar en v. (v) (Jur),** to come in to force, to come into effect, to be introduced; **estar en v. (v),** to be in force, to be in effect; **v. híbrido,** hybrid vigour, hybrid vigor (US).
vigueta, f. joist; **anclaje (m) de v.,** joist anchor.
villa, f. villa (casa), small town (pueblo), borough (municipalidad); **v. miseria (LAm),** squatter settlement, shanty town.
vinilo, m. vinyl.
viña, f. vineyard, vine.
viñador, m. vine-grower.
viñatero, m. vine-grower.
viñedo, m. vineyard.
viñero, m. vine grower, viticulturist.
violación, f. violation, breach; **v. de la ley,** breach of the law.
viral, a. viral.
virología, f. virology.
virulencia, f. virulence.
virulento/a, a. virulent.

virus, m. y m.pl. virus; **v. del mosaico**, mosaic virus; **v. informático**, computer virus.

vísceras, f.pl. viscera.

viscosidad, f. viscosity; **coeficiente (m) de v.**, coefficient of viscosity; **v. cinemática**, kinematic viscosity; **v. dinámica**, dynamic viscosity.

viscosímetro, m. viscosimeter.

viscosola, a. viscous.

Viseense, m. Visean.

visibilidad, f. visibility.

visible, a. visible; **reservas (f.pl) visibles (Com)**, visible reserves.

visión, f. sight, vision; view (vista); **v. corta**, short sight, shortsightedness; **v. de larga**, long sight, longsightedness.

vista, f. view; sight, eyesight, vision (Anat); look, gaze (mirada); **v. aérea**, aerial view; **v. de pájaro**, bird's eye view; **v. panorámica**, panoramic view.

visual, a. visual.

visualización, f. visualization.

visualizar, v. to visualize.

vitamina, f. vitamin; **v. C**, vitamin C, ascorbic acid.

vitelinola, a. vitelline, yolk; **saco (m) v.**, yolk sac.

viticultor, m. vine-grower, viticulturist.

vítreola, a. vitreous.

vitrificación, f. vitrification.

vitrinita, f. vitrinite.

vitriolo, m. vitriol.

vitroclasto, m. vitroclast.

vivero, m. garden centre, garden center (US), nursery.

vivienda, f. housing; **viviendas (PRico)**, shanty town.

vivíparola, a. viviparous.

vivisección, f. vivisection.

vivola, a. living, live; **en vivo**, live (transmisión); **organismo (m) v.**, living organism.

vocabulario, m. vocabulary.

voladura, f. blowing-up, explosion.

volandera, f. millstone, grindstone (para cereales); washer (Ing, anillo).

volante, f. flywheel.

volar, v. to fly (p.e. aves); to blow up, to explode (de explosiones).

volátil, a. volatile (Quím).

volatilidad, f. volatility.

volatilizar, v. to volatilize.

volcán, m. volcano; **v. en actividad**, active volcano; **v. en escudo**, shield volcano; **v. extinto**, extinct volcano.

volcánicola, a. volcanic; **tapón (m) v.**, volcanic plug.

volquete, m. dump truck.

voltaicola, a. voltaic.

voltaje, f. voltage (potencial eléctrica).

voltamperímetro, m. voltammeter, voltameter (US).

voltamperio, m. volt-ampere.

voltímetro, m. voltmeter.

voltio, m. volt (unidad de capacidad eléctrica).

volumen, m. volume; **v. bombeado**, pumpage, amount pumped; **v. crítico**, critical volume; **v. de negocios (Com)**, turnover; **v. de ventilación pulmonar (Med)**, tidal volume; **v. molar**, molar volume; **v. molecular gramo**, gram molecular volume.

volumétricola, a. volumetric.

vomitar, v. to vomit, to be sick.

vómito, m. vomit.

vorágine, f. whirlpool, vortex.

vórtice, m. vortex (pl. vórtices).

vuelco, m. overturn (e.g. of a car); dump, upsetting; **v. de cinta (Comp)**, tape dump.

vuelta, f. turn (gira); revolution (revolución); road bend (en la carretera); **dar vueltas (v)**, to spiral.

vulcanismo, m. vulcanism, volcanism.

vulcanización, f. vulcanization.

vulcanología, f. vulcanology.

vulnerabilidad, f. vulnerability.

vulnerable, a. vulnerable.

W

wadi, m. (Arab) wadi; intermittent watercourse.
warfarina, f. warfarin.
watio, m. watt (unidad de consumo eléctrico).
wavellita, f. wavellite.
Wenlockense, m. Wenlockian.
Westfaliense, m. Westphalian.
windsurf, m. sailboarding, windsurfing.

wireline, m. wireline; **testificación (f) geofísica con w.**, wireline logging.
witherita, f. witherite.
wolframita, f. wolframite.
wollastonita, f. wollastonite.
WWF, m. WWF; **Fondo (m) Mundial para la Naturaleza**, World Wildlife Fund.

X

xantoproteicola, a. xanthoproteic; **test (m) x. (Biol, Quím)**, xanthoprotein test.
xenoblásticola, a. xenoblastic.
xenoblasto, m. xenoblast.
xenolíticola, a. xenolithic.
xenolito, m. xenolith.

xenomórficola, a. xenomophic.
xenón, m. xenon (Xe).
xerofita, f. xerophyte.
xeroserie, f. xerosere.
xilema, m. xylem.
xileno, m. xylene.

Y

yacer, v. to rest on, to lie on.
yacimiento, m. deposit, vein (Geol); **y. mineral**, mineral deposit.
yarda, f. yard (medida de longitud, equiv. a 0,9144 metros).
yate, m. yacht.
yedra, f. ivy.
yema, f. bud, shoot; gemma (Bot, de hepáticas); yolk (de huevo); **y. terminal/axilar/lateral (Bot)**, terminal/axillary/lateral bud.

yermo, m. wasteland, wilderness.
yermola, a. uninhabited (despoblado).
yesca, f. kindling, tinder; punk (US).
yesera, f. gypsum deposit.
yeso, m. plaster (p.e. de paredes).
yoduración, f. iodine disinfection.
yterbio, m. ytterbium (Yb).
yuca, f. yucca.

Z

zafiro, m. sapphire.
zahorí, m. water diviner.
zamuro, m. turkey vulture, turkey buzzard (US).
zanahoria, f. carrot.
zanca, f. stringer (Con).
zancado, m. adult salmon after spawning.
Zancliense, m. Zanclian.
zancuda, f. wader (de aves).
zancudo, m. mosquito (pl. mosquitos, mosquitoes).
zángano, m. drone (insecto).
zanja, f. soakaway, drain, ditch; **z. a cielo abierto**, open ditch; **z. con tubos de barro cocido**, tile drain; **z. de avenamiento rellena de grava**, french drain; **z. de drenaje**, drainage ditch; **z. de impermeabilización**, cut-off trench; **z. de oxidación**, oxidation ditch; **z. de préstamos (Ing)**, borrow pit; **z. de recarga**, recharge ditch.
zanjadora, f. trenching machine.
zanjar, v. to ditch, to dig a ditch or trench (abrir una zanja).
zanjero, m. trenching, ditching.
zapapico, m. pickaxe, mattock.
zapata, f. shoe, track (de oruga), washer (arandela), lintel (Con); **z. de freno**, brake shoe.
zarcillo, m. holdfast, hapteron (de algas); tendril (de plantas); hoe (herramienta).
zarpa, f. claw, paw (un animal); weighing anchor.
zarzal, m. bramble, brambles (pl); blackberry bush; thicket.
zarzamora, f. blackberry.
zenit, m. zenith.
zeolita, f. zeolite.
zincho, m. clamp; **z. ajustable**, adjustable clamp.
zócalo, m. plinth, insular shelf (Geol); square (US) (plaza); sill, plinth; shelf (Geol).
zoisita, f. zoisite.
zona, f. zone, area; **z. abisal**, abyssal zone; **z. aerobia**, aerobic zone; **z. afótica**, aphotic zone; **z. anaerobia**, anaerobic zone; **z. árida**, arid zone; **z. climática**, climate zone; **z. de abrigo de la lluvia**, rain shadow; **z. de abscisión**, abscission zone; **z. de bajas presiones**, trough of low pressure; **z. de captación**, catchment area; **z. de cizalla**, shear zone; **z. de desarrollo**, development area; **z. de drenaje**, drainage area; **z. de falla**, fault zone; **z. de fractura**, fracture zone; **z. de gran radioactividad**, radioactive hotspot; **z. de humedad del suelo**, soil moisture zone; **z. de hundimientos**, collapse zone; **z. de la cuenca**, zona of capture, catchment area; **z. de meteorización**, zone of weathering; **z. de oscilación de las mareas**, intertidal zone; **z. de raiz**, root zone; **z. de recarga**, recharge area; **z. de saturación**, zone of saturation; **z. desnuclearizada**, nuclear-free zone; **z. de subducción**, subduction zone; **z. de tiendas**, shopping centre (center, US); **z. de transición**, transition zone; **z. edificada**, built-up area; **z. eufótica**, euphotic zone; **z. eulitoral**, eulittoral zone; **z. fótica**, photic zone; **z. fronteriza**, frontier area; **z. industrial**, industrial site; **z. intermareal**, intertidal zone; **z. intermediaria**, buffer zone; **z. intertidal**, intertidal zone; **z. litoral**, littoral zone; **z. mareal**, tidal zone, intertidal zone, shore; **z. marginal (LAm)**, squatter settlement; **z. postal**, postal area; **z. semiárida**, semi-arid zone; **z. tampón (Ecol)**, buffer zone; **z. tectonizada**, tectonic zone; **z. templada**, temperate zone; **z. triturada**, shattered zone; **z. urbana**, urban zone; **z. verde**, green belt, green area, green park.
zonificación, f. zonification.
zoogeografía, f. zoogeography.
zoología, f. zoology.
zooplancton, m. zooplankton.
zoospora, f. zoospore.
zopilote, m. turkey vulture, turkey buzzard.
zorrero, m. (Col) waste tip scavenger.
zozobra, f. overturn, overturning (of a ship).
zubia, f. area with numerous streams.
zumaya, f. night bird.
zunchar, v. to fasten with a metal band.